Fernando A. Ponce, Abigail Bell (Eds.)

ICNS 4
Proceedings of the Fourth International Conference on Nitride Semiconductors

Part B

Fernando A. Ponce, Abigail Bell (Eds.)

Proceedings

Part B

Editors:
Fernando A. Ponce, Department of Physics and Astronomy, Arizona State University, USA
e-mail: ponce@asu.edu
Abigail Bell, Department of Physics and Astronomy, Arizona State University, USA
e-mail: abell1@imap.asu.edu

Cover:
With kind permission of M. Stutzmann and O. Ambacher, Walter Schottky Institute, Technical University of Munich, FRG

> This book was carefully produced. Nevertheless, editors, authors and publisher do not warrant the information contained therein to be free of errors. Readers are advised to keep in mind that statements, data, illustrations, procedural details or other items may inadvertently be inaccurate.

Library of Congress Card No.: applied for

A catalogue record for this book is available from the British Libary.

Die Deutsche Bibliothek – CIP Caraloguing-in-Publication-Data
A catalogue record for this publication is available from Die Deutsche Bibliothek

1st edition
ISBN 3-527-40347-7

© WILEY-VCH Verlag Berlin GmbH, 2002
Printed on acid-free paper.
All rights reserved (including those of translation in other languages). No part of this book may be reproduced in any form – by photoprinting, microfilm, or any other means – nor transmitted or translated into machine language without written permission from the publishers. Registered names, trademarks, etc. used in this book, even when not specifically marked as such, are not to be considered unprotected by law.
Composition, printing, bookbinding: Druckhaus „Thomas Müntzer" GmbH, D-99947 Bad Langensalza.
Printed in the Federal Republic of Germany.

Contents

Growth of indium nitride

MOCVD Growth of High-Quality InN Films and Raman Characterization of Residual Stress Effects
E. Kurimoto, H. Harima, A. Hashimoto, and A. Yamamoto   1–4

Growth Temperature Dependences of MOVPE InN on Sapphire Substrates
A. Yamamoto, Y. Murakami, K. Koide, M. Adachi, and A. Hashimoto.   5–8

Anisotropic Superconductivity of InN Grown by Molecular Beam Epitaxy on Sapphire (0001)
T. Inushima, V.V. Vecksin, S.V. Ivanov, V.Y. Davydov, T. Sakon, and M. Motokawa   9–12

Polarity of High-Quality Indium Nitride Grown by RF Molecular Beam Epitaxy
Y. Saito, Y. Tanabe, T. Yamaguchi, N. Teraguchi, A. Suzuki, T. Araki, and Y. Nanishi   13–16

Study of Epitaxial Relationship in InN Growth on Sapphire (0001) by RF-MBE
T. Yamaguchi, Y. Saito, K. Kano, T. Araki, N. Teraguchi, A. Suzuki, and Y. Nanishi   17–20

Growth and Characterization of InN Heteroepitaxial Layers Grown on Si Substrates by ECR-Assisted MBE
T. Yodo, H. Ando, D. Nosei, and Y. Harada.   21–26

A Novel Two-Step Method for Improvement of MOVPE Grown InN Film on GaP(111)B Substrate
A.G. Bhuiyan, A. Yamamoto, and A. Hashimoto   27–30

Phase Separation of $Al_{1-x}In_xN$ Grown at the Resonance Point of Nitrogen-ECR Plasma
K. Murano, T. Inushima, Y. Ono, T. Shiraishi, S. Ohoya, and S. Yasaka . . .   31–34

Spatial Variation of Luminescence of InGaN Alloys Measured by Highly-Spatially-Resolved Scanning Cathodoluminescence
F. Bertram, S. Srinivasan, L. Geng, F.A. Ponce, T. Riemann, J. Christen, S. Tanaka, H. Omiya, and Y. Nakagawa   35–39

InGaN thin films

A Comparison of Rutherford Backscattering Spectroscopy and X-Ray Diffraction to Determine the Composition of Thick InGaN Epilayers
S. Srinivasan, R. Liu, F. Bertram, F.A. Ponce, S. Tanaka, H. Omiya, and Y. Nakagawa   41–44

Local Structure Analysis of $Ga_{1-x}In_xN$ Alloy Using Extended X-Ray Absorption Fine Structure Measurements
T. Miyajima, Y. Kudo, K.-Y. Liu, T. Uruga, T. Asatsuma, T. Hino, and T. Kobayashi   45–48

Indium Surface Segregation during Growth of (In,Ga)N/GaN Multiple Quantum Wells by Plasma-Assisted Molecular Beam Epitaxy
P. Waltereit, O. Brandt, K.H. Ploog, M.A. Tagliente, and L. Tapfer   49–53

Temperature-Independent Stokes Shift in an $In_{0.08}Ga_{0.92}N$ Epitaxial Layer Revealed by Photoluminescence Excitation Spectroscopy
H. KUDO, K. MURAKAMI, H. ISHIBASHI, R. ZHENG, Y. YAMADA, and T. TAGUCHI . . 55–58

Depth Resolved Studies of Indium Content and Strain in InGaN Layers
S. PEREIRA, M.R. CORREIA, E. PEREIRA, K.P. O'DONNELL, C. TRAGER-COWAN, F. SWEENEY, E. ALVES, A.D. SEQUEIRA, N. FRANCO, and I.M. WATSON 59–64

InGaN quantum wells: growth and carrier dynamics

Carrier Dynamics in Group-III Nitride Low-Dimensional Systems: Localization versus Quantum-Confined Stark Effect (*invited*)
P. LEFEBVRE, T. TALIERCIO, S. KALLIAKOS, A. MOREL, X.B. ZHANG, M. GALLART, T. BRETAGNON, B. GIL, N. GRANDJEAN, B. DAMILANO, and J. MASSIES 65–72

Pressure Dependence of Piezoelectric Field in InGaN/GaN Quantum Wells
G. VASCHENKO, D. PATEL, C.S. MENONI, N.F. GARDNER, J. SUN, W. GÖTZ, C.N. TOMÉ, and B. CLAUSEN 73–76

Piezoelectric Field-Induced Quantum-Confined Stark Effect in InGaN/GaN Multiple Quantum Wells
C.Y. LAI, T.M. HSU, W.-H. CHANG, K.-U. TSENG, C.-M. LEE, C.-C. CHUO, and J.-I. CHYI 77–80

Carrier Dynamics in InGaN/GaN SQW Structure Probed by the Transient Grating Method with Subpicosecond Pulsed Laser
K. OKAMOTO, A. KANETA, K. INOUE, Y. KAWAKAMI, M. TERAZIMA, G. SHINOMIYA, T. MUKAI, and SG. FUJITA 81–84

Temperature Dependence and Reflection of Coherent Acoustic Phonons in InGaN Multiple Quantum Wells
Ü. ÖZGÜR, C.-W. LEE, and H.O. EVERITT 85–89

InGaN/GaN Quantum Well Microcavities Formed by Laser Lift-Off and Plasma Etching
P.R. EDWARDS, R.W. MARTIN, H.-S. KIM, K.-S. KIM, Y. CHO, I.M. WATSON, T. SANDS, N.W. CHEUNG, and M.D. DAWSON 91–94

Comparative Study of InGaN/GaN Structures Grown by MOCVD Using Various Growth Sequences
A.V. SAKHAROV, A.S. USIKOV, W.V. LUNDIN, A.F. TSATSULNIKOV, RU-CHIN TU, SUN BIN YIN, and JIM Y. CHI 95–98

Growth and Characterization of InGaN/GaN Multiple Quantum Wells on Ga-Polarity GaN by Plasma-Assisted Molecular Beam Epitaxy
X.Q. SHEN, T. IDE, M. SHIMIZU, F. SASAKI, and H. OKUMURA 99–102

InGaN quantum wells: optical properties

Optical Studies on AlGaN/InGaN/GaN Single Quantum-Well Structures under External Strains
E. KURIMOTO, M. TAKAHASHI, H. HARIMA, H. MOURI, K. FURUKAWA, M. ISHIDA, and M. TANEYA 103–106

Phonon and Photon Emission from Optically Excited InGaN/GaN Multiple Quantum Wells
A.V. AKIMOV, S.A. CAVILL, A.J. KENT, N.M. STANTON, T. WANG, and S. SAKAI . . 107–110

Contents

Dual Contribution to the Stokes Shift in InGaN–GaN Quantum Wells
T.J. OCHALSKI, B. GIL, P. BIGENWALD, M. BUGAJSKI, A. WOJCIK, P. LEFEBVRE,
T. TALIERCIO, N. GRANDJEAN, and J. MASSIES 111–114

Spectroscopy and Modeling of Carrier Recombination in III–N Heterostructures
P.M. SWEENEY, M.C. CHEUNG, F. CHEN, A.N. CARTWRIGHT, D.P. BOUR,
and M. KNEISSL 115–119

Two-Component Photoluminescence Decay in InGaN/GaN Multiple Quantum Well Structures
SHIH-WEI FENG, YUNG-CHEN CHENG, CHI-CHIH LIAO, YI-YIN CHUNG, CHIH-WEN LIU,
CHIH-CHUNG YANG, YEN-SHENG LIN, KUNG-JENG MA, and JEN-INN CHYI 121–124

Time-Resolved Photoluminescence Study of InGaN MQW with a p-Contact Layer
T. KURODA, R. SASOU, A. TACKEUCHI, H. SATO, N. HORIO, and C. FUNAOKA . . . 125–128

Photoluminescence Excitation Spectroscopy of MBE Grown InGaN Quantum Wells and Quantum Boxes
M.E. WHITE, K.P. O'DONNELL, R.W. MARTIN, C.J. DEATCHER, B. DAMILANO,
N. GRANDJEAN, and J. MASSIES 129–132

Photoluminescence Excitation Spectroscopy of $In_xGa_{1-x}N$/GaN Mulitple Quantum Wells with Various In Compositions
C. SASAKI, M. IWATA, Y. YAMADA, T. TAGUCHI, S. WATANABE, M.S. MINSKY,
T. TAKEUCHI, and N. YAMADA 133–136

Temperature Dependent Optical Properties of InGaN/GaN Quantum Well Structures
P. HURST, P. DAWSON, S.A. LEVETAS, M.J. GODFREY, I.M. WATSON, and G. DUGGAN   137–140

Energy Diagram and Recombination Mechanisms in InGaN/AlGaN/GaN Heterostructures with Quantum Wells
A.E. YUNOVICH and V.E. KUDRYASHOV 141–145

InGaN quantum wells: inhomogeneities

Relation between Structural Parameters and the Effective Electron–Hole Separation in InGaN/GaN Quantum Wells
N.A. SHAPIRO, H. FEICK, N.F. GARDNER, W.K. GÖTZ, P. WALTEREIT, J.S. SPECK,
and E.R. WEBER 147–151

Spatial Inhomogeneity of Photoluminescence in InGaN Single Quantum Well Structures
A. KANETA, G. MARUTSUKI, K. OKAMOTO, Y. KAWAKAMI, Y. NAKAGAWA,
G. SHINOMIYA, T. MUKAI, and SG. FUJITA 153–156

Optical Characterization of InGaN/GaN MQW Structures without In Phase Separation
B. MONEMAR, P.P. PASKOV, G. POZINA, T. PASKOVA, J.P. BERGMAN, M. IWAYA,
S. NITTA, H. AMANO, and I. AKASAKI 157–160

Phase Separation in InGaN Epitaxial Layers
A.N. WESTMEYER and S. MAHAJAN 161–164

Structural and Optical Characteristics of InGaN/GaN Multiple Quantum Wells with Different Growth Interruption
H.K. CHO, J.Y. LEE, N. SHARMA, J. HUMPHREYS, G.M. YANG, and C.S. KIM 165–168

Characterisation of Optical Properties in Micro-Patterned InGaN Quantum Wells
K.-S. KIM, P.R. EDWARDS, H.S. KIM, R.W. MARTIN, I.M. WATSON, and M.D. DAWSON   169–172

Indium Distribution within $In_xGa_{1-x}N$ Epitaxial Layers: A Combined Resonant Raman Scattering and Rutherford Backscattering Study
R. CORREIA, S. PEREIRA, E. PEREIRA, E. ALVES, J. GLEIZE, J. FRANDON, and M.A. RENUCCI   173–177

Cathodoluminescence Investigations of Interfaces in InGaN/GaN/Sapphire Structures
M. GODLEWSKI, E.M. GOLDYS, K.S.A. BUTCHER, M.R. PHILLIPS, K. PAKULA, and J.M. BARANOWSKI   179–182

Multi-Emission from InGaN/GaN Multi-Quantum Wells Grown on Hexagonal GaN Microstructures
CHI SUN KIM, YOUNG KUE HONG, CHANG-HEE HONG, EUN-KYUNG SUH, HYUNG JAE LEE, MIN HONG KIM, HYUNG KOUN CHO, and JEONG YONG LEE   183–186

Quantum dots

Uniform Array of GaN Quantum Dots in AlGaN Matrix by Selective MOCVD Growth
K. TACHIBANA, T. SOMEYA, S. ISHIDA, and Y. ARAKAWA   187–190

Self-Assembled Growth of GaN Quantum Dots Using Low-Pressure MOCVD
M. MIYAMURA, K. TACHIBANA, T. SOMEYA, and Y. ARAKAWA   191–194

On Phonon Confinement Effects and Free Carrier Concentration in GaN Quantum Dots
M. KUBALL, J. GLEIZE, S. TANAKA, and Y. AOYAGI   195–198

Temperature Dependent Photoluminescence of MBE Grown Gallium Nitride Quantum Dots
J. BROWN, C. ELSASS, C. POBLENZ, P.M. PETROFF, and J.S. SPECK   199–202

Isoelectronic doping

The Transition from Blue Emission in As-Doped GaN to GaNAs Alloys in Layers Grown by Molecular Beam Epitaxy
C.T. FOXON, S.V. NOVIKOV, Y. LIAO, A.J. WINSER, I. HARRISON, T. LI, R.P. CAMPION, C.R. STADDON, and C.S. DAVIS   203–206

Spatially Resolved Cathololuminescence Study of As Doped GaN
A. BELL, F.A. PONCE, S.V. NOVIKOV, C.T. FOXON, and I. HARRISON   207–211

On the Origin of Blue Emission from As-Doped GaN
I. HARRISON, S.V. NOVIKOV, T. LI, R.P. CAMPION, C.R. STADDON, C.S. DAVIS, Y. LIAO, A.J. WINSER, and C.T. FOXON   213–217

The Influence of As on the Optimum Nitrogen to Gallium Ratio Required to Grow High Quality GaN Films by Molecular Beam Epitaxy
C.T. FOXON, S.V. NOVIKOV, R.P. CAMPION, Y. LIAO, A.J. WINSER, and I. HARRISON   219–222

Temperature Dependence of the Miscibility Gap on the GaN-Rich Side of the Ga–N–As System
S.V. NOVIKOV, T. LI, A.J. WINSER, R.P. CAMPION, C.R. STADDON, C.S. DAVIS, I. HARRISON, and C.T. FOXON   223–225

Contents

The Influence of Arsenic Incorporation on the Optical Properties of As-Doped GaN Films Grown by Molecular Beam Epitaxy Using Arsen
S.V. Novikov, T. Li, A.J. Winser, C.T. Foxon, R.P. Campion, C.R. Staddon, C.S. Davis, I. Harrison, A.P. Kovarsky, and B.Ja. Ber 227–229

Effect of Isoelectronic In-Doping on Deep Levels in GaN Grown by MOCVD
H.K. Cho, C.S. Kim, Y.K. Hong, Y.-W. Kim, C.-H. Hong, E.-K. Suh, and H.J. Lee 231–234

Structural Properties of GaN Grown by Pendeo-Epitaxy with In-Doping
Young Kue Hong, Chi Sun Kim, Hung Sub Jung, Chang-Hee Hong, Min Hong Kim, Shi-Jong Leem, Hyung Koun Cho, and Jeong Yong Lee 235–238

Non-Monotonous Behaviour of In-Doped GaN Grown by MOVPE with Nitrogen Carrier Gas
A. Yamamoto, T. Tanikawa, K. Ikuta, M. Adachi, A. Hashimoto, and Y. Ito . . 239–242

Dilute nitride alloys

A Perspective of $GaAs_{1-x}N_x$ and GaP_xN_{1-x} as Heavily Doped Semiconductors (*invited*)
A. Mascarenhas, Yong Zhang, and M.J. Seong 243–252

Evolution of Electron States with Composition in GaAsN Alloys
P.R.C. Kent and A. Zunger 253–257

Phonon Modes of $In_xGa_{1-x}As_{1-y}N_y$ Measured by Far Infrared Spectroscopic Ellipsometry
G. Leibiger, V. Gottschalch, and M. Schubert 259–262

MOCVD Growth of InN_xAs_{1-x} on GaAs Using Dimethylhydrazine
A.A. El-Emawy, H.-J. Cao, E. Zhmayev, J.-H. Lee, D. Zubia, and M. Osiński . . 263–267

Spectroscopic Ellipsometry Study on the Electronic Structure near the Absorption Edge of GaAsN Alloys
H. Yaguchi, S. Matsumoto, Y. Hijikata, S. Yoshida, T. Maeda, M. Ogura, D. Aoki, and K. Onabe 269–272

Photoluminescence Study on Temperature Dependence of Band Gap Energy of GaAsN Alloys
H. Yaguchi, S. Kikuchi, Y. Hijikata, S. Yoshida, D. Aoki, and K. Onabe 273–277

Phonon Modes and Critical Points of GaPN
G. Leibiger, V. Gottschalch, R. Schwabe, G. Benndorf, and M. Schubert . . . 279–282

Raman Characterization of MBE Grown (Al)GaAsN
A. Hashimoto, T. Kitano, K. Takahashi, H. Kawanishi, A. Patane, C.T. Foxon, and A. Yamamoto 283–286

Electronic Structure of Heavily and Randomly Nitrogen Doped GaAs Near the Fundamental Band Gap
Yong Zhang, S. Francoeur, A. Mascarenhas, H.P. Xin, and C.W. Tu 287–291

Point defects and impurities in GaN

Defect-Related Donors, Acceptors, and Traps in GaN (*invited*)
D.C. Look 293–302

Passivation and Doping due to Hydrogen in III-Nitrides
S. Limpijumnong and C.G. Van de Walle 303–307

Capture Kinetics of Electron Traps in MBE-Grown n-GaN
A. Hierro, A.R. Arehart, B. Heying, M. Hansen, J.S. Speck, U.K. Mishra,
S.P. DenBaars, and S.A. Ringel 309–313

Reduction of Defects in GaN on Reactive Ion Beam Treated Sapphire by Annealing
D. Byun, J. Shin, S. Cho, J. Kim, S.J. Lee, C.H. Hong, G. Kim, and W.-K. Choi . . 315–318

Comparative Study on the Optical Properties of Eu:GaN with Tb:GaN
H. Bang, S. Morishima, Z. Li, K. Akimoto, M. Nomura, and E. Yagi 319–323

Implantation Induced Defect States in Gallium Nitride and Their Annealing Behaviour
A. Krtschil, A. Kielburg, H. Witte, A. Krost, J. Christen, A. Wenzel,
and B. Rauschenbach 325–329

Annealing Behaviour of GaN after Implantation with Hafnium and Indium
K. Lorenz, F. Ruske, and R. Vianden 331–335

Magnetic Properties of Mn- and Fe-Implanted p-GaN
N. Theodoropoulou, M.E. Overberg, S.N.G. Chu, A.F. Hebard,
C.R. Abernathy, R.G. Wilson, J.M. Zavada, K.P. Lee, and S.J. Pearton 337–340

Outgoing Multiphonon Resonant Raman Scattering in Be- and C-Implanted GaN
W.H. Sun, S.J. Chua, L.S. Wang, X.H. Zhang, and M.S. Hao 341–344

Doping of GaN with magnesium

Influence of Dopants on Defect Formation in GaN (*invited*)
Z. Liliental-Weber, J. Jasinski, M. Benamara, I. Grzegory, S. Porowski,
D.J.H. Lampert, C.J. Eiting, and R.D. Dupuis 345–352

Observation of Mg-Rich Precipitates in the p-Type Doping of GaN-Based Laser Diodes
M. Hansen, L.F. Chen, J.S. Speck, and S.P. DenBaars 353–356

Activation of p-Type GaN with Irradiation of the Second Harmonics of a Q-Switched Nd:YAG Laser
Yung-Chen Cheng, Chi-Chih Liao, Shih-Wei Feng, Chih-Chung Yang,
Yen-Sheng Lin, Kung-Jeng Ma, and Jen-Inn Chyi 357–360

Time- and Temperature-Resolved Photoluminescence of GaN:Mg Epitaxial Layers Grown by MOVPE
A.I. Gurskii, I.P. Marko, E.V. Lutsenko, V.N. Pavlovskii, V.Z. Zubialevich,
G.P. Yablonskii, B. Schineller, O. Schön, and M. Heuken 361–364

Spatial Fluctuations and Localisation Effects in Optical Characteristics of p-Doped GaN Films
E.M. Goldys, M. Godlewski, E. Kaminska, A. Piotrowska, and K.S.A. Butcher 365–369

Characterization of Mg-Doped GaN Micro-Crystals Grown by Direct Reaction of Gallium and Ammonia
S.H. Lee, K.S. Nahm, E.K. Suh, and M.H. Hong 371–373

Activation of Mg Acceptor in GaN:Mg with Pulsed KrF (248 nm) Excimer Laser Irradiation
Dong-Joon Kim, Hyun-Min Kim, Myung-Geun Han, Yong-Tae Moon,
Seonghoon Lee, and Seong-Ju Park 375–378

Contents

Analysis of Time-Resolved Donor–Acceptor-Pair Recombination in MBE and MOVPE Grown GaN:Mg
S. STRAUF, S.M. ULRICH, P. MICHLER, J. GUTOWSKI, T. BÖTTCHER, S. FIGGE, S. EINFELDT, and D. HOMMEL 379–383

Investigation of Defect Levels in Mg-Doped GaN Schottky Structures by Thermal Admittance Spectroscopy
N.D. NGUYEN, M. GERMAIN, M. SCHMEITS, B. SCHINELLER, and M. HEUKEN 385–389

Low-Temperature Activation of Mg-Doped GaN with Pd Thin Films
I. WAKI, H. FUJIOKA, M. OSHIMA, H. MIKI, and M. OKUYAMA 391–393

Role of extended defects

Threading Dislocations and Optical Properties of GaN and GaInN (*invited*)
T. MIYAJIMA, T. HINO, S. TOMIYA, K. YANASHIMA, H. NAKAJIMA, T. ARAKI, Y. NANISHI, A. SATAKE, Y. MASUMOTO, K. AKIMOTO, T. KOBAYASHI, and M. IKEDA 395–402

Mosaicity of GaN Epitaxial Layers: Simulation and Experiment
R. CHIERCHIA, T. BÖTTCHER, S. FIGGE, M. DIESSELBERG, H. HEINKE, and D. HOMMEL 403–406

Correlation of Defects and Local Bandgap Variations in GaInN/GaN/AlGaN LEDs
F. HITZEL, A. HANGLEITER, S. BADER, H.-J. LUGAUER, and V. HÄRLE 407–410

Energetic Calculation of Coincidence Grain Boundaries with a Modified Stillinger-Weber Potential
J. CHEN, G. NOUET, and P. RUTERANA 411–414

Structure Characterization of (Al,Ga)N Epitaxial Layers by Means of X-Ray Diffractometry
J. KOZŁOWSKI, R. PASZKIEWICZ, and M. TŁACZALA 415–418

Optical behavior of nitrides

Optical Micro-Characterization of Complex GaN Structures (*invited*)
J. CHRISTEN and T. RIEMANN 419–424

Optical Properties of III-Nitride Ternary Compounds
A. BALDANZI, E. BELLOTTI, and M. GOANO 425–428

Investigation of Refractive Index and Optical Propagation Loss in Gallium Nitride Based Waveguides
E. DOGHECHE, P. RUTERANA, G. NOUET, F. OMNES, and P. GIBART 429–432

Extremly Slow Relaxation Process of a Yellow-Luminescence-Related State in GaN Revealed by Two-Wavelength Excited Photoluminescence
J.M. ZANARDI OCAMPO, N. KAMATA, W. OKAMOTO, K. YAMADA, K. HOSHINO, T. SOMEYA, and Y. ARAKAWA 433–436

Infrared Ellipsometry – a Novel Tool for Characterization of Group-III Nitride Heterostructures for Optoelectronic Device Applications
M. SCHUBERT, A. KASIC, S. EINFELDT, D. HOMMEL, U. KÖHLER, D.J. AS, J. OFF, B. KUHN, F. SCHOLZ, and J.A. WOOLLAM 437–440

Electronic Defect States Observed by Cathodoluminescence Spectroscopy at GaN/Sapphire Interfaces
X.L. SUN, S.H. GOSS, L.J. BRILLSON, D.C. LOOK, and R.J. MOLNAR 441–444

Scanning Tunneling Luminescence Studies of Nitride Semiconductor Thin Films under Ambient Conditions
    S.K. MANSON-SMITH, C. TRAGER-COWAN, and K.P. O'DONNELL   445–448

Raman Scattering and Photoluminescence of Mg-Implanted GaN Films
    LIANSHAN WANG, SOO JIN CHUA, and WENHONG SUN   449–452

Effects of Indium Segregation and Well-Width Fluctuations on Optical Properties of InGaN/GaN Quantum Wells
    A. SOLTANI VALA, M.J. GODFREY, and P. DAWSON   453–456

First-Principles Calculations of Optical Properties of AlN, GaN, and InN Compounds under Hydrostatic Pressure
    B. ABBAR, B. BOUHAFS, H. AOURAG, G. NOUET, and P. RUTERANA   457–460

Near K-Edge Absorption Spectra of III–V Nitrides
    K. FUKUI, R. HIRAI, A. YAMAMOTO, H. HIRAYAMA, Y. AOYAGI, S. YAMAGUCHI,
    H. AMANO, I. AKASAKI, and T. TANAKA   461–465

Excitons

Internal Structure of Free Excitons in GaN
    P.P. PASKOV, T. PASKOVA, P.O. HOLTZ, and B. MONEMAR   467–470

Experimental and Theoretical Tools for the Study of Exciton Properties versus Disorder in Nitride-Based Quantum Structures
    B. GIL, M. ZAMFIRESCU, P. BIGENWALD, G. MALPUECH, and A. KAVOKIN   471–474

Comparison of Exciton–Biexiton with Bound Exciton–Biexciton Dynamics in GaN: Quantum Beats and Temperature Dependence of the Acoustic-Phonon Interaction
    K. KYHM, R.A. TAYLOR, J.F. RYAN, T. AOKI, M. KUWATA-GONOKAMI,
    B. BEAUMONT, and P. GIBART   475–479

Micro-Photoluminescence Spectroscopy of Exciton–Polaritons in GaN with the Wave Vector **k** Normal to the *c*-Axis
    T.V. SHUBINA, T. PASKOVA, A.A. TOROPOV, A.V. LEBEDEV, S.V. IVANOV,
    and B. MONEMAR   481–484

Radiative and Nonradiative Exciton Lifetimes in GaN Grown by Molecular Beam Epitaxy
    G. POZINA, J.P. BERGMAN, B. MONEMAR, B. HEYING, and J.S. SPECK   485–488

The 3.466 eV Bound Exciton in GaN
    B. MONEMAR, W.M. CHEN, P.P. PASKOV, T. PASKOVA, G. POZINA,
    and J.P. BERGMAN   489–492

Exciton Diffusion in GaN Epitaxial Layers
    YU. RAKOVICH, J.F. DONEGAN, A. GLADYSHCHUK, G. YABLONSKII, B. SCHINELLER,
    and M. HEUKEN   493–496

Excitonic Transitions in Homoepitaxial GaN
    G. MARTÍNEZ-CRIADO, C.R. MISKYS, A. CROS, A. CANTARERO, O. AMBACHER,
    and M. STUTZMANN   497–500

Donor and Donor Bound Exciton Spectroscopy in Wurtzite GaN Heterostructures
    M. TEISSEIRE, G. NEU, and C. MORHAIN   501–504

Polarity of hexagonal nitrides

Playing with Polarity (*invited*)
M. STUTZMANN, O. AMBACHER, M. EICKHOFF, U. KARRER, A. LIMA PIMENTA, R. NEUBERGER, J. SCHALWIG, R. DIMITROV, P.J. SCHUCK, and R.D. GROBER 505–512

Investigation of Defects and Polarity in GaN Using Hot Wet Etching, Atomic Force and Transmission Electron Microscopy and Convergent Beam Electron Diffraction
P. VISCONTI, D. HUANG, M.A. RESHCHIKOV, F. YUN, T. KING, A.A. BASKI, R. CINGOLANI, C.W. LITTON, J. JASINSKI, Z. LILIENTAL-WEBER, and H. MORKOÇ .. 513–517

Wetting Behaviour of GaN Surfaces with Ga- or N-Face Polarity
M. EICKHOFF, R. NEUBERGER, G. STEINHOFF, O. AMBACHER, G. MÜLLER, and M. STUTZMANN 519–522

Kinetic Process of Polarity Selection in GaN Growth by RF-MBE
K. XU, N. YANO, A.W. JIA, A. YOSHIKAWA, and K. TAKAHASHI 523–527

V/III Ratio Dependence of Polarity of GaN Grown on GaAs (111)A-Ga and (111)B-As Surfaces by MOMBE
O. TAKAHASHI, T. NAKAYAMA, R. SOUDA, and F. HASEGAWA 529–532

Electron Backscattered Diffraction Patterns from Cooled Gallium Nitride Thin Films
E. SWEENEY, C. TRAGER-COWAN, J. HASTIE, D.A. COWAN, K.P. O'DONNELL, D. ZUBIA, S.D. HERSEE, C.T. FOXON, I. HARRISON, and S.V. NOVIKOV 533–536

Influence of Polarity on Surface Reaction between GaN{0001} and Hydrogen
M. MAYUMI, F. SATOH, Y. KUMAGAI, K. TAKEMOTO, and A. KOUKITU 537–541

A Comparative Study of MBE-Grown GaN Films Having Predominantly Ga- or N-Polarity
F. YUN, D. HUANG, M.A. RESHCHIKOV, T. KING, A.A. BASKI, C.W. LITTON, J. JASINSKI, Z. LILIENTAL-WEBER, P. VISCONTI, and H. MORKOÇ 543–547

Polarity Inversion by Supplying Group-III Source First in MOMBE of GaN/AlN or GaN on GaAs (111)B (As Surface)
F. HASEGAWA, O. TAKAHASHI, T. NAKAYAMA, and R. SOUDA 549–552

Polarization effects

Charge Screening of Polarization Fields in Nitride Nanostructures (*invited*)
A. DI CARLO and A. REALE 553–558

Polarization Effects and UV Emission in Highly Excited Quaternary AlInGaN Quantum Wells
E. KUOKSTIS, JIANPING ZHANG, J.W. YANG, G. SIMIN, M. ASIF KHAN, R. GASKA, and M. SHUR 559–562

Photoluminescence Study of Piezoelectric Polarization in Strained $Al_xGa_{1-x}N$/GaN Single Quantum Wells
V. KIRILYUK, P.R. HAGEMAN, P.C.M. CHRISTIANEN, F.D. TICHELAAR, and P.K. LARSEN. 563–566

Transport properties

GW Self-Energy Correction to the Band Mass of Nitride Semiconductors
M. OSHIKIRI and F. ARYASETIAWAN 567–570

Group-III Nitrides Hot Electron Effects in Moderate Electric Fields
E.A. Barry, K.W. Kim, and V.A. Kochelap 571–574

Temperature Dependent Transport Parameters in Short GaN Structures
A.F.M. Anwar, Shangli Wu, and R.T. Webster 575–578

Simultaneous Impurity-Band and Interface Conduction in Depth-Profiled n-GaN Epilayers
C. Mavroidis, J.J. Harris, K. Lee, I. Harrison, B.J. Ansell, Z. Bougrioua, and I. Moerman 579–583

Band Structure Effects on the Transient Electron Velocity Overshoot in GaN
M. Wraback, H. Shen, E. Bellotti, J.C. Carrano, C.J. Collins, J.C. Campbell, R.D. Dupuis, M.J. Schurmann, and I.T. Ferguson 585–588

Photoconductivity in Porous GaN Layers
M. Mynbaeva, N. Bazhenov, K. Mynbaev, V. Evstropov, S.E. Saddow, Y. Koshka, and Y. Melnik 589–592

High-Field Electron Transport in Nanoscale Group-III Nitride Devices
S.M. Komirenko, K.W. Kim, V.A. Kochelap, and M.A. Stroscio 593–597

Two-dimensional electron gas properties

Electrical Characterization at Cubic AlN/GaN Heterointerface Grown by Radio-Frequency Plasma-Assisted Molecular Beam Epitaxy
T. Kitamura, Y. Ishida, X.Q. Shen, H. Nakanishi, S.F. Chichibu, M. Shimizu, and H. Okumura 599–602

Transport Properties of Two-Dimensional Electron Gases Induced by Spontaneous and Piezoelectric Polarisation in AlGaN/GaN Heterostructures
A. Link, T. Graf, R. Dimitrov, O. Ambacher, M. Stutzmann, Y. Smorchkova, U. Mishra, and J. Speck 603–606

Energy Relaxation by Warm Two-Dimensional Electrons in a GaN/AlGaN Heterostructure
N.M. Stanton, A.J. Kent, S.A. Cavill, A.V. Akimov, K.J. Lee, J.J. Harris, T. Wang, and S. Sakai 607–611

2DEG Characteristics of AlN/GaN Heterointerface on Sapphire Substrates Grown by Plasma-Assisted MBE
K. Jeganathan, T. Ide, S.X.Q. Shen, M. Shimizu, and H. Okumura 613–616

Electron Transport in III–V Nitride Two-Dimensional Electron Gases
D. Jena, I. Smorchkova, A.C. Gossard, and U.K. Mishra 617–619

Investigation for the Formation of Polarization-Induced Two-Dimensional Electron Gas in AlGaN/GaN Heterostructure Field Effect Transistors
H.W. Jang, C.M. Jeon, K.H. Kim, J.K. Kim, S.-B. Bae, J.-H. Lee, J.W. Choi, and J.-L. Lee 621–624

2DEG Mobility in AlGaN–GaN Structures Grown by LP-MOVPE
Z. Bougrioua, J.-L. Farvacque, I. Moerman, and F. Carosella 625–628

Author Index 629–640

Your selection in solid state physics

KOSSEVICH, A.M.
The Crystal Lattice
Phonons, Solitons, Dislocations
1999. 326 pages. Hardcover.
€129.-*/DM252.30/£75.-**
ISBN 3-527-40220-9

MARDER, M.P.
Condensed Matter Physics
2000. XXVI. 896 pages. Hardcover.
€116.57*/DM228.-/£67.95**
ISBN 0-471-17779-2

KOVALENKO, N.P./ KRASNY, Y.P./ KREY, U.
Physics of Amorphous Metals
2001. 296 pages. Hardcover.
€119.-*/DM232.74/£70.-**
ISBN 3-527-40315-9

Physica Status Solidi (a)
2001.
Volume 183-188.
15 issues per year
ISSN 0031-8965

NEW LAYOUT!

SERNELIUS, B.E.
Surface Modes in Physics
2001. 370 pages. Hardcover.
€109.-*/DM213.19/£65.-**
ISBN 3-527-40313-2

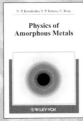

Physica Status Solidi (b)
2001.
Volume 223-228.
15 issues per year
ISSN 0370-1972

NEW LAYOUT!

The discovery of bulk metallic glasses has led to a large increase in the industrial importance of amorphous metals and this is expected to continue. This book is the first to describe the theoretical physics of amorphous metals in a very homogeneous and self-consistent way. It covers the important theoretical development of the last 20 years. The authors are outstanding physicists renowned for their work in the field of disordered systems.
While both theorists and experimentalists interested in amorphous metals will profit from this book, it will also be useful for supplementary reading in courses on solid-state physics and materials sciences.

Crystal Research and Technology
2001. Volume 36.
12 issues per year
ISSN 0232-1300

Das ganze Spektrum der Physik

www.pro-physik.de

Electromagnetic surface modes are present at all surfaces and interfaces between materials of different dielectric properties. These modes have very important effects on numerous physical quantities: adhesion, capillary force, step formation and crystal growth, the Casimir effect etc. They cause surface tension and wetting and they give rise to forces which are important e.g. for the stability of colloids.
This book is a useful and elegant treatment of the fundamentals of surface energy and surface interactions. The concept of electromagnetic modes is developed as a unifying theme for a range of condensed matter physics, both for surfaces and in the bulk. In close relation to the theoretical background, the reader is served with a broad field of applications.

O'HANDLEY, R.C.
Modern Magnetic Materials
Principles and Applications
2000. XXVIII. 740 pages. Hardcover.
€152.36*/DM298.-/£89.50**
ISBN 0-471-15566-7

This book is an essential reference for physicists, electrical engineers, materials scientists, chemists, and metallurgists, and others who work on magnetic materials and magnet design. Focusing on materials rather than the physics of magnetism, it provides a modern, practical treatment of materials that can hold a magnetic field. Cutting-edge topics include nanocrystalline materials, amorphous magnetism, charge and spin transport, surface and thin film magnetism and magnetic recording.

Visit
www.wiley-vch.de

WILEY

John Wiley & Sons Ltd · Baffins Lane · Chichester
West Sussex · PO19 1UD · United Kingdom
Phone +44 (0) 1243 843294 · Fax +44 (0) 1243 843296
email: cs-books@wiley.co.uk, www.wiley.co.uk

Wiley-VCH · P.O. Box 10 11 61 · 69451 Weinheim · Germany
Phone +49 (0) 6201 60 6152 · Fax +49 (0) 6201 60 6184
email: service@wiley-vch.de, www.wiley-vch.de

WILEY-VCH

DEÁK, P./ FRAUENHEIM, T./ PEDERSON, M.R. (eds.)
INCL. CD-ROM
Computer Simulation of Materials at Atomic Level
2000. II. 728 pages. Hardcover.
€179.-*/DM350.09/£105.-**
ISBN 3-527-40290-X

physica status solidi (a)
applied research
2001,
volumes 183-188
15 issues per year
ISSN 0031-8965

physica status solidi (b)
basic research
2001,
volumes 223-228
15 issues per year
ISSN 0370-1972

Editor-in-Chief: Martin Stutzmann,
Technical University Munich

 More speed

NEW: 2 x 15 issues per year

Now published in 15 issues
(6 volumes) in each series (a) and (b).

- physica status solidi is a high-quality, carefully **peer-reviewed** scientific journal
- aiming at a **rapid publication** of important and new results in all areas of modern solid state physics
- coverage is focused on recent trends and materials like semiconductors, superconductors, magnetics, organic substances, nanostructures, computational modelling and others
- both aspects of basic and applied research are covered in the two series **physica status solidi (a)** and **physica status solidi (b)**

Short publication times

Average time between receipt of the final manuscript and print publica-tion was just **73 days** for original papers in 2000. Now, with EarlyView® articles appear electronically up to several weeks ahead of the print edition.

Rapid Research Notes

The latest important results **peer-reviewed** and published on the internet within only two weeks following submission.

See our special 40th anniversary issue with 40 highlight papers from four decades at our homepage
www.physica-status-solidi.com

40 years of physica status solidi
get 40 days of free access at
www.interscience.wiley.com/trial/pss40

More information

NEW: Editor's Choice and Feature Articles

Highlighting particularly relevant or interesting articles with **free publication** of full color figures, prominent positioning, **free online access** and reprints **free of charge.**

New: ContentAlert

Registered users of Wiley InterScience can now receive **tables of contents via e-mail** as soon as a new issue is fully online. This service is offered for all journals on Wiley InterScience, to all registered users, regardless of whether a subscription to a journal exists.

Das ganze Spektrum der Physik
www.pro-physik.de

The place to visit for **InterScience**
Wiley journals online www.interscience.wiley.com

NEW: The innovative reference linking idea allowing a quick jump from journal references to the cited articles

Wiley-VCH, Journals Department,
P.O. Box 10 11 61, 69451 Weinheim,
Germany
Fax: +49 (0) 62 01- 606 172
e-mail: subservice@wiley-vch.de
http://www.wiley-vch.de

WILEY-VCH

MOCVD Growth of High-Quality InN Films and Raman Characterization of Residual Stress Effects

E. Kurimoto[1]) (a), H. Harima (a), A. Hashimoto (b), and A. Yamamoto (b)

(a) Department of Applied Physics, Osaka University, Yamada-oka, Suita, Osaka 565-0871, Japan

(b) Department of Electronics Engineering, Fukui University, Bunkyo, Fukui 910-8507, Japan

(Received July 10, 2001; accepted July 17, 2001)

Subject classification: 63.20.Dj; 71.55.Eq; 78.30.Fs; 81.15.Gh; S7.14

Thick InN films were grown by metal-organic chemical vapor deposition on sapphire c-plane at different growth temperatures, and studied by Raman scattering. Sharp phonon peaks were observed in all samples, showing good crystalline quality. The E_2-phonon modes, which are sensitive to the strain in the c-plane, showed a systematic variation in frequency with the growth temperature. This variation was attributed to thermal strains induced by the difference in thermal expansion coefficient between the InN layer and the sapphire substrate. Frequency of the high-frequency E_2-phonon mode ν (in cm^{-1}) varied with strain ε (in %) as $\nu = 50\varepsilon + 481.5$. Residual strains in InN layers could be evaluated using this empirical formula.

Introduction InN has received much attention for optoelectronic devices to cover continuously the near-infrared to UV region by alloying with other III–V nitrides, GaN and AlN. However, because of difficulty in growing high-quality crystals, basic properties of pure InN are not well investigated when compared to other nitrides. Quite recently, good films have been prepared by molecular beam epitaxy (MBE), and precise phonon frequencies have been extracted using these films [1]. However, for actual device fabrication, a more efficient growth method should also be established. In this work, thick InN layers were grown by metal-organic chemical vapor deposition (MOCVD) and characterized by Raman scattering. It is shown that the crystalline quality is as good as previous MBE films judged from the sharp phonon peaks. The phonon peaks show systematic frequency shifts corresponding to the growth temperature. An empirical linear relationship has been deduced between the phonon frequency and the residual strain.

Experiments Hexagonal InN films were grown by MOCVD on sapphire (0001) substrate at different temperatures in the range of 500–650 °C. Trimethylindium (TMI) and ammonia (NH$_3$) were used as source materials and N$_2$ as the carrier gas. Details of the growth method are given elsewhere [2–4]. Three specimens were selected for this work; Sample A grown at 650 °C with InN layer thickness 0.8 μm, sample B at 600 °C and 0.4 μm, and sample C at 500 °C and 0.2 μm. An X-ray diffraction analysis showed that high-quality, (0001)-oriented films were grown with a good homogeneity. Microscopic Raman scattering was performed at room temperature in the backscattering geometry

---

[1]) Corresponding author; Phone: +81 6 6879 7854, Fax: +81 6 6879 7856, e-mail: eiji@ap.eng.osaka-u.ac.jp

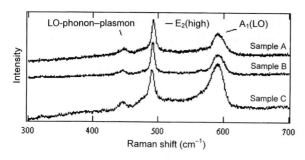

Fig. 1. Observed Raman spectra of InN

using the 514.5 nm Ar$^+$ laser line for excitation. No polarization detection was made. The scattered light was dispersed by a double monochromator with 85 cm focal length and detected by a liquid nitrogen cooled CCD camera. The instrumental width determined by the slit width and the CCD pixel size was about 1.7 cm^{-1}.

Results and Discussion Figures 1 and 2 show the observed spectra from the InN films in the frequency range of 300–700 and 80–100 cm^{-1}, respectively. In these Raman spectra the peaks are assigned as follows: 88 cm^{-1} to E_2 (low frequency) phonon mode, ≈490 cm^{-1} to E_2 (high frequency) phonon mode, ≈590 cm^{-1} to A_1(LO) phonon mode, and ≈450 cm^{-1} to LO-phonon–plasmon coupled mode (lower branch). It is found that the E_2 phonon peaks are both sharp (below 2 cm^{-1} in FWHM for the 88 cm^{-1} mode, and below 5 cm^{-1} for the ~490 cm^{-1} mode). This means that the present MOCVD films have good quality as previous MBE films; Davydov et al. reported, e.g., 6.2 cm^{-1} for the ~490 cm^{-1} mode [1].

Figure 2 shows that the low-frequency E_2 mode shifts to higher frequency in the order of A < B < C. The frequency difference between samples A and C is 0.8 cm^{-1}. Quite interestingly, opposite frequency shift is observed for the high-frequency E_2 mode: As shown in Fig. 3, the phonon peak shifts to higher frequency in the order of C < B < A. The frequency difference between samples A and C is 3 cm^{-1}.

The E_2 modes correspond to atomic oscillations in the c-plane. Therefore, the mode frequency, especially the high-frequency mode, is sensitive to lattice strains in the c-plane as confirmed in hexagonal GaN [5]. Since this mode shifts to higher frequency

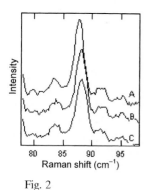

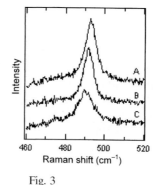

Fig. 2

Fig. 3

Fig. 2. Expansion figure of low-frequency E_2 mode

Fig. 3. Expansion figure of high-frequency E_2 mode

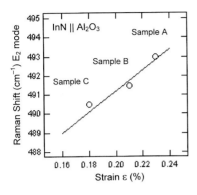

Fig. 4. E_2 peak frequency plotted against residual strain

with the rise of growth temperature, it is suggested that thermal strains induced by the difference in expansion coefficient between the substrate and InN layer are responsible for the shifts. Strains induced by the lattice mismatch can be discarded here, because of the layer thickness much larger than the critical thickness for coherent growth. The unique behavior of the low-frequency E_2 mode, i.e., negative shift with strain, has already been reported in GaN for hydrostatic pressure [6]. In the following, we will analyze only the high-frequency mode which shows clearer shifts.

Residual thermal strains ε (%) in InN layers are written as

$$\varepsilon = (T_g - T_r)(\alpha_{\text{sapphire}} - \alpha_{\text{InN}}) \times 100, \tag{1}$$

where T_g, T_r, α_{sapphire} and α_{InN} are the growth temperature, room temperature, thermal expansion coefficient of sapphire substrate and that of InN, respectively. Putting $\alpha_{\text{sapphire}} = 7.5 \times 10^{-6}$ (K^{-1}) and $\alpha_{\text{InN}} \approx 4 \times 10^{-6}$ (K^{-1}) [7], we obtained $\varepsilon = 0.23$, 0.21, 0.18% for sample A, B, and C, respectively.

Figure 4 plots the E_2 peak frequency against the residual thermal strain ε. The observed frequency shows a linear variation with a slope of approximately 50 cm^{-1} per unit strain (%); $\nu = 50\varepsilon + 481.5$ (cm^{-1}). Let us compare our result with those of MBE films: Davydov et al. [1] observed the high-frequency E_2 mode at 488 cm^{-1}, slightly lower than ours. Since Davydov et al. treated relatively thick films, thermal strains are the probable cause of the frequency shift as our case. If we refer to Fig. 4 and use Eq. (1), a growth temperature of $T_g \approx 400$ °C is deduced. This would be a standard growth temperature for MBE method, which uses lower values compared with CVD. Then, it is found that the frequency as 488 cm^{-1} in MBE-grown InN film is a reasonable value and the relationship that we present is a common one.

Conclusion In this work, thick InN films were grown by MOCVD on sapphire c-plane at different growth temperatures, and studied by Raman scattering. From observed Raman spectra, it was found that the E_2 phonon peaks were both sharp, suggesting that the present MOCVD films had as good quality as previous MBE films, and the peak frequency showed a systematic variation with the growth temperature. This variation in frequency was induced by the thermal strain in InN films, and we showed a relationship between E_2 peak frequency and residual strain as $\nu = 50\varepsilon + 481.5$. Using this relationship, residual strain in InN layers could be evaluated.

References

[1] V. Yu. Davydov, V. V. Emtsev, I. N. Goncharuk, A. N. Smirnov, V. D. Petrikov, V. V. Mamutin, V. A. Vekshin, S. V. Irannov, M. B. Smirnov, and T. Inushima, Appl. Phys. Lett. **75**, 3297 (1999).
[2] A. Yamamoto, T. Shin-ya, T. Sugiura, and A. Hashimoto, J. Cryst. Growth **189/190**, 461 (1998).

[3] M. ADACHI, Y. MURAKAMI, A. HASHIMOTO, and A. YAMAMOTO, Proc. Internat. Workshop on Nitride Semiconductors, IPAP Conf. Ser. **1**, 339 (2000).
[4] M. ADACHI, A. SEKI, A. HASHIMOTO, and A. YAMAMOTO, Proc. Internat. Workshop on Nitride Semiconductors, IPAP Conf. Ser. **1**, 347 (2000).
[5] F. DEMANGEOT, J. FRANDON, M. A. RENUCCI, O. BRIOT, B. GIL, and R. L. AULOMBARD, Solid State Commun. **100**, 207 (1996).
[6] P. PERLIN, C. J. CARILLON, J. P. ITIE, A. S. MIGUEL, I. GRZEGORY, and A. POLIAN, Phys. Rev. B **45**, 83 (1992).
[7] H. MORKOÇ, S. STRITE, G. B. GAO, M. E. LIN, B. SVERDLOV, and M. BURNS, J. Appl. Phys. **76**, 1363 (1994).

Growth Temperature Dependences of MOVPE InN on Sapphire Substrates

A. Yamamoto[1]), Y. Murakami, K. Koide, M. Adachi, and A. Hashimoto

Department of Electrical and Electronics Engineering, Fukui University, 3-9-1 Bunkyo, Fukui 910-8507, Japan

(Received June 23, 2001; accepted August 14, 2001)

Subject classification: 68.55.Jk; 73.61.Ey; 81.15.Aa; S7.14

This paper reports growth temperature dependences of MOVPE InN on sapphire substrates. Surface morphology of grown InN is markedly dependent on growth temperature. An InN film with a fibrous columnar structure with small grains is grown at temperatures lower than 550 °C. At 630–650 °C, a continuous film with enhanced two-dimensional growth is obtained. In the temperature range 500–600 °C, the growth rate of InN is found to be increased with increasing growth temperature and shows a saturation against the increase in TMI supply. At 630–650 °C, such a saturation is not found and a growth rate as high as 0.8 µm/h is obtained by increasing TMI supply. At a growth temperature lower than 600 °C, carrier concentration shows a marked NH_3/TMI molar ratio dependence, while at 630–650 °C it is independent on NH_3/TMI molar ratio. In order to obtain InN with a carrier concentration of $(3-5) \times 10^{19}$ cm^{-3}, an NH_3/TMI molar ratio as low as 9×10^3 is adequate at around 650 °C, while that of 1.8×10^5 is needed at 500 °C. This is due to the higher decomposition rate of NH_3 at 650 °C. Atmosphric pressure growth gives better electrical properties than reduced pressure (76 Torr) growth, suggesting that effective NH_3 decompostion rate at 760 Torr is higher than that at 76 Torr. The difference in carrier concentration between reduced-pressure grown and atmospheric-pressure grown InN films becomes small with increasing growth temperature.

Introduction Indium nitride (InN) has attracted much attention because of its excellent electron transport properties [1], as well as the constituent compounds of InGaN and InAlN. In order to realize novel devices of InN and its alloys, growth of high quality InN film is highly required. However, high quality InN films, comparable to GaN films, have not yet been obtained. Growth reaction for MOVPE-InN is thought to be restricted by a low decomposition rate of NH_3 [2], because of the relatively low (∼500 °C) growth temperature for InN. Although a higher growth temperature is expected to result in a higher decomposition rate of NH_3, it can also bring about thermal decomposition or thermal etching of grown InN. Therefore, growth temperature seems to be one of the most critical factors in MOVPE growth of InN using NH_3. Recently, a higher mobility has been repoerted for MOVPE-InN on GaN [3] and MBE-InN on α-Al_2O_3 [4]. However, major causes for such high mobilities have not yet been known.

In this paper, we report a systematic study on the growth temperature dependences of MOVPE-grown InN. Surface morphology, growth rate and electrical properties have been studied for InN films grown on α-Al_2O_3 substrates at a temperature between 500 and 650 °C, with parameters of TMI supply, NH_3/TMI molar ratio, and growth pressure.

---

[1]) Corresponding author; Phone: +81 776 27 8566; Fax: +81 776 27 8749;
e-mail: yamamoto@kyomu1.fuee.fukui-u.ac.jp

Experimental Epitaxial films of InN are grown on α-Al$_2$O$_3$(0001) substrate at 500–650 °C at a pressure of 76 or 760 Torr, using a MOVPE apparatus with a horizontal reactor. As reactant sources, TMI and NH$_3$ are used and N$_2$ as a carrier gas. NH$_3$/TMI (V/III) molar ratio is varied from 9×10^3 to 1.8×10^5. Prior to the InN growth, the substrate is heated at 950 °C for 10 min in H$_2$ and then nitrided in the flowing NH$_3$ at 900 °C for 30 min. The grown films are characterized with a scanning electron microscope (SEM), atomic force microscopy (AFM), and Hall measurements.

Results and Discussion Surface morphology of grown InN is markedly dependent on growth temperature. Figure 1[2]) shows AFM images for InN films grown on α-Al$_2$O$_3$(0001) substrate at a different temperatures. When the growth is done at a temperature less than 550 °C, a grown film has a surface with many small grains, as shown in Fig. 1a. This is a characteristic feature of InN films with a columnar fibrous structure. At 600–650 °C, a continuous film with enhanced two-dimensional growth is obtained as shown in Figs. 1b and c. For the film grown at 650 °C, many pits are formed on the enlarged grain surface. Such pits seem to be formed by thermal etching during the growth because of the high temperature. Thus, high temperature growth is very effective to enhance grain growth and/or two-dimensional growth of InN. It is found that growth rate of InN is also markedly dependent on growth temperature. Figure 2 shows growth temperature dependence of growth rate of InN for a different TMI supply. One can see that growth rate is increased with increasing growth temperature in the range 500–630 °C, while it is independent of growth temperature at a temperature higher than 630 °C. One can see that, only when the growth is performed at 630–650 °C, growth rate is proportional to TMI supply. The increase in growth rate with increasing growth temperature at atemperature less than 630 °C can be explained by taking account that growth rate is limited by NH$_3$ decomposition rate [5]. It shoud be noted that a growth rate as high as 0.8 μm/h is attained by employing a high growth temperature at around 650 °C.

For MOVPE InN, we have previously reported that carrier concentration is reduced and Hall mobility is increased with increasing V/III ratio in the growth atmosphere [6]. Figure 3 shows V/III ratio dependence of carrier concentration for InN films grwon at different temperatures. For a growth temperature lower than 600 °C, a marked V/III ratio dependence of carrier concentration is found. This seems to show that the in-

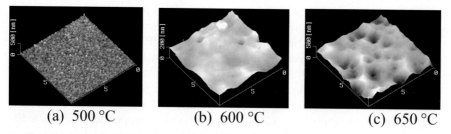

Fig. 1 (colour). Surface morphology (AFM image) for InN films grown at a different temperature. Growth pressure is 0.1 atm

---

[2]) Colour figure is published online (www.physica-status-solidi.com).

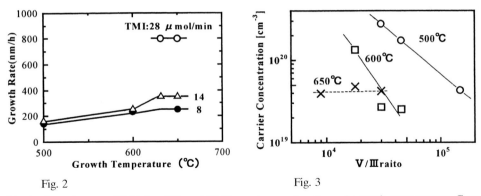

Fig. 2.   Fig. 3

Fig. 2. Growth rate of InN for different TMI supply as a function of growth temperature. Growth pressure is 0.1 atm

Fig. 3. Carrier concentration of InN grown at different teperatures as a function of V/III ratio. Growth pressure is 0.1 atm

crease in V/III ratio brings about the reduction of nitrogen vacancies in the InN by increasing active nitrogen in the growth atmosphre, which results in the improvement of electrical properties. It is notes that carrier concentration is independent of V/III ratio for InN films grown at a temperature around 650 °C. This fact suggests that another mechanism governs the electrical properties of InN films grown at such a high temperature. As described above, both surface morphology and growth rate for the InN clearly shows the change in growth reaction of InN at 600–630 °C. The growth reaction is limited by NH_3 decomposition at a temperature lower than 600 °C, where electrical properties of the grown InN are governed by active nitrogen concentration in the growth atmosphere. At a temperature higher than 630 °C, on the other hand, the growth is limited by TMI supply, where active nitrogen concentration in the growth atmosphere is not a critical factor. At such a high temperature, however, thermal decomposition of the grown InN can govern the electrical properties of InN, showing no V/III ratio dependence of the carier concentration.

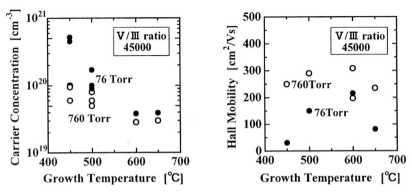

Fig. 4. Comparison of carrier concentration (left part) and Hall mobility (right part) between InN films grown at 76 and 760 Torr

Comparison is also made between electrical properties for InN films grown at a reduced pressure (76 Torr) and atmosphric pressure (760 Torr). Figure 4 shows growth temperature dependence of carrier concentration and Hall mobility for InN films grown at 76 and 760 Torr. Generally, atmosphric pressure growth gives better electrical properties. This means that effective NH_3 decompostion rate at 760 Torr is higher than that at 76 Torr. As shown in Fig. 4a, the difference in carrier concentration between 76 and 760 Torr becomes small with increasing growth temperature. This shows that difference in effective NH_3 decompostion rate becomes small as growth temperature increases. Although carrier concentration for InN grown at 600 and 650 °C is comparable between between 76 and 760 Torr, Hall mobility at 650 °C is lower than that at 600 °C. This may be due to the pit formation on the InN surface shown in Fig. 1c.

Conclusion Growth temperature dependencies of MOVPE InN have been systematically studied. An InN film with a fibrous columnar structure with small grains is grown at temperatures less than 550 °C. At 600–650 °C, a continuous film with enhanced two-dimensional growth is obtained. In the temperature range of 500–600 °C, the growth rate of InN is found to be increased with increasing growth temperature even for a constant TMI supply and shows a saturation against the increase in TMI supply. This is due to the limitation of growth reaction by the NH_3 decomposition. At 630–650 °C, such a saturation is not found for TMI supply up to 28 μmol/min at least and a growth rate as high as 0.8 μm/h is obtained. For a growth temperature lower than 600 °C, carrier concentration is decreased with a marked V/III raio dependence, while it is independent of the V/III ratio at a temperature around 650 °C. It is found that atmospheric pressure growth gives better electrical properties than reduced pressure growth. This means that the effective NH_3 decompostion rate at 760 Torr is higher than at 76 Torr. The difference in carrier concentration between 76 and 760 Torr growth becomes small with increasing growth temperature. Thus, the growth temperature is found to be a predominating factor for surface morphology, growth rate and electrical properties of MOVPE InN. It should be noted that these changes in surface morphology, growth rate and V/III ratio dependence of carrier concentration show the marked change in the growth behavior of MOVPE InN at 600–630 °C.

Acknowledgement This work has been supported in part by CREATE FUKUI of JST (Japan Science and Technology Corporation).

References

[1] S. K. O'Rearly, B. E. Foutz, M. S. Shur, U. V. Bhapker, and L. F. Eastman, J. Appl. Phys. **83**, 826 (1998).
[2] C. R. Abernathy, GaN and Related Materials, Vol. 2, Chap. 2, Ed. St. J. Pearton, Gordon & Breach, New York/London 1997 (p. 11).
[3] S. Yamaguchi, M. Kariya, S. Nitta, T. Takeuchi, C. Wetzel, H. Amano, and I. Akasaki, J. Appl. Phys. **85**, 7682 (1999).
[4] Y. Saito, N. Teraguchi, A. Suzuki, T. Araki, and Y. Nanishi, Jpn. J. Appl. Phys. **40**, L91 (2001).
[5] M. Adachi, Y. Murakami, A. Hashimoto, and A. Yamamoto, Proc. Internat. Workshop Nitride Semiconductors, Nagoya, IPAP Conf. Series **1**, 339 (2000).
[6] A. Yamamoto, T. Shin-ya, T. Sugiura, and A. Hashimoto, J. Cryst. Growth. **189/190**, 461 (1998).

Anisotropic Superconductivity of InN Grown by Molecular Beam Epitaxy on Sapphire (0001)

T. Inushima[1]) (a), V. V. Vecksin (b), S. V. Ivanov (b), V. Y. Davydov (b), T. Sakon (c), and M. Motokawa (c)

(a) Dept. of Electronics, Tokai University, Kitakaname, Hiratsuka, 259-1292 Japan

(b) A. F. Ioffe Physico-Technical Institute, RAS, St. Petersburg, 194021 Russia

(c) Metal Research Institute, Tohoku University, Katahira, Sendai, 980-8577 Japan

(Received June 22, 2001; accepted August 4, 2001)

Subject classification: 71.20.Nr; 74.62.Bf; 81.15.Hi; S7.14

We report a study of anisotropic superconductivity of InN grown by MBE on a sapphire (0001) surface. InN shows a resistivity anomaly below 3.5 K and a superconductivity below 1.5 K. The superconductivity is of the second kind and its H_{c1} and H_{c2} are 0.08 and 0.8 T, respectively, when the magnetic field (**B**) is applied parallel to the c-axis. When $\mathbf{B} \perp c$-axis, H_{c1} and H_{c2} are 0.23 and 2.3 T, respectively. InN demonstrates negative magnetoresistance above H_{c2} and the resistivity decreases 4% at 12 T under the $\mathbf{B} \parallel c$-axis configuration. These results indicate that the superconductivity of InN is anisotropic.

Introduction Among the III-nitride semiconductors, InN is a key material for optical and high temperature device application [1]. In most of the reports so far InN has been grown on a sapphire (0001) substrate and it is an n-type semiconductor with the carrier concentration higher than 10^{19} cm^{-3}. Its crystal structure is hexagonal with the c-axis parallel to the substrate c-axis. The band gap energy was reported in earlier studies to be 1.9 eV [2, 3]. With the recent development of a molecular beam epitaxy (MBE) growth method of InN, the phonon structure of InN has become clearer and we have observed all the six optical phonons of hexagonal InN [4]. The essential point of the phonon structure is the existence of a strong E_2 mode at 87 cm^{-1} which is attributed to the in-phase vibration of In on the c-plane [5]. The improvement of the crystal quality has enabled us to reveal that the band gap energy of InN is ambiguous and can be much smaller than 1.9 eV [6]. Moreover, it was reported that in some InN films the resistivity started to decrease at 3.2 K and was recovered when magnetic field was applied, from which the existence of superconducting elemental indium in the film was suggested [7]. We found similar resistivity anomaly and reported that some InN was superconductive [6].

In this report we present the evidence that pure InN itself has anisotropic superconductivity at 1.5 K.

Experimental Details In this experiment, we compared two samples of InN. These samples were grown on sapphire (0001) substrates by ECR-plasma-assisted MBE [8]. After high temperature nitridation, first a thin (15 nm) InN buffer layer was formed on the substrate at 300 °C. Then after high-temperature annealing at 900 °C, the tempera-

---

[1]) Corresponding author; Phone: +81 46 358 12 11, Fax: +81 46 358 83 20, e-mail: inushima@keyaki.cc.u-tokai.ac.jp

ture was reduced to 470 °C and InN was grown continuously at the growth rate of 80 nm/h for W269 and 125 nm/h for W280. The importance of the high temperature annealing was reported elsewhere [9]. The InN-growth initiation procedure allowed us to get nearly column-free InN films directly on the sapphire surface, which was confirmed by the TEM image [10]. The sample thickness is 400 nm for W269 and 1.0 μm for W280. The carrier concentrations were 2×10^{20} and 5×10^{19} cm^{-3} and their mobilities were 132 and 1700 cm^2 V^{-1} s^{-1} for W269 and W280, respectively, which were obtained from conventional Hall measurements. Being analysed by the use of X'Pert system of Philips Corp., these samples showed a clear hexagonal structure. The major difference between W280 and W269 is that W280 contains a weak metal In phase. In the $2\theta-\omega$ X-ray diffraction spectrum, W269 shows only hexagonal InN. W280, however, has additional weak reflections of (101) and (202) of tetragonal (t) In. Both samples had Raman spectra very similar to those reported before [4] and had a strong E$_2$ mode at 87 cm^{-1}. The precise band gap energies of these samples are difficult to determine due to the weak absorption tail extending to the 2.0 eV region. When we assume that InN has the direct optical transition, then the gap energy is 1.46 and 0.89 eV for W269 and W280, respectively.

The temperature dependence of the resistivity was measured by using a He^3 cryostat. The resistance of the sample was measured by a conventional dc four-probe method. The evaporated Au electrodes on the samples were connected to the outer electrodes by conductive silver paste. The applied current was 0.1 mA. The refrigerator capacity was 1 mW and the resistance of each measured sample was less than 10 Ω so that we could ignore the self-heating effect. For the study of anisotropic critical field, a rotating sample holder was used in a 15 T superconducting solenoid. In these measurements the direction of the current was along the c-plane and normal to the applied magnetic field.

Results and Discussion No sample showed any remarkable temperature dependence of resitivity between 300 and 3.5 K, which indicates that the InN investigated has metallic conductivity. Below 3.5 K, however, these samples showed a resistivity change, which is shown in Fig. 1. In Fig. 1 the ordinate is normalized by each sample's resistivity at 4.2 K. As is shown in the inset of Fig. 1, W280 shows a drastic change of resistivity at 2.5 K, which makes a kink at 2.1 K. Then it becomes less than 10^{-4} at 0.8 K, which is the limit of measuring accuracy. The resistivity of W269 starts to change at 3.5 K, makes the slope of the temperature dependence larger at 2.2 K, then reaches 10^{-4} at 1.5 K. It is obvious that pure InN has a resistivity anomaly at 3.5 K and becomes a superconductor at 1.5 K.

Figure 2 shows the field strength (B) dependence of the resistivity of W280 and W269 at 0.5 K as a function of the angle between the applied field direction and the crystal c-axis. The superconductivity of W280 breaks at 250 gauss and the resistivity reaches the initial value at 0.5 T, but only weak angle dependence is observed. The superconductivity of W269 breaks at 0.08 T, when **B** ∥ c-axis, and at 0.23 T, when **B** ⊥ c-axis. The rise of the resistivity from zero is abrupt, which allows us to define the lower critical field (H_{c1}) unambiguous way, as shown in the inset of Fig. 2. The angle dependence of H_{c1} shown in Fig. 2 can be understood as "resistive broadening" reported for high-T_c superconductors [11]. The origin of the resistive broadening is understood as the low-dimensional fluctuation of the superconductivity on the CuO$_2$ layer in YBCO [12].

The upper critical field (H_{c2}) is defined as the field which brings back the resistivity to the initial value. The resistivity of W269 does not return to the 4.2 K value and

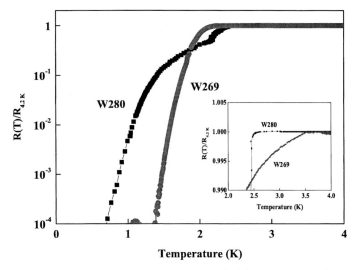

Fig. 1. Temperature dependence of the resistivity of W269 and W280 which is normalized to the value at 4.2 K. The inset shows the detail of the resistivity change near 2.5 K

shows a maximum at 0.8 T for $\mathbf{B} \perp c$-axis and at 2.3 T for $\mathbf{B} \parallel c$-axis, which is demonstrated in Fig. 2. The maximum value is within 0.2% of the value at 4.2 K, which allows us to consider the maximum field as H_{c2} for W269. When $B > H_{c2}$, W269 shows negative magnetoresistance with a resistivity decrease of 4% from the value at 4.2 K at the $\mathbf{B} \parallel c$-axis configuration. When $\mathbf{B} \perp c$-axis, the decrease is 1%. For comparison, the resistivity change of W280 versus B is also shown in Fig. 2. W280 does not show any remarkable angle dependence ($H_{c1} = 250$ Gauss), which indicates that the occurrence of the metallic In phase breaks the superconductivity anisotropy. The resistivity of W280 shows typical magnetoresistance and has a B^2 dependence above 0.5 T (H_{c2}). We could not observe any oscillatory magnetoresistance which is expected in a degenerate semiconductor. The negative magneto-resistance shown in Fig. 2 can be explained by the

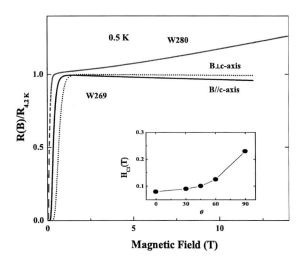

Fig. 2. Magnetic field dependence of the resistivity of W280 and W269 as a function of the angle between the magnetic field and the crystal c-axis. Inset is the angle dependence of H_{c1} of W269, where $\theta = 0$ corresponds to $\mathbf{B} \parallel c$-axis

suppression of the scattering probability of carriers on the c-plane by the external field. When $\mathbf{B} \parallel c$-axis, the effect of suppression is strongest and induces the biggest reduction of the resistivity. To summarize, we conclude that the superconductivity of InN is anisotropic and is produced on the In-plane which is parallel to the sapphire c-plane. If it were due to In droplets on the surface of the samples or In clusters in the buffer layer grown at low temperatures, a drastic change of resistivity should be observed at 3.4 K leading to isotropic superconductivity [13]. We should not completely neglect the explanation that the observed anisotropy of conductivity anomaly is related to indium segregation at columnar boundaries, but it seems to be less probable due to the structural quality of the film revealed by TEM and the absence of any XRD indications of such developed In-enriched network.

When we consider the characteristic parameters of superconducting InN at 0.5 K using the data of W269, the transition temperature (T_c) of InN is 3.5 K and the penetration depth (λ_L) is estimated to be 0.12 μm from the electron density $n = 2 \times 10^{20}$ cm^{-3} and the effective mass of $0.1 m_0$ [14]. The coherent length (ξ_{GL}) obtained from the anisotropic H_{c2} is 20 nm for $\mathbf{B} \parallel c$-axis and 12 nm for $\mathbf{B} \perp c$-axis, which gives the ratio $\kappa = \lambda_L / \xi_{GL} = 6$–$10$. This ratio indicates that InN is a superconductor of the second kind.

Conclusion When InN is grown on a sapphire (0001) plane, it shows a resistivity anomaly at $T_c = 3.5$ K, and anisotropic superconductivity at 1.5 K. The superconductivity is of the second kind and its coherent length is 12–20 nm.

Acknowledgement The authors are grateful to Professor Yao of Tohoku University for his help, and to the Ministry of Education for the Grant-in-Aid for Scientific Research (C) 11650332. The low temperature measurement was carried out at the High Field Laboratory for Superconducting Materials of Tohoku University.

References

[1] S. NAKAMURA, and F. FASOL, The Blue Laser Diode, Springer-Verlag, Berlin 1997 (pp. 69–78).
[2] T. L. TANSLEY and C. P. FOLEY, J. Appl. Phys. **59**, 3241 (1986).
[3] K. OSAMURA, K. NAKAJIMA, Y. MURAKAMI, P. H. SHINGU, and A. OHTSUKI, Solid State Commun. **11**, 617 (1972).
[4] V. Y. DAVYDOV, V. V. EMTSUEV, I. N. GONCHARUK, A. N. SMIRNOV, V. D. PETRIKOV, V. V. MAMUTIN, V. A. VEKSHIN, S. V. IVANOV, M. B. SMIRNOV, and T. INUSHIMA, Appl. Phys. Lett. **75**, 3297 (1999).
[5] T. INUSHIMA, T. SHIRAISHI, and V. YU. DAVYDOV, Solid State Commun. **110**, 491 (1999).
[6] T. INUSHIMA, V. V. MAMUTIN, V. A. VEKSHIN, S. V. IVANOV, T. SAKON, and S. MOTOKAWA, J. Cryst. Growth **227/228**, 481 (2001).
[7] M. MIURA, H. ISHII, A. YAMADA, M. KONAGAI, Y. YAMAUCHI, and A. YAMAMOTO, Jpn J. Appl. Phys. **36**, L256 (1997).
[8] V. V. MAMUTIN, V. A. VECKSHIN, V. YU. DAVYDOV, V. V. RATNIKOV, YU. A. KUDRIAVTSEV, B. YA. BER, V. V. EMTSEV, and S. V. IVANOV, phys. stat. sol. (a) **176**, 373 (1999).
[9] T. V. SHUBINA, V. V. MAMUTIN, V. A. VEKSHIN, V. V. RATNIKOV, A. A. TOROPOV, S. V. IVANOV, M. KARLSTEEN, U. SODERVALL, M. WILLANDER, G. R. POSINA, and B. MONEMAR, phys. stat. sol. (b) **216**, 205 (1999).
[10] V. V. MAMUTIN, T. V. SHUBINA, V. A. VEKSHIN, V. V. RATNIKOV, A. A. TOROPOV, S. V. IVANOV, M. KARLSTEEN, U. SODERVALL, and M. WILLANDER, Appl. Surf. Sci. **166**, 87 (2000).
[11] Y. IYE, T. TAMEGAI, H. TAKEYA, and H. TAKEI, Jpn. J. Appl. Phys. **26**, L1057 (1987).
[12] J. R. CLEM, Phys. Rev. B **43**, 7837 (1991).
[13] D. SHOENBERG, Superconductivity, 2nd Ed., Cambridge University Press, Cambridge 1965.
[14] Y. C. YEO, T. C. CHONG, and M. F. LI, J. Appl. Phys. **83**, 1429 (1998).

Polarity of High-Quality Indium Nitride Grown by RF Molecular Beam Epitaxy

Y. Saito[1]) (a), Y. Tanabe (a), T. Yamaguchi (a), N. Teraguchi (b), A. Suzuki (b), T. Araki (a), and Y. Nanishi (a)

(a) Department of Photonics, Ritsumeikan University, 1-1-1, Noji-Higashi, Kusatsu, Shiga 525-8577, Japan

(b) Advanced, Tech. Res. Labs., Sharp Corp., 2613-1 Ichinomoto, Tenri, Nara 632-8567, Japan

(Received July 2, 2001; accepted July 20, 2001)

Subject classification: 68.55.Jk; 81.15.Hi; S7.14

We have evaluated the polarity of single-crystalline InN using coaxial impact collision ion scattering spectroscopy (CAICISS). The polarity of rf-MBE grown InN on sapphire was found to depend on growth temperature. Nitrogen polarity was observed for low temperature grown InN at 300 °C. On the other hand, high temperature grown InN at 550 °C showed primarily indium polarity. Furthermore, a mixture of indium polarity and nitrogen polarity was observed for two-step grown InN. A GaN cap layer was grown on InN to infer the polarity of InN. From reflection high-energy electron diffraction observations, GaN grown on low temperature grown InN showed a (3×3) reconstruction pattern, whereas GaN grown on high temperature grown InN showed a (2×2) reconstruction pattern. These results suggest that the polarity of InN is much more sensitive to growth temperature compared with other nitride semiconductors.

Introduction InN and its alloys with GaN and AlN are successfully used in short wavelength light emitting diodes and laser diodes [1]. InN, moreover, has the smallest effective mass [2] and the highest electron drift velocity [3] among nitride semiconductors. Therefore, InN is expected to be one of the most promising materials for high speed and high frequency electronic devices. However, due to the extremely high dissociation pressure [4] and the lack of suitable substrate materials the growth of high-quality InN is still very difficult. Radio-frequency plasma-excited molecular beam epitaxial (rf-MBE) growth of high-quality InN with a room-temperature electron mobility of 760 cm^2/Vs has been reported very recently [5]. The growth mechanism responsible for this high-quality InN, however, has not been sufficiently understood. In particular, the polarity of InN, which is a key factor to grow high-quality nitride semiconductors with wurtzite structure, has never been investigated systematically because it has been hard to obtain a sufficient quantity of high-quality InN enough to evaluate polarity. In this paper, we give the first report on the polarity evaluation of InN by coaxial impact collision ion scattering spectroscopy (CAICISS) measurements and reflection high-energy electron diffraction (RHEED) observations.

Experimental Procedure InN films for this study were grown by rf-MBE. The substrate used in this study was c-face sapphire. After the substrate was cleaned by an organic solvent, it was thermally cleaned at 800 °C in vacuum for 10 min. The back-

---

[1]) Corresponding author; Phone: +81 77 561 2884; Fax: +81 77 561 3994;
e-mail: see40063@se.ritsumei.ac.jp

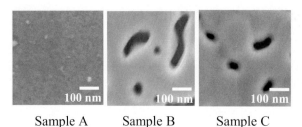

Sample A   Sample B   Sample C

Fig. 1. Surface SEM images of samples A, B and C. The sample A was observed after annealing at 550 °C

ground pressure of the growth chamber was approximately 1×10^{-10} Torr with a liquid nitrogen jacket. Prior to the growth of InN, a nitridation process was carried out at 550 °C for 1 h. We grew three types of InN samples with a film thickness of about 200 nm; sample A was grown at a relatively low temperature of 300 °C, sample B was grown at a relatively high temperature of 550 °C, and sample C was grown at high-temperature on a low temperature-grown buffer layer. Growth conditions of these samples were determined based on in situ RHEED observation. Film thicknesses were determined using a mechanical profile-meter (DECTAK[3]). Scanning electron microscopy (SEM) was used to determine the InN surface morphology. To determine the polarity of the InN, coaxial impact collision ion scattering spectroscopy (CAICISS) analysis was used. The polarity of InN was determined by comparing these results with CAICISS results previously reported for GaN [6, 7]. Precise simulation of the CAICISS spectrum for InN has not yet been carried out. Hence, we grew GaN cap layers on InN samples (A, B and C) with a thickness of about 40 nm to evaluate the polarity of InN; sample I was grown on sample A, sample II was grown on sample B, and sample III was grown on sample C. Here it was assumed that polarity inversion should not occur during GaN growth on InN so that the polarity of GaN should be the same as that of InN. Thus, we studied the polarity of the InN by CAICISS analysis of the GaN grown on top of the InN samples.

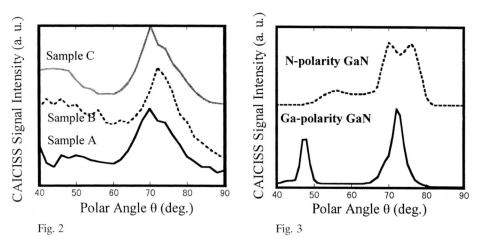

Fig. 2. CAICISS spectra of the InN samples on sapphire

Fig. 3. Simulated CAICISS spectra of GaN

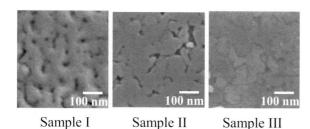

Fig. 4. Surface SEM images of the samples I, II and III

Sample I   Sample II   Sample III

Results and Discussion Figure 1 shows the surface morphologies of samples A, B, and C observed by SEM. We found that the surface morphology of samples B and C had several pits, whereas that of sample A had a few indium (In) droplets. Figure 2 shows CAICISS spectra of samples A, B, and C. The spectrum of the sample A is similar to that obtained for the N-polarity-GaN simulation shown in Fig. 3.

On the other hand, the CAICISS spectrum of sample B had the peak of polar angle at around 72°. This suggests that the polarity of sample B is more similar to In-polarity than N-polarity. The spectrum of sample C is similar to that of sample A. Because the spectrum of sample C has a broad peak of polar angle at approximately 48°, sample C probably has a mixture of In-polarity and N-polarity.

Figure 4 shows the surface morphologies of samples I, II, and III observed by SEM. The surface morphology of all three samples is slightly rough. Figure 5 shows the CAICISS spectra of samples I, II, and III. The spectrum of sample I has a peak at 70°. This spectrum is similar to that obtained by the simulation of N-polarity GaN, which is the same as that of sample A. The spectrum of the sample II has strong peaks around 44° and 72°. This result indicates that the polarity of sample II is most similar to Ga-polarity. However, for Ga-polarity the peak should be at 48° rather than 44°. The reason for the peak shift is not presently understood. The spectrum of sample III has peaks around 70°, 72°, and 76°. This indicates that the polarity of sample III has a mixture of Ga-polarity and N-polarity.

Figure 6 shows the RHEED patterns of samples I, II, and III that were observed at 200 °C. These RHEED patterns show (3 × 3) and (2 × 2) reconstructions on sample I and sample II, respectively. According to the reported relationship between polarity and RHEED reconstruction patterns, [8] these results indicate that sample I has mainly N-polarity, and sample II is mainly Ga-polarity. Sample III did not show any reconstruc-

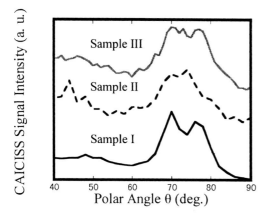

Fig. 5. CAICISS spectra of the GaN samples on InN

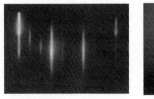 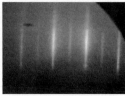 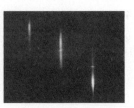

Sample I Sample II Sample III

Fig. 6. RHEED patterns of samples I, II and III. It was observed at around 200 °C after the growth of GaN. (**e** ∥ [11$\bar{2}$0])

tion pattern. These results indicate that the polarity of low-temperature-grown InN layer is mainly N-polarity, that of high-temperature-grown InN layer is mainly In-polarity, and that of two-step grown InN layer is a mixture of In-polarity and N-polarity. In the case of GaN growth, the polarity is probably determined by the polarity of the buffer layer. In the case of InN growth, the polarity of the two-step grown InN has mixed polarity, although the polarity of the buffer layer should have N-polarity. This suggests that it is much easier to change the polarity of InN compared with other nitride semiconductors.

Conclusions The polarity of InN was evaluated by CAICISS analysis and RHEED observations. We found that the polarity of rf-MBE-grown InN on sapphire was very sensitive to growth temperature. The polarity of low temperature-grown InN was mainly N-polarity, that of high temperature grown InN was mainly In-polarity, and that of two-step grown InN was a mixture of In-polarity and N-polarity. Hence, the polarity of InN was sensitive to growth temperature when compared with GaN.

Acknowledgements This work was supported in part by Academic Frontier Promotion project and NEDO Regional Consortium Project.

References

[1] S. Nakamura, M. Senoh, S. Nagahama, N. Iwasa, T. Yamada, T. Matsushita, H. Kiyoku, and Y. Sugimoto, Jpn. J. Appl. Phys. **35**, L74 (1996).
[2] S. N. Mohammad and H. Morkoc, Prog. Quantum Electron. **20**, 361 (1996).
[3] B. E. Fortz, S. K. O'Leary, M. S. Shur, and L. F. Eastman, J. Appl. Phys. **85**, 7727 (1999).
[4] J. B. MacChesney, P. M. Bridenbaugh, and P. B. O'Connor, Mater. Res. Bull. **5**, 783 (1970).
[5] Y. Saito, N. Teraguchi, A. Suzuki, T. Araki, and Y. Nanishi, Jpn. J. Appl. Phys. **40**, L90 (2001).
[6] M. Sumiya, K. Yoshimura, T. Ito, K. Ohtsuka, S. Fuke, K. Mizuno, M. Yoshimoto, H. Koinuma, A. Ohtomo, and M. Kawasaki, J. Appl. Phys. **88**, 1158 (2000).
[7] S. Sonoda, S. Shimizu, Y. Suzuki, K. Balakrishman, J. Shirakashi, and H. Okumura, Jpn. J. Appl. Phys. **39**, L73 (2000).
[8] A. R. Smith, R. M. Feenstra, D. W. Greve, M. S. Shin, M. Skowronski, J. Neugebauer, and J. E. Northrup, Appl. Phys. Lett. **72** 2114 (1998).

Study of Epitaxial Relationship in InN Growth on Sapphire (0001) by RF-MBE

T. Yamaguchi[1]) (a), Y. Saito (a), K. Kano (a), T. Araki (a), N. Teraguchi (b), A. Suzuki (b), and Y. Nanishi (a)

(a) Dept. of Photonics, Ritsumeikan Univ., 1-1-1 Noji-Higashi, Kusatsu, Shiga 525-8577, Japan

(b) Advanced Tech. Res. Labs., Sharp Corp., 2613-1 Ichinomoto, Tenri, Nara 632-8567, Japan

(Received June 25, 2001; accepted August 4, 2001)

Subject classification: 61.10.Nz; 68.55.Jk; 81.15.Aa; 81.15.Hi; S7.14

InN films were grown directly on sapphire (0001) substrates by rf-MBE. In the direct growth of InN on sapphire substrates without nitridation process, InN films have the tendency to form a multi-domain structure due to in-plane rotation and they can have both the epitaxial relationship of $[11\bar{2}0]_{InN} \parallel [11\bar{2}0]_{sapphire}$ and $[10\bar{1}0]_{InN} \parallel [11\bar{2}0]_{sapphire}$. The domain with $[11\bar{2}0]_{InN} \parallel [11\bar{2}0]_{sapphire}$ is mainly observed at relatively low growth temperatures around 520 °C. The domain with $[10\bar{1}0]_{InN} \parallel [11\bar{2}0]_{sapphire}$, however, becomes dominant with increasing the growth temperatures to around 540 °C. The dominant factor in determining the main epitaxial relationship is found to be its growth temperature. However, this is highly sensitive to a small change in growth temperature which should be controlled within 20 °C.

Introduction Group-III nitride semiconductors are attractive materials for optoelectronic devices in visible and ultraviolet regions and for electronic devices that can operate at high-temperature, high-frequency and high-power conditions. In spite of these attentions, study on InN, particularly, has been far behind the other group-III nitride semiconductors such as GaN and AlN because the growth of InN is very difficult, which is attributed to its low dissociation temperature [1]. Recently, epitaxial growth of high quality InN, with good surface morphology and a high electron mobility, was reported [2, 3]. However, the growth mechanism of InN, especially the epitaxial relationship of a-axis between InN films and sapphire substrates, is not fully clarified. When InN films were grown on sapphire (0001) substrates with nitridation process, the epitaxial relationship has been found as $[10\bar{1}0]_{InN} \parallel [10\bar{1}0]_{AlN} \parallel [11\bar{2}0]_{sapphire}$ [4]. On the other hand, when InN films were grown directly on sapphire (0001) substrates, some different results have been obtained. Yamaguchi et al. [5] reported that InN films with the epitaxial relationship of $[10\bar{1}0]_{InN} \parallel [11\bar{2}0]_{sapphire}$ were obtained when they were grown at 450 °C by metalorganic vapor phase epitaxy (MOVPE). On the other hand, Guo et al. [6] reported that InN films, which were grown at 500 °C by microwave-excited metalorganic vapor phase epitaxy (ME-MOVPE), were with $[11\bar{2}0]_{InN} \parallel [11\bar{2}0]_{sapphire}$. In addition, Wakahara et al. [7] reported that InN films grown by ME-MOVPE were with two epitaxial relationships of $[10\bar{1}0]_{InN} \parallel [11\bar{2}0]_{sapphire}$ and $[11\bar{2}0]_{InN} \parallel [11\bar{2}0]_{sapphire}$. It has not been clear yet what has led to these different results on the epitaxial relationship. Various parameters, such as growth method, growth process, substrate temperature and V/III ratio should be considered.

---

[1]) Corresponding author; Phone: +81 77 561 2884, Fax: +81 77 561 3994, e-mail: ro012961@se.ritsumei.ac.jp

In this paper, we discuss the growth temperature dependence of the epitaxial relationship as a study on epitaxial relationship in InN grown directly on sapphire (0001) substrate without nitridation process.

Experimental Procedure InN films were grown directly on sapphire (0001) substrates by radio-frequency plasma-excited molecular beam epitaxy (rf-MBE) for 1 h. Prior to growth, the sapphire substrates were cleaned by organic solvents and thermally cleaned at 800 °C for 10 min in vacuum. Indium cell temperature, rf power and nitrogen flow rate were kept constant throughout this study at 735 °C, 330 W and 2 sccm, respectively. Growth temperatures were varied between 450 and 580 °C. Morphological characteristics of the InN films were investigated by using scanning electron microscopy (SEM; HITACHI S4300SE). X-ray diffraction (XRD; Philips X'Pert MRD) φ-scan measurements were used to investigate the epitaxial relationship between the InN film and the sapphire substrate. Asymmetric $\{11\bar{2}2\}$ reflections of InN and $\{11\bar{2}3\}$ reflections of sapphire were used for them.

Results and Discussion Figure 1 shows the results of XRD φ-scan measurements for the InN films grown at the growth temperature of a) 520 °C, b) 530 °C and c) 540 °C. The solid and broken lines show the peaks for InN and sapphire, respectively. As shown in Fig. 1a, the InN film grown at the growth temperature of 520 °C had three types of domains; (A) domains with non-rotation from the sapphire substrate (i.e. $[11\bar{2}0]_{InN} \parallel [11\bar{2}0]_{sapphire}$), (B) domains with in-plane rotation by 30° from the substrate (i.e. $[10\bar{1}0]_{InN} \parallel [11\bar{2}0]_{sapphire}$) and (C) domains rotated by $\approx 19°$ from the substrate (i.e. $[14\bar{5}0]_{InN} \parallel [11\bar{2}0]_{sapphire}$). The observation of domains rotated by $\approx 19°$ and its formation mechanism were reported elsewhere [8]. Observation of two different domains with epitaxial relationship of $[10\bar{1}0]_{InN} \parallel [11\bar{2}0]_{Sapphire}$ and $[11\bar{2}0]_{InN} \parallel [11\bar{2}0]_{sapphire}$ is in good agreement with the previous report by Wakahara et al. [7]. The main epitaxial relationship of this InN film grown at 520 °C was $[11\bar{2}0]_{InN} \parallel [11\bar{2}0]_{sapphire}$. Similarly, the InN films grown at 530 °C and 540 °C had three types of domains, as shown in Figs. 1b and c. However, the peak intensities of domains with $[11\bar{2}0]_{InN} \parallel [11\bar{2}0]_{sapphire}$ and $[10\bar{1}0]_{InN} \parallel [11\bar{2}0]_{sapphire}$ were almost equivalent in the InN film grown at 530 °C as shown in Fig. 1b. In case of InN film grown at 540 °C as shown in Fig. 1c, the peaks from domains with $[10\bar{1}0]_{InN} \parallel [11\bar{2}0]_{sapphire}$ became dominant.

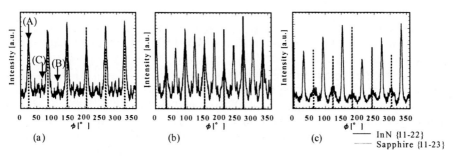

Fig. 1. Profiles of XRD φ-scans for InN $\{11\bar{2}2\}$ and sapphire $\{11\bar{2}3\}$. InN films were grown at a) 520 °C, b) 530 °C and c) 540 °C. The solid and broken lines show the peaks for InN and sapphire, respectively

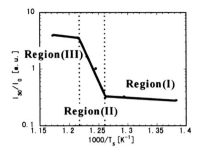

Fig. 2. Plot of the I_{30}/I_0 values as a function of reciprocal of the growth temperature. I_{30} and I_0 were estimated from the average values of the six peak intensities of InN $\{11\bar{2}2\}$

The main epitaxial relationship of the InN films varied with respect to the small difference in growth temperature by 20 °C. These results suggest that InN growth is very sensitive to the growth temperature, especially as for the epitaxial relationship of a-axis.

To evaluate the relationship between $[11\bar{2}0]_{InN}$ ∥ $[11\bar{2}0]_{sapphire}$ and $[10\bar{1}0]_{InN}$ ∥ $[11\bar{2}0]_{sapphire}$ more in detail, we studied the existence ratio I_{30}/I_0; I_{30} and I_0 are the average intensities of the six InN $\{11\bar{2}2\}$ peaks of the domains with $[10\bar{1}0]_{InN}$ ∥ $[11\bar{2}0]_{sapphire}$ and $[11\bar{2}0]_{InN}$ ∥ $[11\bar{2}0]_{sapphire}$, respectively.

Figure 2 shows the plot of the I_{30}/I_0 values as a function of reciprocal of the growth temperature. Three regions with diferent growth temperature dependence of the I_{30}/I_0 values were observed. Typical SEM images of InN surface morphology grown at these three temperature regions are shown in Fig. 3.

In region (I), the values of I_{30}/I_0 were smaller than unity, and almost independent of the growth temperature. It is indicated that the domains with the epitaxial relationship of $[11\bar{2}0]_{InN}$ ∥ $[11\bar{2}0]_{sapphire}$ were mainly observed in this region. However, a small amount of domains with $[10\bar{1}0]_{InN}$ ∥ $[11\bar{2}0]_{sapphire}$ and $[14\bar{5}0]_{InN}$ ∥ $[11\bar{2}0]_{sapphire}$ was still contained. InN films with only $[11\bar{2}0]_{InN}$ ∥ $[11\bar{2}0]_{sapphire}$ could not be grown in such low growth temperature range. It is considered that InN growth in the temperature range of region (I) could not suppress the formation of domains with $[10\bar{1}0]_{InN}$ ∥ $[11\bar{2}0]_{sapphire}$ and $[14\bar{5}0]_{InN}$ ∥ $[11\bar{2}0]_{sapphire}$ completely because of low surface migration of In atoms, leading to the surface morphology with many small grains as shown in Fig. 3a.

In region (II), the values of I_{30}/I_0 decreased exponentially with reciprocal of growth temperature. The main epitaxial relationship changed from $[11\bar{2}0]_{InN}$ ∥ $[11\bar{2}0]_{sapphire}$ to $[10\bar{1}0]_{InN}$ ∥ $[11\bar{2}0]_{sapphire}$. It is confirmed that the epitaxial relationship of the InN films was strongly dependent on the growth temperature in region (II). Surface morphology

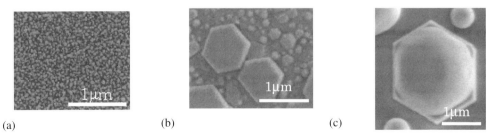

Fig. 3. Typical SEM images of surface morphology of the InN films grown at the growth temperature of a) region I, b) region II and c) region III

showed the formation of large grains with a size of 1 μm, as shown in Fig. 3b. This is considered to be due to the enhancement of In migration with increasing growth temperature.

Again, in region (III), the values of I_{30}/I_0 became almost independent of the growth temperature. Here, domains with the epitaxial relationship of $[10\bar{1}0]_{InN} \parallel [11\bar{2}0]_{sapphire}$ were mainly observed. The size of grains became larger than that obtained in region (II) due to further enhancement of the In migration. In such high growth temperature region, however, the dissociation of InN caused the formation of In droplets as shown in Fig. 3c.

From these results, it is clearly shown that the epitaxial relationship between InN films and sapphire substrates depends on the growth temperature. This does not agree with the previous reports [5–7] about the relationship between the main epitaxial relationship and the growth temperature. It is considered that this might be attributed to the difference of growth methods. The mechanism which determines the main epitaxial relationship is under consideration. However, since this kind of phenomenon was only found in InN growth among group-III nitride semiconductors, it might be due to the smaller difference in lattice mismatch between $[11\bar{2}0]_{InN} \parallel [11\bar{2}0]_{sapphire}$ and $[10\bar{1}0]_{InN} \parallel [11\bar{2}0]_{sapphire}$.

Summary InN films were grown directly on sapphire (0001) substrates by rf-MBE without nitridation process. InN films could be grown with both the epitaxial relationship of $[11\bar{2}0]_{InN} \parallel [11\bar{2}0]_{sapphire}$ and $[10\bar{1}0]_{InN} \parallel [11\bar{2}0]_{sapphire}$ as the main epitaxial relationship of a-axis. The main epitaxial relationship changed from $[11\bar{2}0]_{InN} \parallel [11\bar{2}0]_{sapphire}$ to $[10\bar{1}0]_{InN} \parallel [11\bar{2}0]_{sapphire}$ with increasing the growth temperature from 520 to 540 °C. In addition, the values of I_{30}/I_0 decreased exponentially with the reciprocal of growth temperature. Thus, epitaxial relationships between the InN films and the sapphire substrates are concluded to mainly be determined by the growth temperature.

Acknowledgement This work was supported by the Sasakawa Scientific Reseach Grant from the Japan Science Society, the Academic Frontier Promotion Project and the NEDO's Regional Consortium Project.

References

[1] J. B. MacChesney, P. M. Bridenbaugh, and P. B. O'Connor, Mater. Res. Bull. **5**, 783 (1970).
[2] Y. Saito, T. Yamaguchi, T. Araki, N. Teraguchi, A. Suzuki, and Y. Nanishi, MRS 2000 Fall Meeting, USA, Nov. 27–Dec. 1, 2000 (G11.18).
[3] H. Lu, W. J. Schaff, J. Hwang, H. Wu, W. Yeo, A. Pharkya, and L. F. Eastman, Appl. Phys. Lett. **77**, 2548 (2000).
[4] A. Yamamoto, M. Tsujino, M. Ohkubo, and A. Hashimoto, J. Cryst. Growth **137**, 415 (1994).
[5] S. Yamaguchi, M. Kariya, S. Nitta, T, Takeuchi, C. Wetzel, H. Amano, and I. Akasaki, J. Appl. Phys. **85**, 7682 (1999).
[6] Q.-X. Guo, T. Yamamura, A. Yoshida, and N. Itoh, J. Appl. Phys. **75**, 4927 (1994).
[7] A. Wakahara and A. Yoshida, Appl. Phys. Lett. **54**, 709 (1989).
[8] T. Yamaguchi et al., to be submitted.

Growth and Characterization of InN Heteroepitaxial Layers Grown on Si Substrates by ECR-Assisted MBE

T. Yodo[1]) (a), H. Ando (a), D. Nosei (a), and Y. Harada (b)

(a) Electric Engineering, Osaka Institute of Technology, 5-16-1, Asahi-ku, Ohmiya, Osaka 535-8585, Japan

(b) Applied Physics, Osaka Institute of Technology, 5-16-1, Asahi-ku, Ohmiya, Osaka 535-8585, Japan

(Received June 29, 2001; accepted August 4, 2001)

Subject classification: 61.10.Nz; 68.55.Jk; 78.55.Cr; 81.15.Hi; 81.15.Jj; S7.14

For the first time, we observed strong band-edge photoluminescence at 1.814 eV, and two stronger emissions at 1.880 and 2.081 eV at 8.5 K from the respective 880 nm thick InN heteroepitaxial layers (heteroepilayers) with 10 nm thick InN buffer layers grown on Si(001) and Si(111) substrates by electron cyclotron resonance-assisted molecular beam epitaxy. The former was probably assigned as donor-to-acceptor pair (DAP(α-InN)) emission from wurtzite-InN (α-InN) crystal grains, the latter were assigned as donor bound exciton ($D^0X(\alpha$-InN)) emission, and $D^0X(\beta$-InN) or DAP(β-InN) emission from zincblende-InN (β-InN) crystal grains, respectively. Substrate annealing before growth and the introduction of a buffer layer had strong influences on the crystal structure and crystalline quality of the initial InN heteroepilayers.

Introduction InN is potentially suitable as a nitride-based material for visible light-emitting diodes ranging from red to blue color, by alloying it with GaN (having large band gap) because of the small band gap of InN (1.89 eV at 300 K [1, 2]). Nevertheless, only several researchers [1–10] reported the film growth of wurtzite-InN (α-InN) on glass [1, 2], KBr [2], (0001) sapphire [4–8], (0001) GaN [3] and (111) GaAs [9], and of zincblende-InN (β-InN) on (001) GaAs [3] and (111) GaP [10] because of difficulty of crystal growth. To our knowledge, there were no reports on luminescence from InN heteroepilayers probably because of the very degraded crystalline quality, and only a few experimental results on infrared absorption [1, 2] and Raman scattering [3] have been reported so far in optical properties of α-InN crystal. Furthermore, on β-InN, not only the optical properties but also even experimental values of band-gap energy have not ever been reported and a theoretical value of band-gap energy has merely been reported (it was calculated to be 2.2 eV at 300 K [11]). Moreover, if Si substrates with high conductivities were possible as substrates for InN epitaxial growth in place of sapphire etc., various new visible optical and electrical devices with versatile functions and optoelectronic integrated circuits could be developed in the future. However, to our knowledge, there was one report on growth of InN heteroepilayers on Si substrates [12]. In this paper, we report the growth of InN heteroepilayers on Si(001) and Si(111) substrates, investigate the influences of substrate annealing temperature before growth and the introduction of a buffer layer on the crystal structure and crystalline quality of

---

[1]) Corresponding author; Phone: +81-6-6954-4891, Fax: +81-6-6957-2136, e-mail: yodo@elc.oit.ac.jp

heteroepilayers and report a strong band-edge photoluminescence at 8.5–100 K from the InN heteroepilayers for the first time.

Experimentals The InN heteroepilayers were grown on nominally Si(001) and Si(111) substrates by electron cyclotron resonance-assisted molecular beam epitaxy (ECR-assisted MBE). Before growth, the Si substrates were chemically treated by three different methods: They were only dipped for 30 s in HF(47%), dipped for 5 min in solutions of HF(47%):NH_4F(40%) = 1:7 mixture and following in the respective HF, NH_4F solutions for 5 min (we call it RCA-like method here [13]); or dipped and boiled for 10 min repeatedly in solutions using HF, HNO_3, NH_4OH and HCl mixture (so-called Ishizaka method [14]); thereafter, they were annealed at 850–1250 °C for 5 min without a following substrate nitridation treatment or with it at 800 and 900 °C for 20 min. Then the InN heteroepilayers were grown at 500 °C and at In-K cell temperatures of 840–900 °C under a nitrogen gas flow rate of 2 sccm with 10 nm thick InN buffer layers (grown at 250 °C and an In-K cell temperature of 820 °C) on the pretreated substrates for 8 h, or directly without the introduction of buffer layers for 2 and 3 h while controlling the respective relative intensities of the plasma emission lines at 357 and 391 nm during ECR nitrogen plasma source at micro-wave powers of 300–350 W to 320 and 100 arb. units [15]. The respective relative intensities of the plasma emission lines were controlled with good reappearance by adjusting a three-stab tuner in a micro-wave waveguide, the magnitude of magnetic field and the configuration of the magnet. The 357 nm and 391 nm emission lines are attributed to plasma emissions related to nitrogen radical molecules and nitrogen molecular ions in ECR nitrogen plasma, respectively [15, 16]. The growth rates of the InN heteroepilayers were about 50–110 nm/h, depending on the In-K cell temperature under these growth conditions.

Crystal structure and crystalline quality of the InN heteroepilayers were characterized by X-ray diffraction and photoluminescence (PL) at 8.5–100 K using a 514.5 nm Ar^+ ion laser with a power of 100 mW as the excitation light source, respectively. The resolving power of the monochromator used in the PL system was 0.2 nm in the wavelength range of 520–900 nm. The optical emission spectra of ECR nitrogen plasma during InN growth were monitored in situ through a polychromator. The resolving power of the polychromator was 1 nm in the wavelength range of 240–930 nm.

Results and Discussions Figure 1 shows the θ–2θ X-ray diffraction spectra in the 2θ range of 30–32° of the 200 nm thick InN heteroepilayers grown at 500 °C and at an In-K cell temperature of 880 °C directly without the introduction of buffer layers on Si(111) substrates, chemically treated using the RCA-like method and thereafter annealed at 850–1250 °C for 5 min or without the annealing treatment, but all without the following substrate nitridation treatment. All the InN heteroepilayers exhibited the dominant (0002)α-InN Cu$K_{\alpha1}$ diffractions at around 31.4–31.5° and other weak diffractions at around 31.1–31.2°, probably assigned as (111)β-InN Cu$K_{\alpha1}$ diffraction. While the (0002)α-InN X-ray diffraction intensity dramatically increased up to 1050 °C with an increase of annealing temperature, the (111)β-InN X-ray diffraction appeared from around 900 °C and also abruptly increased up to 1050 °C. The X-ray intensities of these diffractions abruptly decreased at temperatures higher than 1050 °C and finally the very weak (111)β-InN X-ray diffraction was overlapped on the dominant (0002)α-InN X-ray diffraction as a shoulder. Thus, the growth process of β-InN crystal grains would be en-

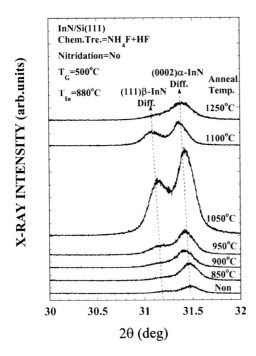

Fig. 1. θ–2θ X-ray diffraction spectra of 200 nm thick InN heteroepilayers without buffer layer on Si(111)

hanced by the substrate annealing before growth although the reason has not yet been known. Moreover, the peak angles of both the (0002)α-InN and (111)β-InN X-ray diffractions gradually shifted toward the lower angle side, particularly the (0002)α-InN peak largely shifted from 31.48° to 31.39° with increasing the annealing temperature from no anneal to 1250 °C, reflecting the increase of compressive strain despite the same growth conditions, although the reason has not yet been known. It is imagined that the substrate surface state after annealing would affect residual strains of respective InN grains in the layer. Although it was not shown here, the respective X-ray diffraction spectra of InN heteroepilayers grown on Si(001) under the same growth conditions as the series of samples in Fig. 1 were also dominated by the (0002)α-InN diffractions, on which the very faint (111)β-InN diffraction was laid as a shoulder. The annealing temperature dependence of X-ray intensities of the (111)β-InN diffraction in InN on Si(001) was very weak compared to those in InN on Si(111), probably reflecting the crystalline quality of the heteroepilayers.

Figures 2 and 3 show the θ–2θ X-ray diffraction spectra in the 2θ range of 30–32° and the X-ray rocking curves on the (0002)α-InN diffractions of the 880 nm thick InN heteroepilayers grown on 10 nm thick InN buffer layers on HF-treated Si(111) and Ishizaka-treated Si(001) substrates, which were annealed at 1200 °C for 5 min before growth, respectively. The 880 nm thick InN heteroepilayer on Si(111) had the dominant (0002)α-InN diffraction at 31.42° and the faint (111)β-InN diffraction at around 31.2° on the shoulder, while the InN heteroepilayer on Si(001) only the dominant (0002)α-InN diffraction at 31.45°. It was concluded that the InN heteroepilayers grown on Si substrates were crystallographically dominated by the (0001)-oriented α-InN crystal including slightly (111)-oriented β-InN grains. From the result of Fig. 3, the sharpest full-width at half-maximum (FWHM) of X-ray rocking curve on the (0002)α-InN diffraction was about 1.3°, which was the best for all the InN heteroepilayers grown on Si in our study, compared to very broad FWHMs of more than 5° in the series of samples in Fig. 1. However, the value was still broader than the FWHMs (37 arcmin [4], 52 arcmin [5], 96 arcsec [6]) of InN heteroepilayers on sapphire substrates reported so far. Furthermore, the FWHM of the 880 nm thick InN heteroepilayer with buffer layer on Si(001) was 3.3°, which was broader than that on Si(111), but sharper than those of the series of samples without introduction of buffer layers in Fig. 1, suggesting the drastic

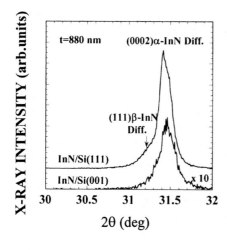

Fig. 2. θ–2θ X-ray diffraction spectra of 880 nm thick InN heteroepilayers with buffer layers on Si(001) and Si(111)

improvement of the crystalline quality due to the introduction of buffer layer, but indicating the still degraded crystalline quality.

Figure 4 shows PL spectra measured at 8.5 K of the same samples as in Fig. 2. The PL spectrum measured at 100 K of only the InN layer on Si(111) is also shown in the figure. The respective 8.5 K PL spectra of InN on Si(001) and Si(111) exhibited a strong band-edge PL emission at 1.814 eV and two stronger emissions at 1.880 and 2.081 eV. The PL emission of 1.814 eV was probably assigned as a donor bound exciton ($D^0X(\alpha\text{-InN})$) or donor-to-acceptor pair ($DAP(\alpha\text{-InN})$) emission considering from the PL peak energy, the band-gap energy of α-InN crystal and the results of X-ray diffraction in Fig. 2. The emission of 1.880 eV is probably assigned as $D^0X(\alpha\text{-InN})$ emission because of the peak energy higher than 1.814 eV, and that of 1.814 eV is assigned not to $D^0X(\alpha\text{-InN})$ but $DAP(\alpha\text{-InN})$ emission. As observed in Fig. 2, since X-ray diffraction of this sample exhibited dominant (0002) α-InN diffraction and very weak (111)β-InN diffraction on the shoulder, only the PL emissions related to α-InN crystal grains would have to be observed. However, the observed PL peak energy was 2.081 eV, which was higher than the band-gap energy of α-InN crystal and lower than that of β-InN crystal. Therefore, we cannot help concluding reasonably that the 2.081 eV PL emission is attributed to be an emission related to β-InN crystal, probably assigned as $D^0X(\beta\text{-InN})$ or $DAP(\beta\text{-InN})$ emission considering from the experimental evidence of the slight inclusion of (111)-oriented β-InN crystal grains based on X-ray diffraction results, although the optical data on β-InN have been little reported so far. However, we here had to assume that the luminescence efficiency of emissions from β-InN grains might be much higher than that from α-InN grains in InN crystal. We have ever reported the similar phenomenon on β-GaN grains in GaN crystal, suggesting that the luminescence efficiency of

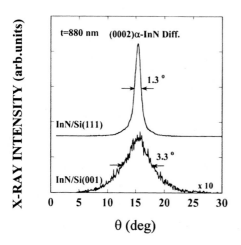

Fig. 3. X-ray rocking curves of the same samples as in Fig. 2

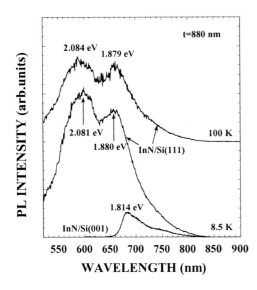

Fig. 4. PL spectra at 8.5 K of the same samples as in Fig. 2

$D^0X(\beta\text{-GaN})$ emission from β-GaN grains was much higher than that from α-GaN grains in GaN crystal [15]. The peak energies of 1.880 and 2.081 eV were little changed by the measurement temperature and the intensities were reduced to a level of half with increasing the measurement temperature from 8.5 to 100 K. The total 8.5 K PL intensity of the 880 nm thick InN heteroepilayer on Si(111) increased by more than six times, compared to that on Si(001). However, we observed no PL emissions from the series of samples in Fig. 1 and from a thinner 330 nm thick InN heteroepilayer grown on Si(111) by shortening only the growth period from 8 to 3 h, maintaining other important growth conditions (such as the introduction of buffer layer) of the emitting sample the same. The PL emission efficiency was much decreased by the degraded quality due to the reduction of film thickness, suggesting that it is strongly affected by the crystalline quality of the InN heteroepilayer. In order to heighten the band-edge PL emission efficiency from the thinner InN heteroepilayers, we are trying to optimize the growth conditions at the present.

Conclusion We observed strong band-edge photoluminescence at 1.814 eV, and two stronger emissions at 1.880 and 2.081 eV at 8.5 K from the respective 880 nm thick InN heteroepilayers with 10 nm thick InN buffer layers grown on Si(001) and Si(111) substrates by ECR-assisted MBE for the first time. The former was probably assigned as DAP(α-InN) emission, the latter as $D^0X(\alpha\text{-InN})$ emission, and $D^0X(\beta\text{-InN})$ or DAP(β-InN) emission, respectively. Substrate annealing temperature before growth and the introduction of low-temperature grown InN buffer layer had strong influences on the crystal structure and crystalline quality of the initial InN heteroepilayers.

Acknowledgment This work was partially supported by a Grant-in-Aid for Scientific Research (No. 13650358) from the Ministry of Education, Science, Sports and Culture of Japan.

References

[1] T. L. TANSLEY and C. P. FOLEY, J. Appl. Phys. **59**, 3241 (1986).
[2] T. L. TANSLEY and C. P. FOLEY, J. Appl. Phys. **60**, 2092 (1986).
[3] G. KACZMARCZYK, A. KASCHNER, S. REICH, A. HOFFMANN, C. THOMSEN, D. J. AS, A. P. LIMA, D. SCHIKORA, K. LISCHKA, R. AVERBECK, and H. RIECHERT, Appl. Phys. Lett. **76**, 2122 (2000).
[4] QI-XIN GUO, TOSHIMI YAMAMURA, AKIRA YOSHIDA, and NOBUO ITOH, J. Appl. Phys. **75**, 4927 (1994).

[5] HAI LU, W. J. SCHAFF, JEONGHYUN HWANG, HONG WU, WESLEY YEO, AMIT PHARKYA, and L. F. EASTMAN, Appl. Phys. Lett. **77**, 2548 (2000).
[6] WEI-KUO CHEN, YUNG-CHUNG PAN, HENG-CHING KIN, JEHN OU, WEN-HSIUNG CHEN, and MING-CHIH LEE, Jpn. J. Appl. Phys. **36**, L1625 (1999).
[7] Y. SAITO, N. TERAGUCHI, A. SUZUKI, T. ARAI, and Y. NANISHI, Jpn. J. Appl. Phys. **40**, L91 (2001).
[8] A. YAMAMOTO, T. SHINYA, T. SUGIURA, and A. HASHIMOTO, J. Cryst. Growth **189**, 461 (1998).
[9] QIXIN GUO, M. NISHIO, H. OGAWA, and A. YOSHIDA, Jpn. J. Appl. Phys. **38**, L490 (1999).
[10] W. E. HOKE, P. J. LEMONIAS, and D. G. WEIR, J. Cryst. Growth **111**, 1024 (1991).
[11] G. POPOVICI, H. MORKOC, and S. N. MOHANNAD, in: Group III Nitride Semiconductor Compounds, Physics and Applications, Ed. B. GIL, Clarendon Press, Oxford 1998 (p. 22).
[12] A. YAMAMOTO, M. TSUJINO, M. OHKUBO, and A. HASHIMOTO, J. Cryst. Growth **137**, 415 (1994).
[13] W. KELN and D. A. PUOTINEN, RCA Rev., pp. 187–206 (1970).
[14] A. ISHIZAKA and Y. SHIRAKI, J. Electrochem. Soc. **133**, 666 (1986).
[15] T. YODO, H. TSUCHIYA, H. ANDO, and Y. HARADA, Jpn. J. Appl. Phys. **39**, 2523 (2000).
[16] R. J. MOLNAR and T. D. MOUSTAKAS, J. Appl. Phys. **76**, 4587 (1994).

A Novel Two-Step Method for Improvement of MOVPE Grown InN Film on GaP(111)B Substrate

A. G. Bhuiyan[1]), A. Yamamoto, and A. Hashimoto

Department of Electrical and Electronics Engineering, Fukui University, 3-9-1 Bunkyo, Fukui 910-8507, Japan

(Received June 23, 2001; accepted August 4, 2001)

Subject classification: 68.37.Ps; 68.55.Jk; 81.15.Aa; S7.14

This paper reports a novel two-step method for improvement of MOVPE grown InN film on GaP(111)B, which includes the growth of a low temperature InN buffer layer and a high temperature epilayer. The additional feature of this two-step method which distinguishes it from the commonly known two-step method is that after growing the low temperature InN buffer layer the temperature is raised to the epitaxial growth temperature while continuing the growth. It is found that a single crystalline InN film with an excellent surface morphology can be grown on GaP(111)B at high temperature ($\sim$600 °C) by this novel two-step method. In contrast, InN film grown by the conventional two-step method is found to be very rough. Differences between these two growth techniques and their influences on the buffer layer and then on main epilayer are also discussed.

Introduction InN has attracted much attention because of its excellent electron transport properties, as predicted theoretically [1, 2]. In addition, InN has gained importance as a constituent compound of InGaN and InAlN ternary alloys, which are currently being used in the heterostructure based electronic device applications. However, unlike the intensively studied GaN, InGaN, and other nitride compounds, InN remains as one of the least studied nitride materials. Sapphire has been widely used as a substrate material for InN. However, a high quality grown film of InN has yet to be reported. To enhance improvement of the grown InN film, we are looking to GaP, as a suitable semiconductor substrate, having smaller lattice mismatch and closer thermal expansion coefficient with InN than that of sapphire. For example, lattice mismatch for InN/GaP(111) is 8%, and for InN/α-Al$_2$O$_3$(0001) is 22.5%. We succeeded to grow a single crystalline InN film on GaP(111)B and found that nitridation of GaP has a poor effect on the grown InN film due to formation of PN$_x$ [3]. We also clearly clarified the significance of the nitridation suppression of GaP and InP substrates and have found that high temperature results in a two-dimensional growth of InN on GaP(111)B [4].

The two-step growth method, which consists of a low temperature buffer layer in the first step and high temperature epilayer in the second step, is now a relatively mature growth technique to obtain good epilayers of III-nitrides [5]. Time of ramping and annealing the buffer have also a large influence on the structure of buffer layer and on the quality of the main epilayer. However, such efforts are yet to be concentrated to obtain good InN epilayers. This paper reports a novel two-step method for improvement of MOVPE grown InN films on GaP(111)B, which includes the growth of a low temperature InN buffer layer and a high temperature epilayer. The additional feature of this two-step method which distinguishes it from the commonly known two-step method is that, after growing the low temperature InN buffer layer the temperature is raised to the epitaxial growth temperature while continuing the growth.

---

[1]) Corresponding author; Fax: +81 776 27 8749; e-mail: e000264@icpc00.icpc.fukui-u.ac.jp

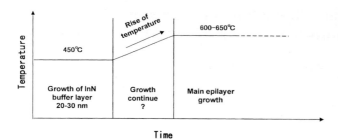

Fig. 1. MOVPE growth of InN on GaP(111)B by a novel two-step method (schematic). An InN buffer layer is grown at low temperature and then the temperature is raised to the desired epitaxial growth temperature while continuing the growth. The InN film is then grown at the desired temperature

Experimental The growth of InN was performed with a conventional MOVPE system with a horizontal reactor. Wafers of GaP(111)B were used as a substrate. The substrates were degreased and then etched by dipping into HCl + HNO$_3$ (3:1). After loading into the reactor the substrates were thermally cleaned at 700 °C for 5 min in the flowing H$_2$. The MOVPE growth of InN was carried out using TMI and NH$_3$ as source materials and N$_2$ as a carrier gas in a reduced pressure of about 0.1 atm. A low temperature InN buffer layer was grown at 450 °C for 20–30 min with a V/III ratio of 7.5×10^4. The temperature was then raised to the desire epitaxial growth temperature while continuing the growth. The epitaxial film of InN was then grown at the temperature range of 600–650 °C with a NH$_3$/TMI molar ratio in the range of $(0.9–7.5) \times 10^4$. The schematic of the InN growth on GaP(111)B by MOVPE is shown in Fig. 1. The thickness of the initial InN layer is estimated to be 20–30 nm and that of total epilayer is 130–400 nm. The InN grown films were evaluated with RHEED and AFM analysis.

Results and Discussion A higher growth temperature is expected to result in a better crystalline quality of the grown film. High temperature growth using a low temperature buffer layer, i.e., two-step growth, has now become a standard method for the III-nitride growth by MOVPE. Time of ramping and annealing the buffer have also a large influence on the structure of buffer layer and on the quality of the main epilayer. However, annealing the InN buffer is thought to be difficult due to low InN dissociation temperature and high equilibrium N$_2$ vapor pressure over the InN film. After growing the low temperature InN buffer layer increasing the temperature to the high temperature (600–650 °C) for the main epitaxial growth is difficult due to decomposition of InN even in any flow environment, like N$_2$ or NH$_3$ flow. For example, InN was not found to be formed if NH$_3$ is exposed to InAs [6] and InP [4] surfaces above 500 °C. Instead, metallic In was found to be formed due to the low decomposition temperature of InN. Guo et al. [7] also reported that if a single crystalline InN film is heated above 550 °C in a N$_2$ flow, the surface undergoes a considerable change, owing to the decomposition and desorption of nitrogen. Therefore, suppression of the InN decomposition during the temperature rising period and shortening of the ramping appear to be very significant for the improvement of the subsequently grown InN film. But in the case of GaN, due to high thermal stability such problems do not occur and the quality of the main epilayer improves by annealing the low-temperature buffer layer.

In order to solve the above-mentioned problem for InN, we are proposing a novel two-step growth technique for the improvement of the grown InN film on GaP(111)B substrate. The additional feature of this novel two-step method is that after growing the low temperature InN buffer layer the temperature is raised to the epitaxial growth

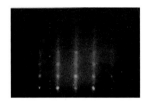 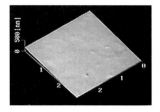

Fig. 2 (colour). RHEED pattern and AFM image of a InN film grown on GaP at 600 °C by the novel two-step method

(a) RHEED   (b) AFM

temperature while continuing the growth. In contrast, in the conventional two-step method after growing the low temperature InN buffer layer, growth is stopped and then the temperature is raised to the epitaxial growth temperature in an N_2 or NH_3 flow environment. Figure 2[2]) shows the RHEED pattern and AFM image of an InN film grown on GaP(111)B at 600 °C by the proposed novel two-step method. The buffer layer was grown at 450 °C and epilayer was at 600 °C. The estimated thickness for the buffer layer is 20 nm and for epilayer is 100 nm. The additional thickness of the InN layer grown during the rise of temperature from 450–600 °C is estimated to be 15 nm. A single crystalline InN film with an excellent surface morphology is grown on GaP(111)B at high temperature (∼600 °C) by the novel two-step method as shown by RHEED pattern and AFM image in Fig. 2. In contrast, if the InN film is grown by the conventional two-step method the film quality becomes very poor. Figure 3[2]). shows the AFM images of InN films grown on GaP(111)B at 600 °C by conventional two-step method with different buffer layer thickness grown at 450 °C. The surface of the InN films grown by the conventional two-step method is very rough as shown in Fig. 3. Considering the additional thickness of the buffer layer for the novel two-step method during the rise of temperature, the thickness of the InN buffer layer for the conventional two-step method was increased from 20 to 30 nm. But no improvement of the grown InN was observed as shown in Fig. 3b. It seems due to that, after growing the low temperature InN buffer layer, when the growth is stopped and then the temperature is increased to the epitaxial growth temperature (∼600 °C) in an N_2 flow environment, the grown InN buffer layer becomes very rough due to the decomposition of InN and results poor effects on the subsequently grown InN. But if the growth is continued during the rise of temperature, the decomposition of InN buffer layer can easily be suppressed due to the continuous growth of InN. Therefore, significant improvement of the InN film grown on GaP(111)B can be obtained by the novel two-step growth tech-

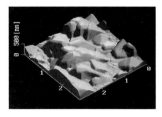 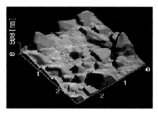

Fig. 3 (colour). AFM images of InN films grown on GaP at 600 °C by the conventional two-step method with different buffer layers

(a) 20 nm   (b) 30 nm

[2]) Colour figures are published online (www.physica-status-solidi.com), where indicated.

nique as explained above. Even a single crystalline InN film with an excellent surface morphology can be grown on GaP(111)B substrate at 650 °C by this novel two-step method.

Conclusion We have studied a novel two-step method for the improvement of MOVPE grown InN film on GaP(111)B and compared it with the conventional two-step method. The additional feature of this two-step method is that, after growing the low temperature InN buffer layer the temperature is raised to epitaxial growth temperature while continuing the growth. A single crystalline InN film with excellent surface morphology can be grown on GaP(111)B at high temperature ($\sim$600 °C) by the proposed novel two-step method, whereas InN films grown by the conventional two-step method are found to be poor. Suppression of the InN decomposition during the temperature rising period is found to be an important factor. The differences between these two growth techniques and their influences on the buffer layer and then on main epilayer are also clarified.

Acknowledgement This work has been supported in part by CREATE FUKUi of JST (Japan Science and Technology Corporation).

References

[1] S. K. O'LEARY, B. E. FOUTZ, M. S. SHUR, U. V. BHAPKAR, and L. F. EASTMAN, J. Appl. Phys. **83**, 826 (1998).
[2] B. E. FOUTZ, S. K. O'LEARY, M. S. SHUR, and L. F. EASTMAN, J. Appl. Phys. **85**, 7727 (1999).
[3] A. G. BHUIYAN, A. HASHIMOTO, A. YAMAMOTO, and R. ISHIGAMI, J. Cryst. Growth **212**, 379 (2000).
[4] A. G. BHUIYAN, A. YAMAMOTO, A. HASHIMOTO, and R. ISHIGAMI, Proc. Internat. Workshop Nitride Semiconductors, Nagoya (Japan), September 24–27, 2000; IPAP Conf. Series **1**, 343 (2000).
[5] I. AKASAKI and H. AMANO, Tech. Dig. Internat. Electron Device Meeting **96**, 231 (1996).
[6] A. YAMAMOTO, T. SHIN-YA, T. SUGIURA, M. OHKUBU, and A. HASHIMOTO, J. Cryst. Growth **189/190**, 476 (1998).
[7] Q. GUO, O. KATO, and A. YOSHIDA, J. Appl. Phys. **73**, 7969 (1993).

Phase Separation of $Al_{1-x}In_xN$ Grown at the Resonance Point of Nitrogen-ECR Plasma

K. Murano (a), T. Inushima[1]) (a), Y. Ono (a), T. Shiraishi (a), S. Ohoya (b), and S. Yasaka (b)

(a) Department of Electronics, Tokai University, Hiratsuka 259-1292, Japan

(b) Kanagawa Industrial Technology Research Institute, Ebina 243-0435, Japan

(Received July 9, 2001; accepted August 4, 2001)

Subject classification: 68.55.Nq; 71.20.Nr; 78.55.Cr; 81.15.Jj; S7.14

We present a novel method to grow $Al_{1-x}In_xN$ by the use of high density active nitrogen species, which are generated at the resonance point of nitrogen-ECR (electron cyclotron resonance) plasma, where metal Al and In are evaporated onto sapphire c-plane simultaneously. By the use of this method, $Al_{1-x}In_xN$ ($x \leq 0.5$) films are grown at a substrate temperature lower than 600 °C. From the X-ray, luminescence, absorption and phonon structure analyses, we conclude that $Al_{1-x}In_xN$ has a phase separation in the small x region and has a trap level at 1.66 eV which originates from the electronic structure of InN.

Introduction Mixed crystal $Al_{1-x}In_xN$ is an attractive material for its wide band gap variation from 6.2 eV of AlN to 1.9 eV of InN, and its application to blue light emitting devices is expected. The attempt to grow $Al_{1-x}In_xN$ so far has been made by the use of reactive sputtering [1, 2] and MOCVD methods [3, 4]. Recently the compositional fluctuation and inhomogeneity of $Al_{1-x}In_xN$ have been reported [5, 6]. The band gap bowing of $Al_{1-x}In_xN$ is especially a controversial issue due to the uncertainty of the band gap energy of InN itself [7]. From the thermodynamical consideration, mixed crystal of $Al_{1-x}In_xN$ has been regarded as an unstable system due to its large lattice mismatches (13.5%) between the constituent compounds and the high volatility of N over InN, and therefore, it is predicted that a compositional immiscibility gap exists [8, 9]. The growth temperature of InN is about 450 °C, while that of AlN is 1100 °C for MOCVD growth and therefore, a new method to grow AlN at much lower temperatures is required for the crystal growth of this system.

Previously we reported the crystal growth of AlN at the resonance point of the electron-cyclotron resonance (ECR) nitrogen plasma and Al was evaporated at the resonance point where high density of active nitrogen was generated [10]. By the use of this method, we deposited AlN onto an Al inter-digital-transducer electrode and fabricated surface acoustic wave filters with a high quality-factor. The growth temperature of this method was lower than 600 °C.

Here we present the optical properties of $Al_{1-x}In_xN$ films ($x \leq 0.5$) grown by the use of this method, where metal Al and In were evaporated onto the sapphire c-plane simultaneously.

Experiments The equipment used for growing $Al_{1-x}In_xN$ was reported elsewhere [10]. Al of 99.999% and In of 99.99% purity were used as the evaporation materials. The In

---

[1]) Corresponding author; Phone: +81 46 358 12 11, Fax: +81 46 358 83 20, e-mail: inushima@keyaki.cc.u-tokai.ac.jp

and Al source crucibles were set at the resonance point of the ECR plasma and the sample holder was set vertically 2 cm above the crucibles. The sample holder should be set within 2 cm from the resonance point (875 Gauss). The crucibles were made of Ta and had Ta covers to protect Al and In from N_2 plasma. A hole of a diameter of 1.5 mm was made in the covers. A 2.45 GHz microwave power of 1.5 kW was used for the generation of N_2 plasma and the deposition pressure was 1×10^{-5} Torr, which was obtained at N_2 flow rate of 30 sccm. Under these deposition conditions, a growth rate of 500 nm/h was obtained on the sapphire (0001) surface. The sample thicknesses were between 400 and 500 nm. The composition ratio was determined by energy dispersive X-ray (EDX) spectroscopy: the accuracy of x was about 2% and the variation of the composition was less than 2% in the area investigated. The content x in $Al_{1-x}In_xN$ was controlled by the heating current of each crucible. In this experiment we did not use any heater to keep the substrate temperature constant, but it was heated by the radiation from the crucible and by the plasma itself, hence the temperature for the crystal growth was 400–500 °C and the deposition temperature was almost constant during the crystal growth. The temperature variation owing to the change of current heating the crucibles was negligibly small. The X-ray data were collected by the conventional $2\theta-\omega$ scan mode using a CuK_α line. The photoluminescence spectra were taken by a 365 nm line of He–Cd laser at room temperature.

Results and Discussion Figure 1 shows the XRD patterns of the $Al_{1-x}In_xN$ film deposited on sapphire (0001). In the X-ray spectra only (0002) and (0004) reflections of $Al_{1-x}In_xN$ are observed, from which it is understood that the c-axis of $Al_{1-x}In_xN$ is perpendicular to the substrate surface. At $x = 0.06$ we observe two peaks of (0002); one is from InN and the other is from AlN. As the In content increases, the (0002) of AlN shifts to the lower angle side, but its angle is larger than that expected from Vegard's law of $Al_{1-x}In_xN$. The peak originating from InN can be observed until $x = 0.2$ at similar position to that of (0002) of InN. At $x = 0.5$, which is the largest composition we obtained in this experiment, only one peak of $Al_{0.5}In_{0.5}N$ is observed at the position expected from Vegard's law.

Figure 2 shows the photoluminescence spectra of $Al_{1-x}In_xN$ at room temperature. As the spectra were obtained by the use of He–Cd laser (365 nm), the luminescence above 3.4 eV could not be observed in Fig. 2. The luminescence consists mainly of two peaks, one observed around 2.3 eV

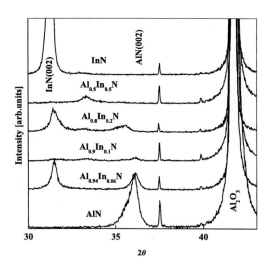

Fig. 1. X-ray diffraction spectra of $Al_{1-x}In_xN$ films for various In content. The alloy composition x is estimated by EDX measurements

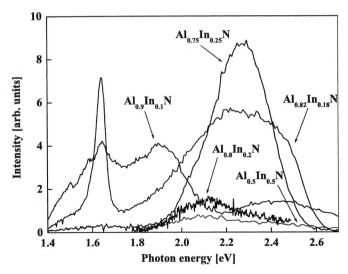

Fig. 2. Photoluminescence spectra of $Al_{1-x}In_xN$ as a function of In content obtained at room temperature

and the other at 1.66 eV. The former is broad and shows weak x dependence and its energy position vs. x is plotted in Fig. 3. The latter is sharp and is sensitive to the position which the laser excites. This peak is observed when $x < 0.2$. When $x > 0.5$, the samples have metallic resistivity and we could not observe any luminescence.

The fundamental absorption spectra of $Al_{1-x}In_xN$ are obtained from the transmission and the reflection spectra at room temperatures. From the extrapolation of the square of the absorption coefficient, the band gap energy (E_g) is determined, which is shown

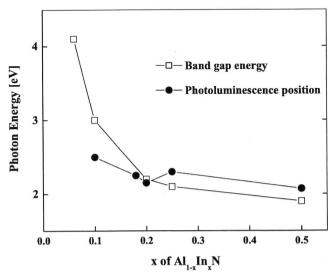

Fig. 3. Band gap energy of $Al_{1-x}In_xN$ as a function of In content. The peak position of the photoluminescence taken at the higher energy side is also plotted

in Fig. 3 together with the energy position of the luminescence peak. As is seen in Fig. 3, the band gap energy of $Al_{1-x}In_xN$ decreases steeply in the small x region and almost saturates at $x \geq 0.2$. At $x \geq 0.2$, the discrepancy between the luminescence position and the band gap energy disappears.

The band structure of $Al_{1-x}In_xN$ was calculated based on the ab initio calculation and it was concluded that $Al_{1-x}In_xN$ is a direct-band-gap material and E_g shifts smoothly from 6.1 to 1.9 eV [11]. On the contrary, experimentally obtained band gap bowing was not so simple. Peng et al. [2] reported that at $x \approx 0.1$, $E_g \approx 3.0$ eV and at $x > 0.2$, E_g had weak x dependence as we observed in this experiment. Yamaguchi et al. [6] studied the relationship between the optically determined band gap energy and the luminescence peak position at $0.1 < x < 0.6$, and it is essentially the same as that we show in Fig. 3. Moreover they reported the convergence of the luminescence position to 1.66 eV in the large x region. It is obvious that there is a trap level at 1.66 eV which originates from the electronic structure of InN.

As is seen in Fig. 1, $Al_{1-x}In_xN$ contains an InN phase in the film at the small x region. When we take the reflection spectra of these films with the incident and reflected light beams nearly normal to the film surface, we see two optical phonons at 595 cm^{-1} and 912 cm^{-1}, which are E_1(LO) phonons of AlN and InN, respectively. At present, we do not have a clear idea of the crystal morphological coexistence of these two phases in $Al_{1-x}In_xN$, but it is obvious that InN phase is involved in $Al_{1-x}In_x$ at small x region. This result is understood that when the In content increases, the crystal quality becomes worse, which obscures the occurrence of phase separation.

The phase separation of $Al_{1-x}In_xN$ has not been reported yet, but that of $Ga_{1-x}In_xN$ was reported [12], and superlattice-like diffraction spots were mentioned. Further investigation is needed with $Al_{1-x}In_xN$.

Conclusion We presented a novel technique to grow AlN by the use of a high density active nitrogen species at low temperatures. Obtained $Al_{1-x}In_xN$ has a phase separation at the small x region.

References

[1] K. Kubota, Y. Kobayashi, and K. Fujimoto, J. Appl. Phys. **66**, 2984 (1989).
[2] T. Peng, J. Piprek, G. Qui, J. O. Olowolafe, K. M. Unruh, C. P. Swann, and E. F. Schubert, Appl. Phys. Lett. **71**, 2439 (1997).
[3] Q. Guo, H. Ogawa and A. Yoshida, J. Cryst. Growth **146**, 462 (1995).
[4] K. S. Kim, A. Saxler, P. Kung, and K. Y. Lim, Appl. Phys. Lett. **71**, 800 (1997).
[5] S. Yamaguchi, M. Kariya, S. Nitta, T. Takeuchi, C. Wetzel, H. Amano, and I. Akasaki, Appl. Phys. Lett. **73**, 830 (1998).
[6] S. Yamaguchi, M. Kariya, S. Nitta, T. Takeuchi, C. Wetzel, H. Amano, and I. Akasaki, Appl. Phys. Lett. **76**, 876 (2000).
[7] T. Inushima et al., J. Cryst. Growth **227/228**, 481 (2001).
[8] T. Matsuoka, Appl. Phys. Lett. **71**, 105 (1997).
[9] T. Takayama, M. Yuri, K. Ito, T. Baba, and S. Harris Jr., Jpn. J. Appl. Phys. **39**, 5057 (2000).
[10] K. Murano, T. Inushima, Y. Wakasugi, H. Kondo, S. Ohoya, and V. Yu. Davydov, IPAP Conf. Ser. **1**, 190 (2000).
[11] A. F. Wright, and J. S. Nelson, Appl. Phys. Lett. **66**, 3465 (1995).
[12] A. Wakahara, T. Tokuda, X. Z. Dang, S. Noda, and A. Sasaki, Appl. Phys. Lett. **71**, 906 (1997).

Spatial Variation of Luminescence of InGaN Alloys Measured by Highly-Spatially-Resolved Scanning Catholuminescence

F. Bertram[1])*) (a), S. Srinivasan (a), L. Geng (a), F. A. Ponce (a),
T. Riemann (b), J. Christen (b), S. Tanaka (c), H. Omiya (c), and
Y. Nakagawa (c)

(a) Department of Physics, Arizona State University, Tempe, AZ 85287-1504, USA

(b) Institute of Experimental Physics, Otto-von-Guericke University, Magdeburg, Germany

(c) Nichia Corporation, Tokushima-ken 774-8601, Japan

(Received June 21, 2001; accepted August 4, 2001)

Subject classification: 61.72.Qq; 78.60.Hk; S7.14

Cathodoluminescence (CL) measurements were performed on a set of thick InGaN layers covering systematically a wide range of indium concentrations ($x = 0.03-0.20$). These thick InGaN layers are exceptionally specular for low indium concentrations ($x < 0.1$), while some degree of microscopic roughness is observed for $x > 0.1$. While in CL mappings the size of the areas with constant emission wavelength decreases with indium content, a similar change of domain size is observed by AFM. For low indium content, statistical fluctuations of the local indium concentration lead to a Gaussian broadening of a single emission line. In contrast, for $x > 0.1$ phase separation results in a multimodal distribution of the peak wavelength, leading to additional low-energy peaks in CL overview spectra. In highly spatially resolved CL measurements we correlate these low-energy emissions to characteristic structural defects.

Introduction InGaN is the active material used for light-emitting diodes (LED) and laser diodes (LD) in the ultraviolet and blue wavelength region. It is surprising that LEDs grown on sapphire exhibit high quantum efficiency even at room temperature in spite of a large number of threading dislocations acting as nonradiative recombination centres. In the recent literature the structural dependence of optical properties of InGaN have been extensively investigated by CL, micro-photoluminescence (μ-PL) and scanning near field optical microscopy (SNOM). Here, the CL is mainly used to show the very inhomogeneous InGaN emission to proof a strong localization [1–3]. Only a few authors were able to directly correlate specific emission with structural features [4, 5].

In this paper thick epitaxial InGaN layer were under study. The study of pure InGaN epilayers allows to focus on the optical properties exclusively related to the InGaN alloy system, avoiding the effects of quantum confinement, thickness fluctuations, and lattice-mismatched induced strain in InGaN/GaN QW structures. We directly correlate the CL emission with structural features observed at the sample surface by AFM and SEM.

---

[1]) Corresponding author; Phone: +49 391 67 11259/12606; Fax: +49 391 67 11130; e-mail: frank.bertram@physik.uni-magdeburg.de
*) Permanent address: Institute of Experimental Physics, Otto-von-Guericke University, Universitätsplatz 2, D-39106 Magdeburg, Germany.

Experimental Set-Up The nominal 80 nm thick InGaN layers were grown by MOVPE on 4 μm thick GaN using a *c*-plane sapphire substrate. The average indium concentration varies systematic over a wide range ($x = 0.03–0.20$). These thick InGaN layers are exceptionally specular for low indium concentrations $x < 0.1$. Some degree of microscopic roughness was observed for $x > 0.1$. The average indium content of the layers was determined by X-ray diffraction (XRD) measurements of the lattice parameters in a $\theta/2\theta$ scan taking into account the effect of biaxial stress. These results agree with the compositions derived from Rutherford Backscattering Spectroscopy (RBS).

The CL measurements were performed on two different systems. The statistical analysis was performed using a self-built CL system in a modified JEOL 6400 scanning electron microscope [6]. The other CL system used in this study was a commercial MonoCL2 of Gatan. The kinetic energy of the electrons used for the CL excitation was reduced to 3 keV for highly resolved CL images, as compared to 5 keV for the standard CL work presented here. At 3 keV the centre of excitation is about $d_{exc} = 30$ nm below the surface and the diameter of the generation volume of secondary electrons and finally the electron–hole pairs, the Bethe range, is largely reduced to about $R_B = 90$ nm ($d_{exc} = 60$ nm and $R_B = 170$ nm, respectively, for 5 keV) [7]. The samples were studied at 4 K and the beam current was varied between 6 and 300 pA.

Results and Discussion In order to correlate microstructure with spatial dependence of the luminescence, average CL overview spectra over large areas were compared with high-spatial-resolution CL measurements. CL overview spectra were taken from samples with average indium concentration ranging from $x = 0.03–0.20$. As expected, the peak corresponding to the InGaN shifts to lower emission energies with increasing indium content. For low x the InGaN emission consists of a single peak from a homogeneous InGaN matrix. For higher indium concentration ($x > 0.1$) additional, clearly separated two to three peaks can be found on the low energy side of the main peak. These additional peaks are about one order of magnitude less in intensity compared to the main InGaN peak. The full width at half maximum (FWHM) of the main peak varies significantly with indium content. At first, line width increases linearly with x due to statistical fluctuations of alloy composition. With the appearance of the additional low energy peaks the FWHM of the main peak remains constant for $x > 0.1$.

The lateral fluctuations of the local emission wavelength at 4 K of two samples ($x = 0.03$ and 0.17) are plotted in the wavelength images of Figs. 1a[2]) and c, respectively. The lateral fluctuations of the luminescence clearly increase for higher indium incorporation. For low average x, relatively small fluctuations occur on a scale of several microns. In contrast, for higher indium content much larger changes are observed on a sub-micron scale. To quantify the lateral fluctuations of the peak position, i.e. the lateral x fluctuation, we have performed a statistical analysis of Figs. 1a and c. Histograms of the CL wavelength images (51200 pixels each) are plotted in Figs. 1b and d, directly quantifying the impact of indium content on optical homogeneity and quality. For the sample with lower indium content the statistical distribution is sharp and has a Gaussian shape resembling a perfectly random fluctuation. For higher indium content the mean value of the histogram shifts to lower energies and broadens, and additional peaks on the low-energy side of the main peak can be found. The distribution of wave-

---

[2]) Colour figure is published online (www.physica-status-solidi.com).

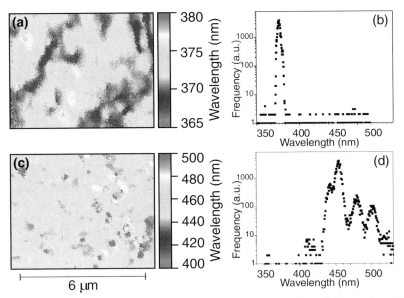

Fig. 1 (colour). CL wavelength images and histograms of the InGaN samples with $x = 0.03$ and 0.17

lengths changes from single-mode Gaussian to a multimodal distribution for higher [In]. This implies that for higher indium concentrations ($x > 0.1$) specific longer wavelength positions, i.e. specific high [In], occur more frequently. The highest CL intensity appears from a homogeneous InGaN matrix at $\lambda = 450$ nm. In addition, a number of laterally clearly separated domains emitting strongly red-shifted luminescence are found in this homogeneous matrix. To summarize, the low-energy peaks in the overview spectra caused by accumulation of sharp single lines appearing at specific longer wavelengths. This is proven by the fact that the peak positions in the histogram Fig. 1d match with the peaks in the overview spectrum. The CL intensity of these low-energy peaks is about one order of magnitude lower than of main peak of the homogeneous InGaN matrix.

In AFM images a strong impact of indium content on the surface morphology is seen. For low indium concentrations large hexagonal hillocks (lateral size of 1.3 μm) with some small craters are obtained. For higher indium content the hillocks lose their hexagonal shape. The size of the craters increases from 50 to 200 nm. For $x > 0.1$ the RMS roughness measured by AFM jumps to a seven times higher value, but the mean lateral feature size decreases linearly with [In]. The reason for this jump in roughness lays in the appearance of additional surface features for $x > 0.1$. The lateral size of these defects varies between 500 and 800 nm and the mean height is 15 nm, pining out of the main surface matrix like needles. For even higher x only the density of these needles increases.

Characteristic features are also visible in the SEM image (Fig. 2a) like in the AFM: craters and needles. The needles are marked with arrows. The big circle in the images assigns three craters and one needle. Most of the craters show a specific SEM contrast: a black hole in the centre with a bright contrast around. A thickness enhancement around the craters leads to a plateau. An AFM measurement gives an averaged height of 3 nm for the plateaus.

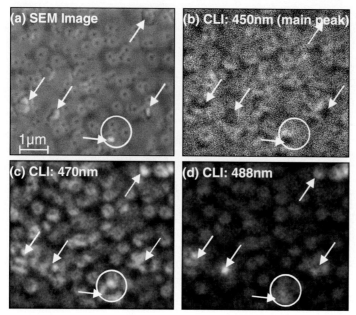

Fig. 2. SEM image together with a set of monochromatic images from InGaN sample with $x = 0.12$. In the SEM image craters and spikes are visible. Arrows mark the spikes and the big circle marks one spike and three craters

To find out the structural origin of the luminescence monochromatic images were taken from an InGaN sample with $x = 0.12$. The three monochromatic images (CLI) in Fig. 2 represent certain spectral regions in the CL overview spectrum and histogram: the main peak is the from InGaN matrix (450 nm), the first (470 nm) and second (488 nm) peaks from the low energy side of the main peak. The CLI of the main peak (InGaN matrix) shows an almost homogeneous intensity distribution (Fig. 2b). At the position of the needles dark spots appear in this CLI. In the longer wavelength interval the luminescence is exclusively detected from the positions of craters and needles. Not all craters show bright luminescence in the longer wavelength region, but all needles. Furthermore, the highest CL intensity in longer wavelength region originates from the needles. In particular, in the CLI at $\lambda = 470$ nm the luminescence is emitted by the plateau around the crater (a ring with a hole in the middle is visible), while in the CLI at $\lambda = 488$ nm high CL intensity is concentrated at the centre of the crater.

Conclusion The spatial variation of InGaN alloys with different composition was measured with highly-spatially-resolved cathodoluminescence microscopy. The dependence on indium composition of the homogeneity of luminescence was obtained. The low energy shoulder of the InGaN peak in the CL overview spectra for $x > 0.1$ was correlated to structural defects like craters and needles. The luminescence change over a crater was analysed. For each alloy composition preferential wavelength positions were found.

Acknowledgements This work was partially supported by ONR (N00014-00-1-0133).

References

[1] S. Chichibu, K. Wada, and A. Nakamura, Appl. Phys. Lett. **71**, 2346 (1997).
[2] S. Chichibu, T. Sota, K. Wada, and S. Nakamura, J. Vac. Sci. Technol. B **16**, 2204 (1998).
[3] S. Chichibu, K. Wada, J. Muellhaeuser, O. Brandt, K. H. Ploog, T. Mizutani, A. Setoguchi, R. Nakai, M. Sugiyama, H. Nakanishi, K. Korii, T. Deuchi, T. Sota, and S. Nakamura, Appl. Phys. Lett. **76**, 1671 (2000).
[4] D. Cherns, S. J. Henley, and F. A. Ponce, Appl. Phys. Lett. **78**, 2691 (2001).
[5] X. H. Wu, C. R. Elsass, A. Abare, M. Mack, S. Keller, P. M. Petroff, S. P. DenBaars, J. S. Speck, and S. J. Rosner, Appl. Phys. Lett **72**, 692 (1998).
[6] F. Bertram, T. Riemann, J. Christen, A. Kaschner, A. Hoffmann, C. Thomsen, K. Hiramatsu, T. Shibata, and N. Sawaki, Appl. Phys. Lett. **74**, 359 (1999).
[7] T. M. Levin, G. H. Jessen, F. A. Ponce, and L. J. Brillson, J. Vac. Sci. Technol. B **16**, 2545 (1999).

A Comparison of Rutherford Backscattering Spectroscopy and X-Ray Diffraction to Determine the Composition of Thick InGaN Epilayers

S. Srinivasan (a), R. Liu (a), F. Bertram (a), F.A. Ponce[1]) (a), S. Tanaka (b), H. Omiya (b), and Y. Nakagawa (b)

(a) Department of Physics and Astronomy, P.O. Box 871504, Arizona State University, Tempe AZ-85287-1504, USA

(b) Nichia Chemical Industries, Tokushima-ken 7748601, Japan

(Received June 26, 2001; accepted August 4, 2001)

Subject classification: 61.10.Kw; 61.18.Bn; 68.55.Nq; 85.40.Xx; S7.14

In this paper, we report the measurements of indium composition of thick InGaN epilayers by X-ray diffraction (XRD) and Rutherford backscattering spectroscopy (RBS). In order to account for the biaxial stress in the InGaN epilayers, we determined both a and c lattice parameters in a $\theta/2\theta$ scan. Indium composition was determined by simultaneous application of Vegard's law to both lattice parameters and by considering the relationship between the lattice parameters under strain. These composition values are compared with values determined by RBS. The value of elastic constants used in these calculations is critical and we show that by careful choice we can obtain a good correlation between the XRD and RBS measurements.

Introduction The variation of the optical properties of the ternary $In_xGa_{1-x}N$ alloy with indium composition has recently evoked tremendous interest [1, 2]. By varying the indium composition it is possible to achieve light emission in the entire visible region [3]. For a complete understanding of this behavior it is important to determine to a high accuracy the indium composition of these alloys. Due to the difficulties involved in growing thick InGaN films [4] most of the work done on these matertials has been on quantum wells (QWs). It has typically been very difficult to determine accurately the indium composition in InGaN QWs. Rutherford backscattering spectroscopy (RBS) has not been a viable technique for studying buried QW layers. In this work, we have studied relatively thick InGaN films where both RBS and X-ray diffraction (XRD) can be used effectively in combination.

Typically, the indium composition of InGaN epilayers has been determined by XRD measurement of the c lattice parameter by a symmetric scan of (0002) peaks followed by the application of Vegard's law [5]. However, in the presence of lattice strain, such an approach could result in erroneous measurements of the composition. We have tried to overcome this problem by measuring both lattice parameters a and c followed by a simultaneous application of Vegard's law to both while also accounting for the biaxial strain in the lattice.

In order to compare these results with another independent technique we have performed RBS. By a careful choice of elastic constants we have found a good correlation between the composition values determined by the two techniques.

---

[1]) Corresponding author; Phone: +1 480 727 6260; Fax: +1 480 965 7954; e-mail: ponce@asu.edu

Experimental Details The samples studied were InGaN epitaxial layers grown by metallorganic chemical vapor deposition. The layers were grown on c-plane sapphire using an AlGaN buffer layer, followed by 4 μm GaN and 100 nm InGaN films.

XRD $\theta/2\theta$ scans were done using a BEDE D[1] Scientific System in four-bounce mode. The asymmetric scans for ($10\bar{1}5$) peaks were performed at glancing incidence. An Enhanced Dynamic Range (EDR) detector was used to detect the weak ($10\bar{1}5$) signal.

RBS was carried out in a dedicated set-up consisting of a 1.7 MV General Ionex Tandetron tandem ion accelerator. The backscattered ions were detected by a silicon surface barrier detector. The pulse height analysis was done by means of a multichannel analyzer.

Results and Discussion Figure 1 shows typical XRD spectra from $\theta/2\theta$ scans. The horizontal axis has been modified using Bragg's law to represent the lattice spacing (d). The scale for the InGaN peak has been magnified to reveal the low-intensity peaks. In Fig. 1a, the (0002) peaks obtained by a symmetric scan are shown. The GaN peak occurs at a $d = 2.5921$ Å. The InGaN peak for (0002) occurs at $d = 2.6555$ Å. For a hexagonal unit cell the interplanar spacing of a family ($hkil$) of planes is given by

$$\frac{1}{d_{hkl}^2} = \frac{4}{3}\frac{h^2 + k^2 + hk}{a^2} + \frac{l^2}{c^2}, \tag{1}$$

where a and c are the lattice parameters. Note that the indices obey the relation $i = -(h + k)$. Using Eq. (1), c is determined to be 5.311 Å.

The geometry of the prism planes in InGaN makes it impossible to determine a directly from a symmetric scan. But it can be calculated from asymmetric $\theta/2\theta$ scans like, in our case, ($10\bar{1}5$) reflections. A prior knowledge of c from a symmetric scan is required for this calculation. The XRD spectrum from an asymmetric scan of ($10\bar{1}5$) peaks is shown in Fig. 1b. The GaN peak occurs at 0.9710 Å and the InGaN peak occurs at 0.9922 Å. Using Eq. (1), a was determined to be 3.209 Å.

The indium composition was then determined using the relation [6]

$$x = \frac{ac(1+v) - ac_0^{GaN} - a_0^{GaN}c}{ac_0^{GaN} - ac_0^{InN} - a_0^{InN}cv + a_0^{GaN}cv}, \tag{2}$$

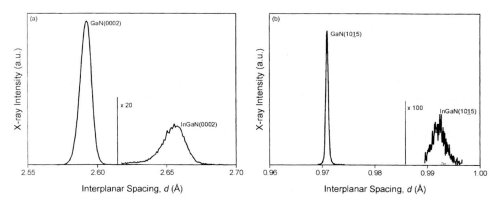

Fig. 1. X-ray diffraction spectra from $\theta/2\theta$ scans. a) Symmetric scan of (0002) peaks and b) asymmetric scan of ($10\bar{1}5$) peaks. The horizontal scale has been modified according to Bragg's law. The scale on the vertical axis has been selectively adjusted to magnify the low intensity InGaN peaks

where $\nu = -\varepsilon_c/\varepsilon_a$ (ε_c and ε_a being the strains along c and a axes, respectively; the lattice parameters of relaxed GaN and InN are designated with the subscript 0). This relationship was derived from a simultaneous application of Vegard's law to both lattice parameters and accounting for strain in the lattice through ν. In a planar biaxially strained wurtzite crystal the relationship between the strains simplifies to $\nu = 2c_{13}/c_{33}$. A wide range of values has been reported in the literature for the elastic constants c_{13} and c_{33} [7] leading to a confusion over the proper choice of ν. The correct choice is critical for an accurate determination of composition. Of the values found in the literature [8, 9], $\nu = 0.6$ best fits our data.

For the current sample, the indium composition was calculated to be 0.167. The overall error in these measurements is estimated to be ± 0.016. Most of this error comes from the poor accuracy in the measurement of a. Better accuracy can be achieved by using higher a-index planes like $(20\bar{2}5)$ or $(11\bar{2}4)$.

The compositions measured by XRD were compared to those determined by RBS. This technique does not require calibrated standards and gives composition values independent of lattice strain and chemical state. Therefore, it is highly suited to the purpose of measuring the composition of strained InGaN epilayers. However, for the study of buried quantum well layers, it has not been a very viable tool to determine composition. We have been able to use RBS to study our thick epilayers and have verified the XRD measurements with these results.

Figure 2 shows the RBS spectrum of the same sample. By fitting a simulated spectrum to the experimental spectrum the indium composition was determined to be 0.17. For a heavy element like indium an accuracy of ± 0.005 is estimated for these measurements.

Figure 3 compares the indium composition determined by the two techniques. The straight line represents equal composition measured from RBS and XRD. It is striking that the XRD measurements are very close to the RBS composition values. As mentioned above, the choice of $\nu = 0.6$ gives the best correlation between the XRD and RBS data. The deviation is within the limits of error estimated above. Better correlation can be obtained by using higher index planes for X-ray analysis.

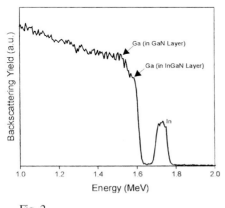
Fig. 2

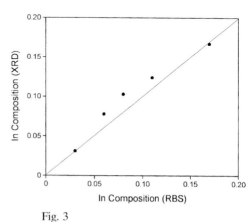
Fig. 3

Fig. 2. Random RBS spectrum of an $In_{0.17}Ga_{0.83}N$ sample

Fig. 3. Indium composition determined by XRD vs. that determined by RBS. The straight line represents equal composition

Conclusions In summary, we have performed XRD and RBS to determine the indium compositions of a set of 100 nm thick InGaN epitaxial films. For XRD measurements, we have accounted for biaxial stress by a simultaneous application of Vegard's law to both lattice parameters and by considering the relationship between the lattice parameters under strain. With a careful choice of the elastic constants used for the calculations, the XRD composition values were found to closely match the compositions determined by RBS. An important finding of this work, verified by the good agreement between XRD and RBS, is that the value $\nu = 0.6$ is the proper value for $In_xGa_{1-x}N$ alloys in the composition range $0 < x < 0.2$.

Acknowledgement This work was supported by ONR (N00014-00-1-0133).

References

[1] C. G. VAN DE WALLE, M. D. MCCLUSKEY, C. P. MASTER, L. T.ROMANO, and N. M. JOHNSON, Mater. Sci. Eng. B **59**, 274 (1999).
[2] K. P. O'DONNELL, R. W. MARTIN, C. TRAGER-COWAN, M. E. WHITE, K. ESONA, C. DEATCHER, P. G. MIDDLETON, K. JACOBS, W. VAN DER STRICHT, C. MERLET, B. GIL, A. VANTOMME, and J. F. W. MOSSELMANS, Mater. Sci. Eng. B **82**, 194 (2001).
[3] F. A. PONCE and D. P. BOUR, Nature **386**, 351 (1997).
[4] T. MATSUOKA, N. YOSHIMOTO, T. SAKAKI, and A. KATSUI, J. Electron. Mater. **21**, 157 (1992).
[5] R. SINGH, D. DOPPALAPUDI, T. D. MOUSTAKAS, and L. T. ROMANO, Appl. Phys. Lett. **70**, 1089 (1997).
[6] H. ANGERER, D. BRUNNER, F. FREUDENBERG, O. AMBACHER, M. STUTZMANN, R. HÖPLER, T. METZGER, E. BORN, G. DOLLINGER, A. BERGMAIER, S. KARSCH, and H.-J. KÖRNER, Appl. Phys. Lett. **71**, 1504 (1997).
[7] C. KIESELOWSKI, J. KRÜGER, S. RUVIMOV, T. SUSKI, J. W. AGER III, E. JONES, Z. LILIENTAL-WEBER, M. RUBIN, E. R. WEBER, M. D. BREMSER, and R. F. DAVIS, Phys. Rev. B **54**, 17745 (1996).
[8] A. POLIAN, M. GRIMSDITCH, and I. GRZEGORY, J. Appl. Phys. **79**, 3343 (1996).
[9] K. KIM, W. R. L. LAMBRECHT, and B. SEGALL, Phys. Rev. B **53**, 16310 (1996).

Local Structure Analysis of $Ga_{1-x}In_xN$ Alloy Using Extended X-Ray Absorption Fine Structure Measurements

T. Miyajima[1]) (a), Y. Kudo (b), K.-Y. Liu (b), T. Uruga (d), T. Asatsuma (a), T. Hino (c), and T. Kobayashi (a)

(a) *Core Technology Development Center, Core Technology & Network Company, Sony Corporation, 4-14-1 Asahi-cho, Atsugi, Kanagawa 243-0014, Japan*

(b) *Environment & Analysis Tech. Dept., Sony Corporation, 4-16-1 Okata, Atsugi, Kanagawa 243-0021, Japan*

(c) *Sony Shiroishi Semiconductor Inc., 3-53-2 Shiratori, Shiroishi, Miyagi 989-0734, Japan*

(d) *Japan Synchrotron Radiation Research Institute, Mikazuki-cho, Hyogo 679-5198, Japan*

(Received July 1, 2001; accepted July 14, 2001)

Subject classification: 61.10.Ht; 68.55.Jk; 81.05.Ea; 81.15.Gh; S7.14

We investigated the local atomic structure around In atoms of MOCVD-grown $Ga_{1-x}In_xN$ alloy ($0.01 \leq x \leq 0.21$) using extended X-ray absorption fine structure (EXAFS) measurements of the In K-edge. For $x \leq 0.16$, the In–Ga and In–In atomic distances were constant at 3.22–3.30 Å and 3.25–3.30 Å, respectively, and close to the Ga–Ga atomic distance in the ideal wurtzite GaN. On the other hand, the In–N bond length was constant at 1.85–2.21 Å, which is close to that in the ideal wurtzite InN. These results suggest that the internal strain of the $Ga_{1-x}In_xN$ alloy is relaxed by the bond angle changing of In–N–Ga or In–N–In. The coordination numbers of the second-nearest neighbor In of In were higher than that estimated from random GaInN alloy, in which In atoms randomly occupy the cation sublattice of GaInN alloy. This behavior can be explained by the segregation of In atoms starting even at $x = 0.01$. For $x = 0.21$, the interatomic distance and the coordination number could not be accurately estimated using our simple model, in which single interatomic distances of In–Ga and In–In were assumed.

Introduction $Ga_{1-x}In_xN$ alloy has been used for the emitting layer of a GaN-based light-emitting diode (LED) and laser diode (LD), however it has a serious problem in that phase separation is occasionally observed in an alloy with a high indium content of more than 20% [1, 2] and the optical properties affect the characteristics of a green and amber LED [3] and a blue LD [4]. Phase separation can be ascribed to the large lattice-constant difference between GaN and InN, but the cause of phase separation is not well understood. We report here the results of our investigation of the local atomic structure of $Ga_{1-x}In_xN$ alloy using the fluorescence-detecting extended X-ray absorption fine structure (EXAFS) measurements.

EXAFS is a powerful tool for characterizing the local atomic structure of an alloy semiconductor [5] and has been applied to $Ga_{1-x}In_xN$ alloy [6–9]. It is difficult, however, to analyze the accurate local structure around the In atom because high brightness X-ray with an energy higher than $E = 27.9$ keV is needed to excite In K-edge. We

---

[1]) Corresponding author; Phone: +81 46 230 5089; Fax: +81 46 230 5775; e-mail: takao.miyajima@jp.sony.com

performed EXAFS measurements on $Ga_{1-x}In_xN$ alloy using synchrotron radiation from an 8 GeV storage ring of SPring-8, which can generate a high brightness X-ray with an energy of several tens of keV.

Experimental Samples of $Ga_{1-x}In_xN$ films and a $Ga_{1-x}In_xN$/GaN single quantum well (SQW) were grown on (0001) sapphire substrates using metal-organic chemical vapor deposition (MOCVD). The $Ga_{1-x}In_xN$ film consisted of an 1 μm thick GaN underlying layer, a 0.18 μm thick $Ga_{1-x}In_xN$ layer ($0.01 \leq x \leq 0.21$), and a 10 nm thick GaN capping layer. The $Ga_{1-x}In_xN$/GaN SQW consisted of a 2 μm thick GaN underlying layer, 3 nm thick $Ga_{0.92}In_{0.08}N$:Si well, and 100nm-thick GaN:Si capping layer. The concentration of Si was 1×10^{19} cm^{-3}. The indium content in the film was determined using X-ray diffraction (XRD) measurements, by assuming coherent growth of $Ga_{1-x}In_xN$ on GaN layer. The accuracy of this estimation was confirmed by Rutherford back-scattering spectroscopy (RBS) measurements.

EXAFS measurements were carried out on the beam-lines of BL16B2 and BL01B1 at SPring-8. The X-ray generated from the storage ring was monochromated by two Si(311) crystals and irradiated on the sample. Fluorescence X-rays of the In K-edge were obtained from the sample and detected using a one- or a 19-element Ge solid-state detector.

Results and Discussion Figures 1 and 2 show the intensities of In K-edge fluorescence X-ray for $Ga_{1-x}In_xN$ films and $Ga_{1-x}In_xN$/GaN SQW as a function of incident photon energy. These intensities were normalized by the intensity of the incident X-ray, which was detected using an ionization chamber. Clear In K-edge EXAFS oscillations are observed beyond the photon energy of 27.92 keV in all samples. This EXAFS measurement system has a high ability for analyzing a small amount of In in a single quantum well. Intensities beyond the photon energy of 27.92 keV increase with increasing In content, but there is no linear relationship because the intensity depends not only on the In content but also on the sample thickness and size.

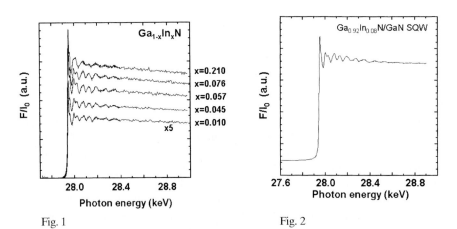

Fig. 1. The intensity of In K fluorescence X-ray for $Ga_{1-x}In_xN$ films as a function of incident photon energy. The intensity for $Ga_{0.99}In_{0.01}N$ was multipled by 5

Fig. 2. The intensity of In K fluorescence X-ray for $Ga_{0.92}In_{0.08}N$/GaN SQW as a function of incident photon energy

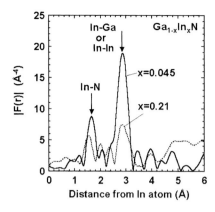

Fig. 3. The absolute values $|F(r)|$ of In K-edge radial structure functions for $Ga_{1-x}In_xN$ of $x = 0.045$ and 0.21

Figure 3 shows the absolute values $|F(r)|$ of In K-edge radial structure functions which were obtained by the Fourier transform of k^3-weight EXAFS oscillations [10] for $Ga_{1-x}In_xN$ films. The two peaks at $r = 1.6$ Å and 2.8 Å correspond to the interatomic distance from In to the first nearest neighbor atom of N, and the second nearest neighbor atom of In or Ga, respectively. The peak position is 0.1–0.5 Å shorter than the real interatomic distance because of the phase shift [10].

The interatomic distances and coordination numbers can be obtained by least-square fitting using the REX2000 package software (Rigaku Corporation). As shown in Fig. 4, the interatomic distances of In–N, In–In and In–Ga in $Ga_{1-x}In_xN$ films do not strongly depend on the In content. The In–N distance is 2.06–2.13 Å, which is close to the 2.169 Å [11] estimated from bulk InN. Below an In content of $x = 0.16$, the In–In and In–Ga interatomic distances are 3.21–3.30 Å, which is close to the 3.189 Å of the Ga–Ga distance in bulk GaN [12]. Therefore, the interatomic distance of In–N in $Ga_{1-x}In_xN$ alloy maintains the bonding character of bulk InN, and the interatomic distances of In–In and In–Ga in $Ga_{1-x}In_xN$ alloy maintain the bonding character of bulk GaN. We believe that the internal strain can be relaxed by changing the bond angle of Ga–N–In and In–N–In. This characteristic was experimentally observed in $Ga_{1-x}In_xN$ alloy grown by molecular beam epitaxy by Morishima et al. [8] and Jeffs et al. [7], and was theoretically predicted by Saito and Arakawa [13] and Takayama et al. [14].

All samples of $Ga_{1-x}In_xN$ alloy have larger coordination numbers of the second-nearest neighbor In atom of In atom than the ideal random alloy, in which In atoms randomly occupy the cation sublattice, as shown in Fig. 5. This means that In atoms are easily segregated around In atoms, and that alloy ordering or clustering is present in $Ga_{1-x}In_xN$ with even a low In content of $x = 0.01$.

At an In content of $x = 0.21$, the interatomic distance of In–In and In–Ga and the coordination numbers of the second-nearest neighbor atom of In could not be accurately estimated using our simple model, in which single interatomic distances of In–Ga and In–In were as-

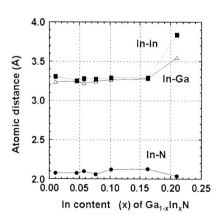

Fig. 4. The interatomic distances of In–N, In–In and In–Ga in $Ga_{1-x}In_xN$ films as a function of In content

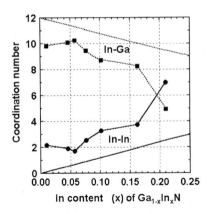

Fig. 5. The coordination numbers of In (●) and Ga (■) second nearest neighbor atom of In in $Ga_{1-x}In_xN$ films as a function of In content. The linear (solid and dotted) lines show the coodination numbers of the In and Ga second nearest atom of In for the ideal random alloy of $Ga_{1-x}In_xN$

sumed. This suggests that a more complicated model is needed for estimating the above parameters. We believe that In compositional spatial fluctuation or phase separation [3, 4] could have occurred in the $Ga_{1-x}In_xN$ film of $x = 0.21$.

Summary $Ga_{1-x}In_xN$ alloy ($0.01 \leq x \leq 0.21$) were investigated using extended X-ray absorption fine structure (EXAFS) measurements of In K-edge. The interatomic distance of In–N is close to that in bulk InN, and the interatomic distance of In–Ga or In–In is close to the Ga–Ga in bulk GaN. We believe that the internal strain can be relaxed by changing the bond angle of In–N–In or In–N–Ga. The coordination numbers of the second-nearest neighbor In of In are larger than those estimated from random alloy. This behavior can be explained by the segregation of In atoms starting even at $x = 0.01$. For $x = 0.21$, the interatomic distance and the coordination number could not be accurately estimated using our simple model. We believe that In compositional spatial fluctuation or phase separation could have occurred in the $Ga_{1-x}In_xN$ film.

Acknowlegements The authors would like to thank F. Nakamura and T. Asano for growing samples. They also thank Dr. S. Kawado, Dr. M. Itabashi, Dr. M. Ikeda and Dr. O. Kumagai for their encouragement during this work.

References

[1] R. SINGH, D. DOPPALAPUDI, T. D. MOUSTAKAS, and L. T. ROMANO, Appl. Phys. Lett. **70**, 1089 (1997).
[2] Y. NARUKAWA, Y. KAWAKAMI, M. FUNATO, SZ. FUJITA, SG. FUJITA, and S. NAKAMURA, Appl. Phys. Lett. **70**, 981 (1997).
[3] T. MUKAI, M. YAMADA, and S. NAKAMURA, Jpn. J. Appl. Phys. **38**, 3976 (1999).
[4] S. NAKAMURA, M. SENOH, S. NAGAHAMA, N. IWASA, T. MATSUSHITA, and T. MUKAI, Appl. Phys. Lett. **76**, 22 (2000).
[5] J. C. MIKKELSEN, JR. and J. B. BOYCE, Phys. Rev. Lett. **49**, 1412 (1982).
[6] A. V. BLANT, T.S. CHENG, N. J. JEFFS, C. T. FOXON, C. BAILEY, P. G. HARRISON, A. J. DENT, and J. F. W. MOSSELMANS, Mater. Sci. Eng. B **50**, 38 (1997).
[7] N. J. JEFFS, A. V. BLANT, T.S. CHENG, C. T. FOXON, C. BAILEY, P. G. HARRISON, J. F. W. MOSSELMANS, and A. J. DENT, Mater. Res. Soc. Symp. Proc. **512**, 519 (1998).
[8] S. MORISHIMA, H. SASAKI, S. ENDO, T. MARUYAMA, K. AKIMOTO Y. KITAJIMA, and M. NOMURA, in: Proc. Internat. Symp. Blue Laser and Light Emitting Diodes, Chiba 1998, Ohmsha Ltd., Tokyo 1998 (p. 242).
[9] K. P. O'DONNELL, R. W. MARTIN, M. E. WHITE, J. F. W. MOSSELMANS, and Q. GUO, phys. stat. sol. (b) **216**, 216 (1999).
[10] D. C. KONINGSBERGER and R. PRINS (Eds.): X-ray Absorption: Principles, Applications, Techniques of EXAFS, SEXAFS and XANES, John Wiley & Sons, New York 1988.
[11] C-Y. YEH, Z. W. LU, S. FROYEN, and A. ZUNGER, Phy. Rev. B **46**, 10086 (1992).
[12] T. DETCHPROHM, K. HIRAMATSU, K. ITO, and I. AKASAKI, Jpn. J. Appl. Phys. **31**, L1454 (1992).
[13] T. SAITO and Y. ARAKAWA, Phys. Rev. B **60**, 1701 (1999).
[14] T. TAKAYAMA, M. YURI, K. ITO, T. BABA, and J. S. HARRIS, JR., J. Appl. Phys. **88**, 1104 (2000).

Indium Surface Segregation during Growth of (In,Ga)N/GaN Multiple Quantum Wells by Plasma-Assisted Molecular Beam Epitaxy

P. Waltereit[1])*) (a), O. Brandt (a), K. H. Ploog (a), M. A. Tagliente (b), and L. Tapfer (b)

(a) Paul-Drude-Institut für Festkörperelektronik, Hausvogteiplatz 5–7, D-10177 Berlin, Germany

(b) Centro Nazionale Ricerca e Sviluppo Materiali (PASTIS-CNRSM), Strada Statale 7 Appia km 712, I-72100 Brindisi, Italy

(Received June 18, 2001; accepted August 4, 2001)

Subject classification: 68.35.Dv; 78.67.De; 81.15.Hi; S7.14

We investigate the synthesis of (In,Ga)N/GaN multiple quantum wells by plasma-assisted molecular beam epitaxy. For metal-stable growth, reflection high-energy electron diffraction and X-ray diffraction reveal massive In surface segregation which is directly confirmed by In depth profiles recorded by secondary-ion mass-spectrometry. These profiles exhibit a top-hat In distribution and are thus indicative of a zero order segregation mechanism instead of a first order process as observed for other materials systems. The segregation of In during metal-stable growth results in quantum wells with smooth interfaces but larger width than intended, and thus causes blue-shifted transition energies and poor quantum efficiency. This unexpected blue-shift may be the reason for the frequent conclusion that the theoretical polarization fields of Bernardini et al. [Phys. Rev. B **56**, R10024 (1997)] are too large for (In,Ga)N. Being in possession of the (at least approximately) correct structural parameters, we find the theoretical fields of Bernardini et al. for (In,Ga)N to be in very satisfactory agreement with the experimental data. Reduction of In segregation by N-stable conditions is possible but inevitably results in rough interfaces.

Introduction (In,Ga)N/GaN multiple quantum wells (MQWs) are widely used in light emitting diodes (LEDs) and injection laser diodes (LDs), but only little insight into the actual structural properties of these MQWs, in particular regarding the formation of their interfaces, has been gained yet. For bulk (In,Ga)N, phenomena such as In surface segregation [1–4], In bulk segregation [5, 6] and complete phase separation [7] are reported. For (In,Ga)N/GaN MQWs, these phenomena are expected to occur as well, but are difficult to distinguish from the impact of the concurrently existing internal electrostatic fields within these structures.

The grower has to pay attention to several phenomena for the synthesis of (In,Ga)N/GaN MQWs. First, the formation enthalpies for InN and GaN are very different, resulting in strong In surface segregation on the growth front [1]. Most reports (see, e. g., Yoshimoto et al. [1]) on the actual segregation mechanism of (In, Ga)N are restricted to the conclusion that the solid phase composition of (In,Ga)N may not follow the gas phase composition linearly. In one work [2], a segregation energy of 0.48 eV was ex-

---

[1]) Corresponding author; Phone: +1 805 893 8523; Fax: +1 805 893 8971; e-mail: patrick@engineering.ucsb.edu
*) Present address: Materials Department, University of California, Santa Barbara, CA 93106, USA.

tracted, which is very large indeed and should lead to a large segregation length in (In,Ga)N/GaN multilayer structures. Second, the vapor pressure of InN is rather large compared to the vapor pressure of GaN resulting in low incorporation rates of In in the (In,Ga)N alloy [8]. Third, a miscibility gap is predicted for (In,Ga)N leading to spinodal decomposition [5]. The generally accepted strategy to circumvent these problems is the use of relatively low growth temperatures (compared to GaN growth) and high V/III flux ratios (significantly greater than unity) [1–3, 9]. Both measures are reported to result in an improved In incorporation rate as evaporation as well as segregation of In are suppressed. However, the use of large V/III flux ratios for the growth of (In,Ga)N wells implies even larger V/III flux ratios during the GaN barrier growth. This surface stoichiometry is contradictory to the commonly employed Ga-stable growth of GaN [10–12], as N-rich surface stoichiometries lead to a roughening of the GaN growth front.

Experimental The samples are grown on on-axis Si-face SiC(0001) by plasma-assisted molecular beam epitaxy (MBE) in a modified Riber-32 system equipped with conventional effusion cells for Ga and In and an EPI rf N-plasma source. Approximately 1 µm thick GaN(0001) buffer layers are grown prior to the deposition of the actual MQW structure. Details of substrate preparation and buffer growth may be found elsewhere [12]. The (In,Ga)N/GaN MQWs are deposited at a substrate temperature of 580 °C, a plasma power of up to 250 W and a N_2 flow of 0.7 sccm, yielding a N-limited growth rate well in excess of 1 µm/h. Growth is monitored in-situ by reflection high-energy electron diffraction (RHEED). After growth, the samples are characterized by high-resolution triple-axis θ–2θ X-ray diffraction (HRXRD) scans using Cu K_{α_1} radiation in conjunction with dynamical diffraction simulations, secondary-ion-mass-spectrometry (SIMS) employing 5.5 keV MCs^+ ions (M standing for metal) and photoluminescence (PL) using the 325 nm line of a He–Cd laser at an excitation density of 0.1 W cm^{-2}.

Results and Discussion During the growth of a MQW period, we notice a characteristic sequence in RHEED. During QW growth a (1 × 1) pattern is observed, which gains in intensity during barrier growth until finally a transition to a threefold reconstruction along the [$1\bar{1}00$] azimuth is detected. The time delay for the appearance of the surface reconstruction depends on the specific surface stoichiometries during well and barrier deposition. Large V/III ratios lead to short time delays and vice versa. As large V/III ratios are known to suppress In segregation in (In, Ga)N growth, it is tempting to explain the time delay by the presence of a floating In layer that is consumed during barrier deposition.

The presence of In segregation for metal-rich grown wells is seen in HRXRD in conjunction with dynamical simulation (Fig. 1). In XRD, only the integrated In content is determined and the kinematical analysis based on the nominal deposition times yields the nominal MQW parameters. The simulation based on these parameters fails to agree with the data. Interestingly, excellent agreement is obtained after kinematical analysis with effective times (sum of nominal deposition time and delay time taken from RHEED) to account for In segregation.

SIMS supports this approach (Fig. 2). For the metal-stable grown wells (Fig. 2a) the actual e^{-1} well width (8.5 nm) is three times larger than the nominal one (3 nm). In contrast, segregation is strongly reduced by nitrogen-stable conditions (Fig. 2b) but this

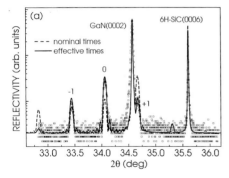
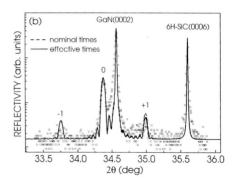

Fig. 1. High-resolution XRD θ–2θ scans around the symmetric GaN(0002) reflection of (In,Ga)N/GaN MQWs. The wells are grown under a) metal-stable and b) N-stable flux conditions. The solid lines represent simulations based on dynamical diffraction theory. Satellite peaks of $(\pm n)$-th order are labelled '$\pm n$'

strategy roughens the growth front leading to poor interface quality. Usually, segregation is a first order process such that a constant fraction of the excess atomic species is incorporated resulting in an asymmetrically shaped composition profile with exponential leading and trailing edges. For metal-stable grown wells (Fig. 2a) the shape of the In profile is symmetric with an almost constant In content and leading and trailing edges of 5 nm/decade and 4 nm/decade, respectively. Such symmetric top-hat' profiles with edges corresponding to the SIMS depth resolution evidence an In incorporation rate at the solubility limit-here around 15%-resulting in constant In contents over the entire well width. Incorporation of In does *not* depend on the amount of floating In on the sample surface and is therefore a zero order process instead of a first order process.

PL spectra at 5 K (Fig. 3a) show clear differences for the metal-stable and the N-stable grown wells. The ratio of the PL intensity of the N-stable and the metal-stable grown wells is around 7 due to the larger spatial separation of electron and hole wavefunctions in the wider metal-stable grown wells. Such a spatial separation results in an increase in radiative lifetime and consequently a reduction in quantum efficiency which is even more pronounced at 300 K. Furthermore, the PL linewidth is 175 meV for the

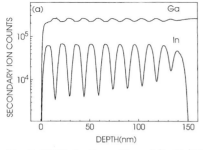
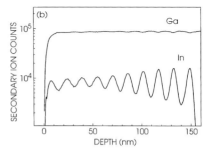

Fig. 2. SIMS depth profiles of (In,Ga)N/GaN MQWs grown under identical conditions except for the In flux resulting in a) metal-stable and b) nitrogen-stable growth conditions during the 8 s well deposition. The barriers were grown under nitrogen-stable conditions for 50 s. The nominal well and barrier thicknesses are 3 nm and 12 nm, respectively. Note that the metal-stable grown wells are 8.5 nm wide due to massive In segregation

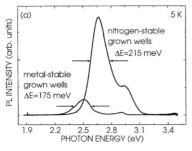
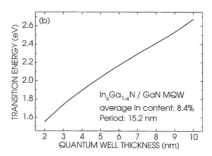

Fig. 3. a) 5 K PL spectra of the two MQW samples under investigation. Note the significantly reduced quantum efficiency and linewidth of the wider and smoother metal-stable grown wells with respect to the N-stable grown wells. b) Calculated transition energy of the metal-stable grown wells for different well widths but identical average In content and period

smooth metal-stable grown wells whereas a value of 215 meV is found for the rough N-stable grown wells. This broadening is a consequence of the rough interfaces of the N-stable grown wells.

We compare experimental and theoretical energies based on self-consistent Schrödinger-Poisson calculations for the metal-stable grown wells (well-width fluctuations of the N-stable grown wells result in a substantial red-shift due to carrier localization). For these metal-stable grown wells, the experimentally determined energy (2.50 eV) is in very good agreement with the theoretical value (2.45 eV) based on the structural parameters obtained from effective deposition times. In contrast, the nominal deposition times lead to grossly underestimated transition energies (1.69 eV). The transition energy critically depends on the actual well width (Fig. 3b) and is therefore a very sensitive indicator for In segregation. Hence, accurate determination of structural parameters is neccessary for a sensible comparison of experiment and simulation. Neglecting In segregation leads to the often stated, but incorrect conclusion that the theoretical polarization fields are overestimated by a factor of two.

Conclusion Summarizing and concluding, the growth of (In,Ga)N/GaN MQWs is strongly affected by In surface segregation. Indium segregation in these structures was shown to be a zero order process instead of the more commonly observed first order process, and thus yields a top-hat In distribution. Neglecting In segregation during metal-stable growth results in quantum wells with poor electron–hole wavefunction overlap and low quantum efficiency since the well width is much larger than the intended one. Furthermore, In surface segregation has a strong impact of PL transition energies. If segregation is correctly taken into account the experimentally observed energies are in satisfactory agreement with those calculated using the polarization fields of Bernardini et al. [13]. N-stable conditions reduce In surface segregation but inevitably deliver rough interfaces.

References

[1] N. Yoshimoto, T. Matsuoka, T. Sasaki, and A. Katsui, Appl. Phys. Lett. **59**, 2251 (1991).
[2] B. Yang, O. Brandt, B. Jenichen, J. Müllhäuser, and K. H. Ploog, J. Appl. Phys. **82**, 1918 (1997).
[3] M. L. O'Steen, F. Fiedler, and R. J. Hauenstein, Appl. Phys. Lett. **75**, 2280 (1999).

[4] N. Duxbury, U. Bangert, P. Dawson, E. J. Trush, W. V. der Stricht, K. Jacobs, and I. Moerman, Appl. Phys. Lett. **76**, 1600 (2000).
[5] I. Ho and G. B. Stringfellow, Appl. Phys. Lett. **69**, 2701 (1996).
[6] S. Chichibu, T. Azuhata, T. Sota, and S. Nakamura, Appl. Phys. Lett. **69**, 4188 (1996).
[7] D. Doppalapudi, S. N. Basu, K. F. Ludwig, and T. D. Moustakas, Appl. Phys. Lett. **84**, 1389 (1998).
[8] T. Nagatomo, T. Kuboyama, H. Minamino, and O. Omoto, Jpn. J. Appl. Phys. **28**, 1334 (1989).
[9] J. M. van Hove, P. P. Chow, A. M. Wowchak, J. J. Klaassen, R. Hickman, and C. Polley, J. Vac. Sci. Technol. B **16**, 1286 (1998).
[10] T. Zywietz, J. Neugebauer, and M. Scheffler, Appl. Phys. Lett. **73**, 487 (1998).
[11] E. J. Tarsa, B. Heying, X. H. Wu, P. Fini, S. P. DenBaars, and J. S. Speck, J. Appl. Phys. **82**, 5472 (1997).
[12] O. Brandt, R. Muralidharan, P. Waltereit, A. Thamm, A. Trampert, H. von Kiedrowski, and K. H. Ploog, Appl. Phys. Lett. **75**, 4019 (1999).
[13] F. Bernardini, V. Fiorentini, and D. Vanderbilt, Phys. Rev. B **56**, R10024 (1997).

Temperature-Independent Stokes Shift in an $In_{0.08}Ga_{0.92}N$ Epitaxial Layer Revealed by Photoluminescence Excitation Spectroscopy

H. Kudo[1]), K. Murakami, H. Ishibashi, R. Zheng, Y. Yamada, and T. Taguchi

Faculty of Engineering, Yamaguchi University, 2-16-1 Tokiwadai, Ube, Yamaguchi 755-8611, Japan

(Received June 24, 2001; accepted August 4, 2001)

Subject classification: 78.40.Fy; 78.55.Cr; S7.14

Optical properties of an $In_{0.08}Ga_{0.92}N$ epitaxial layer have been studied by means of photoluminescence excitation spectroscopy. The photoluminescence spectrum of the $In_{0.08}Ga_{0.92}N$ epitaxial layer was composed of two emission components with an energy separation of 40 meV. Photoluminescence excitation measurements allowed us to observe a clear peak due to the absorption of InGaN and to investigate the temperature dependence of the Stokes shift. At 100 K, the Stokes shifts of the higher and lower energy components were estimated to be 44 and 79 meV, respectively. The Stokes shifts were well consistent with the energy shifts expected from the polaron interaction. The absorption peaks for both the higher and lower energy components were located at the same energy position. Furthermore, the Stokes shift of the higher energy component was not dependent on temperature and indicated a constant value up to room temperature.

Introduction Recent remarkable progress on GaN-based semiconductors has led to a realization of high-brightness blue/green light-emitting diodes (LEDs) [1]. Very recently, high-brightness ultraviolet (UV) LEDs with an external quantum efficiency of 24% have been demonstrated [2]. These achievements enabled us to open a path toward the realization of semiconductor lighting [3]. However, in contrast to the remarkable progress of the optical devices, the radiative recombination mechanism of $In_xGa_{1-x}N$ ternary alloys used as the active light-emitting medium is not fully understood in spite of a considerable amount of research [4, 5].

Photoluminescence excitation (PLE) spectroscopy is one of the powerful tools for understanding the degree of carrier localization. However, PLE spectra of $In_xGa_{1-x}N$ ternary alloys reported to date showed no clear absorption peaks, and indicated a broadened absorption tail [6]. For the reason mentioned above, it has been difficult to identify the Stokes shift precisely.

In this work, we have investigated the temperature dependence of PLE spectra of an $In_{0.08}Ga_{0.92}N$ epitaxial layer. We observed the clear peak due to the absorption of InGaN up to room temperature (RT) in the PLE spectra, and performed a detailed investigation on the temperature dependence of Stokes shift.

Experimental Procedures The $In_xGa_{1-x}N$ epitaxial layer used in this work was grown on a (0001) sapphire substrate by means of metalorganic chemical vapor deposition method, following the deposition of a 3 μm thick GaN buffer layer. The thickness and In mole ratio of the $In_xGa_{1-x}N$ epitaxial layer were 0.1 μm and $x = 0.08$, respectively.

---

[1]) Corresponding author; Phone: +81 836 85 9407; Fax: +81 836 85 9401;
e-mail: c2426@stu.cc.yamaguchi-u.ac.jp

PLE measurement was performed using a dye laser pumped by a Xe–Cl excimer laser (308 nm) as an excitation source. The repetition rate and the pulse width were 100 Hz and 20 ns, respectively.

Results and Discussion Figure 1 shows the photoluminescence (PL; solid line) and PLE (full circles) spectra at 4 K taken from the $In_{0.08}Ga_{0.92}N$ epitaxial layer mentioned above. There appears a single emission band in the PL spectrum at about 3.225 eV with a band width of 47 meV. Besides the dominant emission band, there exists a longitudinal optical (LO)-phonon replica. The detected photon energy of the PLE spectrum was tuned at the PL peak position of 3.225 eV. The main peak located at 3.290 eV in the PLE spectrum is attributed to the absorption of InGaN. This observation of a clear peak allows us to evaluate the difference in peak energy, defined as Stokes shift, between the PL and the PLE spectra, and the Stokes shift is estimated to be 65 meV. The higher energy peaks located around 3.5 eV originate from the contribution of carriers excited in GaN. This experimental result indicates that the carriers generated in GaN play an important role in the radiative recombination of InGaN. The inset of Fig. 1 shows the enlarged PLE spectra of the $In_{0.08}Ga_{0.92}N$ epitaxial layer around the absorption edge of GaN. In the PLE spectrum, three clear peaks are observed at 3.490, 3.497 and 3.520 eV. These three peaks are unambiguously attributed to the A free exciton (E_X^A), B free exciton (E_X^B) and the C free exciton (E_X^C) resonances.

We have reported so far that the PL spectrum of the $In_{0.08}Ga_{0.92}N$ epitaxial layer was composed of two emission components by means of temperature dependent PL measurements [7]. The two emission components were observed clearly in the PL spectrum with an energy separation of about 40 meV at 100 K. In order to investigate the excited states of the two emission components, we performed PLE measurements at 100 K. Figure 2 shows the excitation energy dependence of PL (solid lines) and PLE (full and open circles) spectra of the $In_{0.08}Ga_{0.92}N$ epitaxial layer at 100 K. The PL spectra were obtained under excitation energies from 3.53 (curve a) to 3.38 eV (curve e). The detected photon energies of the PLE spectra were tuned at the PL peak positions of the higher and lower energy components, which were obtained by fitting the experimental curve using the Gaussian function (dashed lines). When the excitation energy is tuned above the band-gap energy of GaN, no notable change is observed for the PL spectra (curve a). However, compared to the lower energy component, the luminescence intensity of the higher energy component decreases dramatically when the excitation energy is tuned below the band-gap

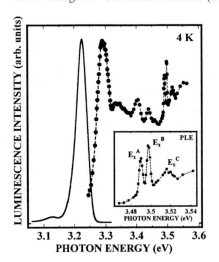

Fig. 1. PL (solid lines) and PLE (full circles) spectra at 4 K taken from an $In_{0.08}Ga_{0.92}N$ epitaxial layer. The detected photon energy is tuned at the PL peak position of 3.225 eV. The inset shows the enlarged PLE spectra around the absorption edge of GaN

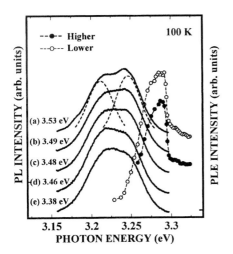

Fig. 2. Excitation-energy dependence of PL (solid line) and PLE (full and open circles) spectra of the $In_{0.08}Ga_{0.92}N$ epitaxial layer at 100 K. The detected photon energies of the PLE spectra are tuned at the PL peak positions of the higher and lower energy components, which are obtained by fitting the experimental curve using the Gaussian function (dashed lines)

energy of GaN (curves b to d). This experimental result indicates that the carriers generated in the GaN play an important role in the radiative recombination of the higher energy component. In the PLE spectra, clear peaks for both the higher and lower energy components are observed, and the difference in peak energy between the PL and the PLE spectra (Stokes shift) of the higher and lower energy components are estimated to be 44 and 79 meV, respectively. The experimentally obtained Stokes shifts are well consistent with the energy shifts ($a_e \hbar \omega = 44.2$ meV, $a_h \hbar \omega = 76.5$ meV) expected from the theoretical approach based on the polaron interaction [8, 9]. Therefore, the contribution of the electron (hole)–polaron state to the recombination process of the higher (lower) energy component is proposed. Furthermore, the peaks in the PLE spectra for both the higher and lower energy components are located at the same energy position. This experimental result indicates that the carriers corresponding to the two emission components are relaxed from the same excited state.

We have also investigated the temperature dependence of PLE spectra of the $In_{0.08}Ga_{0.92}N$ epitaxial layer from 4 to 290 K. Figure 3 shows the temperature dependence of Stokes shift for the higher energy component. The Stokes shift of the higher energy component is not dependent on temperature, and indicates almost a constant value up to RT. If the contribution of the localized carriers to the radiative recombination process is dominant, the Stokes shift should be decreased with temperature owing to the thermal population of the carriers from localized to extended states. Therefore, this experimental result cannot be explained by the effect of carrier localization due to In compositional fluctuation, and strongly supports our proposed recombination model based on the polaron interaction.

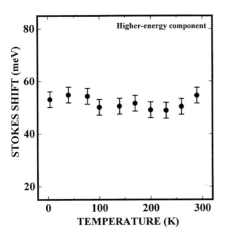

Fig. 3. Temperature dependence of Stokes shift of the higher energy component taken from an $In_{0.08}Ga_{0.92}N$ epitaxial layer. The Stokes shift represents the difference in peak energy between the PL and PLE spectra

Conclusions We have investigated the optical properties of an $In_{0.08}Ga_{0.92}N$ epitaxial layer by means of PLE spectroscopy. The PL spectrum of the $In_{0.08}Ga_{0.92}N$ epitaxial layer was composed of two emission components with an energy separation of 40 meV. PLE measurements allowed us to observe a clear peak due to the absorption of InGaN, and to investigate the temperature dependence of Stokes shift. At 100 K, the Stokes shifts of the higher and lower energy components were estimated to be 44 and 79 meV, respectively. The Stokes shifts of the higher and lower energy components were well consistent with the energy shifts expected from the polaron interaction. In addition, the absorption peaks for both the higher and lower energy components were located at the same energy position. One of the most important experimental results was that the Stokes shift of the higher energy component was not dependent on temperature, and indicated a constant value up to RT.

Acknowledgements We acknowledge the financial support of "The light for the 21st century" National project from METI/NEDO/JRCM. The sample used in this study was provided by Nichia Chemical Industries.

References

[1] S. Nakamura and G. Fasol, The Blue Laser Diode, Springer-Verlag, Berlin/Heidelberg/New York 1997.
[2] K. Tadatomo, H. Okagawa, Y. Ohuchi, T. Tsunekawa, Y. Imada, M. Kato, and T. Taguchi, Jpn. J. Appl. Phys. **40**, L583 (2001).
[3] T. Tamura, T. Setomoto, and T. Taguchi, J. Lumin. **87–89**, 1180 (2000).
[4] W. Shan, B. D. Little, J. J. Song, Z. C. Feng, M. Schurman, and R. A. Stall, Appl. Phys. Lett. **69**, 3315 (1996).
[5] Y. Narukawa, Y. Kawakami, M. Funato, Sz. Fujita, Sg. Fujita, and S. Nakamura, Appl. Phys. Lett. **70**, 981 (1997).
[6] Y. H. Cho, J. J. Song, S. Keller, M. S. Minsky, E. Hu, U. K. Mishra, and S. P. DenBaars, Appl. Phys. Lett. **73**, 1128 (1998).
[7] H. Kudo, H. Ishibashi, R. S. Zheng, Y. Yamada, and T. Taguchi, phys. stat. sol. (b) **216**, 163 (1999).
[8] J. T. Devreese, Polarons in Ionic Crystals and Polar Semiconductors. North-Holland Publ. Co., Amsterdam 1972.
[9] R. S. Zheng, T. Taguchi, and M. Matsuura, J. Appl. Phys. **87**, 2526 (2000).

Depth Resolved Studies of Indium Content and Strain in InGaN Layers

S. Pereira[1]) (a, b), M. R. Correia (a), E. Pereira (a), K. P. O'Donnell (b),
C. Trager-Cowan (b), F. Sweeney (b), E. Alves (c), A. D. Sequeira (c),
N. Franco (c), and I. M. Watson (d)

(a) Departamento de Física, Universidade de Aveiro, P-3810-193 Aveiro, Portugal

(b) Department of Phys. and Appl. Physics, University of Strathclyde, Glasgow, UK

(c) I.T.N., Departamento de Física, E.N.10, P-2686-935 Sacavém, Portugal

(d) Institute of Photonics, University of Strathclyde, Glasgow, UK

(Received July 26, 2001; accepted August 4, 2001)

Subject classification: 61.10.Nz; 61.18.Bn; 68.65.Ac; 78.60.Hk; 78.66.Fd; S7.14

A depth resolved study of optical and structural properties in wurtzite InGaN/GaN bilayers grown by MOCVD on sapphire substrates is reported. Depth resolved cathodoluminescence (CL), Rutherford backscattering spectrometry (RBS) and high resolution X-ray diffraction (HRXRD) were used to gain an insight into the composition and strain depth profiles. It is found that both quantities can vary considerably over depth. Two representative samples are discussed. The first shows a CL peak shift to the blue when the electron beam energy is increased. Such behaviour conforms to the In/Ga profile derived from RBS, where a linear decrease of the In mole fraction from the near surface (≈ 0.20) down to the near GaN/InGaN interface (≈ 0.14) region was found. The other sample discussed shows no depth variations of composition. However, the strain changes from nearly pseudomorphic, close the GaN interface, to an almost relaxed state close to the surface. This discrete variation of strain over depth, originates a double XRD and CL peak related to InGaN.

1. Introduction Group-III nitride epilayers, grown by metalorganic chemical vapour deposition (MOCVD), are currently a major topic of research due to their widespread use in light emitting diodes (LEDs) and laser diodes [1]. Commercial InGaN-based LEDs, developed over the last decade, enjoy unrivalled performance in the UV, blue and green spectral regions. Structural properties of InGaN materials, such as those at the heart of these devices, have been studied intensively. Some key topics under discussion include the role of composition, strain, ordering and phase separation on the optical properties [2–4]. Misinterpretations regarding the effects of strain and composition have generated some systematic uncertainties in the literature. In this work we study selected InGaN films, grown in commercial nitride reactors, which display the main characteristics to be discussed here. Using a combination of depth resolved CL, RBS and HRXRD we address the problem of composition and strain variations over depth in InGaN epilayers.

2. Experimental Details The samples studied are nominally undoped wurtzite InGaN/GaN bilayers, grown by metal organic chemical vapour deposition (MOCVD) on Al_2O_3 substrates in an AIX2000/2400HT CVD reactor. Samples, cooled to a temperature of

---

[1]) Corresponding author; Tel.: +351 918178065; Fax: +351 234 424 965; e-mail: spereira@fis.ua.pt

approximately 25 K in a closed cycle helium cryorefrigerator, were excited on their front faces with a variable energy electron beam. Emitted light was collected from the sample edges in a 90° geometry and analysed using an Oriel InstaSpec™ cooled two-dimensional CCD array mounted at the output focal plane of a Chromex 0.5m monochromator. HRXRD characterisation was performed using a double-crystal diffractometer. A flat Ge (444) monochromator and horizontal divergence slits with widths of 100 μm and a height of 2 mm select Cu $K_{\alpha 1}$ radiation. The instrumental resolution limit is about 30". RBS measurements were performed with a 1 mm collimated beam of 2.0 MeV ^{4}He$^+$ ions at currents of about 5 nA. The backscattered particles were detected at 160° and close to 180°, with respect to the beam direction, using silicon surface barrier detectors located in the standard Cornell geometry.

3. Results and Discussion

3.1 Composition pulling effects in InGaN

In-depth information on the optical properties of solids can be achieved by using an electron beam of variable energy to excite the luminescence. This powerful depth-profiling technique to semiconductor layer properties is based on the fact that the rate of energy loss of an electron beam in a solid depends on its incidence energy [5]. Figure 1 shows the InGaN related low temperature CL emission from the InGaN/GaN epilayer under study at electron beam energies ranging from 0.5 to 20 keV. As can be observed, the CL peak shifts to higher energies with increasing electron kinetic energy at the surface. However this only occurs up to energies of ≈7 keV. It can be noticed that the shift progressively decreases in magnitude until about 7 keV, and thereafter a small shift to lower energies is verified from ≈7 to 9 keV. The CL spectral peak remains practically unchanged for higher voltages. The overall behaviour of the CL peak position indicates that the InGaN composition is not uniform over depth, and the In content is higher near the surface region.

In order to confirm this hypothesis, a RBS study was performed on the sample. This allows an accurate determination of the In mole fraction free from the effects of strain, with a depth resolution of about 5–10 nm under optimised experimental conditions. Additionally this technique requires no standards, as what is measured is the ratio of In/Ga signals. The random RBS spectrum from this sample is shown in Fig. 2. This was acquired with a sample tilt of 50°, with respect to the incident beam, in order to extract the depth profile with an improved depth resolution. Further tilt was not considered to avoid merging of In and Ga related peaks reducing the certainty in the analysis due to the impossibility to collect both In and Ga signals independently. In a random RBS performed on a target with a constant

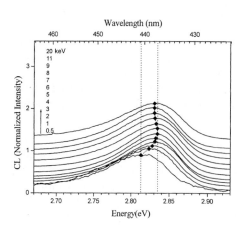

Fig. 1. InGaN related CL emission acquired at electron energies ranging from 0.5 to 20 keV. Spectra were normalised and vertically displaced for clarity

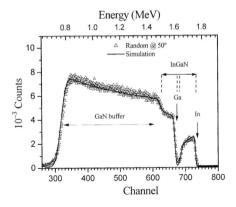

Fig. 2. Random (50° tilt) and simulated RBS spectra. Vertical arrows indicate the scattering energies of the different chemical elements. Horizontal arrows indicate the depth locations

chemical composition the backscattering yield, at a constant bombarding energy, from deeper sample regions is higher ($\propto 1/E^2$) due to the energy loss of the penetrating beam. In Fig. 2 this effect can be observed in the energy window corresponding to the Ga in GaN buffer layer. Regarding the In in the InGaN film, this dependence is not verified, and it can be noticed that the In signal decreases as the beam penetrates deeper in the layer. Accordingly, the Ga signal increases at a faster rate than would be expected in a layer with uniform composition. This is a qualitative indication that the In content is in fact decreasing over depth. For a detailed quantitative analysis, simulation of the RBS spectra was performed using RUMP [6]. A multilayer model consisting of a 75 nm thick InGaN film, with a linearly decreasing In content from about 0.20 near surface, to about 0.14 close to the GaN interface region, followed by a 465 nm thick GaN buffer provides an excellent fit to experiment, as shown in Fig. 2.

The CL results in Fig. 1 can now be further clarified with respect to the compositional profile extracted from the RBS analysis. Monte-Carlo (MC) simulations of the electron beam energy dissipation in solids provide an invaluable tool in the interpretation of the CL spectra [7]. Results of the simulations for this sample are represented in Fig. 3. As it can be observed, for electron energies from 1 keV to about 7 keV, the region of maximal energy deposition (where most electron–hole pairs are created) progressively moves from the near surface region to the InGaN/GaN interface. This point of maximum excitation produces the maximum in the CL spectrum. Further electron energy increase tends to broaden the excitation profile within the InGaN layer. There-

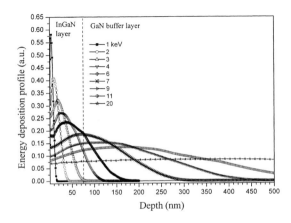

Fig. 3. Electron beam energy deposition curves calculated using Monte-Carlo simulations of the electron trajectories for a 75 nm thick $In_{0.17}Ga_{0.73}N$ layer grown on top of a GaN buffer layer

fore, at this stage the situation changes from one that favours the near interface region, to one where a more uniform excitation over depth is attained. Consequently, a more evenly averaged contribution of the different depth regions to the CL spectrum is achieved, corresponding to a reversal of the shift at around 8–9 keV. Once a uniform excitation over depth is reached for voltages over about 10 keV, an increase in the electron beam energy is not expected to result in any further shifts, as experimentally observed up to 20 keV.

3.2 Depth variations of strain in InGaN The presence of two peaks in the InGaN diffraction profile has often been interpreted as an indication of partial phase segregation, and attributed to the presence of "micro-regions" with different In contents [8–10]. It is worth mentioning that these layers typically exhibit a double InGaN related luminescence peaks [11]. Composition estimations in those sub-regions frequently rely on direct assumption of Vegard's law, which states that the lattice constants of a *relaxed* ternary compound should scale linearly with x, the In mole fraction, between those of the binaries. Here we briefly discuss one of those samples.

An HRXRD reciprocal space map (RSM) around the (105) reflection clearly shows two InGaN-related components, as shown in Fig. 4. Two values for the InGaN lattice constants $c_{\text{InGaN}}(1) = 5.255$ Å and $c_{\text{InGaN}}(2) = 5.286$ Å are calculated from the peak positions in Fig. 4. Let us consider the two peaks as "belonging to" regions (1) and (2). Under direct Vegards' law assumption, each measured lattice parameter, $c_{\text{InGaN}}(1,2)$, would lead to values of indium mole fractions of $x(1) = 0.129$ and $x(2) = 0.180$. However, biaxially strained wurtzite structures suffer distortion of the hexagonal unit cell. In consequence, the procedure discussed above tends to overestimate the In content of layers that are not fully relaxed, as previously discussed by some of the present authors [12].

From the RSM presented in Fig. 4 the in-plane lattice constants (a) of the hexagonal structure can also be determined. It is noticeable that one of the InGaN components is nearly aligned with the GaN diffraction maximum along Q_x, the reciprocal space vector in the plane of the layers. From this, it is apparent that such sub-region, InGaN(2), has practically the same in-plane lattice constant as the underlying GaN buffer layer. This region has a *higher* out-of-plane lattice constant $c_{\text{InGaN}}(2)$, as seen by the peak position relative to Q_z. On the other hand, the other diffraction peak corresponds to a larger value of $a = a_{\text{InGaN}}(1)$ and a smaller value of $c = c_{\text{InGaN}}(1)$. Specifically, the lattice constants $a_{\text{InGaN}}(1) = 3.220$ and $a_{\text{InGaN}}(2) = 3.186$ are determined.

When both InGaN lattice constants a and c are known, the sample strain can be taken into ac-

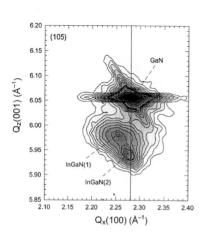

Fig. 4. Reciprocal space map of the (105) plane in the InGaN/GaN layer. The abscissa (Q_x) and the ordinate (Q_z) are proportional to $2/(\sqrt{3}a)$ and $5/c$, respectively

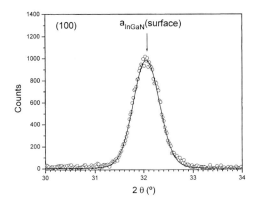

Fig. 5. In-plane X-ray diffraction scattering peak from the (100) InGaN planes near the surface fitted to a Voigt function

count in estimates of composition, as described in detail in Ref. [13]. Solving Poissons' equation for both sub-regions considered yields the values of $x(1) = 0.118$ and $x(2) = 0.119$. The RBS determined composition in this sample is $x = 0.119$ and is found to be uniform over depth. It is rather interesting to notice that if the strain in each sub-region is considered, the value of x obtained is actually the same, within the error limits of such an estimate (± 0.01), and excellent agreement with the RBS determined value is achieved.

In order to depth-locate the regions contributing to the double XRD peak in this layer, in-plane X-ray diffraction was used to measure the InGaN lattice constants only for the first few tens of nanometers of material near the surface. This can be accomplished by making use of the small penetration of the beam when its incident angle is close to the critical angle ($\approx 0.3°$) for reflectivity. The X-rays penetrating into the surface region of the sample can be used to scatter from planes approximately normal to the surface plane. The limit of the penetration for this type of geometry is governed by the path that the X-rays have to take [14].

The (100) diffraction peak acquired under these conditions is shown in Fig. 5. From the maximal position one can calculate the lattice constant a_{InGaN}(surface) directly. A value of a_{InGaN}(surface) = 3.220 Å, identical to $a_{InGaN}(1)$, is determined. This result demonstrates that the two diffraction components, observed in Fig. 4 originate from different depths of the InGaN layer. The relaxed region is close to the surface and the pseudomorphic part near the GaN interface, as it would be expected. Therefore, the explanation for the origin of the double XRD peak and also the double CL peak observed in this sample (not shown) is a discrete strain variation over depth. Nevertheless, despite being under different strains both regions contributing to the double XRD and CL peak have identical In content.

4. Conclusion We have shown that both composition and strain can vary over depth in InGaN layers. A gradient of the In content was verified along the growth direction. We have also demonstrated that a double peak in the InGaN XRD profile can be completely unrelated to phase segregation. Split XRD components may in fact represent the same In content. In the case studied here, in-plane XRD diffraction shows that the InGaN strain varies with depth. The layer is almost pseudomorphic near the GaN interface and partially relaxed close to the surface. Depth resolved CL confirms that both In content gradients and strain variations over depth influence the luminescence properties. Consequently, great care must be taken when relating structural and optical properties if these are evaluated at different depth scales. Further work needs to be performed to understand the influence of the growth conditions in these properties.

Acknowledgements Sérgio Pereira acknowledges financial support from Fundação para a Ciência e Tecnologia (SFRH/BD/859/2000) and useful discussions with Dr. Patricia Kidd (XRD) and Dr. Paul Edwards (MC simulations).

References

[1] B. GIL (Ed.), Group III Nitride Semiconductor Compounds, Physics and Applications. Series on Semiconductor Science and Technology, Vol. 6, Oxford Science Publications, Oxford 1998.
[2] K. P. O'DONNELL, phys. stat. sol. (a) **183**, 117 (2001).
[3] M. E. AUMER, S. F. LEBOEUF, S. M. BEDAIR, M. SMITH, J.Y. LIN, and H. X. JIANG, Appl. Phys. Lett. **77**, 821 (2000).
[4] A. F. WRIGHT, K. LEUNG, and M. VAN SCHILFGAARDE, Appl. Phys. Lett. **78**, 189 (2001).
[5] C. TRAGER-COWAN, A KEAN, F. YANG, B. HENDERSON, and K. P. O'DONNELL, Physica B **185**, 319 (1993).
[6] L. R. DOOLITTLE, Nucl. Instrum. Methods Phys. Res. B **9**, 344 (1985).
[7] E. NAPCHAN and D. B. HOLT, Inst. Phys. Conf. Ser., No. 87, Eds. A. G. CULLIS et al., IOP Publ., Bristol 1987 (p. 733).
[8] N. A. EL-MASRY, E. L. PINER, S. X. LIU, and S. M. BEDAIR, Appl. Phys. Lett. **72**, 40 (1998).
[9] YONG-TAE MOON, DONG-JOON KIM, KEUN-MAN SONG, IM-HWAN LEE, MIN-SU YI, DO-YOUNG NOH et al., phys. stat. sol. (b) **216**, 167 (1999).
[10] D. RUDLOFF, phys. stat. sol. (b) **216**, 315 (1999).
[11] S. PEREIRA, M. R. CORREIA, E. PEREIRA, K. P. O'DONNELL, R.W. MARTIN, M. WHITE, E. ALVES, A. D. SEQUEIRA, and N. FRANCO, presented in E-MRS 2001, Strasbourg, to be published.
[12] S. PEREIRA, M. R. CORREIA, T. MONTEIRO, E. PEREIRA, E. ALVES, A. D. SEQUEIRA, and N. FRANCO, Appl. Phys. Lett. **78**, 2137 (2001).
[13] S. PEREIRA, M. R. CORREIA, E. PEREIRA E. ALVES L. C. ALVES C. TRAGER-COWAN, and K. P. O'DONNELL, Mater. Res. Soc. Symp. Proc., Vol. 639, b 3.52.1 (2001).
[14] P.F. FEWSTER, X-Ray Scattering from Semiconductors, Imperial College Press, UK 2001 (ISBN 1-86094-159-1).

Carrier Dynamics in Group-III Nitride Low-Dimensional Systems: Localization versus Quantum-Confined Stark Effect

P. Lefebvre (a), T. Taliercio (a), S. Kalliakos (a), A. Morel (a),
X.B. Zhang (a), M. Gallart (a), T. Bretagnon (a), B. Gil (a),
N. Grandjean (b), B. Damilano (b), and J. Massies (b)

(a) Groupe d'Etude des Semiconducteurs, CNRS, Université Montpellier II,
Case Courrier 074, F-34095 Montpellier Cedex 5, France

(b) Centre de Recherche sur l'Hétéro Epitaxie et ses Applications, CNRS,
Rue Bernard Grégory, F-06560 Valbonne, France

(Received June 20, 2001; accepted June 26, 2001)

Subject classification: 73.21.Fg; 73.21.La; 78.47.+p; 78.55.Cr; 78.67.De; 78.67.Hc; S7.14

Continuous-wave and time-resolved optical spectroscopy is used to examine a variety of InGaN/GaN quantum-well and quantum-box samples, grown by molecular beam epitaxy. The results are analyzed in order to clarify the respective influences of electric fields and of carrier localizations on radiative recombinations. The coupling of electron–hole pairs with LO-phonons is also studied in detail, from careful analysis of the size-dependent intensities of LO-phonon replica. From our attempt of modelling the Huang-Rhys factor, S, for excitons in these systems, we conclude that the observed optical recombinations are rather those of electrons and holes separately localized on different potential fluctuations.

Introduction All efficient light-emitting devices based on group-III nitride semiconductors involve low-dimensional systems, such as InGaN/GaN [1–3] or GaN/AlGaN [4] quantum wells (QWs). An intense research activity has been devoted recently to radiative recombinations in these artificial materials. For InGaN-based nanometric layers, large Stokes shifts have been readily assigned to strong carrier localization, possibly in self-formed quantum dots [5, 6]. This idea was supported by observations of In-rich nanoclusters in InGaN layers. Although this clustering may depend on the growth conditions, the random distribution of In atoms induces rather strong potential fluctuations, in the volume of this ternary alloy. These fluctuations, together with large carrier effective masses in group-III nitrides, induce the localization of electron and hole wave-functions on nanometric scales, which prevents their efficient capture by nonradiative centers, related to the high density of threading dislocations. This localization was invoked to interpret (i) the strong Stokes shift between photoluminescence (PL) lines and the absorption onset of InGaN epilayers and QWs and (ii) the rather long PL decay times observed in these systems.

Now, for all QWs based on nitrides in their wurtzite phase, including InGaN/GaN ones, electric fields of several hundred kV/cm are present, if the growth axis is (0001). Indeed, this symmetry permits spontaneous and piezoelectric polarizations which are specially strong for group-III nitrides [7, 8]. These fields reduce drastically the oscillator strength for the ground-state optical transition, as evidenced [9, 10] by very large, size-dependent, recombination times. They are also partly at the origin of the Stokes shift observed in QWs [11].

What are the respective effects of carrier localization and of electric fields on the Stokes shift and on recombination dynamics? Concerning carrier localization, should we consider an exciton localized as an entity at some places in the QW plane? Or should we, instead, regard optical recombinations in these systems as those of independently localized electrons and holes? In this communication, we address these questions by analyzing the ensemble of experimental and theoretical results obtained on InGaN/GaN QWs and QBs. In particular, we use measurements of the size-dependent coupling of electron–hole pairs with LO-phonons to propose a microscopic view of carrier localization in these systems.

Experimental Details The samples were all grown by molecular beam epitaxy on c-plane sapphire substrates, covered by GaN buffer layers deposited at 800 °C, at a rate of 1 µm/h. The temperature was decreased down to 550 °C for the growth of the InGaN layer. The samples consist of InGaN/GaN QWs or single planes of QBs, with In compositions of ≈0.15–0.20, as deduced from X-ray diffraction and from reflection high-energy electron diffraction (RHEED).

One of our samples is a graded-width InGaN/GaN QW [12]: for growing the InGaN layer, the rotation of the sample holder was stopped, inducing a gradient of the QW width around an average value of 4.5 nm. The width was measured to vary between 2.0 and 5.5 nm by Transmission Electron Microscopy (TEM). The 18 mm long sample was cut along the vertical symmetry plane of the substrate. In the growth chamber, the In and Ga effusion cells lie symmetrically on both sides of this vertical plane so that the fluxes of Ga and In atoms vary but keep a constant ratio along the vertical axis, yielding a constant In composition.

For the other samples, the holder was normally rotated, providing the homogeneity in layer width and composition. The width of the QWs and the 2D–3D (Stranski-Krastanov) growth mode transition for QBs were monitored by RHEED. The QBs are 3D islands with typical dimensions of 5–10 nm in diameter and 2 nm in height. The latter is a growing function of the deposition time. The typical areal density of the QBs is of a few 10^{11} cm^{-2}.

Time-resolved (TR) PL experiments were conducted with a conventional setup including a streak camera. The excitation was provided by 2 ps laser pulses obtained by frequency-doubling the radiation from a Ti:sapphire mode-locked cavity. The produced photons at 3.2 eV allowed us to create electron–hole pairs in the only InGaN layer. The repetition rate was adjusted between 8 kHz and 82 MHz by using an acousto-optic modulator, avoiding quasi-continuous excitation of slowly decaying PL signals. None of the decays measured by us could be fit using single exponentials. Instead, all these decays appeared to have similar shapes, although on really different time scales [12]. These shapes are somehow comparable to stretched exponentials, but they are not exactly. For this reason, we characterize each decay by the delay, τ_{10}, after which the maximum PL signal is divided by 10.

Electric Field versus Localization Figure 1[2]) displays the time-integrated PL spectra recorded from three different spots on the graded-width InGaN/GaN QW sample. These spots are shown on the photograph (inset) taken during a continuous-wave (cw) PL experiment where the excitation was provided by a source placed near the edge of

---

[2]) Colour figure is published online (www.physica-status-solidi.com).

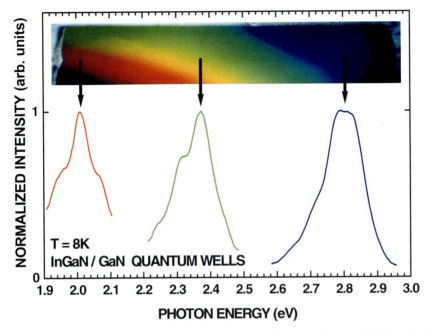

Fig. 1 (colour). PL spectra taken from three spots of a graded-width InGaN/GaN QW. The inset shows a color photograph taken during a PL experiment with indirect UV excitation

the sample. As the well width goes from ≈2.0 nm to ≈5.5 nm, the PL emission goes continuously from the ultraviolet to the red with a constant linewidth of ≈0.12 eV. The PL decay time, τ_{10}, plotted against the PL peak energy in Fig. 2, also varies continuously, but in a nearly exponential way, over more than four orders of magnitude. Figure 2 also displays equivalent data extracted from the literature [5, 11, 13–28] for a variety of samples embedding InGaN/GaN QWs, and for our other QW and QB samples. There is a general trend: an InGaN/GaN system emitting UV light will show a decay time of a few nanoseconds, whereas another one emitting red light will rather decay on a time scale of tens of microseconds. The scattering of data arises mainly from the

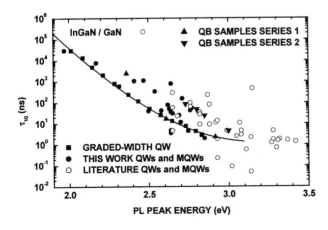

Fig. 2. Collection of PL decay times from the literature and from our measurements on InGaN/GaN QWs and QBs. For the points from literature, the value of τ_{10} was estimated from the given exponential decay time, τ, by $\tau_{10} = \tau \ln(10)$. The solid curve shows the result of our calculation for our graded-width QW

variety of In compositions and of physical systems represented: single or multiple QWs, p–i–n structures. It is now understood, indeed, that all these situations provide different values for the electric field inside the QW, even for a constant size and composition [29–34]. The different growth methods represented may also change the correlation between τ_{10} and the PL energy, by changing the nano-texture of the InGaN, or the surface properties of the sample [33, 34].

Nevertheless, if all parameters are kept constant, apart from the well width, like for our graded-width QW, we obtain a smooth correlation. Moreover, this correlation is almost perfectly described by the calculated variation of the inverse of the oscillator strength of excitons versus transition energy, when varying the well width between 2.0 and 5.5 nm (solid curve in Fig. 2). This calculation was made using the envelope function approximation, including a longitudinal electric field in the QW of 2.45 MV/cm. We thus basically explain the observed variation of τ_{10} by the quantum-confined Stark effect. The quite large value for the field is discussed in detail in a previous work [12]. We simply wish to mention that this value agrees totally with the Stokes shift measured experimentally on this sample [35]. We have checked this point by calculating the whole interband absorption spectrum for our InGaN/GaN QWs [12]. For the narrower wells, emitting UV light, we found that the ground-state transition, e1h1, is fairly allowed and the sharp absorption onset of the system corresponds to this e1h1 transition. Then, the Stokes shift is strictly due to localization effects and it was measured at 0.12 eV, i.e. essentially comparable to the PL linewidth. In fact both Stokes-shift and linewidth correspond to the existence of some distribution of energies for electrons and holes, at various localization sites. Now, increasing the well width, e1h1 becomes less and less allowed, and the absorption onset involves multiple transitions between excited states. By the way, the absorption onset remains nearly at a constant energy but it changes in shape with the well width: it is sharp, at the e1h1 transition, for narrow wells and it becomes smoother for wider wells. On the other hand, due to the quantum confined Stark effect, the PL energy decreases almost linearly with well width. In fact we have shown [12] that this PL energy was simply the e1h1 energy reduced by a constant value of ≈ 0.12 eV, which accounts for localization effects. The rest of the measured Stokes shift simply arises from the quantum-confined Stark effect.

From these results, we learn that, in these conditions of constant composition and nano-texture, the distribution of energy levels near some average value is fairly constant across the sample. This explains the constant linewidth and the constant localization-induced Stokes shift. The electric field effects come "on top" of these localization effects: they rather control the average energy of the emission and the corresponding decay time.

At this stage, we remark that all the above comments for InGaN/GaN QWs are also strictly valid for QB planes. Indeed, none of our experimental results really allows us to distinguish between QWs and QBs. For instance, some of the points in Fig. 2 were obtained on QBs. The overall dependence of PL energy and decay time versus the QB height is similar to that obtained for QWs versus well width. This is illustrated by the PL spectra in Fig. 3, taken from InGaN/GaN QBs of varying heights and by the nearly exponential variation of τ_{10} (see Fig. 2). We have also shown [36] that there is no difference, either, on the temperature (T) dependence of PL energies, intensities and τ_{10}-value, for comparable QWs and QBs. In particular, the so-called s-shaped [37] variation of the PL peak energy versus T is assigned to the thermally activated transfers between

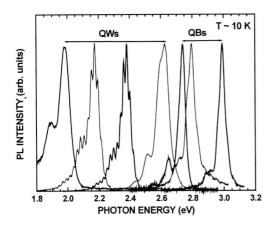

Fig. 3. Continuous-wave PL spectra for InGaN/GaN QWs and QBs of various vertical sizes. Several phonon replica are observed for each spectrum, with a spacing of 91 meV

different localization centers. The similarities observed for QWs and QBs for this behavior and for the other ones led us to conclude that carrier localization in InGaN quantum systems occurs on scales which are much smaller than the size of our QBs themselves, i.e. a few nanometers. In other words, the optical properties of InGaN/GaN QB planes are essentially equivalent to those of QWs of similar vertical size. As an illustration, Fig. 3 shows cw PL spectra taken from other QWs with similar compositions as the QBs, but wide enough to produce lower energy emission. This possibility of maintaining 2D growth above the threshold of the 2D–3D transition will be detailed in another paper. The common feature of all samples in Fig. 3 is that they all presented rather narrow linewidths, allowing us to observe series of LO-phonon replica.

Size Dependence of the LO-Phonon Coupling Although details of emission spectra in Fig. 3 can be different, like the amplitude of some interference fringes due to differences in the multi-layered samples, the general trends are clear. LO-phonon replica are systematically observed, with a constant spacing of ≈91 meV, close to the energy of the LO-phonon in wurtzite GaN. Also, the lower the energy of the zero-phonon line, the larger the relative intensities of the replica. These intensities are related by the relation $I_n = I_0 S^n/n!$, where n is the number of phonons involved and S is the so-called Huang-

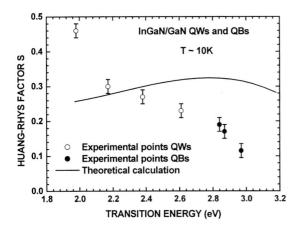

Fig. 4. Plot of the S-factors measured from the spectra of Fig. 3. The solid curve shows the S-factor calculated for excitons in InGaN/GaN QWs, by a variational approach

Rhys coupling factor which accounts for the strength of the Fröhlich interaction between recombining electron–hole pairs and the polar phonons. From this relation we can measure the S-factor for all our samples by plotting $I_n \times n!$ against n ($n = 0, 1, 2, \ldots$) in a logarithmic scale. Although we have observed, by this method, slight deviations from the above equation, we can fit the measured points to extract the average S-factor for a given sample. Figure 4 displays the S-factors measured for QBs (solid symbols) and QWs (open symbols) against their PL energy for $n = 0$. We obtain a clear monotonous decrease of S versus PL energy, i.e. an increase of S with the vertical size of the system.

In order to interpret this result, we have tried to model this variation by calculating the S-factor for excitons in InGaN/GaN QWs (QBs are treated as QWs, as explained above). To do so, we have used the following variational wave function for excitons: $\Phi(\mathbf{r}) = f_e(z_e) f_h(z_h) \Psi(\varrho)$, where $f_e(z_e)$ and $f_h(z_h)$ are the respective envelope functions for electrons and holes in the QW, accounting for a longitudinal electric field of 2.45 MV/cm. The two-dimensional hydrogenic function, $\Psi(\varrho) = N \exp(-\varrho/\lambda)$ accounts for the in-plane relative motion of the electron–hole pair, bound by Coulomb interaction. N is a normalization factor and the variational parameter, λ, is simply the average extension of this relative motion (pseudo-Bohr radius). We have calculated S within the adiabatic approximation [38, 39], in which only the Fröhlich interaction of the exciton with LO phonons is considered since it is the strongest one among all interactions. The very small dispersion of Fröhlich interaction versus the angle with respect to the c-axis of the wurtzite crystal is neglected [40]. The result of our calculation, shown in Fig. 4 by a solid curve, has clearly not a monotonic dependence on the transition energy. In fact, increasing the well width increases the electron–hole dipole along the growth axis, thus increasing the S-factor, which is the dominant effect for the narrower wells. But, simultaneously, increasing the well width also weakens the Coulomb attraction between the electron and the hole. As a result, the pseudo-Bohr radius, λ, along the well plane grows significantly with the well width. This effect, tending to reduce S, becomes dominant for sufficient well width. This explains why S decreases when decreasing the transition energy below ≈ 2.8 eV. The overall calculated variation is much smaller than that measured experimentally.

Again, these results give us a deeper insight into the microscopic aspects of electron–hole recombinations in InGaN QWs and QBs. Indeed, we come to the conclusion that these recombinations are really not those of free or localized excitons with an in-plane relative motion mainly ruled by the Coulomb attraction between the electron and the hole. This picture is incompatible with the observed steady increase of the S-factor with the vertical size of the quantum system. Instead, we propose the following picture: extremely localized potential fluctuations are present in the InGaN layers, due to the statistical distribution of In atoms and, possibly to some clustering (not compulsory). The electron and the hole are pushed apart towards the interfaces, so that they can be attracted to different potential minima along the plane. By the way, this possibility of localization at different sites along the plane is amplified for wider QWs because more fluctuations are present in this larger volume. The consequence of this situation is that the electron and the hole are localized separately: the coupling of the pair with polar phonons is not that of a hydrogenic exciton; rather, it resembles the coupling one may have for a donor–acceptor pair (DAP). The short-range potential fluctuations are a priori isoelectronic traps, but the electric field introduces a discrimination between them. Those near one interface act like donors, because they capture electrons, and those near the other interface act like acceptors, by capturing holes. We cannot do any accurate model because many physical parameters are lacking. But we can remember that the strength of coupling with LO-phonons for DAP is simply a growing function of the average distance R between donor and acceptor ions [41]. This is exactly what we have observed, a minimum for R being simply the distance between localization centers along the growth axis, i.e. essentially a growing function of the well width (or box height).

Conclusion We have clarified the respective influences of carrier localization and of internal electric fields on optical recombination processes in InGaN/GaN QWs and QBs, by using continuous and time-resolved PL experiments. Carrier localization provokes a Stokes shift basically similar to the PL linewidth, but its role is crucial to prevent nonradiative carrier captures. By pushing the wave functions of electrons and holes towards either sides of the QW or QB, the electric fields can have a large impact on the Stokes shift, due to the quantum-confined Stark effect, especially for wide QWs. We have shown, too, that the measured coupling factor of electron–hole pairs with LO-phonons increases steadily with the QW width (or the box height). It is impossible to conciliate this behavior with the picture of an exciton having an hydrogen-like in-plane relative motion of the electron–hole pair. Instead, our results are rather compatible with electrons and holes localized separately at deep potential fluctuations, near either sides of the system.

Ackowledgements We acknowledge support of the European Commission, through the "CLERMONT" Research Training Network, under contract N° HPRN-CT-1999-000132. This work is also supported by the French Ministry of Education, Research and Technology within the "BOQUANI" and "NANILUB" research programs.

References

[1] S. NAKAMURA, J. Cryst. Growth **202**, 290 (1999).
[2] S. NAKAMURA and G. FASOL, The Blue Laser Diode, Springer-Verlag, Berlin 1997.
[3] I. AKASAKI, H. AMANO, K. ITOH, N. KOIDE, and K. MANABE, Inst. Phys. Conf. Ser. **129**, 851 (1993).
[4] J. HAN, M.H. CRAWFORD, R.J. SHUL, J.J. FIGIEL, M. BANAS, L. ZHANG, Y.K. SONG, H. ZHOU, and A.V. NURMIKKO, Appl. Phys. Lett. **73**, 1688 (1998).
[5] Y. NARUKAWA, Y. KAWAKAMI, SZ. FUJITA, SG. FUJITA, and S. NAKAMURA, Phys. Rev. B **55**, R1938 (1997).
[6] K.P. O'DONNELL, R.W. MARTIN, and P.G. MIDDLETON, Phys. Rev. Lett. **82**, 237 (1999).
[7] F. BERNARDINI, V. FIORENTINI, and D. VANDERBILT, Phys. Rev. B **56**, R10026 (1997).
[8] F. BECHSTEDT, U. GROSSNER, and J. FURTHMÜLLER, Phys. Rev. B **62**, 8003 (2000).
[9] J.S. IM, H. KOLLMER, J. OFF, A. SOHMER, F. SCHOLZ, and A. HANGLEITER, Phys. Rev. B **57**, R9435 (1998).
[10] P. LEFEBVRE, J. ALLÈGRE, B. GIL, H. MATHIEU, P. BIGENWALD, N. GRANDJEAN, M. LEROUX, and J. MASSIES, Phys. Rev. B **59**, 15363 (1999).
[11] E. BERKOWICZ, D. GERSHONI, G. BAHIR, E. LAKIN, D. SHILO, E. ZOLOTOYABKO, A.C. ABARE, S.P. DENBAARS, and L.A. COLDREN, Phys. Rev. B **61**, 10994 (2000).
[12] P. LEFEBVRE, A. MOREL, M. GALLART, T. TALIERCIO, J. ALLÈGRE, B. GIL, H. MATHIEU, B. DAMILANO, N. GRANDJEAN, and J. MASSIES, Appl. Phys. Lett. **78**, 1252 (2001).
[13] C.K. SUN, S. KELLER, G. WANG, M.S. MINSKY, J.E. BOWERS, and S.P. DENBAARS, Appl. Phys. Lett. **69**, 1936 (1996).
[14] E.S. JEON, V. KOSLOV, Y.K. SONG, A. VERTIKOV, M. KUBALL, A.V. NURMIKKO, H. LIU, C. CHEN, R.S. KERN, C.P. KUON, and M.G. CRAFORD, Appl. Phys. Lett. **69**, 4194 (1996).

[15] M.S. MINSKY, S.B. FLEISCHER, A.C. ABARE, J.E. BOWERS, E.L. HU, S. KELLER, and S.P. DENBAARS, Appl. Phys. Lett. **72**, 1066 (1998).
[16] H. KOLLMER, J.S. IM, S. HEPPEL, J. OFF, F. SCHOLZ, and A. HANGLEITER, Appl. Phys. Lett. **74**, 82 (1999).
[17] A. HANGLEITER, J.S. IM, H. KOLLMER, S. HEPPEL, J. OFF, and F. SCHOLZ, MRS Internet J. Nitride Semicond. Res. **3**, 15 (1998).
[18] Y.H. CHO, G.H. GAINER, A.J. FISCHER, J.J. SONG, S. KELLER, U.K. MISHRA, and S.P. DENBAARS, Appl. Phys. Lett. **73**, 1370 (1998).
[19] K.C. ZENG, M. SMITH, J.Y. LIN, and H.X. JIANG, Appl. Phys. Lett. **73**, 1724 (1998).
[20] T.J. SCHMIDT, Y.H. CHO, G.H. GAINER, J.J. SONG, S. KELLER, U.K. MISHRA, and S.P. DENBAARS, Appl. Phys. Lett. **73**, 1892 (1998).
[21] T.J. SCHMIDT, Y.H. CHO, G.H. GAINER, J.J. SONG, S. KELLER, U.K. MISHRA, and S.P. DENBAARS, Appl. Phys. Lett. **73**, 560 (1998).
[22] Y.H. CHO, J.J. SONG, S. KELLER, M.S. MINSKY, E. HU, U.K. MISHRA, and S.P. DENBAARS, Appl. Phys. Lett. **73**, 1128 (1998).
[23] J. ALLÈGRE, P. LEFEBVRE, S. JUILLAGUET, W. KNAP, J. CAMASSEL, Q. CHEN, and M.A. KAHN, MRS Internet J. Nitride Semicond. Res. **2**, 34 (1997).
[24] S.F. CHICHIBU, A.C. ABARE, M.S. MINSKY, S. KELLER, S.B. FLEISCHER, J.E. BOWERS, E. HU, U.K. MISHRA, L.A. COLDREN, S.P. DENBAARS, and T. SOTA, Appl. Phys. Lett. **73**, 2006 (1998).
[25] S.F. CHICHIBU, T. SOTA, K. WADA, and S. NAKAMURA, J. Vac. Sci. Technol. B **16**, 2204 (1998).
[26] S.F. CHICHIBU, A.C. ABARE, M.P. MACK, M.S. MINSKY, T. DEGUCHI, D. COHEN, P. KOZODOY, S.B. FLEISCHER, S. KELLER, J.S. SPECK, J.E. BOWERS, E. HU, U.K. MISHRA, L.A. COLDREN, S.P. DENBAARS, K. WADA, T. SOTA, and S. NAKAMURA, Mater. Sci. Eng. B **59**, 298 (1999).
[27] Y. NARUKAWA, Y. KAWAKAMI, SG. FUJITA, and S. NAKAMURA, Phys. Rev. B **59**, 10283 (1999).
[28] B. MONEMAR, J.P. BERGMAN, J. DALFORS, G. POZINA, B.E. SERNELIUS, P.O. HOLTZ, H. AMANO, and I. AKASAKI, MRS Internet J. Nitride Semicond. Res. **4**, 16 (1999).
[29] M. LEROUX, N. GRANDJEAN, J. MASSIES, B. GIL, P. LEFEBVRE, and P. BIGENWALD, Phys. Rev. B **60**, 1496 (1999).
[30] S.F. CHICHIBU, A.C. ABARE, M.S. MINSKI, S. KELLER, S.B. FLEISCHER, J.E. BOWERS, E. HU, U.K. MISHRA, L.A. COLDREN, S.P. DENBAARS, and T. SOTA, Appl. Phys. Lett. **73**, 2006 (1998).
[31] O. AMBACHER, B. FOUTZ, J. SMART, J.R. SHEALY, N.G. WEIMANN, K. CHU, M. MURPHY, A.J. SIERAKOWSKI, W.J. SCHAFF, L.F. EASTMAN, R. DIMITROV, A. MITCHELL, and M. STUTZMANN, J. Appl. Phys. **87**, 334 (2000).
[32] J.L. SANCHEZ-ROJAS, J.A. GARRIDO, and E. MUÑOZ, Phys. Rev. B **61**, 2773 (2000).
[33] J.P. IBBETSON, P.T. FINI, K.D. NESS, S.P. DENBAARS, J.S. SPECK, and U.K. MISHRA, Appl. Phys. Lett. **77**, 250 (2000).
[34] O. GFRÖRER, C. GEMMER, J. OFF, J.S. IM, F. SCHOLZ, and A. HANGLEITER, phys. stat. sol. (b) **216**, 405 (1999).
[35] B. DAMILANO, N. GRANDJEAN, J. MASSIES, L. SIOZADE, and J. LEYMARIE, Appl. Phys. Lett. **77**, 1268 (2000).
[36] P. LEFEBVRE, T. TALIERCIO, A. MOREL, J. ALLÈGRE, M. GALLART, B. GIL, H. MATHIEU, B. DAMILANO, N. GRANDJEAN, and J. MASSIES, Appl. Phys. Lett. **78**, 1538 (2001).
[37] Y.H. CHO, G.H. GAINER, A.J. FISCHER, J.J. SONG, S. KELLER, U.K. MISHRA, and S.P. DENBAARS, Appl. Phys. Lett. **73**, 1370 (1998).
[38] S. RUDIN and T.L. REINECKE, Phys. Rev. B **41**, 3017 (1990).
[39] M. SOLTANI, M. CERTIER, R. EVRARD, and E. KARTHEUSER, J. Appl. Phys. **78**, 5626 (1995).
[40] B. C. LEE, K. W. KIM, M. DUTTA, and M. A. STROSCIO, Phys. Rev. B **56**, 997 (1997).
[41] A.L. GURSKII and S.V. VOITIKOV, Solid State Commun. **112**, 339 (1999).

Pressure Dependence of Piezoelectric Field in InGaN/GaN Quantum Wells

G. Vaschenko[1]) (a), D. Patel (a), C. S. Menoni (a), N. F. Gardner (b), J. Sun (b), W. Götz (b), C. N. Tomé (c), and B. Clausen (c)

(a) Department of Electrical and Computer Engineering, Colorado State University, Fort Collins, CO 80523-1373, USA

(b) LumiLeds Lighting, 370 W. Trimble Road, San Jose, CA 95131, USA

(c) MST Division, Los Alamos National Laboratory, Los Alamos, NM 87545, USA

(Received June 20, 2001; accepted June 29, 2001)

Subject classification: 77.65.Bn; 77.65.Ly; 78.66.Fd; S7.14

In this work we use the well width dependence of the quantum confined Stark effect to determine the variation of the built-in piezoelectric field in InGaN/GaN quantum wells with applied hydrostatic pressure. We find that the field increases from 1.4 MV/cm at atmospheric pressure to 2.6 MV/cm at 8.7 GPa. An analysis of the strain generated by the pressure suggests that the increase in the field is due to a dramatic dependence of the piezoelectric constants of GaN and InGaN on strain.

Built-in piezoelectric fields (F_{pz}) play an important role in devices based on InGaN/GaN quantum wells (QWs) with wurtzite lattice configuration. To quantify the effect of these fields one needs to know the exact geometry of the device, the strain state, and the piezoelectric constants of the materials forming the quantum wells. While the first two characteristics can be determined quite precisely, the latter one is subject to a significant uncertainty. The experimentally defined values of the piezoelectric constant e_{33} of GaN range from 0.43 C/m^2 [1] to 1.12 C/m^2 [2], with the theoretically predicted values of 0.63–0.73 C/m^2 [3, 4]. For InN the piezoelectric constants are not determined experimentally yet. The predicted value of e_{33} in InN is 0.81 C/m^2 [5]. To even worsen this uncertainty, the piezoelectric constants of InGaN quantum wells are commonly linearly interpolated from the values of the piezoelectric constants of the binaries forming the alloy [6, 7]. This approach neglects the effect of the microscopic strains associated with alloy mixing and the macroscopic strain generated by the lattice mismatch between GaN and InGaN. In this work we show that piezoelectric constants of GaN and InGaN have a dramatic dependence on the macroscopic strain. To reveal this dependence we apply hydrostatic pressure to the QW structures to modify the strain state in the wells and in the barriers. This results in a strikingly large increase of the F_{pz}. An analysis of the strain state shows that the increase of the F_{pz} arises from the nonlinear piezoelectric effect [3, 8], i.e. the strain dependence of the piezoelectric constants.

The QW structures studied in this work were grown by metal organic chemical vapour deposition on (0001) sapphire substrates. Each structure has four identical quantum wells with indium composition of ~15% and GaN barriers. The barriers are 12.5 nm wide and the well widths in the three different samples are 2.5, 3.1, and 3.8 nm respectively. The QW regions are grown on 3.5 μm GaN layers.

---

[1]) Corresponding author; Phone: (970) 491-8426; e-mail: vaschen@engr.colostate.edu

To apply pressure samples were mounted in a standard diamond anvil cell filled with liquid argon. The photoluminescence (PL) was excited with the third harmonic of a femtosecond Ti:sapphire laser at 270 nm. The PL emission was collected in a backscattering geometry and detected with a 0.3 m spectrometer and liquid nitrogen cooled CCD camera. Time-resolved PL measurements were obtained with a fast photomultiplier tube and a digital oscilloscope in accumulation mode with 0.8 ns resolution.

Figures 1a and b show the shifts of the PL peak energy with pressure in the QWs and GaN layers at low and high optical excitation intensity, respectively. In both regimes the QW PL peaks shift with pressure at a much smaller rate than that of GaN. The pressure rate of change (dE/dp) in the QWs varies significantly with well width. At an excitation intensity of 2 W/cm^2 (Fig. 1a) dE/dp increases from a very small value of 1.6 meV/GPa in the 3.8 nm wells to 18.9 meV/GPa in the 2.5 nm wells. Increase of the excitation intensity to 200 W/cm^2 leads to a significant increase in dE/dp, which is more pronounced in the wider wells (Fig. 1b).

Figure 2 shows the QW PL decay time constants in the 2.5 and 3.1 nm well samples. The decay time significantly increases with applied pressure. In the sample with 3.8 nm wells decay times were too long to be accurately measured with our experimental set-up.

The well width and excitation dependent dE/dp, and the pressure dependent decay time behaviours can be accounted for by the gradual increase of F_{pz} with pressure [9]. The field produces a Stark shift of the PL peak [10], which effectively reduces dE/dp with respect to the constant field case. Since the Stark shift is $\sim qF_{pz}L_w$, where q is the electron charge and L_w is the well width, dE/dp is reduced by a larger amount in the wider wells than in the narrow wells. Due to this well width dependence we can obtain the magnitude of F_{pz} at different pressures from the slope of the PL peak energy versus well width (Fig. 3a) [8]. The field magnitudes obtained in such a way are shown in Fig. 3b by solid diamonds. We also calculated the change in the field magnitude with pressure from the Stark shift in each well. This shift was defined as an energy differ-

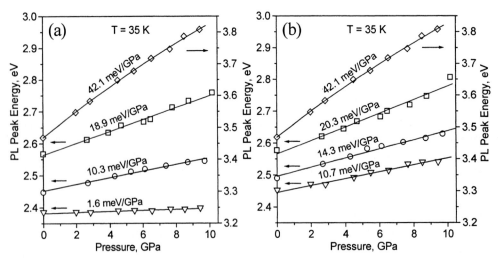

Fig. 1. Pressure dependence of the PL peak energy in 2.5 nm (□), 3.1 nm (○), and 3.8 nm (▽) In$_{0.15}$Ga$_{0.85}$N/GaN QWs and in the GaN layer (◇), at a) 2 and b) 200 W/cm^2 excitation intensity. The lines are the linear fits to the data for the QWs and second-order polynomial for GaN. The numbers above the lines are the slopes of the fit (linear term of the fit for GaN)

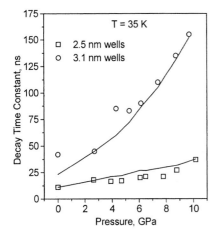

Fig. 2. Pressure dependence of the decay time constants in the InGaN/GaN QWs. The lines show the carrier lifetime change calculated from the experimentally determined F_{pz}

ence between the measured value and that calculated from the InGaN deformation potentials (open symbols in Fig. 3b). The deformation potentials for InGaN were linearly interpolated from the values known for GaN and InN [11]. The strain modified by pressure was calculated using linear elasticity theory [12]. The good agreement obtained between the measured values of piezoelectric field confirms that the small dE/dp are the result of changes in F_{pz} with pressure.

Using the measured values of F_{pz} we calculated the variation of the carrier lifetime with pressure for the QWs of different widths. The lifetime was found as the inverse of the square of the electron–hole wavefunction overlap in the wells with transverse electric field. The results of this calculation normalized to the value of the decay time measured in the 2.5 nm well at atmospheric pressure are shown by the solid lines in Fig. 2.

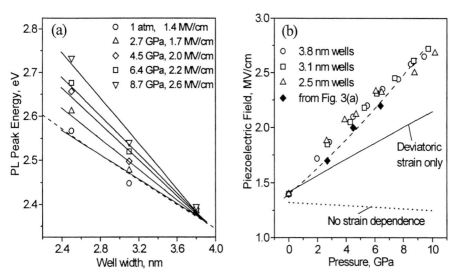

Fig. 3. a) PL peak energy vs. well width at different pressures. The slope of the linear fits yields F_{pz}. The dashed line is the best fit to the atmospheric pressure data with a line obtained by calculating e_1–hh_1 QW transition energy with varied electric field F. The best fit was obtained at $F \approx 1.4$ MV/cm. b) F_{pz} dependence on pressure obtained from a) (solid diamonds), and from the Stark shift of the PL peak (open symbols). The dotted and solid lines are the calculations of F_{pz} using Eqs. (1) and (2) for piezoelectric constants independent of strain and dependent on deviatoric strain, respectively. The dashed line is a polynomial fit to the data points indicated by solid diamonds

The close match between the measured and calculated behaviours unequivocally shows that the small dE/dp in the quantum wells is due to increase in F_{pz} with pressure.

To investigate the origin of the F_{pz} changes with pressure we calculated its values from [6]

$$F_{pz} = L_b(P_b - P_w)/(L_b\varepsilon_w + L_w\varepsilon_b), \qquad (1)$$

where $\varepsilon_{w,b}$ are the permittivities of the InGaN well and GaN barrier, and $L_{w,b}$ are the well and the barrier widths. The piezoelectric polarizations in the well and in the barrier are found as

$$P_{w,b} = e_{33}^{w,b}\varepsilon_{zz}^{w,b}(p) + 2e_{31}^{w,b}\varepsilon_{xx}^{w,b}(p), \qquad (2)$$

where $e_{33}^{w,b}$ and $e_{31}^{w,b}$ are the piezoelectric constants, and $\varepsilon_{xx}^{w,b}$ and $\varepsilon_{zz}^{w,b}$ are the total strains including the strains generated by pressure [12]. The result of the calculation of F_{pz} using Eqs. (1) and (2) with piezoelectric constants independent of strain (Ref. [5]) are shown in Fig. 3b by dotted line. This calculation shows that if strain does not affect the piezoelectric constants, F_{pz} would slightly decrease with pressure. Assuming, instead, that the piezoelectric constants depend on deviatoric (volume conserving) strain, as suggested by Shimada et al. [3], we obtain the increase in F_{pz} shown by the solid line in Fig. 3b, which is smaller than the experimentally measured. The difference between the calculated and measured values arises from the dependence of the piezoelectric constants on the dilatational (volumetric) strain, which is not accounted for in the calculations.

In conclusion, we have determined the piezoelectric field in InGaN/GaN quantum wells of different width as a function of applied pressure. An almost twofold increase of the piezoelectric field at ~9 GPa is shown to be the result of the strong dependence of the piezoelectric constants with total strain, i.e. nonlinear piezoelectric effect. Both, dilatational and deviatoric strain components, contribute to this effect.

Acknowledgements The CSU group acknowledges the support of the National Science Foundation and the Colorado Photonics and Optoelectronics Program.

References

[1] A. D. BYKHOVSKI, V. V. KAMINSKI, M. S. SHUR, Q. C. CHEN, and M. A. KHAN, Appl. Phys. Lett. **68**, 818 (1996).
[2] I. L. GUY, S. MUENSIT, and E. M. GOLDYS, Appl. Phys. Lett. **75**, 4133 (1999).
[3] K. SHIMADA, T. SOTA, K. SUZUKI, and H. OKUMURA, Jpn. J. Appl. Phys. **37**, L1421 (1998).
[4] F. BERNARDINI, V. FIORENTINI, and D. VANDERBILT, Phys. Rev. B **56**, R10024 (1997).
[5] F. BERNARDINI, V. FIORENTINI, and D. VANDERBILT, Phys. Rev. B **63**, 193201 (2001).
[6] V. FIORENTINI, F. BERNARDINI, F. DELLA SALA, A. DI CARLO, and P. LUGLI, Phys. Rev. B **60**, 8849 (1999).
[7] A. D. BYKHOVSKI, B. L. GELMONT, and M. S. SHUR, J. Appl. Phys. **81**, 6332 (1997).
[8] R. ANDRÉ, J. CIBERT, LE SI DANG, J. ZEMAN, and M. ZIGONE, Phys. Rev. B **53**, 6951 (1996).
[9] G. VASCHENKO, D. PATEL, C. S. MENONI, S. KELLER, U. K. MISHRA, and S. P. DENBAARS, Appl. Phys. Lett. **78**, 640 (2001).
[10] D. A. B. MILLER, D. S. CHEMLA, T. C. DAMEN, A. C. GOSSARD, W. WIEGMANN, T. H. WOOD, and C. A. BURRUS, Phys. Rev. B **32**, 1043 (1985).
[11] W. W. CHOW and S. W. KOCH, Semiconductor-Laser Fundamentals, Springer-Verlag, Berlin/Heidelberg/New York 1999.
[12] G. VASCHENKO, D. PATEL, C. S. MENONI, N. F. GARDNER, J. SUN, W. GÖTZ, and C. N. TOMÉ, submitted for publication in Phys. Rev. B.

Piezoelectric Field-Induced Quantum-Confined Stark Effect in InGaN/GaN Multiple Quantum Wells

C. Y. LAI (a), T. M. HSU (a), W.-H. CHANG (a), K.-U. TSENG (a), C.-M. LEE (b), C.-C. CHUO (b), and J.-I. CHYI (b)

(a) Department of Physics, National Central University, Chung-Li 32054, Taiwan

(b) Department of Electrical Engineering, National Central University, Chung-Li 32054, Taiwan

(Received June 21, 2001; accepted July 2, 2001)

Subject classification: 78.65.–j; 78.20.Jq; 78.67.De; S7.14

In this paper, we present an experimental evidence for the piezoelectric field-induced quantum-confined Stark effect (QCSE) on InGaN/GaN quantum wells. The optical transitions of $In_{0.23}Ga_{0.77}N$/GaN p–i–n MQWs were studied by using modulation spectroscopy (electrotransmission ET) at room temperature. Quantum-well-related signals are well resolved in our ET spectra. Clear energy blue shifts in accordance with increasing reversed bias are observed in the ET spectra. The energy blue shift is attributed to the QCSE. The strength of piezoelectric field is found to be 1.9 MV/cm. We also show experimentally how the piezoelectric field affects the energy shift in the strained MQWs.

Introduction In recent years, III-nitride compounds attracted great interest because of their wide band gap and strong bond strength. These properties allowed (Al,In)GaN to be used for full color displays, high brightness blue, green light emitter diodes, high power microwave and high temperature transistor [1]. It is well known that III-nitrides semiconductors have largely pronounced piezoelectric constants [2]. For the MQWs and superlattice system, the strength of the piezoelectric field generated by lattice-mismatch-induced strain can be up to MV/cm. Such strong built-in electric field will modify the band structures and affect the optical and electrical properties of the III-nitride heterostructures. Thus, it is necessary to get a further insight into the effect of piezoelectric field on III-nitride for improving device performance. However, up to now, there are only a few reports on the direct measurement of the piezoelectric field [3].

In this paper, we present an experimental evidence for the piezoelectric field-induced quantum-confined Stark effect (QCSE) on InGaN/GaN quantum wells. The optical transitions of InGaN/GaN p–i–n MQWs ($x < 0.23$) were studied by using modulation spectroscopy (electrotransmission ET) at room temperature. Modulation spectroscopy has been shown to be sensitive to critical point, confinement state transition in the Brillouin zone or microstructure, with the resulting spectrum showing a sharp derivative signal [4–6]. Even at room temperature, due to the derivative-like nature, a large number of sharp signals can be observed. A bias-dependent ET with applied reverse bias varying from 0 up to 20 V was measured, and a direct determination of piezoelectric field was presented in this study.

Experimental Details The sample used in this study was grown on a sapphire (0001) substrate by a low-pressure metal-organic chemical-vapor deposition reactor (MOCVD). The structure consists of a 4 μm n$^+$-GaN (Si: 2×10^{18} cm^{-3}) layer, a 0.3 μm n-type GaN (Si: 5×10^{17} cm^{-3}) layer, ten period of un-doped 20 Å InGaN quantum wells with 70 Å GaN barriers, and a 0.3 μm p-type GaN (Mg: 5×10^{17} cm^{-3}) layer. The nominal indium composition of was ~23%. This sample was processed into mesa devices, on which a semi-transparent electrode was deposited for optical access. The ET experiment was illuminated by a 150 W Xe lamp combined with a 0.25 m monochromator. A dc bias composed with an ac voltage (1 V) was applied to the sample, and the modulated transmission signal was detected by an UV-enhanced silicon detector using the standard lock-in technique.

Results and Discussion Figure 1 shows the room-temperature (T = 300 K) ET spectra. Under zero-bias condition, two well resolved derivative signals A and B are shown in the ET spectra. For the spectrum investigated, the signal A is assigned to the e1–h1 transition, i.e. the absorption between the first electron state (e1) and the first heavy hole state (h1) in MQWs, and the broader at higher energy be near 3.0 eV might be attributed to the forbidden state e1–h2 transition. The forbidden states which come from the huge piezoelectric field break the inversion symmetry of the well and cause that wave function overlap integrals for the forbidden states become significant [7]. The signal B is attributed to the Franz-Keldysh oscillation (FKO) tail of GaN barrier the position of which is smaller than the GaN bandgap energy [8]. The FKO is observed in the electroreflectance spectra for this sample that is not shown here.

To verify the character of the piezoelectric field, the bias-dependent ET spectra were performed at room temperature and are shown in Fig. 2. As shown in Fig. 2, the ET

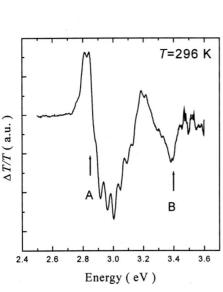

Fig. 1

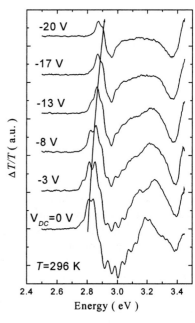

Fig. 2

Fig. 1. Zero-bias ET spectrum for the InGaN/GaN MQWs sample

Fig. 2. Bias-dependent ET spectra for InGaN/GaN MQWs sample at room temperature

signal exhibits an energy blue shift up to about 40 meV with increasing reverse bias from 0 to −20 V. These results are contrary to the typical quantum-confined Stark effect that the optical transition is red shifted by the reverse bias applied [9]. This phenomenon can be explained by the presence of a large internal piezoelectric field in the well. The piezoelectric field is oriented against the p–i–n built-in electric field, and the applied reversed bias will decrease the net electric field in the QW. The band structure of InGaN well will approach flat-band condition and caused an energy blue shift for the QW energy transition. Our result is coinciding with recent work by other groups [3, 10].

In order to estimate the magnitude of the piezoelectric field, the in-well field in piezoelectric MQW incorporated in a p–i–n structure is given simply by

$$E_\mathrm{w} = E_\mathrm{ave} + \frac{L_\mathrm{b}}{L_i} E_\mathrm{piezo},\qquad(1)$$

where E_piezo is the piezoelectric field discontinuity between strained wells and unstrained barriers, L_b is the total length of the barrier and L_i is the total length of the i-th layer. The E_piezo is set as a free parameter for determining the shape of the quantum well. By consideration of the depletion approximation, the electric field in the i-th layer, is a function of position, built-in voltage and applied bias [11]. For simplicity, we set an average electric field E_ave which is define as $E_\mathrm{ave} = \sum E(x_i)/n$, where x_i is the position of the i-th well and n represents the total number of wells embedded in the i-th layer. The material parameters of InGaN were taken by linear extrapolating to those of InN and GaN. After the potential profile is determined by Eq. (1), the confined QW energy state is obtained by a numerical calculation of Schrödinger equation [12]. A ratio of 70:30 for the conduction–valence band offset is adapted, which gave reasonable transition energy for our results.

Figure 3 shows a comparison between the bias dependence of the measured and calculated transition energies for both samples. The solid circles denote the experimental points of the Franz-Keldysh oscillation tail of GaN barrier, which is slightly red-shifted by the reverse bias applied. The solid squares and line denote the experimental point and theoretical calculation of the QWs e1–h1 transitions. The values of E_piezo is set as 1.9 MV/cm,

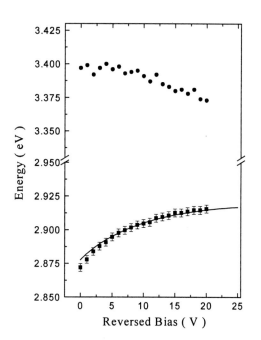

Fig. 3. Calculation of QCSE is compared with experimental data. The solid circles denote experimental points of FKO tail. The solid squares and the line denote experimental points and theoretical calculation of the MQW e1–h1 transitions

which is in good agreement with the experimental data for our sample. For InN mole fractions of 15–16%, $E_{piezo} = 1.2$ MV/cm is reported by Takeuchi et al. [3]. Due to the higher indium composition, our result is in reasonable agreement with the data reported in literature. We found that ET is a useful tool for investigation of the optical character of InGaN/GaN MQWs.

Conclusions In conclusion, we observed ET derivative like signal in both InGaN/GaN p–i–n MQW samples. The variation of transition energies with applied reverse bias has been investigated in InGaN/GaN MQWs. The piezoelectric field is oriented against the built-in electric field and applied reverse bias. The strength of the piezoelectric field of 1.9 MV/cm for the sample, is in reasonable agreement with all the reported results in literature. We have shown experimentally how the piezoelectric field affects the energy shift of strained MQWs. A field of such magnitude is strongly affecting the optical and electronic properties of heterostructure devices. Further investigations on strained InGaN/GaN MQWs are necessary such that one can reliably design and predict the performance of devices.

Acknowledgement This work was supported by the National Science Council of Taiwan under Grant No. NSC 89-2112-M-008-063.

References

[1] S. C. Jain, M. Willander, J. Narayan, and R. Van Overstraeten, J. Appl. Phys. **87**, 965 (2000).
[2] F. Bernardini and V. Fiorentini, Phys. Rev. B **57**, R9427 (1998).
[3] T. Takeuchi, C. Wetzel, S. Yamaguchi, H. Sakai, H. Amano, I. Akasaki, Y. Kaneko, S. Nakagawa, Y. Yamaoka, and N. Yamada, Appl. Phys. Lett. **73**, 1691 (1998).
[4] F. H. Pollak, in: Handbook on Semiconductors, Vol. 2, Ed. M. Balkanski, North-Holland Publ. Co., Amsterdam 1994 (p. 527).
[5] C. Y. Lai, T. M. Hsu, C. L. Lin, C. C. Wu, and W. C. Lee, J. Appl. Phys. **87**, 8589 (2000).
[6] D. E. Aspnes, in: Handbook on Semiconductors, Vol. 2, Ed. T. S. Moss, North-Holland Publ. Co., Amsterdam 1980 (p. 109).
[7] B. K. Laurich, K. Elcess, C. G. Fonstad, J. G. Beery, C. Mailhiot, and D. L. Smith, Phys. Rev. Lett. **62**, 649 (1989).
[8] R. N. Bhattacharya, H. Shen, P. Parayanthal, F. D. Pollak, T. Coutts, and H. Aharoni, Phys. Rev. B **37**, 4044 (1988).
[9] G. Bastard, Wave Mechanics Applied to Semiconductor Heterostructures, Chap. 8, Les Editions de la Physique, Paris 1988.
[10] S. F. Chichibu, T. Azuhata, T. Sota, T. Mukai, and S. Nakamura, J. Appl. Phys. **88**, 5153 (2000).
[11] D. A. B. Miller, D. S. Chemla, T. C. Damen, A. C. Gossard, W. Wiegmann, T. H. Wood, and C. A. Burrus, Phys. Rev. B **32**, 1043 (1985).
[12] A. K. Ghatak, K. Thyagarajan, and M. R. Shenoy, IEEE J. Quantum Electron. **24**, 1524 (1988).

Carrier Dynamics in InGaN/GaN SQW Structure Probed by the Transient Grating Method with Subpicosecond Pulsed Laser

K. Okamoto[1]) (a), A. Kaneta (a), K. Inoue (a), Y. Kawakami (a),
M. Terazima (b), G. Shinomiya (c), T. Mukai (c), and Sg. Fujita (a)

(a) Department of Electronic Science and Engineering, Kyoto University,
Kyoto 606-8501, Japan

(b) Department of Chemistry, Graduate School of Science, Kyoto University,
Kyoto 606-8502, Japan

(c) Nitride Semiconductor Laboratory, Nichia Corporation, 491 Oka, Kaminaka, Anan,
Tokushima 774-8601, Japan

(Received July 16, 2001; accepted July 20, 2001)

Subject classification: 73.63.Hs; 78.55.Cr; 78.67.De; S7.14

Carrier dynamics in GaN and InGaN/GaN SQW structures were observed by using the transient grating (TG) method with sub-picosecond pulsed laser at room temperature. The diffusion coefficients (D) of photo-created carriers were estimated by the decay rate time of TG signals and the photoluminescence (PL) lifetime. It was found that D depends on the emission wavelength (In composition). The relationship between the emission efficiencies and carrier diffusion was considered in terms of the spatial inhomogeneity of In composition.

Recently, InGaN/GaN-based light emitting diodes (LEDs) have been commercialized in ultraviolet (UV), blue, green, and amber spectral region [1, 2]. In particular, the external quantum efficiency (η_{ext}) of about 20% is now achieved in blue (450 nm) LEDs. However, η_{ext} values are still lower for LEDs out of this blue spectral range, but detailed reasons for the reduction of η_{ext} have so far not been elucidated. In this study, we try to elucidate this reason by the viewpoint of carrier dynamics. The transient grating (TG) method which is one of third-order nonlinear spectroscopy, has been used for GaN to detect the nonlinear susceptibility [3], exciton dephasing time [4], time response of scattering [5], or quantum beat [6], etc. It is also a powerful tool to directly detect diffusion processes. By using this method, Haag et al. [7] have measured and reported the carrier diffusion in GaN. We observed the diffusion of heat energy generated by the nonradiative recombination of carriers in GaN and ZnSe by the TG method with nano-second pulsed laser [8, 9]. In this work, we observed carrier diffusion in InGaN/GaN-based SQW structures in UV, blue, green, and amber spectral region by using the TG method with sub-picosecond pulsed laser.

The samples used in this study were grown on a (0001) oriented sapphire (Al_2O_3) substrate by a two-flow metalorganic chemical vapor deposition (MOCVD) technique [10]. The thickness of GaN bulk layer is 4 μm. GaN/InGaN SQW structure is composed of a GaN (1.5 μm), an n-GaN:Si (2.3 μm), an InGaN SQW (3 nm) and a GaN cap (5 nm) layer. LED structure of GaN/InGaN SQW is composed of GaN (30 nm), n-type

---
[1]) Corresponding author; Phone: +81 75 753 7577; Fax: +81 75 753 7579;
e-mail: kokamoto@vbl.Kyoto-u.ac.jp

GaN:Si (5 μm), InGaN SQW (3 nm), p-type AlGaN:Mg (60 nm) and p-type GaN:Mg (150 nm).

For the TG measurement, a mode-locked fiber laser and regenerative amplifier system (Clark) was used. The frequency doubled beam (388 nm) was used as pump beam. Pulse width, power, and repetition rate were 500 fs, 1 mW and 1 Hz, respectively. The pump beam was split into two coherence beams and crossed again in the sample with $\theta = 30°$ and $120°$. The modulation of carrier density (grating) is created in the sample by the interference pattern. The fundamental beam (775 nm) was used as a probe beam with the optical delay unit. The probe beam was partly diffracted, which was detected by a photomultiplier tube (Hamamatsu) and averaged with a boxcar integrator. For the time-resolved photoluminescence (TRPL) measurements, the frequency doubled beam of a mode-locked Ti:sapphire laser (Spectra-Physics; 370 nm) was used for excitation. For the scanning confocal laser microscopy (Tokyo Instruments) was used with the 488 nm line of Ar$^+$ laser. The whole measurements have been performed at room temperature (23 °C).

Figure 1a shows the time profile of the TRPL measurement (A) and the TG measurement with $\theta = 30°$ (B) and $120°$ (C) taken for bulk GaN. These profiles could be fitted by a single exponential function. The PL lifetime (τ_{PL}) and the TG decay time (τ_{TG}) were obtained as $\tau_{PL} = 50$ ps, $\tau_{TG}(30°) = 43$ ps, and $\tau_{TG}(120°) = 16$ ps. The time and spatial behavior of the TG signal intensities were described by the diffraction theory [11] and the diffusion-recombination coupled rate equation of carriers [9]. By solving these equations, the time profile of the TG signals is given by a single exponential function and D is obtained by $1/\tau_{TG} = Dq^2 + 1/\tau_{PL}$, where q is the wave number of grating ($q = 2\pi/\Lambda$). Figure 1b shows the relationship between τ_{TG} and q^2. This plot shows a good linear relationship and D was obtained by the slope as 0.54 cm^{2}s^{-1}. The diffusion length (Λ) of carriers is estimated as $\Lambda = 0.05$ μm by $\Lambda = (D\tau_{PL})^{1/2}$. This value is close to the reported values by cathodoluminescence, $\Lambda = 0.1$ [12], 0.25 [13], or 0.05 μm [14].

In a similar way, τ_{PL} and D values of other samples were measured and plotted against the emission wavelength (In composition) in Fig. 2. It was found that τ_{PL} becomes lager with increasing In composition. A drastic change was observed in the region larger than 450 nm, which is the highest η_{ext} point. D also becomes gradually

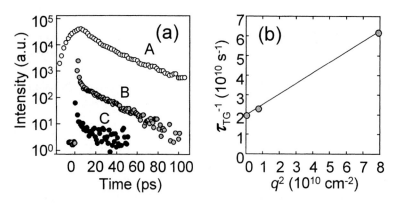

Fig. 1. a) Time profile of the TRPL measurement (A) and TG measurements with $\theta = 30°$ (B) and $120°$ (C) taken for GaN at room temperature. b) Relationship between the TG decay time (τ_{TG}) and the grating constant (q^2) taken for GaN

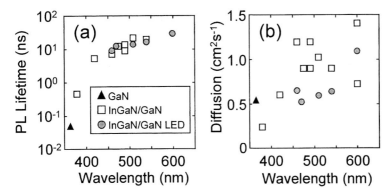

Fig. 2. a) PL lifetimes and b) diffusion coefficients of GaN bulk layer, InGaN/GaN SQW, and InGaN SQW LED structures plotted against the emission wavelength (In composition)

larger with increasing In composition expect lowest value of SQW (380 nm). At the room temperature, τ_{PL} should be nearly equal to the nonradiative recombination lifetime ($\tau_{non\text{-}rad}$) of carriers, because nonradiative recombination is the dominant process of carriers. $\tau_{non\text{-}rad}$ can be described by the product of thermal velocity of carriers (v_{th}), the capture cross section (σ) to the nonradiative recombination center (NRC), and the density (N_{NRC}) of NRC. The increase of $\tau_{non\text{-}rad}$ suggests a decrease of $v_{th}\sigma N_{NRC}$. This fact should be due to the slow carrier diffusion by In fluctuation. In fact, D of SWQ at 380 nm was very small. In this region, carriers should be localized in the fluctuation of In composition, which acts as radiative center (RC) of carriers. This localization contributes to the high η_{ext} value. On the other hand, η_{ext} was reduced for a large amount of In. This fact is often interpreted as the increment of NRC by the degrading crystal properties for a large amount of In. However, Fig. 2 shows that both $\tau_{PL}(\tau_{non\text{-}rad})$ and $D(v_{th})$ become larger with In composition. Thus, $N_{NRC}\sigma$ must become smaller with increasing In content. This fact suggests that the NRC should not increase and so it should not be the origin of the reduction of η_{ext}. Otherwise, the fast diffusion of carriers should be due to the delocalization of carriers, which is a negative factor for η_{ext}.

D should depend on the spatial inhomogeneity of In composition. By using scanning near-field optical microscopy (SNOM), we have been observed the spatial inhomogeneity of PL intensity (In fluctuation) of the InGaN/GaN SQW [15]. Similar inhomogeneity of PL was observed by using scanning confocal laser microscopy (SCLM). Figure 3 shows the SCLM image and the spectrum for LED (540 nm). A PL inhomogeneity within micron scale was observed. This scale of In fluctuation is as long as the diffusion length of carriers. In the scale of recombination pathways, In composition may be homogeneous. Such microscopic homogeneity should be the reason of the fast diffusion of carriers. The PL spectra for each region were also shown in Fig. 3 (right part). It was found that the PL intensities were smaller in the In rich region (longer PL wavelength). This behavior is opposite to the case of UV-blue region and suggests that the In rich region should not act as RC. Another reason of the fast diffusion of carriers in this region may be the effect of a piezoelectric field (PEF). PEF is induced by the large amount of In and fluctuates along the In fluctuation within micron scale. Such a fluctuation of PED causes carrier acceleration in lateral direction.

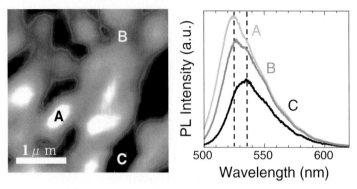

Fig. 3. Image (left part) and PL spectrum (right part) of the scanning confocal laser microscope taken for InGaN/GaN LED structure with emission at 540 nm at room temperature

In conclusion, we propose that the main reason of the reduction of η_{ext} for a large amount of In is not the increment of NRC, but the delocalization of carriers due to fast diffusion. To develop new devices with higher η_{ext} RC in the active layers must be effective.

Acknowledgements The authors would like to thank Mr. J.-K. Choi (Kyoto Univ.) for SCLM measurement. This work was partly supported by the Kyoto University-Venture Business Laboratory Project, Research Foundation for Opto-Science and Technology, Konica Imaging Science Foundation and a Grant-in-Aid for Scientific Research from the Japan Society for the Promotion of Science and Ministry of Education, Science and Culture.

References

[1] T. MUKAI and S. NAKAMURA, Jpn. J. Appl. Phys. **38**, 5735 (1999).
[2] T. MUKAI, M. YAMADA, and S. NAKAMURA, Jpn. J. Appl. Phys. **37**, L1358 (1998).
[3] H. HAAG, P. GILLIOT, R. LÉVY, B. HÖNERLAGE, O. BRIOT, S. RUFFENACH-CLUR, and R. L. AULOMBARD, Appl. Phys. Lett. **74**, 1464 (1999).
[4] S. PAU, J. KUHL, F. SCHOLZ, V. HAERLE, M. A. KHAN, and C. J. SUN, Appl. Phys. Lett. **72**, 557 (1998).
[5] BAHMAN TAHERI, J. HAYS, J. J. SONG, and B. GOLDENBERG, Appl. Phys. Lett. **68**, 587 (1996).
[6] T. AOKI, G. MOHS, M. KUWATA-GONOKAMI, and A. A. YAMAGUCHI, Phys. Rev. Lett. **82**, 3108 (1999).
[7] H. HAAG, B. HÖNERLAGE, O. BRIOT, and R. L. AULOMBARD, Phys. Rev. B **60**, 11624 (1999).
[8] K. OKAMOTO, Y. KAWAKAMI, SG. FUJITA, and M. TERAZIMA, Anal. Sci. **17**, 312 (2001).
[9] K. OKAMOTO, Y. KAWAKAMI, SG. FUJITA, M. TERAZIMA, and S. NAKAMURA, Proc. Internat. Workshop Nitride Semiconductors, Nagoya (Japan), IPAP Conf. Ser. **1**, 540 (2000).
[10] S. NAKAMURA, Jpn. J. Appl. Phys. **30**, L1705 (1991).
[11] H. KOGELMIK, Bell. Syst. Tech. J. **48**, 2909 (1969).
[12] X. ZHANG, P. KUNG, D. WALKER, J. PIOTROWSKI, A. ROGALSKI, A. SAXLER, and M. RAZEGHI, Appl. Phys. Lett. **67**, 2028 (1995).
[13] S. J. ROSNER, E. C. CARR, M. J. LUDOWISE, G. GIROLAMI, and H. I. ERIKSON, Appl. Phys. Lett. **70**, 420 (1997).
[14] T. SUGAHARA, H. SATO, M. HAO, Y. NAOI, S. KURAI, S. TOTTORI, K. YAMASHITA, K. NISHINO, L. T. ROMANO, and S. SAKAI, Jpn. J. Appl. Phys. **37**, L398 (1998).
[15] A. KANETA, T. IZUMI, K. OKAMOTO, Y. KAWAKAMI, SG. FUJITA, Y. NARITA, T. INOUE, and T. MUKAI, Jpn. J. Appl. Phys. **40**, 102 (2001).

Temperature Dependence and Reflection of Coherent Acoustic Phonons in InGaN Multiple Quantum Wells

Ü. Özgür, C.-W. Lee, and H. O. Everitt[1])*)

Department of Physics, Duke University, Durham, NC 27708, USA

(Received July 20, 2001; accepted July 30, 2001)

Subject classification: 63.22.+m; 73.63.Hs; 78.47.+p; 78.67.De; S7.14

Sub-picosecond optical pump–probe techniques were used to generate coherent zone-folded longitudinal acoustic phonons (ZFLAPs) in an InGaN multiple quantum well structure. Differential transmission measurements revealed that carriers injected near the barrier band edge were quickly captured into the quantum wells and generated strong coherent ZFLAP oscillations. Differential reflection measurements were used to explore the acoustic phonon transport and reflection in the multiple quantum well structure.

Electron–phonon interactions limit the dephasing time of electronic or excitonic coherent states, so understanding and controlling the effects of phonons is of fundamental interest [1, 2]. Particularly challenging is the optical investigation of acoustic phonons in bulk semiconductors because the phonon dispersion relation permits only low frequency Brillouin scattering. However, it is well understood that a semiconductor multiple quantum well (MQW) produces zone folding of the acoustic phonon branch so that direct excitation is possible [2–5]. Recently, it has been demonstrated that strong coherent zone-folded longitudinal acoustic phonon (ZFLAP) oscillations can be generated and observed in InGaN MQW structures [6–8]. In this paper, differential transmission measurements describe temperature dependence of the impulsive mechanism for generating these coherent ZFLAP oscillations, and differential reflection measurements reveal the coherent reflection of these phonons from the semiconductor/air interface.

The MQW sample used here was grown by metalorganic chemical vapor deposition (MOCVD) at the University of California at Santa Barbara using a modified two-flow horizontal reactor on double polished c-plane sapphire [9]. It consists of a ten-period, 12 nm per period MQW with 3.5 nm wide $In_{0.15}Ga_{0.85}N$ quantum wells and 8.5 nm wide $In_{0.05}Ga_{0.95}N$:Si barriers. The MQW structure is capped with a 100 nm GaN layer and is grown on a ≈ 2 µm GaN:Si layer. The Si doping concentration in the barriers is $\sim 10^{18}$ cm^{-3}.

Room temperature photoluminescence (PL) and photoluminescence excitation (PLE) spectroscopies were used to identify the energies of the quantum well states and the barriers, respectively [8]. The peak PL emission occurred at 2.99 eV, corresponding to electron–hole recombination from the lowest energy quantum well subband states. The broad PLE absorption peak for emission at 2.99 eV occurred near 3.22 eV, corresponding to electron–hole (e–h) pairs being generated near the MQW barrier band edge. As

---

[1]) Corresponding author; Tel: 919-6602518; Fax: 919-6602525; e-mail: everitt@arl.aro.army.mil
*) Also with U. S. Army Research Office, Research Triangle Park, NC 27709-2211, USA.

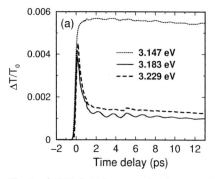 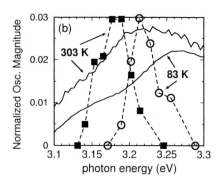

Fig. 1. a) DT data for pump/probe wavelengths near (3.229 eV, 3.183 eV) and below (3.147 eV) the barrier band edge. The feature at 5 ps arises from a pump pulse reflection from the substrate. b) Strength of coherent ZFLAP oscillations as a function of pump/probe energy, normalized with respect to the peak $\Delta T/T_0$ value at that energy (dashed lines), and the PLE signals for 303 K and 83 K (solid lines)

can be seen in Fig. 1b, the PLE signal blueshifts by approximately 40 meV at 83 K as compared with the 303 K data.

Wavelength-degenerate, sub-picosecond, pump–probe differential transmission (DT) has been used to measure the electron capture time [8]. A strongly wavelength-dependent bi-exponential decay of the created carriers (Fig. 1a) indicated that electrons were captured from the 3D barrier states to 2D confined QW states with a time constant between 0.31 and 0.54 ps. Electron capture was most efficient when e–h pairs were generated within 50 meV of the 3.22 eV barrier. The width of the PL signal and region of efficient carrier capture (∼100 meV) suggest sizable material inhomogeneities.

Remarkably strong ZFLAP oscillations in the DT signal are observed (Fig. 1). The damped oscillations always started at the peak of the pump pulse ($t = 0$) and are described by $A \exp(-t/\tau) \cos(2\pi t/P + \pi)$, whose phase term π signifies that the oscillations were always observed to start at a minimum. The oscillation period P is related to the MQW period d as $P = d/v_s$, where v_s is the sound velocity. The ZFLAPs propagate along the c-axis of the wurtzitic InGaN. The oscillations observed in our $d = 12$ nm MQW sample[2]) occurred with a period $P = 1.44$ ps (frequency $f_0 = 0.69$ THz). This yields a sound velocity of 8333 m/s, slightly larger than 7990–8020 m/s [10, 11] for bulk GaN. The characteristic decay time of observed DT oscillations, $\tau = 12$ ps, is approximately equal to the phonon transit time through the MQW region. The coherence time of the ZFLAPs is expected to be much longer than this.

The temperature independent 80 meV span of pump/probe energies over which ZFLAP oscillations are observed corresponds to the region of carrier capture and the width of the barrier absorption band edge (Fig. 1). As the temperature is lowered to 83 K, the amplitude of the ZFLAP oscillations, normalized to the maximum amplitude of the DT signal, was found to be the same as at 303 K. However, the excitation energy which generates the strongest ZFLAP oscillations blueshifts by 40 meV, consistent with the shift observed in the barrier energy measured by PLE at 83 K. The ZFLAP oscillation frequency did not change with temperature.

---

[2]) X-ray measurements of the sample revealed the MQW period to be 12 nm, not the 8 nm originally claimed by the growers and reported in [8].

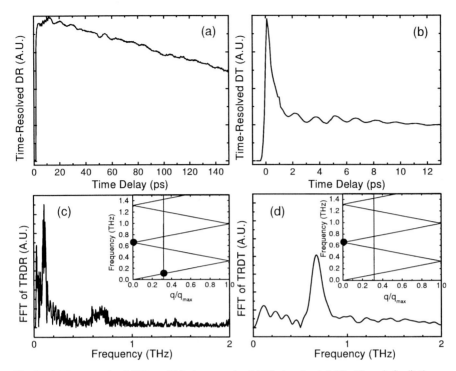

Fig. 2. a) Time-resolved DR and b) time-resolved DT signals at 3.18 eV; and c), d) the corresponding FFT spectra. The inset in c) and d) shows the calculated acoustic phonon dispersion, with the solid line representing the wavevector defined by the laser, and the filled circles corresponding to the FFT peaks

A wavelength-degenerate differential reflection (DR) technique was used to further investigate the coherent acoustic phonons. The DR and DT signals from the MQW sample, excited with pump/probe energy of 3.183 eV, are compared in Figs. 2a and b. In addition to the high frequency oscillations observed in the first 10 ps of both DR and DT data, a low frequency component was observed in the DR data. Figures 2c and d compare the fast Fourier transforms (FFTs) of the DR and DT signals, respectively. The calculated MQW folded acoustic phonon dispersion is shown in the inset, where $q_{max} = \pi/d$. The DT signal has only the previously discussed 0.69 THz frequency component, corresponding to the first order folded frequency at zone center ($q = 0$). The FFT of the DR signal shows an additional frequency component at 0.10 THz, corresponding to the Brillouin-scattered bulk acoustic phonon mode, excited at the wavevector defined by the laser wavelength

$$q = 2\frac{2\pi n}{\lambda}, \qquad (1)$$

$$2\pi f = v_s q = \frac{4\pi n v_s}{\lambda}. \qquad (2)$$

The additional factor of two in Eq. (1) arises from momentum conservation for the reflected photon, and n is the refractive index at the laser wavelength. Higher frequency phonon doublets at this same wavevector, corresponding to the folded branches of the acoustic phonon dispersion, are surprisingly absent. This contrast with the clear observation of phonon doublets in GaAs/AlAs SLs [12] is currently under investigation.

Serendipitously, the absence of phonon doublets dramatically simplifies the analysis of coherent phonon dynamics. For example, in Fig. 2a the magnitude of the low frequency oscillations is clearly observed to decay, then revive near 50 ps delay. To understand this revival, consider that ballistic acoustic phonons travel 400 nm in 50 ps ($v_s = 8000$ m/s). This is roughly the round-trip distance through the MQW region and GaN cap layer. Thus, the 0.10 THz acoustic phonon signal decays as the phonons travel ballistically out of the MQW and into the cap and buffer layers. Those phonons that reach the surface are strongly reflected because of the high acoustic mismatch at the semiconductor/air interface. When these reflected phonons reach the MQW region, the oscillation of the DR signal reappears. Because of the greater phonon travel distance through the buffer layer and the smaller acoustic mismatch between the buffer layer and the substrate, no acoustic phonon reflections were observed from that interface. Measurements of phonon reflections from the substrate/air interface were not attempted.

It is interesting to note that the 0.69 THz ZFLAPs were not observed to reflect from the semiconductor/air interface. Because the decay of the original ZFLAP signal is much too fast to represent phonon decoherence, the coherent ZFLAPs must leak out of the MQW region into the cap and buffer layers. There these formerly $q = 0$ phonons assume a bulk propagating wavevector $q = f/v_s$ and are expected to traverse the cap and buffer layers along with the 0.10 THz phonons. If it is assumed that the coherence time of the high frequency phonons is comparable to that of the low frequency phonons, the absence of reflected 0.69 THz phonons suggests that they scatter and dephase before they can return to the MQW region. Perhaps this occurs at the semiconductor/air interface where the short wavelength 0.69 THz phonons (12 nm) are more susceptible to the surface roughness (≈ 10 nm) than are the longer wavelength 0.10 THz phonons (80 nm).

In summary, spectrally pure, coherent 0.10 THz bulk and 0.69 THz zone folded acoustic phonons were generated using impulsive optical techniques. A single, high frequency acoustic phonon mode was excited using impulsive DT, and it was found that these zone folded acoustic phonons are insensitive to temperature changes between 300 and 83 K. Acoustic phonon modes at both frequencies were simultaneously excited using impulsive DR. It was observed that the bulk phonons travel ballistically through the cap layer and reflect efficiently from the semiconductor/air interface, whereas reflected zone folded phonons were not observed.

Acknowledgements The samples were grown by A. C. Abare, S. Keller, and S. P. DenBaars of the University of California, Santa Barbara. We thank R. Merlin for numerous valuable discussions. We thank A. C. Abare and M. J. Bergmann for the X-ray measurements and acknowledge additional helpful discussions with M. A. Stroscio and H. C. Casey, Jr. This work was supported by ARO grant No. DAAH04-93-D-0002 and by DARPA/ARO grant DAAH04-96-0076.

References

[1] K. SCHWAB, E. A. HENRIKSEN, J. M. WORLOCK, and M. L. ROUKES, Nature **404**, 974 (2000).
[2] A. BARTELS, T. DEKORSY, H. KURZ, and K. KÖHLER, Appl. Phys. Lett. **72**, 2844 (1998).
[3] C. COLVARD et al., Phys. Rev. B **31**, 2080 (1985).

[4] A. V. Kuznetsov and C. J. Stanton, Phys. Rev. Lett. **73**, 3243 (1994).
[5] K. Mizoguchi, M. Hase, S. Nakashima, and M. Nakayama, Phys. Rev. B **60**, 8262 (1999).
[6] C. K. Sun et al., Appl. Phys. Lett. **75**, 1249 (1999).
[7] C. K. Sun, J. C. Liang, and X. Y. Yu, Phys. Rev. Lett. **84**, 179 (2000).
[8] Ü. Özgür et al., Appl. Phys. Lett. **77**, 109 (2000).
[9] S. Keller et al., J. Cryst. Growth **195**, 258 (1998).
[10] C. Deger et al., Appl. Phys. Lett. **72**, 2400 (1998).
[11] M. Yamaguchi et al., J. Appl. Phys. **85**, 8502 (1999).
[12] A. Bartels, T. Dekorsy, H. Kurz, and K. Köhler, Phys. Rev. Lett. **82**, 1044 (1999).

InGaN/GaN Quantum Well Microcavities Formed by Laser Lift-Off and Plasma Etching

P. R. EDWARDS[1]) (a), R. W. MARTIN (a), H.-S. KIM (b), K.-S. KIM (b), Y. CHO (c), I. M. WATSON (b), T. SANDS (c), N. W. CHEUNG (c), and M. D. DAWSON (b)

(a) Department of Physics, University of Strathclyde, Glasgow, G4 0NG, UK

(b) Institute of Photonics, University of Strathclyde, Glasgow, G4 0NW, UK

(c) University of California, Berkeley, California 94720, USA

(Received June 21, 2001; accepted June 28, 2001)

Subject classification: 78.55.Cr; 78.67.De; 81.65.Cf; S7.14

Photoluminescence measurements have been used to investigate InGaN/GaN quantum well microcavities formed between two dielectric Bragg reflectors. Both single and ten-period quantum wells emitting near 420 nm were studied. The structures were formed using a combination of MOCVD growth for the nitride layers, laser lift-off to remove the sapphire substrates and electron-beam evaporation to deposit the mirrors. Room temperature photoluminescence measurements have been used to investigate the cavity modes observed from both plasma etched and unetched microcavities, and half widths as low as 0.6 meV were observed. The cavity modes were visible as dips in measured reflectance spectra and as peaks in the PL. Comparison of the mode wavelengths with simulated reflectivity spectra has allowed the determination of the cavity thickness before and after etching; this has shown the etch-back step to have a degree of control ($\pm 5\%$) necessary for the later fabrication of resonant periodic gain structures.

Introduction Vertical cavity surface-emitting lasers (VCSELs) have significant advantages over their edge-emitting counterparts, but several factors have so far prevented the commercial exploitation of nitride-based systems in such devices. These include the limited difference in refractive index between GaN and compatible semiconductors, hindering the production of the highly efficient distributed Bragg reflectors (DBRs) needed for the cavity mirrors. The use of all-dielectric DBRs has previously been proposed to overcome this problem, in combination with either lateral overgrowth [1] or laser lift-off [2–4] techniques to produce a resonant cavity. This work documents progress in the latter method, and in particular the control of the GaN thickness; this is required in order to ensure that the high gain InGaN quantum wells coincide with antinodes of the cavity electric field [5].

Experimental Both single quantum well (SQW) and multiple quantum well (MQW) samples emitting at ≈ 420 nm were fabricated using an Aixtron 200 series MOVPE reactor. A low-temperature GaN nucleation layer was deposited onto (0001) sapphire substrates, followed by an undoped GaN buffer layer ≈ 1 µm thick. A single layer of InGaN of nominal thickness 2 nm was deposited on the first sample, whilst the other sample was deposited with ten such layers each separated by a ≈ 7 nm barrier of GaN. Both samples were capped with a further ≈ 15 nm of GaN, before the approximate

---

[1]) Corresponding author; Phone: +44 141 548 4369, Fax: +44 141 552 2891, e-mail: paul.edwards@strath.ac.uk

nitride thickness was determined using ex-situ reflectometry measurements. The first DBR was then deposited, consisting of a 10½ period stack of alternate SiO_2 and ZrO_2 $\lambda/4$ layers, terminated top and bottom by SiO_2. These layers were deposited by electron beam evaporation, and designed to give a reflectivity band centred at 420 nm, the peak in the QW luminescence observed in unmirrored samples.

The samples were then bonded by the top surface of the DBR to a silicon wafer in order to support the layers on removal of the sapphire substrate. This substrate removal was carried out by the laser lift-off (LLO) technique [6]. After polishing the bottom surface of the sapphire, pulses from a 248 nm KrF excimer laser are directed through the substrate, resulting in the elemental decomposition of GaN near the GaN/sapphire interface; gentle heating is then sufficient to melt the resulting Ga-rich layer and detach the sapphire.

Before deposition of the top mirror, the samples were etched using an inductively-coupled $Cl_2/BCl_3/Ar$ plasma [7]. This removed the low-quality GaN nucleation layer in addition to allowing control of the GaN thickness. Separate portions of the sample were then etched back by nominally 100 nm and 200 nm; a further unetched control portion was retained from each sample. Residuals from the etching process were removed using an HCl solution. The top DBR, identical to the first, could then be deposited to complete the cavity. Reflectivity measurements were made using a spectrophotometer with $\approx$2 nm resolution. Photoluminescence (PL) spectra were obtained by excitation using a 4 mW HeCd laser (λ = 325 nm) focussed to a spot diameter of $\approx$20 μm, and detection using a spectrometer and CCD array.

Results and Discussion Figure 1a shows the measured reflectivity spectrum of the un-etched SQW sample, and this is compared in Fig. 1b with a simulated reflectivity spectrum of the complete DBR/nitride/DBR cavity structure. This was calculated using the matrix method and published dispersion data for GaN [8], SiO_2 [9] and ZrO_2 [10]. In the simulation the width of the nitride layer has been adjusted slightly from an initial value estimated from reflectometry measurements, in order for the wavelength of the appropriate cavity mode to coincide with that observed at 446 nm.

The measured spectrum shows a marked decease in reflectivity at shorter wavelengths by comparison with the calculated spectrum, as well as less prominent modes in this region. A further simulation, shown in

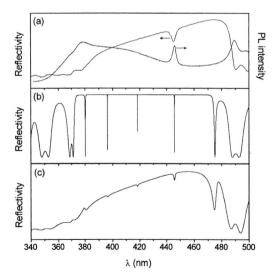

Fig. 1. Reflectivity of the unetched MQW sample as a) measured, b) simulated neglecting absorption and c) simulated with λ^{-4} dependent absorption in the top DBR. The corresponding PL spectrum is also shown

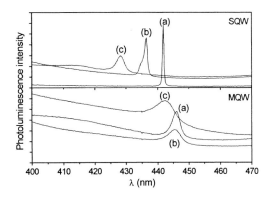

Fig. 2. Room temperature photoluminescence spectra from the SQW and MQW cavities, with unetched, 100 nm etched and 200 nm etched samples labelled (a), (b) and (c), respectively

Fig. 1c, shows that both of these effects can be adequately modelled by assuming an absorption coefficient proportional to λ^{-4} in the top DBR. This is consistent with Rayleigh scattering due to GaN surface roughness prior to the second mirror deposition, on a scale of <100 nm. The similar reflectivity spectra observed for the etched samples (not shown) suggest that this is a result of the LLO process, and not the subsequent etch-back step. Furthermore, as this effect was not observed in a comparable study using simple GaN cavities [4] it is expected that this can be resolved by further process optimisation.

Also shown in Fig. 1a is the measured PL spectrum from the same sample. This shows a peak aligned with the cavity mode observed in the reflectivity, in addition to broader peaks at 370 and 490 nm corresponding to the edges of the high reflectivity band of the DBRs.

The PL data from each of the samples are shown in Fig. 2. Reflectivity simulations were carried out for each of the samples, with the cavity thicknesses again adjusted slightly from their nominal values in order to fit the observed modes. Using this method, total nitride thicknesses of 898, 793 and 686 nm were extracted for the unetched, "100 nm" etched and "200 nm" etched SQW samples, whilst the MQW data yielded values of 910, 819 and 721 nm, respectively. This demonstrates a repeatability of around ±5%.

The spectra in Fig. 2 also show the effect of etching on the width of the cavity mode, with the PL peak clearly broadening with increasing etching depth. The half widths are plotted in Fig. 3, which shows not only the trend with increased etching, but also the fact that the peak in the MQW cavity is consistently 15–25 meV broader than in the SQW device. The panel on the right of the figure shows for comparison the widths of the PL peaks from un-mirrored material, which differ by a similar value.

The narrowest of the cavity modes observed (3.7 meV for the unetched SQW sample) yields a

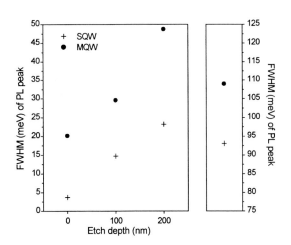

Fig. 3. PL peak widths as a function of etch depth (left) and for unmirrored samples (right)

cavity Q factor of 760, which is comparable with results from similar structures reported by Song et al. [2]. In the absence of multiple modes, the cavity finesse has been estimated by combining the measured peak width with mode spacing values from the simulated reflectivity spectra. This gave a best value of 49, which in turn indicates mirror reflectivities of ≈94% (assuming the mirrors to be equal).

Conclusion Cavity-filtered luminescence has been demonstrated for high-Q InGaN/GaN quantum well cavity structures fabricated using a combination of laser lift-off and e-beam deposition of dielectric mirrors. Inductively-coupled plasma etching has also been shown to be a viable technique for removing low-quality material, and for modifying the nitride thickness with the degree of control necessary for producing resonant periodic gain devices.

Acknowledgement The authors are grateful for the financial support of the U.K. EPSRC.

References

[1] R. W. MARTIN, P. R. EDWARDS, R. PECHARROMAN-GALLEGO et al., phys. stat. sol. (a) **183**, 145 (2001).
[2] Y.-K. SONG, H. ZHOU, M. DIAGNE et al., Appl. Phys. Lett. **74**, 3441 (1999).
[3] Y.-K. SONG, H. ZHOU, M. DIAGNE et al., Appl. Phys. Lett. **76**, 1662 (2000).
[4] R. W. MARTIN, P. R. EDWARDS, H.-S. KIM et al., presented at E-MRS Spring Meeting, June 5–8, 2001, Strasbourg (France).
[5] K. J. EBELING, Semiconductor Quantum Optoelectronics, Institute of Physics Publishing, Bristol 1999 (pp. 295–338).
[6] W. S. WONG, Y. CHO, E. R. WEBER et al., Appl. Phys. Lett. **75**, 1887 (1999).
[7] H.-S. KIM, K.-S. KIM, T. KIM et al, to be submitted to J. Vac. Sci. Technol. (2001).
[8] T. PENG and J. PIPREK, Electron. Lett. **32**, 2285 (1996).
[9] E. D. PALIK, Handbook of Optical Constants of Solids, Academic Press, San Diego 1998.
[10] D. SMITH and P. BAUMEISTER, Appl. Opt. **18**, 111 (1979).

Comparative Study of InGaN/GaN Structures Grown by MOCVD Using Various Growth Sequences

A. V. Sakharov[1]) (a, b), A. S. Usikov (a, b), W. V. Lundin (a, b),
A. F. Tsatsulnikov (a), Ru-Chin Tu (b), Sun Bin Yin (b), and Jim Y. Chi (b)

(a) A. F. Ioffe Physico-Technical Institute, Politekhnicheskaya 26, 194021, St. Petersburg, Russia

(b) Opto-Electronics & Systems Laboratories, Industrial Technology Research Institute, Bldg. 78, 195-8 Sec. 4, Chung Hsing Rd. Chutung, Hsinchu 310, Taiwan

(Received June 22, 2001; accepted July 4, 2001)

Subject classification: 68.65.Fg; 78.55.Cr; 78.60.Fi; 78.67.De; 81.15.Gh; S7.14

In this work we report results of direct comparison of two sets of InGaN/GaN MQW structures grown with and without using of thermocycling (TC) by electroluminescence, photoluminescence and photoluminescence excitation study. For both sets of structures luminescence originates from localized centres, but the density of states differs for structures grown with and without TC. Light emitting diodes (LED) grown with use of TC show nearly two times higher efficiency than the structures with the same design grown without TC.

Introduction III–N-based semiconductors have received growing interest due to their large direct band gap, making them promising materials for UV-to-visible light emitting devices, solar blind UV detectors, and high-power and high-temperature devices. However, due to the lack of ideal substrates for the growth of thin film nitrides, a large number of dislocations and defects are naturally formed in the epitaxial layer to alleviate the lattice mismatch and the strain of postgrowth cooling. Moreover, growth of InGaN/GaN multi-quantum well (MQW) structures is significantly complicated by strong difference in optimal growth conditions for InGaN and GaN. It is well known that there are two approaches for InGaN/GaN growth: a) growth in constant conditions by switching flows and b) growth of InGaN and GaN in different conditions. The second approach includes changing of substrate temperature keeping all other growth regimes of GaN and InGaN constant [1] as well as changing all parameters (such as substrate temperature, gas ambient [2] and reactor pressure) for GaN and InGaN growth. Both these approaches have their advantages and disadvantages. In the first case the growth regime is a compromise between optimal conditions for GaN and InGaN growth; in the second case the difference in growth regimes for GaN and InGaN creates the problem of transient layers. In this work we present the results of direct comparison of these two approaches.

Two sets of InGaN/GaN and InGaN/GaN/AlGaN structures emitting at various wavelengths from 395 to 470 nm designed for photo- and current excitation were grown. The first set of the structures contains InGaN/GaN MQW active regions grown at constant temperature, while for the structures from the second set we have used thermocycling (TC) with different growth temperatures for InGaN well and GaN barrier. Each

---
[1]) Corresponding author; Phone: +7-812-2473182; Fax: +7-812-2473178;
e-mail: val@beam.ioffe.rssi.ru

structure from the first set has an analog in the second set with the same design and nearly same emitting wavelength.

Experiment The InGaN/GaN/AlGaN structures studied in this work were grown in the large-scale AIX2000 MOCVD system on (0001) sapphire substrates employing low-temperature GaN nucleation layer. Structures for optical study consist of 2 μm thick undoped GaN epilayer, 2 μm thick Si-doped GaN epilayer and 15 InGaN/GaN QW active region capped with 50 nm of AlGaN. Growth temperature for active region was chosen to obtain the same emission wavelength for structures grown with and without thermocycling.

For optical studies the samples were mounted on a copper heat sink attached to a wide temperature range close-cycle He cryostat. Conventional photoluminescence (PL) spectra were measured in the back-scattering geometry using a continuous wave He–Cd laser or pulsed N_2 laser as the excitation sources. PL excitation spectra were recorded using light of a 500W Xe lamp passed through monochromator. Structures with p–n junctions were processed as LEDs and electroluminescence (EL) spectra were measured under various driven currents on unmounted chips collecting light from the substrate side.

Results and Discussion Room temperature photoluminescence (PL) spectra of samples grown at different temperature (samples b1115 and b1116 grown without TC, samples b1117 and b1121 with TC) are shown in Fig. 1. The PL spectrum at low excitation densities shows a single relatively broad peak with extended tails on both high and low energy sides. This shape of the spectrum is typical for the structures with In-rich nano-domains, with significant dispersion in size and In concentration. The full width on half-maximum (FWHM) value at low excitation densities is higher for long-wavelength

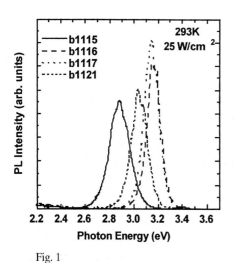
Fig. 1

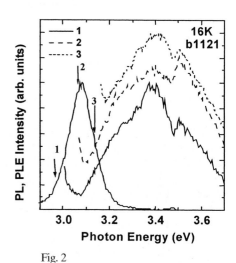
Fig. 2

Fig. 1. Room temperature PL spectra of samples grown with TC (samples b1117, b1121) and without TC (samples b1115, b1116)

Fig. 2. Low temperature PL (left curve) and PLE spectra of a sample grown with TC. Note that the PLE spectra are shifted for clarity. Detection energy is marked by arrows

structures. For short-wavelength structures there is no significant difference between growth with and without TC. The PL intensity is also on the same level, slightly higher for structures grown with TC.

Figure 2 shows PL and PLE spectra recorded at low temperature for structure b1121. The Stokes shift revealed by PLE is about 150 meV, due to formation of local In-rich areas typical for InGaN/GaN system. The shape of PLE spectra practically does not change with registration energy indicating the existence of a continuum spectrum above ≈ 3 eV. PLE spectra recorded for other structures (grown with and without TC) show the similar behaviour indicating the same nature of localized centres. The values of Stokes shift obtained from PLE studies are in good agreement with results published previously [3].

Figure 3 shows the dependence of PL peak position and FWHM on excitation density. The shift of PL peak position even at maximum excitation is smaller than the Stokes shift revealed from PLE, so we can conclude that even at excitation densities of 1 MW/cm^2 recombination comes through localized states. Changes in PL spectra with increasing excitation density are different for structures grown with different approaches; for structures grown with TC with increasing excitation density from 25 W/cm^2 to 3 kW/cm^2 the typical blueshift was smaller than for structures grown without TC, and for structures emitting in UV the increase of excitation density even causes a redshift of PL. Such difference can be explained by the increasing role of recombination via localized states in the case of TC growth regime. Further increase in excitation density produces a blueshift that depends only on emission wavelength and not on growth regime.

The dependence of FWHM on excitation density is shown in Fig. 3b. It should be noted that at excitation density from 25 W/cm^2 to 3 kW/cm^2 for all samples studied, FWHM is nearly the same in the range of 130–160 meV. Further increase of excitation leads to filling of localized states and broadening of PL peak, and at maximal excitation densities FWHM for structures grown with and without TC is comparable.

So, from excitation dependent PL studies we can conclude that the main difference between structures grown with and without TC is in the formation of localized states in the band tail, with higher density of states for structures grown with TC.

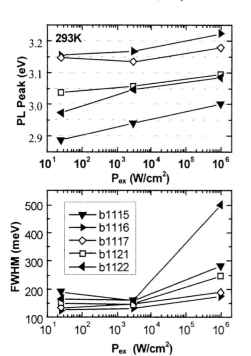

Fig. 3. Dependence of PL peak position and width on excitation density. Samples grown with TC marked by hollow symbols and without TC by solid symbols

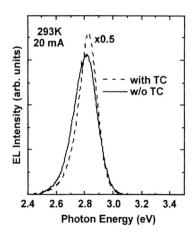

Fig. 4. Room temperature electroluminescence spectra of two LEDs grown with and without TC recorded at 20 mA

Figure 4 shows electroluminescence spectra of two LEDs grown with and without TC. Both structures show similar $I-V$ characteristics, with $V_f \approx 4.5$ V at 20 mA and similar dependence of efficiency on current, but for the structure grown with TC the efficiency was two times higher. We attribute this behaviour to suppression of nonradiative recombination of carriers in GaN barriers grown at higher (in the case of TC) temperature. Moreover, localized states prevent transport of the carriers toward dislocations that can also result in increase in emission efficiency.

Conclusions In summary, we have compared two sets of InGaN/GaN MQW structures grown with and without use of thermocycling by electroluminescence, photoluminescence and photo-luminescence excitation study. It was shown that for both sets of structures luminescence originates from localized centres, but the density of states is different for structures grown with and without TC. Higher density of states for structures grown with TC leads to smaller shift of PL peak position with increasing excitation density, making these structures preferable for high excitation density applications. LED structures grown with use of TC show nearly two times higher efficiency than the structures with same design grown without TC.

Acknowledgement The authors thank the Russian Foundation for Basic Research for financial support of this work.

References

[1] W. V. Lundin, A. V. Sakharov, V. A. Semenov, A. S. Usikov, M. V. Baidakova, I. L. Krestnikov, and N. N. Ledentsov, Proc. of 7th Internat. Symp. Nanostructures: Physics and Technology, St. Petersburg (Russia), June 14–18, 1999 Ioffe Institute 1999 (p. 485).
[2] N. Duxbury, P. Dawson, U. Bangert, E. J. Thrush, W. van der Stricht, K. Jacobs, and I. Moerman, phys. stat. sol. (b) **216**, 355 (1999).
[3] R. Martin, P. G. Middleton, and K. P. O'Donnell, Appl. Phys. Lett. **74**, 263 (1999).

Growth and Characterization of InGaN/GaN Multiple Quantum Wells on Ga-Polarity GaN by Plasma-Assisted Molecular Beam Epitaxy

X. Q. Shen[1]) (a), T. Ide (a), M. Shimizu (a), F. Sasaki (b), and H. Okumura (a)

(a) Power Electronics Research Center, National Institute of Advanced Industrial Science and Technology, Central 2, 1-1-1, Umezono, Tsukuba, Ibaraki 305-8568, Japan

(b) Photonics Research Institute, National Institute of Advanced Industrial Science and Technology, Central 2, 1-1-1, Umezono, Tsukuba, Ibaraki 305-8568, Japan

(Received June 19, 2001; accepted July 7, 2001)

Subject classification: 61.14.Hg; 81.05.Ea; 81.15.Hi; 81.65.Cf; S7.14

InGaN/GaN multiple quantum-wells (MQWs) on Ga-polarity GaN by plasma-assisted molecular-beam epitaxy were grown and characterized. In-situ reflection high-energy-electron diffraction observations and high-resolution X-ray diffraction results indicated that a flat interface and a good periodicity of the InGaN/GaN heterostructures were achieved. Photoluminescence measurements revealed the superior optical properties of InGaN/GaN MQWs emitting from ultraviolet ($\approx$388 nm) to green-yellow ($\approx$528 nm) range with the In composition varying from 0.04 to 0.30. Stimulated-emission features by optical pumping were demonstrated, which implied the high-quality of the MBE-grown InGaN/GaN MQWs.

Introduction GaN and its related alloys (AlGaN, InGaN) have attracted a great deal of attention due to its potential applications in optical and electronic devices [1–4]. At present, the InGaN-based heterostructures are mainly fabricated by the metalorganic chemical vapor deposition (MOCVD) technique. Although molecular-beam epitaxy (MBE) has many advantages over MOCVD, such as low temperature growth and p-type doping without post-annealing [5], there are only limited research works reported on the InGaN-based single and multiple quantum wells (MQWs) fabrication by the MBE technique [6, 7]. We have pointed out that the quality of MBE-grown GaN films on c-face sapphire substrates can be greatly improved by controlling the lattice-polarity of the GaN films during the growth [8], to which Ga-polarity GaN films are preferred. It is well accepted that Ga-polarity GaN films are usually obtained by MOCVD growth, while the lattice orientation of MBE films grown on c-face sapphire substrates is known to be mainly N-polarity or a mixed one [9]. Recently, excellent electron mobility values of MBE-grown GaN films and AlGaN/GaN heterostructures applying the growth on the Ga-polarity mode have been reported [10–12]. We speculate that lattice-polarity control in MBE technique is a key point to approach the high-quality GaN films and try to extend this opinion to InGaN-based heterostructures.

In this paper, we report the growth and characterizations of InGaN/GaN MQWs grown on a Ga-polarity GaN by rf-MBE. In-situ reflection high-energy-electron diffrac-

---

[1]) Corresponding author; Phone: +81-298-61-3373; Fax: +81-298-61-5434; e-mail: xq-shen@aist.go.jp

tion (RHEED) and low-temperature photoluminescence (PL) were used to characterize the surface flatness and the optical properties of the InGaN MQWs.

Experimental The GaN films with Ga-polarity were grown on sapphire c-plane (0001) substrates by rf-MBE. Details of the Ga-polarity GaN film preparation and the confirmation of the lattice-polarity have been published elsewhere [13]. After the growth of the Ga-polarity GaN ($\approx$700 nm) layer at 700 °C, the growth was interrupted and the temperature was decreased to approximately 600 °C for the growth of the InGaN/GaN MQWs and the GaN cap layer. Then, ten periods of InGaN ($\approx$4 nm)/GaN ($\approx$6 nm) pair with different In mole fractions were grown by changing the In flux. At last, the GaN cap layer ($\approx$60 nm) was grown at the same temperature. HRXRD results on the fully relaxed thick InGaN film ($\approx$400 nm) grown at the similar conditions described above have been used to estimate the InN mole fraction in InGaN wells. PL measurements (He–Cd laser as an exciting source) at 4.2 K were carried out to investigate the optical properties of InGaN MQWs. Furthermore, time-resolved PL (TRPL) measurements at 5 K were performed to study the PL peak lifetime and the excitation power dependence, using a fourth-harmonic light from femto-second optical parametric generator and amplifier pumped by a Ti:sapphire regenerative amplifier with 1 kHz repetition.

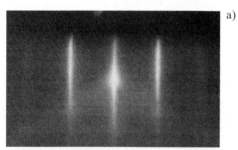

a)

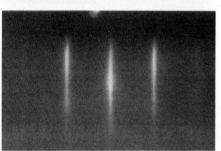

b)

c)

Results and Discussion Figure 1 illustrates the in-situ RHEED observation results during the growth of a Ga-polarity GaN underlayer (Fig. 1a), an InGaN well layer (Fig. 1b) and a GaN barrier layer (Fig. 1c). It is clear that streak RHEED patterns are observed throughout the growth, indicating the two-dimensional growth was maintained during the growth of the wells and barrier layers. In other word, superior surface flatness for each layer was achieved, which is very important for the fabrication of high-quality MQWs. HRXRD measurements were performed and clear satellite peaks up to the third order indicated that high interface quality and periodicity of each InGaN QW were well achieved [14]. From the satellite peak separation, the calculated thickness of InGaN wells and GaN barrier layers using the equation reported by Nakamura and Fasol [4] agrees very well

Fig. 1. In-situ RHEED observations during the growth of a) a GaN underlayer, b) an InGaN well layer and c) a GaN barrier layer

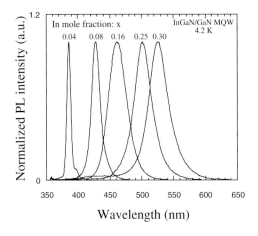

Fig. 2. 4.2 K PL spectra of the InGaN MQWs with In composition varing from 0.04 to 0.30

with the expected ones. From the results, good interface flatness and reproducible periods in the InGaN MQWs by rf-MBE are demonstrated.

4.2 K PL spectra of the MBE-grown InGaN MQW samples with the In composition varying from 0.04 to 0.30 are shown in Fig. 2. PL peak wavelengths from the InGaN MQWs changed from 388 to 528 nm. The PL linewidths corresponding to these samples are comparable to the results in the literature [4]. Intense PL emission from these InGaN MQWs have been recorded. The intensities of the PL emission from the MQWs are comparable to the thick InGaN films, which are ten times thicker than the total thickness of the InGaN wells. This illustrates the high quantum efficiency of luminescence in the MQWs.

TRPL measurements at 5 K were carried out to study the PL peak lifetime and the excitation power dependence. Figure 3 depicts the dependence of the integrated PL intensity of $In_{0.04}Ga_{0.96}N$/GaN MQWs on the excitation power. The inset shows the lifetime of the PL peak at different excitation powers. From Fig. 3, a linear increase followed by a superlinear increase of the integrated PL intensity is clearly seen. The threshold density is ≈ 69 μJ/cm^2. Meanwhile, with respect to the fast component of the peak lifetime being 132 ps below the threshold density point, the peak lifetime above the point is greatly shortened to < 20 ps, which is the resolution limit of our TRPL apparatus. This is a typical stimulated-emission (SE) phenomenon by optical pumping.

This is the first report concerning the SE observation from MBE-grown InGaN/GaN MQWs. It is worth pointing out that all our measured InGaN MQW samples showed the similar SE phenomena, although the threshold density became larger with increasing In composition.

Concerning the reason why the growth by conventional MBE only results in poor

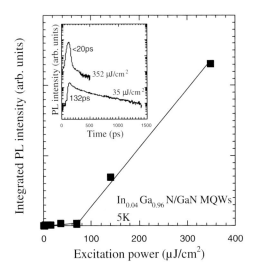

Fig. 3. Dependence of the TIPL intensity of $In_{0.04}Ga_{0.96}N$/GaN MQWs on the excitation power. The inset shows the peak lifetime below and above the threshold density point

quality of InGaN MQWs, we think that the lattice-polarity control of the films plays an essential role. In the case of MOCVD growth, the lattice polarity of GaN underneath is Ga-polarity, while MBE-grown GaN films without giving a special attention to the lattice-polarity usually result in N-polarity [9]. It has been found that the Ga-polarity GaN surface is much more stable than the N-polarity one during the interruption of the MBE growth [15]. We have also found that an InGaN film grown on a N-polarity GaN gives poor quality compared to an InGaN film on a Ga-polarity one [16]. The above two facts clearly reveal the merits of the growth of InGaN on a Ga-polarity GaN. The different lattice orientation results in the difference of the InGaN quality between the conventional MBE and MOCVD techniques. Therefore, we strongly believe that the above opinion provides a future direction to MBE technique for the possibility of exploring high efficient InGaN-based optical devices, which are now mainly explored by MOCVD technique.

Summary In summary, InGaN/GaN MQWs on Ga-polarity GaN were successfully grown on sapphire substrates by rf-MBE. In-situ RHEED observations and HRXRD results indicated that the flat interface and good periodicity of the InGaN/GaN MQWs were achieved. PL measurements revealed the superior luminescence properties of InGaN/GaN MQWs from ultraviolet ($\approx$388 nm) to green-yellow ($\approx$528 nm) range with the In composition varying from 0.04 to 0.30. We demonstrated the SE features by optical pumping of our samples, which showed the high quality of the MBE-grown InGaN/GaN MQWs.

References

[1] S. Strite and H. Morkoc, J. Vac. Sci. Technol. B **10**, 1238 (1992).
[2] S. Nakamura, M, Senoh, N. Iwasa, S. Nagahama, T. Yamada, and T. Mukai, Jpn. J. Appl. Phys. **34**, L1332 (1995).
[3] S. Nakamura, M. Senoh, S. Nagahama, N. Iwasa, T. Yamada, T. Matsushita, Y. Sugimoto, and H. Kiyoku, Appl. Phys. Lett. **70**, 2753 (1997).
[4] S. Nakamura and G. Fasol, The Blue Laser Diode, Springer-Verlag, Berlin/Heidelberg/New York 1997.
[5] T.D. Moustakas and R.J. Molnar, Mater. Res. Soc. Symp. Proc. **281**, 253 (1993).
[6] R. Singh, D. Doppalapudi, and T.D. Moustakas, Appl. Phys. Lett. **69**, 2388 (1996).
[7] B. Damilano, N. Grandjean, J. Massies, L. Siozade, and J. Leymarie, Appl. Phys. Lett. **77**, 1268 (2000).
[8] X.Q. Shen, T. Ide, S.H. Cho, M. Shimizu, S. Hara, H. Okumura, S. Sonoda, and S. Shimizu, Jpn. J. Appl. Phys. **39**, L16 (2000).
[9] E.S. Hellman, MRS Internet J. Nitride Semicond. Res. **3**, 11 (1998).
[10] H. Tang and J.B. Webb, Appl. Phys. Lett. **74**, 2373 (1999).
[11] C.R. Elsass, I. P. Smorchkova, B. Heying, E. Haus, P. Fini, K. Maranowski, J. P. Ibbetson, S. Keller, P.M. Petroff, S. P. Denbaars, U.K. Mishra, and J.S. Speck, Appl. Phys. Lett. **74**, 3529 (1999).
[12] X.Q. Shen, T. Ide, S.H. Cho, M. Shimizu, S. Hara, H. Okumura, S. Sonoda, and S. Shimizu, Jpn. J. Appl. Phys. **40**, L23 (2001).
[13] X.Q. Shen, T. Ide, S.H. Cho, M. Shimizu, S. Hara, H. Okumura, S. Sonoda, and S. Shimizu, J. Cryst. Growth. **218**, 115 (2000).
[14] X.Q. Shen, T. Ide, M. Shimizu, and H. Okumura, J. Appl. Phys. **89**, 5731 (2001).
[15] X.Q. Shen, T. Ide, M. Shimizu, S. Hara, and H. Okumura, Appl. Phys. Lett. **77**, 4013 (2000).
[16] X.Q. Shen, T. Ide, M. Shimizu, S. Hara, and H. Okumura, Jpn. J. Appl. Phys. **39**, L1270 (2000).

Optical Studies on AlGaN/InGaN/GaN Single Quantum-Well Structures under External Strains

E. Kurimoto[1]) (a), M. Takahashi (a), H. Harima (a), H. Mouri (b),
K. Furukawa (b), M. Ishida (c), and M. Taneya (c)

(a) Department of Applied Physics, Osaka University, Suita, Osaka 565-0871, Japan

(b) Electronic Components Laboratories, Sharp Corporation, Tenri, Nara 632-8567, Japan

(c) Advanced Technology Research Laboratories, Sharp Corporation, Tenri, Nara 632-8567, Japan

(Received June 30, 2001; accepted August 4, 2001)

Subject classification: 73.21.Fg; 77.65.Ly; 78.55.Cr; 78.67.De; S7.14

Photoluminescence (PL) spectra from AlGaN/InGaN/GaN single quantum well structures with different In content in the InGaN well layer have been measured with applying biaxial tensile stresses by a central-flexure method for the purpose of studing piezoelectric-field effect on the well layer. The band-edge emission from the InGaN well layer showed linear variations against the applied stress with different coefficients depending on the In content. We obtained, e.g., +17.5 meV/GPa for an InGaN layer with 20% In content. Our result presents a direct evidence for the presence of piezoelectric field which depends on the In content.

Introduction Group III-nitride semiconductors are of much practical interest for optoelectronic devices such as blue and green light emitting diodes and lasers. There have been intensive investigations on the optical and electrical properties in these devices, and some of the devices have been on the market. However, quantum-well structures on which such devices are based [1] still contain several open questions. Especially, the piezoelectric field induced by the strain associated with lattice mismatch between neighboring layers [2] seriously affects the optical properties. Therefore, to understand and control the piezoelectric-field effect is of practical importance. In this work, the piezoelectric field in AlGaN/InGaN/AlGaN single quantum well (SQW) is investigated by a unique method. We have measured the biaxial tensile stress in the c-plane by Raman scattering, and measured photoluminescence (PL) spectra from SQW as a function of applied tensile stress.

Experiments Three types of AlGaN/InGaN/GaN SQW samples were grown on sapphire (0001) substrates by metal-organic chemical vapor deposition (MOCVD). The thicknesses of the different layers are as follows: AlGaN 40 nm, InGaN 3 nm and GaN 2–3 μm. The samples had different In contents in the InGaN well layer; 10% (sample A), 20% (sample B), and 30% (sample C). The biaxial tensile stress was applied to the c-plane of SQW by a central-flexure method [3]. The sapphire substrates were thinned to about 100 μm to bend the samples and apply stresses smoothly. The stress was calibrated using a phonon frequency shift in the GaN layer observed by Raman scattering. Previous papers reported that the E_2-phonon peak of GaN showed a linear frequency

---

[1]) Corresponding author; Phone: +81 6 6879 7854; Fax: +81 6 6879 7856;
e-mail: eiji@ap.eng.osaka-u.ac.jp

shift with the residual stress in the range of about 2–7 cm^{-1}/GPa [4–7], from which we selected 3 cm^{-1}/GPa as the proportional coefficient in this work. The PL measurements were performed at room temperature using an Ar$^+$ laser at 351 nm as excitation source at different tensile stresses. Time-resolved PL measurements were made using a streak camera and a pulse laser excitation at 400 nm obtained by the second harmonic generation of a Ti:sapphire laser.

Results and Discussion Figures 1a and b show the peak energy shift of band-edge emission from the GaN and InGaN layer, respectively, plotted against the applied tensile stress. In Fig. 1a, the GaN peak shows a common variation for all the samples (sample A: circles, B: squares, C: triangles). The peak was positioned at 3.4 eV at zero applied stress and linearly red-shifted with the increase of stress as approximated by the dotted line. The slope was -14 ± 2 meV/GPa. Similar results have been reported in previous papers [4, 8]; Davydov et al. [4] e.g., reported -20 meV/GPa, which is in reasonable agreement with our result. On the contrary, the emission peak of InGaN plotted in Fig. 1b shows very different behavior. The band-edge emission, peaked at 3.2 eV (A), 2.8 eV (B) and 2.6 eV (C) at zero applied stress, shows a red shift for sample A with the increase of applied tensile stress, while the samples B and C show a blue shift.

This result will be explained by the difference of piezoelectric field in the InGaN layer as follows. Since the piezoelectric field is generated by the compressive lattice distortion induced by the lattice mismatch between the GaN and InGaN layer, the piezoelectric field will become stronger with the increase of In content. Thus, the sample C has the strongest field among the tested samples, and is expected to show most clearly a red shift in PL emission peak compared to the absorption edge. This is the so-called quantum confined Stark effect (QCSE), which corresponds to the case II as classified by Chichibu et al. [9] If we apply external tensile strains, the opposite effect, namely, a blue shift in emission peak due to the reduction of piezoelectric field will be expected in the order of strength A < B < C. This was actually observed in the present experiment. The result of sample A, giving only a red shift for external tensile stress as

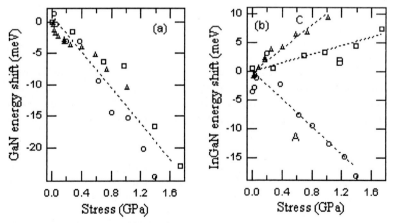

Fig. 1. PL peak energy shift in a) GaN and b) InGaN layers plotted against the applied tensile stress for samples A (circles), B (squares), C (triangles)

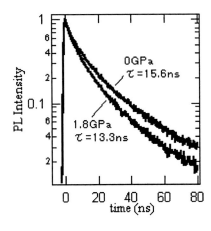

Fig. 2. Time-resolved PL on sample B at different tensile stresses

in the case of GaN layer, shows that the decrease in band gap due to lattice expansion dominates the QCSE contribution.

We have further confirmed the presence of piezoelectric field in In-rich samples by time-resolved PL experiments, considering that the carrier recombination should be suppressed if electron and hole wave functions are spatially separated by the piezoelectric field [9]. For this purpose, we compared the carrier recombination time at different tensile stresses for sample B. The result is shown in Fig. 2. It is found that the carrier recombination time decreases from 16 to 13 ns with the increase in tensile stress from 0 to 1.8 GPa.

We will hereafter evaluate the contribution of piezoelectric field to the energy shift rate of band-edge emission with the tensile stress. The emission energy E_{PL} will be described as a function of biaxial tensile stress p as

$$E_{PL}(p) = E_0 + eF_{pie}(p)\,L + \Delta E_a(p) + \Delta E_q. \tag{1}$$

Here, E_0, F_{pie}, ΔE_a, ΔE_q, e and L are emission energy at zero applied stress, piezoelectric field, band gap variation by lattice expansion, quantum confinement effect, elementary electric charge, and the well width (= 3 nm). The energy shift rate (dE_{PL}/dp) then consists of two components related to piezoelectric field ($dE_{pie}/dp = eL\,dF_{pie}/dp$) and variation of lattice expansion ($d\Delta E_a/dp$),

$$dE_{PL}/dp = dE_{pie}/dp + d\Delta E_a/dp. \tag{2}$$

Let us first estimate the latter part $d\Delta E_a/dp$ for the treated samples. In the case of hydrostatic pressure, $d\Delta E_a/dp$ has been reported as +33 meV/GPa for InN and +40 meV/GPa for GaN [8]. If their ratio is adopted also for the case of biaxial tensile stress, the energy shift coefficient for GaN is −14 meV/GPa (observed in Fig. 1a), and −11.5 meV/GPa for InN. We can then obtain this coefficient for the present ternary alloy samples by interpolation as $d\Delta E_a/dp = -13.8$ meV/GPa (sample A), −13.5 meV/GPa (B) and −13.3 meV/GPa (C).

Using the observed PL energy shift dE_{PL}/dp shown in Fig. 1b, −12 meV/GPa (A), +4 meV/GPa (B) and +10 meV/GPa (C), we can obtain the contribution of piezoelectric field from Eq. (2). The values are $dE_{pie}/dp = +1.8$ meV/GPa (A: 10%), +17.5 meV/GPa (B: 20%) and +23.3 meV/GPa (C: 30%). This result indicates that the effect of piezoelectric field in the band-edge emission energy of InGaN layer drastically increases with the increase of In content above ~10% for the present quantum well structures.

Conclusion We studied photoluminescence properties of AlGaN/InGaN/GaN single quantum wells to investigate the variation of electronic band structure in the InGaN well layer by applying biaxial tensile stresses using a central-flexure method. In the experiment, three types of samples having the same structure but different In contents for the well layer were tested. The band-edge emission peaks for the GaN and InGaN layer were plotted against the applied tensile stress, and it was found that the GaN layer showed commonly a red shift, while the InGaN layers showed different behavior depending on the In content. The result was interpreted by the quantum confined Stark effect due to piezoelectric field induced by the lattice distortion in the InGaN layer. The contribution of piezoelectric field to the energy shift of band-edge emission with tensile stress was evaluated as +1.8 meV/GPa (10% In content), +17.5 meV/GPa (20%) and +23.3 meV/GPa (30%). Our experiment demonstrated that the piezoelectric field effect in the group III-nitride quantum well structures could be directly evaluated by optical means under external strains.

References

[1] S. NAKAMURA, in: Semiconductors and Semimetals, Vol. 50, Eds. J. I. PANKOVE and T. D. MOUSTAKOS, Academic Press, London 1998 (p. 431).
[2] T. TAKEUCHI, S. SOTA, M. KATSURAGAWA, M. KOMIRI, H. TAKEUCHI, and I. AKASAKI, Jpn. J. Appl. Phys. **36**, L382 (1997).
[3] I. I. NOVAK, V. V. BAPTIZMANSKI, and L. V. ZHOGA, Opt. Spectrosc. **43**, 145 (1977).
[4] V. YU. DAVYDOV, N. S. AVERKIEV, I. N. GONCHARUK, D. K. NELSON, I. P. NIKITINA, A. S. POLKOVNIKOV, A. N. SMIRNOV, M. A. JACOBSON, and O. K. SEMCHINOVA, J. Appl. Phys. **82**, 5097 (1997).
[5] T. KOZAWA, T. KACHI, H. KANO, H. NAGASE, N. KOIDE, and K. MANABE, J. Appl. Phys. **77**, 4389 (1995).
[6] C. KISIELOWSKI, J. KRÜGER, S. RUVIMOV, T. SUSKI, J. W. AGER III, E. JONES, Z. L.-WEBER, M. RUBIN, E. R. WEBER, M. D. BREMSER, and R. F. DAVIS, Phys.Rev. B **54**, 17745 (1996).
[7] F. DEMANGEOT, J. FRANDON, M. A. RENUCCI, O. BRIOT, B. GIL and R. L. AULOMBARD, Solid State Commun. **100**, 207 (1996).
[8] S.-H. WEI and A. ZUNGER, Phys. Rev. B **60**, 5404 (1999).
[9] S. F. CHICHIBU, A. C. ABARE, M. S. MINSKY, S. KELLER, S. B. FLEISCHER, J. E. BOWERS, E. HU, U. K. MISHRA, L. A. COLDREN, S. P. DENBAARS, and T. SOTA, Appl. Phys. Lett. **73**, 2006 (1998).

Phonon and Photon Emission from Optically Excited InGaN/GaN Multiple Quantum Wells

A. V. Akimov (a), S. A. Cavill (a), A. J. Kent[1] (a), N. M. Stanton (a), T. Wang (b), and S. Sakai (b)

(a) School of Physics and Astronomy, University of Nottingham, Nottingham, NG7 2RD, UK

(b) Satellite Venture Business Unit, Department of Electrical and Electronic Engineering, University of Tokushima, Minamijosanjima 2-1, Tokushima 770, Japan

(Received June 25, 2001; accepted July 19, 2001)

Subject classification: 63.22.+m; 73.21.Fg; 78.55.Cr; 78.67.De; S7.14

The effect of the well width on both the photoluminescence (PL) and phonon emission in InGaN multiple quantum well (MQW) samples has been investigated. For narrow MQW samples ($w = 1.25$, 2.5 nm) the low temperature PL quantum efficiency is close to unity with the phonon emission being due mainly to carrier relaxation in the QWs. For wider MQWs the PL quantum efficiency is reduced and the intensity of the phonon emission increases. We explain this in terms of non-radiative recombination processes in the QWs which result in phonon emission.

Introduction It has been demonstrated that high quality InGaN/GaN MQWs exhibit strong PL at room temperature, and at low temperatures, the quantum efficiency is close to unity [1, 2]. However, non-radiative processes in InGaN/GaN MQWs and other nitride related nanostructures are not well understood. These non-radiative processes include the relaxation of hot quasiparticles (carriers and excitons) and also non-radiative recombination on defects, which competes with PL reducing the PL quantum efficiency. Evidently non-radiative processes have important practical implications for the performance of opto-electronic devices. Since the various non-radiative processes involve the emission of phonons, techniques which observe the phonons directly give information which is complementary to that obtained by other techniques.

The main aim of the present work is to show that in high quality InGaN/GaN narrow MQW structures the phonon emission comes mostly from the relaxation process whilst for wider MQW structures, the phonon emission is due to both relaxation and non-radiative recombination processes.

Samples and Experimental Technique The samples were grown on (0001) sapphire substrates by atmospheric pressure metal-organic chemical vapor deposition (MOCVD). The details of growth may be found in [1]. The MQW structure consisted of ten $In_{0.13}Ga_{0.87}N$ quantum wells separated by 7.5 nm GaN barriers. Three MQW samples were studied with the thicknesses of quantum wells 1.25 nm, 2.5 nm and 5 nm. The back side of the substrates were polished and superconducting bolometers (40×40 μm^2 active area) were evaporated for detection of the emitted phonons. After polishing, the substrates were between 0.31 mm and 0.37 mm thick depending on the sample.

---

[1]) Corresponding author; e-mail: anthony.kent@nottingham.ac.uk

The experiments were carried out in liquid helium at $T_0 \approx 2$ K, at the superconducting transition temperature of the bolometer. The samples were excited by 10 ns pulses from a frequency-tripled, $\lambda = 355$ nm, Q-switched Nd:YAG laser, focussed to a spot of diameter 50 µm. Excitation power densities in the range 1 kW/cm^2 were used. For characterisation, the PL was detected using a single grating monochromator and avalanche photodiode. The measured PL spectra are identical to earlier work (spectral line width ≈ 45 meV) [1].

Nonequilibrium phonons, emitted as result of hot carrier (exciton) relaxation and non-radiative recombination in the MQW, propagate ballistically through the sapphire substrate and were detected by measuring the transient change in resistance of the superconducting bolometer. The signals were recorded using a high speed digitizer and signal averager with a 5 ns resolution. In our experiments the excitation energy $E_0 = 3.52$ eV $\approx E_{GaN}$ at $T = 2$ K (E_{GaN} is the band gap in the GaN barriers). Therefore, the relaxation path is mainly in the 2D MQWs and so there is negligible phonon emission from the bulk GaN barriers.

The bolometer is sensitive not only to emitted phonons but also to photons, i.e. PL from the InGaN/GaN MQW. In order to extract the "pure" phonon signal we measured the PL pulses separately using a fast UV photodiode. The PL and bolometer measurements were carried out using identical geometries and settings for different samples which allows us to analyze the relative intensity of photon and phonon fluxes in different MQW samples and so obtain information about the PL quantum efficiency.

Images of the phonon emission were made by raster scanning the excitation spot over the surface of the sample using a set of computer controlled galvanometer mirrors.

Experimental Results and Discussion Figure 1 shows the detected bolometer (solid lines) and photodiode (dashed lines) signals in the samples with different MQW width. The signals are normalized to unity for the PL amplitude (peak starting at $t = 0$ ns) in each sample.

The bolometer response to ballistic acoustic phonons shows a sharp rising edge at a time equal to the time of flight between the excitation point and the bolometer. The group velocities are 11000 ms^{-1} and 6000 ms^{-1} for longitudinal (LA) and transverse (TA) polarized phonons, respectively. It is seen that the relative intensity of the phonon

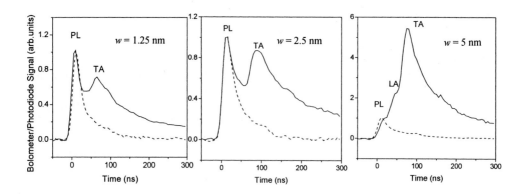

Fig. 1. Bolometer (solid lines) and photodiode (dashed lines) signals

signal increases with increasing MQW width. The phonon contribution is largest in the wide, 5 nm, MQW where the amplitude of the TA signal is 5 times bigger than the amplitude of the PL pulse. Conversly the TA amplitude is smaller than the PL signal in the narrow 1.25 nm MQW. Separate measurements of the PL signal, using the photodiode, show that the PL pulses have the same amplitude in the 1.25 nm and 2.5 nm MQWs, while in the 5 nm MQW the PL amplitude is about 60% of that measured in the other two samples. The relative intensities of the low temperature PL are in good agreement with earlier studies of similar MQWs, where it was shown that the 2.5 nm MQW has a higher quantum efficiency than the wider 5 nm MQW [1]. The temporal evolution of the PL signals measured by the photodiode (dashed curves in Fig. 1) is almost the same for all three MQW samples and is determined by the temporal resolution of the optical and electronic system. In agreement with earlier studies, we obtained that the decay for the 5 nm MQW is slightly longer than for the narrower MQW samples [2]. We explain the results in Fig. 1 in terms of the different PL quantum efficiencies for different MQWs. The total amount of phonon energy, E_{ph}, emitted by hot carriers and excitons may be written as

$$E_{ph} = \Delta E + (1-\eta) E_{PL} , \tag{1}$$

where $\Delta E = E_0 - E_{PL}$ is the excess energy released during the relaxation of hot carriers and excitons. The second term, which includes the PL quantum efficiency η, describes the energy released as a result of non-radiative recombination. If we assume that $\eta = 1$ in the 1.25 nm and 2.5 nm MQW samples and $\eta = 0.6$ in the 5 nm MQW (the value of $\eta = 0.6$ for the 5 nm MQW is used because the detected PL signal was 60% that of the 1.25 and 2.5 nm MQW PL signals), then from Eq. (1), using values for E_{PL} and ΔE (E_{PL} = 3.25, 3.18, 3.01 eV and ΔE = 0.26, 0.33. 0.50 eV for 1.25, 2.5 and 5 nm MQW respectively), we obtain the ratios for total phonon emitted energies as 1/1.3/6.3 for the 1.25 nm, 2.5 nm and 5 nm MQWs respectively. Comparing the detected bolometer and photodiode signals (Fig. 1) we get for the phonon amplitudes the corresponding ratios equal to 1/1.4/5.6 which is in very good agreement with the calculations. The small difference in experimental and calculated ratios may be due to different angular and temporal dependencies of the detected phonon signals for different MQW samples. Thus the results of the analysis based on Eq. (1) leads us to the conclusion that the 1.25 nm and 2.5 nm MQWs have low temperature PL quantum efficiency close to unity as was claimed in previous work on similar samples [1, 2]. Hence in the 1.25 nm and 2.5 nm MQWs the emission of phonons is mainly due to the relaxation of hot carriers and excitons. However in the 5 nm MQW, radiative recombination is reduced whilst non-radiative recombination processes resulting in the emission of acoustic phonons are increased. One possible explanation for this is given in [2]: in InGaN wells the strong piezoelectric polarisation results in band tilting (QCSE). The electron and hole wavefunctions are therefore confined near the edges of the well resulting in a reduced wavefunction overlap. Increasing the well width increases the band tilting and so decreases the e–h wavefunction overlap. Thus a reduction in PL efficiency occurs as the well width increases as is shown experimentally.

Figure 2 shows phonon images of the TA mode for the 1.25 nm and 2.5 nm QWs together with an image from a blank sapphire substrate. In the latter case, the phonons are generated isotropically by optical excitaion of a 100 nm metal film evaporated on the surface of the sapphire. As can be seen from Fig. 2a the emission of acoustic pho-

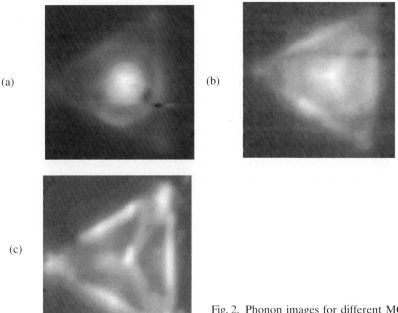

Fig. 2. Phonon images for different MQWs, a) 1.25 nm, b) 2.5 nm; and c) for blank sapphire substrate

nons in the 1.25 nm MQW is anisotropic (the phonons are emitted almost perpendicular to the plane of the MQW) compared with the other two images. The observed anisotropy is in good agreement with theoretical predictions for acoustic phonon emission from narrow QWs [3]. As the QW width increases the amount of $q_\perp$ (component of wavevector perpendicular to the QW plane) allowed by the carrier–phonon selection rules decreases and becomes comparable with the parallel component of wavevector ($q_\parallel$) allowed by the carrier–phonon selection rules. When this occurs ($q_\perp \approx q_\parallel$) the acoustic phonon emission should be more or less isotropic giving an image similar to that of the blank sapphire.

Conclusion We have shown that in 1.25 nm and 2.5 nm InGaN/GaN MQWs, the PL quantum efficiency is close to unity with the phonon emission coming mainly from hot carrier and exciton relaxation. In the wider MQW (5 nm), the non-radiative recombination processes also contribute to phonon emission. This additional path is a consequence of the decrease in overlap of the e–h wavefunctions which reduces the radiative recombination efficiency.

References

[1] T. WANG, D. NAKAGAWA, M. LACHAB, T. SUGAHARA, and S. SAKAI, Appl. Phys. Lett **74**, 3128 (1999).
[2] J. A. DAVIDSON, P. DAWSON, T. WANG, T. SUGAHARA, J. W. ORTON, and S. SAKAI, Semicond. Sci. Technol **15**, 497 (2000).
[3] A. J. KENT, in: Hot Electrons in Semiconductors. Physics and Devices, Ed. N. BALKAN, Claredon Press, Oxford 1998 (p. 81).

Dual Contribution to the Stokes Shift in InGaN–GaN Quantum Wells

T. J. Ochalski[1]) (a), B. Gil (b), P. Bigenwald (c), M. Bugajski (a), A. Wojcik (a), P. Lefebvre (b), T. Taliercio (b), N. Grandjean (d), and J. Massies (d)

(a) Institute of Electron Technology, Al. Lotnikow 32/46, PL-02 668 Warsaw, Poland

(b) Université de Montpellier II, Groupe d'Etude des Semiconducteurs, case courrier 074, F-34095 Montpellier Cedex 5, France

(c) Université d'Avignon et des pays de Vaucluse, 33 rue Pasteur, F-84000 Avignon, France

(d) Centre de Recherche sur l'Hétéro-Epitaxie et Applications, CNRS, Sophia Antipolis, F-06560 Valbonne, France

(Received June 30, 2001; accepted August 4, 2001)

Subject classification: 78.40.Fy; 78.55.Cr; 78.67.De; S7.14

By comparing photoluminescence and photoreflectance spectra taken on a series of InGaN–GaN quantum wells grown under identical conditions except the growth time of the InGaN layers, we could monitor the Quantum Confined Stark Effect (QCSE) without changing the nanotexture of the alloy layers. Our results indicate that, for quantum wells which radiate in the red, the contribution of the QCSE superimposes on the intrinsic localization phenomena of the carriers in the InGaN alloy, and is larger by one order of magnitude. Interpretation of data for samples that emit from the blue to the red can provide only partial conclusions if both localization effects and QCSE are not taken into consideration.

Although InGaN material is extensively used for visible light emitters [1], its physics still remains an enigma. Experimentalists have early reported the existence of a Stokes shift between the photoluminescence (PL) signal collected after that photo-excited electrons and holes recombine, and the signatures of experiments which probe more intrinsic optical processes, e.g. photoconductivity, absorption, PL excitation spectroscopy. There have been many interpretations published to elucidate the mechanism(s) responsible for the effect: (i) The localization of a carrier, or of the interacting electron–hole pair, to spatial regions of the crystal where the local indium composition is larger than the average indium composition of the alloy, is considered [2–5]. (ii) The existence of internal electric field has been evidenced in InGaN-based quantum wells (QWs) where the photoluminescence redshifts compared to its value in micrometer-thick films having the composition of the quantum well layer [6–10]. Unambiguous arguments in favor of that have been obtained from the correlation between the increase of the PL decay time (a probe of the wavefunction overlap rather than of the energy) and the increase of the well widths for series of QWs grown with a constant indium composition [11].

In this paper, we address the dual origin of the Stokes shift in InGaN QWs by using samples that emit light down to the red. The genesis of this investigation is our conclusion that there is, in the literature, a lack of documentation using full series of samples

---

[1]) Corresponding author; Phone: +48 22 548 79 37; Fax: +48 22 847 06 31; e-mail: ochalski@ite.waw.pl

grown under identical conditions so that the QCSE varies, while localization effects due to the alloy nanotexture do not vary. Our samples were grown by Molecular Beam Epitaxy on sapphire substrates. Details about the growth conditions can be found in Ref. [6]. The indium composition, measured by high resolution X-ray diffraction reciprocal mapping, was kept at 18% in the whole set of samples. Again, we only varied the quantum well thickness. The "Stokes shift" was measured by combining room temperature PL and room temperature PR spectroscopy. PL excitation was performed using the 325 nm radiation of the HeCd laser at a pump density of 100 W/cm^2. Photoreflectance set-up used in this work differs somewhat from conventional configuration used for PR measurements at room temperature [12]. The pump beam was generated by a xenon lamp and a 0.35 m monochromator. It was focused on a sample by quartz lens. Such set-up offers certain advantages over laser pump beams. Luminescence, which is inevitably generated in PR measurement, is in all cases a parasitic effect. Thus by reducing the power density of the pump beam, the luminescence is also diminished allowing for clear observation of even weak PR features. Additionally, a possibility of continuous tuning of the pump beam wavelength allows for finding the maximum of modulation strength for each sample. The transition energies were determined as a fit using Aspnes' First Derivative Functional Form (FDFF) model.

Figures 1 and 2 summarize some selected PL and PR spectra. We have chosen to show data corresponding to the long-wavelength regions of the light spectrum for which the investigated effects are the most dramatic. For a 3.5 nm thick QW that radiates at 2.4 eV at room temperature (green light), a reflectance structure is fitted at 2.92 eV

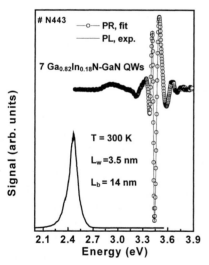
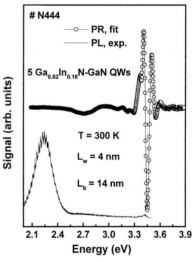

Fig. 1   Fig. 2

Fig. 1. Photoluminescence spectrum (bottom) and photoreflectance spectrum (top) of a $Ga_{0.82}In_{0.18}N$–GaN multiple quantum well. Well width and barrier width are 3.5 nm and 14 nm, respectively

Fig. 2. Photoluminescence spectrum (bottom) and photoreflectance spectrum (top) of a $Ga_{0.82}In_{0.18}N$–GaN multiple quantum well. Well width and barrier width are 4 nm and 14 nm, respectively

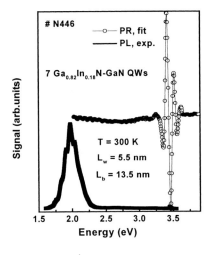

Fig. 3. Photoluminescence spectrum (bottom) and photoreflectance spectrum (top) of a $Ga_{0.82}In_{0.18}N$–GaN multiple quantum well. Well width and barrier width are 5.5 nm and 13.5 nm, respectively

with a broadening of 229 meV as shown in Fig. 1. Data taken on a second (orange) sample are shown in Fig. 2. The reflectance structure of interest is found at 2.76 eV.

Finally we show in Fig. 3 the data taken for a 5.5 nm wide multiple QW for which the PL peaks at 2 eV, at 10 K. For this sample that weakly emits light in the red, one can hardly resolve something, and the automatic fitting gives a broad (900 meV) feature at 2.8 eV. We have fitted the evolution of the room temperature PL peak maximum E_p with well width in our samples and we got E_p (eV) = 3.302 − 0.2618 L_z (nm). This diminution of the PL energy with the well width, we attribute to the QCSE.

This interpretation is reinforced by additional investigation concerning the radiative recombination dynamics in these samples, that provided an internal electric field determined from the data as 2.5–2.6 MV/cm in these samples [13]. Similar treatment performed for the reflectance signal gave E_g(eV) = 3.414 − 0.164 L_z(nm). In the limit case of very thin wells, the energy difference $E_g - E_p$ tends to the localization energy. This result indicates that, for these quantum wells, the Stokes shift increases when the well width increases. To go further, we have made an envelope function determination of the transition energies and band-to-band oscillator strengths.

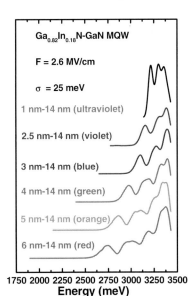

The full set of results is gathered in Fig. 4[2]) – there are plotted, as a function of their energy, the amplitudes of different band-to-band transitions computed for several multiple quantum well designs. All levels are supposed to experience a 25 meV gaussian inhomogeneous broadening and the plot starts in all cases at the energy of the fundamental transition. Due to the QCSE, the oscilla-

Fig. 4 (colour). Evolution of the band to band oscillator strength for $Ga_{0.82}In_{0.18}N$–GaN multiple quantum wells as a function of the well width. A 25 meV broadening was assumed for all transitions. The colour of the photoluminescence is indicated in the figure

---

[2]) Colour figure is published online (www.physica-status-solidi.com).

tor strength of the fundamental transition which is one of the dominant transitions for thin-well samples diminishes when the well width increases. Subsequently, due to the absence of oscillator strength for the low energy transition, one observes a strong Stokes shift between the PL energy and the PR feature for red samples (wide wells) and a weak one for violet light emitters (thin wells). The long-lived fundamental states are a bottleneck for thermalisation of the carriers and a PL signal is detected at their energy. The main band that is detected by PR always corresponds to the energy at which the onset of strong transitions occurs. When increasing the well width, this onset shifts slower than the fundamental energy transition (the PL data). Because the hole mass is extremely heavy, we have a fairly large number of hole states in the case of wide wells which gives the increase of the broadening of the photoreflectance feature with increasing well width.

Acknowledgements This work has been supported by the EU RTN "CLERMONT" program, contract No. HPRN-CT-1999-00132 and by the French Ministry of Education, Research and Technology within the "BOQUANI" and "NANILUB" research programs.

References

[1] S. NAKAMURA and S. F. CHICHIBU (Eds.), Introduction to Nitride Semiconductor Blue Lasers and Light Emitting Diodes, Taylor and Francis, London/New York 2000.
[2] Y. NARUKAWA, Y. KAWAKAMI, M. FUNATO, SZ. FUJITA, SG. FUJITA, and S. NAKAMURA, Appl. Phys. Lett. **70**, 981 (1997).
[3] S. F. CHICHIBU, A. C. ABARE, M. S. MINSKI, S. KELLER, S. B. FLEISCHER, J. E. BOWERS, E. HU, U. K. MISHRA, L. A. COLDREN, S. P. DENBAARS, and T. SOTA, Appl. Phys. Lett. **73**, 2006 (1998).
[4] K. P. O'DONNELL, R. W. MARTIN, and P. G. MIDDLETON, Phys. Rev. Lett. **82**, 237 (1999).
[5] E. BERKOWICZ, D. GERSHONI, G. BAHIR, E. LAKIN, D. SHILO, E. ZOLOTOYABKO, A. C. ABARE, S. P. DENBAARS, and L. A. COLDREN, Phys. Rev. B **61**, 10994 (2000).
[6] B. DAMILANO, N. GRANDJEAN, J. MASSIES, L. SIOZADE, and J. LEYMARIE, Appl. Phys. Lett. **77**, 1268 (2000).
[7] T. TAKEUCHI, H. TAKEUCHI, S. SOTA, H. SAKAI, H. AMANO, and I. AKASAKI, Jpn. J. Appl. Phys. **36**, L177 (1997).
[8] C. WETZEL, T. TAKEUCHI, H. AMANO, and I. AKASAKI, Phys. Rev. B **61**, 2159 (2000).
[9] A. HANGLEITER, J.-S. IM, H. KOLLMER, S. HEPPEL, J. OFF, and F. SCHOLZ, MRS Internet J. Nitride Semicond. Res. **3**, 15 (1998).
[10] K. P. O'DONNELL, R. W. MARTIN, S. PEREIRA, A. BANGURA, M. E. WHITE, W.VAN DER STRICHT, and K. JACOBS, phys. stat. sol. (b) **216**, 141 (1999).
[11] P. LEFEBVRE, A. MOREL, M. GALLART, T. TALIERCIO, J. ALLEGRE, B. GIL, H. MATHIEU, B. DAMILANO, N. GRANDJEAN, and J. MASSIES, Appl. Phys. Lett. **78**, 1252 (2001).
[12] F. H. POLLAK, in: Handbook on Semiconductors, Vol. 2, Ed. M. BALKANSKI, North-Holland, Amsterdam 1994 (p. 527).
[13] P. LEFEBVRE, T. TALIERCIO, A. MOREL, J. ALLEGRE, M. GALLART, B. GIL, H. MATHIEU, B. DAMILANO, N. GRANDJEAN, and J. MASSIES, Appl. Phys. Lett. **78**, 1538 (2001).

Spectroscopy and Modeling of Carrier Recombination in III–N Heterostructures

P. M. Sweeney (a), M. C. Cheung (a), F. Chen (a), A. N. Cartwright[1] (a), D. P. Bour*) (b), and M. Kneissl (b)

(a) Department of Electrical Engineering, 201 Bonner Hall, Box 60-1900, University at Buffalo, State University of New York, NY 14260, USA

(b) Xerox-Palo Alto Research Center, Electronic Materials Laboratory, Palo Alto, CA 94304, USA

(Received June 28, 2001; accepted August 4, 2001)

Subject classification: 78.47.+p; 78.55.Cr; S7.14

Time-resolved temperature dependent PL measurements of InGaN quantum wells are presented. The effects of quantum well-like and localized state emission in similar p–i(MQW)–n structures are discussed. A phenomenological model is presented to explain the observed dynamics.

Introduction The future development of visible optoelectronic devices such as LEDs, lasers, and visible modulators depends critically on a more complete understanding of the growth and emission processes in III–N quantum well structures. Much recent research has incorporated piezoelectricity, indium segregation, localized strain relaxation, quantum well width fluctuations, and spontaneous polarization to explain the complicated emission processes in these materials. Here, we present how time-resolved spectroscopy of nearly identical InGaN quantum well structures provides insight into the role of indium segregation in these structures. Moreover, we discuss how these localized fluctuations result in spatially distributed regions of highly efficient emission and drastic shifts in the emission wavelength, and how time-resolved spectroscopy can be used to estimate the relative percentage of phase segregation.

Indium Segregation and Piezoelectricity The solid phase immiscibility gap in InGaN results from the large difference in interatomic spacing between GaN and InN [1, 2]. One method to reach longer emission wavelengths in III-N heterostructures is to increase the indium concentration of the InGaN layer that is typically used as the quantum well material. High indium concentrations have been difficult to obtain primarily because of the effects on crystalline quality. Attempts to improve epilayer quality by raising the growth temperature have resulted in lower InN concentrations because of the volatility of nitrogen [3]. Moreover, in low temperature growth it is possible to get two separate phases of material within the same quantum well [1, 2, 4]. As a further complication to the emission dynamics in these materials, the observation of piezoelectricity in wurtzite $In_xGa_{1-x}N$/GaN heterostructures [5–7] has changed our understanding of the behavior of these materials under optical or electrical excitation. These piezoelectric fields are quite large (reported from 300 kV/cm to 1.1 MV/cm for 20% indium)

---

[1]) Corresponding author; Phone: (716) 645-3115x1205; Fax: (716) 645-5964; e-mail: anc@eng.buffalo.edu
*) Now at Agilent Laboratories.

[5, 6]. The magnitude of the in-well and barrier fields and the resulting band-structures of InGaN/GaN samples, as well as the accompanying spectral shifts, have been reported [7]. Of course, carrier recombination and transport dynamics in these materials are quite complicated due to the strong interaction between indium segregation and piezo-electricity [8].

Consistent with Stringfellow's theoretical calculation [1, 2], in this paper, we will provide data that is indicative of both a low indium phase and a high indium phase in the same sample. We discuss how these indium rich regions act as highly efficient radiative centers that increase device efficiency by attracting carriers away from non-radiative defect dislocations. To do this, we present the results of time-resolved photoluminescence (PL) of two samples with similar well widths and indium concentrations grown under different conditions (additional measurements have shown similar behavior in a number of other samples). Both samples were grown by organometallic vapor phase epitaxy (OMVPE). Initially, a 4 μm thick layer of GaN:Si was deposited on a c-plane sapphire substrate, followed by the MQW region. The MQW region consists of the GaN barrier and the 2 nm thick $In_{0.2}Ga_{0.8}N$ quantum well region. This barrier and quantum well layer was repeated to achieve the desired number of quantum wells (2 and 20). The last quantum well in each sample was capped with a GaN barrier region followed by a GaN:Mg p-contact layer. For simplicity we refer to these samples as M2 (2 quantum wells) and M20 (20 quantum wells). The frequency doubled, ~200 fs pulses, from a tunable mode-locked Ti:sapphire laser at the repetition rate of 80 MHz served as the excitation source for time-resolved photoluminescence (TRPL). The wavelength of the incident light was varied to effectively pump directly within the quantum well without exciting the higher bandgap barrier material. The resulting PL was spectrally and temporally resolved using a Hamamatsu streak camera with a 20 ps response time.

As can be seen from Fig. 1, the observed spectra for each sample at 15 K is quite different. The M20 spectrum consists of a single emission peak at 412 nm with a spectral width of 67 meV that decays with a time constant of approximately 5 ns. In contrast, M2 is seen to have two distinct peaks, a strong emission peak at 370 nm that decays with a time constant of 800 ps and a second peak at 463 nm that has an exceptionally long lifetime that is beyond the resolution of the interpulse spacing of our Ti:sapphire laser system. A quantum mechanical calculation of the expected excitonic

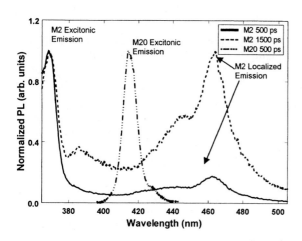

Fig. 1. Normalized time-resolved PL spectra of M2 (at 500 and 1500 ps) and M20 (at 500 ps) at 15 K

emission within the well (including piezoelectricity, the internal p–i–n field, and assuming 20% InGaN) is in excellent agreement with the peak emission energy of sample M20. However, the peak emission energies from sample M2 suggests that sample M2 is not behaving as a traditional quantum well at all. Clearly there are several different recombination states in M2, evidenced from the multiple peaks in the spectrum. The primary peak at short times (370 nm) behaves like a classic quantum well emission and the other very broad peak exhibits a Stokes shifted localized emission. Emission at this low energy (2.68 eV) would require either a much higher indium concentration or the presence of localized states due to phase segregation. The indium concentration within the quantum wells of M2 and M20 have both been estimated to be approximately 20% by X-ray diffraction. Although longer wavelength emission can be explained by several different theories including interface roughness, piezoelectricity, and localized strain, the 200 meV shift from the expected energy, and the broad emission, observed in sample M2 is consistent with a distribution of localized regions of varying size and indium concentration [8, 9]. Moreover, the observed increased lifetime of the low energy emission peak in M2 is consistent with stronger confinement due to localization of carriers in nanometer scale regions [10].

Model These studies are consistent with phase segregation producing regions of radiative recombination centers observed at lower emission energies. In other words, in InGaN the presence of the indium rich regions, which reduces crystalline quality, provides an efficient alternate, radiative trap for carriers in the quantum well [8, 9]. Although space limits full treatment here, Fig. 2 shows an energy level diagram with representative levels for the quantum well emission, for a spatially localized lower energy emission region, and finally a level that represents the defects within the system. For sample M20, there are very few localized low energy emission centers, and the emission spectrum is dominated by a single exponential decay representative of the excitonic emission, and the non-radiative recombination, from a fairly uniform alloy within the quantum well. Sample M2, on the other hand, is observed to have two distinct emission paths. The first (short wavelength) emission is a quantum well-like emission that rapidly decays as the carriers in the conduction band recombine radiatively through excitonic emission and also diffuse in the plane of the quantum well to the regions with lower potentials. Since the in-plane transport process can be very fast, the observed emission decay and the extracted lifetime from this high-energy state is probably more indicative

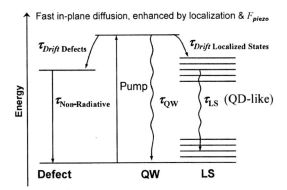

Fig. 2. Simple carrier dynamics model. Efficiency is affected by diffusion and drift to both defect dislocations and indium rich localized regions

of the in-plane transport. The second (long wavelength) emission appears to be strongly localized and has a very long lifetime (>13 ns), which is also representative of the typical emission observed from commercially available long wavelength emitting LED structures [10]. Moreover, this emission is well characterized by a disordered system with a large characteristic confinement potential. This disorder can be explained by phase segregation in M2 resulting in the formation of indium rich regions. Consistent with phase segregation, we should expect regions of high (>20%) indium concentrations and regions of low (<20%) indium concentrations.

The emission spectrum near zero delay (the solid curve in Fig. 1) is extremely useful for estimating the relative areas of the efficient emission centers and the lower indium concentration material. At short times, before diffusion has taken place, carriers should be fairly uniformly distributed locally in the excitation region. Therefore, since the short wavelength emission represents quantum well emission and the long wavelength emission represents localized state emission in indium rich regions, the relative ratio of the area under the curve should be approximately representative of the area of each region. The ratio of the two peaks (quantum well-like emission and quantum dot-like emission) in M2 at short times is roughly 6:1. Based on Stringfellow's calculations [1, 2], a possible explanation of this would be that nearly 86% of the material produced was a true alloy of approximately 12% indium (estimated from a fit of the observed emission energy to the expected excitonic emission energy of this quantum well), while the other 14% is composed of indium rich regions that vary in size and concentration with an estimated average indium concentration of 62%. This is evident in the broad emission at long wavelengths. As diffusion starts to take place, carriers will be attracted to the lower energy localized potential minima, where they will recombine radiatively. In addition, the lifetimes of the longer wavelength emission will be enhanced, not only by the localization, but also due to the spatial separation of carriers resulting from the large piezoelectric fields generated in these indium rich regions.

Summary In this paper, we have presented experimental studies of two InGaN/GaN MQW structures and provided a description of phase-segregation as well as the resulting carrier dynamics. Moreover, we have demonstrated the use of time-resolved photoluminescence to estimate the segregation in these materials. Structural analyses are necessary to verify the accuracy of this simple technique. There are many more complications in these III–N materials that contribute to changes in emission energy, linewidth, and the lifetimes of the carriers. Moreover, the development of composite structures that can accurately control the in-well field and reduce defect densities should result in device geometries for high quality visible LEDs, lasers, and modulators.

Acknowledgements We wish to acknowledge the support of the National Science Foundation CAREER Award, NSF #9733720, and Dr. Colin Wood for his support through the Office of Naval Research Young Investigator Program Award # N00014-00-1-0508, and we acknowledge the useful discussions with Dr. David Kofke at the University at Buffalo.

References

[1] G. B. STRINGFELLOW, J. Electrochem. Soc. **119**, 1780 (1972).
[2] I. HO and G. B. STRINGFELLOW, Appl. Phys. Lett. **69**, 2701 (1996).

[3] T. Matsuoka, Appl. Phys. Lett. **71**, 105 (1997)
[4] T. Takayama, M. Yuri, K. Itoh, T. Baba, and J. S. Harris, Jr., J. Appl. Phys. **88**, 1104 (2000).
[5] S. F. Chichibu, A. C. Abare, M. S. Minsky, S. Keller, S. B. Fleischer, J. E. Bowers, E. Hu, U. K. Mishra, L. A. Coldren, S. P. DenBaars, and T. Sota, Appl. Phys. Lett. **73**, 2006 (1998).
[6] A. Hangleiter, Jin Seo Im, H. Kollmer, S. Heppel, J. Off, and F. Scholz, MRS Internet J. Nitride Semicond. Res. **3**, 15 (1998).
[7] A. N. Cartwright, P. M. Sweeney, T. Prunty, D. P. Bour, and M. Kneissl, MRS Internet J. Nitride Semicond. Res. **4**, 12 (1999).
[8] N. A. Shapiro, P. Perlin, C. Kisielowski, L. S. Mattos, J. W. Yang, and E. R. Weber, MRS Internet J. Nitride Semicond. Res. **5**, 1 (2000).
[9] K. P. O'Donnell, R. W. Martin, and P. G. Middleton, Phys. Rev. Lett. **82**, 237 (1999).
[10] M. Pophristic, F. H. Long, C. Tran, I. T. Ferguson, and R. F. Karlicek, Jr., J. Appl. Phys. **86**, 1114 (1999).

Two-Component Photoluminescence Decay in InGaN/GaN Multiple Quantum Well Structures

SHIH-WEI FENG (a), YUNG-CHEN CHENG (a), CHI-CHIH LIAO (a), YI-YIN CHUNG (a), CHIH-WEN LIU (a), CHIH-CHUNG YANG[1]) (a), YEN-SHENG LIN (b), KUNG-JENG MA (b), and JEN-INN CHYI (c)

(a) Department of Electrical Engineering and Graduate Institute of Electro-Optical Engineering, National Taiwan University, 1, Roosevelt Road, Sec. 4, Taipei, Taiwan

(b) Department of Mechanical Engineering, Chung Cheng Institute of Technology, Tahsi, Taoyuan, Taiwan

(c) Department of Electrical Engineering, National Central University, Taiwan

(Received June 21, 2001; accepted August 4, 2001)

Subject classification: 68.55.Jk; 68.55.Ln; 78.55.Cr; 78.66.Fd; S7.14

Two-component decay of time-resolved photoluminescence (TRPL) intensity in three InGaN/GaN multiple quantum well samples were observed. The first-decay component was attributed to exciton relaxation of free-carrier and localized states; the second-decay one was dominated by the relaxation of localized excitons. The second-decay lifetime was related to the extent of carrier localization or indium aggregation and phase separation. The lifetime of free-carrier states was connected with the defect density. Based on the temperature-dependent data of PL and stimulated emission (SE), the localization energies of the three samples were calibrated to show the consistent trend with the second-decay lifetime and previous material analyses.

Introduction Due to the low miscibility between InN and GaN, indium aggregation and phase separation (formation of either InN or GaN clusters) occurred in InGaN/GaN quantum well (QW) structures [1]. Indium-rich clusters in such a QW structure form the localized states, which may effectively trap carriers for radiative recombination [2]. In this paper, we report several novel optical properties of three InGaN/GaN QW samples. We conducted time-resolved photoluminescence (TRPL) measurements and observed two-component decay of photoluminescence intensity. We also combine the temperature-dependent data of PL and stimulated emission (SE) for calibrating the localization energy, which is regarded as the energy difference between the localized and free-carrier states (both at the ground levels). Note that excitons can exist in both localized and free-carrier states.

The samples were grown in a low-pressure metal-organic chemical vapor deposition reactor. The InGaN/GaN multiple QW samples consisted of five periods of Si-doped InGaN well with 3 nm in thickness. The designated indium compositions of the sample were 11, 16, and 21% (samples A, B, and C, respectively). The Si doping concentration was 10^{18} cm^{-3}. The barrier was 7 nm GaN. In each sample, the QW layers were sandwiched with a 1.5 μm GaN buffer layer on a sapphire substrate and a 50 nm GaN cap layer. The growth temperatures were 1050 and 740 °C for GaN and InGaN, respectively.

We used HeCd (3.815 eV) laser and the fourth harmonic (4.661 eV) of a Q-switch Nd: YAG laser for PL and edge-mode SE measurements, recpectively. The TRPL meas-

---

[1]) Corresponding author; Phone: 886-2-23657624, Fax: 886-2-23652637,
e-mail: ccy@cc.ee.ntu.edu.tw

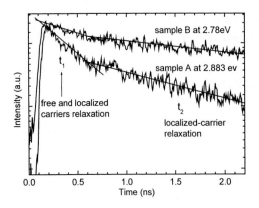

Fig. 1. Typical TRPL profiles and fitting curves

urements were performed with a Hamamatsu streak camera. Frequency-doubled optical pulses of about 100 fs in width were generated from a mode-locked Ti:sapphire laser for excitation. The pump photon energy and power of TRPL were 3.1 eV (400 nm) and 20 mW, respectively.

Experimental Results Two room-temperature TRPL intensity profiles of different samples at two different photon energies are shown in Fig. 1. The two-component decay in either profile can be clearly seen. Although a multiple-component decay feature in wurtzite InGaN samples was reported [3], the decay mechanism in InGaN/GaN QWs has not been fully discussed yet.

We fitted our data with two-step exponential decay to obtain the first-decay time t_1 and second-decay time t_2. The second lifetimes of the three samples are shown as functions of emission photon energy in the upper portion of Fig. 2. The decay time t_2 increases with decreasing photon energy for each of all three samples. This is a typical property of the carrier localization model [4]. The second-decay lifetimes can be regarded to describe the relaxation of localized excitons. Therefore, t_2 can be re-defined as t_L, the localized carrier lifetime. Among the three samples, the increasing trend of t_L with increasing nominal indium content is consistent with, what previously has been reported, that stronger carrier localization resulted in a longer lifetime [5]. On the other hand, t_1 describes the decay rate of simultaneous exciton relaxation of the free-carrier and localized states. In other words, $1/t_1 = 1/t_F + 1/t_L$, where t_F is the lifetime of free-carrier states. Thus, from the last equation we can obtain t_F, which is strongly related to defect recombination. The data of t_F are shown in the lower portion of Fig. 2. Note that on the low energy side, t_F of sample B is larger than those of the other two samples. Hence, it

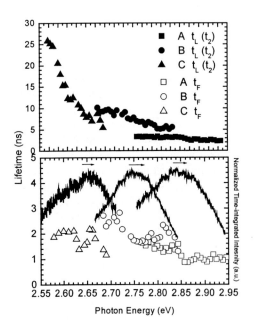

Fig. 2. Decay times t_L (top) and t_F (bottom) as functions of emission photon energy for the three samples at room temperature

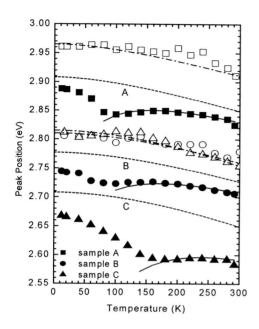

Fig. 3. PL (filled symbols) and SE (empty symbols) spectral peak positions of the three samples as functions of temperature

is believed that sample B is superior to the other two samples in quality. To confirm it, we performed AFM surface scanning. It was found that the surface morphology of sample B was indeed better than samples A and C.

The temperature variations of spectral peak positions of PL and SE are shown with filled and empty symbols, respectively, in Fig. 3. Because SE is supposed to originate from (excited) free-carrier recombination [6], its peak positions, $E_S(T)$, as functions of temperature T, can be assumed following Varshni's equation as [7]

$$E_S(T) = E_S(0) - \alpha T^2/(T + \beta). \tag{1}$$

The three dash-dotted curves stand for the fitting results of SE data with Varshni's equation. Table 1 lists the fitting values of $E_S(0)$, α and β. On the other hand, because PL is supposed to come from the recombination in both free-carrier and localized states, the formula for characterizing its temperature variations is given as follows [8]:

$$E_P(T) = E_P(0) - \alpha T^2/(T + \beta) - \sigma^2/k_B T. \tag{2}$$

With the same values of α and β for fitting the SE data, we can obtain $E_P(0)$ and σ by fitting the high temperature portions of the PL data in Fig. 3 (shown as the solid curves). The fitting results of $E_P(0)$ and σ are also listed in Table 1. Then, keeping the same α, β and $E_P(0)$ values in (2) and ignoring the contribution of the localized-state term, i.e., $\sigma^2/k_B T$, we can reproduce the pure effect of band gap shrinkage of the free-carrier states (ground level), as shown with the short-dashed curves in Fig. 3. As a result, the energy difference between the PL peak position at the lowest temperature

Table 1
Fitting parameter values and fitting results in Fig. 3

	sample A	sample B	sample C
$E_S(0)$ (eV)	2.965	2.810	2.815
α (meV/K)	0.37	0.42	0.67
β (K)	264.6	450	700.7
$E_P(0)$ (eV)	2.908	2.777	2.708
σ (meV)	22	22.3	38.24
localization energy E_L (meV)	21.3	35.4	41.27
distribution energy E_D (meV)	57	43	107

and the intercepted point of the short-dashed curve at 0 K (i.e., $E_P(0)$) roughly corresponds to the energy difference between the localized and free-carrier states (both at the ground levels), i.e., the localization energy, E_L. The calibrated values of E_L are shown in Table 1. We also listed the values of distribution energy, E_D, defined as the difference between the SE peak energy and the presumably ground level of free-carrier states. In other words, $E_D = E_S(0) - E_P(0)$.

Several interesting features can be observed in Table 1. The increasing trend of σ with increasing indium content is reasonable because it is proportional to the degree of carrier localization. This trend is consistent with the high-resolution transmission electron microscopy (HRTEM) images [1]. One can see the trend that a higher nominal indium content leads to a larger localization energy. This trend again is consistent with the extent of composition fluctuation, confirmed with HRTEM and X-ray diffraction (XRD) [1]. Note that the localization energies of samples B and C are closer, compared with sample A. Nevertheless, the σ values of samples A and B are closer, compared with sample C. These results can be explained with the following hypotheses. The number densities of local potential minima (indium-rich clusters), which are supposed to be related to the σ values, of samples A and B are about the same; however, the potential depths of these minima (indium composition contrasts), which are represented by the localization energy E_L, of sample B are generally closer to those of sample C. Meanwhile, the best quality of sample B is confirmed with the values of the lowest distribution energy E_D.

Conclusions In summary, we have reported the two-component decay of TRPL intensity. The first-decay component was attributed to simultaneous exciton relaxation of free-carrier and localized states; the second-decay one was dominated by localized exciton relaxation. The second-decay lifetime represented the extent of carrier localization or indium aggregation and phase separation. The exciton relaxation lifetime of free-carrier states was related to defect density. We also calibrated the localization energy and distribution energy by fitting our PL and SE data with models. The calibrated results were consistent with those of previous HRTEM and XRD observations, TRPL, and AFM surface scanning.

Acknowledgement This research was supported by National Science Council, The Republic of China (Taiwan), under grants NSC 88-2215-E-002-023, NSC 88-2215-E-002-040, NSC 88-2215-E-014-005, and NSC 89-2215-E-002-0036. It was also sponsored by Chung Shan Institute of Science and Technology, Taiwan, R.O.C.

References

[1] Y. S. LIN, K. J. MA, C. HSU, S. W. FENG, Y. C. CHENG, C. C. LIAO, C. C. YANG, C. C. CHOU, C. M. LEE, and J. I. CHYI, Appl. Phys. Lett. **77**, 2988 (2000).
[2] S. CHICHIBU, T. AZUHATA, T. SODA, and S. NAKAMURA, Appl. Phys. Lett. **69**, 4188 (1996).
[3] M. POPHRISTIC, F. H. LONG, C. TRAN, I. T. FERGUSON, and R. F. KARLICEK, JR., J. Appl. Phys. **86**, 1114 (1999).
[4] Y. NARUKAWA, Y. KAWAKAMI, SZ. FUJITA, SG. FUJITA, and S. NAKAMURA, Phys. Rev. B **55**, R1938 (1997).
[5] H. S. KIM, R. A. MAIR, J. LI, J. Y. LIN, and H.X. JIANG, Appl. Phys. Lett. **76**, 1252 (2000).
[6] K. DOMEN, A. KURAMATA, and T. TANAHASHI, Appl. Phys. Lett. **72**, 1359 (1998).
[7] B. MONEMAR, Phys. Rev. B **10**, 676 (1974).
[8] P. G. ELISEEV, P. PERLIN, J. LEE, and M. OSINSKI, Appl. Phys. Lett. **71**, 569 (1997).

Time-Resolved Photoluminescence Study of InGaN MQW with a p-Contact Layer

T. Kuroda[1]) (a), R. Sasou (a), A. Tackeuchi (a), H. Sato (b), N. Horio (b), and C. Funaoka (b)

(a) Department of Applied Physics, Waseda University, Tokyo 169-855, Japan

(b) Stanley Electric Company, Ltd., Edanisi 1-3-1, Aoba, Yokohama 225-0013, Japan

(Received June 21, 2001; accepted August 4, 2001)

Subject classification: 73.61.Ey; 78.47.+p; 78.55.Cr; 78.67.De; S7.14

To clarify the influence of diffused Mg impurities on the carrier recombination in InGaN multiple quantum wells (MQWs), we have performed a systematic study using time-resolved photoluminescence (PL) measurements. It was found that the MQWs with a p-contact layer and the MQWs with a nondoped GaN layer had almost the same carrier lifetime and PL intensity below 200 K. However, the MQWs with the p-contact layer had shorter carrier lifetime and lower PL intensity than the MQWs with the nondoped GaN layer above 200 K. This degradation of the PL for the MQWs with the p-contact layer can be attributed to nonradiative recombination caused by the diffusion of Mg impurities from the p-contact layer into MQWs.

Introduction Recently, a number of developments have been made in the growth techniques of III–V nitride semiconductors. Nitride-based optoelectronic devices such as light-emitting diodes (LEDs) and laser diodes (LDs) have been fabricated in the blue and ultraviolet wavelength regions [1, 2]. An important breakthrough among these achievements is the success of p-type doping in epitaxial GaN [3, 4]. In the fabrication of these devices, Mg has been widely used as an acceptor to achieve electrical p-conductivity. Since Mg acceptors in GaN are considered to be passivated by hydrogen during metalorganic chemical vapor deposition (MOCVD) growth, the as-grown layer is usually activated by thermal annealing to break the Mg–H bonds. Kuroda et al. [5] observed the unwanted diffusion behavior of Mg impurities from the p-type side toward the n-type side by secondary ion mass spectroscopy (SIMS) and electron-beam-induced current (EBIC) measurements.

To clarify the influence of diffused Mg impurities on the carrier recombination in InGaN multiple quantum wells (MQWs), we have performed a systematic study using time-resolved photoluminescence (PL) measurements for the MQWs with a p-contact layer and the MQWs with a nondoped GaN layer. These samples had almost the same carrier lifetime and PL intensity below 200 K. However, the MQWs with the p-contact layer had a shorter carrier lifetime and lower PL intensity than the MQWs with the nondoped GaN layer above 200 K. This degradation of the PL for the MQWs with the p-contact layer can be attributed to nonradiative recombination caused by the Mg impurities diffusing from the p-contact layer into the MQWs.

Experimental The samples used in this study were grown by MOCVD. The GaN nucleation layer was grown on a sapphire substrate. The sample includes a 350 nm thick

---

[1]) Corresponding author; Phone: +81 3 5286 3853; Fax: +81 3 5286 3853; e-mail: tkuroda@mn.waseda.ac.jp

p-contact layer next to the MQWs which consists of five periods of 1.5 nm thick $In_{0.16}Ga_{0.84}N$ quantum wells and 8.0 nm thick $In_{0.03}Ga_{0.97}N$ barriers. The growth temperature, V/III ratio and growth rate of well and barrier layer were 750 °C, 5000, 0.07 nm/s and 900 °C, 15000, 0.2 nm/s, respectively. The Mg concentrations of the p-contact layer were of the order of 10^{19} cm^{-3}. In the well width of 1.5 nm, the quantum-confined Stark effect (QCSE) which causes the longer carrier lifetime and the lower PL energy is estimated to be negligible [6]. Thermal annealing was carried out for Mg activation. A reference sample with a nondoped GaN cap layer instead of the p-contact layer was also grown. We measured the time-resolved PL spectrum using a streak camera with a time resolution of 15 ps. Frequency-doubled optical pulses generated from a Kerr-lens mode-locked Ti:sapphire laser were used for pumping. The laser wavelength was 395 nm, which excites only the well layers. The repetition rate of the laser pulses was 100 MHz. The pumping power was 10 mW.

Results and Discussion PL spectra of the MQWs with p-contact layer and with nondoped GaN layer are shown in Figs. 1a and b, respectively. The PL peak energy is 2.90 eV for the MQWs with the p-contact layer and 2.88 eV for the MQWs with the nondoped GaN layer at 10 K. Figures 2a and b show the PL decay curves, given at PL peak energy (4 nm range) of the MQWs with p-contact layer and with the nondoped GaN layer measured at various temperatures, respectively. The decay curves of the two samples were almost the same below 200 K. However, the decreases of the carrier lifetime and the PL intensity for the MQWs with p-contact layer were larger than those for the MQWs with nondoped GaN layer above 200 K. The carrier lifetime of the MQWs with nondoped GaN layer, as determined by single exponential fitting from the later parts, was 6.9 ns at 300 K. In contrast, for the MQWs with the p-contact layer, the carrier lifetime was shortened to 5.3 ns.

In Fig. 3 we plotted the time-integrated PL intensity and carrier lifetime as a function of temperature. Clearly, the MQWs with the p-contact layer have larger PL degradation than the MQWs with the nondoped GaN layer above 200 K.

These experimental results can be attributed to nonradiative recombination. Since, at low temperature, the radiative recombination lifetime is shorter than nonradiative life-

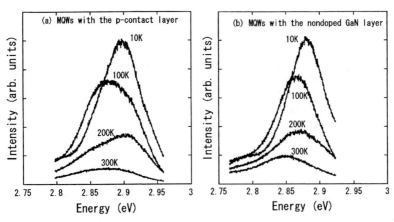

Fig. 1. PL spectra for InGaN MQWs a) with p-contact layer and b) with the nondoped GaN layer for various temperatures

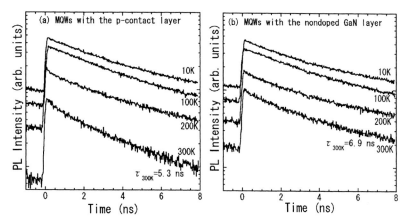

Fig. 2. PL decay curves for a) InGaN MQWs a) with p-contact layer and b) with the nondoped GaN layer for various temperatures.

time, the radiative recombination is dominant. However, with increasing temperature, the radiative recombination lifetime becomes long and the nonradiative recombination lifetime becomes short. As a result, nonradiative recombination is dominant at 300 K [7]. The close similarities of the PL decay curves of the two MQWs below 200 K indicate that the two samples had similar radiative recombination. Above 200 K, the larger degradation of the PL for the MQWs with the p-contact layer indicates that these MQWs had larger nonradiative recombination than the MQWs with the nondoped GaN layer. Kuroda et al. [5] reported that Mg impurities doped in the p-contact layer diffuse heavily into the n-type side when the sample is grown on a sapphire substrate. The appearance of threading dislocations causes this diffusion of the Mg impurities. Since our MQWs were grown on a sapphire substrate, Mg impurities likely diffused into the MQW layers. Therefore, we can conclude that the diffused Mg impurities act as nonradiative recombination centers in MQWs, resulting in the larger degradation of the PL for the MQWs with the p-contact layer.

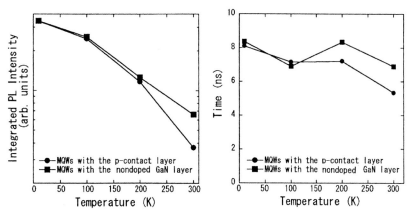

Fig. 3. Temperature dependence of the PL intensity and carrier lifetime for the MQWs with the p-contact layer (●) and with the nondoped GaN layer (■)

Conclusion   To clarify the influence of the Mg impurities diffused from the p-contact layer, we have performed time-resolved PL measurements of InGaN MQWs with the nondoped GaN layer and the MQWs with the p-contact layer. The MQWs with the p-contact layer had shorter carrier lifetime and lower PL intensity than the MQWs with the nondoped GaN layer at room temperature. This degradation of the PL can be attributed to nonradiative recombination, which was caused by the diffusion of Mg impurities from the p-contact layer into the InGaN MQWs.

Acknowledgements   We thank K. Taniguchi and Y. Yamashita of the Stanley Electric Company, Ltd. for the samples. This work was supported in part by the High-Tech Research Center Project from the Ministry of Education, Culture, Sports, Science and Technology. T. Kuroda and A. Tackeuchi are grateful for Waseda University Grant for Special Research Projects.

References

[1] S. Nakamura, T. Mukai, and M. Senoh, Jpn. J. Appl. Phys. **30**, L1998 (1991).
[2] S. Nakamura, M. Senoh, S. Nagahama, N. Iwasa, T. Yamada, T. Matsushita, H. Kiyoku, and Y. Sugimoto, Jpn. J. Appl. Phys. **35**, L74 (1996).
[3] H. Amano, M. Kito, K. Hiramatsu, and I. Akasaki, Jpn. J. Appl. Phys. **28**, L2112 (1989).
[4] S. Nakamura, T. Mukai, M. Senoh, and N. Iwasa, Jpn. J. Appl. Phys. **31**, L139 (1992).
[5] N. Kroda, C. Sasaoka, A. Kimura, A. Usui, and Y. Mochizuki, J. Cryst. Growth **189/190**, 551 (1998).
[6] T. Kuroda, A. Tackeuchi, and T. Sota, Appl. Phys. Lett. **76**, 3753 (2000).
[7] Y. Narukawa, K. Sawada, Y. Kawakami, S. Fujita, S. Fujita, and S. Nakamura, J. Cryst. Growth **189/190**, 606 (1998).

Photoluminescence Excitation Spectroscopy of MBE Grown InGaN Quantum Wells and Quantum Boxes

M. E. White[1]) (a), K. P. O'Donnell (a), R. W. Martin (a), C. J. Deatcher (a), B. Damilano (b), N. Grandjean (b), and J. Massies (b)

(a) Department of Physics and Applied Physics, University of Strathclyde, Glasgow, G4 0NG, Scotland, UK

(b) CRHEA/CNRS, Rue Bernard Gregory, Sophia Antipolis, F-06560 Valbonne, France

(Received June 20, 2001; accepted August 4, 2001)

Subject classification: 78.40.Fy; 78.55.Cr; 78.67.De; 78.67.Hc; S7.14

Photoluminescence excitation (PLE) spectroscopy was carried out to investigate the excitation/emission cycle of MBE grown InGaN quantum structures. Quantum well and Stranski-Krastanov type quantum box samples were chosen that emit from blue to red. The bandgap energy (E_g), determined from the PLE spectrum, was found to decrease concurrently with the detection energy. This indicates that the emission spectrum from the sample is inhomogeneously broadened. A plot of the resultant Stokes shift against detection energy shows a linear trend. Our results agree with those from independent measurements of thermally detected optical absorption (TDOA) in some of the samples. On comparison with absorption and PLE measurements on MOCVD grown InGaN, a difference in the bandgap energies obtained becomes apparent for detection energies below ≈ 2.6 eV.

Introduction InGaN based optoelectronics are continually escalating in their applications. High radiative efficiencies and the potential for blue to red emission make it of substantial commercial interest [1]. However, despite extensive use, there is still much uncertainty regarding the physical processes which produce the optical properties of this material. Here we report on photoluminescence excitation (PLE) spectroscopy to investigate the excitation/emission cycle and measure the optical bandgap of InGaN. PLE has advantages over optical absorption techniques in that it does not rely on the substrate material being transparent at the excitation energy and is not dependent on the thickness of the active layer. From the literature, PLE spectroscopy on InGaN has been confined to samples emitting at energies greater than 2.65 eV, blue emission [2–4]. For a given sample, no variation in E_g on changing the detection energy was reported. In this paper, we present PLE spectroscopy taken on MBE grown quantum well (QW) and Stranski-Krastanov type quantum box (QB) samples that emit from blue to red, 2.86 eV > emission peak > 1.96 eV. We also extract further information by obtaining PLE at different detection energies, selected to cover the emission peak and low energy spectral tail, of each individual sample. These results are compared to absorption and PLE measurements taken on MOCVD grown InGaN.

Experimental Details InGaN QW and Stranski-Krastanov type QB samples were grown using molecular beam epitaxy (MBE), as described previously [5, 6]. Samples were grown on sapphire substrate with a 25 nm GaN nucleation layer and a GaN buf-

---

[1]) Corresponding author; Phone: +44 141 548 3458; Fax: +44 141 552 2891; e-mail: m.white@strath.ac.uk

fer typically a few microns thick. The QW InGaN layers were grown with double the NH$_3$ flux (200 sccm) of the QB InGaN layers to delay the formation of any 3D islands and promote 2D layer-by-layer growth [7]. 2D–3D transitions in the Stranski-Krastanov growth type occur above a critical In composition and are a consequence of the very large compressive strain due to the lattice mismatch of InN and GaN [6]. The InGaN layer thickness (L_z) varied from 1.5 nm to 5.5 nm in the QWs and from 2.5 nm to 3.5 nm in the QBs. A GaN capping layer was deposited on each sample. Low temperature (<25 K) PLE spectroscopy was performed in a closed cycle helium refrigerator. A 1 kW xenon arc lamp coupled with a 0.25 m monochromator was used as the excitation source. The excitation spot was chopped and focussed to a spot of about 1 mm diameter on the sample. A 0.67 m monochromator in conjunction with a cooled PMT was used to select and detect the emission energy.

Experimental Results and Discussion PLE spectra obtained from QW and QB samples were found to have the same general shape as shown in Fig. 1.

The bandgap energy (E_g) was determined for each sample using a fit to the low energy side of the PLE spectrum by the sigmoidal function [8]

$$y(E) = \frac{a_0}{1 + \exp\left(\frac{E_g - E}{\Delta E}\right)}, \quad (1)$$

where ΔE is a broadening parameter, a_0 is a constant and E is the excitation energy at which the intensity, y, is recorded. E_g was found to decrease with the sample emission peak. Furthermore, E_g values from PLE data obtained at different energies within the emission band of an individual sample were found to decrease with the detection energy selected, E_p. Similar behaviour was observed for both the QW and QB samples and indicates that the emission spectrum is inhomogeneously broadened in both cases. Taking the Stokes shift to be the difference between the bandgap energy determined from the PLE measurement and the detection energy selected, a plot of the resultant Stokes shift against detection energy was found to follow a linear trend, as shown in Fig. 2. Values of the Stokes shift between 860 and 100 meV were determined for detection energies in the range of 1.85–2.82 eV. Independent thermally detected optical absorption (TDOA) measurements [7] taken on some of the samples (also plotted in Fig. 2) agree fairly well with these results.

However, on comparison to absorption [9] and PLE [10] measurements taken on InGaN QWs and epilayers grown by metalorganic chemical vapour deposition (MOCVD), a reduc-

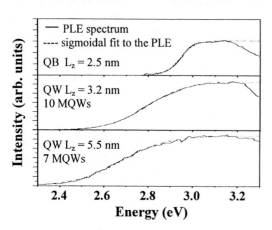

Fig. 1. Selection of PLE spectra from InGaN QW and QB samples

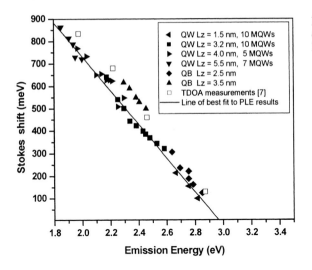

Fig. 2. A plot of Stokes shift against emission detection energy for the MBE samples

tion in the E_g values for a given E_p was observed, see Fig. 3. The difference between E_g determined for the MBE and MOCVD samples becomes apparent for detection energies below ≈2.6 eV. Subsequently, a linear trendline of E_p against E_g given by the equations

$$E_p = 2.98(30)\, E_g - 6.11(90) \tag{2}$$

and

$$E_p = 1.328(21)\, E_g - 1.15(6) \tag{3}$$

can be plotted for the MBE and MOCVD InGaN samples, respectively. Nevertheless, the bandgap energies obtained from PLE on the MOCVD samples, using the method outlined previously, also indicate that the sample emission is inhomogeneously broadened. Strong localisation effects attributed to quantum dots of uniform composition,

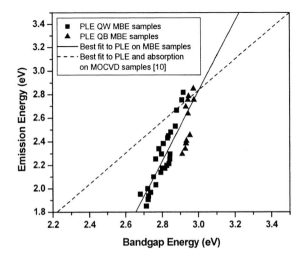

Fig. 3. Bandgap energy against emission detection energy for the InGaN MBE and MOCVD samples

close to pure InN, combined with the accompanying strain induced piezoelectric fields have been presented [8, 11] to explain the linear trend in the absorption measurements as shown in Fig. 3. This argument may be extended to the PLE results of the MBE samples. The different trendlines plotted for the MBE and MOCVD results are likely to be related by the different growth techniques and the resulting piezoelectric fields and strain effects. However, the linear relationship between E_p and E_g in both the MBE and MOCVD samples and the trendline obtained from the MBE samples, irrespective of QW or QB, indicate that the excitation/emission cycle of InGaN is dependent on some form of localisation at a common source.

Conclusion PLE spectroscopy was carried out on MBE grown QW and QB samples that emit from blue to red. Bandgap energy determined from the PLE spectrum was found to decrease linearly with decreasing detection energy. This indicates the emission spectrum is inhomogeneously broadened in both the QW and QB samples. Independent measurements taken by TDOA on some of these samples agree with the large Stokes shifts calculated from the results. On comparison with absorption and PLE measurements on InGaN MOCVD samples, a change in the E_g to E_p relationship was observed for E_p below ≈ 2.6 eV.

References

[1] S. NAKAMURA and G. FASOL, The Blue Laser Diode, Springer-Verlag, Berlin 1997.
[2] H. C. YANG, P. F. KUO, T. Y. LIN, Y. F. CHEN, K. H. CHEN, L. C. CHEN, and J.-I. CHYI, Appl. Phys. Lett. **76**, 3712 (2000).
[3] Y-H KWON, G. H. GAINER, S. BIDNYK, Y. H. CHO, J. J. SONG, M. HANSON, and S. P. DENBAARS, Appl. Phys. Lett. **75**, 2545 (1999).
[4] J. A. DAVIDSON, P. DAWSON, T. WANG, T. SUGAHARA, J. W. ORTON, and S. SAKAI, Semicond. Sci. Technol. **15**, 497 (2000).
[5] B. DAMILANO, N. GRANDJEAN, J. MASSIES, L. SIOZADE, and J. LEYMARIE, Appl. Phys. Lett. **77**, 1268 (2000).
[6] B. DAMILANO, N. GRANDJEAN, S. DALMASSO, and J. MASSIES, Appl. Phys. Lett. **75**, 3751 (1999).
[7] B. DAMILANO, N. GRANDJEAN, J. MASSIES, S. DALMASSO, J. L. REVERCHON, M. CALLIGARO, J. Y. DUBOZ, L. SIOZADE, and J. LEYMARIE, phys. stat. sol. (a) **180**, 363 (2000).
[8] R. W. MARTIN, K. P. O'DONNELL, P. G. MIDDLETON, and W. VAN DER STRICHT, Appl. Phys. Lett. **74**, 263 (1999).
[9] K. P. O'DONNELL, R.W. MARTIN, C. TRAGER-COWAN, M. E. WHITE, K. ESONA, C. DEATCHER, P. G. MIDDLETON, K. JACOBS, W. VAN DER STRICHT, C. MERLET, B. GIL. A. VANTOMME, and J. F. W. MOSSELMANS, Mater. Res. Soc. Symp. Proc. **595**, W11.26.1 (2000).
[10] M. E. WHITE, K. P. O'DONNELL, R. W. MARTIN, S. PEREIRA, C. J. DEATCHER, and I. M. WATSON, Proc. Europ. Mater. Res. Soc., Strasbourg, June 2001.
[11] K. P. O'DONNELL, R. W. MARTIN, and P. G. MIDDLETON, Phys. Rev. Lett. **82**, 237 (1999).

Photoluminescence Excitation Spectroscopy of $In_xGa_{1-x}N/GaN$ Multiple Quantum Wells with Various In Compositions

C. Sasaki[1]) (a), M. Iwata (a), Y. Yamada (a), T. Taguchi (a), S. Watanabe (b), M. S. Minsky (b), T. Takeuchi (b), and N. Yamada (b)

(a) *Department of Electrical and Electronic Engineering, Yamaguchi University, 2-16-1, Tokiwadai, Ube, Yamguchi 755-8611, Japan*

(b) *Agilent Laboratories, 3-2-2 Sakado, Takatsu-ku, Kawasaki 213-0012, Japan*

(Received June 25, 2001; accepted August 4, 2001)

Subject classification: 78.47.+p; 78.55.Cr; S7.14

Luminescence properties of $In_xGa_{1-x}N/GaN$ multiple quantum wells (MQWs) with various In compositions have been studied by means of photoluminescence excitation (PLE) spectroscopy. The clear peak due to the absorption of $In_xGa_{1-x}N$ quantum wells was observed in the PLE spectrum of the MQW sample with $x < 0.01$ at 4 K, and the Stokes shift was estimated to be 63 meV. It was found from temperature-dependent PLE measurements that the Stokes shift was independent of temperature up to 300 K. This result suggests that the large Stokes shift cannot be explained only by the effect of carrier localization due to compositional fluctuation.

Introduction GaN-based semiconductors have been studied for application to optoelectronic devices such as light-emitting diodes (LEDs) and laser diodes (LDs). Blue/green LEDs and purple LDs are commercialized [1, 2]. It is well known that $In_xGa_{1-x}N$-based light-emitting devices reveal high quantum efficiency in spite of large dislocation densities (10^8–10^{10} cm^{-2}) in the device structures. However, the radiative recombination mechanism in the $In_xGa_{1-x}N$ ternary alloys is not fully understood at present in spite of many reports [3–6].

In this paper, we have studied the luminescence properties of $In_xGa_{1-x}N/GaN$ multiple quantum wells (MQWs) with various In compositions by means of temperature-dependent photoluminescence excitation (PLE) spectroscopy. We observed a clear peak due to the absorption of $In_xGa_{1-x}N$ quantum wells in the PLE spectra, and investigated the temperature dependence of Stokes shift.

Experimental Procedure The samples used in the present work were grown by metalorganic vapour phase epitaxy (MOVPE) on (0001) sapphire substrates, following the deposition of a 3.0 μm thick GaN layer. The MQW samples consisted of five periods of 2.0 nm thick $In_xGa_{1-x}N$ well layers separated by 7.5 nm thick GaN:Si barrier layers. The In composition was varied from $x < 0.01$ to $x = 0.22$. The In composition x in quantum wells was estimated by X-ray diffraction measurements by taking biaxial strain into consideration [7]. Photoluminescence (PL) and PLE measurements were performed using a tunable dye laser pumped by a Xe–Cl excimer laser (308 nm). The pulse width and the repetition rate were 20 ns and 100 Hz, respectively. The PL spectra

---

[1]) Corresponding author; Phone: +81-836-85-9407; Fax: +81-836-85-9401; e-mail: c2507@stu.cc.yamaguchi-u.ac.jp

were measured by a liquid nitrogen cooled charge-coupled-devices camera in conjunction with a 50 cm single-grating monochromator.

Results and Discussion Figure 1, curves a to d show the PL and PLE spectra at 4 K taken from $In_xGa_{1-x}N/GaN$ MQW samples with $x < 0.01$, $x = 0.06$, 0.18, and 0.22 under an excitation power density of 48 kW/cm^2, respectively. The luminescence band of the $In_xGa_{1-x}N$ quantum wells is located at 3.312 eV for $x < 0.01$, 3.043 eV for $x = 0.06$, 2.784 eV for $x = 0.18$, and 2.672 eV for $x = 0.22$. The luminescence from the GaN cladding layers is also observed at about 3.480 eV. The detected photon energy in the PLE measurement is tuned at the peak photon energy of each PL spectrum. A clear peak due to the absorption of the $In_xGa_{1-x}N$ quantum wells is observed at 3.375 eV in the PLE spectrum of the MQW sample with $x < 0.01$. Then, the Stokes shift of the sample is estimated to be 63 meV. With increasing In composition x, the absorption edge of the PLE spectra becomes broader. This is due to the increase in In compositional fluctuation. The inflection point around the absorption edge of the PLE spectra is indicated by vertical arrows. The energy position is located at 3.220 eV for $x = 0.06$, 2.987 eV for $x = 0.18$, and 2.941 eV for $x = 0.22$. Then, the Stokes shift is estimated to be 177 meV for $x = 0.06$, 203 meV for $x = 0.18$, and 269 meV for $x = 0.22$. The Stokes shift becomes larger with increasing In composition x.

Next, we performed temperature-dependent PL and PLE measurements. Figure 2, curves a to d show the PL and PLE spectra taken from the MQW sample with $x < 0.01$ at 4, 90, 180, and 300 K under the excitation power density of 48 kW/cm^2, respectively. Interference fringes are observed in the PL spectra at higher temperatures. The absorp-

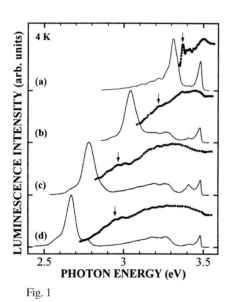

Fig. 1

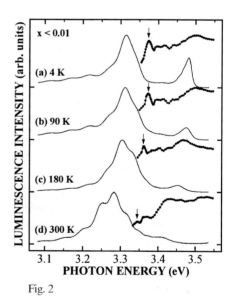

Fig. 2

Fig. 1. PL (solid lines) and PLE (solid circles) spectra at 4 K taken from $In_xGa_{1-x}N/GaN$ MQW samples with (a) $x < 0.01$, (b) $x = 0.06$, (c) 0.18, and (d) 0.22

Fig. 2. Temperature dependence of PL (solid lines) and PLE (solid circles) spectra taken from an $In_xGa_{1-x}N/GaN$ MQW sample with $x < 0.01$ at (a) 4, (b) 90, (c) 180, and (d) 300 K

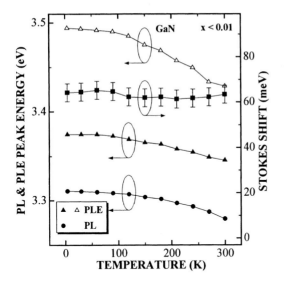

Fig. 3. PL and PLE peak photon energies, Stokes shift of an $In_xGa_{1-x}N/GaN$ MQW sample with $x < 0.01$ as a function of temperature. The open and solid triangles indicate the PLE peak photon energy of the GaN cladding layers and $In_xGa_{1-x}N$ quantum wells, respectively. The solid circles indicate the PL peak photon energy of the $In_xGa_{1-x}N$ quantum wells. The solid squares indicate the Stokes shift

tion peak is clearly observed at lower temperatures. With increasing temperature, the PL and PLE spectra become broader and both of the peak photon energy similarly shift toward the low-energy side. The absorption peak is observed even at 300 K. At 300 K the peak photon energies of the PL and PLE spectra are located at 3.282 and 3.344 eV, respectively. Then, the Stokes shift is estimated to be 62 meV. This indicates that the carriers are strongly localized even at 300 K. Figure 3 shows the PL and PLE peak energies and the Stokes shift as a function of temperature in the MQW sample with $x < 0.01$. The open and solid triangles indicate the PLE peak photon energy of the GaN cladding layers and $In_xGa_{1-x}N$ quantum wells, respectively. The energy shift of the $In_xGa_{1-x}N$ quantum wells with temperature is smaller than that of the GaN cladding layers. The solid circles indicate the PL peak photon energy of the $In_xGa_{1-x}N$ quantum wells. Both the PL and PLE peak photon energies of the $In_xGa_{1-x}N$ well layers similarly shift toward the low-energy side up to 300 K. As a result, the Stokes shift (indicated by the solid squares) is independent of temperature up to 300 K. If the carrier localization originates from compositional fluctuations, the Stokes shift should be decreased with increasing temperature. Therefore, the observed temperature dependence of the Stokes shift cannot be explained by the effect of carrier localization due to compositional fluctuations.

In order to evaluate the degree of carrier localization, we have separately investigated the temperature dependence of time-resolved luminescence of the MQW sample with $x < 0.01$ [8]. At 7 K the PL decay time was dependent on the emission photon energy. The decay time at the higher-energy side of the PL spectrum was fast, and was estimated to be about 0.1 ns. With decreasing the detected photon energy, the decay time became longer, and was estimated to be about 0.65 ns at lower-energy side. This temporal behavior indicated that the carriers were localized at 7 K. On the other hand, the decay time was independent of the emission photon energy at 300 K. The decay time was almost constant over the emission photon energy, and was estimated to be about 0.1 ns. This indicated that carriers were delocalized at 300 K. However, as mentioned above, the Stokes shift was estimated to be about 62 meV at 300 K. These experimental results indicated this Stokes shift cannot be explained by the effect of carrier localization due to compositional fluctuation. Therefore, we must consider another origin of the large Stokes shift. It has been reported recently that GaN and InN have a large electron-phonon coupling energy [9]. The LO-phonon coupling energies of GaN and InN are 92 and 86 meV, respectively. Therefore, the carrier–LO-phonon interaction will play an imporant role in the emission mechanism. Thus, one probable origin of the Stokes shift may be due to the large electron–phonon interaction. This is suggested by the temperature dependence of the Stokes shift.

Conclusions We have studied the luminescence properties of $In_xGa_{1-x}N$/GaN MQWs with various In compositions by means of PLE spectroscopy. The clear peak due to the absorption of $In_xGa_{1-x}N$ quantum wells was observed in the PLE spectrum of the MQW sample with $x < 0.01$. The PL and PLE peak photon energies of the $In_xGa_{1-x}N$ well layers similarly shifted toward the lower-energy side up to 300 K. As a result, the Stokes shift was independent of temperature up to 300 K. These experimental results suggested that this large Stokes shift could not be explained only by the effect of carrier localization due to compositional fluctuation.

Acknowledgement This work was supported by the Japanese National Project "The Light for the 21st Century" from METI/NEDO/JRCM.

References

[1] S. NAKAMURA, M. SENOH, N. IWASA, S. NAGAHAMA, T.YAMADA, and T. MUKAI, Jpn. J. Appl. Phys. **34**, L1332 (1995).
[2] S. NAKAMURA, M. SENOH, S. NAGAHAMA, T. MATSUSHITA, H. KIYOKU, Y. SUGIMOTO, T. KOZAKI, H. UMEMOTO, M. SANO, and T. MUKAI, Jpn. J. Appl. Phys. **38**, L226 (1999).
[3] Y. H. CHO, G. H. GAINER, A. J. FISCHER, J. J. SONG, S. KELLER, U. K. MISHRA, and S. P. DENBAARS, Appl. Phys. Lett. **73**, 1370 (1998).
[4] Y. NARUKAWA, Y. KAWAKAMI, SG. FUJITA, and S. NAKAMURA, Phys. Rev. B **59**, 10283 (1999).
[5] C. WETZEL, S. KAMIYAMA, H. AMANO, and I. AKASAKI, Internat. Workshop Nitride Semiconductors, Nagoya (Japan), IPAP Conf. Ser. **1**, 510 (2000).
[6] H. KUDO, H. ISHIBASHI, R. ZHENG, Y. YAMADA, and T. TAGUCHI, phys. stat. sol. (b) **216**, 163 (1999).
[7] T. TAKEUCHI, H. TAKEUCHI, S. SOTA, H. SAKAI, H. AMANO, and I. AKASAKI, Jpn. J. Appl. Phys. **36**, L177 (1997).
[8] S. WATANABE, M. S. MINSKY, N. YAMADA, T. TAKEUCHI, R. SCHNEIDER, C. SASAKI, M. IWATA, Y. YAMADA, and T. TAGUCHI, Proc. Internat. Workshop Nitride Semiconductors, Nagoya (Japan), IPAP Conf. Ser. **1**, 532 (2000).
[9] R. ZHENG, T. TAGUCHI, and M. MATSUURA, J. Appl. Phys. **87**, 2526 (2000).

Temperature Dependent Optical Properties of InGaN/GaN Quantum Well Structures

P. Hurst[1]) (a), P. Dawson (a), S. A. Levetas (a), M. J. Godfrey (a),
I. M. Watson (b), and G. Duggan (c)

(a) Physics Department, UMIST, PO Box 88, Manchester, M60 1QD, UK

(b) Institute of Photonics, University of Strathclyde, Glasgow, UK

(c) IQE (Europe) Ltd., Cardiff, UK

(Received June 22, 2001; accepted July 2, 2001)

Subject classification: 78.47.+p; 78.55.Cr; 78.67.De; S7.14

We have investigated the variation of the photoluminescence intensity and decay time as a function of temperature of a series of InGaN/GaN quantum well structures in which the number of quantum wells was varied. All the samples exhibited a decrease in photoluminescence intensity and decay time with increasing temperature with the rate of decrease being reduced as the number of quantum wells was increased. We have compared these results with a theoretical model which describes the effects of thermally excited carrier escape and recapture. We find reasonable agreement with the results of the model and the experiments for the samples incorporating only a few quantum wells supporting the idea that thermally excited carrier loss is the main non-radiative recombination path.

Introduction During the last ten years the use of the group III-nitride material system in opto-electronic devices has enabled the fabrication of high brightness LEDs and laser diodes emitting in the blue and green parts of the spectrum [1–5]. However, despite the rapid progress in device development, there are still some significant unanswered questions concerning the recombination mechanisms and carrier dynamics in the InGaN/GaN quantum wells which constitute the active regions of these devices. This is particularly important as photoluminescence spectroscopy and decay time measurements are quite often used in the assessment of the quantum well substructure prior to their inclusion in complete device structures.

In this paper, we report on the temperature dependence of the time integrated photoluminescence intensity and photoluminescence decay time of a series of InGaN/GaN multiple quantum well structures incorporating different numbers of quantum wells.

Experimental Details The samples studied were grown on (0001) sapphire substrates by metal organic chemical vapour deposition (MOCVD). The substrates were initially treated in hydrogen at 1140 °C, followed by the growth of a 25 nm GaN nucleation layer at a temperature of 540 °C and a 1.2 μm thick GaN buffer at a growth temperature of 1140 °C. The undoped quantum well structures were then grown at a temperature of 830 °C and consisted nominally of 7.5 nm thick GaN barriers and 2.5 nm $In_{0.14}Ga_{0.86}N$ wells. Finally, a 15 nm GaN cap layer was deposited at a temperature of 830 °C on each sample. The multiple quantum well structures comprised 2, 3, 5, 10 or 18 quantum wells in the different samples (referred to as 2, 3, 5, 10 and 18QW, respectively).

The optical excitation was provided by pulses from a mode-locked, cavity-dumped laser whose output was frequency doubled to give light with a wavelength of 290 nm.

For the time integrated photoluminescence spectroscopy the excitation was chopped at a frequency ~250 Hz and the signal processed by a lock-in detector. For the time decay measurements, the signal was processed by a time-correlated single photon counting system, such that the minimum decay time measurable was ~400 ps.

Results and Discussion Measurements of the photoluminescence spectra and photoluminescence decay were made on all five samples over the temperature range 6–300 K. The photoluminescence spectra at a temperature of 6 K are dominated by a single emission line centred at an energy of 3.017 eV. The full widths at half-maximum height of the spectra are 100, 92, 67, 55 and 53 meV for the 18, 10, 5, 3 and 2QW samples, respectively. The width of the spectra for 2 and 3QW samples are caused primarily by fluctuations in the In fraction in the quantum wells [6–8], for the samples with greater number of quantum wells the increased width of the spectra are probably due to changes in the average In fraction in the different wells. The integrated photoluminescence intensity as a function of temperature for the five samples is presented in Fig. 1. At a temperature of 6 K the photoluminescence intensity from all the samples is equal within a factor of 2.5 and the decay times τ, characterised by the time it takes for the photoluminescence intensity to fall to $1/e$ of its initial value, from all the samples is 6 ns. However, as the temperature was increased the photoluminescence intensity from all samples decreased, but at differing rates. As can be seen in Fig. 1 the rate at which the intensity decreased over the temperature range becomes progressively larger as the number of quantum wells is decreased. Specifically, the 2QW sample shows a decrease in intensity of 10000, whereas at 300 K the 18QW sample only showed a decrease of a factor of ten. Also, there was a general decrease in the measured τ as the temperature was increased, the decrease being much greater for the samples incorporating only a few wells. It was not possible to measure the decay time of the samples with only a few quantum wells over the whole temperature range as the photoluminescence intensity became too weak.

The reduction in photoluminescence intensity and decay time with increasing temperature indicates that non-radiative recombination plays an increasingly important role as the temperature is increased. Of particular importance to this work are the conclusions reached by Heath [9] and Thucydides [10] who showed that thermionic emission of carriers from single GaAs/AlGaAs quantum wells can be the dominant non-radiative recombination process at high temperatures. Qualitatively, the results presented here could reflect the same phenomena

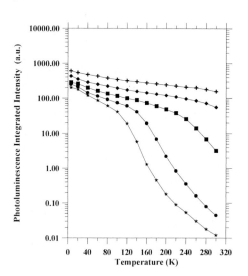

Fig. 1. Photoluminescence integrated intensity versus temperature for the 2QW (★), 3QW (●), 5QW (■), 10QW (♦) and 18QW (+) samples

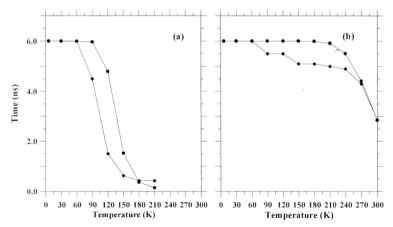

Fig. 2. Measured values of τ (♦) and calculated values of τ_c (●) versus temperature for the a) 2QW and b) 10QW samples

whereby the effects of thermionic emission are reduced as the number of quantum wells is increased due to the increased importance of carrier recapture. On this basis we have used a rate equation model which incorporates the probabilities of carrier capture and emission into and between the quantum wells to calculate the lifetimes τ_c for carriers (either electrons or holes) within the quantum well system. The values for the capture probability and emission rate were calculated as described by Levetas and Godfrey [11]. The effective masses of the electron and hole in the InGaN quantum wells were taken to be $0.22m_0$ and $2.2m_0$, respectively [12–14].

The results of the decay time measurements have been compared directly with the results of the model. The dominant input parameter to enable the results of the model to be compared with the experimental results is the depth of the confining potential. In the foregoing comparison we used an energy barrier of 190 meV: such a depth for the valence band well is in good agreement with our calculations for the hole confined states. In Fig. 2 are shown the the values of τ and τ_c over the temperature range 6–300 K for the 2 and 10QW samples. This set of data is shown as it is representative of the behaviour of the samples with a small number or quantum wells (2, 3, 5QW) and a large number of quantum wells (10, 18QW). As expected, the 2QW sample is particularly susceptible to thermally excited carrier escape which is reflected in the reasonable agreement between the measured values of τ and the calculated values of τ_c. The agreement between the model and experiment for the 10QW sample is less satisfactory over the temperature range 60–210 K where the predicted values of τ_c remain constant but the measured values of τ decrease. This would indicate some other non-radiative process is dominant, such as that associated with the delocalisation of carriers from the In fluctuations that occur in each quantum well [15]. In the temperature regime where the model predicts that thermally excited carrier escape would be important the measured values of τ fall off rapidly in reasonable agreement with the predictions of the model.

Summary We have investigated the temperature dependence of the photoluminescence intensity and decay times of a series of InGaN/GaN quantum well structures grown by MOCVD on sapphire substrates. Measurements of the samples under identical conditions show the general trend that, as the number of quantum wells is increased, the temperature dependent fall off in the photoluminescence intensity and decay time becomes less severe. The results for the samples with few quantum wells are explained in terms of thermionic emission leading to the thermal quenching of the photoluminescence, although the degree of quenching is determined by the number of quantum wells that are available to recapture the thermally excited carriers.

References

[1] S. Nakamura and G. Fasol, The Blue Laser Diode, Springer-Verlag, Berlin 1997.
[2] I. Akasaki and H. Amano, J. Cryst. Growth **163**, 351 (1996).
[3] J. W. Orton and C. T. Foxon, Rep. Prog. Phys. **61**, 1 (1998).
[4] I. Akasaki, S. Sota, H. Sakia, T. Tanika, M. Koike, and H. Amano, Electron. Lett. **32**, 1105 (1996).
[5] B. Gil (Ed.), Group III-Nitride Semiconductor Compounds, Physics and Applications, Clarendon Press, Oxford 1998.
[6] N. Duxbury, P. Dawson, U. Bangert, E. J. Thrush, W. Van Der Stricht, K. Jacobs, and I. Moermann, phys. stat. sol. (b) **216**, 355 (1999).
[7] B. Monemar, J. P. Bergman, J. Dalphors, G. Pozina, B. E. Sernelius, P. O. Holtz, H. Amano, and I. Akasaki, MRS Internet J. Nitride Semicond. Res. **4S1**, G2.5 (1999).
[8] B. Monemar, J. P. Bergman, J. Dalphors, G. Pozina, B. E. Sernelius, P. O. Holtz, H. Amano, and I. Akasaki, MRS Internet J. Nitride Semicond. Res. **4**, 16 (1999).
[9] A. D. Heath, Time Resolved Photoluminescence Studies of Low Dimensional Semiconductors, PhD Thesis, UMIST, 1996.
[10] G. Thucydides, J. M. Barnes, E. Tsui, K. W. J. Barnham, C. C. Phillips, T. S. Cheng, and C. T. Foxon, Semicond. Sci. Technol. **3**, 331 (1996).
[11] S. A. Levetas and M. J. Godfrey, Phys. Rev. B **59**, 10202 (1999); One-Phonon Scattering of Carriers at a Quantum Well, MPhil, UMIST, 1998.
[12] S. N. Mohammad and H. Morkoc, Prog. Quantum Electron. **20**, 361 (1996).
[13] A. S. Barker, Jr. and M. Ilegems, Phys. Rev. B **7**, 743 (1973).
[14] J. S. Im, A. Moritz, F. Steuber, V. Härle, F. Scholz, and A. Hangleiter, Appl. Phys. Lett. **70**, 63 (1997).
[15] E. Berkowicz, D. Gershoni, G. Bahir, E. Lakin, D. Shilo, E. Zolotoyabko, A. C. Abare, S. P. DenBaars, and L. A. Coldren, Phys. Rev. B **61**, 10994 (2000).

Energy Diagram and Recombination Mechanisms in InGaN/AlGaN/GaN Heterostructures with Quantum Wells

A. E. Yunovich[1]) and V. E. Kudryashov

M.V. Lomonosov Moscow State University, Department of Physics, Moscow, Russia

(Received June 21, 2001; accepted July 2, 2001)

Subject classification: 68.65.Fg; 73.21.Fg; 78.60.Fi; S7.14

Electroluminescence spectra of GaN based LEDs are analyzed quantitatively using a model of 2D density of states with band tails. Calculations take into account an energy diagram with given band offsets between AlGaN, InGaN and GaN layers. The model describes spectral shapes with four fitting parameters in a wide range of currents and intensities with a good accuracy. Fluctuations of well thickness (heterointerface roughness), of In content, of charged impurities and piezoelectric fields are discussed for an evaluation of the exponential tail parameter of 2D density of states in the active region. Spectral maxima move with current changes due to the filling of the band tails and redistribution of current carriers between regions with a larger probability of radiative recombination. 2D concentrations and lifetimes of minority carriers in active layers are evaluated from the parameters of the model.

Introduction The problem of recombination mechanisms in LEDs based on InGaN/AlGaN/GaN heterostructures with quantum wells is still in discussion. This paper continues the analysis of electroluminescent spectra of GaN-based LEDs published in [1, 2].

Experimental Results Experimental results were obtained on samples of the green light-emitting diodes HLMP-CE30, based on $In_xGa_{1-x}N/Al_yGa_{1-y}N/GaN$ heterostructures with multiple quantum wells, made in Hewlett Packard laboratories [1, 2]. Electroluminescent (EL) spectra in a wide range of currents and intensities were studied and published in detail in [1, 2].

An example of EL spectra of a green LED in the current range $J = 1\,\mu A$ to 20 mA is shown in Fig. 1[2]). It is possible to divide the spectra into three ranges: (i) At high currents J, the maxima move with J ($\hbar\omega_{max} = 2.35$–2.52 eV), as it was seen in other green LEDs with SQWs and MQWs [1, 2]. This "moving" band, as discussed and will be analyzed below in detail, can be described by a model of 2D density of states with band tails caused by potential fluctuations. (ii) In the range of $0.1\,mA < J < 1\,mA$, the maxima do not depend on J, $\hbar\omega_{max} = 2.35$–2.36 eV (so called "standing band"). This band can be described by a model of radiative recombination in localized tail states formed by inhomogeneities in InGaN layers (clusters or "dots") [3]. (iii) A band on the long-wavelength side, in the range $J \leq 70\,\mu A$, has maxima which move with voltage approximately linearly: $\hbar\omega_{max} \approx eU = 1.92$–$2.04$ eV. This band can be described by a model of the tunnel radiative recombination [1, 2].

---

[1]) Corresponding author; Phone: +7-(095)-9392994; Fax: +7-(095)-9393731; e-mail: yunovich@scon175.phys.msu.su

[2]) Colour figures are published online (www.physica-status-solidi.com), where indicated.

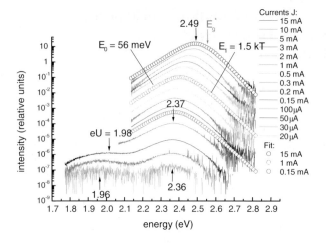

Fig. 1 (colour). Dependence of EL spectra on the current J for Hewlett-Packard green LED HLMP-CE30. Circles show approximation of the spectral form

Typical features of the spectra in the ranges (i) and (ii) are exponential tails on the long-wavelength side: $I \sim \exp(\hbar\omega/E_0)$, slope $E_0 = 55$–60 meV; and on the short-wavelength side: $I \sim \exp(-\hbar\omega/E_1)$, slope $E_1 = 37$–40 meV. Slopes E_0 and E_1 changed only slightly versus current, the value E_1 increased with J for currents $J \geq 10$ mA.

Discussion

Energy diagram of InGaN/AlGaN/GaN p–n heterostructures According to the measurements of LED's dynamical capacitance versus voltage [2] the heterostructure has three following regions of potential fall: space charge regions on p- and n-sides of GaN (of 60–100 nm) and a compensated i-region between them. In the i-region there is a constant electric field E_{max} that leads to nearly linear potential fall. The active layer with quantum wells InGaN and barriers GaN is situated in the i-region. Charged donors Si^+ and acceptors Mg^- create almost parabolic band curvature on both sides of the p–n heterojunction (see Fig. 2). The potential falls on both sides (n- and p-) are inversely proportional to the concentrations of charged impurities. Energy diagram was calculated for acceptor concentration $N_A = 10^{19}$ cm^{-3}; the acceptor ionization energy was assumed to be $\Delta E_A = 160$ meV and the donor concentration N_D (Si) = 10^{18} cm^{-3} ($\Delta E_D = 35$ meV). This diagram does not include piezoelectric and spontaneous polarization on heterointerfaces [4], these effects are assumed in our model as potential fluctuations.

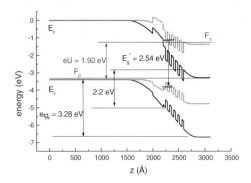

Fig. 2 (colour). Energy diagram of hererostructure InGaN/AlGaN/GaN in the case $N_A = 10^{19}$ cm^{-3}; $\Delta E_A = 160$ meV; $N_D = 10^{18}$ cm^{-3} in equilibrium (solid line) and at forward bias $U = 1.92$ V (dotted line)

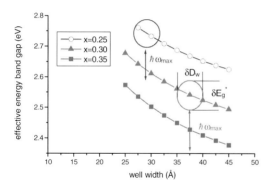

Fig. 3 (colour). Dependence of the effective energy band gap in $In_xGa_{1-x}N$ quantum well on the well width for various In contents: $x = 0.25$ (○), $x = 0.30$ (▲) and $x = 0.35$ (■)

The effective energy bandgap is determined by In content "x" in $In_xGa_{1-x}N$ active layers, quantum size levels of electrons and holes in quantum wells ΔE_{c1} and ΔE_{v1}, deformation strain ΔE_P, piezoelectric fields ΔE_{pe}, random fields of charged impurities ΔE_{DA}, Coulomb interaction of electrons and holes ΔE_{exc}:

$$E_g^* = E_{g(In_xGa_{1-x}N)} + \Delta E_{c1} + \Delta E_{v1} + \Delta E_P + \Delta E_{pe} + \Delta E_{DA} - \Delta E_{exc}. \tag{1}$$

We use the following parameters of $In_xGa_{1-x}N$ and $Al_yGa_{1-y}N$ energy gaps from [3, 4]:

$$E_{g(InGaN)} = xE_{g(InN)} + (1-x)E_{g(GaN)} - b_{(InGaN)}x(1-x) \tag{2}$$

$$E_{g(AlGaN)} = xE_{g(AlN)} + (1-x)E_{g(GaN)} - b_{(AlGaN)}x(1-x) \tag{3}$$

Energy gaps in these equations are $E_{g(GaN)} = 3.4$ eV, $E_{g(InN)} = 1.9$ eV, $E_{g(AlN)} = 6.22$ eV; the bowing factors $b_{(InGaN)} = 3$ eV and $b_{(AlGaN)} = 0.5$ eV for $E_g(x, y)$ are taken from [3, 4].

The dependence of E_g^* on the well width D_w calculated for three values of In content taking into account only three first members on the right side of Eq. (1) is shown in Fig. 3.

Fluctuations of the effective energy gap E_g^* are caused by various terms in Eq. (1). The major part of them is caused by In content fluctuations Δx and changes of quantum size levels ΔE_{c1}, ΔE_{v1} due to the well width roughness ΔD_w. The value $E_g^*(x, D_w)$ in Eq. (1) has derivatives, at $x = 0.3$ and $D_w = 35$ Å, of $\partial E_g^*/\partial D_w = 9.26$ meV/Å and $\partial E_g^*/\partial x = 2.377$ eV, respectively. Let us take ΔD_w of the order of a lattice constant $a_\| = 5.185$ Å and Δx of the order of 1%, than we can estimate the value of $\Delta E_g^* \approx 53$ meV (see Fig. 3).

The diagram in Fig. 2 explains some characteristic energies of the spectra in Fig. 1. The distance between the bottom of conduction band and valence band in the quantum well of InGaN is 2.4 eV. It corresponds to the "standing" spectral band $\hbar\omega = 2.36$ eV. Below bias $U = 2.4$ V, the distance between the Fermi quasi-levels is $\Delta F = F_n - F_p = eU < 2.36$ eV and the radiative recombination occurs between the local levels of electron and hole tails state density close to band edges. At higher voltages all local levels are occupied and the spectrum follows quasi-Fermi levels; then the spectral maximum shifts to short wavelengths with the voltage.

Model of effective combined 2D density of states Electroluminescence spectra of LEDs based on GaN heterostructures are described numerically with a good precision in a wide range of currents and intensities by the model of effective 2D state density [1, 2]:

$$I(\hbar\omega) \sim N^{2D}(\hbar\omega - E_g^*)\, f_c(\hbar\omega, kT, F_n)\, (1 - f_v(\hbar\omega, kT, F_p)). \tag{4}$$

Here f_c, f_v are the electron and hole quasi-Fermi functions close to the band edges:

$$f_c(E) = (1 + \exp((E - F_n)/kT))^{-1},$$

$$1 - f_v(E) = (1 + \exp((F_p - E)/kT))^{-1}, \tag{5}$$

$N^{2D} \sim N_{cv}^{2D} (1 + \exp(\hbar\omega - E_g^*))^{-1}$ is the effective 2D state density with an exponential tail close to the band edges [1, 2]. If we assume that the majority carriers (electrons) fill the states near the band edge, that is $f_c(\hbar\omega, kT, F_n) \approx 1$, then the model has four parameters. The physical meaning of them follows: $E_0 = \Delta E_g^* = 50$–60 meV is the exponential slope on the long-wavelength side corresponding to fluctuations of E_g^*; $E_g^* = 2.5$–2.7 eV is the effective energy bandgap corresponding to a distance between quantum size levels in the wells; $E_1 = mkT = 1.5\,kT$ is the exponential slope of spectra on the short-wavelength side due to the quasi-Fermi functions; and $\Delta F_p = F_p - E_v$ is the quasi-Fermi level of holes.

Approximation curves of the spectra calculated by Eq. (4) are shown in Fig. 1. Theoretical curves fit the experimental ones in a wide range of J with a fairly good accuracy.

The slight changes of parameters E_0 and E_1 with the current were discussed in [1, 2].

An interesting result was received from the fitting parameter ΔF_p. The 2D concentration of minority carriers (holes) injected into the active 2D layer can be calculated from ΔF_p and effective density of 2D states:

$$\delta p_{2D} = N_v^{2D} \exp(\Delta F_p / kT). \tag{6}$$

The δp_{2D} value calculated from this equation is plotted versus current in Fig. 4. We may conclude that in some range of currents ($J > J_1$) the concentration of holes injected into the active 2D layer is proportional to the current:

$$\delta p_{2D} = A\,J = (\tau/eS)\,J. \tag{7}$$

If the area of the structure S is known ($S \approx 10^{-3}$ cm^{-2}) we may estimate the lifetime τ of minority carriers injected into the active 2D layer. The data plotted in Fig. 4 give a value of carrier lifetime $\tau \approx 10^{-8}$ s. This estimation is in good agreement with values determined from carrier recombination kinetics in InGaN quantum wells [5].

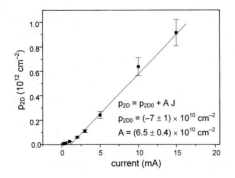

Fig. 4. 2D concentration of holes in the active region of the structure versus the current J

Conclusions 1. The analyzed energy diagram of InGaN/AlGaN/GaN p–n heterostructures and parameters of the model of effective 2D density of states explain specific values of energies in electroluminescent spectra and specific voltages in current–voltage characteristics of LEDs based on these structures.

2. The spectral shape of the LEDs spectral band is described by the model with a good accuracy in a wide range of currents.

3. There is a possibility to evaluate 2D concentration δp_{2D} and lifetime τ of minority carriers (holes) in the active layer from parameters of the model.

References

[1] A. E. Yunovich, V. E. Kudryashov, S. S. Mamakin, A. N. Turkin, A. N. Kovalev, and F. I. Mahyakhin, phys. stat. sol. (a) **176**, 125 (1999).
[2] V. E. Kudryashov, A. N. Turkin, A. E. Yunovich, A. N. Kovalev, and F. I. Manyakhin, Semicond. **33**, 445 (1999).
[3] B. Monemar, J. P. Bergman, J. Dalfors, G. Pozina, B. E. Sernelius, P. O. Holtz, H. Amano, and I. Akasaki, MRS Internet J. Nitride Semicond. Res. **4**, 16 (1999).
[4] F. Bernardini and V. Fiorentini, Phys. Rev. B **57**, R9427 (1998).
[5] Y.-H. Cho, B. D. Little, G. H. Gainer, J. J. Song, S. Keller, U. K. Mishra, and S. P. DenBaars, MRS Internet J. Nitride Semicond. Res. **4S1**, G2.4 (1999).

Relation between Structural Parameters and the Effective Electron–Hole Separation in InGaN/GaN Quantum Wells

N. A. Shapiro[1]) (a), H. Feick (a), N. F. Gardner (b), W. K. Götz (b),
P. Waltereit (c), J. S. Speck (c), and E. R. Weber (a)

(a) Materials Science Division, Lawrence Berkeley National Laboratory and University of California at Berkeley, 1 Cyclotron Road, Berkeley, California 94720, USA

(b) LumiLeds Lighting, San Jose, California 95131, USA

(c) Materials Department, University of California at Santa Barbara, Santa Barbara, California 93106, USA

(Received June 22, 2001; accepted July 11, 2001)

Subject classification: 42.50.Hz; 78.55.Cr; 78.67.De; S7.14

The photoluminescence (PL) of InGaN/GaN quantum well (QW) structures is measured as a function of biaxial strain to study the dependence of the luminescence emission on the built-in electric field. The direction and magnitude of the shift in luminescence energy with strain reveals an effective e–h separation in the well. We have used this method to evaluate the effective e–h separation in a number of structures with varying QW thickness and indium content. Our results show that the e–h separation increases with increasing QW thickness and with increasing indium content.

Introduction InGaN has emerged in the last years as the most important material for short-wavelength optoelectronics. Devices based on this material are already commercially available, yet the nature of the radiative transitions that occur in these devices is still under debate. Two distinct transition types that have been proposed are the recombination of carriers localized in indium-rich nanoclusters [1] and the recombination of carriers separated due to strong built-in electric fields [2].

We recently developed a method to distinguish between these two transition types, by measuring the photoluminescence (PL) from epitaxially grown InGaN/GaN QW structures as a function of externally applied biaxial strain [3]. This has an advantage over using hydrostatic strain because it changes the electric fields in the QW structure. This allows us to investigate the dependence of the luminescence on the built-in electric field.

Experimental For the application of biaxial strain, the sample is used as a window of a pressure cell. The biaxial strain varies linearly with pressure (p) and with the square of the ratio of window diameter to sample thickness. Details of the experimental procedure are outlined in Ref. [3]. The thickness of all samples (substrate + epilayer) was about 400 μm, and the window diameter was 3.175 mm. PL spectra were recorded at room temperature, using a 0.85 m double spectrometer equipped with a GaAs photomultiplier. A 325 nm He–Cd laser was used for excitation. Time-resolved photolumines-

---

[1]) Corresponding author; Phone: (510) 486-5083; Fax: (510) 486-5530; e-mail: shapboy@uclink.berkeley.edu

Table 1
Structural parameters of the studied InGaN QW layers

#	L_w (nm)	L_b (nm)	x (%)	$h\nu$ (eV)
1	2.45	11	8.9	2.97
2	3.02	11	12	2.66
3	3.02	11	15.9	2.43
4	3.36	11	15.4	2.30
5	1.76	11	11	2.95
6	2.71	11	11	2.81
7	3.24	11	11	2.75
8	4.2	7.3	2.5	3.13
9	4.2	7.3	8.5	2.75
10	4.2	7.3	11.9	2.58

cence was analyzed with a ps-resolution streak camera, using 200 fs excitation pulses from a frequency-tripled mode-locked Ti:sapphire laser. The indium content and well and barrier thicknesses of the samples under study were determined by high-resolution X-ray diffraction.

A number of samples of different structures, grown either by MOCVD or MBE, are included in this study. Samples 1–7 are InGaN/GaN MQWs with four periods grown by MOCVD on a sapphire substrate. Samples 1–4 cover the entire wavelength range for common nitride LEDs, from UV to yellow-green. Samples 5–7 are identical except for the QW layer thickness. Samples 8–10 are InGaN/GaN MQWs with 13 periods grown by MBE on top of an MOCVD grown GaN template at identical conditions except for the indium flux. Table 1 lists the structural parameters of all samples.

Results Figure 1 shows the PL peak energy as a function of pressure in the biaxial strain device for sample 3. The peak energies were determined through a Gaussian fit of the QW luminescence line. The luminescence blueshifts linearly with pressure as the pressure is increased to 69 bar, and then it redshifts linearly with no hysteresis as the pressure is decreased back to zero.

Figure 2a shows the slopes of the PL peak energy shift with pressure ($dh\nu_p/dp$) for samples 1–4 as a function of $h\nu_{peak}$. Figure 2b shows the slopes as a function of QW thickness for samples 5–7. Figure 2c shows the slopes as a function of indium content for samples 8–10.

Discussion To interpret our results, we have to identify the physical factors responsible for the direction and magnitude of the PL shift. The energy gap

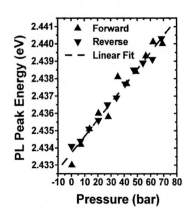

Fig. 1. PL peak energy as a function of pressure in the biaxial strain device (sample 3)

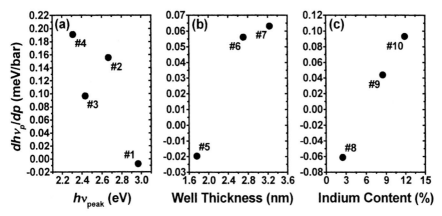

Fig. 2. a) Slope of the PL shifts with pressure ($dh\nu_p/dp$) of samples 1–4 vs. $h\nu_{peak}$. b) $dh\nu_p/dp$ of samples 5–7 vs. the QW thickness. c) $dh\nu_p/dp$ of samples 8–10 vs. the indium content

shrinks linearly with increasing tensile biaxial strain or pressure ($dE_g/dp \sim -c$). This should result in a redshift of the luminescence. The electric field in the well under the application of biaxial strain is

$$E_w = \frac{L_b(P_b - P_w) - \varepsilon_0 \varepsilon_b \left(\dfrac{L_w \Delta P_{pz,w}}{\varepsilon_0(\varepsilon_w - 1)} + \dfrac{L_b \Delta P_{pz,b}}{\varepsilon_0(\varepsilon_b - 1)} \right)}{\varepsilon_0 \varepsilon_b L_w + \varepsilon_0 \varepsilon_w L_b} \qquad (1)$$

where ε_0 is the permittivity of free space, $\varepsilon_w(\varepsilon_b)$ is the QW (barrier) dielectric constant, L_w (L_b) is the thickness of the QW (barrier), and $P_w(P_b)$ is the polarization of the QW (barrier). $P = P_{sp} + P_{pz}$, where $P_{pz} = e_{31}(\varepsilon_{xx} + \varepsilon_{yy}) + e_{33}\varepsilon_{zz}$; P_{sp} and P_{pz} are the spontaneous and piezoelectric polarizations, respectively, and e_{ij} are the electromechanical tensors. $\Delta P_{pz,w}$ ($\Delta P_{pz,b}$) is defined as the change in the piezoelectric polarization in the well (barrier) induced by the applied biaxial strain. We assume zero voltage-drop across one period without external stress but finite voltage drop due to the stress [3]. The influence of the surface depletion field will be discussed elsewhere [4]. The built-in electric field in the well decreases linearly with increasing tensile biaxial strain, the rate depending on the indium content ($dE_w/dp \sim -f(x)$).

The built-in electric field in InGaN/GaN QW structures generates a potential drop across the QW that induces the electron and hole wave functions to shift toward opposite surfaces of the QW. As a result, the radiative transition probability diminishes and the luminescence resulting from the transition is redshifted by $\sim E_w L_r$, where we introduce L_r as the effective e–h separation. Since applying the tensile strain reduces the electric field in the QW, this redshift should also be reduced by $\sim \Delta E_w L_r$, and the luminescence should experience an equivalent blueshift. The total change of peak energy with biaxial strain is therefore given by

$$\frac{dh\nu_p}{dp} \approx \frac{dE_g}{dp} - L_r \frac{dE_w}{dp}. \qquad (2)$$

Since dE_g/dp should be constant and the differences in dE_w/dp between our samples are relatively small (due to small range of x), we expect $dh\nu_p/dp$ should scale roughly with the effective e–h separation (L_r).

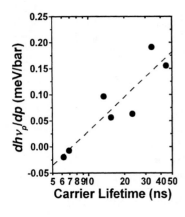

Fig. 3. $dh\nu_p/dp$ of samples 1–7 vs. the lifetime plotted on a logarithmic scale

This hypothesis is supported by measurements of the carrier lifetimes of samples 1–7, the results of which are plotted in Fig. 3 vs. $dh\nu_p/dp$. According to Eq. (2), $dh\nu_p/dp$ should depend linearly on L_r. On the other hand, the relationship between transition probability (carrier lifetime) and carrier separation (L_r) is roughly exponential [5]. In Fig. 3, $dh\nu_p/dp$ indeed increases linearly with the logarithm of the carrier lifetime.

Figure 2b shows an increased $dh\nu_p/dp$ with increasing QW thickness. This is expected because increased well-thickness (L_w) should translate into increased effective e–h \mtheta separation (L_r). However, from our quantitative model (Eq. (1), Ref. [3]) we would have expected a significantly greater increase in $dh\nu_p/dp$ as we progress from sample 5 to 7. One explanation for this might be that with the increased well thickness there is an increased clustering of indium atoms, forming indium-rich nano-clusters. Localization of carriers in these centers would inhibit carrier separation and thus reduce L_r. Another explanation would be that in thinner wells the confinement increases thus reducing the magnitude of $dh\nu_p/dp$ [6]. This would act to increase $dh\nu_p/dp$ in \mtheta thinner wells, making L_r appear larger than it really is.

From Fig. 2c, we see that the $dh\nu_p/dp$ increases with increasing indium content. This is expected because increasing the indium content should increase dE_w/dp due to an increased difference between the piezoelectric constants between the well and barrier layers. However, the increase we observe in $dh\nu_p/dp$ is significantly stronger than our model calculations. We interpret this to mean that in addition to an increase in dE_w/dp, there is also an increase in L_r. This is reasonable because as we increase the indium content, we also increase the strength of the built-in electric field in the QW. The stronger electric field should induce a greater separation between the electrons and holes.

From Fig. 2a, we see that as the QW is designed to emit luminescence at longer wavelength, the effective separation of the carriers also increases. This is consistent with the above analysis because longer wavelengths require thicker wells and/or greater indium content. It is interesting to note that the greatest internal quantum efficiencies tend to be reached for LEDs designed to emit around blue-green (500 nm) [7], where we observe significant blueshifts and large carrier lifetimes. This suggests that the carrier separation is not strongly deleterious to the operation of these devices, even though we would expect that the reduced oscillator strength of the transition would reduce the internal quantum efficiency.

Conclusion We have demonstrated that the application of biaxial strain allows to directly study the nature of the transitions in InGaN/GaN QW structures. It is found that our results can be consistently described in terms of an effective e–h separation, induced by the built-in electric field. This separation depends on the structural parameters of the device. In particular, the separation increases with increasing QW thickness and increasing indium content. We have also observed that the separation increases with decreasing $h\nu_{peak}$. We attribute this observation to the thicker QWs and greater indium content necessary to produce longer wavelengths.

Acknowledgement This work was supported by the Department of Defense MURI program through the Office of Naval Research under Contract No. N00014-99-1-0729 (C. Wood, program manager).

References

[1] Y. NARUKAWA, Y. KAWAKAMI, S. FUJITA, and S. FUJITA, Phys. Rev. B **55**, R1938 (1996).
[2] T. TAKEUCHI, S. SOTA, M. KATSURAGAWA, M. KOMORI, H. TAKEUCHI, H. AMANO, and I. AKASAKI, Jpn. J. Appl. Phys. **36**, L382 (1997).
[3] N. A. SHAPIRO, Y. KIM, H. FEICK, E. R. WEBER, P. PERLIN, J. W. YANG, I. AKASAKI, and H. AMANO, Phys. Rev. B **62**, 16318 (2000).
[4] P. WALTEREIT, J. S. SPECK, N. A. SHAPIRO, H. FEICK, and E. R. WEBER, (unpublished).
[5] E. BERKOWICZ, D. GERSHONI, G. BAHIR, E. LAKIN, D. SHILO, E. ZOLOTOYABKO, A. C. ABARE, S. P. DENBAARS, and L. A. COLDREN, Phys. Rev. B **61**, 10994 (2000).
[6] W. PAUL, Semicond. Semimetals **54**, 1 (1998).
[7] T. MUKAI, M. YAMADA, and S. NAKAMURA, Jpn. J. Appl. Phys. **38**, 3976 (1999).

Spatial Inhomogeneity of Photoluminescence in InGaN Single Quantum Well Structures

A. Kaneta (a), G. Marutsuki (b), K. Okamoto (a), Y. Kawakami (a), Y. Nakagawa (b), G. Shinomiya (b), T. Mukai (b), and Sg. Fujita (a)

(a) Department of Electronic Science and Engineering, Kyoto University, Kyoto 606-8501, Japan

(b) Nitride Semiconductor Laboratory, Nichia Corporation, 491 Oka, Kaminaka, Anan, Tokushima 774-8601, Japan

(Received June 22, 2001; accepted August 4, 2001)

Subject classification: 78.47.+p; 78.55.Cr; 78.67.De; S7.14

Spatial distribution of photoluminescence (PL) spectra has been assessed in an InGaN single quantum well (SQW) structure by means of fluorescence microscopy and scanning near-field optical microscopy (SNOM) under illumination-collection mode. The PL intensity of fluorescence image is uniform at 77 K, but the dark spot areas were extended with increasing temperature. The near-field PL images revealed the variation of both peak energy and intensity in PL spectra according to the probing location with the scale less than a few hundreds nm.

$In_xGa_{1-x}N$-based light emitting diodes (LEDs) are currently commercialized between the near-ultraviolet and amber spectral regions [1–3]. In spite of high threading dislocation density (10^8–10^{10} cm^{-2}), such LEDs exhibit a substantially high external quantum efficiency (η_{ext}) (10–15%) at the emission wavelength of blue region. These phenomena have been understood in terms of two major mechanisms. The first one is that the nonradiative recombination centers (NRC), whose main origins are probably not macroscopic defects but point defects, are suppressed by the substitution of In atoms to Ga sites. The second one is the so-called localization effect where excitons and/or carriers are trapped at deep energy states formed by large alloy fluctuation, so that the pathway to the NRC is hindered very effectively [4–6]. Nevertheless, it has also been reported that η_{ext} of InGaN-based LEDs decreases if the emission wavelength becomes longer than the blue-green region. Although the mechanism accountable for those phenomena has not been clarified yet, the key of this mechanism would be revealed by assessing the correlation between radiative/nonradiative recombination processes and micro/nanoscopic structures.

Several reports have recently appeared on spatial emission mapping measurements in InGaN single quantum wells (SQWs) by employing cathodoluminescence (CL) [7, 8] and scanning near-field optical microscopy (SNOM) [9–14]. It has been found that the lateral size of In-composition fluctuations is about 100 nm, which may be limited by the diffusion length of carriers, and/or by the resolution of the spectroscopy. In this paper, the spatial inhomogeneity of photoluminescence has been investigated in an InGaN SQW structure by using the fluorescence microscope and SNOM under illumination-collection mode. The measurement of this mode leads to high spatial resolution because both photoexcitation and PL probing made by the same fiber tip prevent the spatial resolution from being affected by the diffusion effect of excitons and/or carriers [15].

The sample used in this study was grown on (0001) oriented sapphire (Al_2O_3) substrate by the two-flow metalorganic chemical vapor deposition (MOCVD) technique [16]. The sample is composed of a GaN (1.5 µm), an n-GaN:Si (2.3 µm), an $In_{0.2}Ga_{0.8}N$ SQW (3 nm) and a GaN cap (5 nm) layer. The macroscopic PL peak of this sample was located at 470 nm at room temperature (RT). The fluorescent image was taken at temperatures from 77 K to RT using an optical microscope.

The near-field PL measurements were performed with an NFS-300 near-field spectrometer developed at JASCO Corp. The InGaN laser diode (LD, λ = 400 nm) was used as an excitation source in order to achieve the selective photo-excitation to an InGaN SQW. The optical power of 10 mW was coupled to the probe, and about 10 µW was utilized to illuminate the sample through the probe. PL collected by the probe (illumination-collection) was introduced into the 50 cm monochromator. The PL signal was detected by using a charge-coupled device (CCD) detector.

Figure 1 shows fluorescence images of a blue InGaN SQW taken in the range from 77 K to RT. The emission intensity was almost uniform at 77 K, the area as well as the density of dark spots were enhanced with increasing temperature. This is probably because the pathway to NRC was activated in addition to the increase of capture cross section to NRC [17].

Micro-PL and micro-time-resolved PL spectra of an InGaN SQW at 77 K and RT are shown in Figs. 2a and b, respectively. In each figure, PL spectra and decay profiles labeled (1)–(5) indicate the probing positions in the fluorescence image. The PL intensity and PL decay time fluctuate according to the location of the probing area reflecting that internal quantum efficiency of PL is inhomogeneously varied within the active layer. However, the detailed structures were blurred by the diffraction limit. Therefore, the PL mapping technique was developed using SNOM under an illumination-collection mode.

Near-field PL image of peak intensities and peak wavelengths were taken at an InGaN SQW structure by scanning 4 µm × 4 µm area with the mapping interval of 0.1 µm. The aperture size of the probe was about 0.3 µm in diameter. The spatial inhomogeneity with an area of a few hundreds nm was observed as shown in Fig. 3. Similar results have also been observed in other ternary alloys such as InGaAs [15] or GaNAs

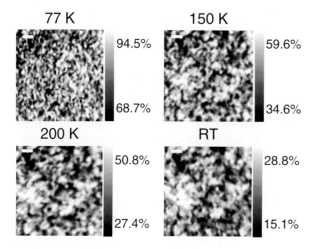

Fig. 1. The fluorescence images of an InGaN SQW at 77 K to RT. The white bars in the images indicate 5 µm scale

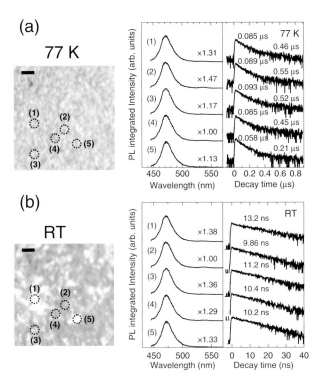

Fig. 2. Micro-PL and micro-time resolved PL spectra of InGaN SQW at a) 77 K and b) RT. The excitaion spot size and excitation power density are $2\,\mu m \times 2\,\mu m$ and $9.375\,\mu J/cm^2$, respectively. The bars in the fluorescence images indicate 5 µm. Dotted circles labelled with numbers indicate excitation positions in the fluorescence images

[18]. Spatial inhomogeneity of the PL integrated intensity indicate that the density of the defects and/or the NRC are distributed spatially. The results of Fig. 3 show a clear spatial correlation between PL integrated intensity and peak wavelength. The strong PL intensity regions correspond to long PL peak wavelength regions. These results suggest spatially inhomogeneous distribution of In alloy composition of the well layer. It would be interesting how such spatially inhomogeneous distribution changes with temperature as well as In composition of well layer of various SQW-LED samples. Such approach is now in progress.

We have studied a spatial inhomogeneity of PL in an InGaN SQW structure using fluorescence microscopy and SNOM under illumination-collection mode. The emission intensity of fluorescence images was almost uniform at 77 K, the area as well as the

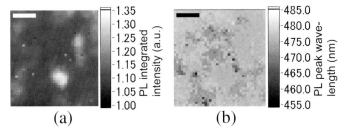

Fig. 3. a) Near-field PL image mapped with the PL integrated intensity and b) peak wavelength at RT. The bars in the images indicate 1 µm scale. The excitation power density is 1.11×10^3 W/cm^2 under continuous wave (cw) condition

density of dark spots were enhanced with increasing temperature. The near-field PL mapping image obtained in this investigation revealed the variation of both peak and intensity in PL spectra according to the probing location with the area less than a few hundreds nm.

Acknowledgements The authors would like to thank Mr. Y. Narita for his kind advice concerning the near-field spectroscopy. They are also grateful to Dr. R. Micheletto for valuable comments for the instrumentation. This work was partly supported by the Kyoto University-Venture Business Laboratory Project, the Kawakami Memorial Foundation and by a Grant-in-Aid for the special area research project of Photonics based on wavelength integration and manipulation from the Ministry of Education, Science, Sports and Culture, Japan.

References

[1] S. Nakamura, M. Senoh, N. Iwasa, and S. Nagahama, Jpn. J. Appl. Phys. **34**, L797 (1995).
[2] S. Nakamura, M. Senoh, N. Iwasa, S. Nagahama, T. Yamada, and T. Mukai, Jpn. J. Appl. Phys. **34**, L1332 (1995).
[3] T. Mukai, H. Narimatsu, and S. Nakamura, Jpn. J. Appl. Phys. **37**, L479 (1998).
[4] Y. Narukawa, S. Saijou, Y. Kawakami, Sg. Fujita, T. Mukai, and S. Nakamura, Appl. Phys. Lett. **74**, 558 (1999).
[5] S. Chichibu, T. Azuhata, T. Sota, and S. Nakamura, Appl. Phys. Lett. **69**, 4188 (1996).
[6] Y. Narukawa, Y. Kawakami, M. Funato, Sz. Fujita, and Sg. Fujita, Appl. Phys. Lett. **70**, 981 (1997).
[7] T. Sugahara, M. Hao, T. Wang, D. Nakagawa, Y. Naoi, K. Nishino, and S. Sakai, Jpn. J. Appl. Phys. **37**, L1195 (1998).
[8] S. Chichibu, K. Wada, and S. Nakamura, Appl. Phys. Lett. **71**, 2346 (1997).
[9] P. A. Crowell, D. K. Young, S. Keller, E. L. Hu, and D. D. Awschalom, Appl. Phys. Lett. **72**, 927 (1998).
[10] A. Vertikov, M. Kuball, A. V. Nurmikko, Y. Chen, and S.-Y. Wang, Appl. Phys. Lett. **72**, 2645 (1998).
[11] A. Vertikov, A. V. Nurmikko, K. Doverspike, G. Bulman, and J. Edmond, Appl. Phys. Lett. **73**, 493 (1998).
[12] A. Vertikov, I. Ozden, and A. V. Nurmikko, Appl. Phys. Lett. **74**, 850 (1999).
[13] D. K. Young, M. P. Mack, A. C. Abare, M. Hansen, L. A. Coldren, S. P. DenBaars, E. L. Hu, and D. D. Awschalom, Appl. Phys. Lett. **74**, 2349 (1999).
[14] A. Kaneta, T. Izumi, K. Okamoto, Y. Kawakami, Sg. Fujita, Y. Narita, T. Inoue, and T. Mukai, Jpn. J. Appl. Phys. **40**, 102 (2001).
[15] T. Saiki, K. Nishi, and M. Ohtsu, Jpn. J. Appl. Phys. **37**, 1638 (1998).
[16] S. Nakamura, Jpn. J. Appl. Phys. **30**, L1705 (1991).
[17] T. Someya and Y. Arakawa, Jpn. J. Appl. Phys. **38**, L1216 (1999).
[18] K. Matsuda, T. Saiki, M. Takahashi, A. Moto, and S. Takagishi, Appl. Phys. Lett. **78**, 1508 (2001).

Optical Characterization of InGaN/GaN MQW Structures without In Phase Separation

B. Monemar[1]) (a), P. P. Paskov (a), G. Pozina (a), T. Paskova (a),
J. P. Bergman (a), M. Iwaya (b), S. Nitta (b), H. Amano (b), and I. Akasaki (b)

(a) Department of Physics and Measurement Technology, Linköping University,
S-581 83 Linköping, Sweden

(b) Department of Materials Science and Engineering, Hi Tech Research Center,
Meijo University, 1-501 Shiogamaguchi, Tempaku-ku, Nagoya, 468 Japan

(Received June 21, 2001; accepted August 4, 2001)

Subject classification: 78.55.Cr; 78.60.Hk; 78.67.De; S7.14

Photoluminescence and cathodoluminescence spectroscopies are used to investigate the properties of the band edge emission of InGaN/(In)GaN multiple quantum well (MQW) structures which do not show evidence of phase separation in high resolution electron microscopy. The data still show a clear low energy peak in the spectra, about 0.1 eV below the main exciton peak. Possible interpretations of this second peak are discussed.

1. Introduction The InGaN alloy system is of paramount importance for optical emitter devices with applications in the visible to near UV region [1]. A property of this alloy system that is by now documented both theoretically and experimentally is the instability of the random alloy over a very large range of compositions. A prediction is that at a growth temperature of 800 °C the random alloy is unstable in the composition range $0.04 < x < 0.88$ [2]. A separation of the material into two phases with drastically different In composition is then expected in this range. The presence of In rich inclusions has been evidenced in several cases recently, from both optical and structural studies [3, 4]. The properties of the InGaN alloy and related QW structures have been shown to vary substantially with growth conditions, as judged from the literature data [3–6].

In this work we have studied specific $In_{0.12}Ga_{0.88}N/(In)GaN$ MQW structures grown by MOVPE, and optimized to give high uniformity of the average In composition in the QW structures. We have also studied areas with different dislocation density in structures prepared by the mass transport growth technique [7]. The observed presence of a low energy peak in PL spectra in addition to the expected excitonic near band edge emission is discussed.

2. Samples and Experimental Procedure A nominally undoped GaN layer of thickness 7 μm was grown by MOVPE on a (0001) sapphire substrate with AlN buffer layer. Micrometer-sized trenches (lines) on the GaN surface were patterned in the GaN buffer layer by reactive ion etching. After that the structure was annealed at 1100 °C in a MOVPE reactor in ammonia with nitrogen gas atmosphere. It results in removal (etching) of some GaN material in the original surface areas, and a lateral overgrowth of GaN in the dry-etched areas. Finally, three InGaN quantum wells of width 35 Å were grown on the top of such a GaN structure. The barriers of thickness 105 Å were InGaN

---

[1]) Corresponding author; Phone: +46 13 281765, Fax: +46 13 142337, e-mail: bom@ifm.liu.se

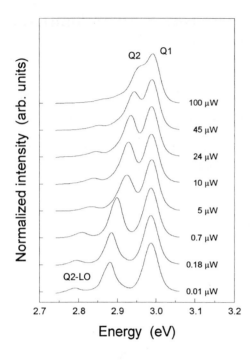

Fig. 1. Photoluminescence spectra of an $In_{0.12}Ga_{0.88}N/(In)GaN$ MQW sample obtained at 2 K at a series of different excitation intensities

with a small In composition, doped with Si to about 10^{18} cm^{-3}.

The micro-PL measurements were performed with cw optical excitation (λ = 266 nm). Cathodoluminescence (CL) was measured using a standard Leo 1500 Gemini scanning electron microscope (SEM) with a MonoCL2 system and a cooled GaAs photomultiplier.

3. Photoluminescence Spectra and Transient Behaviour

The PL spectra from the MQW samples studied in this work depend on excitation intensity, as demonstrated in Fig. 1. At the lowest excitation (about 0.3 W/cm^2) the PL linewidth of the main peak Q1 is about 40 meV. The PL linewidth of excitons in this materials system is influenced by both the alloy broadening and the interface roughness. The alloy broadening contribution in the case of $In_{0.12}Ga_{0.88}N$ is expected to be < 15 meV [8]. The interface roughness is of the order one monolayer, leading to a broadening in the range 20–25 meV, as judged from the best data reported from the AlGaN/GaN MQW systems, where the interface roughness effects dominate [9]. We interpret the Q1 peak as a localized QW exciton transition, in a QW which has a certain electron filling due to transfer of donor electrons from the lightly Si doped barriers. The S-shaped temperature dependence of the spectral position confirms this interpretation. A localization energy of about 18 meV is deduced from the Arrhenius plot in Fig. 2. The transient PL data reveal a PL decay time of about 5 ns at 2 K, which agrees with the expected radiative lifetime of excitons in a 3 nm InGaN/GaN QW [5].

A weaker low energy peak Q2 is observed about 0.1 eV below the main PL

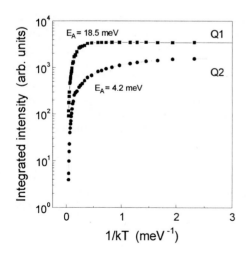

Fig. 2. Arrhenius plots of the temperature dependence of the photoluminescence intensities of the Q1 and Q2 peaks

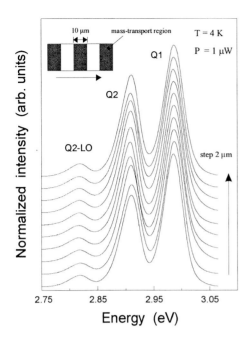

Fig. 3. Variation of the photoluminescence spectra with threading dislocation density, obtained at 2 K by scanning the focused laser beam across the mass transport region (see inset). No significant variation is observed

peak (Fig. 1). This peak is due to a separate PL transition, i. e. not an LO replica of Q1. The Q2 transition is increased in strength with increasing excitation intensity, and shifts to higher energy (Fig. 1). With increasing temperature the low energy PL spectrum disappears, leaving only the Q1 transition at room temperature. The temperature dependence of the PL intensity of the Q2 peak shows the presence of a small localization energy, about 4 meV (Fig. 2). The PL decay is very slow for the Q2 peak, >100 ns at 2 K [7].

In Fig. 3 is shown a set of micro-PL spectra obtained at 4 K with an excitation spot size <2 µm. The laser spot is moved across the low dislocation stripe regions where the mass transport overgrowth was conducted in the buffer layer. Clearly there is no measurable spectral difference in the Q1 or Q2 transitions with dislocation density, only a slight reduction in PL intensity with an order of magnitude increase of the dislocation density (from the low 10^8 cm^{-2} range to the low 10^7 cm^{-2} range).

4. Cathodoluminescence Data CL topographs were obtained for selected areas and selected emission wavelengths in the InGaN MQW PL region. As seen in Fig. 4 there is dark contrast present from threading dislocations [7]. There is also some broader area contrast present in the pictures. This contrast does not vary across the InGaN MQW recombination spectral range, and is presumably related to surface or interface

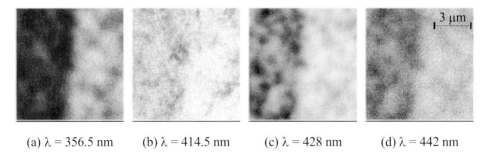

(a) λ = 356.5 nm   (b) λ = 414.5 nm   (c) λ = 428 nm   (d) λ = 442 nm

Fig. 4. Cathodoluminescence spectra at 4 K of the MQW region, showing monochromatic pictures in the energy region of the Q1 and Q2 transitions

variations. No difference in contrast is seen for the spectral ranges corresponding to the Q1 and Q2 peaks, respectively (Fig. 4). These CL data do not give any information on possible few nm scale inhomogeneities.

5. Discussion The samples studied in this work were part of a large series of samples grown under similar conditions. For many of these samples TEM data were obtained with atomic resolution. None of these samples showed any evidence of phase separation, the InGaN material was homogeneous. The analysis of the optical data therefore should be concentrated on physical models excluding phase separation (or nm size In rich inclusions [3, 4, 6]).

One possibility is that the potential across the MQW structure is not flat, so that the different QWs have different properties due to a different internal field in each QW [7]. This would allow a different decay time for different QWs, as observed for Q1 and Q2. Another possibility would be that there exist two kinds of isolated areas with only weak thermal connection in the same QW. The low energy transition could then be due to a domain with a higher electric field, possible due to the presence of polarization fields that are not passivated at the outer surface. If such effects occur, they seem to disappear at elevated temperatures, since the Q2 peak is not observed at room temperature in our samples. A third model involving self trapping effects in InGaN [10] cannot be ruled out by the present data. More work is needed to settle the reason for the strange observation of two PL transitions in these QW systems.

References

[1] S. Nakamura and G. Fasol, The Blue Laser Diode, Springer Verlag, Berlin/Heidelberg/New York 1997.
[2] T. Saito and Y. Arakawa, Phys. Rev. B **60**, 1701 (1999).
[3] Y. Narukawa, Y. Kawakami, S. Fujita, S. Fujita, and S. Nakamura, Phys. Rev. B **55**, R1938 (1997).
[4] S. Chichibu, T. Azuhata, T. Sota, and S. Nakamura, Appl. Phys. Lett. **69**, 4188 (1997).
[5] E. Berkowicz, D. Gershoni, G. Bahir, Lakin, D. Shilo, E. Zolotoyabko, C. Abare, S. P. Denbaars, and L. A. Coldren, Phys. Rev. B **61**, 10994 (2000).
[6] T. Wang, D. Nakagawa, M. Lachab, T. Sugahara, and S. Sakai, Appl. Phys. Lett. **74**, 3128 (1999).
[7] G. Pozina, J. P. Bergman, B. Monemar, M. Iwaya, S. Nitta, H. Amano, and I. Akasaki, Appl. Phys. Lett. **77**, 1638 (2000).
[8] S. M. Lee and K. K. Bajaj, J. Appl. Phys. **73**, 1788 (1993).
[9] N. Grandjean, J. Massies, and M. Leroux, Appl. Phys. Lett. **74**, 2361 (1999).
[10] H. Kudo, H. Ishibashi, R. Zheng, Y. Yamada, and T. Taguchi, phys. stat. sol. (b) **216**, 163 (1999).

Phase Separation in InGaN Epitaxial Layers

A. N. Westmeyer[1]) and S. Mahajan

Department of Chemical and Materials Engineering and
Center for Solid State Electronics Research,
Arizona State University, Tempe, AZ 85287-6006, USA

(Received June 21, 2001; accepted July 4, 2001)

Subject classification: 61.14.Dc; 64.75.+g; 68.55.Nq; S7.14

Epitaxial layers of InGaN were deposited by metalorganic chemical vapor deposition on a GaN layer/sapphire substrate in order to ascertain compositional variation in the GaN–InN system. Samples were examined with In contents from $x = 0.09$ to 0.31. Plan-view images obtained by transmission electron microscopy reveal a domain structure within which the composition is modulated. Satellites appear around the fundamental reflections in the diffraction pattern. The spacing between the satellite and the reflection can be related to the wavelength of the modulations. Equations are derived for modulations in the $[10\bar{1}0]$ and $[11\bar{2}0]$ directions. Modulations were measured in the $[10\bar{1}0]$ direction and found to be about $\lambda = 3.2$ nm for all samples. Strain energy considerations explain the observation of modulations along different directions.

Introduction GaN and its alloys have attracted considerable attention for optoelectronic devices. Emission over the entire visible spectrum can be obtained by suitable alloying. InGaN has provided the active layer in blue laser diodes [1]. It was predicted [2] to undergo spinodal decomposition, which researchers [3] have found indeed to be the case. Non-circular spots in the diffraction patterns have been described as resulting from a multicomposition structure [4, 5]. Although this is true, the means by which the spots obtain this shape have not yet been elucidated. Analysis of the spots can provide important information about the microstructure.

Experimental Details A series of samples was grown by metalorganic chemical vapor deposition (MOCVD). Basal-plane (0001) sapphire was used for a substrate. On this was grown a low-temperature GaN nucleation layer followed by a 2 μm high-temperature GaN buffer layer, both in a H_2 ambient. Three InGaN samples were then grown to a thickness of about 140 nm in a N_2 ambient at nominal temperatures of 850, 880, and 910 °C (although a temperature calibration had not been done for this reactor, results from other reactors suggest that the actual temperature is about 65 °C less than the nominal temperature). All other growth conditions were identical between the samples. The precursors were trimethylgallium (TMGa), trimethylindium (TMIn), and ammonia (NH_3), with flow rates of 24.6, 30.4, and 134 μmol/min, respectively. The resultant V/III ratio was 2437. The In content x was determined by the method of Schuster et al. [6]. This procedure involves measuring the strained lattice parameters, calculating the relaxed lattice parameters, and applying Vegard's law to the relaxed values. Confirmation was provided by Rutherford Backscattering Spectroscopy (RBS). The samples

---

[1]) Corresponding author; Phone: +1 480 965 7649; Fax: +1 480 965 0037;
e-mail: qwerty@cmu.edu

Table 1
Comparison between methods of determining the In content

sample	temperature (°C)	x by Vegard	x by Schuster	x by RBS	relaxation
A	850	0.37	0.31	0.345	0.88
B	880	0.25	0.21	0.22	0.86
C	910	0.13	0.09	0.09	0.30

were examined by transmission electron microscopy (TEM) along the [0001], [10$\bar{1}$0], and [11$\bar{2}$0] zone-axes with a JEOL 4000EX.

Results and Discussion Table 1 summarizes the results of the In content determination. Several trends are apparent. First, the choice of temperature has a significant impact on the resultant value of the In content. The incorporation of In decreases sharply with temperature. Second, consideration of strain is necessary to provide an accurate measure of the In content. Simply applying Vegard's law to the strained (i.e., measured) lattice parameters results in an overestimation. The procedure of Schuster et al. [6] provides a value consistent with that from RBS, which is not sensitive to strain effects. And third, the degree of relaxation increases with In content. This result is consistent with trends in critical thicknesses [7]. Films with lower In contents are strained less, so they require greater thicknesses to build up sufficient strain energy to initiate relaxation mechanisms. Sample C is mostly pseudomorphic to the underlying GaN layer, whereas samples A and B are mostly relaxed.

Figure 1 is a dark-field plan-view image of sample A. Numerous domain boundaries can be observed. They are oriented along several different types of directions, including $\langle 10\bar{1}0\rangle$, $\langle 11\bar{2}0\rangle$, and others. A similar domain structure has been observed before in InGaAsP [8]. Each domain defines a region in which the composition is modulated along a particular direction.

The existence of these domains results in a diffraction effect, as seen in Fig. 2a. The spots are non-circular, with satellites oriented in the directions of the modulations. One spot with particularly clear satellites is enlarged in Fig. 2b. Multiple satellites are present because the diffraction pattern was taken from several domains. Also, the satellites are aligned along several different directions, indicating that no particular direction is favored for the modulations. This result is consistent with the isotropy of Young's modulus within the basal plane of hexagonal crystals [9].

Periodic Composition Modulations Periodic modulations in composition result in satellites around the fundamental reflections in a diffraction

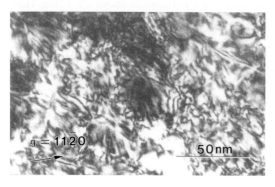

Fig. 1. Dark-field plan-view image of sample A

Fig. 2. a) [0001] zone-axis diffraction pattern of sample A and b) enlargement of the $(1\bar{1}20)$ spot

pattern. This phenomenon was first observed by Daniel and Lipson [10, 11] in Cu_4FeNi_3. McDevitt [12] related the spacing between the satellite and fundamental reflection to the wavelength of the modulations resulting from phase separation in the zinc-blende InGaAsP system. The wurtzite crystal structure has non-orthogonal unit cell axes, so the relations to determine the wavelength need to be adjusted accordingly. The starting point is Bragg's law, which can be written squared as

$$\frac{\lambda^2}{d^2} = \lambda^2 \left(\frac{4}{3} \frac{h^2 + hk + k^2}{a^2} + \frac{l^2}{c^2} \right) = 4 \sin^2 \theta. \qquad (1)$$

The direction of the modulations determines the orientation of the satellite. The mathematics is simplified if a specific direction is chosen at this point. For modulations along the $[10\bar{1}0]$ direction, a small distortion δh occurs, while k and l remain constant. By differentiating Eq. (1), this distortion can be stated as

$$\lambda^2 (2h + k) \, \delta h = 6a^2 \sin \theta \cos \theta \, \delta\theta. \qquad (2)$$

Daniel and Lipson [10, 11] defined a term Q as the breadth of the satellites, or equivalently the period of unit cells in the direction of the modulations. In the $[10\bar{1}0]$ direction,

$$Q = 1/\delta h. \qquad (3)$$

The period of the modulations, or modulation wavelength λ, is obtained by multiplying Q by the distance across the unit cell in the direction of the modulations. In the $[10\bar{1}0]$ direction,

$$\lambda = Qa = a/\delta h. \qquad (4)$$

A small change in the diffraction angle $\delta\theta$ can be related to a small displacement in the diffraction pattern δr by

$$\frac{\tan \theta}{\delta \theta} = \frac{r}{\delta r}. \qquad (5)$$

The final equation is obtained by taking the ratio of Eq. (1) to Eq. (2) for the case in which $l = 0$ (i.e., a [0001] zone-axis diffraction pattern) and suitably applying Eqs. (4) and (5),

$$\lambda = \frac{1}{2} \frac{2h + k}{h^2 + hk + k^2} \frac{r}{\delta r} a. \qquad (6)$$

This procedure can be repeated for modulations in other directions. In the $[11\bar{2}0]$ direction, distortions δh and δk occur, with $\delta h = \delta k$, while l remains constant. Equation (3) is modified to $1/Q = \delta h + \delta k$ and Eq. (4) to $\lambda = Q\,a = a/(\delta h + \delta k)$. The final equation for this direction is

$$\lambda = \frac{3}{4} \frac{h+k}{h^2 + hk + k^2} \frac{r}{\delta r} a. \qquad (7)$$

For samples A, B, and C, the modulation wavelengths in the [10$\bar{1}$0] direction were (3.2 ± 1.3), (3.1 ± 1.3), and (3.6 ± 1.3) nm, respectively. These values were averaged from measurements of six or seven different locations on the diffraction pattern. The driving force for phase separation increases with In content up to $x = 0.5$ [2], so samples with higher x are expected to have smaller λ. This trend did not appear in these measurements, presumably due to the significant deviations of the values.

The existence of phase separation in sample C is inconsistent with the phase diagram [2]. Two explanations can account for this. First, the phase diagram was calculated, so its correspondence to experiment is not perfect. And second, the phase separation observed here occurred during growth; i.e., it was a surface phenomenon. The phase diagram, however, represents bulk equilibrium.

Conclusions In summary, InGaN epitaxial layers were grown on GaN/sapphire composites. Transmission electron microscopy revealed that the microstructure consisted of domains. Diffraction patterns contained satellites around the fundamental reflections. These satellites are the result of periodic composition modulations within the domains. The equations were derived relating the spacing of the satellites to the wavelength of the modulations.

Acknowledgements The authors gratefully acknowledge the support of this work through DOD-MURI Grant No. F49620-95-I-0447, MRSEC Grant No. 001081900, and Sandia National Labs.

References

[1] S. Nakamura, M. Senoh, S. Nagahama, N. Iwasa, T. Matsushita, and T. Mukai, Appl. Phys. Lett. **76**, 22 (2000).
[2] I. Ho and G. B. Stringfellow, Appl. Phys. Lett. **69**, 2701 (1996).
[3] I. P. Soshnikov, V. V. Lundin, A. S. Usikov, I. P. Kalmykova, N. N. Ledentsov, A. Rosenauer, B. Neubauer, and D. Gerthsen, Semiconductors **34**, 621 (2000).
[4] N. A. El-Masry, E. L. Piner, S. X. Liu, and S. M. Bedair, Appl. Phys. Lett. **72**, 40 (1998).
[5] E. L. Piner, N. A. El-Masry, S. X. Liu, and S. M. Bedair, Mater. Res. Soc. Symp. Proc. **482**, 125 (1998).
[6] M. Schuster, P. O. Gervais, B. Jobst, W. Hösler, R. Averbeck, H. Riechert, A. Iberl, and R. Stömmer, J. Phys. D **32**, A56 (1999).
[7] R. People and J. C. Bean, Appl. Phys. Lett. **49**, 229 (1986).
[8] S. Mahajan, B. V. Dutt, H. Temkin, R. J. Cava, and W. A. Bonner, J. Cryst. Growth **68**, 589 (1984).
[9] J. F. Nye, Physical Properties of Crystals, Oxford University Press, Oxford 1985 (p. 144).
[10] V. Daniel and H. Lipson, Proc. R. Soc. London A **181**, 368 (1943).
[11] V. Daniel and H. Lipson, Proc. R. Soc. London A **182**, 378 (1944).
[12] T. L. McDevitt, Ph. D. Thesis, Carnegie Mellon University, Pittsburgh 1990.

Structural and Optical Characteristics of InGaN/GaN Multiple Quantum Wells with Different Growth Interruption

H. K. Cho (a), J. Y. Lee[1]) (a), N. Sharma (b), J. Humphreys (b), G. M. Yang (c), and C. S. Kim (c)

(a) Department of Materials Science and Engineering, Korea Advanced Institute of Science and Technology, 373-1 Gusong-dong, Yusong-gu, Daejon 305-701, Korea

(b) Department of Materials Science and Metallurgy, University of Cambridge, Pembroke Street, Cambridge, CB2 3QZ, UK

(c) Department of Semiconductor Science & Technology and Semiconductor Physics Research Center, Chonbuk National University, Duckjin-Dong, Chunju 561-756, Korea

(Received June 7, 2001; accepted August 4, 2001)

Subject classification: 61.10.Kw; 68.37.Lp; 68.65.Fg; 78.55.Cr; 78.67.De; 81.15.Gh; S7.14

InGaN/GaN multiple quantum wells (MQWs) grown with various growth interruptions between InGaN well and GaN barrier by metalorganic chemical vapor deposition were investigated using photoluminescence, high-resolution transmission electron microscopy (HRTEM), and energy filtered transmission electron microscopy (EFTEM). The luminescence intensity of the MQWs with growth interruptions is abruptly reduced compared to that of the MQW without growth interruption. Also, as the interruption time increases the peak emission shows a continuous blue shift. We found that the higher intensity and lower energy emission of the MQW grown without interruption is caused by the recombination of excitons localized from indium clustering regions. Evidence of indium clustering is directly observed by indium ratio map of MQWs and indium composition measurements along an InGaN well using EFTEM.

Introduction It has been reported that strong emission from InGaN/GaN MQWs was caused by the recombination of excitons localized in indium clustering regions [1]. Direct evidence for indium clustering within MQWs has been reported based on dark contrast features observed in transmission electron microscopy (TEM) [2] or composition measurements using scanning TEM (STEM)/energy-dispersive X-ray (EDX) [1] analysis. However, dark contrast images in an InGaN layer can also be observed in TEM due to ion milling induced damages during TEM sample preparation because of the weaker InN bond. Also, because of the very small size of indium clusters (a few nm), the limited resolution of the techniques that have been used, and the spread of the incident electron beam within the specimen, much of the X-ray signals in EDX/STEM are generated from the surrounding areas. Therefore, up to this moment it has been difficult to measure the exact indium composition and reveal the true indium distributions in nanometer scale clusters due to technical limitations. In this paper, we show direct quantitative evidence from EFTEM that the high intensity emission from the MQW without growth interruption is correlated with strong indium clustering effect in the InGaN wells.

---

[1]) Corresponding author; Phone: +82 42 869 4216; Fax: +82 42 869 4276; e-mail: jylee@mail.kaist.ac.kr

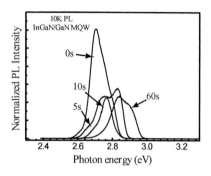

Fig. 1. 10 K PL spectra for InGaN/GaN MQWs with different growth interruption times

Experimental MQWs composed of six periods of InGaN/GaN were grown at 750 °C after growing a 2 µm thick Si-doped GaN layer at 1130 °C. The indium composition of the $In_xGa_{1-x}N$ well layer of the MQW grown without interruption is expected to be around 30–33%. The thicknesses of the wells and barriers are 15 Å and 85 Å, respectively. All MQW samples studied here were grown under the same growth conditions except for the growth interruption time between the InGaN well and GaN barrier which is varied between 0 and 60 s. HRTEM and EFTEM experiments were carried out in a JEOL 2000EX microscope operated at 200 kV and a Philips CM300 field emission gun TEM with a Gatan Imaging Filter, respectively.

Results Figure 1 shows the PL spectra measured at low temperature (10 K) on the InGaN/GaN MQWs with different growth interruption times. As the interruption time increases the peak emission shows a continuous blue shift. Besides the shift of the emission peak, the luminescence intensity of the samples with growth interruptions (5–60 s) is abruptly reduced compared to that of the sample grown without interruption. These data show that growth interruption during the growth of InGaN/GaN MQWs highly affects the emission energy and intensity. We reported that the V-defects originating from stacking faults result in different growth rates of the GaN barriers, change the period thickness of the superlattice, and finally induce the broadening and multiple emission peaks of PL [3]. All specimens used in this study have many V-defects ($>10^9$ cm^{-2}) originating from stacking faults due to the increased misfit strain and show the broad and multiple emission peaks as shown in Fig. 1.

Figure 2 shows cross-sectional HRTEM images of samples with interruption times of 0, 5, 10, and 60 s, respectively. As shown in Fig. 2a, for no growth interruption the $In_xGa_{1-x}N$ well layer exhibits a strong lateral variation in contrast, which may suggest the presence of indium clusters in the QWs. For the sample with interruption time of more than 5 s, the degree of strain contrast rapidly decreases while the well thickness remains unchanged up to an interruption time of 30 s. For even longer interruption times the thickness of the InGaN wells decreases down to about two monolayers as shown in Fig. 2d. From the HRTEM results, we deduce

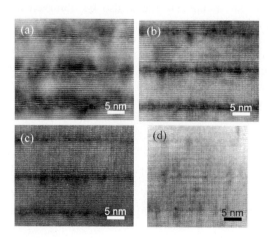

Fig. 2. Cross-sectional HRTEM images obtained from MQWs grown with growth interruption of a) 0 s, b) 5 s, c) 10 s, and d) 60 s

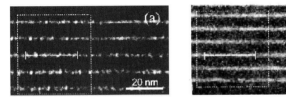

Fig. 3. EFTEM indium ratio maps of the MQWs with a) no interruption (five QWs) and b) 5 s interruption (six QWs)

that growth interruption results in strain relaxation and indium desorption due to thermal annealing.

EFTEM analysis was performed to highlight the chemical composition variations and measure the indium composition of MQWs (Fig. 3). Indium ratio maps using EFTEM are well suited for the quantification of the indium distribution due to the elimination of diffraction effects and thickness variation effects which are generated from the sample preparation process [4]. Figures 3a and b are indium ratio maps of the MQWs with no interruption (five QWs) and 5 s interruption (six QWs), respectively. For no growth interruption, the indium clustering can be clearly observed from the strong white contrast in Fig. 3a showing the indium-rich regions. For 5 s interruption, however, the $In_xGa_{1-x}N$ wells show an even indium distribution indicating an apparent lack of indium clustering.

The accurate indium compositions across (Figs. 4a and b) and along (Figs. 4c and d) the different InGaN/GaN MQW structures (i.e., 0, 5, and 30 s interruption time) are calculated based on the assumption that the nitrogen composition is the same in GaN and InGaN [5]. As shown in Figs. 4a and b (taken from the regions marked with the dashed rectangles in Fig. 3), the average indium composition in the InGaN well across the MQW remained unchanged for 0 and 5 s interruption time. For the interruption time of more than 30 s, however, the indium composition is reduced due to desorption caused by the relative weakness of the In–N bond compared to Ga–N [4]. To determine the indium distribution within an $In_xGa_{1-x}N$ well, line scans were taken along an

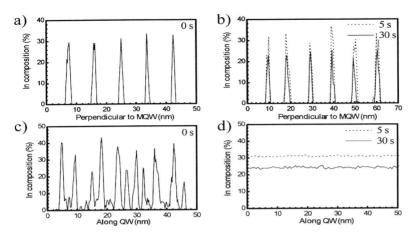

Fig. 4. The indium composition profiles a), b) across the InGaN/GaN MQWs (region marked with dashed rectangle in Fig. 3); and c), d) along an $In_xGa_{1-x}N$ well layer (horizontal line in Fig. 3) using EFTEM. Interruption time as indicated

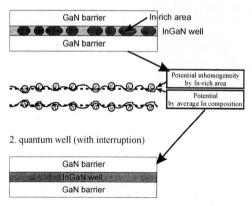

Fig. 5. Schematic model of the presence of the indium cluster and carrier localization observed in InGaN/GaN MQWs

$In_xGa_{1-x}N$ well. For the samples with growth interruption, the indium compositions are nearly constant along an $In_xGa_{1-x}N$ well (Fig. 4d). The average indium compositions in the InGaN well of samples with 5 and 30 s interruption time are determined at 31 and 24%, respectively. For no growth interruption, however, the indium composition fluctuations between 5 and 40% are clearly observed, which provides direct evidence of the strong indium clustering.

Several research groups have reported that for samples with indium composition more than 15–20% and InGaN well thickness less than 3 nm, indium clustering dominates the luminescence properties compared to the piezoelectric field effect, and the presence of indium clustering with high indium composition results in high luminescence efficiency [6]. The thickness and indium composition of the InGaN well of our samples with no interruption and 5 s interruption time is 15 Å and 31%, respectively. Therefore, we expect that although the average indium composition in the InGaN QWs remains unchanged, the presence of indium clusters in the wells results in dominant radiative recombination of excitons in those indium-rich regions as shown in Fig. 5, causing the emission intensity to decrease and the peak energy to increase when the growth between InGaN wells and GaN barriers is interrupted from 0 to 5 s.

Conclusions We have investigated the effect of growth interruptions on the optical and structural properties of InGaN/GaN MQWs. We provide direct evidence of indium clustering along an $In_xGa_{1-x}N$ well layer using EFTEM in the MQW structure grown without growth interruption. We found that for the MQW grown without interruption, indium clustering leads to higher emission intensity and lower peak energy compared to the MQWs grown with interruption.

Acknowledgement This work has been supported by the Ministry of Science and Technology of Korea through the NRL Program and the BK 21 Program in KOREA.

References

[1] Y. Narukawa, Y. Kawakami, M. Funato, Sz. Fujita, Sg. Fujita, and S. Nakamura, Appl. Phys. Lett. **70**, 981 (1997).
[2] Y. S. Lin, K. J. Ma, C. Hsu, S. W. Feng, Y. C. Cheng, C. C. Liao, C. C. Yang, C. C. Chou, C. M. Lee, and J. I. Chyi, Appl. Phys. Lett. **77**, 2988 (2000).
[3] H. K. Cho, J. Y. Lee, C. S. Kim, G. M. Yang, N. Sharma, and C. J. Humphreys, J. Cryst. Growth **231**, 28 (2001).
[4] H. Chen, R. M. Feenstra, J. E. Northrup, T. Zywietz, and J. Neugebauer, Phys. Rev. Lett. **85**, 1902 (2000).
[5] N. Sharma, P. Thomas, D. Tricker, and C. Humphreys, Appl. Phys. Lett. **77**, 1274 (2000).
[6] N. A. Shapiro, P. Perlin, C. Kisielowski, L. S. Mattos, J. W. Yang, and E. R. Weber, MRS Internet J. Nitride Semicond. Res. **5**, 1 (2000).

Characterisation of Optical Properties in Micro-Patterned InGaN Quantum Wells

K.-S. Kim[1]) (a), P.R. Edwards (b), H.-S. Kim (a), R.W. Martin (b), I.M. Watson (a), and M.D. Dawson (a)

(a) Institute of Photonics, University of Strathclyde, Glasgow, G4 0NW, UK

(b) Department of Physics, University of Strathclyde, Glasgow, G4 0NG, UK

(Received June 21, 2001; accepted August 4, 2001)

Subject classification: 78.55.Cr; 78.67.De; 85.60.Jb; S7.14

Optical resonance modes of micro-scale patterned InGaN quantum wells including disks and rings of various diameters (0.5–20 μm) have been investigated. For the observation of resonant modes, well-defined features and striation-free sidewalls were essential. Each of the patterned structures displayed resonant modes superimposed on the main InGaN photoluminescence peak. The large number of such modes observed has allowed analysis of the mode spacing as a function of both wavelength and diameter. Comparison with a simple model has confirmed that these are "whispering gallery" modes. No evidence of radial modes was observed. Microrings showed fewer optical resonance modes than microdisks. In addition, the number of observable modes correlated with the width of microring structures.

Introduction Semiconductor microdisk lasers are of interest for micro-scale, low-threshold laser devices. Their lasing modes approximate to whispering gallery (WG) modes, for which the high reflectivity at the curved disk boundary gives low-threshold operation [1, 2]. It has been expected that AlGaInN structure nitrides on sapphire can confine light strongly in such structures, because of the large discontinuity in refractive index at the boundary between the cavity and surrounding air [3]. In order to fabricate novel optical microstructures that strongly confine the optical fields, a detailed analysis of the optical modes in the structure is needed. In this work, we fabricated a range of different micro-scale disks and rings, and characterised the optical modes.

Experiment GaN/InGaN multi-quantum well (MQW) structures were grown on sapphire (0001) substrates by metal-organic chemical vapour deposition in an Aixtron 200-series reactor. The structures consist of a 1 μm thick undoped GaN buffer layer, a MQW region, consisting of ten periods of undoped 2 nm/7 nm thick $In_{1-x}Ga_xN$/GaN ($x \approx 0.1$), followed by a 15 nm thick GaN capping layer. [4] Micro-scale disks and rings with diameters covering range of 1–20 μm were fabricated using either photolithography or electron-beam lithography. The samples were etched into the substrates using a Cl_2-based inductively coupled plasma, with plasma conditions optimised to produce vertical side walls. The morphology was observed using scanning electron microscopy (SEM). Low-temperature (12 K) photoluminescence (PL) spectroscopy was carried out on individual disk or ring features using a low-intensity 325 nm HeCd laser focused through a UV microscope to a 20 μm spot size.

---

[1]) Corresponding author; Phone: +44 141 553 4120, Fax: +44 141 552 1575, e-mail: ki.sung.kim@strath.ac.uk

Fig. 1. SEM images of InGaN/GaN MQW micro-scale structures. a) Microring with an outer diameter of 6 μm and inner diameter of 2.5 μm fabricated by standard photolithography. b) Microdisk with a diameter of 2 μm made by e-beam lithography

Results and Discussions In order to investigate the effect of edge irregularities on optical mode behaviour, two lithography methods were employed. All other patterning procedures were identical.

Figure 1a shows a SEM image of the sample patterned by using photolithography, with an estimated resolution of 250 nm. Compared to the results in Fig. 1a, the circularity and sidewall morphology in micro-size features were improved when e-beam lithography was adopted, as shown in Fig. 1b. A further factor affecting the edge roughness was found to be the crystal quality of the nitride layers. SEM images of the micro-scale patterns fabricated with lower quality material have shown that the etching process has left striations on the sidewalls.

Figure 2[2]) shows PL spectra from microdisks fabricated by photolithography. Optical cavity modes around the main InGaN peak were not observed, indicating that irregularity of the sidewall disturbed the propagation of optical modes. Despite the absence of optical modes in the PL spectrum, some interesting features were noted. The full width at half maximum (FWHM) values of the main MQW peaks narrowed significantly compared with unpatterned samples. In addition, micro-scale patterned samples with sub-10 μm diameters show optical mode behaviour in the region of GaN emission peak. Phonon replica features in InGaN/GaN MQW spectra were observed in the case

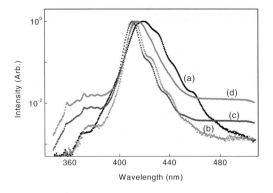

Fig. 2 (colour). Low-temperature PL spectrum for microdisks fabricated by photolithography. a) InGaN/GaN samples prior to microstructure fabrication. Microdisks with diameters of b) 20 μm, c) 8 μm, d) 2 μm

---

[2]) Colour figures are published online (www.physica-status-solidi.com), where indicated.

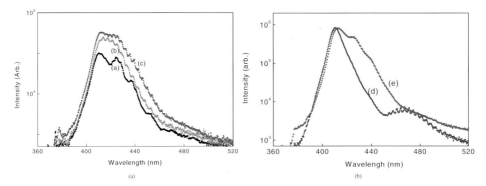

Fig. 3 (colour). Low temperature PL spectra obtained from microdisks with diameters of a) 2 µm, b) 6 µm, c) 8 µm, d) 10 µm, and e) 20 µm, all fabricated by e-beam lithography

of samples larger than 8 µm in diameter, but were not observable in spectra from smaller samples.

Figure 3 shows PL spectra from micro-disks with various diameters (2–20 µm) fabricated by e-beam lithography. In contrast to the results shown in Fig. 2, optical cavity modes were observed around the main MQW peak from all samples. In microdisks with larger diameters, higher mode densities within the emission spectra were observed. From the criterion for formation of WG modes in a circularly symmetric cavity, the mode spacing at a given wavelength is approximately inversely proportional to the radius [3].

PL spectra from disk and ring samples with a 6 µm outer diameter are shown in Fig. 4, together with calculated WG mode positions (vertical solid lines). Assuming a refractive index of 2.65, appropriate to a wavelength of 410 nm [5], the simulated mode positions results are in good agreement with those observed. Microring structures showed fewer resonant modes than the microdisk. In the case of the 1 µm width ring, no optical modes could be clearly observed (line (a) in Fig. 4). An increase in ring width led to more observable optical modes, as illustrated by line (b) in Fig. 4. The calculated WG mode positions in Fig. 4 ignore dispersion, and therefore the calculated mode positions are not well matched with the measured ones at longer wavelengths.

Conclusion Optical modes in a range of micro-scale structures (disks and rings) have been characterised. The WG modes were observed only for those structures with good

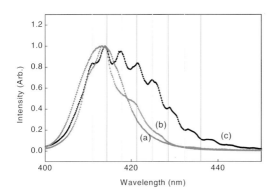

Fig. 4 (colour). PL spectra from a) a ring of outer diameter 6 µm and inner diameter 4 µm, b) a ring of outer diameter 6 µm and inner diameter 2.4 µm, and c) a disk of diameter 6 µm. The vertical solid lines represent the calculated WG modes

edge definition. Disks of a larger diameter display more WG modes and narrower inter-mode spacing. The adoption of microring type structure changed the optical modes compared to microdisks, with fewer and less pronounced modes observed.

References

[1] LORD RAYLEIGH, The Problem of the Whispering Gallery, in: Scientific Papers, Vol. 5, Cambridge University, Cambridge 1912 (pp. 617–620).
[2] S.L. McCALL, A.F.J. LEVI, R.E. SLUSHER, S.J. PEARTON, and R.A. LOGAN, Appl. Phys. Lett. **60**, 289 (1992).
[3] K.C. ZENG, L. DAI, J.Y. LIN, and H.X. JIANG, Appl. Phys. Lett. **75**, 2563 (1999).
[4] R. PECHARROMAN-GALLEGO, P.R. EDWARDS, R.W. MARTIN, and I.M. WATSON, Proc. E-MRS Spring 2001 Meeting, submitted.
[5] E. EJDER, phys. stat. sol. (a) **6**, 445 (1977).

Indium Distribution within $In_xGa_{1-x}N$ Epitaxial Layers: A Combined Resonant Raman Scattering and Rutherford Backscattering Study

R. Correia[1]) (a), S. Pereira (a, b) E. Pereira (a), E. Alves (c), J. Gleize (d), J. Frandon (d), and M. A. Renucci (d)

(a) Departamento de Física, Universidade de Aveiro, Aveiro, Portugal

(b) Department of Physics, University of Strathclyde, Glasgow, UK

(c) Departamento de Física, Instituto Tecnológico e Nuclear, Sacavém, Portugal

(d) Laboratoire de Physique des Solides, CNRS, UMR 5477, Université Paul Sabatier, F-31062 Toulouse Cédex 4, France

(Received July 27, 2001; accepted August 1, 2001)

Subject classification: 63.20.–e; 63.22.+m; 78.30.Fs; S7.14

We report a Raman spectroscopy study performed at room temperature on $In_xGa_{1-x}N$ epitaxial layers grown by metalorganic chemical vapour deposition on top of GaN/sapphire (0001) substrates. Resonant Raman measurements have been performed using excitation energies of 2.34, 2.54 and 3.02 eV. A shift of the $A_1(LO)$ phonon mode frequency was observed under different excitation energies. This is interpreted with respect to the composition and strain variations within the sample. The $A_1(LO)$ phonon line shape is analysed using a Spatial Correlation Model (SC). The structural parameters were determined by Rutherford backscattering spectrometry (RBS).

Introduction The $In_xGa_{1-x}N$ material system has played a considerable role in the development of light emitting devices operating in the visible spectral region [1].

Raman spectroscopy has proven to be an informative tool in the investigation of semiconductors with the wurtzite (WZ) crystal structure. For the case of the $In_xGa_{1-x}N$ alloy, it has been theoretically [2] and experimentally [3–6] demonstrated that it shows one-mode behavior, i.e. exhibits only one set of longitudinal optical (LO) and transverse optical (TO) phonons, whose frequencies vary almost linearly with compositional changes. Due to the lack of good quality samples for a wide range of composition, the $A_1(LO)$ phonon frequency dependence on the indium content x is not yet well-established and different behaviors can be found in the literature [4–6]. In some cases the observed asymmetrical band shape leads to an uncertainty on the phonon frequency at the Brillouin center zone. Furthermore, the state of strain of the samples is not always properly taken into account.

In this work, we study an InGaN sample by RBS and resonant Raman scattering measurements. The low energy tail associated with the $A_1(LO)$ phonon line shape is analysed using the Spatial Correlation Model (SC) [7]. The Raman scattering enhancement is discussed assuming that resonance is achieved when the excitation energy is nearly the band gap E_g.

---

[1]) Corresponding author; Phone: +351 234 370943; Fax: +351 234 424965; e-mail: rcorreia@fis.ua.pt

Experimental Details The nominally undoped $In_xGa_{1-x}N$ heterostructure studied was grown by low-pressure metalorganic chemical vapour deposition (MOCVD) on a GaN/(0001) sapphire substrate. Raman scattering measurements were carried out at room temperature (RT) using a "Dilor" micro and macro-Raman spectrometer working in backscattering geometry, with the incident and scattered light propagating parallel to the c-axis. Wavelengths of 413, 488 and 530 nm were used as excitation lines. With the present micro-Raman set up, the diameter of the spot on the sample surface is about 1 µm. The RBS measurements were performed with a 1 mm collimated beam of 2 MeV $^4He^+$. The detection angle of the backscattering particles, with respect to the beam direction, was 160°, and the energy resolution was about 13 keV.

Results and Discussion Figure 1 shows the random and aligned RBS spectra of the sample in study. In a random RBS spectrum from a target with a constant chemical composition, at a constant bombarding energy, the backscattering yield from deeper sample regions is higher ($\propto 1/E^2$) due to the energy loss of the penetrating beam. This effect can be observed in the energy window corresponding to Ga in the random spectrum of the GaN buffer layer. In the InGaN film layer it can be noticed that the In related signal decreases as the beam penetrates deeper in the layer until about channel 675, where the In and Ga signals start to overlap. This is a qualitative indication that x is decreasing over depth. In fact a detailed quantitative analysis of the RBS spectrum using RUMP [8] reveals that the experimental spectrum can only be reproduced if a composition decrease over depth is considered. A model consisting of a 285 nm thick InGaN layer, with a variation from about $x = 0.33$ near the surface to about $x = 0.05$ close to the InGaN/GaN interface, followed by a thick GaN buffer provides the best fit to experimental data, as shown in Fig. 1. Considering the thickness and the indium mole fraction of each model sub-layer, an average composition of $x = 0.23$ is estimated.

Figure 2[2]) presents the Raman spectra of the InGaN sample. In the backscattering configuration, $z(xx)\underline{z}$, Raman scattering by $A_1(LO)$ and the E_2 phonon modes are symmetry allowed via deformation potential mechanism. Using the same geometry for the polar crystals, under resonant excitation a drastic enhancement of light scattering by LO phonons due to strong electron–phonon Fröhlich interactions is also observed. The presence of E_2^{GaN} in Fig. 2 is a good indication that all the InGaN film is being probed by each of the excitation energies. A significant enhancement of the relative intensity of the $A_1^{InGaN}(LO)$ compared to the E_2^{GaN} is observed, indicating that in InGaN the scat-

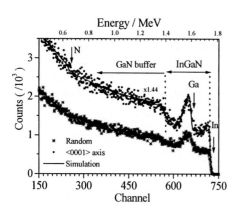

Fig. 1. Random, aligned and simulated RBS spectra from an InGaN/GaN/sapphire (0001) sample

---

[2]) Colour figures are published online (www.physica-status-solidi.com), where indicated.

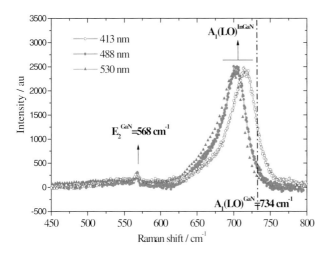

Fig. 2 (colour). RT Raman spectra measured at different excitation energies

tering by Fröhlich mechanisms was promoted. The shift of the $A_1(LO)$ phonon mode to lower frequencies, when the excitation energy decreases, is evident. This shift can be explained by composition and/or strain variations within the sample. An asymmetrical $A_1(LO)$ band shape with a nearly exponential band tail to lower frequencies is observed. This asymmetry can be analysed using a spatial correlation model (SC) [7] based on finite phonon correlation length, related to **q**-vector relaxation, induced by the microscopic nature of the alloy disorder. In this model the Raman intensity $I(\omega)$ at a frequency ω can be written as

$$I(\omega) \propto \int_0^{0.5} \exp\left(\frac{-q^2 l^2}{4}\right) \frac{d^3 q}{[\omega - \omega(q)]^2 + \left(\frac{\Gamma_0}{2}\right)^2} , \qquad (1)$$

with q expressed in units of $2\pi/c$, where c is the measured lattice constant by XRD, l the spatial correlation length, Γ_0 the width of the intrinsic Raman line and $\omega(q)$ is the phonon dispersion curve. The $A_1(LO)$ phonon mode propagates along the c-axis, thus it is reasonable to assume a relationship based on the one-dimensional linear-chain model [9] given by

$$\omega^2(\mathbf{q}) = C + [C^2 - D[1 - \cos(\pi q)]]^{0.5} , \qquad (2)$$

where C and D are related to the frequencies of the $A_1(LO)$ phonon mode, ω^Γ and ω^A at $|\mathbf{q}| = 0$ and $|\mathbf{q}| = \pi/c$, respectively. To solve Eq. (1), a spherical Brillouin zone and an isotropic phonon dispersion curve are assumed. The quantities l, Γ_0, D and C are treated as variable parameters to fit the experimental data. The fit result obtained for 3.02 eV is shown in Fig. 3. In spite of the large number of variable parameters, the values obtained for ω^Γ and ω^A are in good agreement with those expected for an one-mode behavior alloy, considering the values for the binary compounds GaN [10] ($\omega^\Gamma = 734$ cm^{-1}, $\omega^A = 690$ cm^{-1}) and InN [11] ($\omega^\Gamma = 586$ cm^{-1}; $\omega^A = 584.7$ cm^{-1}), respectively. The small value obtained for the spatial correlation length, $l = 11.3$ Å, suggests a lack of long-range order in this alloy system. This is in good agreement with the large value for $\chi_{min} = 78\%$ obtained from RBS measurements along (0001) direction. χ_{min} is

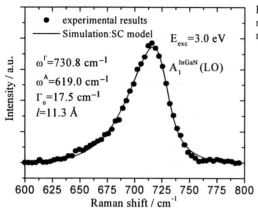

Fig. 3. Experimental data of the $A_1(LO)$ phonon at different excitation energies and the SC model fit

defined as the ratio between aligned and random backscattering yields. A large value of χ_{min} means a poor crystal quality along the considered direction. Furthermore, the value of the phonon frequency ω^Γ is used to determine the composition and strain of InGaN regions accounting to the resonance effect. Considering that this is achieved when the excitation energy coincides with E_g, and that E_g, varies simultaneously with strain and composition, both parameters should be taken into account. Moreover, the $A_1(LO)$ phonon frequency is also sensitive to these quantities. This can be described by the linear system of equations

$$\begin{cases} E_g = E_g(x)^{\text{relax}} + 15.4\varepsilon_{zz}, \\ \omega^\Gamma = \omega(x)^{\text{relax}} + [(-2 \times 685\varepsilon_{xx}) + (-997\varepsilon_{zz})], \\ \varepsilon_{zz} = \dfrac{-2C_{13}(x)}{C_{33}(x)}\varepsilon_{xx}, \end{cases} \quad (3)$$

where $E_g(x)^{\text{relax}}$ is the band gap dependence on x for relaxed material determined in previous work [12], ω^Γ the phonon frequency obtained by the SC model, $C_{i,j}(x)$ are elastic constants linearly interpolated from the binary values [13] and ε_{xx} and ε_{zz} the parallel and perpendicular strain components, respectively. In the absence of better knowledge, calculations were performed assuming that InGaN follows the strain correction on band gap [14] and phonon frequency [15] reported for GaN. Several reported phonon frequency dependencies on the composition $\omega^{\text{relax}}(x)$ were considered. The third equation in (3) is the Poisson ratio connecting perpendicular and parallel strain components. The system (3) was solved for $E_g = 3.02$ and 2.54 eV. Using either a Vegards' like $\omega^{\text{relax}}(x)$ dependence or the those experimentally determined in Ref. [4, 5], the results for $(x, \varepsilon_{xx}, \varepsilon_{zz})$ give, for a larger indium mole fraction, a higher value for compressive strain. This would be expected while the lattice mismatch is accommodated by the elastic strain. A totally different conclusion would be taken if the expression for $\omega^{\text{relax}}(x)$ established by D. Behr et al. [6] was used. In this case, larger values of x would correspond to more relaxed regions. If this were directly connected to the RBS layer model, the relaxed InGaN material would lie near the surface of the sample, contrary to the conclusion drawn from the $\omega(x)$ variations given in Refs. [4, 5]. However, great care should be taken when comparing the results given by a local technique such as resonant Raman scattering with a technique witch laterally averages the indium content over 1 mm.

Conclusions The results indicate that the SC model can account for the asymmetric broadening of the A_1(LO) phonon Raman line shape. Based on this interpretation, further work, using Raman spectroscopy, should be done to obtain valuable insights into the microscopic nature of InGaN alloy disorder. From this study it is also clear that, due to the poorly established A_1(LO) phonon frequency dependence on x, the interpretation of the resonant Raman results is far from a definitively conclusion.

References

[1] B. GIL (Ed.), Group III-Nitride Semiconductor Compounds, Physics and Applications. Series on Semiconductor Science and Technology 6, Oxford Sci. Publ., 1998.
[2] H. GRILLE, CH. SCHNITTLER, and F. BECHSTEDT, Phys. Rev. B **61**, 6091 (2000).
[3] H. HARIMA, E. KURIMOTO, Y. SONE, S. NAKASHIMA, S. CHU, A. ISHIDA, and H. FUJIYASU, phys. stat. sol. (b) **216**, 785 (1999).
[4] D. ALEXSON, L. BERGMAN, R. J. NEMANICH, M. DUTTA, M. A. STROSCIO, C. A. PARKER, S. M. BEDAIR, N. A. EL-MASRY, and F. ADAR, Appl. Phys. Lett. **89**, 798 (2001).
[5] R. CORREIA, S. PEREIRA, T. MONTEIRO, E. PEREIRA, and E. ALVES, Mater. Res. Soc. Symp. Proc. **639**, G6. 10 (2001).
[6] D. BEHR, R. NIEBUHR, H. OBLOH, J. WAGNER, K. H. BACHEM, and U. KAUFMANN, Mater. Res. Soc. Symp. Proc. **468**, 213 (1997).
[7] P. PARAYANTHAL and F. H. POLLAK, Phys. Rev. Lett. **52**, 1822 (1984).
[8] L. R. DOOLITTLE, Nucl. Instrum. Methods B **9**, 344 (1985).
[9] O. MADELUNG, Springer Ser. Solid State Sci. **2**, 135 (1978).
[10] H. SIEGLE, G. KACZMARCZYK, L. FILIPPIDIS, A. P. LITVINCHUK, A. HOFFMANN, and C. THOMSEN, Phys. Rev. B **55**, 7000 (1997).
[11] V. Y. DAVIDOV, V. V. EMTSEV, I. N. GONCHARUK, A. N. SMIRNOV, V. D. PETRIKOV, V. V. MAMUTIN, V. A. VEKSHIN, S. V. IVANOV, M. B. SMIRNOV, and T. INUSHIMA, J. Appl. Phys. **75**, 3297(1999).
[12] S. PEREIRA, R. CORREIA, T. MONTEIRO, E. PEREIRA, E. ALVES, A. D. SEQUEIRA, and N. FRANCO, Appl. Phys. Lett. **78**, 2137 (2001).
[13] A. F. WRIGHT, J. Appl. Phys. **82**, 2833 (1997).
[14] A. SHIKANAI, T. AZUHATA, T. SOTA, S. CHICHIBU, A. KURAMATA, K. HORINO, and S. NAKAMURA, J. Appl. Phys. **81**, 417 (1997).
[15] F. DEMANGEOT, J. FRANDON, M. R. RENUCCI, H. S. SANDS, D. N. BATCHELDER, O. BRIOT, and S. RUFFENACH-CLUR, Solid State Commun. **109**, 519 (1999).

Cathodoluminescence Investigations of Interfaces in InGaN/GaN/Sapphire Structures

M. Godlewski (a), E.M. Goldys[1]) (a), K.S.A. Butcher (a), M.R. Phillips (b), K. Pakula (c), and J.M. Baranowski (c)

(a) Division of Information and Communication Sciences, Macquarie University, Sydney 2109 NSW, Australia

(b) Microstructural Analysis Unit, University of Technology, Sydney 2009 NSW, Australia

(c) Warsaw University, Warsaw, Hoza 69, Poland

(Received June 19, 2001; accepted August 4, 2001)

Subject classification: 61.72.Ff; 73.20.Hb; 78.60.Hk; S7.14

Scanning electron microscopy and cathodoluminescence (CL) in spot and depth-profiling modes were used to evaluate the in-plane and in-depth uniformity of light emission from InGaN/GaN quantum well (QW) structures. The structures were grown by MOCVD on sapphire with a low-temperature (LT) GaN buffer. Depth-profiling CL investigations were used to identify the observed CL emissions, which show a complicated in-depth evolution. The influence of a LT GaN buffer on the structural and optical properties of the GaN/sapphire interface is discussed.

Introduction Most of the presently produced GaN-based opto-electronic structures are grown on strongly mismatched sapphire. Noticeable improvements of electrical and optical properties of GaN epilayers grown with the use of LT AlN or GaN buffer layers at GaN/sapphire interface have been achieved. These thin layers (50–100 nm thick in the case of AlN and less than 30 nm for GaN), with a very high dislocation density, reduce detrimental effects related to lattice mismatch, relax strain at interface and also supply nucleation centres for the growth of improved quality GaN epilayers. Not surprisingly, the problem of optimised AlN, AlGaN or GaN buffer layers is one of the most studied in GaN technology. In this work we study the properties of the GaN/sapphire interface in two InGaN QW structures.

Experiment The two InGaN/GaN quantum well structures were grown by metal-organic chemical vapour deposition (MOCVD) on (0001) sapphire with a LT GaN buffer layer. The active part of the structures, grown on top of a 3 μm thick GaN layer, consisted of an InGaN quantum well (QW) doped with silicon to the level of 10^{18} cm^{-3} and covered with a 20 nm wide GaN cap layer. The two samples studied had different In fractions in the QW region and also the width of the QW varied, they were identical otherwise. In the first sample, labelled as #1, the In fraction was about 6%, as determined from the growth conditions, and the InGaN QW was about 5 nm wide. Sample #2 was grown with a 1 nm wide InGaN QW and with 3% In. Cathodoluminescence (CL) spectra and scanning electron microscopy (SEM) images were taken in a JEOL35C scanning electron microscope with a MonoCL2 CL system by Oxford Instruments and de-

---

[1]) Corresponding author; Phone: +61 2 9850 8902; Fax: +61 2 9850 8115; e-mail goldys@ics.mq.edu.au

tected using a Hamamatsu R943-02 Peltier cooled photomultiplier. The spectra were not corrected for the system response. Charging effects were carefully minimised.

Results We concentrate on the emission coming from the sapphire/GaN interface region of the structure probed at accelerating voltages in the range of 30–35 kV. In addition to a sharp and dominant "edge" CL, we observed the emergence of two defect-related [1] GaN emissions (see Fig. 1), namely the yellow luminescence (YL), which is observed at low accelerating voltages, and the blue luminescence (BL). The latter apparently comes from the interface region of the film. The fact that yellow emission comes from the upper layer of the structure is likely to be due to strain conditions in this region of the structure. The GaN and InGaN layers are lattice mismatched and thus the QW region is characterised by a lower structural quality. Following this idea, a strong YL should also be excited from the interface region of the structure, as we reported earlier [2]. Instead, in the examined structures we observed another defect-related emission coming from the interface region. In addition to a dominant "edge" band, the blue band is observed under preferential excitation of the GaN/sapphire region of the films. The origin of various CL emissions can be identified by studying the CL in the spot mode, and by comparing the CL spectra excited from the region of a flat surface of the films and at micro-holes. Three bright CL emissions are observed in large discontinuities present in the film. These emissions originate from the underlying GaN buffer, directly covering the sapphire substrate (Fig. 2). The first one is a very strong emission line with the maximum at about 3.7 eV, i.e., above the GaN band-gap energy. This line we relate to the band edge ("edge") CL emission in AlGaN due to intermixing of Al from the sapphire within the GaN buffer. The AlGaN has about 10% Al fraction, as we estimate from the known band gap of AlGaN alloys [3]. The second is an excitonic "edge" emission of GaN, but red-shifted and much broader than that observed in a strain relaxed film. The third is a relatively broad and strong BL band, with the maximum at about 2.95–3.0 eV.

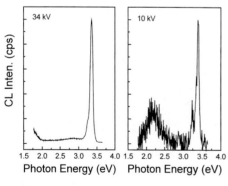
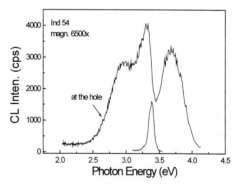

Fig. 1

Fig. 2

Fig. 1. Comparison of the CL spectra taken at 10 kV and 34 kV accelerating voltages (sample #1). The latter spectrum comes from the interface region of the structure

Fig. 2. Spot-mode CL spectra at 20 kV detected from the region of a large hole. For comparison we also show the "edge" GaN CL (lower curve) observed from the film excited outside of the large hole, also at 20 kV

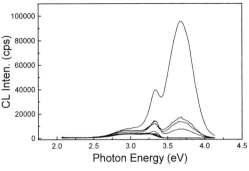 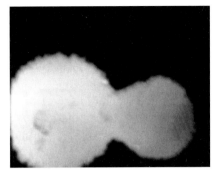

Fig. 3.   Fig. 4.

Fig. 3. In-plane homogeneity of CL in spot-mode experiments measured in the region of a large discontinuity, measured at 20 kV

Fig. 4. Scanning CL image of in-plane dependence of the AlGaN-related CL emission taken at the same region as Fig. 3, at 360× magnification (image size 250 × 300 μm^2)

We further studied the in-plane properties of the CL bands (shown in in Fig. 2) by exciting the CL from different regions of a large discontinuity (see Fig. 3), and by performing scanning CL investigations (Fig. 4). At the detection wavelength set to the band assigned to AlGaN emission we observe a very bright emission, which definitely comes from the discontinuity. The same is true in the case of BL, which also comes from the interface region of the structure. As the donor–acceptor pair recombination origin of this band is well known [1], the latter result indicates donor and acceptor accumulation at the interface. The spot-mode CL data shown in Fig. 3, and also the CL spectrum shown in Fig. 2, indicate that the interface region is strained and contains a large density of defects. The "edge" GaN emission is strongly broadened and red-shifted compared to high quality thick GaN films (see comparison in Fig. 2). Moreover, the CL measured from different regions of the discontinuity fluctuates strongly in intensity. Two of the emissions (AlGaN-related and BL) are highly inhomogeneous and originate from the hole region only.

Finally, using the advantages available in the depth-profiling CL studies we verified the in-plane homogeneity of the "edge" CL in GaN excited at different depths in the structure. In this experiment we took the scanning CL images by setting the detection at the "edge" GaN CL excited at different accelerating voltages, within a continuous region of a thick GaN layer. A comparison of the scanning CL spectra taken at 10 kV with those at 30–35 keV enables us to follow the variations of film quality at various depths of the structure and the corresponding variations in optical properties. At 10 keV the electron beam excites the region of the InGaN quantum well, while at 30–35 keV the interface region is probed as well. We selected for this study a region of the sample with some small micro-defects. The depth-profiling CL directly visualises a direct link between the in-plane homogeneity of the "edge" GaN CL and structural quality of the sample. In the region where the granular growth mode dominates, that is for the emission excited from the interface region of the film, strong spatial variations of the "edge" emission are observed. With increasing distance from the interface (i.e. at lower voltages) the effect becomes less and less pronounced and a relatively homogeneous "edge" CL is observed from the upper region of the GaN film.

Discussion Two mechanisms were proposed earlier to explain the mechanism of efficient strain relaxation at the GaN/sapphire interface by thin LT-GaN buffer. The first assumed that a mixed cubic–hexagonal structure of the buffer may play an important role in stress relaxation at the interface [4]. This model was based on the observation of cubic GaN islands, surrounded by hexagonal GaN in the region of the LT-GaN buffer [4]. The second model assumed an enhanced stress relaxation caused by Al inter-diffusion from the sapphire to the LT-GaN buffer, which results in formation of AlN and/or AlGaN micro-crystals in the region close to the GaN/sapphire interface [5]. The present study clearly supports the second mechanism of stress relaxation at the LT-GaN buffer. The CL emission from the interface region clearly contains GaN- and AlGaN-related bands, with the latter observed at energies above the GaN band gap energy. Since the interface region of the sample is strained, we cannot precisely estimate the Al fraction in AlGaN micro-crystallites. The rough estimate, from the spectral position of the relevant CL band, suggests about 10% of Al.

The scanning depth-profiling CL investigations show variations of the sample uniformity with increasing distance from the interface. A granular microstructure of the LT-GaN buffer results in strong fluctuations of the CL intensity at the GaN "edge" CL. We already reported such an interlink between the sample morphology and the CL intensity [2]. With increasing distance from the interface the CL intensity fluctuations become less pronounced and finally a homogeneous emission is observed, which reflects a smooth surface of our GaN/InGaN structures.

We also report the observation of a very pronounced acceptor-related BL from the interface region of the structure, which indicates an enhanced doping at the interface region. Oxygen is the most likely donor species in this region of the sample [5]. The origin of acceptor centres is less clear. Carbon is certainly one of the potential candidates for acceptor contamination at the interface region. Intrinsic acceptor-like centres (vacancies) should also be abundant in the highly-defected interface region of the structures. The presence of a strong acceptor-related CL from the buffer region of the samples, may help explain why the yellow emission of GaN is not observed at the interface. This emission is not observed, or is weak in the case of p-type GaN [6]. The YL is however coming from the upper region of GaN film, close to the InGaN QW, formed at the lower growth temperature necessary for InGaN deposition, and is also strained due to the GaN-to-InGaN lattice mismatch. In consequence, the GaN layer close to the QW region is of lower structural quality, resulting in the YL, which is known to anti-correlate with sample quality [1].

References

[1] T.L. TANSLEY, E.M. GOLDYS, M. GODLEWSKI, B. ZHOU, and H.Y. ZUO, in: Optoelectronic Properties of Semiconductors and Superlattices, Series Ed. M.O. MANASREH, Vol. 2: GaN and Related Materials, Ed. S.J. PEARTON, Gordon and Breach Publ., New York/London 1997 (pp. 233–293), and references therein.
[2] M. GODLEWSKI, E.M. GOLDYS, M.R. PHILLIPS, R. LANGER, and A. BARSKI, Appl. Phys. Lett. **73**, 3686 (1998).
[3] T. MATSUOKA, see [1] (pp. 53–83).
[4] T. ONITSUKA, T. MARUYAMA, K. AKIMOTO, and Y. BANDO, J. Cryst. Growth **189/190**, 295 (1998).
[5] SHU-YOU LI and JING ZHU, J. Cryst. Growth **203**, 473 (1999).
[6] S.K. ESTREICHER and D.E. BOUCHER, see [1] (pp. 171–199).

Multi-Emission from InGaN/GaN Multi-Quantum Wells Grown on Hexagonal GaN Microstructures

Chi Sun Kim (a), Young Kue Hong (a), Chang-Hee Hong[1] (a), Eun-Kyung Suh (a), Hyung Jae Lee (a), Min Hong Kim (b), Hyung Koun Cho (c), and Jeong Yong Lee (c)

(a) Department of Semiconductor Science and Technology and Semiconductor Physics Research Center, Chonbuk National University, Chonju 561-756, Korea

(b) Department of OE Team, Device and Materials Laboratory, LG Electronics Institute of Technology, Seoul 137-724, Korea

(c) Department of Materials Science and Engineering, KAIST, Taejon 305-701, Korea

(Received June 22, 2001; accepted July 17, 2001)

Subject classification: 68.37.Hk; 68.37.Lp; 68.65Fg; 78.55.Cr; 78.67.De; 81.15.Gh; S7.14

GaN hexagonal microstructures with InGaN/GaN multi-quantum wells were fabricated by using the selective metal organic chemical vapor deposition technique on dot-patterned GaN/sapphire(0001) substrates. The shape of these GaN microstructures was strongly related to the formation of self-limited (0001) facet with various growth conditions, which affect the Ga diffusion and surface reaction on the GaN surface. InGaN/GaN multi-quantum well structures on hexagonal GaN microstructures were selectively grown and characterized by scanning electron microscopy, photoluminescence, cathodoluminescence and transmission electron microscopy measurements. Blue emission was obtained from six ($1\bar{1}01$) facets as well as the flat (0001) facet, while yellow luminescence was observed at lateral overgrown region through the six ($1\bar{1}01$) side-walls.

Introduction The selective metalorganic chemical vapor deposition(MOCVD) growth technique of GaN-based materials can be a useful method for the development of semiconductor microstructure devices and optoelectronic devices [1–3]. Recently, nanostructures, such as InGaN quantum wires, dots and wells, grown on three-dimensional microstructures have been achieved by using selective MOCVD [4–6]. These active structures of LED or LD have been widely studied for increasing photon confinement and high efficiency on patterned GaN/sapphire substrates. Multi-emissions from various facet structures of the GaN or InGaN/GaN hexagonal structures may be used to apply to much higher performance optical devices and new functional devices. In order to achieve these devices selectively grown nanostructures, understanding the growth mechanism and the characteristics of the grown layers will be an important issue.

In this work, the growth of GaN hexagonal microstructures has been systematically investigated depending on the growth pressure, growth temperature and fill factor grown on 5.5 μm size dot-patterned GaN(0001)/sapphire with 100 nm thick SiO_2 films. Also, blue InGaN/GaN three quantum wells (3QWs) were selectively grown on hexagonal GaN microstructures and characterized by scanning electron microscope (SEM), photoluminescence (PL), cathodoluminescence (CL) and transmission electron microscopy (TEM) measurements.

---

[1]) Corresponding author; Phone:+82-63-270-2831; Fax:+82-63-270-3585; e-mail: chhong@moak.chonbuk.ac.kr

Experiment The GaN microstructures were grown on 2 μm thick GaN(0001) films by selective MOCVD. The 250 Å thick GaN nucleation layer was grown on sapphire c-plane substrates. The 100 nm thick SiO_2 mask was deposited by plasma enhanced chemical vapor deposition. Patterning of the mask was carried out by conventional photolithography and BOE etching to form dot-patterned windows with a circle of 5.5 μm in diameter. The selective MOCVD growth was carried out by varying the growth pressure from 75 to 750 Torr with a constant growth temperature of 950 °C. The growth temperature was varied from 950 to 1150 °C with a constant growth pressure of 75 Torr. The fill factor was ranged from 0.04 to 0.25. The flow rates of trimethylgallium and ammonia were fixed at 112 μmol/min and 3.5 slm as the precursors of Ga and N, respectively. In the second step, the InGaN/GaN 3QW structure was grown on the GaN microstructure with partially flat (0001) facets. The growth temperature for InGaN/GaN 3QWs was 750 °C. Trimethylindium and N_2 gas was used as In source material and main carrier gas, respectively. SEM, PL, CL and TEM measurements were employed for the characterization of the grown layers.

Results and Discussion Figures 1a to e show top-view SEM images of GaN microstructures grown on dot-patterned GaN/sapphire substrates depending on the growth pressure at 950 °C and the growth temperature at 75 Torr for the growth time of 20 min. The fill factor of all samples is fixed at 0.16. At a relatively high growth pressure in Figs. 1a and b, the formation of island-like pyramidal structures with no flat (0001) facet were observed on the (0001) surface. However, at a lower growth pressure of 75 Torr, partially flat (0001) facets and tiny reverse pyramidal pits, which have six $(1\bar{1}01)$ facets, were observed as shown in Fig. 1c. It was also found that the width of the top and bottom plane was increased while the height was decreased at relatively lower growth pressures. These phenomena can be explained by the Ga diffusion length, which becomes longer with decreasing the growth pressure. More details are shown in Ref. [7]. When the two-dimensional growth became dominant, the top flat (0001) facet was widened. In Figs. 1c to e, as with increasing the growth temperature, the part of flat (0001) facet became wider on the top surface and finally the GaN hexagonal pyramid with the completely flat (0001) surface was obtained [8, 9]. It is because the growth mode on the top (0001) surface of the GaN is quickly changed two-dimensionally at the initial growing step.

InGaN/GaN 3QWs on the hexagonal GaN microstructure as shown in Fig. 1c and on the typical planar structure as a reference were grown at 950 °C and 75 Torr. Figure 2 shows PL intensity and peak position of the blue InGaN/GaN 3 QWs grown on the

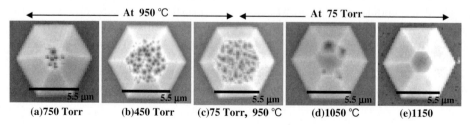

Fig. 1. SEM images of GaN microstructures grown by varying the growth pressure (parts a to c) at 950 °C and by varying the growth temperature (parts c to e) at 75 Torr

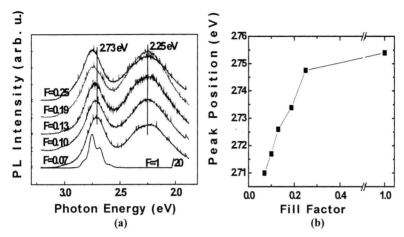

Fig. 2. a) PL intensity and b) peak position of the blue InGaN/GaN 3QWs grown on the GaN microstructure as function of fill factor

GaN microstructure as a function of fill factor. The PL spectra exhibit two strong emission peaks, one with blue emission centered around 2.73 eV and the other with yellow emission centered around 2.26 eV. The intensity of yellow luminescence (YL) was over 80% comparing to that of blue emission. Thus, it looks white-emitted pixel owing to multi-emission. The blue emission peak was blue-shifted from 2.71 to 2.754 eV as the fill factor increased, whereas yellow emission peak was not affected. It seems that In incorporation or growth rate of the quantum wells was slightly increased because of enhanced incoming In and Ga adatoms from wider mask region with decreasing the fill factor. However, YL is related with more intrinsic characteristics of the selective MOCVD than the geometry of the mask. Figures 3a and b show SEM images of the microstructures before and after the growth of the InGaN 3QWs, respectively, and CL images taken at 368 (c), 438 (d) and 564 nm (e), respectively, after the growth. The scale bars in Fig. 3 indicate the width of GaN window. Blue emission by CL was obtained from the full-sized microstructure while the yellow luminescence was strongly observed at lateral epitaxial overgrown (LEO) region through the six $(1\bar{1}01)$ sidewalls. Comparing Figs. 3a, b and e, this YL was not emitted from entire $(1\bar{1}01)$ facets. In this lateral overgrown region, band-to-band emission was also blue-shifted as shown in Fig. 3c (dark region). It seems that this phenomenon comes from the compressive-strained LEO region where columnar GaN was converted into single-crystallized GaN $[10\bar{1}1]$. However, more detailed studies should be needed. Figures 4a and b are high-resolution TEM

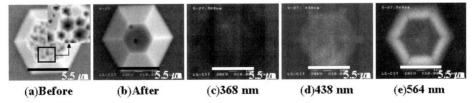

Fig. 3. SEM images of the microstructures a) before and b) after the growth of InGaN 3QWs and the CL images taken at c) 368, d) 438 and e) 564 nm after the growth

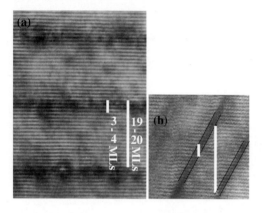

Fig. 4. High-resolution TEM images of InGaN 3QWs grown on the a) reference sample and b) structure as shown in Fig. 3b

images of InGaN 3 QWs grown on the reference sample and on the hexagonal structure as shown in Fig. 3b. The thickness of InGaN quantum wells of both two samples is nearly the same, three to four monolayers along $\langle 0001 \rangle$ direction. This result indicates that blue emission can come from (0001) and also $\{1\bar{1}01\}$ facets.

Conclusion The selective growth mechanism of the hexagonal GaN microstructures was clearly understood by observing change of the facet structure with different growth conditions. Blue InGaN/GaN 3QWs were fabricated on the hexagonal GaN microstructure with partially flat (0001) facets. Two strong PL emission peaks concerning white emission were observed, one with blue emission centered around 2.73 eV and the other with yellow emission centered around 2.26 eV. Blue emission was observed on the entire GaN microstructure with flat (0001) facets and $(1\bar{1}01)$ facets, while YL was obtained at LEO region through the six $(1\bar{1}01)$ side-walls.

Acknowledgements This work has been supported by the project of the Brain Korea 21(BK21) through Semiconductor Physics Research Center (SPRC) at Chonbuk National University.

References

[1] T. ZHELEVA, O. H. NAM, M. D. BREMSER, and R. F. DAVIS, Appl. Phys. Lett. **71**, 2472 (1997).
[2] T. AKASAKA, Y. KOBAYASHI, S. ANDO, N. KOBAYASHI, and M. KUMAGAI, J. Cryst. Growth **189**, 72 (1998).
[3] K. C. ZENG, J. Y. LIN, and H. X. JIANG, Appl. Phys. Lett. **74**, 1227 (1999).
[4] S. BIDNYK, B. D. LITTLE, Y. H. CHO, J. KRASINSKI, J. J. SONG, W. YANG, and S. A. MCPHERSON, Appl. Phys. Lett. **73**, 2242 (1998).
[5] D. KAPOLNEK, R. D. UNDERWOOD, B. P. KELLER, S. KELLER, S. P. DENBAARS, and U. K. MISHRA, J. Cryst. Growth **170**, 340 (1997).
[6] S. ANDO, T. HONDA, and N. Kobayashi, Jpn. J. Appl. Lett. **32**, L104 (1992).
[7] C. S. KIM, Y. K. HONG, K. S. KIM, C.-H HONG, K. Y. LIM, G. M. YANG, and H. J. LEE, IPAP Conf. Series **1**, 296 (2000).
[8] S. KITAMURA, K. HIRAMATSU, and N. SAWAKI, Jpn. J. Appl. Phys. **34**, L1184 (1995).
[9] G. B. STRINGFELLOW, Organometallic Vapor-Phase Epitaxy: Theory and Practice, 2nd ed., Academic Press, San Diego 1999 (Chap. 6, p. 309).
[10] X. LI, O. W. BOHN, and L. L. COLEMAN, Appl. Phys. Lett. **75**, 4049 (1999).
[11] F. A. PONCE, MRS Bull. **22**, 51 (1997).

Uniform Array of GaN Quantum Dots in AlGaN Matrix by Selective MOCVD Growth

K. Tachibana[1]), T. Someya, S. Ishida, and Y. Arakawa

Research Center for Advanced Science and Technology and Institute of Industrial Science, University of Tokyo, 4-6-1 Komaba, Meguro-ku, Tokyo, 153-8505, Japan

(Received June 26, 2001; accepted August 4, 2001)

Subject classification: 68.65.Hb; 78.55.Cr; 78.66.Fd; 78.67.Hc; 81.10.Bk; 81.15.Gh; S7.14

We demonstrate GaN quantum dots (QDs) embedded in an AlGaN matrix, using metalorganic chemical vapor selective deposition, on a uniform array of hexagonal pyramids of GaN. The hexagonal pyramids have clear $\{1\bar{1}01\}$ side facets and their radius of curvature at the tops is not larger than 10 nm, which indicates that very sharp tops are realized. Intense photoluminescence was observed from GaN QDs at a peak energy of 3.61 eV at room temperature.

Introduction The quantum dot (QD) structure is very interesting from the viewpoint of device applications as well as physical properties. The QD lasers have the superior characteristics such as suppression of the temperature dependence of threshold current [1] and a lower threshold current density [2], compared to quantum well (QW) lasers. In nitride semiconductors, stimulated emission was observed from GaN self-assembled dots [3]. Moreover, the lasing action of InGaN self-assembled QD lasers was demonstrated at room temperature under optical excitation [4]. To realize the predicted characteristics of QD lasers, the uniformity of the QDs is needed. Selectively-grown QDs have the advantage of uniformity and control of the position because the selective growth is performed on a substrate patterned by lithography. In previous works, InGaN QDs were formed on a Si/GaN/sapphire substrate patterned by focused ion beam etching, and cathodoluminescence was investigated at 80 K for a few tens of QD structures [5]. Investigated was the single spectroscopy of selectively-grown InGaN QDs formed on hexagonal pyramids of GaN on a SiO_2/GaN/sapphire substrate, using the microphotoluminescence (micro-PL) with a spatial resolution of a few hundred nanometers [6].

In this paper, we investigate selectively-grown GaN QDs embedded in an $Al_{0.2}Ga_{0.8}N$ matrix formed on hexagonal pyramids of GaN on a SiO_2/GaN/sapphire substrate, using metalorganic chemical vapor deposition (MOCVD) with selective growth. The hexagonal pyramids have very sharp tops, with their radius of curvature at the tops not larger than 10 nm. Intense PL was observed from GaN QDs even at room temperature.

Fabrication of GaN Quantum Dots The sample was grown, using atmospheric-pressure two-flow MOCVD with a horizontal quartz reactor. A 2.6 µm thick GaN layer was grown at 1071 °C after a 25 nm thick GaN nucleation layer was deposited on a (0001)-oriented sapphire substrate. 40 nm of SiO_2 was deposited, using rf sputter. A grid-like pattern was formed, with the period 4 µm and square openings of side length 2 µm,

---

[1]) Corresponding author; Phone: +81-3-5452-6098; ext. 57590; Fax: +81-3-5452-6247; e-mail: tachi@iis.u-tokyo.ac.jp

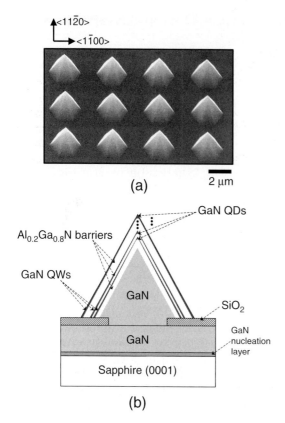

Fig. 1. a) SEM bird's-eye-view of a uniform array of hexagonal pyramids including GaN QDs in $Al_{0.2}Ga_{0.8}N$ matrix at the tops. b) Schematic of GaN QDs in $Al_{0.2}Ga_{0.8}N$ matrix formed on hexagonal pyramids of GaN on a SiO_2/GaN/sapphire substrate

prepared by conventional photolithography and a buffered HF solution. The selective growth was performed using MOCVD again. The uniform array of hexagonal pyramids of GaN was realized at 945 °C, confirmed by scanning electron microscopy [7]. Then, the selective growth of twenty periods of GaN/$Al_{0.2}Ga_{0.8}N$ multiple quantum wells followed at 1040 °C. The growth times for GaN QW and $Al_{0.2}Ga_{0.8}N$ barrier materials were such as to give 1.4 and 3.2 nm thicknesses, respectively, in planar growth.

Figure 1a shows an SEM bird's-eye-view of a uniform array of hexagonal pyramids of GaN including GaN QDs in $Al_{0.2}Ga_{0.8}N$ matrix at the tops. The hexagonal pyramids have clear $\{1\bar{1}01\}$ side facets. The radius of curvature at the tops was not larger than 10 nm, from the SEM cross-sectional image of one pyramid structure after cleaved. This means that very sharp tops are realized. It is considered that GaN QDs in $Al_{0.2}Ga_{0.8}N$ matrix are formed at the tops on hexagonal pyramids of GaN, schematically shown in Fig. 1b, as demonstrated in the GaAs [8] or InGaN [6] system. The lateral size of the QDs is thought to be comparable to the radius of curvature at the tops.

Optical Properties of Selectively-Grown GaN Quantum Dots The PL measurement was carried out from 5 to 300 K, excited by an excimer laser (ArF) with peak energy of 6.42 eV and repetition rate of 100 Hz. The collected light was dispersed by a 0.3 m monochromator and detected by a liquid nitrogen-cooled charge-coupled device camera. Figure 2 shows the PL spectra of selectively-grown GaN QDs. All the spectra were shown on the same scale. Two peaks from GaN bulk were observed; 3.34 and 3.42 eV at 300 K. The lower peak energy of 3.34 eV is due to hexagonal pyramids of GaN and the higher peak energy of 3.42 eV is due to the GaN layer under SiO_2, from the cathodoluminescence study of selectively-grown GaN hexagonal pyramids [9]. There were no clear PL peaks observed above 3.48 eV at 5 K (or 3.42 eV at 300 K) in the PL spectra of bulk $Al_{0.2}Ga_{0.8}N$, without GaN QDs, on hexagonal pyramids of GaN. Moreover, as the bandgap energy of $Al_{0.2}Ga_{0.8}N$ at room temperature is 3.9 eV [10], the PL peak of 3.61 eV at 300 K in Fig. 2 is due to GaN QDs.

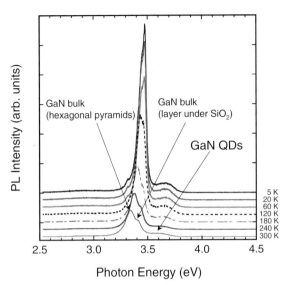

Fig. 2. PL spectra of selectively-grown GaN QDs, excited by an excimer laser (ArF) with excitation energy of 6.42 eV. The spectra were measured from 5 to 300 K

The PL intensity of GaN QDs is smaller than that of GaN bulk, because the volume of GaN QDs is much smaller than that of GaN bulk. The PL peak intensity of GaN QDs decreases above 180 K, although it decreases more slowly than that of GaN bulk. This indicates that there exists stronger confinement of carriers in the QDs, however, the effect of the suppression of non-radiative recombination in the QDs becomes relatively small at higher temperature because the carriers are captured into the non-radiative centers such as the defects around the QDs. Further investigation is needed to realize much stronger confinement of carriers in the QDs.

Conclusions We demonstrate selectively-grown GaN QDs embedded in an AlGaN matrix, formed on a uniform array of hexagonal pyramids of GaN. The radius of curvature of hexagonal pyramids at the tops is not larger than 10 nm, which indicates that very sharp tops are realized. A clear peak of PL spectra is observed from GaN QDs even at room temperature.

Acknowledgements The authors thank M. Nishioka, K. Hoshino, and M. Miyamura, University of Tokyo, for technical support with the MOCVD system. This work was partly supported by the Grant-in-aid for COE Research from Ministry of Education, Culture, Sports, Science and Technology (#12CE2004) and Research for the Future Program of the Japan Society for the Promotion of Science (JSPS) (Project No. JSPS-RFTF96P00201). One of the authors (K.T.) acknowledges JSPS Research Fellowships for Young Scientists for partial financial support.

References

[1] Y. Arakawa and H. Sakaki, Appl. Phys. Lett. **40**, 939 (1982).
[2] M. Asada, Y. Miyamoto, and Y. Suematsu, IEEE J. Quantum Electron. **22**, 1915 (1986).
[3] S. Tanaka, H. Hirayama, Y. Aoyagi, Y. Narukawa, Y. Kawakami, Sz. Fujita, and Sg. Fujita, Appl. Phys. Lett. **71**, 1299 (1997).
[4] K. Tachibana, T. Someya, Y. Arakawa, R. Werner, and A. Forchel, Appl. Phys. Lett. **75**, 2605 (1999).

[5] J. Wang, M. Nozaki, M. Lachab, Y. Ishikawa, R. S. Qhalid Fareed, T. Wang, M. Hao, and S. Sakai, Appl. Phys. Lett. **75**, 950 (1999).
[6] K. Tachibana, T. Someya, S. Ishida, and Y. Arakawa, Appl. Phys. Lett. **76**, 3212 (2000).
[7] K. Tachibana, T. Someya, S. Ishida, and Y. Arakawa, J. Cryst. Growth **221**, 576 (2000).
[8] Y. Nagamune, M. Nishioka, S. Tsukamoto, and Y. Arakawa, Appl. Phys. Lett. **64**, 2495 (1994).
[9] F. Bertram, J. Christen, M. Schmidt, K. Hiramatsu, S. Kitamura, and N. Sawaki, Physica E **2**, 552 (1998).
[10] T. Takeuchi, H. Takeuchi, S. Sota, H. Sakai, H. Amano, and I. Akasaki, Jpn. J. Appl. Phys. **36**, L177 (1997).

Self-Assembled Growth
of GaN Quantum Dots Using Low-Pressure MOCVD

M. Miyamura[1]), K. Tachibana, T. Someya, and Y. Arakawa

Research Center for Advanced Science and Technology and Institute of Industrial Science, University of Tokyo, 4-6-1 Komaba, Meguro-ku, Tokyo 153-8904, Japan

(Received July 1, 2001; accepted July 14, 2001)

Subject classification: 68.65.Hb; 81.07.Ta; 81.15.Gh; 85.35.Be; S7.14

We have demonstrated the self-assembled growth of GaN quantum dots (QDs) on an AlN layer by low-pressure metalorganic chemical vapor deposition. The dependence of the QD density on the GaN coverage was investigated. The density of QDs increased as the GaN coverage was higher and it reached 6×10^8 cm^{-2} when the amount of GaN deposited was around 2.5 monolayers (ML). The average diameter and height of the QDs were 35 and 1.6 nm, respectively. It was found that the formation of the QDs was also much affected by the growth temperature.

1. Introduction Nitride semiconductors have been much attractive from the viewpoint of short-wavelength emitting devices. For example, the violet laser diodes using InGaN multiple quantum well (MQW) structure as the active layer were demonstrated at room temperature [1]. On the other hand, several attempts have been made to realize UV-light emitters using nitride semiconductor system, such as AlGaN MQW light emitting diodes [2, 3]. One of the major obstacles for the realization of UV-light emitting devices is the relatively poor crystalline quality of AlGaN active layers. In order to overcome this problem, the quantum dot (QD) structure is promising since QDs contain no dislocations. Moreover, QD structure has been theoretically predicted to enhance the characteristics of lasers significantly [4]. By these reasons, some researches about the GaN QDs have been conducted. Daudin et al. [5] demonstrated GaN/AlN QDs grown by molecular beam epitaxy taking advantage of Stranski-Krastanov (S-K) growth mode. The growth of GaN/AlGaN QDs using "anti-surfactants" by metalorganic chemical vapor deposition (MOCVD) and stimulate emission from these QDs were also reported [6, 7]. As another way to fabricate GaN or InGaN QDs, the selective growth method was proposed [8]. But there are no reports on the self-assembled growth of GaN QDs without any "anti-surfactants" by MOCVD.

In this paper, we demonstrate GaN self-assembled QDs grown on a very flat surface of AlN by low-pressure MOCVD. The formation of GaN QDs is explained mainly by the 2.5% lattice mismatch of the GaN/AlN system. First, the growth conditions of an AlN buffer layer and GaN QDs are introduced. We investigated the dependence of the QD density on the GaN coverage. Finally, the influence of the growth temperature on the formation of GaN QDs is discussed.

2. Experimental Procedures The growth of AlN and GaN was carried out using two-flow MOCVD with a horizontal quartz reactor under the reactor pressure of 200 Torr.

---

[1]) Corresponding author; Phone: +81-3-5452-6098, ext. 57590; Fax: +81-3-5452-6247;
e-mail: miyamura@iis.u-tokyo.ac.jp

As group III sources, trimethylgallium (TMG) and trimethylaluminum (TMA) were used. NH_3 was used as a group V source. We grew an AlN layer on a (0001)-oriented 6H-SiC substrate at 1080 °C. During the growth of AlN layer, the flow rate of TMA was 8.0 μmol/min, and that of NH_3 was 2.5 standard liters per minute (slm) with carrier gases of H_2 and N_2 at 1.5 and 1.5 slm, respectively. XRD was used to characterize the AlN layer grown on a (0001)-oriented 6H-SiC substrate. Clear fringe peaks were observed. Its thickness was estimated to be 75 nm by intervals between each fringe peak. The root-mean-square roughness of the AlN surface was 0.17 nm measured by contact mode AFM. From these results, it was confirmed that the AlN layer showed nice crystalline quality and a quite flat surface. This is very important for the fabrication of GaN QDs at the following steps.

After the AlN layer was grown, the growth temperature was reduced to 975 °C to grow GaN. The flow rate of TMG was 3.1 μmol/min, and that of NH_3 was 0.8 slm with carrier gases of H_2 and N_2 at 1.2 and 2.5 slm, respectively. The V/III ratio was about 11500 and the growth rate was 5.0×10^{-2} monolayers (ML)/s. The GaN coverage was controlled by changing the growth time of GaN. The growth time was systematically varied from 20 to 50 s. After the deposition of GaN, each sample was kept under the same condition for 30 s as growth interruption and then cooled down to room temperature. The surface morphologies of GaN were observed using AFM.

3. Results and Discussion Figure 1 shows AFM images of GaN surface. The growth times of GaN layer for each sample were 20 and 50 s, respectively.

When the growth time was 20 s, only monolayer steps were observed. The GaN QDs were formed by deposition of over 20 s. The dependence of the QD density on the GaN coverage is shown in Fig. 2. The density of the QDs increased as the GaN coverage was higher, and it reached 6×10^8 cm^{-2} when the growth time was 50 s. The average diameter and height of the QDs were 35 and 1.6 nm at this time, respectively.

The formation of the QDs is mainly due to the 2.5% lattice mismatch of the GaN/AlN system. The critical thickness is estimated to be around 1.0 ML, what corresponds to a growth time of 20 s. This value is, however, relatively small compared with that

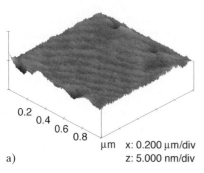
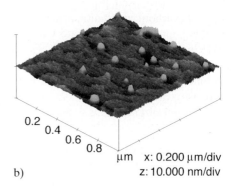

Fig. 1. AFM images of GaN surface morphology. The growth times of GaN layer were a) 20 and b) 50 s, respectively. The image scale is 1×1 $μm^2$

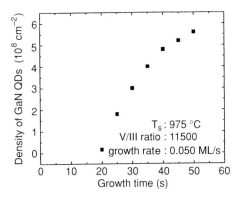

Fig. 2. Dependence of GaN QD density on the growth time

expected by the lattice mismatch of 2.5%. Since we estimated the GaN coverage by calibrating the growth rate from XRD measurement for GaN/AlN superlattice structures, one possible explanation of the discrepancy is that the estimated growth rate of GaN cannot accurately applied to the initial stage of GaN deposition. Another possibility is that the wetting layer is locally thick (2 or 3 ML) and the QDs are formed on such disk-lie wetting layer. As result, the actual critical thickness of the wetting layer is much thicker than 1 ML. This phenomenon was also observed in InGaAs/InAs QDs [9].

Next, we investigate the dependence on the growth temperature. Figure 3 shows AFM images of GaN surface morphology, which were grown at 960 and 990 °C, respectively.

Note that the QDs grown at 990 °C have larger diameter than those grown at 960 °C. This can be explained that increasing temperature made the migration of supplied Ga atoms enhanced, and larger dots were formed. Thus, the formation of QDs was much affected by the growth temperature, and this behavior was similar to other QDs grown by the S-K mode, such as InAs QDs.

4. Conclusions We have successfully grown GaN self-assembled QDs on quite a smooth AlN layer by low-pressure MOCVD. The average diameter and height of the QDs was 35 and 1.6 nm, respectively. The density of GaN QDs increased up to 6×10^8 cm^{-2} as the GaN coverage was higher. We also investigated the dependence on the growth temperature. It was found that the formation of the QDs was much affected by the growth temperature.

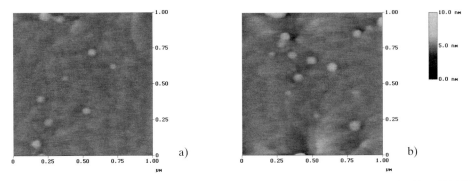

Fig. 3. AFM images of GaN surface morphology. The growth temperatures were a) 960 and b) 990 °C. The image scale is 1×1 μm^2

Acknowledgements The authors gratefully acknowledge M. Nishioka, S. Ishida, and K. Hoshino, University of Tokyo, for technical support with the MOCVD system. This work was supported in part by COE Research from Ministry of Education, Culture, Sports, Science and Technology (#12CE2004) and the Research for the Future Program of the Japan Society for the Promotion of Science (Project No. JSPS-RFTF96P00201).

References

[1] S. Nagahama, N. Iwasa, M. Senoh, T. Matushita, Y. Sugimoto, H. Kiyoku, T. Kozaki, M. Sano, H. Matumura, H. Umemoto, K. Chocho, and T. Mukai, Jpn. J. Appl. Phys. **39**, L647 (2000).
[2] T. Nishida and N. Kobayashi, Proc. 3rd Internat. Conf. Nitride Semiconductors; phys. stat. sol. (a) **176**, 45 (1999).
[3] N. Otsuka, A. Tsujimura, Y. Hasagawa, G. Sugahara, M. Kume, and Y. Ban, Proc. Internat. Workshop Nitride Semiconductors, Nagoya (Japan) 2000; IPAP, Tokyo 2000 (p. 837).
[4] Y. Arakawa and H. Sakaki, Appl. Phys. Lett. **40**, 939 (1982).
[5] B. Daudin, F. Widmann, G. Feuillet, Y. Samson, M. Arlery, and J. L. Rouvière, Phys. Rev. B **56**, R7069 (1997).
[6] S. Tanaka, S. Iwai, and Y. Aoyagi, Appl. Phys. Lett. **69**, 4096 (1996).
[7] S. Tanaka, H. Hirayama, Y. Aoyagi, Y. Narukawa, Y. Kawakami, Sz. Fujita, and Sg. Fujita, Appl. Phys. Lett. **71**, 1299 (1997).
[8] K. Tachibana, T. Someya, S. Ishida, and Y. Arakawa, Appl. Phys. Lett. **76**, 3212 (2000).
[9] M. Kitamura, M. Nishioka, R. Schur, and Y. Arakawa, J. Cryst. Growth **170**, 563 (1997).

On Phonon Confinement Effects and Free Carrier Concentration in GaN Quantum Dots

M. Kuball[1]) (a), J. Gleize (b), Satoru Tanaka (c), and Y. Aoyagi (d)

(a) H. H. Wills Physics Laboratory, University of Bristol, Bristol BS8 1TL, United Kingdom

(b) Laboratoire de Physique des Solides, Université Paul Sabatier, F-31062 Toulouse, France

(c) Research Institute for Electronic Science, Hokkaido University, Sapporo 060-0812, Japan

(d) The Institute of Physical and Chemical Research (RIKEN), Saitama 351-0198, Japan

(Received June 26, 2001; accepted August 4, 2001)

Subject classification: 68.37.Ps; 78.30.Fs; 78.67.De; 81.07.Ta; S7.14

Self-assembled GaN quantum dots (QDs) grown on $Al_{0.15}Ga_{0.85}N$ using Si as anti-surfactant were investigated by resonant Raman scattering. Phonons of GaN QDs of different sizes were probed selectively by using laser excitation energies of 3.53 and 5.08 eV. Phonon confinement effects were evidenced in GaN QDs of 2–3 nm height. Resonant Raman scattering on GaN grown on $Al_{0.23}Ga_{0.77}N$ after the deposition of an increasing amount of Si anti-surfactant, i.e., the morphological transition from a GaN quantum well (2D) to GaN quantum dots (0D), was also investigated.

GaN quantum dot (QD) structures with high photoluminescence (PL) efficiencies have recently been fabricated using the Stranski-Krastanov method [1] and using a Si anti-surfactant [2]. Tanaka et al. [2] demonstrated that the deposition of a small amount of Si anti-surfactant onto AlGaN during a growth interruption inhibits the normal two-dimensional step-flow growth of the subsequent GaN deposition and GaN QDs form. This allows the formation of GaN QDs onto low aluminum content AlGaN layers (lattice mismatch between GaN and $Al_{0.15}Ga_{0.85}N$ < 0.2%) where lattice mismatch is not large enough for the Stranski-Krastanov method. Only a few studies on vibrational properties of GaN QDs have been reported in literature [3], although vibration properties play an important role for thermal and optical properties including phonon-assisted optical transitions. We report here on the resonant Raman investigation of self-assembled GaN QDs grown on AlGaN.

GaN quantum dots were formed on a 0.6 μm thick $Al_{0.15}Ga_{0.85}N$ layer by supplying tetraethylsilane (TESi) prior to the deposition of GaN in a conventional horizontal-type metal-organic chemical vapor deposition (MOCVD) reactor at 1080 °C (25 sccm TESi for 20 s, GaN QDs: 5 sccm trimethylgallium (TMG)/2 slm NH_3 for 60 s). The $Al_{0.15}Ga_{0.85}N$ layer was grown on a 2–3 nm thick AlN buffer layer on a 6H-SiC(0001) substrate. Ammonia (NH_3), trimethylaluminum (TMA), and trimethylgallium (TMG) were used as precursors for the GaN and AlGaN growth. A reference sample was grown under similar conditions except that no TESi was supplied: a continuous GaN layer of 0.3 μm thickness was

---
[1]) Corresponding author; Phone: +44 117 928 8734; Fax: +44 117 925 5624; e-mail: Martin.Kuball@bristol.ac.uk

formed on the $Al_{0.15}Ga_{0.85}N$. To investigate the effect of TESi on the GaN quantum dot formation, a second set of samples was grown on a 0.8 μm thick $Al_{0.23}Ga_{0.77}N$ layer on 6H-SiC, keeping the TESi (10 sccm) and TMG (10 sccm)/NH_3 (2 slm) flow constant and varying the exposure time [4]: 10 s-GaN/3 s-TESi, 10 s-GaN/5 s-TESi, 10 s-GaN/10 s-TESi, 3 s-GaN/30 s-TESi. All samples were characterized by atomic force microscopy (AFM). Raman spectra were recorded in backscattering $z(x,\cdot)\bar{z}$ geometry with unpolarized detection from the top surface of the samples. The experiments were performed using a DILOR macro-Raman spectrometer with the Ar-laser line at 3.53 eV and using a Renishaw micro-Raman spectrometer with the frequency doubled Ar-laser line at 5.08 eV as excitation source with 2–3 cm^{-1} spectral resolution.

The inset of Fig. 1[2]) shows an AFM image of the GaN QDs grown on $Al_{0.15}Ga_{0.85}N$. The sample contains two different kinds of GaN dots: large dots of about 40 nm height and 300–400 nm diameter as well as small dots of about 2–3 nm height and 5–10 nm diameter (located between the large GaN dots). Figure 1 displays Raman spectra recorded under 3.53 eV excitation. The Raman spectrum of the continuous GaN layer grown on $Al_{0.15}Ga_{0.85}N$ is shown for comparison. The fundamental transitions of bulk GaN and $Al_{0.15}Ga_{0.85}N$ were estimated to be 3.42 and 3.66 eV [5]. Under 3.53 eV excitation, both spectra show Raman peaks at 736 cm^{-1} located on top of a GaN PL peak, attributed to first-order Raman scattering from the $A_1(LO)$ phonons of GaN, resonantly enhanced under the ultraviolet excitation. No difference in $A_1(LO)$ phonon frequency was detectable between the GaN QD sample and the continuous GaN layer. For the large GaN dots in this sample, phonon confinement effects are expected to be smaller than the detection limit in our experiments. We conclude that the $A_1(LO)$ Raman signal obtained under 3.53 eV excitation originates from the large GaN dots. Note that Si (a common donor for GaN) was used to initiate the GaN dot formation. Information on the free carrier concentration can be obtained from the $A_1(LO)$ frequency since $A_1(LO)$ phonons interact with collective excitations of free carriers (plasmons) [6]. The $A_1(LO)$ frequency

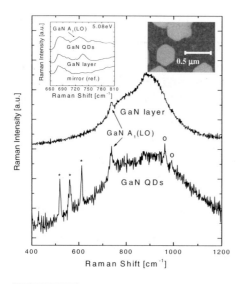

Fig. 1. Raman spectra of GaN QDs grown on $Al_{0.15}Ga_{0.85}N$ and of a continuous GaN layer grown on $Al_{0.15}Ga_{0.85}N$, obtained under 3.53 eV excitation. Asterisks and circles mark laser plasma lines and the Raman signal of the 6H-SiC substrate, respectively. The inset shows Raman spectra obtained under 5.08 eV excitation. A reference spectrum recorded on an aluminum mirror illustrates the system response function under 5.08 eV excitation. Also shown as inset (colour) is an AFM image of the sample

---

[2]) Colour figure are published online (www.physica-status-solidi.com), where indicated.

(736 cm^{-1}) found for both samples is shifted with respect to undoped GaN (734 cm^{-1}) by only 2 cm^{-1}. No large electron concentration is therefore present in the GaN dots despite the use of Si as anti-surfactant. From the measured A$_1$(LO) frequency we estimate the free carrier concentration to be at maximum in the upper 10^{17} cm^{-3}-range.

The inset in Fig. 1 shows Raman spectra recorded under 5.08 eV excitation. A clear GaN A$_1$(LO) phonon frequency shift between the GaN QD sample (727 cm^{-1}) and the continuous GaN layer (736 cm^{-1}) is visible. This is in contrast to the results obtained under 3.53 eV excitation (Fig. 1). For the continuous GaN layer the results obtained under 3.53 eV and 5.08 eV excitation agree. Consequently, the origin of the A$_1$(LO) Raman peak for the GaN QD sample obtained under 3.53 eV and 5.08 eV excitation has to be different. This sample contains only two regions that contain GaN: the 40 nm high dots and the 2–3 nm high dots. The large GaN dots were probed under 3.53 eV excitation, i.e., the only remaining region of the sample that contains GaN are the 2–3 nm high GaN QDs. Phonon confinement effects which shift the A$_1$(LO) phonon frequency towards lower wavenumbers are expected to occur for GaN QDs of 2–3 nm height. The observed frequency shift of 9 cm^{-1} is consistent with simple phonon confinement models [7].

Figure 2[2]) shows AFM images of GaN grown on Al$_{0.23}$Ga$_{0.77}$N. Sample A is a quantum well (QW) (about 5–10 nm thick). Increasing the amount of Si anti-surfactant prior to the GaN deposition results in a transition from a GaN QW (A) to GaN QDs (D). GaN QDs of 5–10 nm height and $\approx$60 nm diameter are present in sample D. Figure 3 shows Raman spectra recorded under 3.53 eV excitation. Visible is the GaN A$_1$(LO) Ra-

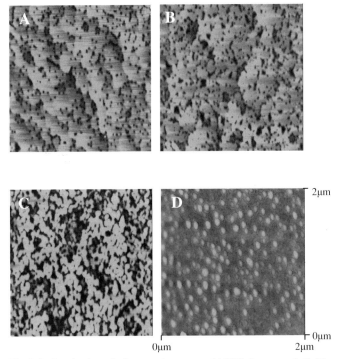

Fig. 2 (colour). Atomic force microscopy (AFM) images of GaN grown on Al$_{0.23}$Ga$_{0.77}$N after the deposition of an increasing amount of Si anti-surfactant. A) 10 s-GaN/3 s-TESi, B) 10 s-GaN/5 s-TESi, C) 10 s-GaN/10 s-TESi, D) 3 s-GaN/30 s-TESi

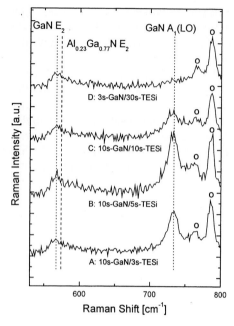

Fig. 3. Raman spectra of GaN grown on $Al_{0.23}Ga_{0.77}N$ after the deposition of an increasing amount of Si anti-surfactant, obtained under 3.53 eV excitation. The circles mark the Raman signal of the 6H-SiC substrate. The location of the E_2 (high) phonon frequency of unstrained GaN (567 cm^{-1}) and $Al_{0.23}Ga_{0.77}N$ (574 cm^{-1}) is indicated

man signal at 734 cm^{-1} resonantly enhanced under the 3.53 eV excitation and located on top of a GaN PL signature. The GaN $A_1(LO)$ Raman intensity decreases from sample A/B to C with C exhibiting a morphology between QW and QD (Fig. 2). No significant difference in $A_1(LO)$ Raman intensity was found between sample A and B (also similar AFM images (Fig. 2)). A weak shoulder is present around the GaN $A_1(LO)$ phonon frequency in the Raman spectrum of sample D, related to the GaN QDs. The decreasing GaN $A_1(LO)$ Raman signal from sample A to C (same TMG/NH$_3$, but different TESi exposure) illustrates the decreasing amount of GaN deposited onto the AlGaN with increasing Si coverage. Si masks areas of the AlGaN surface which finally results in GaN QD formation for sample D. Also visible in the Raman spectra is the $Al_{0.23}Ga_{0.77}N$ E_2 (high) Raman mode at 568 cm^{-1} (intensity independent of GaN surface coverage). The small discrepancy between measured (568 cm^{-1}) and expected $Al_{0.23}Ga_{0.77}N$ E_2 (high) phonon frequency (574 cm^{-1}) is likely to be related to strain and possibly to an Al composition somewhat lower than the nominal composition of $Al_{0.23}Ga_{0.77}N$.

In conclusion, phonon confinement effects with a decrease in $A_1(LO)$ phonon frequency as large as 9 cm^{-1} were evidenced in 2–3 nm high GaN QDs. We estimate the free carrier concentration in the GaN QDs to be at maximum in the upper 10^{17} cm^{-3}-range despite the use Si as anti-surfactant – a common donor for GaN – for the GaN QD growth. The transition from QW to QDs with increasing Si surface coverage was evidenced by Raman scattering.

References

[1] F. WIDMANN, B. DAUDIN, G. FEUILLET, Y. SAMSON, J. L. ROUVIÈRE, and N. PELEKANOS, J. Appl. Phys. **83**, 7618 (1998).
[2] S. TANAKA, S. IWAI, and Y. AOYAGI, Appl. Phys. Lett. **69**, 4096 (1996).
[3] J. GLEIZE, F. DEMANGEOT, J. FRANDON, M. A. RENUCCI, M. KUBALL, F. WIDMANN, and B. DAUDIN, phys. stat. sol. (b) **216**, 457 (1999).
[4] S. TANAKA, I. SUEMUNE, P. RAMVALL, and Y. AOYAGI, phys. stat. sol. (b) **216**, 431 (1999).
[5] D. BRUNNER, H. ANGERER, E. BUSTARRET, F. FREUDENBERG, R. HÖPLER, R. DIMITROV, O. AMBACHER, and M. STUTZMANN, J. Appl. Phys. **82**, 5090 (1997).
[6] H. HARIMA, H. SAKASHITA, and S. NAKASHIMA, Mater. Sci. Forum **264–268**, 1363 (1998).
[7] M. KUBALL, H. MOKHTARI, D. CHERNS, J. LU, and D. I. WESTWOOD, Jpn. J. Appl. Phys. **39**, Pt. 1, 4753 (2000).

Temperature Dependent Photoluminescence of MBE Grown Gallium Nitride Quantum Dots

J. Brown[1]), C. Elsass, C. Poblenz, P. M. Petroff, and I. S. Speck

Materials Department, University of California at Santa Barbara, Santa Barbara, CA, USA

(Received June 22, 2001; accepted August 8, 2001)

Subject classification: 73.21.La; 78.67.Hc; 81.07.Ta; 81.15.Hi; S7.14

We report on the growth and optical properties of gallium nitride quantum dots (QDs) grown by plasma-assisted molecular beam epitaxy. We have observed strong photoluminescence (PL) from the QDs from 8 to 750 K. Atomic force microscopy studies demonstrate that the QDs have diameters of (30 ± 5) nm and heights of (3 ± 1) nm. PL from the quantum dots was compared to that of a gallium nitride growth template film to unambiguously demonstrate the contribution of the QDs to the spectra. Integrated PL intensity was observed to remain strong well above 300 K, and we attribute the decrease in the quantum dot PL at higher temperatures to phonon-mediated carrier ionization of deep level.

Introduction Nitride optoelectronic structures offer emission across the visible spectrum and into the ultraviolet. Due to their wide bandgaps and thermal stability, nitride devices have the possibility of high temperature operation. Nitride quantum dots (QDs) have been investigated through molecular beam epitaxy (MBE), metalorganic chemical vapor deposition (MOCVD), and by theoretical studies [1–4]. In this work, we report on temperature dependent photoluminescence (PL) of GaN QDs grown in an AlN matrix.

A strong motivation for investigating nitride QDs is that the large bandgaps and band offsets in the group III-nitrides, in comparison with the group III-arsenides, should dramatically reduce the thermal ionization of QD confined carriers into the barrier material [4, 5]. In group III-arsenide QDs, carrier thermal ionization into the barrier was shown to be a dominant process at temperatures above 150 K [6]. In this work, phonon-mediated non-radiative carrier recombination is proposed as a dominant mechanism for the thermally induced loss of luminescent recombination above 300 K in GaN/AlN QDs.

Experiments and Results Plasma-assisted MBE has been used to grow GaN quantum dots in an AlN matrix. The following structure was grown: sapphire (0001) substrates with 0.3 μm MOCVD-grown GaN were used as templates for a 30 nm GaN buffer, followed by a partially relaxed 0.3 μm AlN buffer layer grown at 750 °C. The AlN was grown under metal-rich conditions near the cross-over to form metal droplets on the surface [7]. After deposition of each AlN layer, a 60 s delay under N flux was utilized to ensure the incorporation of any residual Al, followed by a 60 s delay with all shutters closed. Next, a stack of ten layers of GaN QDs was grown at 750 °C in the Stranski-Krastanov (S-K) growth mode [8]. To ensure adequate Ga incorporation, the Ga/N

---

[1]) Corresponding author; e-mail: jsbrown@engineering.ucsb.edu

flux ratio was set such that the GaN growth took place just inside the Ga droplet regime [7]. After opening both the Ga and N shutters for 15 s (8 Å of GaN growth), the shutters were then closed and the appearance of islands at the surface was observed in the RHEED transition ((1 × 1) changed from streaky to spotty) from 2D to 3D during a 20 s interval. After island formation, a 30 nm AlN capping layer was deposited to form QDs. Finally, a layer of GaN islands was grown on the surface and left uncapped for subsequent atomic force microscopy (AFM) studies.

Tapping mode AFM and temperature dependent PL from 8 to 750 K were used for sample characterization. Three separate structures were characterized in this work: the MOCVD-grown GaN on sapphire template; a partially relaxed (0.3 μm) AlN layer on the MOCVD-grown GaN on sapphire template; and the QD heterostructure described above.

The PL experiments were performed by optical pumping with a 325 nm He–Cd laser (cw). Low (8–300 K) temperature measurements were carried out with the sample attached to a liquid He cooled cryostat, while high (300–750 K) temperature measurements were performed with the same excitation power by replacing the cryostat with a hot plate.

The uncapped GaN islands exhibited a diameter of (30 ± 5) nm and a height of (3 ± 1) nm as shown in Fig. 1a. Figure 1b shows a reference 0.3 μm partially relaxed AlN film on MOCVD-grown GaN/sapphire template, which clearly shows cracking. Figure 1c is a $1 \times 1 \ \mu m^2$ scan of the same AlN reference film exhibiting spiral hillocks between the cracks, providing a locally continuous stressor film for QD nucleation.

The low temperature PL spectrum (50 K) shows a sharp peak at 3.50 eV near the band edge of GaN attributed to the bulk GaN buffer and a very broad (453 meV FWHM) peak centered at 2.76 eV which is attributed to the GaN QD as shown in Fig. 2a[2]). Theoretical studies including strain and polarization-related electric field effects predict a QD recombination of approximately 2.85 eV for 3 nm GaN/AlN QD height in reasonable agreement between the measured and calculated QD emission for the size of QDs grown for these experiments [4]. The disagreement between the model and experimental results for the QD PL include the lack of QD size resolution of the AFM, as well as the absence of a constraining AlN cap on GaN islands on the surface increasing the probability of QD ripening after the growth. Comparison of the QD PL with the growth template PL at low temperature in Fig. 2a indicates that the 2.2 eV

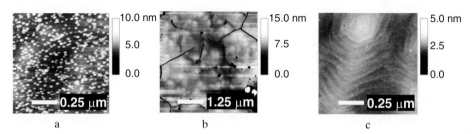

Fig. 1. Tapping mode AFM. a) GaN islands on AlN surface of ten layer GaN QD structure; b) $5 \times 5 \ \mu m^2$ scan: 300 nm partially relaxed AlN on MOCVD-grown GaN template; c) $1 \times 1 \ \mu m^2$ scan: 300 nm partially relaxed AlN, exhibiting step-flow growth between large scale surface fissures

---

[2]) Colour figure is published online (www.physica-status-solidi.com).

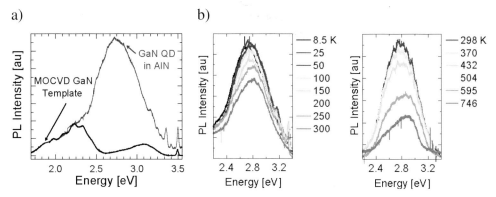

Fig. 2 (colour). a) Low temperature (50 K) PL (in arb. units) of GaN QD in AlN matrix and MOCVD-grown GaN on sapphire template. b) PL vs. temperature: low temperature (8.5–300 K) and high temperature (298–746 K) PL of GaN QD in AlN matrix

yellow luminescence is the origin of the low energy shoulder of the QD peak. The donor–acceptor emission, centered at 3.1 eV, was not observed above 100 K, allowing the observed emission at 2.76 eV to be attributed to GaN QDs.

Temperature dependent PL from 8.5 to 746 K was observed in the GaN/AlN QD heterostructure, as shown in Fig. 2b. The QD PL intensity was observed to remain strong well above room temperature. A blue shift of the QD emission with increasing temperature was observed, as shown in Fig. 2b, indicating a preferential loss of carriers from larger QDs.

Excluding the template yellow luminescence, an Arrhenius plot of the integrated luminescence intensity of the QD emission is shown in Fig. 3a. An activation energy for exciton concentration under steady state conditions as a function of temperature of (46 ± 2) meV was determined from the Arrhenius plot [6, 9].

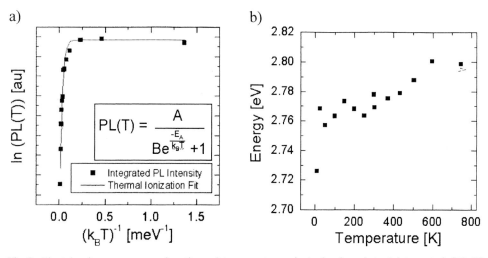

Fig. 3. Photoluminescence as a function of temperature. a) Arrhenius plot of integrated QD PL intensity. b) QD emission energy vs. temperature

Due to the small value of the experimentally determined activation energy, in contrast with the relatively large thermal ionization barriers predicted for this structure [4], the PL decrease mechanism is proposed to be phonon mediated non-radiative recombination. In this process, deep levels in the AlN barrier are thermally ionized, creating non-radiative recombination centers for QD carriers. Polarization-related electric fields cause carrier separation in GaN/AlN QDs, in which electrons are localized near the QD tops and holes are localized in the wetting layer immediately below the QD [4]. The spatial separation of carriers is proposed to enhance the loss of carriers from the QDs to ionized traps, preferentially in larger QDs because carrier separation is expected to be proportional to QD size. The proposed mechanism provides a qualitative explanation for the observed blue shift of QD PL with increasing temperature. Further work is underway to verify this model for the loss in QD luminescence at high temperature.

Conclusion In conclusion, GaN/AlN QDs have been shown to display unique high temperature luminescent behavior. The spontaneous and strain induced polarization complicates the analysis of carrier loss mechanisms in GaN/AlN QD heterostructures, but the promise of high temperature optoelectronics applications necessitates further investigation of these phenomena.

Acknowledgements The authors gratefully acknowledge the support of ARO-DARPA and AFOSR (D. Johnstone, Program Manager).

References

[1] F. WIDMANN, B. DAUDIN, G. FEUILLET, Y. SAMSON, J. L. ROUVIÈRE, and N. PELEKANOS, J. Appl. Phys. **83**, 7618 (1998).
[2] B. DAMILANO, N. GRANDJEAN, F. SEMOND, J. MASSIES, and M. LEROUX, Appl. Phys. Lett. **75**, 962 (1999).
[3] S. TANAKA, S. IWAI, and Y. AOYAGI, Appl. Phys. Lett. **69**, 4096 (1996).
[4] A. D. ANDREEV and E. P. O'REILLY, Phys. Rev. B **62**, 15851 (2000).
[5] A. RIZZI, R. LANTIER, F. MONTI, H. LUTH, F. DELLA SALA, A. DI CARLO, and P. LUGLI, J. Vac. Sci. Technol. B **17**, 1674 (1999).
[6] S. FAFARD, S. RAYMOND, G. WANG, R. LEON, D. LEONARD, S. CHARBONNEAU, J. L. MERZ, P. M. PETROFF, and J. E. BOWERS, Surf. Sci. **361/362**, 778 (1996).
[7] B. HEYING, R. AVERBECK, L. F. CHEN, E. HAUS, H. RIECHERT, and J. S. SPECK, J. Appl. Phys. **88**, 1855 (2000).
[8] D. J. EAGLESHAM and M. CERULLO, Phys. Rev. Lett. **64**, 1943 (1990).
[9] G. BACHER, H. SCHWEIZER, J. KOVAC, and A. FORCHEL, Phys. Rev. B **43**, 9312 (1991).

The Transition from Blue Emission in As-Doped GaN to GaNAs Alloys in Layers Grown by Molecular Beam Epitaxy

C. T. Foxon[1]) (a), S. V. Novikov*) (a), Y. Liao**) (a), A. J. Winser (a, b),
I. Harrison (b), T. Li (a), R. P. Campion (a), C. R. Staddon (a),
and C. S. Davis (a)

(a) *School of Physics and Astronomy, University of Nottingham, Nottingham, NG7 2RD, UK*

(b) *School of Electrical and Electronic Engineering, University of Nottingham, Nottingham, NG7 2RD, UK*

(Received July 2, 2001; accepted August 1, 2001)

Subject classification: 78.55.Cr; 81.05.Ea; 81.15.Hi; S7.14

The transition from As-doped GaN showing strong blue emission ($\sim$2.6 eV) at room temperature to the formation of $GaN_{1-x}As_x$ alloys for films grown by molecular beam epitaxy was investigated. This study demonstrates that with increasing N to Ga ratio there is first an increase in the intensity of blue emission at about 2.6 eV and then a transition to the growth of $GaN_{1-x}As_x$ alloy films. Several possible models, which can explain how this might occur are presented.

1. Introduction Recently there has been considerable interest in As-doped GaN ([1–5] and references in [5]). There are two main reasons, which motivate such investigations; a large negative bowing in the energy band gap for GaNAs solid solutions and strong blue emission at room temperature from arsenic doped GaN. However, what causes the transition from As-doped GaN showing blue emission to a GaNAs alloy is not presently understood. The aim of this paper is to study the transition from As-doped GaN showing blue emission to the formation of GaNAs alloys.

2. Experimental Details The samples used in this study were grown by plasma-assisted molecular beam epitaxy (PA-MBE) on sapphire substrates, full details of the growth procedures used can be found elsewhere [5]. All the samples discussed here were grown at 800 °C. The N/Ga ratio was changed by two methods. In the first instance the Ga and As fluxes were kept constant and the amount of active nitrogen was changed. The amount of active nitrogen is determined by the intensity of the plasma, which in turn is measured using the optical emission detector signal (OED). In a second case, the As and N fluxes were kept constant and the Ga flux was varied over a wide range.

3. Results and Discussion Using the first method, the influence of stoichiometry on the optical properties of As-doped GaN films was studied. With increasing N to Ga ratio, three effects are observed; the intensity of the blue emission increases, the peak

---

[1]) Corresponding author; Phone: +44-115-9515138; Fax: +44-115-9515184;
e-mail: C.Thomas.Foxon@nottingham.ac.uk
*) On leave from the Ioffe Physico-Technical Institute, St. Petersburg, Russia.
**) On leave from the University of Science and Technology of China, Hefei 230026, China.

at 3.2 eV disappears at the high N to Ga ratio and the position of the band edge peak shifts to lower energy implying that we are forming a GaNAs alloy. The observation that it is necessary to have a high N/Ga flux ratio to promote the growth of the alloy is at first surprising. With increasing N/Ga ratio, one might expect to increase the density of Ga vacancies in the lattice and thus increase the possibility of incorporating As on the Ga sublattice as predicted theoretically [3]. Theoretically more N-rich conditions are predicted to raise the formation energy of As_N and lower the formation energy of As_{Ga} [3], which is directly contrary to the present observations that one needs nitrogen rich conditions to form the alloy.

To investigate this in more detail, the second method was used to give larger variations in the N:Ga ratio by changing the Ga flux. Figures 1a and b show the variation in band edge and blue emission intensity as a function of Ga flux. It is evident that there are two regions of Ga flux. At high Ga fluxes, both the band edge and blue emission intensities are strong and as shown in Fig. 1c there is no significant shift in the energy

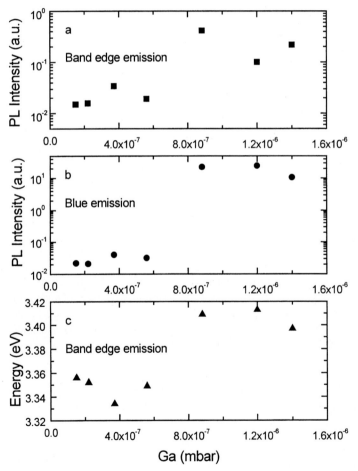

Fig. 1. Variation in intensity of the a) band edge emission and b) blue emission as a function of the Ga flux, and c) shift in the position of the band edge peak as a function of the Ga flux

of the band edge emission. On the other hand, at lower Ga fluxes the intensities of the band edge and blue emission intensities decrease by several orders of magnitude for blue emission and by an order of magnitude for the band edge emission. At the same time there is a significant shift in the position of the band edge emission to lower energy indicating the formation of a $GaN_{1-x}As_x$ alloy with a significant arsenic content. It is interesting to note that there is no significant difference in the position of the band edge emission for the samples grown with the low Ga fluxes, which in turn indicates an alloy of constant composition. Finally, it is important to recognise that the two effects are related, as the blue emission decreases, the alloy formation takes place.

At high N/Ga ratios a completely different PL spectrum is observed. All evidence for blue emission at 2.6 eV disappears and the position of the band to band recombination shifts to much lower energy. The most likely explanation for this shift is the formation of a $GaN_{1-x}As_x$ alloy. From this shift in energy, using theoretical predictions of the band gap bowing in this alloy system [2], the As mole fraction in the $GaN_{1-x}As_x$ alloy is estimated to be $x \sim 0.2\%$. For growth at 800 °C, this is similar to the best results previously reported at 750 °C.

There is a progressive shift in the energy of the band edge emission with increasing N/Ga ratio. This implies that with a constant arsenic flux, increasing N/Ga ratio, i.e. in more nitrogen rich conditions, the alloy composition changes from $x \sim 0$ to $\sim 0.2\%$. In Fig. 1, the change in N:Ga ratio is rather large and this progressive shift if not so obvious, but combining all the information in both sets of data it is clear that the transition occurs over a finite N:Ga flux ratio. Thus, contrary to theoretical prediction [3], the expermental results discussed above show that in order to form an alloy of $GaN_{1-x}As_x$, it is necessary to have high nitrogen to gallium flux ratios.

X-ray diffraction studies also show several general trends with increasing N/Ga ratio. In samples grown with a low N/Ga ratio, a small peak due to [111] oriented cubic GaAs is observed in addition to the major peak corresponding to [0001] oriented wurtzite GaN. This indicates that at low N/Ga ratios, phase separation occurs. As the N/Ga ratio is increased the amount of GaAs decreases and can no longer be detected. The lattice parameter for the wurtzite GaN is at first constant within experimental error and then increases significantly. This suggests that As is initially incorporated predominantly on the Ga site, which does not significantly alter the lattice parameter, because the covalent radii of As and Ga are similar (the covalent radii of Ga, As and N are 0.126, 0.118 and 0.07 nm, respectively). With increasing N/Ga ratio, the As is incorporated on the N site, thus forming the $GaN_{1-x}As_x$ alloy and increasing the lattice parameter. From the shift in lattice parameter and assuming Vegard's law applies, the maximum alloy composition achieved in this study is estimated to be $x \sim 0.2\%$, in good agreement with the estimates from PL.

Several possible models can be proposed to explain this surprising result based on previous experience of GaN growth kinetics. A first explanation for this data may come from the fact that it is well established that under Ga-rich conditions a thin layer of Ga (1–3 monolayers) exists on the surface during growth, but under N-rich conditions this is probably absent. In addition, at sufficiently high Ga-rich conditions Ga droplets are observed on the surface. The above data supports the hypothesis that to obtain GaN films which show strong blue emission excess Ga is required on the surface during growth, but to form alloys it is necessary to increase the N:Ga ratio and thus avoid the presence of excess Ga on the surface. This absence or presence of excess Ga may in turn lead to several possible models consistent with the data. A first possibility to consider is that films of opposite polarity are obtained under Ga-rich or N-rich conditions and thus the incorporation of As in each polarity will differ. A second possibility is that the incorporation rate of impurities and the stoichiometry of the GaN will be different in the regions grown through a Ga surface liquid compared to the areas of the surface on which there is no Ga present and thus growth occurs at a semiconductor–vacuum interface.

An alternative model may be based on the premise that if the mechanisms controlling alloy growth are similar to those in other III–V materials, there should be significant concentrations of Ga and N vacancies at the growth temperature. As atoms will initially occupy the vacancies on the Ga-sublattice, due to the size of the As atom being similar to Ga and much larger than N. With increasing N/Ga ratio an increase in the number of Ga vacancies is expected at high temperature leading to an initial increase in the density of As atoms on the Ga-sublattice and thus increasing the intensity of the blue emission at 2.6 eV. When the solubility limit for arsenic on the Ga-sublattice is achieved, further increase in the N/Ga ratio results in arsenic being incorporated either onto the N-sublattice or phase separation into As-doped GaN and N-doped GaAs occurs. However, as experimentally demonstrated above, with increasing N/Ga ratio, the tendency to phase separation decreases and the additional As is thus incorporated on the N-sublattice promoting the growth of the alloy. It is also possible that during cooling As which is initially on the Ga-sublattice will form a $GaN_{1-x}As_x$ alloy. However, in alloy samples one might expect to still see blue emission due to As on the Ga site, but this is not observed experimentally, which shows that the real process is more complicated than this simple model.

In conclusion, this study demonstrates that with increasing N to Ga ratio there is a transition from As-doped GaN films showing strong blue emission at 2.6 eV to the growth of $GaN_{1-x}As_x$ alloy films of narrower band gap.

Acknowledgements This work was undertaken with support from EPSRC (GR/M67438) and two of us A.J.W. and C.S.D. would like to thank EPSRC for their studentships.

References

[1] J. I. PANKOVE and J. A. HUTCHBY, J. Appl. Phys. **47**, 5387 (1976).
[2] S. SAKAI, Y. UETA, and Y. TERAUCHI, Jpn. J. Appl. Phys. **32**, 4413 (1993).
[3] C. G. VAN DE WALLE and J. NEUGEBAUER, Appl. Phys. Lett. **76**, 1009 (2000).
[4] T. MATTILA and A. ZUNGER, Phys. Rev. B **58**, 1367 (1998).
[5] S. V. NOVIKOV, A. J. WINSER, I. HARRISON, C. S. DAVIS, and C. T. FOXON, Semicond. Sci. Technol. **16**, 103 (2001).

Spatially Resolved Cathodoluminescence Study of As Doped GaN

A. Bell[1]) (a), F. A. Ponce (a), S. V. Novikov (b), C. T. Foxon (b), and I. Harrison (c)

(a) Department of Physics and Astronomy, Bateman Physical Sciences Center F-wing, Arizona State University, Tempe, AZ, 85287-1504, USA

(b) School of Physics and Astronomy, University of Nottingham, Nottingham, NG7 2RD, UK

(c) School of Electrical and Electronic Engineering, University of Nottingham, Nottingham, NG7 2RD, UK

(Received June 21, 2001; accepted August 2, 2001)

Subject classification: 73.20.Hb; 78.55.Cr; 78.60.Hk; S7.14

The introduction of arsenic (As) into GaN to produce a group-V ternary alloy has been of much recent interest, mostly because of the prospect of reducing the GaN bandgap. We have performed a systematic study of the role of As in GaN grown by molecular beam epitaxy (MBE). The As content of this series of samples varies from 3.4×10^{17} to 4.2×10^{18} cm^{-3}. The data are presented to show how As effects the optical properties of GaN. Our focus is on the nature of the strong luminescence band found at $\sim$475 nm. The intensity of the GaN near bandedge emission is shown to decrease and the 475 nm emission to increase with As content. This is attributed to the large As atoms disrupting the GaN lattice and creating defects or stacking faults that act as non-radiative centers. We have used scanning electron microscopy (SEM) and cathodoluminescence (CL) to investigate the spatial uniformity of the $\sim$475 nm emission in these materials and show that the luminescence is inhomogeneous indicating arsenic segregation.

1. Introduction There has been a growing interest in the development of the GaNAs alloy system in recent years [1–7]. The prospect of reducing the bandgap of GaN by the addition of As has a major influence on the research being conducted. The GaNAs alloy system has a bandgap range stretching from the UV into the IR and the large mismatch, of more than 20%, in the lattice constants of GaN and GaAs has been shown theoretically to lead to a limited miscibility and a strong, composition-dependent bowing of the bandgap [1]. The limited miscibility makes the incorporation of As into GaN difficult, however, the large bowing parameter will lead to a substantial bandgap change when relatively small amounts of As are incorporated into GaN.

In 1976, Pankove and Hutchby [2] conducted a study of implanted GaN. One of the implant elements that were studied was As. A characteristic luminescence band at 2.58 eV was observed only in the As implanted samples. This band has been observed in GaNAs layers grown by metal organic chemical vapor deposition (MOCVD) at 2.58 eV [3] and 2.5 eV [4] and layers grown by MBE at $\sim$2.6 eV [5]. In a theoretical paper, the nature of this luminescence band has been attributed to As occupying a Ga site (As$_{Ga}$) in the GaN lattice [1]. It was shown that the large difference in the covalent radii of N (at 0.75 Å) and As (at 1.20 Å) compared to Ga (at 1.26 Å), makes it is possible for As to occupy a Ga site. The transition level for the As$_{Ga}$ (2+/+) donor was

---

[1]) Corresponding author; Phone: 480 965 0138; Fax: 480 965 7954; e-mail: abigail.bell@asu.edu

Table 1

The As flux used during growth and As concentration found using SIMS are shown for each sample

sample ID	run #	As flux (mbar)	As conc. (10^{17} cm^{-3})
1	MS-115	1×10^{-9}	5.4
2	MS-117	1×10^{-7}	9.3
3	MS-120	5×10^{-7}	7.1
4	MS-119	1×10^{-6}	9.0
5	MS-122	3.5×10^{-6}	18.5
6	MS-223	7.9×10^{-6}	16.4
7	MS-118	1.3×10^{-5}	42.4

shown to be 2.7 eV and for the (+/0) donor state, 2.2 eV. Since As on a nitrogen site (As$_N$) creates donor levels that occur at 0.11 eV and 0.31 eV above the valence band, the ~2.6 eV luminescence was attributed to As$_{Ga}$.

Arsenic doped GaN layers have also been shown to enhance the ~3.28 eV donor-to-acceptor pair (DAP) luminescence [6], which was attributed to As-associated isoelectronic centers.

2. Experimental Details In an effort to further understand the luminescence processes that occur in As-doped GaN, spatially resolved cathodoluminescence (CL) was carried out on As-doped GaN. The As-doped GaN samples were grown on sapphire substrates, at 800 °C, in a home-made MBE reactor, which has been described in detail elsewhere [7]. The active nitrogen species were generated using an Oxford Applied Research CARS25 rf activated plasma source. The As$_2$ flux was generated using a purpose built As source and was varied for different samples. Prior to growth, the sapphire substrate was nitrided at 800 °C for 30 min. The thickness of the samples is estimated to be between 1.0 and 1.1 μm.

The As concentration of the samples was found using secondary ion mass spectroscopy (SIMS) with an As implanted standard. The error in the As concentration was estimated to be approximately 100%, however, the relative As concentrations is more accurate than that. The As flux used during growth and As concentration in the films are displayed in Table 1.

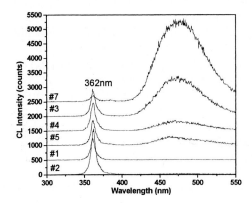

Fig. 1. Room temperature CL spectra of As-doped GaN with $V_{accel} = 3$ keV, $I_{beam} = 170$ pA and magnification ×10000. The ~475 nm emission appears to compete with the near band-edge emission

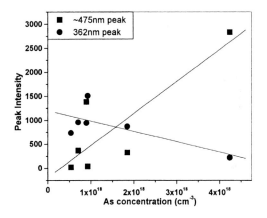

Fig. 2. Peak intensities versus As concentration, taken from the room temperature CL spectra

Scanning electron microscopy (SEM) and CL measurements were carried out using a JSM 6300 microscope. The sample was cooled to 4.2 K using liquid helium. The room temperature CL spectra were taken and the results were related to the As concentration. An SEM image was taken at a high magnification and spatially resolved CL was performed on the features found.

3. Results and Discussion

Figure 1 shows the room temperature CL spectra of samples 1–5 and 7. The acceleration voltage (V_{accel}) was 3 kV, the beam current (I_{beam}) was 170 pA and the magnification (mag) was ×10000 (i.e. scan width ∼20 μm). The GaN bandedge emission occurs at 362 nm (3.43 eV) and the As defect related emission band is observed at ∼475 nm (2.61 eV). Since the As concentration in these samples is in the range of parts per million, we observe no emission related to the GaNAs alloy bandedge. The intensity of the bandedge emission appears to get stronger as the ∼475 nm emission decreases. This would suggest that the two are competing processes. The position of the near 475 nm emission band is approximately the same for all samples. The intensities of the GaN bandedge emission and the ∼475 nm emission are plotted versus As concentration in Fig. 2. The trend shown in this figure is that the bandedge emission intensity decreases with increasing As concentration and that the ∼475 nm emission intensity increases with As concentration. The increase in the As emission correlates with the theory that this emission is somehow related to As. A possible explanation for the decrease in the GaN bandedge emission is that when As is introduced into the GaN lattice, even in small concentrations, the large As atoms distort the GaN lattice, creating dislocations or stacking faults that act as non-radiative centers. As the As concentration is increased, the number of non-radiative centers is increased and hence the GaN bandedge emission intensity will decrease.

SEM images and CL spectra of sample 6 were taken at 4.2 K, with $V_{accel} = 5$ keV, $I_{beam} = 120$ nA at a magnification of ×40 K (i.e. the horizontal width is 5 μm) and a slit width of 1 mm. Figure 3 shows the SEM image of the sample. Light contrast features, measuring ∼1 μm across were observed. Fig. 4 shows the CL spectra taken over the whole region shown in the SEM image and at a magnification of ×300 K in the positions marked 1–4. The spectra taken over the whole area exhibits emission at 379 nm (3.27 eV) and a broad band at ∼475 nm. The 379 nm emission could be due to bandedge emission from zinc blende GaN. This seems unlikely since the room temperature emission of sample 6 is consistent with wurtzite GaN. It is more likely to be the donor–acceptor pair (DAP) commonly found in GaN [6]. The

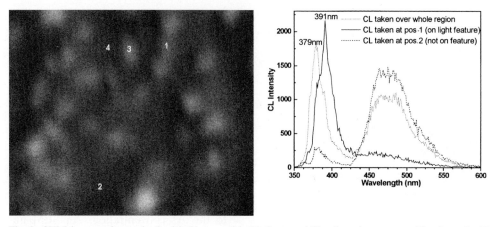

Fig. 3. SEM image of sample 6 with $V_{accel} = 5$ keV, $I_{beam} = 120$ nA and at a magnification of x40 K. The horizontal width is 5 μm

Fig. 4. CL spectra of sample 6 taken at a magnification of $40000\times$ over the whole region shown in Fig. 3. and at a magnification of $300000\times$ at positions 1 and 2 shown in Fig. 3. $V_{accel} = 5$ keV and $I_{beam} = 120$ nA

4 K spectra of the other samples exhibit the wurtzite bandedge and the 379 nm emission which is often accompanied by the first and second LO phonon replicas which are a common feature of DAP emission. Thus, we attribute the emission to a DAP.

The CL spectra taken from the light feature at position 1, exhibits strong emission at 391 nm (3.17 eV) and the intensity of the broad emission ∼475 nm is reduced. The 391 nm emission is redshifted in comparison to the spectra taken over the whole region, which would suggest this is a region of high As concentration in which the GaNAs alloy is forming. The spectra taken from the dark region at position 2 exhibits weak emission in the region of the GaN bandedge and strong 475 nm emission. Similar results were obtained for positions 3 and 4. This inhomogeneous emission would suggest that the As is segregated across the sample.

4. Conclusions In conclusion, as the As concentration in GaN is increased, the GaN near bandedge emission intensity decreases. This is explained as being caused by nonradiative centers, which are created by disruption to the GaN lattice by the large As atoms. The introduction of As into GaN increases the intensity of the ∼475 nm emission. This adds strength to the theory that the ∼475 nm emission is As related. An SEM/CL study showed that the CL emission is inhomogeneous and that in ∼1 μm regions, the GaNAs alloy is formed.

Acknowledgements We are grateful for support from the Office of Naval Research (contract No. N00014-00-1-0133) and from Durel Corporation. The growth of the material was undertaken with support from UK EPSRC (GR/M67438).

References

[1] J. Neugebauer and C. G. Van de Walle, Phys. Rev. B **51**, 10568 (1995).
[2] J. I. Pankove and J.A. Hutchby, J. Appl. Phys. **47**, 5387 (1976).
[3] X. Li, S. Kim, E. E. Reuter, S. G. Bishop, and J. J. Coleman, Appl. Phys. Lett. **72**, 1990 (1998).
[4] A. F. Tsatsulnikov, B. Ya. Ber, A. P. Kartashova, Yu. A. Kudryavtsev, N. N. Ledentsov, V. V. Lundin, M. V. Maksimov, A. V. Sakharov, A. S. Usikov, Zh. I. Alferov, and A. Hoffmann, Semiconductors **33**, 728 (1999).
[5] S. V. Novikov, A. J. Winser, I. Harrison, C. S. Davis, and C. T. Foxon, Semicond. Sci. Technol. **16**, 103 (2001).
[6] S. R. Jin, M. Ramsteiner, H. T. Grahn, K. H. Ploog, Z. Q. Zhu, D. X. Shen, A. Z. Li, P. Metev, and L. J. Guido, J. Cryst. Growth **212**, 56 (2000).
[7] C. T. Foxon, T. S. Cheng, S. V. Novikov, N. J. Jeffs, O. H. Hughs, Yu. V. Melnik, A. E. Nikolaev, and V. A. Dmitriev, Surf. Sci. **421**, 377 (1999).

On the Origin of Blue Emission from As-Doped GaN

I. Harrison (a), S. V. Novikov (b), T. Li (b), R. P. Campion (b), C. R. Staddon (b), C. S. Davis (b), Y. Liao (a), A. J. Winser (a, b), and C. T. Foxon (b)

(a) School of Electrical and Electronic Engineering, University of Nottingham, Nottingham, NG7 2RD, UK

(b) School of Physics and Astronomy, University of Nottingham, Nottingham, NG7 2RD, UK

(Received June 21, 2001; accepted June 28, 2001)

Subject classification: 61.10.Nz; 68.37.Ps; 68.55.Jk; 78.55.Cr; S7.14

As-doped GaN films have been grown by plasma-assisted molecular beam epitaxy and their properties investigated using atomic force microscopy, X-ray diffraction and photoluminescence (PL) spectroscopy. The structural properties of the As-doped GaN films improve with increasing sample thickness. The room temperature PL is dominated by a strong blue emission band, exhibiting multiple peaks centered at 2.6 eV. The number of peaks increases monotonically with sample thickness. From this we conclude that the multiple peaks in the blue emission band of As-doped GaN samples arise mainly from optical interference effects. However, the possibility of several transitions involving As being responsible for the blue emission process cannot be excluded.

Group III-nitride semiconductors are now widely studied in the literature owing to their potential applications for light emitting devices in the visible part of the spectrum and for high-power, high-frequency, high-temperature electronic device applications. The present range of devices use the AlGaInN material system, but more recently alloys with mixed group V elements have been increasingly studied for a variety of applications. The Ga–N–As materials system is extremely interesting for several reasons. N-doped GaAs is being increasingly studied due to the large negative bowing in the band gap [1–5]. This allows one to achieve a significant shift in the band gap for comparatively small amounts of nitrogen in the alloy, with potential applications for long wavelength opto-electronic devices. However, the large difference in lattice parameters between GaN and GaAs and the large miscibility gap makes it difficult to grow films of high quality [1–5].

Much less attention has been focussed on As-doped GaN due to the lower solubility of As in GaN [2–4]. One interesting observation, with potential device applications, is the report of blue luminescence at low temperature in As-doped GaN prepared by ion implantation and in films grown by metalorganic vapour phase epitaxy [6, 7]. Recently we reported strong blue luminescence at room temperature in As-doped GaN samples grown by Plasma-Assisted Molecular Beam Epitaxy (PA-MBE) [8–10]. As-doped GaN films might, therefore, be a suitable replacement for InGaN layers in opto-electronic devices based on growth by MBE.

In the room temperature photoluminescence (PL) spectrum from our As-doped GaN samples grown by PA-MBE [8–10] several peaks are observed in the near band edge region at about 3.2 and 3.4 eV together with strong blue emission at about 2.6 eV. With increasing As flux during growth, the intensity of the band edge features decrease and the blue emission intensity increases [8]. With increasing nitrogen to gallium ratio, however, the intensities of both band edge emission and blue emission increase in direct proportion to one another [10]. In all of our As-doped samples, the blue emission band is observed with a maximum intensity around 2.6 eV, but the structure of the emission varies from sample to sample. Our result suggests that As may be incorporated on the Ga site [10] and therefore acts as a double donor [11] and several transitions may be involved in the blue emission process. On the other hand, we observe systematic changes in the position of the peaks with changing growth conditions [8, 9]. In this paper, we will investigate the origin of the blue emission in As-doped GaN by studying the structural, electrical and optical properties of the material using X-ray diffraction, atomic force microscopy (AFM) and PL spectroscopy for As-doped GaN samples of different thickness.

The samples used in this study were grown by PA-MBE on sapphire substrates, full details of the growth procedures used can be found elsewhere [8–10]. All the samples discussed here were grown at 800 °C, where our previous studies have shown that strong blue photoluminescence is observed in the As-doped GaN samples [8–10]. For the present samples, an As_2 flux of 4×10^{-6} mbar (Beam Equivalent Pressure, BEP) and a Ga flux of 8×10^{-7} mbar BEP were used. The nitrogen flux was 2×10^{-5} mbar (BEP) with a plasma source operated at 450 W. All the growth conditions remained constant and only the growth time was varied as shown in Table 1.

The thickness of each of the As-doped GaN samples was determined by measuring the optical reflectivity as a function of wavelength in range 500–1100 nm, using a refractive index of 2.4. Figure 1 shows the variation in sample thickness with growth time. As expected there is a linear relation between the thickness and growth time, indicating that the growth rate remained constant.

The surface morphology of the samples was investigated using Atomic Force Microscopy (TopoMetrix Explorer 2000) and all layers were analysed in the contact mode. The samples showed a monotonic increase in subgrain size from about 200 nm to approximately 500 nm with increasing sample thickness. The structural properties of the samples were investigated by X-ray diffraction using a Philips X'Pert MRD diffractometer. A variety of scan modes including symmetric θ–2θ scans, pole plots with asymmetric reflections and reciprocal lattice scans were employed to determine the various phases present, their relative concentrations and symmetry properties, and Ω-scans were used to estimate the mosaic spread of the sub-grains. Table 1 summarises the results of this structural study. As shown in Fig. 2, the structural quality of the As-doped GaN improves with sample thickness. From the full width at half maximum (FWHM) of the Ω-scans, we have deduced the size of the hexagonal GaN subgrains. As shown in Table 1, the size of the subgrains in-

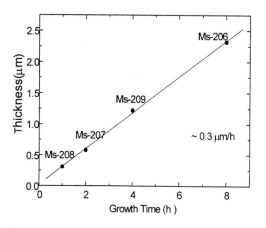

Fig. 1. Dependence of the thickness of As-doped GaN samples on growth time for constant As, Ga and active N fluxes

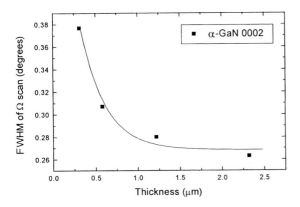

Fig. 2. Dependence of the X-ray FWHM Ω-scans on sample thickness

creases with increasing sample thickness, in good quantitative agreement with the AFM data.

The X-ray studies also show that in all our As-doped GaN samples we have a mixture of (0001) oriented hexagonal α-GaN, (111) oriented cubic β-GaN and (111) oriented GaAs. As shown in Table 1, the proportion of cubic GaN decreases rapidly with film thickness indicating that the cubic regions are close to the sapphire–epilayer interface.

From the X-ray intensity of the GaAs peaks relative to GaN, we can conclude that GaAs proportion is at about 0.03% of the total content of the film. The fraction is roughly constant for all four films. We can estimate the size of the GaAs crystallites to be roughly constant at about 60 nm. Our previous studies indicated the presence of a significant As surface concentration on the As-doped GaN samples [9]. However, because the GaAs fraction is roughly constant with film thickness, we can conclude that small GaAs crystallites are present throughout the layer. The average size of the GaAs crystallites in our samples it too large, to expect there to be quantum dot like behaviour from such inclusions.

The optical properties of the samples were investigated using room temperature photoluminescence. The excitation source was a Kimmon He–Cd laser operating at 325 nm. The maximum power of the laser at the sample surface was 9.4 mW. Further experiment details have been previously published [12]. For all samples, we see both near band edge and blue emission. The intensity of the blue emission is much larger than the band edge emission as previously reported [8–10]. Figure 3 shows the structure of the blue band region of the PL spectra as a function of film thickness. The

Table 1
Sample details. The Ω-scan used the (0002) peak from α-GaN

sample number	Ms-208	Ms-207	Ms-209	Ms-206
growth time (h)	1	2	4	8
thickness (μm)	0.31	0.58	1.22	2.32
FWHM of Ω-scan	0.3767°	0.3072°	0.2801°	0°2
β-GaN cubic content (%)	23.3	15.5	9.2	4.9
GaAs size (nm)	53.6	66.6	49.5	69.8
α-GaN size (nm)	289	383	454	520

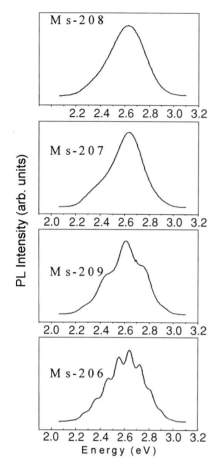

Fig. 3. Blue band PL spectra from As-doped GaN samples of increasing thickness

intensity of blue emission varies with sample thickness, but for convenience we have normalised the curves. For all samples the blue emission is centred around 2.6 eV as we have previously observed [8–10]. For the thinnest samples, only one peak is observed, but for thicker samples multiple peaks can be resolved. The important point here is that the number of peaks observed increases monotonically with sample thickness indicating that the key factor involved in determining the structure of the blue band is film thickness and that most of the peaks observed are due to optical interference effects. The blue emission is very broad, and so the possibility of several transitions being involved cannot be completely excluded.

In conclusion, As-doped GaN films of different thickness grown by PA-MBE have been studied and the structural quality of the films improves monotonically with sample thickness. There is clear evidence for the existence of a small fraction of cubic GaN at the substrate–epilayer interface and for small inclusions of GaAs in the bulk. The PL at room temperature is dominated by strong blue emission centred at 2.6 eV. Multiple peaks in blue band of the PL spectrum centred at 2.6 eV are observed in our As-doped GaN samples. The number of peaks in the blue band increases monotonically with sample thickness and arise from optical interference effects. However, we cannot exclude the possibility of several transitions involving As being responsible for the blue emission process.

Acknowledgements This work was undertaken with support from EPSRC (GR/M67438) and two of us AJW and CSD would like to thank EPSRC for their studentships.

References

[1] D. SCHLENKER, T. MIYAMOTO, Z. PAN, F. KOYAMA, and K. IGA, J. Cryst. Growth **196**, 67 (1999).
[2] S. SAKAI, Y. UETA, and Y. TERAUCHI, Jpn. J. Appl. Phys. **32**, 4413 (1993).
[3] M. WEYERS, M. SATO, and H. ANDO, Jpn. J. Appl. Phys. **31**, L853 (1992).
[4] K. IWATA, H. ASAHI, K. ASAMI, R. KUROIWA, and S. GONDA, Jpn. J. Appl. Phys. **37**, 1436 (1998).
[5] M. KONDOW, T. KITATANI, S. NAKATSUKA, M. C. LARSON, K. NAKAHARA, Y. YAZAWA, M. OKAI, and K. UOMI, IEEE J. Sel. Top. Quantum Electron. **3**, 719 (1997).
[6] J. I. PANKOVE and J. A. HUTCHBY, J. Appl. Phys. **47**, 5387 (1976).

[7] X. Li, S. Kim, E. E. Reuter, S. G. Bishop, and J. J. Coleman, Appl. Phys. Lett. **72**, 1990 (1998).
[8] A. J. Winser, S. V. Novikov, C. S. Davis, T. S. Cheng, C. T. Foxon, and I. Harrison, Appl. Phys. Lett. **77**, 2506 (2000).
[9] C. T. Foxon, S. V. Novikov, T. S. Cheng, C. S. Davis, R. P. Campion, A. J. Winser, and I. Harrison, J. Cryst. Growth **219**, 327 (2000).
[10] S. V. Novikov, A. J. Winser, I. Harrison, C. S. Davis, and C. T. Foxon, Semicond. Sci. Technol. **16**, 103 (2001).
[11] C. G. Van de Walle and J. Neugebauer, Appl. Phys. Lett. **76**, 1009 (2000).
[12] D. J. Dewsnip, A. V. Andrianov, I. Harrison, J. W. Orton, D. E. Lacklison, G. B. Ren, S. E. Hooper, T. S. Cheng, and C. T. Foxon, Semicond. Sci. Technol. **13**, 500 (1998).

The Influence of As on the Optimum Nitrogen to Gallium Ratio Required to Grow High Quality GaN Films by Molecular Beam Epitaxy

C. T. Foxon[1]) (a), S. V. Novikov[*]) (a), R. P. Campion (a), Y. Liao[**]) (a), A. J. Winser (a, b), and I. Harrison (b)

(a) School of Physics and Astronomy, University of Nottingham, Nottingham, NG7 2RD, UK

(b) School of Electrical and Electronic Engineering, University of Nottingham, Nottingham, NG7 2RD, UK

(Received May 29, 2001; accepted August 1, 2001)

Subject classification: 68.37.Ps; 68.55.Ln; 78.55.Cr; 81.05.Ea; 81.15.Hi; S7.14

We have studied the influence of arsenic on the growth and optical properties of GaN films grown by plasma-assisted molecular beam epitaxy (PA-MBE) under different stoichiometric conditions. For GaN films grown without arsenic, there is an optimum N to Ga flux ratio, which results in films showing a good morphology and the highest room temperature band edge luminescence efficiency. However, with arsenic present, the highest blue and band edge luminescence efficiency is observed in GaN at a higher N to Ga ratio. The onset of GaNAs alloy formation is associated with a change in surface morphology. This suggests that arsenic is acting as an isoelectronic surfactant during the growth of GaN films by PA-MBE, changing the morphology, the optical properties and nature of the electrically active defects.

Introduction Gallium nitride and its related alloys are now widely used for electronic and optoelectronic devices. Several studies have shown that the properties of GaN films grown by molecular beam epitaxy (MBE) depend critically on the growth conditions. Growth temperature is an important parameter, but the N to Ga ratio also plays a critical role in determining GaN crystal quality, the electrical properties and device performance [1–3].

It is now well established that surfactants influence the properties of III–V compounds grown by MBE. We have previously demonstrated that for GaN films grown by plasma-assisted MBE (PA-MBE), arsenic has a very significant influence on both the morphology of the films and on their optical properties [4]. In addition to the normally observed band edge emission at about 3.4 eV, arsenic doped GaN films show strong blue luminescence (centred around 2.6 eV) at room temperature. In this paper we discuss in detail the influence of arsenic on the nitrogen to gallium ratio required to obtain optimum morphology and optical properties for GaN films grown by PA-MBE.

Results and Discussion The two sets of samples used in this study were grown by PA-MBE at 800 °C on sapphire substrates, full details of the growth procedures used can be found elsewhere [4]. One set of GaN layers were grown with an intentional arsenic flux of about 4×10^{-6} mbar (beam equivalent pressure, BEP) and for comparison a

---

[1]) Corresponding author; Fax: +44-115-9515184, e-mail: C.Thomas.Foxon@nottingham.ac.uk
[*]) On leave from: Ioffe Physical-Technical Institute, St. Petersburg, Russia.
[**]) On leave from: University of Science and Technology of China, Hefei 230026, P.R. China.

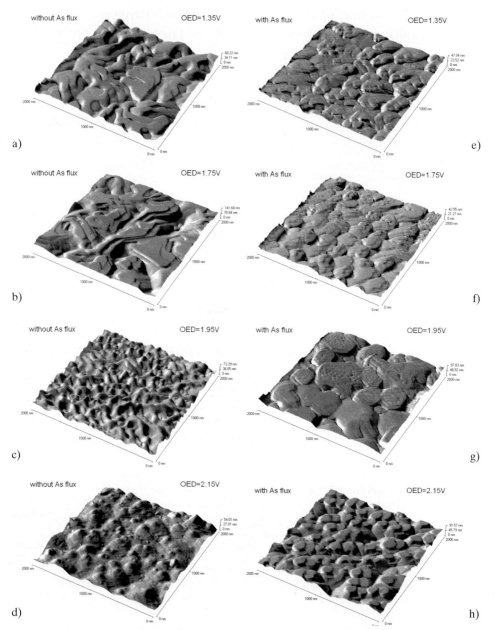

Fig. 1. (colour) Atomic force microscope images of the surface of GaN films grown by PA-MBE under different N to Ga flux ratios: a–d) without arsenic flux; e–h) with arsenic flux. OED values during growth were as follows: a, e) 1.35 V; b, f) 1.75 V; c, g) 1.95 V and d, h) 2.15 V

similar set of samples were grown without arsenic flux. However, due to the previous use of arsenic, there is a background pressure of arsenic in the MBE chamber of about 10^{-9} mbar. The intensity of the plasma is monitored using an optical emission detector

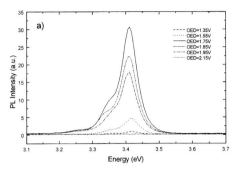 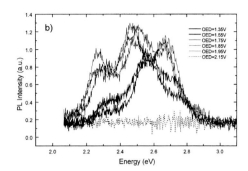

Fig. 2. Room temperature a) near band edge and b) deep PL emission from GaN films grown without an arsenic flux for different N to Ga flux ratios

(OED). The OED signal is proportional to the amount of active nitrogen coming from the source.

Figure 1[2]) shows AFM images for the two sets of samples, grown with and without an arsenic flux for the same N to Ga ratios. In both cases there is a progressive change in the active N to Ga ratio, from Ga rich to more N rich growth conditions. In the case of the films grown with arsenic, the transition occurs at a measurably higher N to Ga ratio. This indicates that the arsenic is having a significant influence on the growth kinetics and surface morphology of the GaN films.

The room temperature photoluminescence from the set of films grown without an intentional arsenic flux is shown in Fig. 2. At the lowest N to Ga ratio, the band edge emission at 3.4 eV is relatively weak as shown in Fig. 2a. As we approach stoichiometric growth, by increasing the OED, the intensity of the band edge emission increases to a maximum value. At the point where the morphology changes from a flat terrace structure to columnar growth (Figs. 1b to c), the intensity of the band edge luminescence starts to decrease. At the highest N to Ga ratio, no photoluminescence is detected at room temperature. At the same time, as shown in Fig. 2b, we also see significant differences in the spectrum of the deep emission above and below the optimum OED value. This possibly indicates that there may be different types of defect

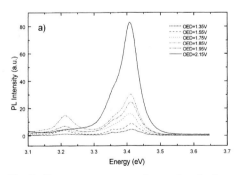 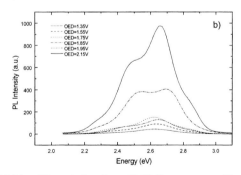

Fig. 3. Room temperature a) near band edge and b) blue PL emission from GaN films grown with an arsenic flux for different N to Ga flux ratios

[2]) Colour figure is published online (www.physica-status-solidi.com).

levels introduced under Ga and N-rich growth conditions. However, the major point is that optimum photoluminescence at room temperature is obtained in films grown under slightly gallium rich conditions, this is similar to the situation for optimum transport properties where the highest mobilities are also obtained under slightly Ga-rich conditions [3].

For films grown with arsenic a quite different behaviour is observed. As shown in Fig. 3, with increasing N to Ga ratio, there is a progressive increase in the intensity of the band edge emission at about 3.4 eV and of the blue emission at about 2.6 eV. However, for the band edge emission at 3.2 eV we see an initial increase in intensity and at a certain N to Ga ratio the intensity then decreases abruptly. The nature of the emission at 3.2 eV will be discussed elsewhere. Finally the energy of the band edge peak at about 3.4 eV shifts progressively to lower energy with increasing N to Ga ratio, indicating the onset of GaNAs alloy formation. At the highest N to Ga ratio studied here, the red shift in band edge emission starts to be significant and we also see the biggest change in surface morphology, as shown in Fig. 1h. Because the lattice parameters of GaAs and GaN are very different, we may speculate that this change in morphology may be associated with the strain introduced into the lattice due to the onset of alloy formation.

Our results suggest that arsenic is acting as an isoelectronic surfactant during the growth of GaN films by PA-MBE, changing the growth kinetics, the surface morphology, the optical properties and nature of the electrically active defects.

Acknowledgements This work was undertaken with support from EPSRC (GR/M67438) and one of us (AJW) would like to thank EPSRC for a studentship.

References

[1] S. H. Cho, K. Hata, T. Maruyama, and K. Akimoto, J. Cryst. Growth **173**, 260 (1997).
[2] M. A. Sanchez-Garcia, E. Calleja, E. Monroy, F. J. Sanchez, F. Calle, E. Munoz, and R. Beresford, J. Cryst. Growth **183**, 23 (1998).
[3] B. Heying, R. Averbeck, L. F. Chen, E. Haus, H. Riechert, and J. S. Speck, J. Appl. Phys. **88**, 1855 (2000).
[4] S. V. Novikov, A. J. Winser, I. Harrison, C. S. Davis, and C. T. Foxon, Semicond. Sci. Technol. **16**, 103 (2001).

Temperature Dependence of the Miscibility Gap on the GaN-Rich Side of the Ga–N–As System

S. V. Novikov[1])[*] (a), T. Li (a), A. J. Winser (a, b), R. P. Campion (a), C. R. Staddon (a), C. S. Davis (a), I. Harrison (b), and C. T. Foxon (a)

(a) School of Physics and Astronomy, University of Nottingham, Nottingham, NG7 2RD, UK

(b) School of Electrical and Electronic Engineering, University of Nottingham, Nottingham, NG7 2RD, UK

(Received June 29, 2001; accepted July 8, 2001)

Subject classification: 68.55.Nq; 81.05.Ea; 81.15.Hi; S7.14

We have investigated the temperature dependence of the transition from single phase films of $GaN_{1-x}As_x$ to phase separated layers, which show regions of hexagonal [0001] oriented GaN, cubic [111] oriented GaAs and hexagonal [0001] oriented $GaN_{1-x}As_x$. We see a strong temperature dependence of the arsenic flux at which GaAs inclusions are first observed. Finally the intensity of blue emission observed in As-doped GaN samples decreases strongly with decreasing growth temperature.

Introduction There is now considerable interest in the N-rich side of the GaNAs system [1–15]. There are two main reasons, which motivate such studies; the large negative bowing in the band gap for GaNAs solid solutions and strong blue room temperature emission from arsenic doped GaN. However, data on the miscibility gap at this end of the phase diagram is largely absent from the literature.

$GaN_{1-x}As_x$ alloys at the N-rich end of the phase diagram have been grown by metalorganic vapour phase epitaxy (MOVPE) [8, 9] and by molecular beam epitaxy (MBE) [6, 7, 10]. For both techniques, it is difficult to obtain a high concentration of As in the alloy before phase separation occurs. The highest concentrations reported in MBE layers are $x \sim 0.26\%$ [6, 7] grown at 750 °C and $x \sim 1\%$ [10] at 500 °C, respectively. This suggests that the solubility limit may be a function of temperature.

Several authors have reported blue emission from As-doped GaN. The first observations were from As ion implanted layers of GaN [1], similar results have been obtained later in arsenic doped GaN films grown by MOVPE [9, 11, 12]. Recently we have demonstrated very strong blue emission at room temperature in As-doped GaN layers grown by MBE on sapphire substrates using both arsenic dimers and tetramers [13, 14]. The intensity of the blue emission at about 2.6 eV from the As-doped samples is more than an order of magnitude stronger than the band edge emission in undoped GaN samples and can be seen in normal lighting.

The aim of this paper is to discuss the temperature dependence of the miscibility gap in N-rich side of the Ga-N-As system and the influence this has on the properties of the films.

---

[1]) Corresponding author; Phone: +44-115-9515193; Fax: +44-115-9515184; e-mail: Sergei.Novikov@nottingham.ac.uk

[*]) On leave from the A.F. Ioffe Physico-Technical Institute, St. Petersburg, Russia.

Experimental Details The samples used in this study were grown by plasma-assisted molecular beam epitaxy (PA-MBE) on sapphire substrates, full details of the growth procedures used can be found elsewhere [13–15]. The samples discussed here were grown in the temperature range 500–800 °C. For all films we have kept the N/Ga ratio constant and changed the arsenic flux over a wide range from a background level of approximately 10^{-9} to about 2×10^{-5} mbar. The resulting films have been studied by optical microscopy, atomic force microscopy, Auger electron spectroscopy, X-ray diffraction and photoluminescence spectroscopy (PL).

Results and Discussion At all growth temperatures, with a sufficient arsenic flux, we observe phase separation in the GaNAs films. Figure 1 shows a typical example of such phase separation in films grown at the N-rich end of the Ga-N-As system observed using X-ray diffraction. In this particular case the films were grown at 700 °C. In the absence of an arsenic flux GaN films are obtained without any phase separation. At low arsenic fluxes a $GaN_{1-x}As_x$ alloy film is obtained with a low concentration of arsenic as shown in Fig. 1a. With increasing arsenic flux we observe a transition to phase separated material. In this case in detailed X-ray diffraction studies we see three peaks corresponding within experimental error to [0001] oriented hexagonal GaN, [111] oriented cubic GaAs and [0001] oriented hexagonal $GaN_{1-x}As_x$. From thermodynamic considerations it is clear that the GaAs inclusion may contain some dilute concentration of N, which cannot be resolved within the accuracy of our measurements due to the broad nature of the spectra. Figure 1b shows a typical example for such phase-separated material around the GaN peak position. For $GaN_{1-x}As_x$ alloys showing no phase separation (Fig. 1a grown with an arsenic flux of 6×10^{-8} mbar) the arsenic content estimated from the X-ray peak position is rather small $x = 0.003$. However, in the phase-separated material (Fig. 1b grown with a higher arsenic flux of 5×10^{-7} mbar) the $GaN_{1-x}As_x$ alloy content is much higher $x = 0.013$, about four to five times. In calculating the arsenic content, we have assumed the films are fully relaxed, which may not be strictly true, nevertheless there is clearly a significant difference in arsenic content.

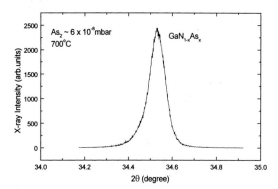

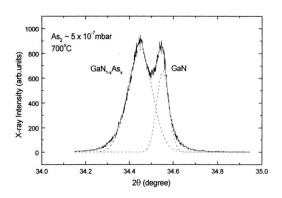

Fig. 1. X-ray θ–2θ study of a dilute $GaN_{1-x}As_x$ alloy showing a) a single peak and b) phase separation into two peaks

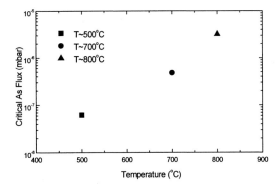

Fig. 2. Temperature variation in critical arsenic flux at which phase separated layers containing GaAs inclusions first occurs

Using X-ray diffraction, we can also detect GaAs inclusions in films grown with a sufficiently high arsenic flux at all substrate temperatures. However, as shown in Fig. 2 the flux at which we first observe GaAs inclusions depends strongly on the growth temperature, increasing with increasing temperature. This suggests that the lifetime of arsenic on the surface may be a crucial factor involved in determination of this process.

As we have previously reported [13–15], we observe strong blue emission at room temperature from As-doped GaN films grown at 800 °C. As the substrate temperature is reduced, the intensity of this blue emission decreases and cannot be observed in samples grown at 500 °C for all arsenic fluxes we have studied.

Acknowledgements This work was undertaken with support from EPSRC (GR/M67438) and two of us AJW and CSD would like to thank EPSRC for their studentships.

References

[1] J. I. PANKOVE and J. A. HUTCHBY, J. Appl. Phys. **47**, 5387 (1976).
[2] S. SAKAI, Y. UETA, and Y. Terauchi, Jpn. J. Appl. Phys. **32**, 4413 (1993).
[3] C. G. VAN DE WALLE and J. NEUGEBAUER, Appl. Phys. Lett. **76**, 1009 (2000).
[4] T. MATTILA and A. ZUNGER, Phys. Rev. B **58**, 1367 (1998).
[5] T. MATTILA and A. ZUNGER, Phys. Rev. B **59**, 9943 (1999).
[6] K. IWATA, H. ASAHI, K. ASAMI, R. KUROIWA, and S. GONDA, Jpn. J. Appl. Phys. **37**, 1436 (1998).
[7] R. KUROIWA, H. ASAHI, K. ASAMI, S. J. KIM, K. IWATA, and S. GONDA, Appl. Phys. Lett. **73**, 2630 (1998).
[8] M. GERASIMOVA, B. GAFFEY, P. MITEV, L. J. GUIDO, K. L. CHANG, K. C. HSIEH, S. MITHA, and J. SPEAR, MRS Internet J. Nitride Semicond. Res. **4S1**, G3.44 (1999).
[9] S. YOSHIDA, T. KIMURA, J. WU, J. KIKAWA, K. ONABE, and Y. SHIRAKI, MRS Internet J. Nitrides Semicond. Res. **5S1**, W3.41 (2000).
[10] Y. ZHAO, F. DENG, S. S. LAU, and C. W. TU, J. Vac. Sci. Technol. B **16**, 1297 (1998).
[11] X. LI, S. KIM, E. E. REUTER, S. G. BISHOP, and J. J. COLEMAN, Appl. Phys. Lett. **72**, 1990 (1998).
[12] A. F. TSATSUL'NIKOV, I. L. KRESTNIKOV, W. V. LUNDIN, A. V. SAKHAROV, A. P. KARTASHOVA, A. S. USIKOV, ZH. I. ALFEROV, N. N. LEDENTSOV, A. STRITTMATTER, A. HOFFMANN, D. BIMBERG, I. P. SOSHNIKOV, D. LITVINOV, A. ROSENAUER, D. GERTHSEN, and A. PLAUT, Semicond. Sci. Technol. **15**, 766 (2000).
[13] C. T. FOXON, S. V. NOVIKOV, T. S. CHENG, C. S. DAVIS, R. P. CAMPION, A. J. WINSER, and I. HARRISON, J. Crystal Growth **219**, 327 (2000).
[14] A. J. WINSER, S. V. NOVIKOV, C. S. DAVIS, T. S. CHENG, C. T. FOXON, and I. HARRISON, Appl. Phys. Lett. **77**, 2506 (2000).
[15] S. V. NOVIKOV, A. J. WINSER, I. HARRISON, C. S. DAVIS, and C. T. FOXON, Semicond. Sci. Technol. **16**, 103 (2001).

The Influence of Arsenic Incorporation on the Optical Properties of As-doped GaN Films Grown by Molecular Beam Epitaxy Using Arsenic Tetramers

S. V. Novikov[1]) (a, b), T. Li (a), A. J. Winser (a, c), C. T. Foxon (a),
R. P. Campion (a), C. R. Staddon (a), C. S. Davis (a), I. Harrison (c),
A. P. Kovarsky (b), and B. Ja. Ber (b)

(a) School of Physics and Astronomy, University of Nottingham, Nottingham, NG7 2RD, UK

(b) Ioffe Physical-Technical Institute, St.Petersburg, 194021, Russia

(c) School of Electrical and Electronic Engineering, University of Nottingham, Nottingham, NG7 2RD, UK

(Received June 29, 2001; accepted July 17, 2001)

Subject classification: 68.55.Ln; 78.55.Cr; 81.05.Ea; 81.15.Hi; S7.14

We have studied the influence of the incorporation of As on the optical properties of As-doped GaN layers grown by plasma-assisted molecular beam epitaxy (PA-MBE) using arsenic tetramers. The doping level of arsenic was determined by secondary ion mass spectrometry. The arsenic concentration is uniform throughout the layers. There is a sub-linear dependence of the arsenic incorporation on the flux with a log–log slope of about 0.1. The photoluminescence from the As-doped GaN films consists of UV excitonic emission at 3.4 eV, UV emission at 3.2 eV and a strong blue band centred at 2.6 eV. The intensity of the blue band centred at 2.6 eV increases more rapidly with arsenic flux than the concentration of arsenic in the bulk, and has a log–log slope of about 0.49. This suggests an approximately fourth-power dependence of the intensity of the blue emission on the concentration of arsenic in the GaN films.

Introduction During the last few years there has been considerable theoretical and experimental interest in arsenic doped GaN [1–4]. There are several reasons, which motivate such investigations – a large negative bowing in the energy band gap for GaNAs solid solutions, the influence of arsenic as a surfactant during the growth of GaN and strong blue emission at room temperature from arsenic doped GaN. The aim of this paper is to understand the concentration of arsenic in GaN films, which leads to dramatic changes in their optical properties giving rise to strong blue emission.

Experimental Details The samples used in this study were grown by PA-MBE on sapphire substrates, full details of the procedures used can be found elsewhere [4]. The samples for this study were grown at 800 °C, where our previous studies have shown that strong blue photoluminescence is observed in the As-doped GaN samples [4]. For the present samples all the growth conditions remained constant, except for the As_4 flux, which was varied from 10^{-9} to 10^{-5} mbar (beam equivalent pressure, BEP). A Ga flux of 8×10^{-7} mbar (BEP) was used. The nitrogen flux was 2×10^{-5} mbar (BEP) with a plasma source operated at 450 W. The structural properties of the samples were

---

[1]) Corresponding author; Phone: +44-115-9515193, Fax: +44-115-9515184,
e-mail: Sergei.Novikov@nottingham.ac.uk

investigated by X-ray diffraction using a Philips X'Pert MRD diffractometer. The optical properties of the samples were investigated using room temperature photoluminescence. The As-doped GaN films were analysed for As, C, and O using a CAMECA IMS-4f secondary ion mass spectrometer (SIMS). For quantitative analysis, properly implanted GaN standards were used for all elements.

Results and Discussion We have studied the arsenic incorporation in GaN films grown by PA-MBE. SIMS spectrum from the As-doped GaN samples show that the As concentration is constant with depth. The concentration of arsenic increases monotonically with increasing As_4 flux as shown in Fig. 1. Surprisingly, we observe a change by a factor of only about three in the arsenic concentration for a four-decade change in the arsenic flux. The rate of increase of concentration with flux has a log–log slope of approximately 0.1.

The photoluminescence from the As-doped GaN films consists of UV excitonic emission at 3.4 eV, UV emission at 3.2 eV and a strong blue band centred at 2.6 eV. Figure 1 also shows the dependence of the intensity of the blue PL peak at 2.6 eV on the arsenic flux. The important point to note is that the slope of the line for the blue intensity versus arsenic flux (0.49) is significantly greater than the slope of the As-concentration versus arsenic flux (0.1). This implies that the blue PL intensity depends very strongly on the As concentration in the GaN lattice, an approximately fourth-power dependence, which may relate to the use of As_4 as a source for arsenic.

In Fig. 1, the SIMS concentration corresponds to the total amount of arsenic in the GaN. From SIMS, we cannot easily distinguish between the location of As in the two sub-lattices of GaN. Our X-ray studies demonstrate that a small amount of GaAs is observed for the two highest arsenic fluxes, as shown in Fig. 2. For all the other samples, no GaAs is detectable. From X-ray studies we can also estimate the size of the GaAs inclusions in the most heavily doped GaN samples being about 50 nm. It is also interesting to observe that the dependence of the As concentration on As flux changes abruptly when we first detect As inclusions, as shown in Fig. 2. Therefore, we may tentatively suggest that

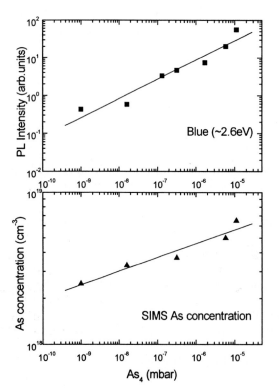

Fig. 1. Dependence of the As concentration determined by SIMS and PL intensity at 2.6 eV (blue emission) on the arsenic flux in As-doped GaN films grown by PA-MBE

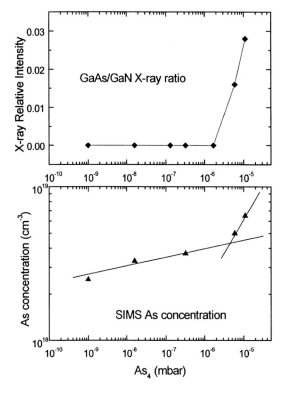

Fig. 2. Dependence of the ratio of GaAs to GaN as measured from X-ray studies and arsenic concentration determined by SIMS on the arsenic flux

the dependence on arsenic flux of arsenic incorporation into the GaN lattice is even lower than 0.1.

The low value for the dependence of the As incorporation into GaN as a function of arsenic flux suggests an unusual growth mechanism. One possible model to explain this behaviour is to suggest that arsenic is incorporated from a limited number of surface sites, whose density depends only very weakly on the arsenic flux. This model is consistent with our previous observations by Auger electron spectroscopy, which also suggested that the surface concentration of arsenic did not depend on the arsenic flux.

The intensity of the blue emission at 2.6 eV depends more strongly on the arsenic flux than the concentration of As incorporated into the GaN. This in turn means that the PL intensity varies very strongly with the concentration of As in the GaN lattice. If As is simply incorporated as a donor on the Ga site, we would not expect such a strong dependence on the amount of As in the GaN. It is possible, therefore, that in addition to this direct influence on the PL of the arsenic incorporated into the lattice, the existence of the arsenic flux may change the stoichiometry of the GaN films grown using PA-MBE.

Acknowledgements This work was undertaken with support from EPSRC (GR/M67438) and two of us (A.J.W. and C.S.D.) would like to thank EPSRC for their studentships.

References

[1] J. I. Pankove and J. A. Hutchby, J. Appl. Phys. **47**, 5387 (1976).
[2] C. G. Van de Walle and J. Neugebauer, Appl. Phys. Lett. **76**, 1009 (2000).
[3] T. Mattila and A. Zunger, Phys. Rev. B **58**, 1367 (1998).
[4] S. V. Novikov, A. J. Winser, I. Harrison, C. S. Davis, and C. T. Foxon, Semicond. Sci. Technol. **16**, 103 (2001) and references therein.

Effect of Isoelectronic In Doping on Deep Levels in GaN Grown by MOCVD

H. K. Cho (a), C. S. Kim (a), Y. K. Hong (a), Y.-W. Kim (b), C.-H. Hong[1]) (a), E.-K. Suh (a), and H. J. Lee (a)

(a) Semiconductor Physics Research Center and Department of Semiconductor Science and Technology, Chonbuk National University, Chonju 561-756, Korea

(b) Frederick Seitz Materials Research Laboratory, University of Illinois at Urbana-Champaign, 104 S. Gordwin, Urbana, IL 61801, USA

(Received June 21, 2001; accepted August 15, 2001)

Subject classification: 68.55.Ln; 71.55.Eq; S7.14

We have investigated the effect of isoelectronic In doping on deep levels in GaN films grown by metalorganic chemical vapor deposition via deep level transient spectroscopy. Deep level E2 0.5 eV below the conduction band was observed in both undoped GaN and In-doped GaN. However, with increasing In mole flow rate, the trap concentration of E2 decreases sharply from 2.3×10^{14} to 2.27×10^{13} cm^{-3}, nearly an order of magnitude in reduction, comparing to undoped GaN. This might be due to the dislocation pinning and/or bending effect. Therefore, In doping in GaN growth can effectively decrease the dislocation density and suppress the formation of deep levels E2.

Introduction Heteroepitaxy of high-quality of GaN films on the substrate (typically a sapphire) is the key issue concerning GaN-related devices, such as high brightness blue and green light emitting diodes, UV detectors and blue laser diodes [1, 2]. A high density of threading dislocations is inevitable due to the large lattice mismatch and difference between the thermal expansion coefficient of GaN and the substrate. Such defects can limit the performance of devices and also introduce deep level traps associated with non-recombination centers. In order to reduce the dislocation density, many growth techniques based on lateral epitaxial overgrowth (LEO) have been introduced [3–5]. However, there still exist some problems such as remaining dislocations from the (0001) facet of GaN seed layer to the surface and a crystallographic tilt in the adjacent regions by using dielectric mask. Recently, it was found that isoelectronic In doping has definite effects of controlling the density of dislocations and improving the GaN film quality. Some experimental reports on the characteristics of In-doped GaN films investigated from a viewpoint of optical [6–8], electrical [9, 10], and growth properties [11] have been published. However, there are few reports about the behavior of deep level defects with isoelectronic In doping. In this paper, we report the effect of isoelectronic In doping on deep levels in GaN using deep level transient spectroscopy (DLTS).

Experiment The GaN materials used in this study were grown in a horizontal-type reactor. The substrate for heteroepitaxial growth was a (0001) oriented sapphire. The growth conditions for the In-doped sample were the same as that for the undoped one. Prior to growth, 250 Å thick GaN buffer layers were deposited at 560 °C. Afterwards, 2 μm thick GaN layers were typically grown on the buffer layers at 1130 °C. The Ga

---

[1]) Corresponding author; Phone: +82-63-270-2831; Fax: +82-63-270-3585; e-mail: chhong@moak.chonbuk.ac.kr

and N precursors were trimethygallium and ammonia. For isoelectronic In doping, tremethylindium (TMIn) diluted with hydrogen was used at four different flow rates of 0–28 μmol/min. The carrier concentrations of these samples were measured using van der Pauw method. High resolution cross-sectional TEM was also performed.

Schottky diodes were fabricated by vacuum evaporating Au onto n-GaN through a shadow mask with 0.5 mm diameter. Prior to the deposition the sample was first cleaned with trichloroethane, acetone and methanol and then boiled in 1% HF solution for 10 min to remove oxide films in GaN. The ohmic contacts were subsequently applied by evaporating Al. Capacitance–voltage (C–V) and DLTS data using a Bio-Rad DL 4600 DLTS system were obtained. DLTS measurements were performed at a test frequency of 1 MHz and a test signal of 100 meV. It was taken at different temperature ranges from 100 to 450 K. A 10 ms wide pulse at 0 V was applied to fill the electron traps in GaN. The output signal was integrated and analyzed with the rate windows ranging from 20 to 1000 s^{-1}.

Results and Discussion The DLTS spectra of undoped and In-doped GaN films are shown in Fig. 1a[2]). In case of In-doped GaN, the In mole flow rates are 7, 14, 28 μmol/min (these are labeled as samples A, B, C, respectively). The emission rate window was 1000 s^{-1}. One distinct level at 378 K, labeled E2 is observed on undoped GaN. The spectrum is induced by the electron emission from deep level to the conduction band. The thermal activation energy from each deep level is determined from an Arrhenius plot as shown in the inset of Fig. 1b. The slope of each set of data yields the activation energy of E2 being 0.50 eV for the high-temperature peak.

Generally, two deep levels can be found in the undoped GaN, E1 and E2 [12–14]. In our case, the deep level E1 trap was strongly related to the buffer growth conditions [15]. Under optimum growth condition, only peak E2 level was detected in both undoped GaN and In-doped GaN. However, with increase of In mole flow rate, the intensity of deep level E2 decreased as shown in Fig. 1a.

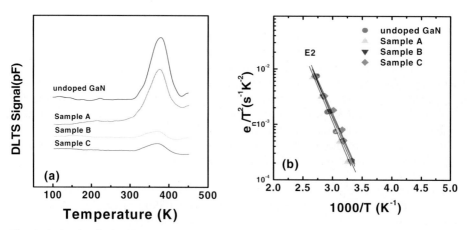

Fig. 1 (colour). a) DLTS spectra of undoped GaN and isoelectronic In-doped GaN samples, b) Arrhenius plot ot the detected deep level E2 in undoped GaN and isoelectronic In-doped GaN; sample A (f_{TMIn} = 7 μmol/min), sample B (f_{TMIn} = 14 μmol/min), sample C(f_{TMIn} = 28 μmol/min)

---

[2]) Colour figures are published online (www.physica-status-solidi.com), where indicated.

Table 1

Results of deep leel transient spectroscopy measurements for undoped GaN and isoelectronic In-doped GaN

sample	n (cm^{-3})	μ_H (cm^2/Vs)	N_d-N_a (cm^{-3})	E_a (eV)	N_t (cm^{-3})	σ (cm^2)
undoped	5.58×10^{16}	208	8.27×10^{16}	0.5 ± 0.04	2.3×10^{14}	4.65×10^{-17}
sample A	7.96×10^{16}	264	6.58×10^{16}	0.5 ± 0.04	1×10^{14}	4.94×10^{-17}
sample B	1.39×10^{17}	547	9.03×10^{16}	0.5 ± 0.04	2.5×10^{13}	4.67×10^{-17}
sample C	1.53×10^{17}	550	1.68×10^{17}	0.5 ± 0.04	4.6×10^{13}	1.87×10^{-17}

For In-doped GaN with In mole flow rate of 2.5 µmol/min (sample A), only one peak E2 level is observed, the same deep level as was detected in undoped GaN. As In mole flow rate is increased to 5, 10 µmol/min (samples B, C, respectively), only the E2 level is detected, but the intensity of this deep level E2 is decreased, as shown in Fig. 1, comparing to that of undoped GaN and sample A.

For undoped GaN, the trap concentration and capture cross section of deep level E2 is 2.3×10^{14}, 4.65×10^{-17} cm^2, respectively. In case of In-doped GaN, the trap concentrations of the deep level E2 which is located 0.5 eV below the conduction band and their capture cross sections are 1×10^{14} cm^{-3} 4.94×10^{-17} cm^2; 2.5×10^{13} cm^{-3}, 4.67×10^{-17} cm^2, 4.6×10^{13} cm^{-3}, 1.87×10^{-17} cm^2 for samples A, B, C, respectively. The detailed characteristics are listed in Table 1. The exact origin of this deep level remains unclear so far. Hacke et al. [13] and Hasse et al. [14] reported that the deep level E2 is a nitrogen antisite point defect (N_{Ga}). Also, Jenkins and Dow [16] by tight bonding calculations suggested that the deep level is associated with N_{Ga} and is at a location of ~0.52 eV below the conduction band. This level agrees well with our result as shown in Fig. 1b.

It is known that the impurity–dislocation reaction is dependent on the species and concentration of impurities as well as the core structure of dislocation. Therefore, impurity precipitates formed on a dislocation can play an important role in the strength of dislocation pinning.

The dislocation pinning and bending by In doping was found from an analysis of TEM images (not shown here). In case of GaAs and InP materials, isoelectronic doping atoms cause a dramatic reduction in the dislocation density by pinning of dislocations at In atoms due to the large radius of In as compared to that of Ga [17]. In our case, such a pinning action shows that feature and also may withhold the generation of nitrogen antisite point defects. This occurs because isoelectronic In atoms can occupy normally vacant Ga sites along the dislocation core sites in the growing films, thereby lowering the density of Ga vacancy-related defects. The trap concentration of deep level E2 in In-doped GaN samples B and C decreases sharply, nearly an order of magnitude in reduction as shown in Table 1. Therefore, the trap concentration of E2, which is associated with nitrogen antisite, can be reduced by adding In.

Summary We have employed deep level transient spectroscopy to study the effect of isoelectronic In doping on deep levels in GaN. The trap concentration of deep level E2 located 0.5 eV below the conduction band can be effectively reduced with increase of isoelectronic In doping. The dislocation pinning and bending by In doping can generally withhold the generation of point defects, including the native defects and also reduce the dislocation density. Therefore, In doping in GaN can effectively suppress the formation of deep levels E2 which is related to nitrogen antisite.

Acknowledgements This work has been supported by the project of the Brain Korea 21 (BK21) through the Semiconductor Physics Research Center (SPRC) at Chonbuk National University.

References

[1] S. NAKAMURA, M. SENOH, S.-I. NAGAHAMA, T. MATSUHITA, K. KIYOKI, Y. SUGIMOTO, T. KOZAKI, H., and Y. SUGINOTO, Jpn. J. Appl. Phys. **35**, 74 (1996).
[2] S. NAKAMURA, M. SENOH, S. NAGAHAMA, N. IWASA, T. YAMADA, T. MATSUSHITA, H. KIYOKU, UMEMOTO, M. SANO, and T. MUKAI, Jpn. J. Appl. Phys. (Part 2) **38**, L1966 (1999).
[3] S. NAKAMURA, Jpn. J. Appl. Phys. **30**, L1705 (1991).
[4] A. SAKAI, H. SUNAGAWA, and A. ISUI, Appl. Phys. Lett. **71**, 2259 (1997).
[5] K. J. LINTCHICUM, T. GEHRKE, D. B. TOMSON, K. M. TRACY, E. P. CARSON, T. P. SMITH, T. S. ZHELEVA, C. A. ZORMAN, M. MEHREGANY, and R. F. DAVIS, MRS J. Internet I. Nitride Semicond. Res. **4S1**, G4.9 (1999).
[6] X. Q. SHEN and Y. AOYAGI, Jpn. J. Appl. Phys. **38**, L14 (1999).
[7] C. K. SHU, J. OU, H. C. LIN, W. K. CHEN, and M. C. LEE, Appl. Phys. Lett. **73**, 641 (1998).
[8] H. KUMANO, K.-I. KOSHI, S. TANAKA, I. SUEMUNE, X.-Q. SHEN, P. RIBLET, P. RAMVALL, and Y. AOYAGI, Appl. Phys. Lett. **75**, 2879 (1999).
[9] X. Q. SHEN, P. RAMVALL, P. RIBLET, and Y. AOYAGI, Jpn. J. Appl. Phys. **38**, L411 (1999).
[10] H. M. CHUNG, W. C. CHUANG, Y. C. PAN, C. C. TSAI, M. C. LEE, W. H. CHEN, and W. K. CHEN, Appl. Phys. Lett. **76**, 897 (2000).
[11] F. WIDMANN, B. DAUDIN, G. FEUILLET, N. PELEKANOS, and J. L. ROUVIERE, Appl. Phys. Lett **73**, 2642 (1998).
[12] W. GOTZ, N. M. JOHNSON, H. AMANO, and I. AKASAKI, Appl. Phys. Lett. **65**, 463 (1994).
[13] P. HACKE, T. DETCHPROHM, K. HIRAMATSU, and N. SAWAKI, J. Appl. Phys. **76**, 304 (1994).
[14] D. HASSE, M. SCHMID, W. KUMER, A. DOMEN, V. HARLE, F. SCHOLZ, M. BURKARD, and H. SCHWEIZER, Appl. Phys. Lett. **69**, 2525 (1996).
[15] H. K. CHO, K. S. KIM, C.-H. HONG, and H. J. LEE, J. Cryst. Growth **223**, 38 (2001).
[16] D. W. JENKINS and J. D. DOW, Phys. Rev. B **39**, 3317 (1989).
[17] D. I. BARRETT, S. MCGUIGAN, H. M. HOBGOOD, G. W. ELDRIDGE, and R. N. THOMAS, J. Cryst. Growth **70**, 179 (1984).

Structural Properties of GaN Grown by Pendeo-Epitaxy with In-Doping

Young Kue Hong (a), Chi Sun Kim (a), Hung Sub Jung (a),
Chang-Hee Hong[1]) (a), Min Hong Kim (b), Shi-Jong Leem (b),
Hyung Koun Cho (c), and Jeong Yong Lee (c)

(a) Department of Semiconductor Science and Technology and Semiconductor Physics Research Center, Chonbuk National University, Chonju 561-756, Korea

(b) Department of OE Team, Device & Materials Laboratory LG Electronics Institute of Technology, Seoul 137-724, Korea

(c) Department of Materials Science and Engineering, Korea Advanced Institute of Science and Technology, Taejon 305-701, Korea

(Received June 21, 2001; accepted August 4, 2001)

Subject classification: 61.10.Nz; 61.72.Lk; 68.37.Lp; 68.37.Ps; 68.55.Ln; S7.14

We have studied the effect of isoelectronic In-doping on the structural properties of GaN grown by pendeo-epitaxy. From an analysis of cross-sectional transmission electron microscopy (TEM) images, the threading dislocation originating from the (0001) facet of GaN seed layer, thereafter propagating onto the top surface of regrown GaN layer, were reduced due to isoelectronic In-doping, which could enhance vacancy trapping. In addition, threading dislocations in the coalescence region were not observable. These results indicate that these dislocations are bent or terminated in the boundary of coalesced region. Also, the crystalline quality was improved from the results of high resolution X-ray diffraction and TEM measurements.

Introduction High luminescence blue light emitting diodes and InGaN multi-quantum-well laser diodes have been successfully achieved on sapphire substrates [1–3]. Although they show a high performance of devices, it is still difficult to control a high dislocation density of GaN epitaxial layers. Therefore, the reduction of defect density and stress in heteroepitaxial growth is one of remaining main issues, especially for GaN-based lasers. The significant progress in defect density reduction in GaN layers has been achieved using lateral epitaxial overgrowth and pendeo-epitaxy technique [4–10]. The pendeo-epitaxy GaN is a new approach of selective epitaxial growth dominated by the growth from sidewalls of rectangular stripes. This process allows the growth of uniformly low defect density material, especially in the lateral growth region from (11$\bar{2}$0) facets toward [1$\bar{1}$00] direction. However, dislocations originating from (0001) facet of the GaN seed layer and propagating to the surface cannot be reduced. Also, the use of a dielectric mask can cause a crystallographic tilt in the adjacent regions and the formation of boundaries at the interface of coalescence region of two growth fronts [11]. Recently, it was found that isoelectronic In atoms cause a dramatic reduction of dislocation density by pinning of dislocations [12]. In this study, we present the results of pendeo-epitaxy of GaN layers without the use of a dielectric mask to avoid crystallographic tilt and also descibe the effect of isoelectronic In-doping to reduce dislocations on GaN seed layers.

---

[1]) Corresponding author; Phone +82-63-270-2831; Fax : +82-63-270-3585;
e-mail: chhong@moak.chonbuk.ac.kr

Experimental Procedure The pendeo-epitaxial GaN has been grown on a 2.0–2.5 μm thick GaN seed layer grown by low pressure metalorganic chemical vapor deposition. The patterned GaN seed layer was formed along the $\langle 1\bar{1}00 \rangle$ GaN direction using conventional photolithography technique and dry etching process. The widths of seed and spacing region were 4.5 μm and 8 μm, respectively. For surface cleaning, $H_2SO_4:H_2O_2:DI = 3:1:1$, HF and then solvent cleaning (trichloroethylene, acetone, and methanol) were employed. The growth temperature was ranging from 1130 to 1160 °C and the growth pressure from 80 to 400 mbar. The flow rates of trimethylgallium, NH_3 and trimethylindium during the growth were 75–150 μmol/min, 3.5 l/min and 1–10 μmol/min, respectively. The crystallographic tilting in the GaN was determined by HRXRD measurements. The morphology and defect microstructures were investigated using atomic force microscopy (AFM) and TEM measurements.

Results and Discussion Figure 1[2]) shows AFM images for the samples of undoped and In-doped pendeo-epitaxial GaN. It is known that threading dislocations with screw component terminated at the surface are seen as dark pits in AFM images. In the vertical growth regions, different surface features of undoped and In-doped pendeo-epitaxial GaN were observed. In case of In-doped pendeo-epitaxial GaN, the dark pit density is reduced by above one order of magnitude comparing to undoped pendeo-epitaxial GaN. Also the interface of coalescence region of two growth fronts can not be distinguished. The rms roughnesses of non-doped and In-doped pendeo-epitaxial GaN were 0.223 nm and 0.144 nm, respectively. This result indicates that the surface roughness of GaN in the vertical grown region is improved by isoelectronic In-doping.

Figures 2a and b show cross-sectional TEM images of undoped and In-doped pendeo-epitaxial GaN, respectively. For a comparative analysis, two spacious samples were divided into three parts:

I) The vertical growth region (T1 and T2). In case of the undoped sample, the threading dislocation density of the regrown epilayer was almost equal to that of the underlying GaN seed layer. That indicates that the threading dislocations have propagated from the GaN (0001) seed facet to the top surface. On the other hand, in case of the

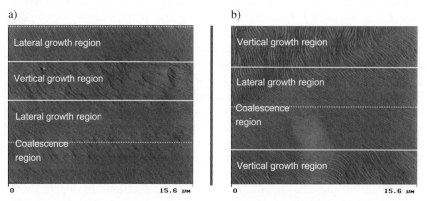

Fig. 1 (colour). AFM images for a) undoped and b) In-doped pendeo-epitaxial GaN samples

---

[2]) Colour figures are published online (www.physica-status-solidi.com), where indicated.

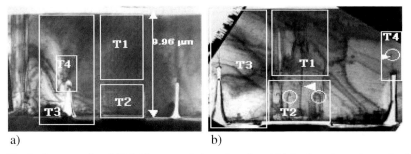

Fig. 2. Cross-sectional TEM images of a) undoped and b) In-doped pendeo-epitaxial GaN

In-doped sample, some of the threading dislocations having propagated from (0001) seed facet were bent toward ⟨1$\bar{2}$10⟩ or disappeared and did not propagate toward the surface (as shown by the circled areas at T2 in Fig. 2b). The threading dislocation density was at least decreased by fifty percent. This suggests that the nature of isoelectronic In-doping has a significant effect on dislocation distribution. Isoelectronic In atoms which are bigger than Ga atoms can cause a reduction in the dislocation density by pinning dislocation. This can enhance vacancy trapping by In, and therefore hinder dislocation source nucleation.

II) The lateral overgrown region (T3). For both undoped and In-doped pendeo-epitaxial GaN, there were no threading dislocations originated at the GaN/sapphire interface in the lateral overgrown region. Preliminary analysis of the areas of (1$\bar{1}$00) faceted lateral growth revealed only a few threading dislocations or stacking faults which formed parallel to the (0001) plane. These dislocations did not subsequently propagate to the (0001) surface of the overgrown GaN layers. Therefore, the dislocation density is significantly reduced in the regrown areas of both samples.

III) The coalescence region (T4). In this region, there are dislocation structures in the boundary formed at the coalescence fronts under our growth conditions. For the undoped sample, these dislocations propagated almost along the ⟨0001⟩ direction. By con-

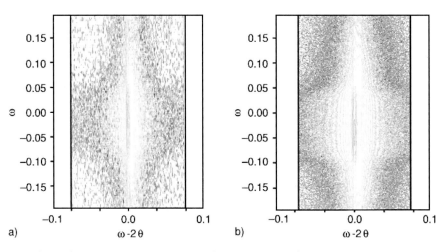

Fig. 3 (colour). Reciprocal space maps of a) In-doped and b) undoped pendeo-epitaxial GaN

strast, for the In-doped sample, they disappeared or were bent during the growth. This might be strongly related to the vacancy trapping and/or the dislocation pinning by isoelectronic In-doping as mentioned before.

Figure 3 shows reciprocal space maps of In-doped and undoped pendeo-epitaxy GaN measured perpendicular to the stripe axis ($\langle 1\bar{1}00 \rangle$, $\phi = 90°$) near (0002) reflection. They do not show the crystallographic tilt because of no use of a dielectric mask. The d spacing related to the strain can be determined from the spreading along the x-axis ($\omega - 2\theta$). It was found that the d spacing was more broadened in the undoped sample than in the In-doped one. This means that the in-plane compressive strain is decreased in the In-doped pendeo-epitaxial GaN.

Summary The effect of isoelectronic In-doping on structural properties of pendeo-epitaxial GaN grown by MOCVD was studied using TEM, HRXRD and AFM measurements. The density of threading dislocations originating from the (0001) template facet was reduced due to dislocation pinning or bending effect, which is associated with isoelectronic In-doping. In addition, threading dislocations in the coalescence region were not observable. From an analysis of high resolution X-ray reciprocal maps, in-plane compressive strain is decreased in the In-doped pendeo-epitaxial GaN.

Acknowlegement This work has been supported by the project of the Brain Korea 21 (BK21) through the Semiconductor Physics Research Center (SPRC) at Chonbuk National University.

References

[1] M. HARADA, Nikkei Electronics Asia, News/Focus, Dec. 30, 1997.
[2] Proc. Second Internat. Conf. on Nitride Semiconductors, INCS'97, Oct. 27–31, 1997, Tokushima, Japan, and references therein.
[3] S. NAKAMURA, M. SENOH, S. NAGAHAMA, N. IWASA, T. YAMADA, and H. KIYOKU, Jpn. J. Appl. Phys. **36**, L1059 (1997).
[4] Y. KATO, S. KITAMURA, K. HIRAMATSU, and N. SAWAKI, J. Cryst. Growth **144**, 133 (1994).
[5] O. H. NAM, M. BREMSER, B. WARD, R. NEMANICH, and R. DAVIS, Jpn. J. Appl. Phys. **36**, L532 (1997).
[6] A. SAKAI, H. SUNAKAWA, and A. USUI, Appl. Phys. Lett. **73**, 481 (1998).
[7] T. GEHRKE, K. J. LINTHICUM, D. B. THOMSON, P. RAJAGOPAL, A. D. BATCHELOR, and R. F. DAVIS, MRS Internet J. Nitride Semicond. Res. **4S1**, G3.2 (1999).
[8] K. J. LINTHICUM, T. GEHRKE, D. B. THOMSON, K. M. TRACY, E. P. CARLSON, T. P. SMITH, T. S. SMITH, T. S. ZHELEVA, C. A. ZORMAN, M. MEHREGANY, and R. F. DAVIS, MRS Internet J. Nitride Semicond. Res. **4S1**, G4.9 (1999).
[9] K. J. LINTHICUM, T. GEHRKE, D. B. THOMSON, K. M. TRACY, E. P. CARLSON, T. P. SMITH, T. S. SMITH, and R. F. DAVIS, Appl. Phys. Lett. **75**, 196 (1999).
[10] D. B. THOMSON, T. GEHRKE, K .J. LINTHICUM, P. RAJAGOPAL, P. HARTLIEB, T. S. ZHELEVA, and R. F. DAVIS, MRS Internet J. Nitride Semicond. Res. **4S1**, G3.37 (1999).
[11] IG-HYEON KIM, C. SONE, OK-HYUN NAM, YONG-JO PARK, and TAEIL KIM, Appl. Phys. Lett. **75**, 4109 (1999).
[12] S. YAMAGUCHI, M. KARIYA, S. NITTA, H. AMANO, and I. AKASAKI, Jpn. J. Appl. Phys. **39**, 2385 (2000).

Non-Monotonous Behavior of In-Doped GaN Grown by MOVPE with Nitrogen Carrier Gas

A. Yamamoto[1]) (a), T. Tanikawa (a), K. Ikuta (a), M. Adachi (a), A. Hashimoto (a), and Y. Ito (b)

(a) Department of Electrical and Electronics Eng. Fukui University, 3-9-1 Bunkyo Fukui 910-8507, Japan

(b) Wakasa Wan Energy Research Center, 64-52-1 Nagatani, Tsuruga, Fukui 910-0192, Japan

(Received June 23, 2001; accepted August 14, 2001)

Subject classification: 68.55.Jk; 73.50.Dn; 73.61.Ey; 81.15.Kk; S7.14

This paper reports the In-doping effects in MOVPE grown GaN using a N_2 carrier gas. Electrical and optical properties of the In-doped GaN are found to show non-monotonous behavior, when the In source (TMI) supply is increased. Carrier concentration in the In-doped GaN is monotonously decreased with increasing TMI supply, while Hall mobility and PL intensity ratio between exciton emission and yellow deep emission (I_{EX}/I_{YL}) are decreased in the region of low TMI supply (TMI/TEG <0.5). In this region, tensile stress in the grown GaN estimated from the exciton emission peak energy is also increased with increasing TMI/TEG ratio. For TMI/TEG ratio in the range of 0.5–1.5, on the other hand, they show very normal behavior (improvement by In doping). Such anomalous behavior for mobility, I_{EX}/I_{YL} and tensile stress is discussed from the viewpoints of impurity hardening of GaN, as well as stress relaxation by both crack formation and In doping.

Introduction A significant improvement in the quality of the epitaxially grown compound semiconductors has been attained by using isoelectronic doping or a surfactant species [1]. Indium is the most widely studied surfactant in the epitaxial growth of GaN. The improvement in the crystalline quality, electrical and optical properties, and surface morphology of the In-doped GaN have been reported for MOVPE [2–5] and MBE [6, 7]. The roles of the In surfactant in the MOVPE growth of GaN are more significant when the N_2 carrier gas is used in comparison with the H_2. This is due to that the migration of the GaN species on the growing surface is less enhanced and, therefore, a higher density of defects is included, when N_2 carrier gas is used. In addition to the favorable influences mentioned above, an unfavorable effect of In doping has also been reported; Huang et al. [5] reported that the recombination lifetime for the MOVPE GaN was sharply decreased by In-doping. The reason for this has not yet been clarified.

In this paper, we report a systematic study on the In-doping effects in the MOVPE growth of GaN using N_2 carrier gas. As many authors reported previously, carrier concentration in the In-doped GaN is monotonously decreased with increasing TMI supply. For TMI/TEG molar ratio in the range of 0–0.5, however, Hall mobility and PL intensity ratio between exciton emission and yellow deep emission (I_{EX}/I_{YL}) are decreased in spite of In doping. Tensile stress in the grown GaN estimated from the exciton emission peak energy is also found to be increased in this TMI/TEG molar ratio region. For

---
[1]) Corresponding author; Phone: +81 776 27 8566, Fax: +81 776 27 8749, e-mail: yamamoto@kyomu1.fuee.fukui-u.ac.jp

TMI/TEG ratio in the higher range (0.5–1.5), on the other hand, mobility, PL intensity ratio and stress show very normal behavior (improvement by In doping). Such anomalous behavior in the electrical and optical properties of the In-doped GaN is discussed from the viewpoint of impurity hardening and residual stress relaxation by In doping and crack formation in the grown films.

Experimental GaN films of about 2 μm thickness are grown at 1000 °C on sapphire substrates at a pressure of 0.1 atm using MOVPE apparatus with a horizontal reactor. 20 nm thick GaN grown at 550 °C is used for a buffer layer. TEG, TMI and NH_3 are used as source materials and N_2 as carrier gas for TEG and TMI. The TMI/TEG molar ratio is varied from 0 to 2. The composition of In included in the grown GaN, estimated by X-ray diffraction taking account of the residual strain, is less than 1% even for TMI/TEG molar ratio 2. Surface morphology of the grown GaN is evaluated with an atomic force microscope (AFM). Electrical properties are measured with Hall measurements. Photoluminescence spectrum is measured at room temperature with a He–Cd laser as excitation source. Number of cracks introduced in the grown films is evaluated with an optical microscope (Nomarski contrast).

Results and Discussion Figure 1 shows the TMI/TEG ratio dependence of carrier concentration and Hall mobility for the In-doped GaN films. As seen in Fig. 1a, the carrier concentration decreases monotonously when increasing the TMI/TGE ratio up to 2. In contrast with the carrier concentration change in Fig. 1a, Hall mobility is decreased for TMI/TEG molar ratio below 0.5 as shown in Fig. 1b. After showing the lowest value at a TMI/TEG molar ratio of ≈0.25, the mobility is then increased with increasing TMI/TEG molar ratio and reached to about 400 cm^2/Vs at a molar ratio around 1.5. Photoluminescence (PL) spectrum is measured at room temperature for the GaN films grown with a different TMI/TEG molar ratio. For PL studies, samples with an average carrier concentration for each TMI/TEG molar ratio level (see Fig. 1a) were used. As a general feature of PL spectrum, the exciton emission at about 365 nm is enhanced and yellow deep emission at 550 nm is depressed by doping with In. Figure 2 shows the TMI/TEG molar ratio dependence of the intensity ratio between the exciton emission and the yellow emission (I_{EX}/I_{YL}). One can see that the intensity ratio I_{EX}/I_{YL} also shows non-monotonous behavior as the TMI/TEG molar ratio is increased. Similar to

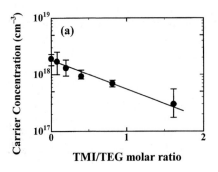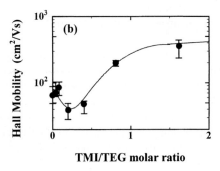

Fig. 1. a) Carrier concentration and b) Hall mobility of GaN layers as a function of TMI/TEG molar ratio

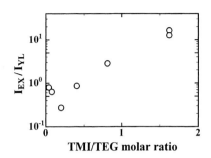

Fig. 2. Intensity ratio between the exciton emission and the yellow emission, I_{EX}/I_{YL}, as a function of TMI/TEG molar ratio

the mobility change with TMI/TEG molar ratio, I_{EX}/I_{YL} also shows a minimum at TMI/TEG molar ratio ≈ 0.25.

How can we explain the anomalous behavior of mobility and I_{EX}/I_{YL} at TMI/TEG molar ratio 0–0.5? Figure 3 shows the exciton peak energy as a function of TMI/TEG molar ratio. Since the stress-free GaN is reported to have an exciton emission at 3.41 eV, one can see that all the films prepared here have a tensile stress. It should be pointed out that, in spite of the increase in the TMI/TEG molar ratio, tensile stress in the GaN films is increased for a TMI/TEG molar ratio 0–0.5. At a TMI/TEG molar ratio higher than 0.5, tensile stress is decreased as expected from the previous report [3]. We found that the number of cracks in the GaN films is drastically changed in this TMI/TEG molar ratio range. In Fig. 3, the number of cracks observed with an optical microscope (Nomarski contrast) is also plotted. Crack density is drastically decreased in the TMI/TEG molar ratio range 0–0.1 and gradually decreased at 0.1–0.8. Since the introduction of cracks in the grown films is due to the tensile stress, the decrease of crack density seems to show a decrease in tensile stress. Unlike the expectancy, however, stress is rather increased in the TMI/TEG molar ratio range 0–0.5. Therefore, there should be another mechanism in the reduction of crack density in the TMI/TEG molar ratio range 0–0.5. It is known that a solute atom with a dissimilar size from a matrix atom in a crystal behaves like an elastic inclusion and depresses dislocation motion [8], which is called "impurity hardening". It is reasonable to consider that, in the TMI/TEG molar ratio range 0–0.5, the hardening effect causes the decrease of crack density, which results the increase in tensile stress. Yamaguchi et al. [9] reported the strain relief by In doping for MOVPE GaN grown with H_2 or N_2 carrier gas. In their case, however, the increase in residual stress by In doping was not observed for the samples grown with N_2 carrier. The major difference between their case and ours seems to be the stress relaxation by crack formation in our samples. The decrease of both tensile stress

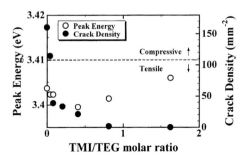

Fig. 3. Exciton peak energy and crack density as a function of TMI/TEG molar ratio

and number of cracks in the TMI/TEG molar ratio higher than 0.5 may be due to the original In-doping effects reported by many authors.

From the above discussion, it can be concluded that the deterioration of Hall mobility and I_{EX}/I_{YL} at a TMI/TEG molar ratio around 0.2–0.5 are closely related to the increase in tensile stress in this region. The unfavorable effect of In-doping [5] may be also related to this. Further investigation will be needed to clarify the mechanism involved in the anomalous behavior of Hall mobility and I_{EX}/I_{YL}. It may be reasonable to consider that the reduction of stress in the grown films is significant in order to obtain a high mobility and a high exciton emission for the heteroepitaxial GaN films.

Conclusion A systematic study has been made on the In-doping effects of MOVPE GaN grown using a N_2 carrier gas. As many authors reported previously, the carrier concentration is monotonously decreased with increasing the supply. For TMI/TEG molar ratio within the range 0–0.5, however, Hall mobility and PL intensity ratio (I_{EX}/I_{YL}) are decreased and the tensile stress is increased as the TMI/TEG ratio is increased. The increase in tensile stress and the decrease of crack density in this TMI/TEG region are due to the hardening effects of the In-doped GaN. Such deterioration of the electrical and optical properties by In doping seems to be closely related to the increase in tensile stress in this TMI/TEG ratio region. For TMI/TEG ratio in the range 0.5–1.5, on the other hand, the electrical and optical properties are improved and tensile stress is reduced as the TMI/TEG ratio increases.

References

[1] H. BENKING, P. NAROZNY, and N. EMEIS, Appl. Phys. Lett. **47**, 828 (1985).
[2] C. K. SHU, J. OU, H. C. LIN, W. K. CHEN, and M. C. LEE, Appl. Phys. Lett. **73**, 641 (1998).
[3] S. YAMAGUCHI, M. KARIYA, S. NITTA, H. AMANO, and I. AKASAKI, Appl. Phys. Lett. **75**, 4106 (1999).
[4] G. POZINA, J. P. BERGMAN, B. MONEMAR, S. YAMAGUCHI, H. AMANO, and I. AKASAKI, Appl. Phys. Lett. **76**, 3388 (2000).
[5] Y. H. HUANG, C. K. SHU, W. C. LIN, K. C. LIAO, C. H. CHUANG, M. C. LEE, W. H. CHEN, W. K. CHEN, and Y. Y. LEE, in: Proc. Internat. Workshop on Nitride Semiconductors, Nagoya 2000, IPAP Conf. Ser. **1**, 610 (2001).
[6] H. KUMANO, K. HOSHI, S. TANAKA, I. SUEMUNE, X. Q. SHEN, P. RIBELT, P. RAMVALL, and Y. AOYAGI, Appl. Phys. Lett. **75**, 2879 (1999).
[7] X. Q. SHEN, P. RAMVALL, P. RIBELT, Y. AOYAGI, K. HOSI, S. TANAKA, and I. SUEMUNE, J. Cryst. Growth **209**, 396 (2000).
[8] G. YACOB, Proc. Semi-Insulating III–V Materials, Evian 1982, Shiva Publishing Ltd., Kent 1982 (p. 2).
[9] S. YAMAGUCHI, M. KARIYA, T. KASHIMA, S. NITTA, K. KOSAKI, Y. YUKAWA, H. AMANO, and I. AKASAKI, Phys. Rev. B **64**, 035318 (2001).

A Perspective of GaAs$_{1-x}$N$_x$ and GaP$_x$N$_{1-x}$ as Heavily Doped Semiconductors

A. Mascarenhas[1]), Yong Zhang, and M. J. Seong

National Renewable Energy Laboratory, 1617 Cole Blvd. Golden, CO 80401, USA

(Received August 31, 2001; accepted September 1, 2001)

Subject classification: 71.55.Eq; 72.20.Jv; 78.30.Fs; 78.55.Cr; S7.14

The behavior of GaAs$_{1-x}$N$_x$ and GaP$_x$N$_{1-x}$ is contrasted with that of Al$_x$Ga$_{1-x}$As and In$_x$Ga$_{1-x}$As with respect to irregular and regular alloy behavior. It is proposed that GaAs$_{1-x}$N$_x$ and GaP$_x$N$_{1-x}$ behave as heavily nitrogen doped semiconductors rather than dilute nitride alloys and that their abnormal or irregular alloy behavior is associated with impurity band formation that manifests itself in the giant bowing and poor transport properties characteristic of these materials.

Introduction Semiconductor heterostructures constitute the building blocks of several electronic and photonic devices such as solar cells, diode lasers, light emitting diodes, and heterojunction bipolar transistors. The heterostructure components and their stacking geometry are often utilized for tailoring features for improving device performance. Although the choice of semiconductors for synthesizing a heterostructure for a particular device application is guided by the specific electrical or optical design requirements, this choice is also guided by the ability to grow the required stack of semiconductor layers epitaxially. Thus the semiconductor lattice size in addition to its electronic bandstructure is an important criterion for heterostructure design. The requirement of a specified bandstructure feature and lattice size is most often fulfilled by the use of substitutional alloys, thus circumventing the discreet and finite repertoire of available elemental and compound semiconductors. They have thus been the subject of extensive studies as regards their alloy thermodynamics and electronic properties. Recently, the phenomenon of giant band gap "bowing" that has been observed in several III–V dilute nitride alloys offers the exciting promise of increasing the flexibility in choice of semiconductor band gaps available with specified lattice constants. However, the poor electrical transport properties that these materials exhibit, seriously limit their usefulness. In this paper, these limitations will be discussed from the perspective of impurity band phenomena.

Background The lattice size of a substitutional semiconductor alloy A$_{1-x}$B$_x$C has experimentally been found to be well approximated by the concentration weighted average of that of its constituents AC and BC, which is usually referred to as Vegard's rule or law [1]. Also, the band gap is found to be close to but generally lower than the concentration weighted average of that of the constituent compounds and this deviation well described in terms of a bowing coefficient which is usually quite small (about 1 eV) [2]. Thus the band gap and lattice size of the alloy covers the range of values between that for the two end point constituent compounds provided

---

[1]) Corresponding author; Phone: +01 303 384 6608; Fax: +01 303 384 6481; e-mail: amascar@nrel.gov

that they are not immiscible in this range. But in spite of the variety of band gaps and lattice sizes that are made possible by alloying semiconductors, there have been many situations encountered over the past few decades, where although it is possible to obtain an alloy having the desired lattice constant, the alloy does not have the required band gap. Such frustrations impose severe penalties in device design. An example of this is the technologically important field of vertical cavity surface emitting lasers (VCSELs) used for fiber optic communications [3]. The difficulty with Bragg mirror stacks of quaternary alloys grown on InP substrates lies in their poor thermal conductivity. Although high reflectivity AlAs/GaAs Bragg mirror stacks can readily be grown on GaAs substrates, it is not possible to find an alloy for the active region of the VCSEL, with a band gap near 0.8 or 0.95 eV and which is lattice matched to GaAs. To grow such VCSELs, it has thus been necessary to resort to complicated techniques which often involve "lift-off" and "fusion bonding" steps in the laser fabrication process [3]. Another alloy constraint example is the quadruple junction solar cell where the optimal heterostructure for realizing a 40% efficient solar cell would be comprised of light absorbing semiconductor active regions grown epitaxially with the sequence of band gaps 0.67 eV/1.05 eV/1.42 eV/1.9 eV on a Ge substrate [4]. GaInP$_2$, GaAs and Ge are ideally suited lattice matched semiconductors for the top, next to top, and bottom cells, but there existed no material lattice matched to GaAs with a band gap of 1.05 eV until quite recently [4].

Giant Bandgap "Bowing" Phenomena About a decade ago, Weyers et al. [5] succeeded in incorporating almost 1% N into GaAs using OMVPE (organo metallic vapor phase epitaxy) techniques and observed a surprising lowering of the band gap by about 140 meV which is contrary to what is expected by Vegard's rule. This rekindled interest in a subject that is over three decades old, namely, the N isoelectronic trap. Research from the 60's to the 80's had established that N behaves as an isoelectronic trap in GaP and GaAs [6–8]. Thermodynamic calculations [9] predicted that N was insoluble in both these materials and in the early work on bulk crystals and LPE material it was only possible to achieve N doping levels less than 10^{19} which allowed for the identification of N induced trap levels in these semiconductors [6–8] as shown in Fig. 1. The ability to incorporate over an order of magnitude more N (>1%) into these semiconductors is attributed to the use of non-equilibrium growth techniques such as OMVPE and MBE (Molecular Beam Epitaxy). The experiments of Weyers et al. [10] were soon repeated by several groups around the world, confirming that the band gap of GaAs was lowered by about 180 meV with just 1% of N incorporation into GaAs, and by 400meV with just 3% N. GaAs$_{1-x}$N$_x$ began to be viewed as an alloy and the anomalous large lowering of its band gap to be described as a giant band gap "bowing" (the bowing parameter varies with x but is <20 eV) [11]. These results were followed by a flurry of research activity after the suggestion by Kondow et al. [12] about the possibility of realizing 1.3 μm lasers using GaAs$_{1-x}$N$_x$ active regions grown on GaAs substrates. At around the same time it became evident that GaAs$_{1-x}$N$_x$ could also be used to provide the 1.05 eV band gap material lattice matched to GaAs for realizing the 40% efficient quadruple junction solar cell and several research groups raced towards this goal [4]. It thus appeared that the anomalous and giant "bowing" of the band gap in GaAs$_{1-x}$N$_x$ offered a way out of the semiconductor alloy constraints that limited the design of some technological important devices.

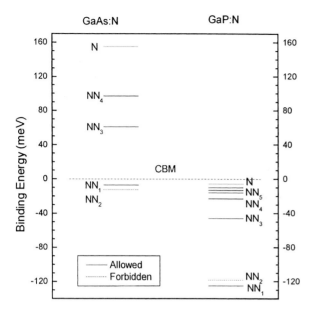

Fig. 1. Nitrogen induced trap levels in GaAs and GaP

Soon it was evident, however, that the giant band gap "bowing" in GaAs$_{1-x}$N$_x$ came with a caveat. The electron mobilities in this material were severely degraded and several efforts to improve this using rapid thermal annealing resulted only in marginal improvements, thus dampening the initial excitement of using this material for both solar cells and lasers [13]. In the remainder of this paper, the exciting research during the past few years on understanding the physical mechanisms underlying the phenomenon of the anomalous giant band gap "bowing" will be discussed from the viewpoint of determining whether the limitations of the GaAs$_{1-x}$N$_x$ system can be overcome.

Heavy Doping and Impurity Bands Using photoluminescence studies at low temperatures and high pressures on GaAs samples doped with, 0.01% N, Wolford et al. [7] and Liu et al. [8] had determined that the isolated N impurity forms a resonant level 150–180 meV above the conduction band edge of GaAs. Almost a decade later, in an effort to understand the origin of the giant "bowing" phenomenon, Perkins et al. [14] used electro-modulated reflectivity studies to investigate the giant "bowing" phenomenon in heavily N doped GaAs, and directly observed a level (denoted as E_+) above the conduction band edge whose existence was inferred by Shan et al. [15] using photoluminescence studies under high pressure. Employing a simple two level repulsion model, Shan et al. assumed that the giant band gap "bowing" in GaAs$_{1-x}$N$_x$ arose from level repulsion between the isolated N resonant level and the GaAs conduction band edge. But soon thereafter, Zhang et al. [16] demonstrated a very unusual behavior for the conduction band effective mass in GaAs:N. Since the carrier lifetimes in GaAs:N are too short to be able to measure the effective mass with conventional techniques that involve cyclotron resonance, this was determined by fabricating GaAs/GaAs:N quantum wells with several well widths for each value of N doping, and inferring the effective mass from the confinement induced shifts in the QW ground state transition energies. The results obtained revealed that the conduction band effective mass exhibited a

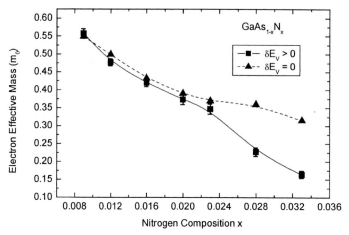

Fig. 2. Variation of conduction band effective mass in GaAs:N (from Ref. [16])

very abrupt increase by almost an order of magnitude on going from 0.01 to 0.1% N and then gradually decreased towards the normal GaAs value as the N doping approached a value of about 3% as shown in Fig. 2. This surprising behavior was a clue to the presence of an impurity band in GaAs:N induced by the heavy N doping. Subsequent low temperature photoluminescence and selective excitation of photoluminescence studies by Zhang et al. [17] showed the evolution of the discreet N trap levels in GaAs into a N induced impurity band. Recently, peculiarities of the N resonant levels in GaAs:N have been dramatically demonstrated in resonant Raman scattering studies by Cheong et al. [18]

The physics of the N isoelectronic trap in GaP and $GaAs_xP_{1-x}$ was extensively studied in the 1960's and 70's because of the potential for use of these materials for light emitting diodes (LED's), however, because of the N solubility problem analogous to that for GaAs:N, the doping levels studied were limited to <0.1%. This work has recently been reviewed by Zhang and Ge [19]. In the early 1990's it became possible to incorporate much larger amounts of N into GaP using non-equilibrium growth techniques such as MBE, and Baillargeon et al. [20] and Liu et al. [21] revealed the existence of a large red shift of the absorption edge or a band gap reduction in GaP doped with about 1% N, suggesting the formation of an N induced impurity band in this material as well. Almost a decade later, research by Xin et al. [22] showed that the absorption edge in GaP:N for N concentrations exceeding 1% appeared to have the energy dependence characteristic of a direct gap semiconductor ($\sqrt{E - E_g}$ dependence), renewed the interest in this material for LED technologies. The evolution of the conduction band edge in GaP:N from the merger of bands formed from the N trap levels in this material is quite apparent in Fig. 3.

As mentioned earlier, the caveat with both the nitrogen doped GaAs and GaP was the poor carrier mobilities. Both these are a natural consequence of the peculiar nature of the conduction band minimum in GaAs:N and GaP:N. The conduction band edge in these nitrogen doped materials evolves out of the formation of an impurity band. As the conduction band effective mass studies in GaAs have indicated, the increase in curvature (see Fig. 4) of the impurity band with increased N doping causes the effective

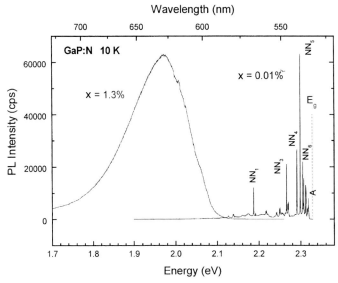

Fig. 3. Low temperature photoluminescence spectra showing evolution of conduction band edge in GaP from nitrogen induced trap levels (from Zhang et al. [30])

mass to decrease from its abnormally high value in the flat impurity band that is characteristic of low N doping concentrations, and that is reminiscent of hopping transport between impurity levels that overlap only weakly. The impurity bands which evolve from the NN trap levels, broaden with heavy nitrogen doping and merge with the bottom of the conduction band, thus giving rise to the phenomenon of giant band gap "bowing". A question that thus emerges in this unusual situation, is whether one should view $GaAs_{1-x}N_x$ as an abnormal alloy or rather as a heavily doped semiconductor GaAs:N [16, 17, 23].

Irregular Alloys When two semiconductors AC and BC are mixed, this typically results in the formation of a disordered alloy $A_xB_{1-x}C$ if the physical properties of A and B do not differ greatly from each other, and in that case the properties of the alloy change smoothly from those of BC to those of AC as x is changed from 0 to 1. Such is the case for $Al_xGa_{1-x}As$ and $In_xGa_{1-x}As$ for example. The semiconductors AC and BC must of course be miscible for a range of x which is mostly true when the properties of A and B do not differ greatly. This contrasts

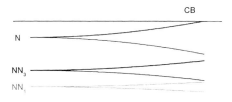

Fig. 4. Schematic showing evolution of impurity bands from different N bound states in GaP

with the n or p-type charge doping of semiconductors, where the solubility of the donor or acceptor in the host is limited (as for e.g. GaAs:Si and GaAs:Zn). In some situations the differences between alloys and doped semiconductors is not so clear. Heavy n-type doping of the order of 10^{19} (or 0.1% dopant concentration) causes a nearly 200 meV band gap reduction in GaAs which results from impurity band formation [24], whereas, a 0.1% N doping in GaAs results in a band gap reduction of less than 20 meV [10]. However, one does not speak of GaAs:Si as an alloy and so the question arises as to why is GaAs:N being referred to as a $GaAs_{1-x}N_x$ alloy? The reason for this is that N is an isoelectronic impurity in III–V alloys and thus does not result in charge doping. There are two types of isoelectronic impurities: those that do not give rise to bound states (such as GaAs:Al or GaAs:In) and those that do give rise to bound states in the host. If the isoelectronic impurity generates bound states located in the band gap (either through isolated centers or pairs) such as is the case for GaAs:N and GaP:N, then with heavy doping the impurity levels associated with these bound states evolve into impurity bands that broaden and merge with the conduction band edge, and this merger gets manifested as a giant band gap "bowing". The formation of an impurity band in heavily n or p-type doped semiconductors has been well studied together with its associated Mott metal-insulator transition [25]. Although the formation of impurity bands in heavily isoelectronically doped alloys like GaAs:N and GaP:N is relatively new, the phenomenon was actually observed in the $CdS_{1-x}Te_x$ system for concentrations $x > 10^{-4}$ almost thirty five years ago [26]. More insight into the common physical relationships between alloys, heavily charge doped semiconductors and heavily isoelectronically doped semiconductors is provided by a scaling rule recently observed by Zhang et al. [23]. As shown in Fig. 5, the band gap reduction in a doped semiconductor

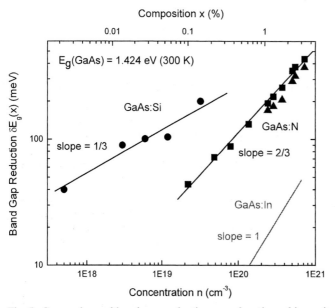

Fig. 5. Comparison of band gap reduction as a function of impurity concentration for three typical systems. Data for a) GaAs:Si (n-type doping) are from Ref. [24], b) GaAs:In (regular alloy) are from Laurenti et al. [32], and c) GaAs:N are from Ref. [23]

is observed to follow the scaling rule

$$\delta E_g(x) = \beta x^\alpha, \tag{1}$$

where x is the mole fraction of the dopant. For alloys like $In_xGa_{1-x}As$, the scaling exponent α is very close to unity. For heavily doped p-type semiconductors $\alpha \approx 1/3$. The physics underlying the $x^{1/3}$ scaling rule is simply that the bandwidth of the impurity band or the band gap reduction is proportional to the electron–electron interaction, and this interaction is proportional to the average impurity separation. For isoelectronically doped GaAs:N the scaling exponent $\alpha \approx 2/3$. The significance of the scaling exponent α being close to 2/3 lies in that it confirms that the band gap reduction in heavily doped GaAs:N is primarily due to the formation of an impurity band associated with nitrogen pair bound states. This is the primary reason for the irregular or abnormal behavior of these alloys as will be discussed below.

Physics of Isoelectronic Traps Because of the difference in valence between the dopant atom and the host atom that it replaces, a non iso-electronic donor (acceptor) atom donates an electron (hole) to the conduction (valence) band of the host crystal. The Coulomb potential of the resulting ionized donor (acceptor) atom varies with distance as r^{-1} and generates a shallow donor (acceptor) bound state. In contrast, for isoelectronic traps that are generated by isoelectronic impurities such as N in GaAs or GaP, it is the difference in electronegativity, size, and pseudopotential between the isoelectronic impurity and the host atom it replaces that generates the trap state [27]. Such traps are characterized by a potential that varies with distance much faster than r^{-1} [28]. The potential well created by the isoelectronic trap is therefore much steeper than that created by the non-isoelectronic donor (acceptor) and because of this an electron (hole) trapped around the isoelectronic impurity atom is localized much more tightly around it. This spatial localization of electrons (holes) by isoelectronic traps smears out the electronic eigenstates in k-space causing them to be delocalized in the Brillouin zone. This proves advantageous in allowing radiative transitions from these states to the band edges, thus enabling light emission from indirect gap semiconductors like GaP. As discussed above, heavy N doping in GaP and GaAs leads to impurity band formation, red-shifts in the photoluminescence, and the giant band gap "bowing" phenomenon. However, the spatial localization around the isoelectronic traps that generate the impurity bands is precisely what disadvantageously affects the carrier mobility. This is the caveat with heavily isoelectronically doped semiconductors wherein the properties of the resulting alloy are irregular. Evidently, the very success in incorporating large amounts of insoluble isoelectronic dopants using non-equilibrium growth techniques is what leads to "irregular alloy" behavior.

Resonance Raman Studies A powerful technique for probing the structure of the impurity bands is resonant Raman scattering, dealing with the interaction between lattice vibrations and the intermediate electronic states involved. If the intermediate electronic state is a Bloch state with a well-defined momentum $\mathbf{k}$, only zone-center (Γ) phonons with momentum $\mathbf{q} = 0$ are involved in the light scattering process, in the dipole approximation, and the phonon line shape is not expected to change near the resonance. However, if the intermediate electronic state originates from a strongly localized deep-impurity level so that its wavefunction is delocalized in k-space, phonons with $\mathbf{q} \neq 0$

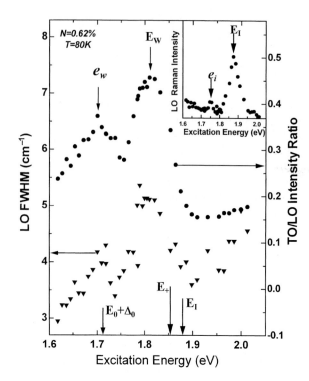

Fig. 6. Resonance profile for the FWHM of the LO(Γ) phonon, left axis, and TO(Γ)/LO(Γ) intensity ratio, right axis, for the $x = 0.62\%$ sample. The resonance profile of the LO(Γ) phonon intensity as a function of excitation energy is shown in the inset

become allowed near resonance and, as a result, the phonon line should show broadening near resonance. The changes in the Raman spectrum near resonance thus provide strong clues to the nature of the intermediate electronic state. The resonance Raman profile for the full width at half maximum (FWHM) of the LO(Γ) phonon and TO(Γ)/LO(Γ) intensity ratio for $x = 0.62\%$ sample exhibits distinct double maxima, labelled e_W and E_W, as illustrated in Fig. 6. The resonance profile of the LO(Γ) phonon intensity as a function of excitation energy also shows two maxima, labeled e_i and E_I, located near $E_0 + \Delta_0$ and E_+, respectively, as displayed in the inset of Fig. 6. It is important to note that the line width of the LO(Γ) phonon in pure GaAs was observed to be almost constant for the excitation energy from 1.55 to 2.0 eV, which encompassed both $E_0 + \Delta_0$ of pure GaAs and E_+ of GaAs$_{1-x}$N$_x$ for $x < 3\%$ at $T = 300$ K [18]. The similar anomalous double resonance maxima were more distinctly observed in the resonance profile of the line width of the 2LO(Γ) phonon for the same $x = 0.62\%$ sample as shown in Fig. 7.

The distinct double resonance of the LO(Γ) phonon line width near $E_0 + \Delta_0$ and E_+ for the $x = 0.62\%$ sample strongly suggest that the electronic states for $E_0 + \Delta_0$ transition have quite similar characteristics as those involved in E_+ transition. It is interesting to note that nitrogen incorporation hardly affects the valence band structure of GaAs whereas it changes the GaAs conduction bands drastically since all the nitrogen induced states (N$_X$ and NN pairs) are located near the conduction band edge but are very far from the valence bands of GaAs. Since the valence bands are hardly perturbed upon dilute incorporation of N into GaAs, the anomalous RRS behavior near $E_0 + \Delta_0$ of GaAs$_{1-x}$N$_x$ originates from the nitrogen-induced change in the nature of the conduc-

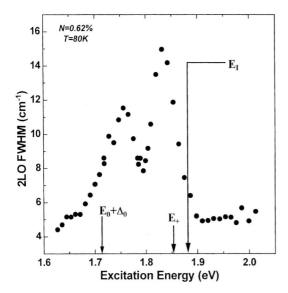

Fig. 7. Resonance profile of the FWHM of the 2LO(Γ) phonon for the $x = 0.62\%$ sample. E_I represents the excitation energy position where the LO(Γ) intensity is maximum

tion band edge at Γ. Therefore, the similarity in the line width resonance near the $E_0 + \Delta_0$ and E_+ transitions is a strong indication that the conduction band minimum around Γ does contain significant contributions from various k-components other than Γ, just as for E_+. It is well known that NN pair states as well as the isolated nitrogen resonant state (N_X) are strongly localized due to the short-range potential produced by the isoelectronic nitrogen atoms [7, 8]. The E_+ state, whose origin can be traced back to the isolated nitrogen resonant state (N_X), contains in its wavefuntion significant amount of k-components from other than the Brillouin-zone center (Γ) [29]. The significant non-Γ components in the electronic states involved in the $E_0 + \Delta_0$ transition are expected to be introduced by NN-pair and cluster states as they form impurity bands that merge with GaAs conduction band minimum at Γ [23, 30, 31].

Conclusion A description of the electronic properties that result from substitution of isovalent impurities which do not generate bound states in a host semiconductor such as GaAs or GaP can be obtained quite satisfactorily within the framework of conventional semiconductor alloy models. In contrast to this, the electronic properties that result from the substitution of non-isovalent dopant impurities which generate shallow bound states in a host are quite well described by conventional models for shallow dopants. In the case of substitutional N impurities in GaAs or GaP, N although isovalent generates bound states in the host, posing the question: should $GaAs_{1-x}N_x$ and $GaP_{1-x}N_x$ be described as conventional alloys or as heavily doped semiconductors? The abnormal or irregular behavior of these materials together with the results of a scaling rule are used to argue that they are better described as heavy isoelectronically doped semiconductors rather than as alloys.

Acknowledgements This research was supported by the U.S. Department of Energy under contract No. DE-AC36-83CH10093 and by the NREL DDRD program No. 0659.0004.

References

[1] L. VEGARD, Z. Phys. **5**, 17 (1921).
[2] J. A. VAN VECHTEN and T. K. BERGSTRESSER, Phys. Rev. B **1**, 3351 (1970).
[3] K. S. GIBONEY, L. B. ARONSON, and B. E. LEMOFF, IEEE Spectr. **35**, 43 (1998).

[4] T. WHITAKER, Compound Semicond. **4**, 32 (1998).
[5] M. WEYERS, M. SATO, and H. ANDO, Jpn. J. Appl. Phys. **31**, L853 (1992).
[6] D. G. THOMAS and J. J. HOPFIELD, Phys. Rev. **150**, 680 (1966).
[7] D. J. WOLFORD, J. A. BRADLEY, K. FRY, J. THOMPSON, and H. E. KING, Gallium Arsenide and Related Compounds 1982, Inst. Phys. Conf. Ser. No. 65, Ed. G. E. STILLMAN, The Institute of Physics, Bristol 1983 (p. 477).
D. J. WOLFORD, J. A. BRADLEY, K. FRY, and J. THOMPSON, Proc. 17th Internat. Conf. Physics of Semiconductors, Eds. J. D. CHADI and W. A. HARRISON, Springer-Verlag, New York 1984 (p. 627).
[8] X. LIU, M.-E. PISTOL, L. SAMUELSON, S. SCHWETLICK, and W. SEIFERT, Appl. Phys. Lett. **56**, 1451 (1990).
[9] I. HO and G. B. STRINGFELLOW, J. Cryst. Growth **178**, 1 (1997).
[10] H. GRÜNING, L. CHEN, T. H. HARTMANN, P. J. KLAR, W. HEIMBRODT, F. HÖHNSDORF, J. KOCH, and W. STOLZ, phys. stat. sol. (b) **215**, 39 (1999).
[11] J. NEUGEBAUER and CH. G. VAN DE WALLE, Phys. Rev. B **51**, 10568 (1995).
[12] M. KONDOW, K. UOMI, A. NIWA, T. KITANI, S. WATAHIKI, and Y. YAZAWA, Jpn. J. Appl. Phys. **35**, 1273 (1996).
[13] S. R. KURTZ, A. A. ALLERMAN, C. H. SEAGER, R. M. SIEG, and E. D. JONES, Appl. Phys. Lett. **77**, 400 (2000).
[14] J. D. PERKINS, A. MASCARENHAS, Y. ZHANG, J. F. GEISZ, D. J. FRIEDMAN, J. M. OLSON, and S. R. KURTZ, Phys. Rev. Lett. **82**, 3312 (1999);
DOE Next Generation Photovoltaics Team Meeting Abstracts, Denver, Colorado, September 11, 1998.
[15] W. SHAN, W. WALUKIEWICZ, J. W. AGER III, E. E. HALLER, J. F. GEISZ, D. J. FRIEDMAN, J. M. OLSON, and S. R. KURTZ, Phys. Rev. Lett. **82**, 1221 (1999);
DOE Next Generation Photovoltaics Team Meeting Abstracts, Denver, Colorado, September 11, 1998.
[16] Y. ZHANG, A. MASCARENHAS, H. P. XIN, and C. W. TU, Phys. Rev. B **61**, 7479 (2000).
[17] Y. ZHANG, A. MASCARENHAS, J. F. GEISZ, H. P. XIN, and C. W. TU, Phys. Rev. B **63**, 85205 (2001).
[18] H. M. CHEONG, Y. ZHANG, A. MASCARENHAS, and J. F. GEISZ, Phys. Rev. B **61**, 13687 (2000).
[19] YONG ZHANG and W.-K. GE, J. Lumin. **85**, 247 (2000).
[20] J. N. BAILLARGEON, K. Y. CHENG, G. E. HOFLER, P.J. PEARAH, and K. C. HSIEH, Appl. Phys. Lett. **60**, 2540 (1992).
[21] X. LIU, S. G. BISHOP, J. N. BAILLARGEON, and K. Y. CHENG, Appl. Phys. Lett. **63**, 208 (1993).
[22] H. P. XIN, C. W. TU, Y. ZHANG, and A. MASCARENHAS, Appl. Phys. Lett. **76**, 1267 (2000).
[23] Y. ZHANG, A. MASCARENHAS, H. P. XIN, and C. W. TU, Phys. Rev. B **63**, 161303(R) (2001).
[24] H. YAO and A. COMPAAN, Appl. Phys. Lett. **57**, 147 (1990).
[25] B. I. SHKLOVSKII and A. L. EFROS, Electronic Properties of Doped Semiconductors, Springer-Verlag, Berlin 1984, and references therein.
[26] A. C. ATEN, J. H. HAANSTRA, and H. DE VRIES, Philips Res. Rev. **20**, 395 (1965).
[27] V. K. BAZHENOV and V. I. FISTUL, Sov. Phys. – Semicond. **18**, 843 (1984).
A. N. PIKHTIN, Sov. Phys. – Semicond. **11**, 245 (1977).
[28] R. A. FAULKNER, Phys. Rev. **175**, 991 (1968).
[29] M. J. SEONG, A. MASCARENHAS, and J. F. GEISZ, Appl. Phys. Lett. **79**, 1297 (2001).
[30] Y. ZHANG, B. FLUEGEL, A. MASCARENHAS, H. P. XIN, and C. W. TU, Phys. Rev. B **62**, 4493 (2000).
[31] P. R. C. KENT and A. ZUNGER, Phys. Rev. Lett. **86**, 2613 (2001).
[32] J. P. LAURENTI, P. ROENTGEN, K. WOLTER, K. SEIBERT, H. KURZ, and J. CAMASSEL, Phys. Rev. B **37**, 4155 (1998).

Evolution of Electron States with Composition in GaAsN Alloys

P. R. C. Kent[1]) and A. Zunger

National Renewable Energy Laboratory, 1617 Cole Boulevard, Golden, CO 80401, USA

(Received June 25, 2001; accepted June 28, 2001)

Subject classification: 71.15.Dx; 71.20.Nr; 71.55.Eq; S7.14

We present a theory of the evolution of the electronic structure of GaAsN alloys, from the dilute impurity limit to the fully formed alloy. Using large scale empirical pseudopotential calculations, we show how substitutional nitrogen forms Perturbed Host States (PHS) inside the conduction band whereas small nitrogen aggregates form localized Cluster States (CS) in the band gap. By following the evolution of these states with increasing nitrogen composition we develop a model that explains many of the experimentally observed phenomena, including high effective masses, Stokes shift in emission versus absorption, and anomalous pressure dependence.

Introduction GaAsN is a special class of semiconductor alloy: due to the difference in the properties of the host and impurity atoms, bound states (or "cluster states", CS) are created near the fundamental gap, leading to discontinuous or rapidly changing optical properties with increasing nitrogen composition. The change in the electronic structure with nitrogen composition has been popularly described theoretically by approaches [1, 2] that ignore the evolution of the CS with nitrogen composition and the (statistical) fluctuations in nitrogen content present in even perfectly random samples. However, recent photoluminescence (PL) [3], pressure [4], and theoretical [5, 6] evidence demonstrate that fluctuations in nitrogen content, including the random formation of pairs, triplets and clusters of nitrogen atoms, must be explicitly included for accurate modeling and understanding of these materials.

We present fully atomistic pseudopotential calculations of the evolution of the GaAsN conduction band edge (of primary importance in PL and absorption, and applications) for purely random alloys, and calculations of the energy levels induced by prototype clusters of several nitrogen atoms. Combining these results we present a description of the conduction band edge structure of current grown material that explains many of the experimentally observed phenomena, including enhanced effective mass [7, 8], Stokes shift between emission and absorption, and anomalous pressure dependence [1].

Methodology To study the role of nitrogen in GaAs we use a supercell approach, where substitutional nitrogen atoms are placed in a large supercell. We relax all atomic positions using a valence force field method, and solve the Schrödinger equation for this periodically-repeated supercell using the plane-wave pseudopotential method with high-quality empirical pseudopotentials (EPM). Our method [5, 6] is nearly identical to that of Bellaiche et al. [9], except that we use improved pseudopotentials [10], larger supercells for better statistics, and we analyze our results in greater detail.

---

[1]) Corresponding author; e-mail: paul_kent@nrel.gov

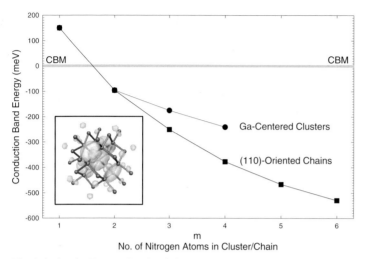

Fig. 1 (colour). Energy levels of Ga-centered nitrogen clusters and (110) directed nitrogen chains in GaAs, calculated in 4096 atom cells. Inset: wave function isosurface of cluster state at an GaN$_4$ cluster, showing strong nitrogen localization

Properties of Single Nitrogen Impurities and Small Nitrogen Clusters in GaAs The fundamental physics of dilute nitride impurities in GaAs is characterized by the formation of nitrogen localized near band gap CS. Historically, only the $a_1(N)$ level, resonant 150–180 meV above the conduction band minimum (CBM) has been identified [11], but small clusters of nitrogen atoms are expected to create other levels. The CS result from the differences in atomic size and orbital energies between the nitrogen and arsenic atom it substitutes. Our empirical pseudopotential calculated $a_1(N)$ level is at $E_c + 150$ meV and $E_c + 180$ meV for 4096 and 13824 atom cells, respectively, in close agreement with experiment.

To consider the role of small nitrogen aggregates formed during growth, we have considered a number of prototypical clusters: pairs, triplets, clusters of multiple nitrogens around a single gallium, and directed chains of nitrogen atoms [6]. In Fig. 1[2]), we show the calculated energy levels for a Ga-centered tetrahedron with its four vertices occupied by As$_{4-p}$N$_p$, with $0 \leq p \leq 4$. Note that $p = 1$ corresponds to an isolated impurity, and $p = 2$ to a first nearest neighbor N–N pair. We see that the levels become deeper as p increases, consistent with the fact that on an absolute scale the CBM of GaN is ~0.5 eV below that of GaAs. The induced CS are highly nitrogen localized, evidenced by the wave function isosurface (inset). We also considered, in Fig. 1, extended [110]-oriented chains of increasing length, motivated by the comparatively deep nature of even a [110]-oriented pair ($p = 2$, above). Consisting of 3, 4, 5 etc. nitrogen atoms we observed that each additional atom in the chain produced successively deeper levels. In general we find that an increased local concentration of nitrogen atoms, of any orientation, induces deep, dipole allowed levels. Small nitrogen aggregates therefore can contribute to below bandgap PL even at low impurity concentrations.

---

[2]) Colour figure is published online (www.physica-status-solidi.com).

Evolution of GaAsN Alloy Properties Nitrogen introduces two types of states in the dilute limit: (i) The Perturbed Host states (PHS) residing within the continuum such as $a_1(X_{1c})$, $a_1(L_{1c})$, and $a_1(\Gamma_{1c})$, and (ii) the Cluster States (CS) residing inside (or near) the band gap, e.g. the pair and higher cluster states (Fig. 1). We next address the question of how the PHS and CS evolve as the nitrogen composition increases.

We perform calculations as a function of nitrogen concentration, by randomly distributing up to 20 nitrogen atoms onto the anion sites of GaAs in a 1000 atom supercell, and 13824 atom supercells for convergence checks. We relax the atomic positions and calculate the electronic structure, repeating this for 15 randomly selected configurations at each composition to establish a statistically representative sample. The ensuing energy levels are then collected and analyzed for their degree of localization, by computing for each level ψ_i the distance $R_\alpha^{(i)}$ from the α-th nitrogen site at which 20% of the wave function is enclosed. Through this measure we have classified each level as either "localized" or "quasi-localized". This polymorphous approach describes the evolution of alloy states in an unbiased manner as alloy fluctuations are fully retained: specific CS and PHS are not assumed (e.g. as in isomorphous models [1]).

Figure 2 depicts the spectral dependence of the average localization $\sum_\alpha 1/R_\alpha^{(i)}$ for localized and quasi-localized levels of GaAsN. Panel a shows the resonant localized single-impurity $a_1(N)$ state, located within the conduction band, and selected pair, triplet and quadruplet (GaAs(N$_3$) and Ga(N$_4$)) cluster states, appearing inside the band gap. These wave functions are highly localized. Panel b shows the more extended per-

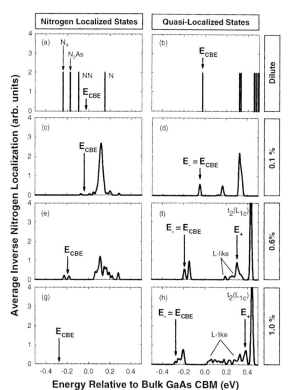

Fig. 2. Spectral dependence of average nitrogen localization for: left panel: nitrogen localized "cluster states" and right panel: quasi-localized "perturbed host states" of GaAsN for selected nitrogen compositions. The vertical arrows show the position of the alloy conduction band edge E_{CBE}

turbed X, L, and Γ host states, and the edge of the conduction band, denoted by the arrow "E_{CBE}" (also called "E_-" [1]). As the nitrogen concentration increases, Figs. 2d, f, h show that the edge E_{CBE} of the conduction band minimum (vertical arrow) moves rapidly to lower energies, due to anti-crossing and repulsion with higher energy members of the PHS. At the same time, the energy of the CS is pinned and remains fixed, as these highly localized states do not strongly interact with each other. Indeed, the wave functions of the CS do not change with composition. At the same time, the $t_2(L_{1c})$ band appears constant in energy, at $E_c + 0.4$ eV, while the upper edge of the PHS (also called "E_+" [1]) appears for $x \sim 0.6\%$ and moves up in energy as x_N increases. This broad band represents mostly delocalized or weakly localized a_1 perturbed host states.

As the edge of the PHS moves rapidly to lower energies ("optical bowing") this broad band of states sweeps past the discrete CS one-by-one. At a critical composition x_c (which depends on the degree of randomness in the samples), the deepest CS is overtaken by the moving PHS. Near x_c, the conduction band minimum is an "amalgamated state" formed from both semi-localized (Fano-resonance like) states and more delocalized states of comparable energy. The duality of semi-localized and delocalized states at the conduction band edge is responsible for many of the anomalous optical properties of dilute nitride alloys, discussed below and elsewhere [5, 6].

Conduction Band Edge Structure By considering our calculated random alloy results in combination with our results for small impurity clusters in the dilute, the current experimental data and conduction band edge structure may be interpreted as follows:

(i) The conduction band edge is formed from the delocalized PHS and some localized CS.

(ii) The low energy side of the band edge is dominated by the low energy CS, while the higher energy end consists of the more extended PHS, resulting in a Stokes shift between emission (from CS) and absorption (into PHS), exciton localization and long exciton lifetimes [12].

(iii) The effective mass at the bottom of the conduction band is enhanced as heavy, non-Γ, character is mixed in.

(iv) The pressure coefficient of the alloy is strongly reduced from the bulk [1], due to the weaker pressure dependence of the localized CS at the band edge, which couple only weakly with the extended PHS [6].

(v) The long, low energy PL tail results from low energy below gap CS, due to clusters of nitrogen atoms. These states are populated in PL by sufficiently mobile excitons finding the low energy CS. The tail CS may be due to [110]-oriented chains, or other energetically unfavorable clusters locked-in during growth. Sufficient concentrations (cross-section) of these states may also be detected in absorption. Optimized annealing of GaAsN samples will therefore give an overall blue-shift of PL and narrowing of the peak, as low-energy but unfavorable clusters are removed.

Conclusion We have presented a polymorphous theory of the evolution of the electronic structure of GaAsN alloys with nitrogen composition. Current GaAsN samples exhibit substantial alloy fluctuations resulting in many below gap cluster states, as evidenced by our calculations and broad PL seen experimentally.

Acknowledgements This work was supported by the US DOE through SC-BES-OER.

References

[1] W. Shan et al., Phys. Rev. Lett. **82**, 1221 (1999).
[2] A. Lindsay and E. P. O'Reilly, Solid State Commun. **112**, 443 (1999).
[3] Y. Zhang et al., Phys. Rev. B **63**, 085205 (2001).
[4] M. S. Tsang et al., Appl. Phys. Lett **78**, 3595 (2001).
[5] P. R. C. Kent and A. Zunger, Phys. Rev. Lett. **86**, 2613 (2001).
[6] P. R. C. Kent and A. Zunger, Phys. Rev. B **64**, 115208 (2001).
[7] Y. Zhang, A. Mascarenhas, H. P. Xin, and C. W. Tu, Phys. Rev. B **61**, 7479 (2000).
[8] P. N. Hai et al., Appl. Phys. Lett. **77**, 1843 (2000).
[9] L. Bellaiche, S.-H. Wei, and A. Zunger, Phys. Rev. B **56**, 10233 (1997).
[10] T. Mattila, S.-H. Wei, and A. Zunger, Phys. Rev. B **60**, R11245 (1999).
[11] D. J. Wolford, J. A. Bradley, K. Fry, and J. Thompson, in: Proc. 17th Internat. Conf. Physics of Semiconductors, Springer-Verlag, New York 1984 (p. 627).
[12] I. A. Buyanova et al., Appl. Phys. Lett. **75**, 3781 (1999).

Phonon Modes of $In_xGa_{1-x}As_{1-y}N_y$ Measured by Far Infrared Spectroscopic Ellipsometry

G. Leibiger[1]) (a), V. Gottschalch (b), and M. Schubert (c)

(a) Center for Microelectronic and Optical Materials Research,
and Department of Electrical Engineering, University of Nebraska-Lincoln,
Lincoln, NE 68588, USA

(b) Faculty of Chemistry and Mineralogy, Semiconductor Chemistry Group,
University of Leipzig, Linnéstraße 3, D-04103 Leipzig, Germany

(c) Faculty of Physics and Earthscience, Solid State Optics Group, University of Leipzig,
Linnéstraße 5, D-04103 Leipzig, Germany

(Received June 29, 2001; accepted July 30, 2001)

Subject classification: 63.20.Dj; 63.20.Pw; 78.30.Fs; S7.14

We study the phonon properties of compressively strained $In_xGa_{1-x}As_{1-y}N_y$ ($x < 0.13$, $y < 0.03$) single layers for wavenumbers from 100 to 600 cm^{-1} using far infrared spectroscopic ellipsometry. The intentionally undoped InGaAsN layers were grown pseudomorphically on top of undoped GaAs buffer layers deposited on Te-doped (001) GaAs substrates by metal-organic vapor phase epitaxy. The InGaAsN layers show a two-mode phonon behaviour in the spectral range from 100 to 600 cm^{-1}. We detect the transverse GaAs and GaN sublattice phonon modes at wavenumbers of about 267 and 470 cm^{-1}, respectively. The polar strength f of the GaN sublattice resonance changes with nitrogen composition y and with the biaxial strain ε_{xx} resulting from the lattice mismatch between InGaAsN and GaAs. This effect can be used to derive the nitrogen and indium content of the InGaAsN layers combining the observed f-dependence with results from high-resolution double-crystal X-ay diffractometry.

Introduction The quaternary InGaAsN alloy has attracted increasing interest in the past few years due to its potential application for long wavelength emitters [1]. However, there are still properties, such as the phonon behaviour, which are unknown yet. In Refs. [2] and [3] we reported on optical and infrared-optical properties of GaAsN/GaAs and GaAsN/InAs/GaAs superlattices, and GaAsN single layers using spectroscopic ellipsometry (SE). In this study we investigate the phonon properties of $In_xGa_{1-x}As_{1-y}N_y$ ($x < 0.13$, $y < 0.03$) single layers using far-infrared spectroscopic ellipsometry (FIRSE).

Experimental The InGaAsN layers ($d \sim 450$ nm) were grown pseudomorphically by MOVPE at growth temperatures ranging from 560 to 600 °C on undoped GaAs buffer layers ($d \sim 350$ nm) which were deposited on Te-doped (001) GaAs substrates. Trimethylindium, trimethylgallium, 1,1-dimethylhydrazine, and arsine were used as precursors. A GaAs-cap layer of about 30 nm thickness was grown on top of the InGaAsN layers. The indium and nitrogen compositions, which are given in Table 1, were determined by FIRSE and high-resolution X-ray diffractometry (HRXRD) as described in Ref. [4]. All samples were measured at room temperature by spectroscopic ellipsometry for wavenumbers from 100 to 600 cm^{-1} at 50° angle of incidence.

---

[1]) Corresponding author; Phone: +001(402)472-1964; Fax: +001(402)472-7987;
e-mail: pge97jrk@studserv.uni-leipzig.de

Table 1
In and N concentrations of our $In_xGa_{1-x}As_{1-y}N_y$ layers

sample	A	B	C	D	E
x	0.09	0.11	0.11	0.12	0.09
y	0.013	0.019	0.022	0.024	0.029

Ellipsometry Ellipsometry can determine the complex dielectric function ε and thickness d of a thin-film sample by comparing the measured data with best-fit model calculations. For modelling the dielectric function of InGaAsN we used the same parametric models that we successfully applied to GaAsN [3]. For a detailed description of the applied models as well as for a short introduction into ellipsometry we refer to Refs. [2] and [3].

Results and Discussion In Fig. 1 we present the experimental (dashed lines) and modeled (solid lines) Ψ spectra of our InGaAsN/GaAs samples. The Ψ spectra were calculated using the FIR-MDF introduced in Section 3, and the best-fit model parameters are given in Table 1. For the Te-doped GaAs substrates we assumed an effective mass parameter of $0.073m_e$. The values of the GaAs substrate free-carrier concentrations N and mobilities μ obtained from the FIRSE data analysis vary between 3.5×10^{17} and 4.3×10^{17} cm^{-3}, and between 2570 and 2884 cm^2/Vs, respectively. Note that the phonon frequencies of the substrate, the buffer and the cap layer were set to the values of unstrained GaAs ($\omega_{LO1} = 291.3$ cm^{-1}, $\omega_{TO1} = 267.7$ cm^{-1}). Note further that we assumed the same high-frequency dielectric constant parameter (ε_∞) for all substrates and GaAs layers, the same broadening parameter for all substrates (γ_{sub}), and for all buffer and cap layers (γ_{buffer}). As a result of the multiple-sample analysis we obtain $\varepsilon_\infty = (12.3 \pm 0.3)$, $\gamma_{sub} = (3.1 \pm 0.1)$ cm^{-1}, and $\gamma_{buffer} = (2.5 \pm 0.1)$ cm^{-1}.

The plasma edge of the Te-doped GaAs substrates dominates the spectra for wavenumbers below 200 cm^{-1}. All InGaAsN layers show a two-mode phonon behaviour. The reststrahlenband of the GaAs phonons is located between the GaAs-TO and LO frequencies ($\omega_{TO1} \approx 267$ cm^{-1}, $\omega_{LO1} \approx 291$ cm^{-1}, respectively). We further detect the GaN sublattice resonance at wavenumbers of about 470 cm^{-1} which is discussed in more detail further below. We do not detect any InN or InAs related sublattice reso-

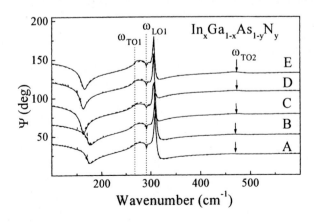

Fig. 1. Experimental (dashed lines) and calculated (solid lines) Ψ spectra of InGaAsN/GaAs in the FIR spectral range

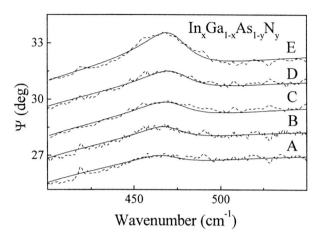

Fig. 2. Experimental (dashed lines) and calculated (solid lines) Ψ spectra of InGaAsN/GaAs in the vicinity of the GaN sublattice resonance

nances. Thus, the phonon properties of our InGaAsN layers are very similar to the phonon properties of GaAsN [2, 3]. The InN phonon migth be absent because of the relatively low nitrogen and indium compositions. The absence of the InAs phonon is well known from FIR reflection studies of Ga-rich $Ga_{1-x}In_xAs$ compounds ($x \leq 0.2$) [5, 6]. Brodsky et al. [5] explained this observation with the lack of a frequency gap between the acoustical and optical branches in the phonon dispersion spectrum of GaAs. Yamazaki et al. [6] calculated the oscillator strength of the InAs sublattice resonance on the basis of a modified cluster model. The authors found that the calculated oscillator strength is too small to be detected experimentally for $x \leq 0.2$.

Figure 2 presents a detail of Fig. 1 containing the structures due to the GaN sublattice resonance. The TO-frequencies ω_{TO2} increase from 469 cm^{-1} ($y = 0.013$) to 473 cm^{-1} ($y = 0.029$). In general, one expects a blueshift with increasing nitrogen composition in view of a linear interpolation between the TO-frequencies of the local mode of nitrogen in GaAs ($\omega_{TO2}^{loc} = 469$ cm^{-1}) [7] and cubic GaN ($\omega_{TO2}^{GaN} = 553$ cm^{-1}) [8] and

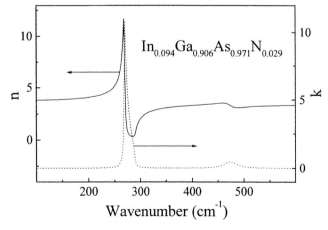

Fig. 3. Optical functions n and k of $In_{0.094}Ga_{0.906}As_{971}N_{0.029}$ layers in the FIR spectral range calculated using the FIR-MDF and best-fit parameters obtained in this work

accounting for compressive layer strain [9]. However, due to the different indium concentrations, and thus change of the layer strain in our InGaAsN layers we cannot compare the obtained frequencies ω_{TO2} with each other directly. Instead, the amplitude f ($f \equiv (\omega_{LO2} - \omega_{TO2})/\omega_{TO2}$) of the GaN sublattice resonance together with the results of HRXRD can be used to calculate nitrogen and indium concentrations in InGaAsN, which is described in Ref. [4].

Figure 3 presents the optical constants n and k of the $In_{0.094}Ga_{0.906}As_{0.971}N_{0.029}$ layer in the FIR spectral range calculated using the FIR-MDF (Eq. (2)), and the best-fit model parameters given in Table 1. These spectra are very similar to the spectra of GaAsN with comparable N content, which were shown in Refs. [2] and [3].

Conclusion In summary, we investigated the phonon properties of MOVPE-InGaAsN layers grown on (001) Te-GaAs substrates in the far ($100~\text{cm}^{-1} \leq \omega \leq 600~\text{cm}^{-1}$) infrared spectral range using spectroscopic ellipsometry. We present model dielectric functions, which can be used to calculate the optical functions $n(E)$ and $k(E)$. We observed a two-mode phonon behaviour similar to the ternary alloy GaAsN. The GaAs and GaN sublattices are resonant at TO frequencies of $\omega_{TO1} \cong 267~\text{cm}^{-1}$ and $\omega_{TO2} \cong 470~\text{cm}^{-1}$, respectively. InAs and InN phonons could not be detected. The amplitude of the GaN phonon, together with high-resolution X-ray diffraction measurements, can be used to calculate nitrogen and indium concentrations in InGaAsN layers.

Acknowledgements This work is supported by the Deutsche Forschungsgemeinschaft under grant Go 629/4-1, and by the National Science Foundation under contract DMI-9901510. The Center for Microelectronic and Optical Materials Research at University of Nebraska further supported this work.

References

[1] M. Kondow, K. Uomi, K. Hosomi, and T. Mozume, Jpn. J. Appl. Phys. **33**, L1056 (1994).
[2] J. Šik, M. Schubert, G. Leibiger, V. Gottschalch, and G. Wagner, J. Appl. Phys. **89**, 294 (2001).
[3] G. Leibiger, V. Gottschalch, B. Rheinländer, J. Šik, and M. Schubert, J. Appl. Phys. **89**, 4927 (2001).
[4] G. Leibiger, V. Gottschalch, and M. Schubert, submitted to J. Appl. Phys.
[5] M. H. Brodsky and G. Lucovsky, Phys. Rev. Lett. **21**, 990 (1968).
[6] S. Yamazaki, A. Ushirokawa, and T. Katoda, J. Appl. Phys. **51**, 3722 (1980).
[7] V. Riede, H. Neumann, H. Sobotta, R. Schwabe, W. Seifert, and S. Schwetlick, phys. stat. sol. (a) **93**, K151 (1986).
[8] J. W. Orton and C. T. Foxon, Rep. Prog. Phys. **61**, 1 (1998).
[9] B. Jusserand and M. Cardona, in: Ligth Scattering in Solids V, Eds. M. Cardona and G. Güntherodt, Springer-Verlag, New York 1988.

MOCVD Growth of InN_xAs_{1-x} on GaAs Using Dimethylhydrazine

A. A. El-Emawy, H.-J. Cao, E. Zhmayev, J.-H. Lee, D. Zubia, and M. Osiński[1])

Center for High Technology Materials, University of New Mexico,
1313 Goddard SE, Albuquerque, New Mexico 87106-4343, USA

(Received July 22, 2001; accepted August 4, 2001)

Subject classification: 68.37.Ps; 68.37.Vj; 68.65.Fg; 81.05.Ea; 81.15.Gh; S7.14; S7.15

InNAs/GaAs multiple-quantum-well samples were grown by MOCVD on (100) n⁺-GaAs substrates at 500 °C and 60 Torr using uncracked dimethylhydrazine (DMHy). Quantum well layers were grown using trimethylindium, tertiarybutylarsine, and 95–97.5% of DMHy in the vapor phase, while GaAs buffer, barrier, and cap layers were grown using trimethylgallium and arsine. The crystalline quality and solid phase composition were evaluated using high-resolution X-ray diffraction analysis. The nitrogen content in InNAs wells was determined to be 18%. Surface morphology was investigated by atomic force microscopy (AFM) and field emission microscopy (FEM). Photoluminescence measurements confirm that the bandgap energy of InNAs is significantly lower than that of InAs. The peak emission wavelength of ∼6.5 μm at 10 K is the longest reported so far for dilute nitride semiconductors.

Introduction The InN_xAs_{1-x} alloy is a very promising material for mid-infrared (2–5 μm) emitters and detectors. With a single exception of recently reported successful growth of InNAs by MOCVD with plasma cracked ammonia source [1], all studies of InNAs growth utilized plasma-source MBE [2–4] and related techniques, such as gas-source MBE [5, 6]. In this paper, we report a successful MOCVD growth of InNAs by using for the first time the uncracked dimethylhydrazine (DMHy) as nitrogen source.

As with other dilute nitrides, the bandgap of InN_xAs_{1-x} was predicted to shrink with increasing nitrogen content x [7, 8], although so far this has been confirmed experimentally only for compositions $x < 6\%$ [6]. The InN_xAs_{1-x} alloy can be lattice-matched to GaAs when $x = 38\%$ [2]. The large band offset between this material and GaAs barriers makes it particularly attractive for reducing temperature sensitivity of mid-IR lasers. Since the solubility limit of nitrogen in InAs is among the highest among binary III–V compounds (three orders of magnitude higher than in GaAs, for example) [9], we considered it feasible to attempt the low temperature growth using DMHy.

Growth Conditions and Procedures All samples were grown in a Thomas Swan vertical MOCVD reactor at 60 Torr. Arsine and trimethylgallium were used during the growth of GaAs buffer, barrier, and cap layers. InNAs films were grown using trimethylindium, tertiarybutylarsine and DMHy as source materials. Before the growth, epi-ready (100) n⁺-GaAs substrates were deoxidized at 760 °C for 5 min. 500 nm GaAs buffer layer was grown at 680 °C. Subsequently, the growth temperature T_g was lowered to 500 °C for the growth of InNAs/GaAs multiple quantum well (MQW) structures and the GaAs capping layer.

---

[1]) Corresponding author; Phone: +1 505 272 7812; Fax: +1 505 272 7801;
e-mail: osinski@chtm.unm.edu

The growth rate was kept constant at 7.5 and 1.25 Å/s for GaAs and InNAs films, respectively. The mole fraction of DMHy to total group-V sources in the vapor phase was maintained between 0.95 and 0.975, corresponding to DMHy flow rate of 600–900 sccm.

X-Ray Analysis of InNAs/GaAs MQW Structures Crystalline quality, composition, and thicknesses of grown layers were investigated by high-resolution X-ray diffraction (HRXRD) Philips triple axis X-ray apparatus. Figure 1 shows $\omega-2\theta$ scans of (004) reflections for two 15-period MQW structures grown under otherwise identical conditions, except for the presence of DMHy during the well layer growth. Sample DE217 contained InNAs/GaAs quantum wells, with the well layers grown under 600 sccm flow rate of DMHy. Sample DE218 was grown under the same conditions as sample DE217, but the well layers contained only InAs. Comparison of the two scans shown in Fig. 1 clearly reveals improvement in quality of the sample containing nitrogen, which we attribute to the reduced lattice constant mismatch between InNAs and GaAs. The full width at half maximum (FWHM) of the $m = 0$ InNAs peak (top scan in Fig. 1) is 65 arcsec, which together with a clearly resolved spectrum displaying up to six diffraction orders indicates high crystalline quality of the sample. The FWHM of the $m = -3$ satellite peak of sample DE217 is 115 arcsec, which compares favorably to 280 arcsec in $InN_{0.06}As_{0.94}$/InGaAsP MQWs grown on InP [6]. The satellite peaks wash out in the case of InAs/GaAs MQWs (sample DE218), revealing poorer interface flatness and partial relaxation.

Measured HRXRD spectra of sample DE217 were analyzed using Bede Scientific Instruments RADS dynamic simulation software, assuming cubic InN and InAs lattice parameters. As shown in Fig. 2, nitrogen composition of 18% resulted in a very good fit between the measured and simulated spectra. Using the same simulation, the well and barrier layer thicknesses were determined as 2.5 nm and 375 nm, respectively.

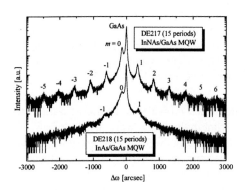

Fig. 1

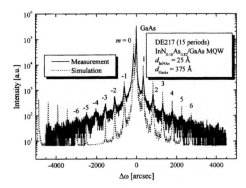

Fig. 2

Fig. 1. HRXRD spectra of 15-period MQW samples with InNAs wells (top scan) and InAs wells (bottom scan)

Fig. 2. Measured (solid line) and simulated (dotted line) HRXRD spectra of 15-period InNAs/GaAs MQW sample

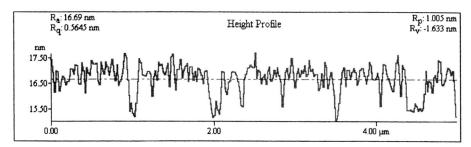

Fig. 3. AFM scan of a sample DE220 consisting of 2.5 nm InNAs/37.5 nm GaAs/2.5 nm InNAs structure grown on GaAs

Microscopic Analysis of InNAs/GaAs Samples The surface morphology of InNAs layers was investigated using Park Scientific Instruments Autoprobe CP atomic force microscope (AFM) and JEOL JSM-6400F high resolution field-emission scanning electron microscope (FEM). Figure 3 shows an AFM scan. The average surface roughness as determined by AFM was 0.5 nm, indicating complete coverage of GaAs buffer layer and high uniformity of the InNAs layer. Figure 4 shows a cross-sectional FEM micrograph revealing the quantum wells.

Photoluminescence (PL) Measurements In order to measure PL spectra, three sets of triple-quantum-well InNAsGaAs samples were grown at different nitrogen flow rates (600 sccm for samples DE208 and DE209, 900 sccm for sample DE210) and with different well thicknesses (2.5 nm for sample DE208, and 5 nm for samples DE209 and DE210). Each sample had a 0.25 μm GaAs cap layer at the top. HRXRD analysis showed that with increasing well thickness, the satellite peaks became broader, indicating partially relaxed films. With increasing nitrogen flow, almost no change in the HRXRD spectra was observed, which we attribute to saturation limit of DMHy decomposition at the low T_g of 500 °C.

The PL setup included a Coherent Innova 300 Ar-ion laser source, with combined 514 and 488 nm lines pumping a Schwartz Electro-Optics Titan-ML tunable Ti:sapphire laser, a Nicolet Magna-IR 760 Fourier transform spectrometer, and a Nicolet MCT-B cooled detector. The samples were placed in a cryostat and cooled to 10 K. The output power from the argon-ion laser was 7.5 W, while the output power incident on the sample from the Ti:sapphire laser tuned to 890 nm was 170 mW, illuminating a spot with ~1 mm diameter.

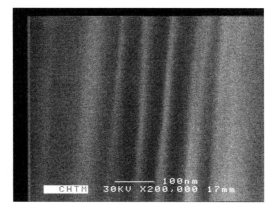

Fig. 4. Cross-sectional FEM micrograph of sample DE233 consisting of a three-period 5 nm InNAs/37.5 nm GaAs quantum well grown on GaAs. The bright part on the right-hand side corresponds to sample surface

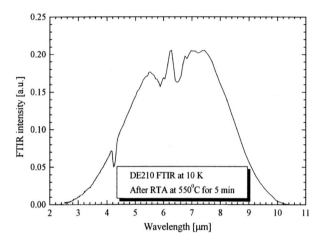

Fig. 5. PL spectrum of sample DE210 containing a triple quantum well 5 nm InNAs/37.5 nm GaAs structure

Initial experiments with as-grown samples resulted in no detected PL signal. The samples were then annealed at 550 °C for 5 min using a rapid thermal annealer, and the PL measurements were repeated. Figure 5 shows the PL spectrum from sample DE210 that emitted intense PL signal after annealing. The peak emission wavelength was about 6.5 μm (190 meV), and the FWHM of the PL spectrum was about 3.5 μm (110 meV).

The dip at 4.26 μm is due to CO_2 band absorption, while the broad minima near 5.9 and 6.5 μm are caused by water vapor absorption in the ambient atmosphere. The maximum at 6.3 μm coincides with water vapor transmission window.

Recently, Tit and Dharma-Wardana [8] reported tight-binding calculations of InNAs band structure, considering various possible arrangements of group-V atoms around a single In atom. For 17.6% nitrogen content, they predict a bandgap ranging from 8 meV for maximally nitrogen-rich clusters to 251 meV for maximally As-rich clusters. This wide range of bandgap energies is consistent with our observation of broad PL emission spectrum shown in Fig. 5. It is also very promising for widely tunable mid-IR lasers.

Conclusions We have successfully grown InNAs/GaAs multiple-quantum-well structures by MOCVD using uncracked dimethylhydrazine (DMHy). The crystalline quality and solid phase composition were evaluated by high-resolution X-ray diffraction. Nitrogen content in InNAs wells was determined to be 18%. Surface morphology was investigated by atomic force microscopy (AFM) and field emission microscopy (FEM). AFM and FEM micrographs indicate high quality of quantum-well structures. PL measurements confirm that the bandgap energy of InNAs is significantly lower than that of InAs. The peak emission wavelength of ∼6.5 μm is the longest reported so far for dilute nitride semiconductors.

Acknowledgement Support from the U.S. Air Force Office of Scientific Research under the Optoelectronics Research Center program is gratefully acknowledged.

References

[1] H. Naoi, D. M. Shaw, Y. Naoi, G. J. Collins, and S. Sakai, J. Cryst. Growth **222**, 511 (2001).
[2] Y. C. Kao, T. P. E. Broekaert, H. Y. Liu, S. Tang, I. H. Ho, and G. B. Stringfellow, MRS Symp. Proc. **423**, 335 (1996).
[3] S. Sakai, T. S. Cheng, T. C. Foxon, T. Sugahara, Y. Naoi, and H. Naoi, J. Cryst. Growth **189/190**, 471 (1998).
[4] R. Beresford, K. S. Stevens, and A. F. Schwartzman, J. Vac. Sci. Technol. B **16**, 1293 (1998).
[5] J.-S. Wang and H.-H. Lin, J. Vac. Sci. Technol. B **17**, 1997 (1999).
[6] J.-S. Wang, H.-H. Lin, L.-W. Song, and G.-R. Chen, J. Vac. Sci. Technol. B **19**, 202 (2001).
[7] T. Yang, S. Nakajima, and S. Sakai, Jpn. J. Appl. Phys. **36**, L320 (1997).
[8] N. Tit and M. W. C. Dharma-Wardana, Appl. Phys. Lett. **76**, 3576 (2000).
[9] I.-H. Ho and G. B. Stringfellow, J. Cryst. Growth **178**, 1 (1997).

Spectroscopic Ellipsometry Study on the Electronic Structure near the Absorption Edge of GaAsN Alloys

H. Yaguchi[1]) (a), S. Matsumoto (a), Y. Hijikata (a), S. Yoshida (a), T. Maeda (b), M. Ogura (b), D. Aoki (c), and K. Onabe (c)

(a) Department of Electrical and Electronic Systems, 255 Shimo-Okubo, Saitama-shi, Saitama 338-8570, Japan

(b) National Institute of Advanced Industrial Science and Technology, 1-1-1 Umezono, Tsukuba, Ibaraki 305-8568, Japan

(c) Department of Advanced Materials Science, The University of Tokyo, 7-3-1 Hongo, Bunkyo-ku, Tokyo 113-8656, Japan

(Received July 14, 2001; accepted July 18, 2001)

Subject classification: 61.10.Nz; 71.20.Nr; 78.20.Ci; 78.66.Fd; S7.14

Spectroscopic ellipsometry has been used to investigate the electronic structure near the fundamental absorption edge of GaAsN alloys grown by metalorganic vapor phase epitaxy. The fundamental absorption edge is clearly observed in the imaginary part of the dielectric function and shifts to lower energies with increasing N concentration. In addition, the absorption structure is observed near the E_0 gap energy of GaAs even in GaAsN alloys. This unequivocally shows that the fundamental absorption edge of GaAsN is not shifted from the E_0 gap of GaAs but newly formed by the N incorporation. Thus, the formation of the narrowest band gap of GaAsN alloys is found to be completely different from that of conventional compound semiconductor alloys, such as AlGaAs and GaAsP.

Introduction Recently, III–V–N alloys are intensively studied since these semiconductor alloys have promising potential for optoelectronic device applications due to their unique electronic and optical properties, such as a huge band gap bowing [1–6], anomalous increase in the refractive index [7], and so on. Particularly, (In)GaAsN is expected as a material for long-wavelength semiconductor lasers used in the optical fiber communications, because their large conduction band offset to (Al)GaAs realizes the sufficient electron confinement and leads to high characteristic temperature of the semiconductor lasers. Although a lot of studies have been made to clarify the mechanism of the huge band gap bowing, one of the most interesting properties of III–V–N alloys, it has not yet been clarified. Previously, we have measured complex dielectric functions of GaAsN alloys using spectroscopic ellipsometry to investigate higher-energy band gaps [8], and have found that the optical transition probability at the E_1 gap decreases with increasing N concentration and that the lowest conduction band is formed by contribution of the L-point state. In the present study, we have focused our attention on the electronic structure near the fundamental absorption edge.

Experimental The samples used in this study were $GaAs_{1-x}N_x$ alloys ($x = 0$, 1.0%, and 2.3%) grown on GaAs (001) substrates by metalorganic vapor phase epitaxy. Tri-

---

[1]) Corresponding author; Phone: +81 48 858 3841; Fax: +81 48 858 3841; e-mail: yaguchi@opt.ees.saitama-u.ac.jp

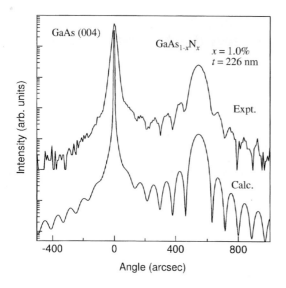

Fig. 1. (004) X-ray diffraction curve of GaAs$_{1-x}$N$_x$ alloy

methylgallium, arsine and 1,1-dimethylhydrazine were used as Ga, As, and N sources, respectively. The N concentration and layer thickness of GaAsN alloys were determined using X-ray diffraction.

Figure 1 shows an example of results of the (004) X-ray diffraction curve of GaAsN alloy taken using Cu Kα_1 radiation. Upper and lower curves represent the experimental and calculated diffraction curves, respectively. The lower diffraction curve was calculated based on dynamical X-ray diffraction theory [9]. From this X-ray diffraction measurement the N concentration and the layer thickness were estimated to be 1.0% and 226 nm, respectively. The spectroscopic ellipsometry measurements (Sopra, model GESP-5) were carried out in the wavelength range of 700–1200 nm at room temperature. The light was incident at 75° on the samples. For a GaAs epitaxial layer grown simultaneously with GaAsN layers, the thickness of the native oxide layer was 3 nm. Thus, we assumed the same thickness as on GaAs. The complex dielectric functions, $\varepsilon_1 + i\varepsilon_2$, of GaAsN alloys were numerically determined using the Newton-Raphson method for the sample structure determined from the X-ray diffraction measurements. We have also measured photoluminescence at room temperature using an Ar ion laser as the excitation source to compare with the results of spectroscopic ellipsometry.

Results and Discussion Figure 2 shows the complex dielectric functions, ε_1 and ε_2 of GaAs$_{1-x}$N$_x$ alloys ($x = 0\%$, 1.0%, 2.3%). The real part of the dielectrics functions, ε_1, is shifted by +0.5 and +1.0 for $x = 1.0\%$ and 2.3%, respectively.

The absorption edge is clearly observed in the imaginary part of the dielectric functions, ε_2, of GaAsN alloys and is found to shift to lower energies with increasing N concentration. The peak corresponding to the absorption edge is also observed in the real part of the dielectric functions. In addition, it is worth noting that the absorption structure was clearly observed near the E_0 gap energy of GaAs even in both the real and imaginary parts of the dielectric functions of GaAsN alloys. This unequivocally shows the fundamental absorption edge of GaAsN alloy is not shifted from the E_0 gap of GaAs but newly formed by the N incorporation. Therefore, the nature of the fundamental absorption edge, or the formation of the narrowest band gap of GaAsN alloys is found to be completely different from that of conventional compound semiconductor alloys, such as AlGaAs and GaAsP, but similar to GaPN [3]. The anomalous increase in the refractive index observed in InGaAsN alloys [7] can be also explained by the fact that the fundamental absorption edge is newly formed by the N incorporation.

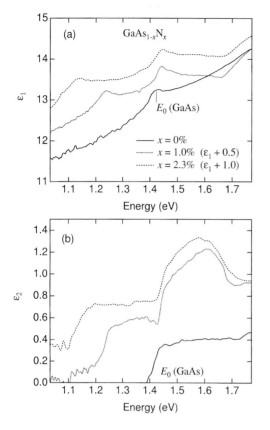

Fig. 2. Complex dielectric functions of GaAsN alloys: a) real part, b) imaginary part

The transition energies of GaAsN alloys were estimated from the complex dielectric functions using model dielectric functions [10]. We treated both fundamental absorption edge and the optical transition near the E_0 gap of GaAs, which we name the "E_0" gap transition hereafter, as three-dimensional M_0 critical points.

Figure 3 shows the fundamental absorption edge energy and the "E_0" gap energy estimated from the analysis using model dielectric functions. The PL peak energy at room temperature is also plotted in this figure.

The fundamental absorption edge energy decreases with increasing N concentration and agrees well with the photoluminescence peak energy at room temperature. This energy shift is also in good agreement with the results for the absorption measurements [5]. On the other hand, the "E_0" gap energy is almost constant but slightly increases with increasing N concentration. Since the GaAsN layers used in this study were coherently grown on GaAs, the gap energy is affected by the strain. For example, the E_0 gap energy of GaAs decreases due to its negative volume deformation potential under the tensile strain. However, an increase in the "E_0" gap is found in spite of the tensile strain of the GaAsN layers. This indicates that the energy shift of the "E_0" gap is not due to the strain effect but mainly due to the intrinsic alloy effect.

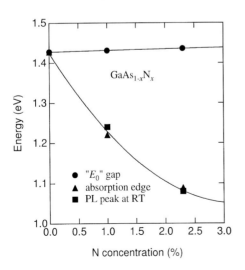

Fig. 3. Absorption edge and the "E_0" gap energies as a function of N concentration

Conclusions The complex dielectric functions near the fundamental absorption edge of GaAsN alloys were studied using spectroscopic ellipsometry. The fundamental absorption edge was clearly observed in the imaginary part of the dielectric functions and the peak corresponding to the absorption edge is also observed in the real part of the dielectric functions. The fundamental absorption edge energy decreases with increasing N concentration and agrees well with the PL peak energy at room temperature. In addition, the absorption structure was clearly observed near the E_0 gap energy of GaAs even in GaAsN alloys. This shows that the fundamental absorption edge of GaAsN is not shifted from the E_0 gap of GaAs but newly formed by the N incorporation.

Acknowledgement This work was partly supported by the Nippon Sheet Glass Foundation for Materials Science and Engineering.

References

[1] J. N. BAILLARGEON, K. Y. CHENG, G. E. HOFLER, P. J. PEARAH, and C. HSIEH, Appl. Phys. Lett. **60**, 2540 (1992).
[2] S. MIYOSHI, H. YAGUCHI, K. ONABE, R. ITO, and Y. SHIRAKI, Appl. Phys. Lett. **63**, 3506 (1993).
[3] H. YAGUCHI, S. MIYOSHI, G. BIWA, M. KIBUNE, K. ONABE, Y. SHIRAKI, and R. ITO, J. Cryst. Growth **170**, 353 (1997).
[4] W. G. BI and C. W. TU, Appl. Phys. Lett. **70**, 1608 (1997).
[5] K. UESUGI and I. SUEMUNE, Jpn. J. Appl. Phys. **36**, 1572 (1997).
[6] K. ONABE, D. AOKI, J. WU, H. YAGUCHI, and Y. SHIRAKI, phys. stat. sol. (a) **176**, 231 (1999).
[7] T. KITATANI, M. KONDOW, K. SHINODA, Y. YAZAWA, and M. OKAI, Jpn. J. Appl. Phys. **37**, 753 (1998).
[8] S. MATSUMOTO, H. YAGUCHI, S. KASHIWASE, T. HASHIMOTO, S. YOSHIDA, D. AOKI, and K. ONABE, J. Cryst. Growth **221**, 481 (2000).
[9] M. A. G. HALIWELL, M. H. LYONS, and M. J. HILL, J. Cryst. Growth **68**, 523 (1987).
[10] S. ADACHI, Phys. Rev. **B35**, 7454 (1984).

Photoluminescence Study on Temperature Dependence of Band Gap Energy of GaAsN Alloys

H. Yaguchi[1]) (a), S. Kikuchi (a), Y. Hijikata (a), S. Yoshida (a), D. Aoki (b), and K. Onabe (b)

(a) Department of Electrical and Electronic Systems, Saitama University, 255 Shimo-Okubo, Saitama-shi, Saitama 338-8570, Japan

(b) Department of Advanced Materials Science, The University of Tokyo, 7-3-1 Hongo, Bunkyo-ku, Tokyo 113-8656, Japan

(Received July 16, 2001; accepted July 19, 2001)

Subject classification: 78.55.Cr; 78.66.Fd; S7.14

We have studied the temperature dependence of photoluminescence (PL) spectra of GaAsN alloys. The PL peak energy shift due to the temperature change decreases with increasing N concentration of GaAsN alloys. The localized state emission partly contributes to the decrease in the PL peak energy shift. In addition, the small PL peak energy shift at high temperatures is due to the reduction in the temperature dependence of the band gap energy. From the analysis using the Bose-Einstein statistical expression, the average phonon energy is much larger than that expected from the linear interpolation between GaAs and GaN, indicating that the interaction between electrons and phonons localized at N atoms plays an important role in the reduction of the temperature dependence of the band gap energy of GaAsN alloys.

1. Introduction

Recently, III–V–N alloys have been receiving much attention because of their unique electronic and optical properties, and thus they are expected to be novel materials for optoelectronic device applications. Particularly, (In)GaAsN alloy is a promising material for long-wavelength semiconductor lasers with superior characteristic. As for optical-fiber communications, the stability against temperature during operation of the semiconductor lasers is desirable, i.e. the output power stability and the wavelength stability. Large conduction band offsets between (In)GaAsN and (Al)GaAs leads to an output power stability against temperature, and indeed a high characteristic temperature for the InGaAsN semiconductor laser is obtained [1]. On the other hand, the temperature dependence of the band gap energy determines the wavelength stability. It is reported that the temperature dependence of the band gap energy of GaAsN alloy is smaller than GaAs from absorption measurements [2] and is explained in terms of band anti-crossing model [3]. In addition, photoluminescence measurements revealed that the emission related to localized states results in the reduction of the PL peak energy shift due to the temperature change [4]. In this study we have measured photoluminescence (PL) spectra of GaAsN alloys to examine in detail the temperature dependence of the band gap energy and the influence of localized states.

---

[1]) Corresponding author; Phone: +81 48 858 3841; Fax: +81 48 858 3841; e-mail: yaguchi@opt.ees.saitama-u.ac.jp

2. Experimental

The samples used in this study were GaAsN alloys the N concentration of which ranged from 0 to 3.1% on GaAs (001) substrates grown by metalorganic vapor phase epitaxy (MOVPE). The sources for Ga, As, and N were trimethylgallium, arsine, and 1,1-dimethylhydrazine, respectively. Owing to the optimization of the MOVPE growth, highly luminescent samples have been successfully obtained even without post annealing [4]. The PL measurements were carried out at 10 to 300 K using an Ar ion laser (488 nm, 10 mW) as the excitation source.

3. Results and Discussion

Figure 1 shows the temperature dependence of the PL spectra of GaAsN alloy with the N concentration of 0.97%. Emission related to localized states is observed at low temperatures and vanishes with increasing temperature. On the other hand, a new PL peak appears at the higher energy side and becomes dominant with increasing temperature. The PL peak which is dominant at high temperatures shifts to lower energies with increasing temperature. This indicates that the PL observed at high temperatures is the band edge emission and that the peak shift is related to the temperature dependence of the band gap energy.

The temperature dependence of the PL peak energy of GaAsN alloys is shown in Fig. 2. At low temperatures, except GaAs, the PL peak energy rapidly decreases with increasing temperature and the peak intensity also rapidly decreases, and at the end the lower energy emission vanishes. As can be seen from this figure, the PL peak energy shift is reduced as the N concentration increases, which partly results from the influence of the localized states in GaAsN alloys. However, the PL peak shift at high temperatures also decreases with increasing N concentration. This indicates that the temperature dependence of the band gap energy of GaAsN alloys decreases with increasing N concentration. Thus, the PL peak energies at high temperatures were analyzed by the Bose-Einstein statistical expression [5],

$$E_{PL} = E_B - a_B \left[\frac{2}{\exp(\Theta_B/T) - 1} + 1 \right] + a k_B T, \quad (1)$$

where a_B represents the electron–phonon interaction strength, Θ_B is

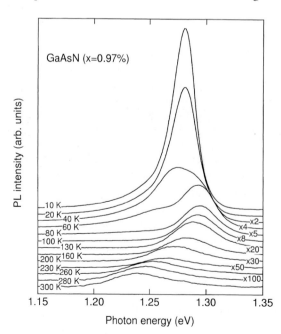

Fig. 1. Temperature dependence of photoluminescence spectra of GaAs$_{1-x}$N$_x$ ($x = 0.97\%$)

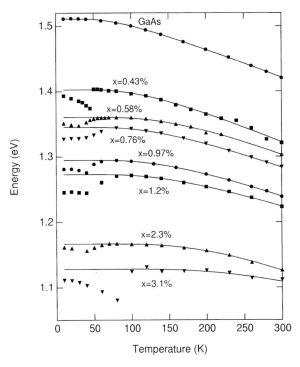

Fig. 2. Temperature dependence of photoluminescence peak energy of GaAsN alloys

the average phonon energy, and α is the parameter reflecting the shape of the joint density of states, which is assumed to be 1/2 here. T is the temperature and k_B is Boltzmann constant. Equation (1) was fitted to the PL peak energies at high temperatures where the localized states and exciton is less influenced. The solid curves shown in Fig. 2 are obtained from this analysis. The band gap energy shift due to temperature change can be also estimated.

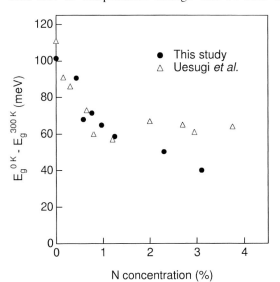

Figure 3 shows the band gap energy differences between 0 and 300 K as a function of N concentration. The N concentration dependence of the band gap shift of GaAsN alloys with temperature changed between 25 and 297 K estimated from absorption measurements by Uesugi et al. [2] is also

Fig. 3. Band gap energy differences between 0 and 300 K as a function of N concentration

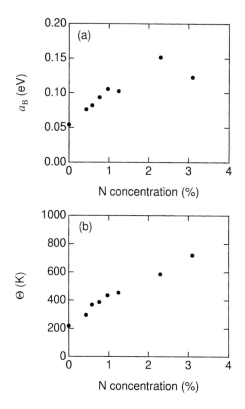

Fig. 4. Electron–phonon interaction strength and average phonon energy obtained from the analysis using the Bose-Einstein statistical expression

shown in this figure. For N concentration of 1%, our results are in good agreement with the results by Uesugi et al. However, the band gap energy shift still decreases with increasing N concentration for $x > 1\%$ in our results.

Figure 4 shows the electron–phonon interaction strength and the average phonon energy obtained from the analysis using the Bose-Einstein expression. Both the electron–phonon interaction strength and the average phonon energy are found to increase with increasing N concentration. However, the average phonon energy obtained from the analysis is much larger than that estimated from the linear interpolation between the longitudinal optical (LO) phonon energies of GaAs (293 cm^{-1}) and cubic GaN (741 cm^{-1}) [6]. For example, the average phonon energy of GaAs$_{1-x}$N$_x$ ($x = 2.3\%$) is estimated as 410 cm^{-1} in this study. This suggests that the interaction between electrons localized at N atoms interact preferentially with the localized Ga–N vibration mode with higher frequency. In fact, it is reported that GaN-type LO$_2$ phonon energy is 473 cm^{-1} [7] for GaAs$_{1-x}$N$_x$ alloys ($x = 2$–3%). Thus, the interaction between electrons localized at N atoms and the Ga–N localized vibration is considered to play an important role in the reduction of the temperature dependence of the band gap energy of GaAsN alloys.

4. Conclusions

We studied the temperature dependence of the photoluminescence of GaAsN alloys. The PL peak energy shift due to the temperature change decreases with increasing N concentration of GaAsN alloys. The decrease in the PL peak energy shift is partly due to the influence of the localized state emission which is dominant at low temperatures. In addition, the reduction in the temperature dependence of the band gap energy is responsible for the small PL peak energy shift. The Bose-Einstein statistical expression is used to analyze the temperature dependence of the band gap energy, and both the electron–phonon interaction strength and the average phonon energy are found to increase with increasing N concentration. The average phonon energy is much larger than that expected from the linear interpolation between the phonon energies of GaAs and GaN. This indicates that the preferential interaction of electrons localized at N atoms interact with the localized Ga–N vibration mode with higher energy leads to the reduction of the temperature dependence of the band gap energy of GaAsN alloys.

Acknowledgements This work was partly supported by Nippon Sheet Glass Foundation for Materials Science and Engineering.

References

[1] M. Kondow, T. Kitatani, K. Nakahara, and T. Tanaka, Jpn. J. Appl. Phys. **38**, L1355 (1999).
[2] K. Uesugi, I. Suemune, T. Hasegawa, T. Akutagawa, and T. Nakayama, Appl. Phys. Lett. **76**, 1285 (2000).
[3] I. Suemune, K. Uesugi, and W. Waulukiewicz, Appl. Phys. Lett. **77**, 3021 (2000).
[4] K. Onabe. D. Aoki, J. Wu, H. Yaguchi, and Y. Shiraki, phys. stat. sol. (a) **176**, 231 (1999).
[5] L. Vina, S. Logothetidis, and M. Cardona, Phys. Rev. B **30**, 1979 (1984).
[6] H. Yaguchi, J. Wu, B. Zhang, Y. Segawa, H. Nagasawa, K. Onabe, and Y. Shiraki, J. Cryst. Growth **195**, 323 (1998).
[7] A. M. Mintairov, P. A. Blagnov, V. G. Melehin, N. N. Faleev, J. L. Merz, Y. Qiu, S. A. Nikishin, and H. Temkin, Phys. Rev. B **56**, 15836 (1997).

Phonon Modes and Critical Points of GaPN

G. Leibiger[1]) (a, b), V. Gottschalch (b), R. Schwabe (c), G. Benndorf (d), and M. Schubert (a, d)

(a) Center for Microelectronic and Optical Materials Research
and Department of Electrical Engineering, University of Nebraska-Lincoln,
Lincoln, NE 68588, USA

(b) Faculty of Chemistry and Mineralogy, University of Leipzig, Linnéstraße 3,
D-04103 Leipzig, Germany

(c) Institute for Surface Modification e.V., Permoserstraße 15, D-04318 Leipzig, Germany

(d) Faculty of Physics and Earthscience, University of Leipzig, Linnéstraße 5,
D-04103 Leipzig, Germany

(Received July 13, 2001; accepted August 4, 2001)

Subject classification: 63.20.Dj; 63.20.Pw; 71.20.Nr; 78.30.Fs; 78.66.Fd; S7.14

Spectroscopic ellipsometry in the mid-infrared and near-infrared to vacuum-ultra violet spectral range is employed to study the phonon properties and critical points of a $GaP_{0.977}N_{0.023}$ layer grown on GaP. We observe a two-mode phonon behaviour, i.e., a GaN-like and a GaP-like phonon. We detect six critical-point structures, which we assign as E_0^{dir}, E_0', E_1, E_1', E_2^1 and E_2^2 transitions. We observe a blueshift of the direct band gap E_0^{dir} and of the E_1 transition in contrast to the redshift of the photoluminescence peak and the absorption tail, which we observed on the same sample. The critical-point energies E_0', E_1', and E_2 do not show any significant composition dependence.

Introduction Nitrogen-doped GaP has been used for decades as active material in green light emitting diodes. With the availability of modern non-equilibrium growth techniques, metastable GaNP alloys, with nitrogen concentrations above the impurity limit can be grown successfully [1]. Similar to GaN_yAs_{1-y}, incorporation of only a few percent of nitrogen in GaN_yP_{1-y} leads to a strong redshift of the fundamental band-gap energy [2]. The underlying microscopic picture of this effect is still in discussion [3–6]. However, up to now no report on the observation of critical-point transitions, with energies larger than that of the photoluminescence-emission energy has been published. Likewise, only little is known about the phonon properties of GaNP. Kent et al. [6] performed large-supercell pseudopotential calculations and predicted a negligible y dependence of the E_1 transition. Buyanova et al. [7] reported on Raman measurements of $GaP_{1-y}N_y$ ($y \leq 0.03$) layers. They observed a two-mode phonon behaviour, i.e., the GaP-like ($\omega_{LO1} \approx 401$–399 cm^{-1}) and the GaN-like ($\omega_{LO2} \approx 495$–503 cm^{-1}) longitudinal optical phonons. Recently, spectroscopic ellipsometry (SE) was used as novel technique for precise determination of phonon modes and critical points in GaN_yAs_{1-y} [8, 9]. SE for mid-infrared (mir) and near-infrared (nir) to vacuum-ultraviolet (vuv) wavelengths are used in this study to investigate phonon modes and critical points of $GaP_{0.977}N_{0.023}$, respectively.

---

[1]) Corresponding author; Phone: +001(402)472-1964; Fax: +001(402)472-7987;
e-mail: pge97jrk@studserv.uni-leipzig.de

Experimental The GaP$_{0.977}$N$_{0.023}$ layer (thickness ~350 nm) was grown on a (001) GaP substrate at a growth temperature of 650 °C using low-pressure (p_{tot} = 50 mbar) metal-organic vapor phase epitaxy. Trimethylgallium, phosphine, and dimethylhydrazine were used as precursors. A GaP-buffer layer of about 150 nm thickness was deposited prior to the GaP$_{0.977}$N$_{0.023}$ layer. The N-concentration was determined by high-resolution X-ray diffraction. The SE spectra were recorded at room temperature with a spectral resolution of 2 cm^{-1} at 50° angle of incidence for wavenumbers from 300 to 600 cm^{-1} and with a resolution of 5 meV at 75° angle of incidence for photon energies from 0.75 to 8.3 eV.

Ellipsometry can determine the complex dielectric function ε and thickness d of a thin-film sample by comparing the measured data with best-fit model calculations. The standard ellipsometric parameters are defined by Ψ and Δ. They are related to the complex reflectance ratio ϱ

$$\varrho \equiv r_p/r_s = \tan \Psi \exp i\Delta, \quad (1)$$

where r_p and r_s are the reflection coefficients for light polarized parallel (p), and perpendicular (s) to the plane of incidence, respectively. The pseudodielectric function $\langle \varepsilon \rangle$ is a common representation of the ellipsometric data Ψ and Δ [10]. In both spectral ranges, the dielectric functions of the GaP substrate and the buffer layer were obtained by analyzing the two reference samples, i.e., the GaP substrate and the substrate with the buffer layer deposited only. The mir-dielectric function of GaP$_{0.977}$N$_{0.023}$ is described as a factorized form of the sum of harmonic oscillators with Lorentzian broadening. In the nir-vuv spectral range we applied Adachi's composite model, using Lorentzian lineshape functions for the description of all critical points above the E_1 transition. For a detailed description of the applied model dielectric functions we refer to Ref. [8] and references therein.

Results and Discussion Figure 1 shows experimental and calculated Ψ spectra of our GaNP sample and of a GaP reference sample. The GaP phonon band between the transverse (TO) and longitudinal optical (LO) frequencies ω_{TO1} and ω_{LO1} dominates both spectra. Whereas the GaP-like TO frequency does not shift, we observe a redshift of the GaP-like LO frequency by 2 cm^{-1} (from 403 to 401 cm^{-1}) in GaP$_{0.977}$N$_{0.023}$ compared to GaP. This effect is due to the tensile strain within the layer, and results in a step-like

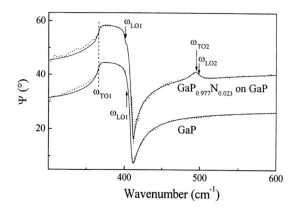

Fig. 1. Experimental (dotted lines) and calculated (solid lines) Ψ spectra of GaP$_{0.977}$N$_{0.023}$/GaP. The upper spectrum has been shifted by an amount of 14 for better readability

structure on the high-frequency side of the GaP phonon band of the GaPN sample. A second phonon band arises in the upper spectrum in Fig. 1, which we attribute to the GaN-like lattice vibration. The observed TO frequency, $\omega_{TO2} = (495.0 \pm 0.6)$ cm^{-1}, agrees well with the well-known local mode frequency of nitrogen in N-doped GaP, which indicates that the effects of alloying (resulting in a blueshift because $\omega_{TO}^{\beta\text{-GaN}} = 553$ cm^{-1}) and tensile strain (resulting in a redshift) cancel each other. The splitting of the TO- and LO frequencies of about 3 cm^{-1} is a measure of the polar strength of the GaN-like phonon. The resulting LO frequency ($\omega_{LO2} = (498.3 \pm 0.5)$ cm^{-1}) agrees well with the results of Raman measurements reported in Ref. [7].

In Fig. 2 we show the second derivatives of the imaginary part of the pseudodielectric function of the GaP$_{0.977}$N$_{0.023}$ sample and of the GaP reference sample in the nir-vuv spectral range. All arrows and dotted lines mark critical-point (CP) transition energies, which, together with the corresponding amplitude and broadening parameters (not shown here), result from our best-fit analysis. In contrast to the well-known redshift of the initially indirect band gap, we observe a blueshift of the direct band gap E_g^{direct} from about 2.77 eV for GaP to about 3.02 eV for GaP$_{0.977}$N$_{0.023}$. The interference oscillations in the low-energy part of the upper spectrum reach beyond the direct band gap of GaP up to E_g^{direct} of GaP$_{0.977}$N$_{0.023}$, which makes the observed blueshift most apparent. The attenuation of the interference oscillations indicates, however, that there is an absorption tail below E_g^{direct}. This absorption tail was also observed in a transmission spectrum, taken from the same sample, and extends to longer wavelengths compared to that of the GaP. This is in agreement with a redshift of both, low- and room-temperature photoluminescence spectra, taken from the same sample. We also observe a blueshift of the E_1 transition from about (3.675 ± 0.002) eV to (3.709 ± 0.006) eV for GaP$_{0.977}$N$_{0.023}$. This blueshift, which is due to the dominating influence of alloying ($E_1^{\beta\text{-GaN}} \approx 7$ eV), was also observed for GaAsN alloys using SE [9]. The CP's E_1, E_1', and E_0', are due to transitions along the Δ direction and at the Γ-point of the first Brillouin zone (BZ), respectively [11, 12]. The CPs E_2^1 and E_2^2 have been tentatively assigned by Zollner et al. [11] to transitions along the Δ direction and at the P-point (P = (0.75, 0.25, 0.25) $2\pi/a$) of the BZ [11]. We do not observe any significant shift of the E_0' ((4.82 ± 0.01) eV), E_1' ((6.74 ± 0.01) eV), E_2^1 ((5.11 ± 0.02) eV), and E_2^2 ((5.26 ± 0.02) eV) transitions, which could indicate, that the effects of alloying and tensile strain cancel each other. However, exact calculation of both effects is difficult,

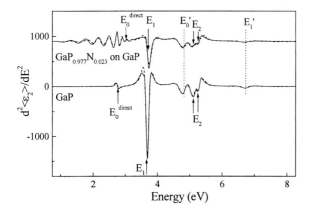

Fig. 2. Second derivative of the imaginary part of the pseudodielectric function of the GaP$_{0.977}$N$_{0.023}$ sample and of a GaP reference sample. The dashed (solid) lines refer to the experimental (calculated) spectra. The upper spectrum has been shifted by an amount of 900 for better readability

because there is a lack of both, critical-point energies of β-GaN and deformation potentials of GaP and β-GaN.

Conclusion In summary, we have determined phonon properties and critical points of a MOVPE grown $GaP_{0.977}N_{0.023}$ layer using mir and nir-vuv spectroscopic ellipsometry. We assign a GaP- and GaN-like TO and LO phonon, and observe no shift of both TO modes. In the nir-vuv spectral range, we detect six critical-point structures and assigned them as E_0^{dir}, E_0', E_1, E_1' E_2^1 and E_2^2 transitions adopting the standard zinkblende notation. We observe a blueshift of the direct band gap E_0^{dir} and of the E_1 transition and no significant composition dependence of the critical-point energies E_0', E_1', and E_2.

Acknowledgements This work is supported by the Deutsche Forschungsgemeinschaft under grant Go 629/4-1, and by the National Science Foundation under contract DMI-9901510. The Center for Microelectronic and Optical Materials Research at University of Nebraska further supported this work.

References

[1] S. MIYOSHI, H. YAGUCHI, K. ONABE, R. ITO, and Y. SHIRAKI, Appl. Phys. Lett. **63**, 3506 (1993).
[2] J. N. BAILLARGEON, K. Y. CHENG, G. E. HOFLER, P. J. PEARAH, and K. C. HSIEH, Appl. Phys. Lett. **60**, 2540 (1992).
[3] W. SHAN, W. WALUKIEWICZ, K. M. YU, J. WU, J. W. AGER III, E. E. HALLER, H. P. XIN, and C. W. TU, Appl. Phys. Lett. **76**, 3251 (2000).
[4] Y. ZHANG, B. FLUEGEL, A. MASCARENHAS, H. P. XIN, and C. W. TU, Phys. Rev. B **62**, 4493 (2000).
[5] L. BELLAICHE, S.-H. WEI, and A. ZUNGER, Appl. Phys. Lett. **70**, 3558 (1997).
[6] P. R. C. KENT and A. ZUNGER, Phys. Rev. Lett. **86**, 2613 (2001).
[7] I. A. BUYANOVA, W. M. CHEN, E. M. GOLDYS, H. P. XIN, and C. W. TU, Appl. Phys. Lett. **78**, 3959 (2001).
[8] G. LEIBIGER, V. GOTTSCHALCH, B. RHEINLÄNDER, J. ŠIK, and M. SCHUBERT, J. Appl. Phys. **89**, 4927 (2001).
[9] G. LEIBIGER, V. GOTTSCHALCH, B. RHEINLÄNDER, J. ŠIK, and M. SCHUBERT, Appl. Phys. Lett. **77**, 1650 (2000).
[10] D. E. ASPNES, The Accurate Determination of Optical Properties by Ellipsometry, in: Handbook of Optical Constants of Solids, Vol. 1, Ed. E. D. PALIK, Academic Press, New York 1998 (p. 89).
[11] S. ZOLLNER, M. GARRIGA, J. KIRCHER, J. HUMLIČEK, M. CARDONA, and G. NEUHOLD, Phys. Rev. B **48**, 7915 (1993).
[12] D. E. ASPNES, C. G. OLSON, and D. W. LYNCH, Phys. Rev. B **12**, 2527 (1975).

Raman Characterization of MBE Grown (Al)GaAsN

A. Hashimoto[1]) (a), T. Kitano (a), K. Takahashi (b), H. Kawanishi (b),
A. Patanè (c), C. T. Foxon (c), and A. Yamamoto (a)

(a) Department of Electrical and Electronics Engineering, Fukui University,
3-9-1 Bunkyo, Fukui 910-8507, Japan

(b) Advanced Techno. Res. Lab., Sharp Co. 2613-1 Ichinomoto, Tenri, Nara, Japan

(c) School of Physics and Astronomy, University of Nottingham, Nottingham NG7 2RD, UK

(Received July 11, 2001; accepted August 31, 2001)

Subject classification: 78.30.Fs; 78.66.Fd; 81.15.Hi; S7.14

Raman characterization of MBE grown (Al)GaAsN layers has been performed in order to investigate the structure of the lattice. Several new Al–N related Raman modes appear in the spectra. The Raman results clearly indicate that the nitrogen atoms in the film form mainly Al–N bonds instead of Ga–N bonds. An extremely strong, broad deep luminescence is observed in the low temperature photoluminescence spectrum, which is related to the nitrogen incorporation. The luminescence suggests that, because the electronic states are strongly coupled to the (Al)GaAsN lattice, localized electron transitions occur from the excited state to the ground state of the deep centres.

Introduction The (Al)GaAsN quaternary system has not been studied as extensively as other narrower band gap III–V–N systems such as GaAsN and InGaAsN. An early theoretical study of the (Al)GaAsN quaternary system, based on a quantum dielectric method predicted that the (Al)GaAsN alloy system would have large direct band gaps over the whole composition range [1]. More recent ab initio relativistic pseudopotential calculations suggest that completely opposite results such as band-gap collapse will be observed [2, 3]. Experimentally, several works are beginning to address this issue [4] and the results have supported the theoretical predictions of band-gap narrowing. However, the spatial distribution of the N atoms in the lattice structure has not been established for the (Al)GaAsN system, in spite of the great influence the N distribution might have on the fundamental properties of the alloy such as the band gap. Raman spectroscopy is a sensitive technique, which enables one to study the local structure of impurity atoms incorporated in the lattice and this method also enables one to observe deviations from long-range order induced by the incorporation of foreign atoms.

In this paper, we report, for the first time, Raman characterization of the (Al)GaAsN system in order to investigate the lattice structure via the phonon properties. The (Al)GaAsN Raman spectra are compared with the Raman spectra of GaAsN and the GaInAsN alloys in order to investigate the spatial distribution of the N atoms. The low temperature photoluminescence properties of the (Al)GaAsN layers are also discussed.

Experimental A conventional MBE system equipped with an ammonia (NH_3) source and effusion cells of gallium (Ga), aluminium (Al) and arsenic (As_4) was used for the

---

[1]) Corresponding author; Phone: +81 776 27 8565; Fax: +81 776 27 8749;
e-mail: hasimoto@kyomu1.fuee.fukui-u.ac.jp

experiments. NH_3 is an excellent N source, since it is carbon free and high purity gas is commercially available. Details of the epitaxial growth have been published elsewhere [4]. The incorporation efficiency of N from NH_3 during the growth of GaAsN was extremely low at normal growth temperature, but the incorporation of N was enhanced by supplying an additional aluminium flux to the surface during the growth process. The surface reaction of uncracked NH_3 was strongly enhanced by the presence of Al on the surface and the N incorporation in the (Al)GaAsN layers was proportional to the Al composition [4]. In this study the sample thickness was about 1.0 μm. An undoped buffer layer of 0.5 μm thick AlGaAs and a 0.2 μm thick GaAs layer were grown on the (001) n-type GaAs substrates ($n = 2 \times 10^{18}$ cm^{-3}), prior to the growth of the active films.

Microscopic Raman scattering characterization was carried out at room temperature in backscattering geometry using a 632.8 nm He–Ne laser. Typical molar fractions of Al and N in the films were 0.3 and 0.0032, respectively. The incident laser power was about 3 mW and the measurements were made in the range of 100 to 800 cm^{-1} with a resolution of about 1 cm^{-1} wave numbers. Low temperature photoluminescence measurements were carried out at 5 K using a 514.5 nm Ar$^+$ laser. Excitation power dependence was measured at 5 K using excitation powers from 0.1 to 100 mW.

Results and Discussion A typical Raman spectrum from an $Al_{0.3}Ga_{0.7}As_{0.9968}N_{0.0032}$ layer is shown in Fig. 1. It is well known that, for the conventional AlGaAs system, the Raman spectra show two-mode behaviour with GaAs-like and AlAs-like phonon modes. This two-mode behaviour is also clearly observed in the (Al)GaAsN Raman spectrum as shown in Fig. 1, GaAs-like LO (278.9 cm^{-1}), GaAs-like TO (263.2 cm^{-1}), AlAs-like LO (375.1 cm^{-1}) and AlAs-like TO (364.6 cm^{-1}) modes were clearly resolved. An acoustic local phonon (AL_1) mode, which appears in AlGaAs layers due to the motion of As atoms about a Ga atom on an Al site, was also observed at ∼250 cm^{-1} [5, 6]. Second-order Raman peaks, such as the GaAs-like 2TO mode (a broad spectrum peaked at ∼531.0 cm^{-1}), the GaAs-like 2LO mode (556.6 cm^{-1}), the AlAs-like 2TO mode and the 2LO AlAs-like mode (a broad spectrum peaked at 547.4 cm^{-1}) were also observed. Most crucially, several new peaks were observed clearly at 397.9, 651.9 and 676.7 cm^{-1}. We believe that the Raman peak at 651.9 cm^{-1} corresponds to the AlN-like TO mode. The sharp and strong peak at 397.9 cm^{-1} may correspond to the local vibrational mode (LVM) of an Al–N bond stretch in the AlGaAs crystal. The peak at 676.7 cm^{-1} corresponds to a second-order phonon of the peak at 339.1 cm^{-1} and we believe that the Raman peak at 339.1 cm^{-1} corresponds to the

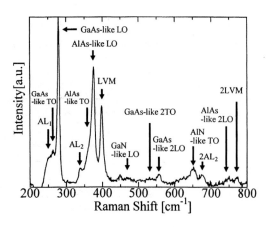

Fig. 1. Typical Raman spectrum from an $Al_{0.3}Ga_{0.7}As_{0.9968}N_{0.0032}$ layer

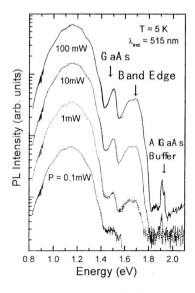

Fig. 2. Excitation power dependence of low temperature PL spectra of AlGaAsN at 5 K

acoustic local phonon (AL$_2$) mode, which is due to the motion of As atoms about an Al atom on a Ga site in the AlGaAs lattice. The Raman peak at 772.6 cm^{-1} corresponds to a second-order phonon of the LVM at 397.9 cm^{-1}. Unfortunately, we could not assign the peak at 448.7 cm^{-1}, however, the broad spectral structure around 475 cm^{-1}, which is the GaN related LO mode in the GaAsN Raman spectra, may be related to this peak. Polarized Raman measurements revealed that all Raman peaks for the (Al)GaAsN systems obey the zinc-blende Raman selection rules, in contrast to the case for the GaAsN system [7, 8].

The appearance of the new, rather strong Al–N related Raman modes clearly indicates that most of the N atoms incorporated in the (Al)GaAsN lattice form Al–N bonds instead of Ga–N bonds. Moreover, the polarized Raman results indicate that the distribution of the N atoms is completely different from the GaAsN case in which the N atoms form spontaneously ordered clusters [7, 8].

Typical low-temperature (5 K) PL spectra are shown in Fig. 2 as a function of excitation power. Several peaks were observed in the PL spectra. The highest energy peak at about 1.89 eV corresponds to luminescence from the Al$_{0.3}$Ga$_{0.7}$As buffer layer due to the high power excitation. The broad luminescence peaked at 1.7 eV can be assigned to the near band-edge emission of the (Al)GaAsN grown layer, because this peak remains at room temperature and becomes the main luminescence peak in the spectrum. Although the peak observed at 1.52 eV is usually assigned to a peak from the GaAs substrate, the total thickness (1.5 μm) of the (Al)GaAsN and (Al)GaAs layers is too thick and the intensity of the luminescence is too strong for this to come from direct excitation of the substrate under the PL conditions used for the measurements. Therefore, we cannot exclude the possibility that this luminescence comes from the (Al)GaAsN layer, which may imply that the (Al)GaAsN grown layer contains small GaAs domains. However, we also cannot rule out the possibility that this GaAs emission comes from the buffer layer and is excited indirectly by higher energy luminescence from the AlGaAs and/or the (Al)GaAsN.

The broad PL peak at 1.1 eV was mainly observed at 5 K. This feature has many shoulders, which can be decomposed as shown in Fig. 3. The peaks are separated by about 60 meV and/or about 120 meV and are multi-phonon replicas. The spectral features, including the multi-phonon replica of the GaN related optical (∼60 meV) mode and the AlN LO (∼113 meV) mode, suggest that localized electron transitions occur from the excited state to the ground state of the deep centres. Because the electronic states couple strongly with the (Al)GaAsN lattice, the defect related luminescence is similar to that observed in AlGaAs. The deep centres are thus probably related to N defects.

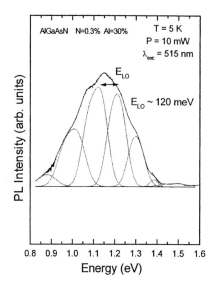

Fig. 3. Typical PL spectrum consisting of multi-phonon lines

No excitation power dependence of the peak is observed in contrast to the case for GaAsN. This indicates that the optical properties of the N atoms in (Al)GaAsN are very different from those in GaAsN in which the N atoms form the spontaneously ordered clusters of GaAs and GaN.

Conclusions In conclusion, we report the first Raman characterization for the (Al)GaAsN system to investigate the lattice structure by studying the phonon properties. We have observed several new Raman modes related to Al–N bonds. We also have investigated the optical properties of the layers by low temperature PL measurements. An extremely strong and broad deep luminescence, which is related to N incorporation, appears in the PL spectra. The spectral features strongly indicate that there are localized electron transitions occurring from the excited state to the ground state of the deep centres and that these electronic states strongly couple with the (Al)GaAsN lattice. The Raman and PL results strongly suggest that most of the N atoms form bonds with Al atoms and that spontaneously ordered clusters do not occur in the (Al)GaAsN system.

References

[1] D. P. MUNICH and R. F. PIERRET, Solid-State Electron. **30**, 901 (1987).
[2] Y. M. GU, T. PANG, C. CHEN, E. G. WANG, C. S. TING, D. M. BYLANER, and L. KLEINMAN, Phys. Rev. B **54**, 784 (1996).
[3] C. CHEN, E. G. WANG, Y. M. GU, D. M. BYLANER, and L. KLEINMAN, Phys. Rev. B **57**, 753 (1998).
[4] K. TAKAHASHI, Y. TOMOMURA, H. IKEDA, and H. KAWANISHI, Appl. Phys. Lett. **78**, 364 (2001).
[5] H. KAWAMURA, R. TSU, and L. ESAKI, Phys. Rev. Lett. **29**, 397 (1972).
[6] B. JUSSERAND and J. SAPRIER, Phys. Rev. B **24**, 194 (1981).
[7] A. M. MINTAIROV, P. A. BLAGNOV, V. G. MELEHINN. N. FALEEV, J. L. MERZ, Y. QIU, S. A. NIKISHIN, and H. TEMKIN, Phys. Rev. B **56**, 836 (1997).
[8] A. HASHIMOTO, T. FURUHATA, T. KITANO, A. K. NGUYEN, A. MASUDA, and A. YAMAMOTO, J. Cryst. Growth, to be published.

Electronic Structure of Heavily and Randomly Nitrogen Doped GaAs near the Fundamental Band Gap

Yong Zhang[1]) (a), S. Francoeur (a), A. Mascarenhas (a), H. P. Xin (b), and C. W. Tu (b)

(a) National Renewable Energy Laboratory, 1617 Cole Blvd. Golden, CO 80401, USA

(b) Department of Electrical and Computer Engineering, University of California, San Diego, La Jolla, CA 92093, USA

(Received June 21, 2001; accepted July 12, 2001)

Subject classification: 71.55.Eq; 78.55.Cr; S7.14

On increasing the nitrogen doping concentration in GaAs, states associated with isolated, paired and clustered (i.e., more complex configurations) nitrogen atoms sequentially appear, with their energy levels being resonant for the isolated center and most of the pairs and becoming bound for a couple of pairs and clusters. At a nitrogen mole concentration of $x \sim 0.1\%$, the shallow nitrogen bound states have merged with the GaAs band edge, which effectively gives rise to a band gap reduction, but the deeper nitrogen bound states persist as discrete levels. We study the behavior of nitrogen at this "transition" concentration, using various techniques (photoluminescence under selective excitation, electroreflectance, and Raman scattering), in order to gain insight into the large band gap reduction observed at all nitrogen concentrations. The validity of a few existing models proposed for explaining the large band gap reduction will be briefly discussed.

Introduction Isolated nitrogen in GaAs is well known to form a resonant state 150–180 meV above the GaAs conduction band edge [1–4]. Nitrogen pairs are found to form either resonant or bound states, depending on their configurations [4]. In addition to these nitrogen impurity states which appear at rather low nitrogen doping levels, deeper and also nitrogen related bound states have been observed more recently at somewhat higher nitrogen doping level ([N] > 10^{18} cm^{-3}) [5, 6]. Some of these newly observed bound states persist as discrete levels up to a nitrogen doping level near 0.1%. These bound states are more likely to be associated with nitrogen clusters [6, 7] than nitrogen pairs [5]. On further increasing nitrogen concentration, only a broad emission band, with its peak energy red-shifting proportionally to the nitrogen concentration, can be observed in a photoluminescence measurement [8]. The large band gap reduction in GaAs$_{1-x}$N$_x$ due to the nitrogen incorporation has been observed for nearly a decade [9], and its scaling rule has recently been accurately measured as $\delta E_g(x) = \beta x^\alpha$ (eV), with scaling exponent $\alpha = 0.667$ ($\approx 2/3$) and $\beta = 4.1$ [10]. The existence of such a scaling rule supports the impurity band model [11] in which the formation of an impurity band of nitrogen pair bound states is the primary mechanism for the large band gap reduction. However, there exist several other controversial views over this issue [12–14]. In this paper, we report a spectroscopy study on a GaAs$_{1-x}$N$_x$ sample with $x = 0.1\%$. At this composition, we are able to simultaneously investigate the behavior for two types of nitrogen induced states: those forming a continuous spectrum and those remaining as discrete states. We would like to point out that even those nitrogen induced

---

[1]) Corresponding author; Phone: +01 303 384 6617; Fax: +01 303 384 6655; e-mail: yzhang@nrel.gov

states belonging to the continuous spectrum do not behave like extended states in GaAs. At least to some extent, they remained spatially localized.

Sample and Experiment The $GaAs_{1-x}N_x$ sample was grown by gas-source molecular beam epitaxy on a semi-insulating (001) GaAs, using a rf nitrogen radical beam source with a mixture of N_2 and Ar in a ratio of 1:9. The growth temperature was 420 °C and the growth rate was 0.8 μm/h. The epilayer thickness is nominally 4000 Å, with a 2000 Å GaAs buffer. The N concentration was determined by high-resolution X-ray rocking curve measurement and theoretical dynamical simulation to be 0.1%. Photoluminescence (PL) spectra were taken at 10 K, using both above band gap and near band gap selective excitation. Electroreflectance (ER) was measured at 80 K.

Results Figure 1 shows a PL spectrum (using above band gap excitation) and an ER spectrum. The dominant PL peaks appears at energies significantly lower than the band gap. Note that even ER was measured at 80 K, the band gap shift between 80 and 10 K is expected to be minimal (a few meV). In addition to the multiple peaks in the low energy region, there is a weak PL peak at 1.473 eV which is very close to the band gap. In fact, this peak is also close to the energy of a transition labeled as NN_A in Ref. [5]. A comparison of the PL and ER spectrum indicates the coexistence of discrete and continuously distributed states introduced by nitrogen doping.

Figure 2 shows the PL spectra obtained under selective excitations. For any excitation energy below the band gap, we always observe a sharp transition, labeled as NN′, at 1.1–1.3 meV below the excitation energy. NN′ most of time is found to be followed by a TA phonon replica of 8.6 meV. Besides this moving peak, we also observe a few other peaks (NN_B, NC, and NN_D) whose peak positions remain stationary on varying the excitation energy. Here NN_B and NN_D are labeled as in Ref. [5], but they are reasonably to be interpreted as excitons bound to nitrogen clusters as for NC, although the exact configurations are not known at this time. A similar N

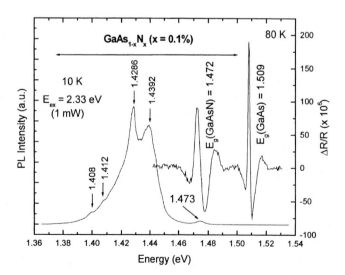

Fig. 1. Photoluminescence (left) and electroreflectance (right) spectrum

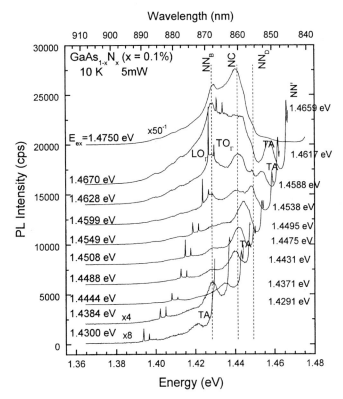

Fig. 2. Selective excitation PL spectra with different excitation energies indicated on the left. The peak energies for the NN' transition are shown on the right. The spectra are vertically shifted for clarity

cluster related transition has also been observed in GaP:N [15]. The tunable below band gap excitation or absorption reveals that the continuous spectrum actually extends into an energy region at least 40 meV below the band gap, which means that the energy spectrum of nitrogen induced states is comprised of closely spaced (thus, practically continuous) states and well separated discrete states. The continuous spectrum can be understood as nitrogen pair states under the random perturbation of nearby nitrogen atoms which do not form isolated clusters. Very similar phenomenon has been observed in GaP:N [15]. The red shift between the excitation energy and NN' can be explained as follows: the excitons created at the excitation energy tend to transfer to nearby trapping centers of local energy minima within their radiative lifetime [16]. These observations lead to the conclusion that although the nitrogen induced states form a continuous spectrum energetically, they retain a certain degree of spatial localization. When the exciation energy is tuned just above the band gap, alongside with a large enhancement in the PL intensity (as shown in the topmost spectrum of Fig. 2), we find that not only the NN' peak disappears but also the Raman lines (LO_Γ and TO_Γ) become invisible. In fact, the disappearence of the NN' peak is indicative of that the states above the band gap are less localized than those below the band gap.

Discussions In recent papers [7, 10, 11, 15], we have pointed out and demonstrated the close similarity in various aspects between the heavily nitrogen doped III–V semiconductors and heavily n- or p-type doped semiconductors where one also observes a very large band gap reduction together with effects due to the random nature of doping [17]. Here we summarize three possible cases, which we have discussed previously at different occasions, for the impurity band formation in nitrogen doped III–V semiconductors. The first case is that bare electrons bound to nitrogen centers directly couple with each other. The second case is unique to the systems discussed here. Note that usually the experimentally significant is not the bare electron bound state but the exciton bound state. The impurity band formation may originate from the coupling between the exciton bound to nitrogen centers. This situation is in fact analogous to the case of the impurity band formation for the acceptor in semiconductors, recalling that the nitrogen bound exciton was classified as acceptor-like [18]. With the excitonic effect taken into account, the inter-center coupling may become more probable than fully relying on the coupling of the bare electron states, since the radius of the hole bound state of the bound exciton is significantly larger than that of the electron bound state. The third case is unique to the random distribution of impurities. In fact, this is a well discussed situation for the impurity band in literature [17]. Essentially, in a random system, the impurity centers can have closely spaced energy levels, but each center remains spatially localized, which effectively forms a continuous spectrum. All the three cases discussed here may occur in $GaP_{1-x}N_x$ and $GaAs_{1-x}N_x$, depending on the nitrogen concentration and on which center is considered. To theoretically model these systems, one faces two difficulties: the first is to accurately describe the very localized impurity potential, and the second is to properly simulate the random structure. For instance, the impurity potential used in a recent calculation [19] was able to match the energy of the resonant isolated nitrogen state reasonably well in its energy, but yielded nitrogen pair states that very poorly match experimental results, which makes the conclusion arrived questionable. Also, the supercell size used for simulating the random nitrogen doped GaAs was not quite sufficient for properly modeling the inter-pair interaction for a relatively low doping level of $x \sim 0.1\%$ [19]. Thus, it still remains to be confirmed both experimentally and theoretically whether the states just above the band edge are due to the impurity band formation of nitrogen pair states [7] or the GaAs host states being perturbed and pushed down [19].

Acknowledgements The work at NREL was supported by the U.S. DOE under contract No. DE-AC36-83CH10093 and by the NREL DDRD under program No. 0659.0004, and the work at UCSD was partially supported by Midwest Research Institute under subcontract No. AAD-9-18668-7 from NREL.

References

[1] D. J. WOLFORD, J. A. BRADLEY, K. FRY, J. THOMPSON, and H. E. KING, in: Inst. Phys. Conf. Ser. No. 65, Ed. G. E. STILLMAN, The Institute of Physics, Bristol 1983 (p. 477).
D. J. WOLFORD, J. A. BRADLEY, K. FRY, and J. THOMPSON, in: Proc. 17th Internat. Conf. Physics of Semiconductors, Eds. J. D. CHADI and W. A. HARRISON, Springer-Verlag, New York 1984 (p. 627).
[2] X.-S. ZHAO, G.-H. LI, H.-X. HAN, Z.-P. WANG, R.-M. TANG, and R.-Z. CHE, Acta Phys. Sin. **33**, 588 (1984) [Chin. Phys. **5**, 337 (1985)].
[3] M. LEROUX, G. NEU, and C. VÈRIÉ, Solid State Commun. **58**, 289 (1986).

[4] X. Liu, M.-E. Pistol, L. Samuelson, S. Schwetlick, and W. Seifert, Appl. Phys. Lett. **56**, 1451 (1990).
X. Liu, M.-E. Pistol, and L. Samuelson, Phys. Rev. B **42**, 7504 (1990).
[5] T. Makimoto, H. Saito, T. Nishida, and N. Kobayashi, Appl. Phys. Lett. **70**, 2984 (1997).
T. Makimoto, H. Saito, and N. Kobayashi, Jpn. J. Appl. Phys. **36**, 1694 (1997).
[6] H. Güning, L. Chen, Th. Hartmann, P. J. Klar, W. Heimbrodt, F. Höhnsdorf, J. Koch, and W. Stolz, phys. stat. sol. (b) **215**, 39 (1999).
[7] Y. Zhang, A. Mascarenhas, J. F. Geisz, H. P. Xin, and C. W. Tu, Phys. Rev. B **63**, 85205 (2001).
[8] S. Francoeur, G. Sivaraman, Y. Qin, S. Nikishin, and H. Temkin, Appl. Phys. Lett. **72**, 1857 (1998).
[9] M. Weyers, M. Sato, and H. Ando, Jpn. J. Appl. Phys. **31**, L853 (1992).
[10] Y. Zhang, A. Mascarenhas, H. P. Xin, and C. W. Tu, Phys. Rev. B **63**, R161303 (2001).
[11] Y. Zhang, A. Mascarenhas, H. P. Xin, and C. W. Tu, Phys. Rev. B **61**, 7479 (2000).
[12] W. Shan, W. Walukiewicz, J. W. Ager III, E. E. Haller, J. F. Geisz, D. J. Friedman, J. M. Olson, and S. R. Kurtz, Phys. Rev. Lett. **82**, 1221 (1999).
[13] E. D. Jones, N. A. Modline, A. A. Allerman, S. R. Kurtz, A. F. Wright, S. T. Tozer, and X. Wei, Phys. Rev. B **60**, 4430 (1999).
[14] T. Mattila, S.-H. Wei, and A. Zunger, Phys. Rev. B **60**, R11245 (1999).
[15] Y. Zhang, B. Fluegel, A. Mascarenhas, H. P. Xin, and C. W. Tu, Phys. Rev. B **62**, 4493 (2000).
[16] J. H. Collet, J. A. Kash, D. J. Wolford, and J. Thompson, J. Phys. C **16**, 1283 (1983).
[17] B. I. Shklovskii and A. L. Efros, Electronic Properties of Doped Semiconductors, Springer-Verlag, Berlin 1984.
[18] J. J. Hopfield, D. G. Thomas, and R. T. Lynch, Phys. Rev. Lett. **17**, 312 (1966).
[19] P. R. C. Kent and A. Zunger, Phys. Rev. Lett. **86**, 2613 (2001).

Defect-Related Donors, Acceptors, and Traps in GaN

D. C. Look[1])

Semiconductor Research Center, Wright State University, Dayton, OH 45435
and Materials and Manufacturing Directorate, Air Force Research Laboratory,
Dayton, OH 45433, USA

(Received July 5, 2001; accepted July 12, 2001)

Subject classification: 61.72.Ji; 61.72.Lk; 71.55.Eq; 72.20.Fr; S7.14

Point defects have been created in GaN by various types of irradiation: electrons (1 and 2.5 MeV, and the spectrum from ^{90}Sr); protons (0.15 MeV, 2 MeV, and 24 GeV); He ions (5.4 MeV); γ-rays (^{60}Co); and sputtering and e-beam deposition of metals. They have been studied by temperature-dependent Hall-effect measurements (T-Hall), deep-level transient spectroscopy (DLTS), optically detected magnetic resonance (ODMR), positron annihilation spectroscopy (PAS), and photoluminescence (PL). Confirmed defect energies, and firm or tentative defect assignments, are as follows: T-Hall (donor at 0.06 eV, V_N); DLTS (electron trap at 0.18 eV (thermal 0.06 eV), V_N; electron trap at 0.9 eV, N_I or Ga_I–X); ODMR (Ga_I and Ga_I–X); PAS (V_{Ga}); PL (0.85 eV band with 0.88 eV zero-phonon line, O_N or O_N–Ga_I; 0.93 eV band; 3.37 eV line; 3.39 eV line). Many of these defect signatures have also been observed in *as-grown* GaN. Dislocations, of the threading-edge type, are found to be acceptor-like in n-type GaN.

Introduction Since the beginnings of GaN materials development, defects have been assumed to strongly affect the observed electrical and optical properties. For example, the high carrier concentration observed in the earliest hydride-vapor-phase-epitaxial (HVPE) growth was assumed to arise from N-vacancy (V_N) donors [1]. This assignment was reasonable in light of the propensity for N to leave the growing GaN surface at high growth temperatures. Then, as high-resolution microscopy became available, the heavy dislocation structure present in lattice-mismatched growth, mainly GaN on Al_2O_3, began to emerge as a fundamental issue that needed to be considered. Perhaps the most perplexing aspect of the dislocation problem was the fact that light-emitting diodes (LEDs) and laser diodes (LDs) actually worked, given the expected high recombination probability because of the dislocations [2]. Beginning in the mid-1990s, a much fuller understanding of point defects and dislocations began to develop from *ab-initio* theoretical calculations, based on the local-density approximation. One point that emerged here was that self-compensation can be a major force in a wide-bandgap material, since the lattice can often lower its energy by several eV by creating defect donors and acceptors, to compensate p-type and n-type material, respectively [3]. From these considerations, V_N donors might be expected in p-type GaN, and V_{Ga} acceptors in n-type GaN. Moreover, it was suggested that threading-edge dislocations would attain acceptor-like character in n-type GaN, by forming V_{Ga} along their cores, and donor-like behavior in p-type GaN, by forming V_N [4].

The question naturally arises, how do we test some of these ideas? It is first necessary to obtain "fingerprints" of various defects, by applying relevant characterization techniques. If we were dealing with impurity fingerprints, we would simply dope with

---

[1]) Phone: 937-255-1725; Fax: 937-255-3374; e-mail: david.look@wpafb.af.mil

the particular impurity. For point defects, this is much harder, but can sometimes be accomplished with high-energy electron irradiation [5]. For dislocations, it is usually necessary to accept what nature has produced, although stress can sometimes be used to create them. In this paper, we will attempt to outline what is actually known about the electrical and optical properties of various point defects and dislocations. The techniques which have proved to be most useful for fingerprinting point defects are the temperature-dependent Hall effect (T-Hall), deep level transient spectroscopy (DLTS), positron annihilation spectroscopy (PAS), photoluminescence (PL), and optically detected magnetic resonance (ODMR). For dislocations, transmission electron microscopy (TEM) is usually the technique of choice, for both qualitative and quantitative analysis.

Effects of Electron Irradiation To create point defects, in our laboratory, we have used high-energy (0.7–2.0 MeV) electrons from a van de Graaff accelerator. Other groups have used electrons of energy 0.3–2.5 MeV, from van de Graaff accelerators; 0.7 MeV (mean) Compton electrons from ^{60}Co γ-rays; and the 0.25–2.3 MeV electron spectrum from ^{90}Sr decay. Irradiation with other particles will be discussed later, but mainly, the same types of defects are produced.

Most of the energy loss in high-energy electron bombardment occurs from electron–electron, rather than electron–nucleus, collisions. Such e–e collisions limit the electron range in GaN to about 0.7 mm for 1 MeV electrons, for example. For epitaxial layers, of thickness 100 μm or less, the electrons lose very little energy in traversing the sample. If a relativistic electron of energy E makes a *direct* hit on a nucleus, it will transfer a maximum energy E_m given by

$$E_m = \frac{2E(E + 2m_e c^2)}{Mc^2} = \frac{2.147 \times 10^{-9} E(E + 1.022 \times 10^6)}{A} \qquad (1)$$

where m_e and M are the electron and ion masses, respectively, A is the atomic weight, and the energies are in eV. The threshold energy E_{th} necessary to produce an atomic displacement is then just given by the condition $E_m = E_d$, where E_d is the displacement energy. Van Vechten has estimated E_d values of 32.5 and 24.3 eV for N and Ga displacements, respectively, and it can be shown from Eq. (1) that these values would lead to threshold energies $E = 0.18$ and 0.51 MeV, for production of V_N–N_I and V_{Ga}–Ga_I Frenkel pairs, respectively. It can also be shown that the respective N and Ga production rates at, say, 1 MeV would be $\tau_N = 0.27$ cm^{-1} and $\tau_{Ga} = 0.69$ cm^{-1}. (Note that the units of production rate are (displaced atoms per cm^3)/(bombarding electrons per cm^2) = cm^{-1}.) In fact, it turns out that the actual production rate of a shallow donor and trap is about 1 cm^{-1}, which suggests that the E_d values may be somewhat lower than van Vechten's estimates. For example, an E_d of 11 eV for N displacement would give $\tau_N = 1$ cm^{-1}.

Temperature-Dependent Hall Effect (T-Hall) T-Hall measurements constitute the standard method of determining donor N_D and acceptor N_A concentrations in semiconductor materials [6]. The temperature dependences of both the Hall mobility μ_H and the carrier concentration n (assumed n-type) are fitted to determine N_D, N_A, and E_D, the donor activation energy. The mobility, for elastic scattering processes, can be calculated from $\mu_H = e\langle\tau^2\rangle/m^*\langle\tau\rangle$, where $\langle\tau\rangle$ denotes an average of the relaxation time $\tau(E)$

over electron energy E. The relaxation *rate* $\tau^{-1}(E)$ has contributions from various scattering mechanisms,

$$\tau^{-1}(E) = \tau_{ac}^{-1}(E) + \tau_{pe}^{-1}(E) + \tau_{po}^{-1}(E) + \tau_{ii}^{-1}(E) + \tau_{dis}^{-1}(E) \qquad (2)$$

in which acoustical-mode lattice vibrations scatter electrons through the deformation potential (τ_{ac}) and piezoelectric potential (τ_{pe}); optical-mode vibrations through the polar potential (τ_{po}); ionized impurities and defects through the screened Coulomb potential (τ_{ii}); and charged dislocations, also through the Coulomb potential (τ_{dis}). The strengths of these various scattering mechanisms depend upon certain lattice parameters, such as dielectric constants and deformation potentials, and extrinsic factors, such as donor, acceptor, and dislocation concentrations, N_D, N_A, and N_{dis}, respectively. For the irradiated samples of interest here, N_{dis} is not important, and the only fitting parameter is N_A, since the ionized defect/impurity density is given by $N_{ion} = 2N_A + n \approx 2N_A + n_H$, where n_H is measured in the experiment. In reality, since polar-optical scattering is not elastic, we often use a more accurate fitting scheme for μ_H versus T [6].

To determine N_D and E_D we must solve the charge-balance equation (CBE),

$$n + N_A = \frac{N_D}{1 + n/\phi_D}, \qquad (3)$$

where $\phi_D = (g_0/g_1) N_C' \exp(\alpha_D/k) T^{3/2} \exp(-E_{D0}/kT)$. Here, g_0/g_1 is a degeneracy factor (= 1/2 for an s-state), $N_C' = 2(2\pi m_n^* k)^{3/2}/h^3$, where h is Planck's constant, E_D is the donor energy, k is the Boltzmann constant, and E_{D0} and α_D are defined by $E_D = E_{D0} - \alpha_D T$. If more than one donor exists within a few kT of the Fermi energy, then equivalent terms are added on the right-hand side of Eq. (3).

Recently we have applied the above analyses to μ_H versus T and n_H versus T data for a very pure free-standing HVPE GaN layer grown by Samsung [7]. The 300 K and peak mobilities of this sample were 1245 and 7400 cm^2/Vs, respectively, and the fitted donor and acceptor concentrations were 6.7×10^{15} and 1.7×10^{15} cm^{-3}, respectively. Interestingly, the value of N_A is very close to the V_{Ga} concentration measured by PAS in similar material [8]. In fact, V_{Ga} is often the dominant acceptor in undoped GaN, over a wide range of acceptor concentrations [9]. After 1 MeV electron irradiation, both N_D and N_A increase, each by about 1 cm^{-3} for each bombarding electron per cm^2, giving a production rate of about 1 cm^{-1} [10]. From various considerations, it has been argued that the donor is likely an N vacancy V_N, and the acceptor an N interstitial N_I [10]. Also determined from the experiment is the donor activation energy $E_D \approx 0.06$ eV. This value is compatible with the theoretical conclusion that V_N should be a shallow donor [11]. Also, theory predicts N_I to have a deep-acceptor state [11], consistent with the Hall data. If we are indeed seeing only N-sublattice damage, then the absence of Ga-sublattice damage is a mystery. These points will be discussed later.

Deep Level Transient Spectroscopy (DLTS) DLTS is a technique capable of determining electron and hole trap parameters: concentration, activation energy, and capture cross section [6]. In its common form, a reverse-biased Schottky barrier or p–n junction is subjected to a forward-bias pulse in order to flood the depletion region with electrons (or holes), and thus temporarily fill the traps in that region. Upon returning to the original reverse bias, the temporarily trapped electrons or holes will be re-emitted.

The experiments to be discussed here involve a Schottky barrier on n-type material, which will have a capacitance C immediately before the pulse, and $C - \Delta C$ immediately afterwards. To first order, the trap concentration N_T is given by $N_T = -2N_D(\Delta C/C)$, where N_D is the net shallow donor concentration. As time proceeds after the filling pulse, the capacitance will return to its original value C, usually in an exponential manner. In the box-car technique, this exponential is sampled at two times, t_1 and t_2, to effect a "rate window" defined by $r = \ln(t_2/t_1)/(t_2 - t_1)$. As temperature is swept upwards, the emission rate e_n of a particular trap increases, according to $e_n = CT^2 \exp(-E/kT)$, where C is a constant involving the capture cross section, and E is an activation energy, consisting of the trap energy E_T plus the cross-section barrier height E_σ, if non-zero (i.e., $E = E_T + E_\sigma$). The DLTS "spectrum" then consists of a series of peaks due to different traps N_{Ti}, with each peak occurring at the temperature for which $e_{ni} = r$.

In GaN, two common defect-related traps, designated here as ED and AD, are nearly always observed, no matter what type of irradiation is used [12–27]. For the rate window set at $r = 4 \text{ s}^{-1}$, traps ED and AD have DLTS signal peaks at about 100 and 400 K, respectively. Other traps have also been seen, as outlined in Table 1, but virtually all workers have seen ED and AD. Trap ED was first reported by Fang et al. [13] to have an energy E of 0.18 eV, but later analysis by Polenta et al. [20] revealed that it consisted of two overlapping traps, each with a thermal energy component E_T of 0.06 eV, and capture-cross-section components E_σ of 0 and 0.05 eV, respectively. Indeed, Goodman et al. [26] have recently found that ED consists of three components, but their source of electrons (0.25–2.3 MeV, ^{90}Sr) was different from that used by Fang et al. and Polenta et al. (1 MeV, van de Graaff). The combined production rate of all of the components is about 1 cm^{-1} [26], the same as that determined from the T-Hall analysis [10], and since the thermal energies are also the same (0.06 eV), it is almost certain that a common defect (probably V_N) is being observed. The different DLTS trap components could perhaps correspond to different separations of the Frenkel-pair components, V_N and N_I [28]. The existence of several components in ED also can explain the slight differences in overall peak position, 0.13–0.20 eV, seen by various workers (cf. Table 1).

It is important to emphasize that the various components of trap ED appear not only in irradiated samples, but also in as-grown [25, 29, 30], ion-implanted [14], and contact-metal-deposited [17, 18] samples. In MBE layers [25, 29, 30], the concentration of trap ED increases with decrease of N flux during growth, again supporting its identification with V_N.

One other trap, AD at 0.9 eV, also is commonly produced by electron irradiation and other forms of irradiation or processing [14, 17, 19, 21–25, 27, 30], and also consists of multiple components [27]. Furthermore, just as is the case with trap ED, trap AD is produced in ion-implanted samples [14], and in samples subjected to e-beam evaporation of metals [17]. Finally, it even occurs in as-grown HVPE and MBE samples [25, 30]. Interestingly, in as-grown samples, traps AD and ED seem to be anticorrelated in intensity, whereas in irradiated samples, they usually appear together. These observations would be consistent if ED is related to V_N, as conjectured above, and AD to N_I, because then they both would be expected as a result of irradiation (Frenkel-pair production), but only one of them in as-grown material, due to stoichiometry considerations. However, further studies are needed on this issue.

Table 1
Experimentally observed defects in GaN

species	experimental energy (eV)	thermal energy (eV)	type of irradiation	observation technique	ID	references
donors	0.06	0.06	e⁻(1 MeV)	T-Hall	V_N	Look, 1997 [10]
acceptors		deep	e⁻(1 MeV)	T-Hall	N_I	Look, 1997 [10]
		deep	e⁻(2 MeV)	Positron annih.	V_{Ga}	Saarinen, 2001 [8]
electron traps	0.06, 0.10, 0.20, 0.27		e⁻(^{90}Sr)	DLTS	V_N	Goodman, 2001 [26]
	0.12, 0.16	0.08	H⁺(24 GeV)	DLTS	V_N	Castaldini, 2000 [24]
	0.13, 0.16, 0.20		H⁺(2 MeV)	DLTS		Auret, 1999 [15]
	0.15, 0.95		γ-ray(^{60}Co)	DLTS		Shmidt, 1999 [19]
	0.18	0.06	e⁻(1 MeV)	DLTS	V_N	Fang, 1998 [13]; Polenta, 2000 [20]
	0.19, 0.92		Ru(e-beam)	DLTS		Auret, 1999 [17]
	0.20, 0.30, 0.40, 0.45		Au(sputtered)	DLTS		Auret, 1999 [18]
	0.20, 0.78, 0.95		He⁺(5.4 MeV)	DLTS		Auret, 1998 [14]
	0.2, 0.3, 0.45, 0.75, 0.95		H⁺(0.15 MeV)	DLTS		Polyakov, 2000 [22]
	0.52, 0.59, 0.90		H⁺(24 GeV)	DLTS		Castaldini, 2000 [24]
	0.59, 0.82		γ-ray(^{60}Co)	DLTS		Wang, 2000 [23]
	0.67		N²⁺(0.3 MeV)	DLTS		Haase, 1996 [12]
	0.9		e⁻(^{90}Sr)	DLTS		Goodman, 2001 [27]
	0.9		e⁻(1 MeV)	DLTS		Fang, 2000 [25]
hole traps	0.25, 0.6, 0.9		H⁺(0.15 MeV)	DLTS		Polyakov, 2000 [22]
paramag. centers		deep	e⁻(2.5 MeV)	ODMR	Ga_I^{2+}-X	Linde, 1997 [32]
		deep	e⁻(2.5 MeV)	ODMR	Ga_I^{2+}	Chow, 2000 [35]
radiative centers	0.85 (0.88 ZPL), 0.93		e⁻(2.5 MeV)	PL		Linde, 1997 [32]; Buyanova, 1998 [33]
	0.85 (0.88 ZPL)		e⁻(2.5 MeV)	PL	O_N, O_N-X	Chen, 1998 [34]
	3.37		e⁻(2.5 MeV)	PL		Buyanova, 1998 [33]
	3.37		e⁻(2 MeV)	PL		Jones, 2001 [36]
	3.39		e⁻(2 MeV)	PL		Jones, 2001 [36]

Positron Annihilation Spectroscopy (PAS) Positrons injected into defect-free GaN are annihilated by the core atomic electrons in a mean time of 160–165 ps. However, if there are negatively charged vacancies present, some of the positrons will become trapped at those locations, and will have longer lifetimes, because of the reduced electron density at vacancies. In the case of GaN, Ga vacancies (but not N vacancies) would be expected to fill this role, and indeed, PAS has been used to identify and quantify V_{Ga}-related defects [31]. For example, it has been shown that 2 MeV electrons produce V_{Ga} centers at a rate of about 1 cm^{-1} in bulk, semi-insulating GaN [8]. Moreover, comparisons of V_{Ga} concentrations with acceptor concentrations N_A in a series of undoped, n-type HVPE GaN samples, with N_A ranging from 10^{15} to 10^{19} cm^{-3}, show that $[V_{Ga}] \approx N_A$, to within experimental error [8, 9]. Thus, it appears that V_{Ga}, and not any impurity, is the dominant acceptor in HVPE GaN, and probably other types of undoped GaN, also. Indeed, theory predicts that V_{Ga} centers should be abundant in n-type GaN [3].

Photoluminescence (PL) Two, infrared PL bands, at roughly 0.85 and 0.93 eV, are produced by 2.5 MeV electrons [32–35]. Both bands are broad, but the former has a sharp zero-phonon line (ZPL) at 0.88 eV and accompanying phonon structure. Some workers have proposed that the lower-energy band with the ZPL is much like the well-known O_P band in GaP, and, in the GaN case, involves a transition between a deep ground state of O_N (at $E_C - 0.90$ eV) and an excited state of O_N (at $E_C - 0.02$ eV), along with associated phonon side bands [34]. However, there are several apparent problems with this model. First of all, it is difficult to determine the role played by the irradiation. That is, if no displaced lattice atoms are directly involved in the 0.88 eV center, then it would seem that the irradiation would have to be "uncovering" the center, perhaps by causing the Fermi level E_F to drop. However, the 0.88 eV PL line is observed even when E_F remains relatively constant during the irradiation [36]. A second problem involves the ground state of O_N, which must be deep in this model, but which should be shallow to account for the well-known introduction of shallow donors in O-doped GaN [9]. And a final problem is introduced by a recent theoretical study, which has concluded that isolated O_N cannot account for observed local-vibrational-mode (LVM) spectra [37]. However, this same study [37] has suggested that an O_N–Ga_I complex is consistent with both the LVM spectra and the existence of a deep electronic ground state. Thus, it seems that a large share of the observed data involving O in GaN can be explained by O_N shallow-donor centers, and O_N–Ga_I deep centers. However, it remains to be seen whether or not an O_N–Ga_I complex can account for the 0.88 eV ZPL.

The high-energy PL band (0.95 eV) also has a definite (although probably indirect) involvement with native defects, as discussed below. Interestingly, both the 0.85 and 0.95 eV PL bands are very weak, or even nonexistent, after 1 MeV irradiation [36], so that it is difficult to correlate the PL results with the Hall and DLTS results.

Another PL peak observed after irradiation occurs at 3.37 eV [33, 36]. This peak has also been observed in as-grown GaN on lattice-mismatched substrates, especially near the highly dislocated interface region, and has been attributed to excitons bound to strongly localized defects [38]. This scenario is consistent with highly localized Ga_I-related defects, discussed in the next section. However, copper sample holders can also exhibit an emission near 3.37 eV, so care is needed in the interpretation of this line. A final peak, at 3.39 eV, is produced by 2 MeV electrons [36]. This peak has an annealing behavior similar to that of the shallow donor and deep acceptor found from the Hall-effect analysis.

Optically Detected Magnetic Resonance (ODMR) A change in spin state, as induced by resonant rf radiation, can sometimes affect a PL transition, either directly, or indirectly, through a competing process. If this is the case, then a sensitive optical emission experiment can be used to detect the magnetic resonance. At room temperature, 2.5 MeV electron irradiation produces 0.85 and 0.95 eV PL bands, as mentioned above, and four new paramagnetic centers, L1–L4, as catalogued by Chow et al. [35]. The magnetic resonances of L1, L3, and L4 are detected via the 0.95 eV PL band, and L2, via the 0.85 eV band [32, 35]. However, irradiation at 4 K produces only the 0.95 eV PL band, and only one ODMR center, L5, which is interpreted as the isolated, doubly-charged Ga interstitial Ga_I^{2+} [35]. Upon annealing near room temperature, L5 begins to disappear, and L2, along with the 0.85 eV PL band, begins to appear in 1:1 correspondence. Thus, L2 is believed to be a Ga_I complex, formed by the migration of the interstitial. Centers L3 and L4 are also believed to be Ga_I related, and along with L5, have highly localized wave functions. Thus, L3–L5 are deep centers, and are evidently different than the shallow 0.06 eV donor and electron trap observed by Hall-effect and DLTS measurements, respectively.

Dislocation Studies Recently, threading-edge dislocations (TEDs) have been shown to be negatively charged in n-type GaN, behaving as a line charge with a linear charge density of about 1e per c-lattice distance, 5.185 Å [39]. Thus, these dislocations are acceptor-like, and theoretical studies are consistent with this picture, suggesting that the dislocation cores in n-type GaN may contain Ga vacancies [4, 40], or V_{Ga}–O_n complexes, with $n = 1$–3 [41]. It might be assumed that simple V_{Ga} centers along the core would have a charge of -3 each, but it turns out that electron–electron repulsion reduces the charge to about -1 each, for typical material with $n \sim 1 \times 10^{17}$ cm^{-3} [40]. For HVPE GaN on Al_2O_3, the interface region has a very high TED density, $N_{dis} \sim 10^{10}$–10^{11} cm^{-2}, and an even higher misfit dislocation density, $N_{mis} > 10^{12}$ cm^{-2}, as deduced by transmission electron microscopy (TEM) measurements [9]. Also, from secondary-ion mass spectroscopy (SIMS) measurements, $[O] > 10^{19}$ cm^{-3} in this region; from PAS, $[V_{Ga}] > 10^{19}$ cm^{-3}; and from Hall-effect and electrochemical C–V (ECV) measurements, $n > 10^{19}$ cm^{-3} [9]. The picture that emerges here is that the donors are O, and the acceptors are V_{Ga}, but the V_{Ga} in this case are probably associated with the dislocations, either as isolated centers along the core, or as V_{Ga}–O complexes. Interestingly, O is the dominant donor only in the interface region, since SIMS and T-Hall data show that Si takes over in the bulk region. However, PAS and T-Hall measurements show that V_{Ga} is the dominant acceptor everywhere, both in the interface and bulk regions [9].

Discussion A summary of the observed defects in GaN is presented in Table 1. However, the reader should be warned that the characteristic energies measured by one particular experimental technique cannot necessarily be directly compared with those of another technique. For example, the donor energy $E_D = 0.06$ eV, determined from T-Hall measurements [10], is effectively the donor ground-state energy, with respect to the conduction band, at $T = 0$. On the other hand, the DLTS trap energies represent the quantity $(E_T + E_o)$ at $T = 0$ [6], and the PL energies can result from many different types of electron or hole transitions, at the temperature of the actual measurement. Thus, it is important to avoid simple comparisons of energies. However, certain conclusions can still be drawn from the data, as discussed below.

All of the different types of irradiation listed in Table 1 produce a relatively shallow center, $E_T = 0.06$ eV from T-Hall data, and $E_T + E_\sigma = 0.13\text{--}0.20$ eV, from DLTS. Part of the spread in the DLTS energies results from the fact that this particular trap usually contains two or more components, and its overall shape is influenced by choice of filling pulse width and other parameters. For at least two of the components, the thermal portion of the energy can be shown be $E_T = 0.06\text{--}0.08$ eV, in agreement with the Hall-effect result [20, 24, 26]. Furthermore, the e-irradiation production rates of the Hall and multiple DLTS centers are equal, within error [20, 26]. This center was originally assigned to V_N, or V_N close-pair complexes [10], and that identification has received some indirect support. For example, a level at $E_C - 0.05$ eV is seen by optical admittance spectroscopy only in acceptor-doped samples, which might be expected to contain compensating donors, such as V_N [42]. Also, a DLTS trap identical to the irradiation trap is seen in MBE samples grown under N-lean conditions [29]. Besides V_N, the only serious candidate for a shallow donor produced by irradiation would be Ga_I; however, so far, only *deep* centers associated with Ga_I (Ga_I^{2+} and $Ga_I^{2+}-X$) have been found after irradiation [32, 35]. Although the Ga_I^0 charge state of the former would theoretically be expected to have a shallow 0/+ transition [11], this species should not even exist after room-temperature irradiation, because it is known that the Ga_I atoms will migrate and form complexes [32, 35]. One of the complexes may be Ga_I-O_N, which theoretically has only deep donor levels [37], as discussed above. In short, V_N is still the best candidate for the dominant shallow donor and trap produced by irradiation.

The other major center produced by most forms of irradiation (cf. Table 1) is a 0.9 eV electron trap, observed by DLTS. It is tempting to associate this center with the one producing the 0.88 eV ZPL, which presumably would have a ground state also at about 0.9 eV [34]. A good candidate might be the Ga_I-O_N complex, which, as discussed above, could be formed by Ga_I migration during irradiation [32, 35], and which is theoretically expected to have a deep donor state [37]. Another candidate would be N_I or N_I-X, since N_I defects are obviously being produced along with V_N. It is reasonable that the acceptor deduced from T-Hall analysis is N_I, because the measured production rates of donors and acceptors are about equal, as expected if they were V_N and N_I, respectively [10]. Also, in some cases at least, the DLTS 0.18 and 0.9 eV traps have about equal intensity [25, 30]. However, these issues are still not settled, and more investigations are needed.

The final defect warranting discussion is V_{Ga}, which has been observed by positron annihilation spectroscopy (PAS), after 2 MeV electron irradiation [8]. As discussed above, V_{Ga} is also found to be the dominant acceptor in many forms of as-grown, n-type GaN [8, 9, 31], as expected from theoretical calculations of defect formation energies [3]. However, in the case of donors, it appears that *impurities* such as Si and O are dominant in as-grown GaN [9], rather than *defects* such as V_N [10] and Ga_I. With regard to traps, it should be noted that those with energies in the range 0.6–0.9 eV will likely have emission rates in the few-Hz regime near room temperature, and thus may lead to low-frequency noise. For example, a trap called "B" by our group, which is often the dominant trap in epitaxial GaN [25, 30], has an emission rate of 6 Hz at 300 K, and may be associated with some noise measurement anomalies in this frequency region, as reported earlier ([43], Fig. 3). Thus, defects are important in GaN, and methods to control them should be a strong consideration in the development of device materials.

Acknowledgements I wish to thank K. Saarinen for permission to quote results prior to publication. Support was received from AFOSR Grant F49620-00-1-0347 and AFRL Contract F33615-00-C-5402.

References

[1] H. P. MARUSKA and J. J. TIETJEN, Appl. Phys. Lett. **15**, 327 (1969).
[2] S. EVOY, H. G. CRAIGHEAD, S. KELLER, U. K. MISHRA, and S. P. DENBAARS, J. Vac. Sci. Technol. B **17**, 29 (1999).
[3] J. NEUGEBAUER and C. G. VAN DE WALLE, Phys. Rev. B **50**, 8067 (1994).
[4] A. F. WRIGHT and U. GROSSNER, Appl. Phys. Lett. **73**, 2751 (1998).
[5] F. AGULLO-LOPEZ, C. R. A. CATLOW, and P. D. TOWNSEND, Point Defects in Materials, Academic Press, New York 1988.
[6] D. C. LOOK, Electrical Characterization of GaAs Materials and Devices, Wiley, New York 1989.
[7] D. C. LOOK and J. R. SIZELOVE, Appl. Phys. Lett. **79**, 1133 (2001).
[8] K. SAARINEN, private communication.
[9] D. C. LOOK, C. E. STUTZ, R. J. MOLNAR, K. SAARINEN, and Z. LILIENTAL-WEBER, Solid State Commun. **117**, 571 (2001).
[10] D. C. LOOK, D. C. REYNOLDS, J. W. HEMSKY, J. R. SIZELOVE, R. L. JONES, and R. J. MOLNAR, Phys. Rev. Lett. **79**, 2273 (1997).
[11] P. BOGUSLAWSKI, E. L. BRIGGS, and J. BERNHOLC, Phys. Rev. B **51**, 17255 (1995).
[12] D. HAASE, M. SCHMID, W. KÜRNER, A. DÖRNEN, V. HÄRLE, F. SCHOLZ, M. BURKARD, and H. SCHWEIZER, Appl. Phys. Lett. **69**, 2525 (1996).
[13] Z.-Q. FANG, J. W. HEMSKY, D. C. LOOK, and M. P. MACK, Appl. Phys. Lett. **72**, 448 (1998).
[14] F. D. AURET, S. A. GOODMAN, F. K. KOSCHNICK, J.-M. SPAETH, B. BEAUMONT, and P. GIBART, Appl. Phys. Lett. **73**, 3745 (1998).
[15] F. D. AURET, S. A. GOODMAN, F. K. KOSCHNICK, J.-M. SPAETH, B. BEAUMONT, and P. GIBART, Appl. Phys. Lett. **74**, 407 (1999).
[16] F. D. AURET, S. A. GOODMAN, F. K. KOSCHNICK, J.-M. SPAETH, B. BEAUMONT, and P. GIBART, Appl. Phys. Lett. **74**, 2173 (1999).
[17] F. D. AURET, S. A. GOODMAN, G. MYBURG, F. K. KOSCHNICK, J.-M. SPAETH, B. BEAUMONT, and P. GIBART, Physica B **273/274**, 84 (1999).
[18] F. D. AURET, W. E. MEYER, S. A. GOODMAN, F. K. KOSCHNICK, J.-M. SPAETH, B. BEAUMONT, and P. GIBART, Physica B **273/274**, 92 (1999).
[19] N. M. SHMIDT, D. V. DAVYDOV, V. V. EMTSEV, I. L. KRESTNIKOV, A. A. LEBEDEV, W. V. LUNDIN, D. S. POLYSKIN, A. V. SAKHAROV, A. S. USIKOV, and A. V. OSINSKY, phys. stat. sol. (b) **216**, 533 (1999).
[20] L. POLENTA, Z.-Q. FANG, and D. C. LOOK, Appl. Phys. Lett. **76**, 2086 (2000).
[21] S. A. GOODMAN, F. D. AURET, F. K. KOSCHNICK, J.-M. SPAETH, B. BEAUMONT, and P. GIBART, Mater. Sci. Eng. B **71**, 100 (2000).
[22] A. Y. POLYAKOV, A. S. USIKOV, B. THEYS, N. B. SMIRNOV, A. V. GOVORKOV, F. JOMARD, N. M. SHMIDT, and W. V. LUNDIN, Solid-State Electron. **44**, 1971 (2000).
[23] C.-W. WANG, B.-S. SOONG, J.-Y. CHEN, C.-L. CHEN, and Y.-K. SU, J. Appl. Phys. **88**, 6355 (2000).
[24] A. CASTALDINI, A. CAVALLINI, and L. POLENTA, J. Phys.: Condens. Matter **12**, 10161 (2000).
[25] Z.-Q. FANG, L. POLENTA, J. W. HEMSKY, and D. C. LOOK, in: Proc. Internat. Semiconducting and Insulating Mater. Conf., IEEE, Piscataway 2000 (p. 35).
[26] S. A. GOODMAN, F. D. AURET, M. J. LEGODI, B. BEAUMONT, and P. GIBART, Appl. Phys. Lett. **78**, 3815 (2001).
[27] S. A. GOODMAN, F. D. AURET, G. MYBURG, M. J. LEGODI, P. GIBART, and B. BEAUMONT, Mater. Sci. Eng. B **82**, 95 (2001).
[28] D. C. LOOK, J. W. HEMSKY, and J. R. SIZELOVE, Phys. Rev. Lett. **82**, 2552 (1999).
[29] Z.-Q. FANG, D. C. LOOK, W. KIM, Z. FAN, A. BOTCHKAREV, and H. MORKOÇ, Appl. Phys. Lett. **72**, 2277 (1998).
[30] D. C. LOOK, Z.-Q. FANG, and L. POLENTA, MRS Internet J. Nitride Semicond. Res. **5S1**, W10.5 (2000).
[31] K. SAARINEN, J. NISSILÄ, P. HAUTOJÄRVI, J. LIKONEN, T. SUSKI, I. GRZEGORY, B. LUCZNIK, and S. POROWSKI, Appl. Phys. Lett. **75**, 2441 (1999).

[32] M. Linde, S. J. Uftring, G. D. Watkins, V. Härle, and F. Scholz, Phys. Rev. B **55**, R10177 (1997).
[33] I. A. Buyanova, Mt. Wagner, W. M. Chen, B. Monemar, J. L. Lindström, H Amano, and I. Akasaki, Appl. Phys. Lett. **73**, 2968 (1998).
[34] W. M. Chen, I. A. Buyanova, Mt. Wagner, B. Monemar, J. L. Lindström, H. Amano, and I. Akasaki, Phys. Rev. B **58**, R13351 (1998).
[35] K. H. Chow, G. D. Watkins, A. Usui, and M. Mizuta, Phys. Rev. Lett. **85**, 2761 (2000).
[36] R. L. Jones, unpublished.
[37] C. J. Fall, R. Jones, P. R. Briddon, and S. Öberg, Mater. Sci. Eng. B **82**, 88 (2001).
[38] C. Wetzel, S. Fischer, J. Krüger, E. E. Haller, R. J. Molnar, T. D. Moustakas, E. N. Mokhov, and P. G. Baranov, Appl. Phys. Lett. **68**, 2556 (1996).
[39] D. C. Look and J. R. Sizelove, Phys. Rev. Lett. **82**, 1237 (1999).
[40] K. Leung, A. F. Wright, and E. B. Stechel, Appl. Phys. Lett. **74**, 2495 (1999).
[41] J. Elsner, R. Jones, M. I. Heggie, P. K. Stich, M. Haugk, Th. Frauenheim, S. Öberg, and P. R. Briddon, Phys. Rev. B **58**, 12571 (1998).
[42] A. Krtschil, H. Witte, M. Lisker, J. Christen, A. Krost, U. Birkle, S. Einfeldt, D. Hommel, A. Wenzel, and B. Rauschenbach, phys. stat. sol. (b) **216**, 587 (1999).
[43] M. E. Levinshtein, S. L. Rumyantsev, D. C. Look, R. J. Molnar, M. Asif Khan, G. Simin, V. Adivarahan, and M. S. Shur, J. Appl. Phys. **86**, 5075 (1999).

Passivation and Doping due to Hydrogen in III-Nitrides

S. Limpijumnong[1])[*]) and C. G. Van de Walle

Xerox Palo Alto Research Center, 3333 Coyote Hill Road, Palo Alto, CA 94304, USA

(Received June 29, 2001; accepted August 4, 2001)

Subject classification: 61.72.–y; 71.55.Eq; S7.14

We have systematically studied the electronic structure and stability of hydrogen in AlN, GaN, and InN, based on first-principles calculations. In GaN and AlN, H is amphoteric and always compensates the prevailing conductivity: in GaN, H^+ is stable for Fermi levels below 2.2 eV, and in AlN, H^+ is stable for E_F below 2.5 eV. In InN, we find that H^+ is stable for *all* Fermi level positions; i.e., H behaves exclusively as a donor. Consequences for controlling the conductivity of InN are discussed.

Introduction Hydrogen is a common impurity in III-nitrides since it is abundantly present in the growth environment of many techniques used to grow nitrides. It is now well known that H passivates acceptors during growth of p-type GaN, necessitating a post-growth anneal to activate the acceptors, for a review, see [1]. The behavior of H in GaN has previously been addressed with first-principles calculations, producing detailed information about the stability and microscopic structure of different charge states [2]. Device applications of nitrides, however, almost always involve alloying with AlN and/or InN. Knowledge of the behavior of H in these other compounds is therefore essential. In this work, we systematically study the behavior of H in the three nitride compounds. The behavior in AlN turns out to be qualitatively similar to GaN: H is amphoteric and always compensates the prevailing conductivity. Hydrogen thus behaves as a donor in p-type GaN or AlN, and as an acceptor in n-type material. InN, however, exhibits a totally different behavior, with H acting exclusively as a donor.

The tendency for H to act as a donor in p-type material forms the driving force for *passivation* of acceptors. Vibrational spectroscopy of acceptor–hydrogen complexes provides a powerful way of monitoring the presence of H and the degree of acceptor activation [3]. A comparison with first-principles calculations then allows for an unambiguous identification of the nature and microscopic structure of the complex. We have recently performed a comprehensive investigation of structures and vibrational properties of acceptor–hydrogen complexes in GaN, taking proper account of anharmonic effects (which are large). The results, including a novel configuration for the Mg–H complex, are discussed elsewhere [4]. In this paper, we focus on the behavior of H as an isolated interstitial impurity.

Methods Our calculations are based on density-functional theory within the local density approximation, using ab initio pseudopotentials with a plane-wave basis set [5]. For GaN and InN we use the so-called "non-linear core correction" (nlcc) [6], with an energy cutoff of 40 Ry. Explicit inclusion of the Ga 3d, and In 4d electrons leads to

---

[1]) Corresponding author; e-mail: SXL57@po.cwru.edu
[*]) Permanent address: School of Physics, Institute of Science, Suranaree University of Technology, Nakhon Ratchasima 30000, Thailand.

very similar results for the configurations studied here, as reported in Ref. [7]. In this study we focus on the zinc-blende phase of the III-nitrides; results for the wurtzite phase (which differs in local environment only beyond third nearest neighbors) are expected to be very similar. Our calculations are performed at the theoretical lattice constants: 4.30 Å for AlN, 4.38 Å for GaN, and 5.10 Å for InN. These are in satisfactory agreement with the experimental values of 4.40, 4.50, and 5.00 Å [8]. Our calculated band gaps are 3.03 eV for GaN, −0.23 eV for InN, and 4.60 eV (direct) and 3.26 eV (indirect) for AlN. As usual for density-functional calculations, these values underestimate the actual gaps: 3.2 eV for GaN [8], 1.8 eV for InN (based on the calculated difference between the zinc-blende and wurtzite gaps [9]), and 5.0 eV for AlN (indirect gap from quasiparticle calculations [10]). The band-gap problem should not affect our conclusions, provided care is taken in the interpretation of the results, as discussed below. For the impurity calculations we used supercells containing 32 and 64 atoms, with at least 24 host atoms relaxed. Brillouin-zone integration was carried out with a $2 \times 2 \times 2$ mesh in the reciprocal unit cell, reduced by symmetry to a set of two irreducible **k** points.

The *formation energy* E^f determines the concentration c of an impurity through the expression: $c = N_{\text{sites}} \exp(-E^f/kT)$. N_{sites} is the number of sites in the lattice (per unit volume) on which the impurity can be incorporated, k is the Boltzmann constant, and T is the temperature. The formation energy of an interstitial H atom in charge state q is defined as

$$E^f(H^q) = E_{\text{tot}}(H^q) - E_{\text{tot}}(\text{bulk}) - \mu_H + qE_F, \qquad (1)$$

where $E_{\text{tot}}(H^q)$ is the total energy of H^q at a specific interstitial site, $E_{\text{tot}}(\text{bulk})$ is the total energy of a bulk supercell of the same dimensions as the one used to perform the impurity calculations, and μ_H is the energy of the hydrogen reservoir, i.e., the H chemical potential. For purposes of displaying our results we choose μ_H to be the energy of an H_2 molecule at $T = 0$. No zero-point energies are included. E_F is the Fermi level. For more information and calculational details we refer to Refs. [5] and [11].

Hydrogen in GaN Our calculations here duplicate those published in Ref. [2]; they were carried out to ensure consistency when comparing with AlN and InN. The comprehensive exploration of the complete energy surface in the earlier work highlighted three interstitial locations where H can be stable or metastable: the bond-center site (BC), the antibonding site near a nitrogen atom (AB_N), or the antibonding site near a gallium atom (AB_{Ga}).

For *positively* charged ($\underline{H^+}$) we find almost equal formation energies at BC and AB_N sites, AB_N being marginally lower. The N–H bond lengths are 1.03 Å for BC and

Table 1

Formations energies (in eV) (at $E_F = 0$) and stable for H in nitride semiconductors

charge state	stable site	AlN	GaN	InN
H^+	AB_N	−1.04	−0.30	−1.11
H^0	AB_{cation}	2.69	3.09	2.03
H^-	AB_{cation}	3.85	4.09	2.55

1.05 Å for AB_N. The AB_{Ga} site is 2.38 eV higher in energy. For *neutral* hydrogen ($\underline{H^0}$), AB_{Ga} and BC are comparable in energy, with AB_N only 0.69 eV higher. Note that the Ga–H bond length is such that the AB_{Ga} almost coincides with the tetrahedral interstitial (T_d) site. For *negatively* charged hydrogen ($\underline{H^-}$), the AB_{Ga} site is strongly preferred; the energy is higher by 2.21 eV at BC and by 3.83 eV at AB_N.

The stable sites and corresponding formation energies are tabulated in Table 1, and the formation energies as a function of E_F are shown in Fig. 1b. H^+ is energetically stable for E_F below $\varepsilon(+/-) = 2.20$ eV, while H^- is more stable for higher Fermi levels. This means that in p-type GaN H^+ is stable, while in n-type GaN H^- is stable, providing the basis for hydrogen's tendency to always act as a compensating center. H^0 is never stable, which is the characteristic of a "negative-U" system. The value of U is equal to $\varepsilon(0/-) - \varepsilon(+/0) = -2.39$ eV. The origin of this large negative U in GaN has been discussed in Ref. [2].

Hydrogen in AlN The lattice constant of AlN is slightly smaller than that of GaN, and the elastic constants slightly larger [8], causing the BC site to be less favorable. We find, in fact, that BC is no longer a local minimum for $\underline{H^+}$: a slight displacement of H off the BC site causes it to relax towards the AB_N site. AB_N is the energetically most stable site, with a N–H bond length of 1.06 Å. AB_{Al} is higher in energy by 2.40 eV; the latter difference is essentially equal to the value in GaN. For $\underline{H^0}$, AB_{Al} is most stable, with H sitting very close to the T_d site. AB_N is 0.40 eV higher in energy, and BC is 0.78 eV higher; again we see (just like in GaN) that the energy differences between these sites are relatively small in the case of H^0. Negatively charged hydrogen ($\underline{H^-}$), finally, prefers to stay close to the cations and strongly prefers AB_{Al}. The Al atoms move toward H, making the Al–H bond length 5% shorter than Al–N. The energy of H^- at BC and AB_N sites is much higher (by 3.42 and 3.50 eV) than at the AB_{Al} site. The formation energies as functions of Fermi level are shown in Fig. 1a.

Overall, the behavior of H in AlN is very similar to the case of GaN with $\varepsilon(+/-)$ now at 2.45 eV, and a slightly larger magnitude of $U = -2.57$ eV. Specifically, the general picture presented in Ref. [2] holds true for AlN, which has an even wider band gap

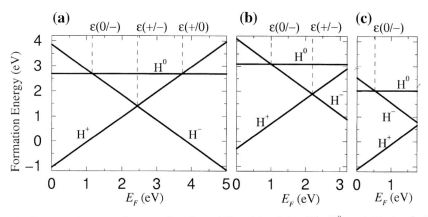

Fig. 1. Formation energies as a function of Fermi level for H^+, H^0, and H^- in a) AlN, b) GaN, and c) InN. $E_F = 0$ corresponds to the valence-band maximum, and formation energies are referenced to the energy of an H_2 molecule

than GaN. The preference for different sites in different charge states can be explained on the basis of Coulomb interactions: H^+ prefers regions of high charge density and thus stays close to the anion (N), favoring AB_N or BC. On the other hand, H^- prefers to stay away from the anion and closer to the cation, making AB_{Ga} or AB_{Al} the preferred site. Both H^+ and H^- exhibit large energy differences between sites close to anions versus cations, while the energy surface for H^0 is much "flatter".

Although the behavior of H in AlN is very similar to GaN, the larger gap of AlN leads to an interesting effect. In GaN, H^- is never quite as stable as H^+, due to an asymmetric placement of $\varepsilon(+/-)$ in the gap (see Fig. 1b). In AlN, however, the larger gap can lead to E_F positions far exceeding $\varepsilon(+/-)$, leading to a higher possible stability of H^-. We therefore predict that H will play a more important role in n-type AlN than in n-type GaN.

Hydrogen in InN The band gap of InN is much smaller than that of AlN and GaN (1.8 eV in the zinc-blende structure), and we will see that this leads to an interesting behavior. For $\underline{H^+}$, the AB_N site is favored again, and BC is higher in energy by only 0.18 eV. At AB_{In}, the total energy is higher by 2.95 eV. For $\underline{H^0}$ and $\underline{H^-}$, great care has to be taken in the interpretation of the density-functional calculations (as discussed previously for GaN in Ref. [7]). We carefully checked the nature of the energy level in the band structure that is occupied with one or two electrons in the H^0 and H^- calculations. For AB_{In}, the corresponding state is indeed localized near the H atom. For AB_N and BC, however, this state turns out to be an *extended* state. Correct calculations for H^0 and H^- should therefore *not* place electrons in this state – the actual hydrogen-induced energy level occurs higher in the band structure. When these considerations are taken into account, AB_N and BC are found to be higher in energy than AB_{In} for both H^0 and H^-. The formation energies as a function of Fermi level for all three charge states are shown in Fig. 1c. Again, U is large and negative ($U = -2.62$ eV), and the value of $\varepsilon(+/-)$ is 1.83 eV. Since the energy gap of InN is smaller than $\varepsilon(+/-)$, the positive charge state is the *only* stable charge state for H in InN, regardless of Fermi-level position. Note that band-gap corrections would *raise* the energy of H^- (the charge state with occupied gap levels) with respect to H^+, thus strengthening our conclusions about the stability of H^+. Therefore, hydrogen always acts as a *donor* in InN.

This behavior is surprising, since it differs from the usual amphoteric character of H in other semiconductors (including GaN and AlN). In InN, H can act as a *source* of n-type conductivity. This is important to keep in mind when growing InN; growth techniques such as metal-organic chemical vapor deposition and hydride vapor phase epitaxy introduce hydrogen from source gases and carrier gases, and even in molecular beam epitaxy H can be present either intentionally (when NH_3 is used as the N source) or unintentionally (as a hard-to-avoid background impurity). We have previously discussed other sources of n-type conductivity in InN [12]: nitrogen vacancies are too high in energy to be a concern, but unintentional impurities such as silicon or oxygen do act as donors and can be easily incorporated. Hydrogen should be added to this list of unintentional donors in InN.

Acknowledgements This work was supported by the Air Force Office of Scientific Research, Contract No. F4920-00-C-0019, monitored by G. Witt. We thank J. E. Northrup, N. Johnson, and J. Neugebauer for useful discussions and suggestions.

References

[1] C. G. VAN DE WALLE and N. M. JOHNSON, in: Gallium Nitride (GaN) II, Eds. J. I. PANKOVE and T. D. MOUSTAKAS, Semiconductors and Semimetals, Vol. 57, Treatise Eds. R. K. WILLARDSON and E. R. WEBER, Academic Press, Boston 1998 (p. 157).
[2] J. NEUGEBAUER and C. G. VAN DE WALLE, Phys. Rev. Lett. **75**, 4452 (1995).
[3] W. GÖTZ et al., Appl. Phys. Lett. **69**, 3725 (1996).
B. CLERJAUD et al., Phys. Rev. B **61**, 8238 (2000).
[4] S. LIMPIJUMNONG, J. E. NORTHRUP, and C. G. VAN DE WALLE, to be published.
[5] M. BOCKSTEDTE, A. KLEY, J. NEUGEBAUER, and M. SCHEFFLER, Comp. Phys. Commun. **107**, 187 (1997).
[6] S. G. LOUIE, S. FROYEN, and M. L. COHEN, Phys. Rev. B **26**, 1739 (1982).
[7] J. NEUGEBAUER and C. G. VAN DE WALLE, Mater. Res. Soc. Symp. Proc. **378**, 503 (1995).
[8] J. H. EDGAR, S. STRITE, I. AKASAKI, H. AMANO, and C. WETZEL (Eds.), Properties, Processing and Applications of Gallium Nitride and Related Semiconductors, EMIS Datareviews Series No. 23, The Institution of Electrical Engineers, London 1999.
[9] C. STAMPFL and C. G. VAN DE WALLE, Phys. Rev. B **59**, 5521 (1999).
[10] A. RUBIO et al., Phys. Rev. B **48**, 11810 (1993).
[11] C. G. VAN DE WALLE, S. LIMPIJUMNONG, and J. NEUGEBAUER, Phys. Rev. B **63**, 245205 (2001).
[12] C. STAMPFL et al., Phys. Rev. B **61**, R7846 (2000).

Capture Kinetics of Electron Traps in MBE-Grown n-GaN

A. Hierro[1]) (a), A. R. Arehart (a), B. Heying (b), M. Hansen (b),
J. S. Speck (b), U. K. Mishra (b), S. P. DenBaars (b), and S. A. Ringel (a)

(a) Department of Electrical Engineering, The Ohio State University, Columbus, OH 43210, USA

(b) Materials and Electrical and Computer Engineering Departments, University of California, Santa Barbara, CA 93016, USA

(Received June 20, 2001; accepted August 4, 2001)

Subject classification: 61.72.Ji; 61.72.Lk; 71.55.Eq; S7.14

The carrier capture kinetics of the E_c–0.59 eV and E_c–0.91 eV electron traps found in molecular beam epitaxy (MBE)-grown n-GaN have been determined by means of deep level transient spectroscopy (DLTS). The 0.59 eV trap does not show the behaviour of either ideal point defects or line defects. In contrast, the 0.91 eV trap displays the kinetics of linearly arranged interacting point defects, which generate a time-dependent local Coulombic potential with a characteristic time constant of ≈ 8.6 μs.

Introduction Even though III–nitrides have already been applied successfully to optoelectronic devices and have shown great potential for microelectronic applications, there are several limiting factors that need be addressed. Among these, electrically active defects that behave as carrier traps in these semiconductors are believed to be a source for $1/f$ noise and other detrimental effects on electronic devices [1]. Understanding the basic electronic properties of these traps and their sources is of great interest and will be valuable for the future understanding of the impact of material properties on device characteristics. We focus here on the carrier capture kinetics of the electron traps observed in MBE-grown n-GaN by means of DLTS.

Experimental Approach MBE was used to grow Si-doped 0.5 μm thick n-GaN layers on an MOCVD template on sapphire substrates. The electron carrier concentration, measured by capacitance–voltage, was $\approx 1 \times 10^{17}$ cm^{-3}. The GaN was grown in the Ga stable regime near transition to Ga droplets [2]. The impact of the Ga/N flux ratio on the traps will be discussed elsewhere. The DLTS experiments were performed using a modified Bio-Rad DL4600 system at –0.5 V reverse bias, 0 V fill bias, and fill pulse times ranging from 0.01 to 100 ms.

Electron Traps in n-GaN The DLTS spectra of MBE as-grown n-GaN shows two electron traps found at E_c–0.59 eV ($E(0.59)$) and E_c–0.91 eV ($E(0.91)$) with capture cross sections of $\approx 2 \times 10^{-15}$ and $\approx 3 \times 10^{-14}$ cm^2, respectively. Photocapacitance analysis also shows several other traps across the rest of the n-GaN bandgap that will be discussed elsewhere. $E(0.59)$ has been previously observed in HVPE- [3], MOCVD- [4], and MBE-grown n-GaN [5]. $E(0.91)$ has been seen in MOCVD-grown n-GaN after He-ion

---

[1]) Corresponding author; Phone: +1-614-292-1721; Fax: +1-614-292-7956; e-mail: hierroa@ee.eng.ohio-state.edu

irradiation [6] and may be related to a trap previously reported in HVPE [7] n-GaN. The fact that both of these levels are also observed in our as-grown films indicates they are likely related to intrinsic defects in n-GaN.

Interacting vs. Non-Interacting Defects In this section we establish the theory behind the approach followed in the next section to characterize the capture kinetics of $E(0.59)$ and $E(0.91)$, which consists of analysing the dependence of the DLTS signal on the fill pulse time. First, let us focus on non-interacting point defects that act as an electron trap, where we can neglect hole capture. During the DLTS fill pulse, where the electron emission rate is negligible compared to the capture rate, and for fill pulse times such that the trap is not saturated, the solution to the rate equation for capture of electrons is given in terms of the duration of the fill pulse, t_p, and can be expressed as [8]

$$n_T(t_p) = N_T + [n_T(0) - N_T] \exp(-c_n t_p), \qquad (1)$$

where n_T, N_T and c_n are the concentration of occupied defect states, the total concentration of states, and the electron capture rate, respectively. Thus, it follows that the trap occupancy depends exponentially on the fill pulse time for non-interacting point defects. However, for closely spaced defects, trapped electrons in turn affect electron capture at neighbouring defects by providing a repulsive time-dependent potential, $\Phi(t)$, against electron capture. The electron capture rate equation is now given by [8]

$$dn_T/dt = c_n(N_T - n_T) \exp(-q\Phi(t)/kT), \qquad (2)$$

where a Boltzmann factor accounts now for the fraction of electrons whose energy is sufficient to overcome the barrier and become trapped. If we assume a linear spacing of the defects, the time dependent Coulombic barrier [9] can be simplified and the solution to Eq. (2) for $t_p \gg \tau$ is [8]

$$n_T(t_p) = c_n \tau N_T \ln(t_p/\tau), \qquad (3)$$

where τ represents a characteristic time required for charge capture to begin influencing $\Phi(t)$. Equation (3) predicts that trap occupancy for a linear arrangement of interacting defects depends logarithmically on fill pulse time. In contrast, Eq. (1) predicts an exponential dependence for non-interacting point defects. Since under constant reverse bias the quiescent capacitance is constant, changes in n_T with t_p are proportional to the DLTS peak height (ΔC), and a plot of ΔC as a function of $\log(t_p)$ should produce a straight line for linearly arranged interacting defects, while non-interacting defects will yield a non-linear curve.

Carrier Capture Kinetics of E(0.59) and E(0.91) Figures 1 and 2 show the DLTS spectra for $E(0.59)$ and $E(0.91)$, respectively, as a function of fill pulse time. Two very different behaviours can be observed for the two levels.

$E(0.59)$ has an occupancy of electrons that depends non-linearly on $\log(t_p)$ and saturates with fill pulse time (Fig. 1). Following the discussion in the previous section, this non-linear behaviour indicates that this defect is not linearly arranged in the crystal. Moreover, for a non-interacting point defect Eq. (1) predicts that $\ln(n_T)$ has to depend linearly on the fill pulse time. Such dependence was not observed for this level either (not shown here for the sake of brevity). We can conclude that $E(0.59)$ does not follow

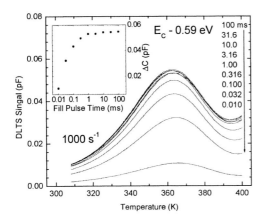

Fig. 1. Dependence of the DLTS signal from $E(0.59)$ on t_p. The inset shows the non-linear behavior of the DLTS peak height with $\log(t_p)$

the capture kinetics of either ideal point or line defects, and a more elaborate model may be needed.

In contrast, $E(0.91)$ displays a logarithmic electron capture rate that does not saturate for the range of fill pulse times used here (Fig. 2). Indeed, this trend closely matches that predicted by Eq. (3) for a linear arrangement of defects. This type of behaviour is consistent with the formation of a localized Coulombic barrier whose height increases with fill pulse time (Fig. 3). As more electrons get captured (local negative charge increases), the barrier height increases, preventing other electrons from being trapped such that the defect states do not fill completely (although for t_p long enough it may eventually saturate [10]).

The fact that the capture kinetics of $E(0.91)$ follows Eq. (3) only indicates that this defect is arranged along a line. This behaviour has been previously reported in MOCVD material [6], and $E(0.91)$ was correlated to an extended defect. However, such capture kinetics can be explained to result from either point defects arranged along a threading dislocation (TD) or from the threading dislocation core itself. The impact that either type of defect arrangement can have on the DLTS signal has been modelled before by Schröter et al. [11]. Since point defects will generate a localized state in the bandgap whereas TD cores will create a band of states, the trap occupancy, which is the solution to Eq. (2) and is proportional to ΔC, will differ. In the case of linearly arranged point defects, the DLTS peak temperature is independent of t_p, whereas in the case of TD cores it shifts to higher temperatures with decreasing t_p. This can be understood if we consider the distribution of states of the TD where the lowest energy states are preferentially filled with electrons for short t_p values. Hence, at lower t_p values the DLTS signal will be detected at progressively higher temperatures. Moreover, in the case of TDs the high temperature side of the DLTS peak

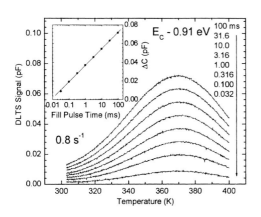

Fig. 2. DLTS signal from $E(0.91)$ as a function of fill pulse time. In contrast to $E(0.59)$, the DLTS peak height is linear with $\log(t_p)$ (shown in the inset)

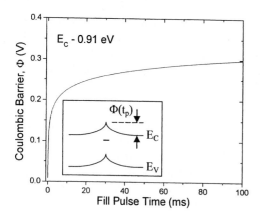

Fig. 3. Dependence of the Coulombic barrier associated with $E(0.91)$ as a funtion of electron capture time

will be aligned for all t_p [11]. Such differences between point defects linearly arranged and TDs have been experimentally observed before ([10] and [9], respectively).

The DLTS signal from $E(0.91)$ clearly shows a constant peak temperature as well as a constant increase in the magnitude of the entire curve for increasing t_p (Fig. 2), consistent with electron trapping at linearly arranged point defects. This assignment to point defects is consistent with previous reports where He-ion irradiation was shown to generate this defect in MOCVD material [6]. Thus, we can conclude that the behaviour of the capture kinetics of $E(0.91)$ closely matches that of point defects arranged along a line, most likely a TD.

Finally, one can extract the time dependence of the potential barrier associated with $E(0.91)$. The potential barrier $\Phi(t_p)$ can be expressed in terms of τ as [8]

$$q\Phi(t_p) = kT \ln(t_p/\tau). \tag{4}$$

Proper rearrangement of Eq. (3) shows that τ can be extracted from a linear fit to n_T vs. $\ln(t_p)$ [8]. From this analysis, a characteristic time constant of ≈ 8.6 μs is obtained for the potential barrier associated with $E(0.91)$. Note that the fill pulse times used in this analysis are much larger than τ, consistent with the assumption used in Eq. (3). When $t_p \ll \tau$ there should be a negligible potential barrier and the capture kinetics should follow those of non-interacting point defects. An estimation of the dependence of the potential barrier height with fill pulse time can then be obtained from Eq. (4), and is shown in Fig. 3.

Conclusion The capture kinetics of electron traps $E(0.59)$ and $E(0.91)$ present in as-grown MBE n-GaN have been determined. These traps present two different behaviours. $E(0.59)$ does not display the behaviour of simple non-interacting point defects or linear defects. In contrast, $E(0.91)$ displays the behaviour of interacting point defects arranged along a line, likely a TD. A local Coulombic potential was determined for this trap as a function of capture time, with a characteristic time for the potential to start affecting carrier capture of ≈ 8.6 μs.

Acknowledgements This work was supported by ONR grant N00014-00-1-0055 (J. Zolper).

References

[1] S. L. RUMYANTSEV, M. S. SHUR, R. GASKA, X. HU, A. KHAN, G. SIMIN, J. YANG, N. ZHANG, S. DENBAARS, and U. K. MISHRA, Electron. Lett. **36**, 757 (2000).
[2] B. HEYING, I. SMORCHKOVA, C. POBLENZ, C. ELSASS, P. FINI, S. P. DENBAARS, U. K. MISHRA, and J. S. SPECK, Appl. Phys. Lett. **77**, 2885 (2000).

[3] P. Hacke, T. Detchprohm, K. Miramatsu, N. Sawaki, K. Tadatomo, and K. Miyake, J. Appl. Phys. **76**, 304 (1994).
[4] W. Götz, N. M. Johnson, H. Amano, and I. Akasaki, Appl. Phys. Lett. **65**, 463 (1994).
[5] C. D. Wang, L. S. Yu, S. S. Lau, E. T. Yu, W. Kim, A. E. Botchkarev, and H. Morkoç, Appl. Phys. Lett. **72**, 1211 (1998).
[6] F. D. Auret, S. A. Goodman, F. K. Koschnick, J-M. Spaeth, B. Beaumont, and P. Gibart, Appl. Phys. Lett. **73**, 3745 (1998).
[7] Z.-Q. Fang, D. C. Look, J. Jasinski, M. Benamara, Z. Liliental-Weber, and R. J. Molnar, Appl. Phys. Lett. **78**, 332 (2001).
[8] P. Omling, E. R. Weber, L. Montelius, H. Alexander, and J. Michel, Phys. Rev. B **32**, 6571 (1985).
[9] T. Figielski, Solid State Electron. **21**, 1403 (1978).
[10] P. N. Grillot, S. A. Ringel, E. A. Fitzgerald, G. P. Watson, and Y. H. Xie, J. Appl. Phys. **77**, 3248 (1995).
[11] W. Schröter, J. Kronewitz, U. Gnauert, F. Riedel, and M. Seibt, Phys. Rev. B **52**, 13726 (1995).

Reduction of Defects in GaN on Reactive Ion Beam Treated Sapphire by Annealing

D. Byun (a), J. Jhin (a), S. Cho (a), J. Kim (a), S. J. Lee (a), C. H. Hong (b), G. Kim (c), and W.-K. Choi (c)

(a) Department of Materials Science, Korea University, 1, 5-Ka, Anam-Dong, Sungbuk-Ku, Seoul, Korea 136-701

(b) Department of Semiconductor Science and Technology, Chonbuk University 664-14, Dukjin-Dong, Dukjin-Gu, Chonju, Korea 561-756

(c) Korea Institute of Science and Technology, PO Box 131, Cheongryang, Seoul, Korea 130-650

(Received July 5, 2001; accepted August 14, 2001)

Subject classification: 68.35.Dv; 68.55.Ln; S7.14

Previous studies showed that reactive ion beam (RIB) pretreatment of sapphire prior to GaN deposition results in the reduction of dislocation density in the GaN film. It was also found that an amorphous phase remained at the interface region after the GaN deposition at high temperature. Annealing was performed to obtain the structural change due to the recrystallization of the remaining amorphous phase, and the effect on the electrical properties of the GaN thin film on RIB treated sapphire (0001) substrate. DCXRD spectra and Hall mobility of the specimen were studied as a function of the annealing time at 1000 °C in N_2 atmosphere. For the annealed specimen, FWHM of DCXRD decreased and the mobility increased. The annealed specimen was compared with a not annealed sample by TEM. A decrease of lattice strain and a reduction of the dislocation density about 56–59% was observed. The present results clearly show that the combination of RIB pretreatment and proper post annealing conditions improve the properties of GaN films grown by MOCVD.

Introduction Sapphire has been the most commonly used substrate for GaN growth but the large differences in thermal expansion coefficient and lattice constant with GaN (about 25.4% and 16.1%, respectively) cause defects [1–3]. There have been various efforts to reduce the defect density. It is reported that the buffer layer helps nucleation of GaN by restraining the isolated island type nucleation during the initial growth stage [3]. In particular, the coexistence of both amorphous and crystallized phases at the interface between substrate and thin film after high temperature growth of GaN on sapphire (0001) pretreated by reactive N_2^+ ion beam (RIB) was reported [4, 5]. There remained a residual amorphous phase at the interface even after the GaN deposition at high temperature. This result suggests the possibility to modify the stress distribution and to reduce the dislocation density by recrystallization of this residual amorphous phase by annealing.

Experimental The sapphire (0001) was irradiated by reactive N_2^+ ions at room temperature with a dose of 1×10^{16} N_2^+ ions/cm^2 at an energy of 800 eV, current of 0.5 μA and 1.6×10^{-4} Torr of reactor pressure, using a 5 cm gridded cold-hollow ion source [6]. Buffer layer and GaN growth were carried out in a Aixtron RF200 horizontal reactor at 300 Torr. Source gases for Ga and N were TMG and NH_3, respectively. H_2 was used as a dilutant and carrier gas for TMG. The GaN thin film samples were annealed by electrical furnace at 1000 °C, atmospheric pressure for various time intervals under N_2. The heating rate was set to 5 °C/min and the layers were cooled down in the furnace to exclude thermal shock from rapid change of temperature. DCXRD spectra of GaN (0002) peak were recorded to analyze the crystalline quality. Hall measurements by van der Pauw method were performed to examine the annealing effect on the electrical property. The lattice strain and the dislocation density was analyzed by cross-sectional TEM (Philips, CM30: LaB_6 filament) at 200 kV ($\lambda = 0.0251$ Å) of acceleration voltage. The high order Laue zone (HOLZ) line from the convergent beam electron diffraction (CBED) technique was analyzed to measure the lattice strain and its distribution in the GaN film. The conventional diffraction contrast images were obtained in bright field (BF) and the weak beam dark field (WBDF) of three different diffraction vectors using two-beam diffraction conditions. The two-beam CBED was used for thickness measurement to calculate the dislocation density from WBDF images.

Results and Discussion It was reported that partial crystalline and amorphous regions coexist at the interface between GaN and RIB pretreated sapphire even after the GaN deposition at 1080 °C [4]. FWHM data of GaN (0002) peak by DCXRD spectra are shown in Fig. 1. A decrease in the FWHM was observed after annealing for 1–2 h. The change of strain distribution in the film by the recrystallization of the residual amorphous region resulted in the improvement of crystallinity. FWHM value was increased when the sample was annealed for 3–5 h, but still was lower than for the sample without heat treatment. This result clearly showed that the crystallinity increased with annealing process.

The change in mobility and carrier concentration is shown in Fig. 2. The mobility first decreased by about 70 cm^2/Vs after annealing for 30 min but then increased to a maximum value of 467 cm^2/Vs for the specimen heat treated for 90 min. But the mobility decreased for samples annealed over 90 min consistent with the FWHM values of DCXRD spectra in Fig. 1. The result of DCXRD spectra and Hall measurement show the improvements in both electrical and crystalline property of GaN thin film via annealing process.

Figure 3 shows the depth profile of three different types of dislocations as a function of the distance from the interface of substrate. Both samples exhibited the same thickness of about 1.6 μm. The specimen without and with 90 min annealing were compared. The three different types of dislocation, a-type ($\mathbf{b} = 1/3[11\bar{2}0]a$: edge dislocation),

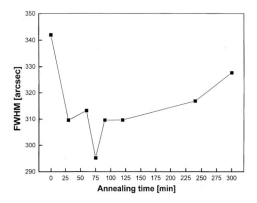

Fig. 1. FWHM of GaN (0002) peak from DCXRD as a function of thermal treatment time

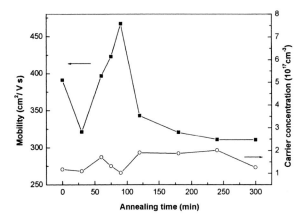

Fig. 2. Hall measurement results of GaN as a function of thermal treatment time. Solid squares represent the Hall mobility and circles represent the carrier concentration

c-type (**b** = [0001]c: screw dislocation) and a + c-type (**b** = 19/30[11$\bar{2}$3]a: mixed dislocation) were identified by the WBDF images of three different diffraction vectors, **g** = [0002], **g** = [1$\bar{1}$01] and **g** = [1$\bar{1}$02] [3, 4]. All the directional vectors of dislocations in the GaN thin film were [0001]. The 90 min annealed sample exhibited a reduction of a-type dislocation density of about 58.6%. The c-type dislocation density was reduced about 57.6%, and a + c-type dislocation density was reduced about 56.8%. The high dislocation density near the interface was significantly reduced, and an overall reduction throughout the film was observed. Also the uniform distribution of defects was obtained via annealing process.

The reason for the high dislocation density of the annealed sample near the surface was the decomposition of GaN at the surface during high temperature annealing. The WBDF image of the annealed sample revealed the decomposed surface of GaN. Therefore, the high density of a + c type dislocations near the surface actually represent the 3D defects created by the decomposition of GaN. The decomposition was restricted near the surface. Excluding the 3D defects near the surface, the densities of all three dislocation types were reduced significantly. It can be suggested that the improvement of crystallinity and mobility is due to this reduction in dislocation density.

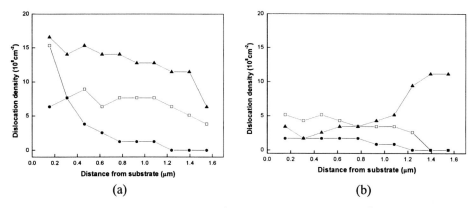

Fig. 3. Depth profile of dislocation density a) before thermal treatment, b) after 1 h 30 min treatment. Three types of dislocation density were presented: a-type (●), c-type (□) and a + c-type (▲)

Conclusions The annealing of a GaN thin film grown on a RIB treated sapphire (0001) substrate by MOCVD was studied. Recrystallization of the partially remaining amorphous phase after MOCVD growth of GaN on RIB treated sapphire caused the reduction of dislocation density and the decrease of lattice strain through thermal treatment. The electrical properties such as Hall mobility and carrier concentration were improved by this increase of crystallinity. This result clearly shows that the GaN thin film properties can be improved by a combination of RIB pretreatment of substrate and proper annealing process, and provides a possible approach to address the high dislocation density problem in GaN epitaxy.

Acknowledgement This work was supported by The Korean Research Foundation (Grant No. 98-0300-08-01-03).

References

[1] C.-Y. Hwang, P. Lu, W. E. Mayo, Y. Lu, and H. Liu, Mater. Res. Soc. Symp. Proc. **326**, 347 (1994).
[2] S. Chadda, M. Pelcynski, K. Malloy, S. Hersee, and H. Liu, Mater. Res. Soc. Symp. Proc. **326**, 353 (1994).
[3] D. Byun, G. Kim, D. Lim, D. Lee, I.-H. Choi, D. Park, and D.-W. Kum, Thin Solid Films **289**, 256 (1996).
[4] H.-J. Kim, D. Byun, G. Kim, and D.-W. Kum, J Appl. Phys. **87**, 7940 (2000).
[5] D. Byun, H.-J. Kim, C.-H. Hong, C.-S. Park, G. Kim, S.-K. Koh, W.-K. Choi, and D.-W. Kum, phys. stat. sol. (a) **176**, 643 (1999).
[6] S. K. Koh, S. K. Song, W. K. Choi, H.-J. Jung, and S. N. Han, J. Mater. Res. **10**, 2390 (1995).

Comparative Study on the Optical Properties of Eu:GaN with Tb:GaN

H. Bang[1]) (a), S. Morishima (a), Z. Li (a), K. Akimoto (a), M. Nomura (b), and E. Yagi (c)

(a) Institute of Applied Physics, University of Tsukuba, 1-1-1 Tennodai, Tsukuba 305-8573, Japan

(b) Photon Factory, Institute of Materials Structure Science, High Energy Accelerator Research Organization, Oho, Tsukuba 305-0801, Japan

(c) The Institute of Physical and Chemical Research (RIKEN), Wako, Saitama, 351-01, Japan

(Received June 24, 2001; accepted August 1, 2001)

Subject classification: 68.55.Ln; 78.30.Fs; 78.55.Cr; S7.14

Sharp luminescence originating from intra-atomic 4f–4f transition from both Eu and Tb doped GaN was observed at 622 nm which can be assigned to $^5D_0$–$^7F_2$ transition of Eu^{3+} and at 545 nm which can be assigned to $^5D_4$–$^7F_5$ transitions of Tb^{3+}. However, the luminescence intensity of Eu^{3+} is two orders of magnitude stronger than that of Tb^{3+} though the content of the rare earth elements is almost the same. The cause of the difference in the luminescence properties was studied based on the defect-related energy transfer model, and the role of the defects observed by Fourier transform infrared (FTIR) spectra was discussed.

Introduction Rare earth ions in inorganic solids generally show sharp and intense luminescence originating from intra-atomic f–f transitions, and the peak position and intensity of the luminescence is hardly affected by chemical environment [1, 2].

There are several reports regarding infrared emission or visible light emission from various rare earth ions (for example, Er^{3+}: infrared and green, Tb^{3+}, Ho^{3+}: green, Tm^{3+}: blue, Sm^{3+}, Pr^{3+}, Eu^{3+}: red) in GaN [3–10], and electroluminescence of red, green and blue color from MIS structure was already demonstrated [9, 11, 12]. Particularly, green to red emission seems to be interesting because of the difficulty of InGaN growth with high content of In. So the rare earth ion doped GaN can be regarded as a promising material in application for novel optical devices.

We have reported the growth of single crystalline Eu or Tb doped GaN by gas source molecular beam epitaxy and showed that the emissions from Eu^{3+} and Tb^{3+} in GaN originating from 4f intra-atomic transition was generated by the host excitation and its intensity and wavelength were extremely stable with temperature [7–9, 13, 15]. However, the emission intensity of Eu^{3+} in GaN is two orders of magnitude stronger than that of Tb^{3+}, though the content of the rare earth ions is almost the same.

In this study, the cause of the difference in the PL properties between Eu doped GaN and Tb doped GaN was studied using defect-related energy transfer model.

Experimental Eu or Tb doped GaN films were grown on sapphire (0001) substrates by gas source molecular beam epitaxy (GSMBE) using uncracked ammonia as nitrogen

---

[1]) Corresponding author; Phone: +81-298-53-6177; Fax: +81-298-55-7440; e-mail: bk993503@s.bk.tsukuba.ac.jp

source. Metallic Ga with 6N purity, Tb and Eu with 3N purity were evaporated from conventional Knudsen effusion cells, and uncracked ammonia gas with 6N purity was introduced to the growth surface through a nozzle of stainless steel tube.

The growth temperatures of Eu or Tb doped GaN were both 700 °C and the cell temperatures of Eu and Tb was 500 and 1300 °C, respectively. The Ga cell temperature and ammonia pressure were kept constant at 950 °C and 2.6×10^{-3} Pa, respectively.

The content of Eu or Tb in GaN was roughly estimated by Rutherford backscattering spectroscopy (RBS) using 2 MeV He ion beam, and found to be about 2% for both cases. PL measurements were performed at 77 K and room temperature using the 325 nm line of a He–Cd laser as excitation source. X-ray diffraction (XRD) was measured with θ–2θ mode using both CuKα_1 and Kα_2 radiation. Fourier transform infrared spectroscopy (FTIR) spectra were measured at room temperature in the spectral region of 400–7000 cm^{-1} for all samples using Jasco FT/IR-300 system.

Results and Discussion Figure 1 shows the X-ray diffraction profiles of Eu doped GaN and Tb doped GaN, compared with that of undoped GaN for reference. The peak positions of Eu doped GaN and Tb doped GaN shifted a little bit to lower angle possibly due to the large atomic radii of Eu and Tb. The values of the full with at half maximum (FWHM) of diffraction peak of Eu doped GaN is almost two times wider than that of undoped GaN, whereas Tb doped GaN shows almost the same value of FWHM as undoped GaN. The structural property of Tb doped GaN is superior to that of Eu.

Figure 2 shows PL spectra, measured at 77 K, of Eu and Tb doped GaN. Luminescence originating from intra-atomic f–f transitions was observed at 622 and 545 nm which can be tentatively assigned as $^5D_0$–$^7F_2$ transition of Eu^{3+} ion, and $^5D_4$–$^7F_5$ transition of Tb^{3+} ion [1, 2, 13].

In case of the luminescence spectra of Eu doped GaN, no luminescence related with GaN was observed, and the intensity of Eu related luminescence was two orders of magnitude larger than that of Tb although the content of Tb is almost the same as that of Eu.

As reported previously [8, 9], the Eu related luminescence is generated through an excitation of GaN as confirmed by photoluminescence excitation spectroscopy, and Tb related luminescence can be considered to be generated also through an excitation of GaN since there is no excited level in Tb^{3+} ion which coincides with the energy value

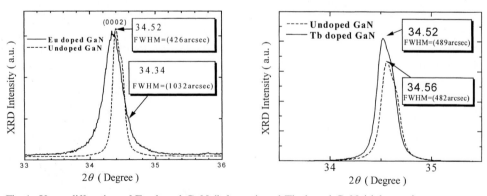

Fig. 1. X-ray diffraction of Eu doped GaN (left part) and Tb doped GaN (right part)

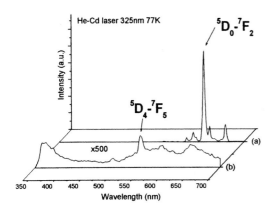

Fig. 2. PL spectra of (a) Eu doped GaN and (b) Tb doped GaN at 77 K

of He–Cd laser (3.81 eV); that is, direct transition from the ground state to an excited state of Tb^{3+} ion hardly occurs by He–Cd laser [1]. Therefore, the large difference in the luminescence intensity between Eu doped GaN and Tb doped GaN may be caused by the difference of energy transfer probability from GaN to the rare earth ions.

The reason for the poor energy transfer probability in Tb doped GaN may not be a degradation of crystal quality, since most of Tb atoms were found to be incorporated into Ga lattice sites just the same as the case of Eu in GaN [14], and the FWHM of X-ray diffraction is even narrower than Eu doped GaN.

There are some proposed models for the energy transfer from the host material to rare earth ion, and among them, defect-related energy transfer model has been widely believed, that is, the transfer process is considered as below; excited electron in the host material is captured at a defect level and the rest of the energy is transferred to the rare earth ions through Auger process, dipole interaction or electron exchange interactions [15–18].

To detect defect levels, FTIR spectra were measured. The FTIR spectra of Tb doped GaN, and Eu doped GaN in the spectral range between 2500 and 3500 cm^{-1} are shown in Fig. 3 together with the sapphire substrate and undoped GaN for reference. Peaks around 2950 cm^{-1} corresponding to 365 meV whose intensity is roughly proportional to the emission intensity from Eu^{3+} [14] were observed. So it may be possible to consider that the defect at 365 meV is related to the energy transfer.

The excited energy levels of Eu^{3+} and Tb^{3+} ions together with the band gap energy of GaN are shown in Fig. 4. The captured electron at the trapped level has an excess energy of 24400 cm^{-1} (3.03 eV) as shown in Fig. 4, and this value is well consistent with the excitation energy of $^7F_2$–$^5D_3$ in Eu^{3+} ion. For Tb^{3+} ion, however, there is no excited level around 24400 cm^{-1}, and this means that the energy transfer from GaN to Tb^{3+} ion can

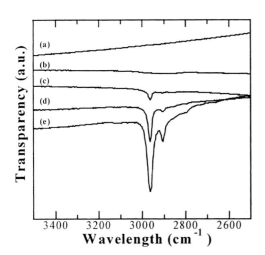

Fig. 3. FTIR spectra of (a) sapphire substrate, (b) undoped GaN, (c) Tb doped GaN, (d) Eu doped GaN (Eu; 0.1%), (e) Eu doped GaN (Eu; 2%)

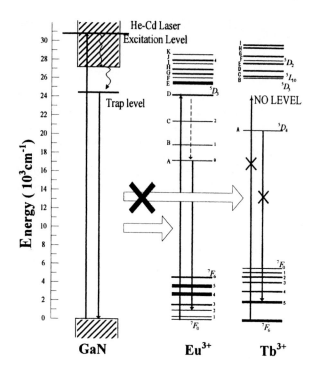

Fig. 4. Schematic model of energy transfer from GaN to Eu^{3+} or Tb^{3+}

hardly occur. These results possibly can be one of the reasons for the remarkable difference in the PL property.

Summary The cause of the remarkable difference in the luminescence intensity between Eu^{3+} and Tb^{3+} doped into GaN was studied by FTIR spectra. The defect level at 365 meV was detected by FTIR measurements, and it is suggested that energy transfer from GaN to rare earth ions proceeds through the defect level. The reason for the poor luminescence of Tb^{3+} in GaN may be concluded to the mismatching of transfer energy.

References

[1] G. H. Dieke, in: Spectra and Energy Levels of Rare Earth Ions in Crystals, Eds. H. M. Crosswhite and H. Crosswhite, Wiley, New York 1968.
[2] M. J. Weber, Phys. Rev. **171**, 283 (1968).
[3] J. D. Mackenzie, C. R. Abernathy, S. J. Pearton, U. Hömmerich, J. T. Seo, R. G. Wilson, and J. M. Zavada, Appl. Phys. Lett. **72**, 2710 (1998).
[4] H. J. Lozykowski, W. Jadwisienczak, and I. Brown, Appl. Phys. Lett. **74**, 1129 (1999).
[5] A. J. Steckl, M. Garter, D. S. Lee, J. Heikenfeld, and R. Birkhahn, Appl. Phys. Lett. **75**, 2184 (1999).
[6] K. Hara, N. Ohatake, and K. Ishi, phys. stat. sol. (b) **216**, 625 (1999).
[7] H. Bang, S. Morishima, T. Maruyama, K. Akimoto, M. Nomura, and E. Yagi, Proc. Internat. Workshop Nitride Semiconductors, IPAP Conf. Series **1**, 494 (2000).
[8] S. Morishima, T. Maruyama, and K. Akimoto, J. Cryst. Growth **378**, 209 (2000).
[9] S. Morishima, T. Maruyama, M. Tanaka, Y. Masumoto, and K. Akimoto, phys. stat. sol. (a) **176**, 113 (1999).
[10] T. Maruyama, H. Sasaki, S. Morishima, and K. Akimoto, phys. stat. sol. (b) **216**, 629 (1999).

[11] D. S. Lee, J. Heikenfeld, R. B. Birkhahn, M. Garter, B. K. Lee, and A. J. Steckl, Appl. Phys. Lett. **76**, 1525 (2000).
[12] A. J. Steckl, M. Garter, D. S. Lee, J. Heikenfeld, and R. B. Birkhahn, Appl. Phys. Lett. **75**, 2184 (1999).
[13] P. A. Lee, P. H. Citrin, P. Eisenberger, and B. Kincaid, Rev. Mod. Phys. **53**, 769 (1981).
[14] S. Morishima, H. Bang, T. Maruyama, and K. Akimoto, see [7] (p. 490).
[15] A. Kudo and T. Sakata, J. Phys. Chem. **99**, 15965 (1995).
[16] H. Kühne, G. Weiser, E. I. Terukov, A. V. Kusnetsov, and K. Kudoyarova, J. Appl. Phys. **86**, 896 (1999).
[17] I. Yassievich, M. Bresler, and O. Gusev, J. Non-Cryst. Solids **226**, 192 (1998).
[18] T. Gregorkiewicz, D. T. Thao, J. M. Langer, H. H. P. Bekman, M. S. Bresler, J. Michel, and L. C. Kimerling, Phys. Rev. B **61**, 5369 (2000).

Implantation Induced Defect States in Gallium Nitride and Their Annealing Behaviour

A. Krtschil[1]) (a), A. Kielburg (a), H. Witte (a), A. Krost (a), J. Christen (a), A. Wenzel (b), and B. Rauschenbach (c)

(a) Institute of Experimental Physics, Otto-von-Guericke-University Magdeburg, PO Box 4120, D-39016 Magdeburg, Germany

(b) Institute of Physics, University of Augsburg, D-86135 Augsburg, Germany

(c) Institute of Surface Modification Leipzig and Institute of Experimental Physics II, University of Leipzig, D-04318 Leipzig, Germany

(Received June 25, 2001; accepted August 4, 2001)

Subject classification: 68.55.Ln; 71.55.Eq; 73.50.Gr; 73.50.Pz; S7.14

Gallium nitride layers grown by metal organic vapour phase epitaxy on sapphire were implanted with different ion species, i.e. silicon, sulfur, and magnesium, and thermally annealed at 1150 °C under nitrogen atmosphere. The impact of this annealing procedure on the resulting deep levels was analyzed by transient and admittance spectroscopy. Several electron traps with thermal activation energies between 200 and 900 meV as well as very deep states at photon energies ranging from 1.8 to 2.5 eV were induced by the implantation process independent of the ion species. After annealing, in general the deep level spectrum shows only minor changes, but an enhancement of the shallower electron traps and a new electron trap for the Si-implanted layers can be observed. These results are explained by a re-arrangement of the induced defects and supported by photoluminescence experiments. Finally, the ineffectiveness of this annealing procedure which is often used by the community to reduce the implantation damage is demonstrated.

Introduction Ion implantation of semiconductor materials is a powerful technique applicable for pattern doping as well as for electrical isolation of distinct device regions. However, the key for success is the control of the implantation damage. The introduced ions commonly generate additional defects which may overcompensate the primary doping effect or which are able to improve the electrical isolation. The most convenient way to reduce the implantation damage is an additional rapid thermal annealing step (RTA) performed at high temperatures (empirical rule: annealing temperature $\approx 2/3$ of the melting point [1]). RTA of GaN at equivalent temperatures (≈ 1650 °C) suffers from a significant material decomposition, i.e. evaporation of nitrogen from the surface. Consequently, the annealing temperature commonly used is reduced to 1150 °C or below. In this report, the impact of this typical RTA treatment on electrically active defects induced by the ion implantation is analyzed.

Experimental The samples were grown by metal organic vapour phase epitaxy on c-axis oriented sapphire with a typical layer thickness of 1 µm and an effective carrier concentration of 2×10^{17} cm^{-3} revealed by C–V measurements. These layers were implanted with 90 keV Mg ions, 100 keV Si ions or 150 keV S ions with an ion dose I_0 of 10^{14} ions per cm^2 and thermally annealed for 15 s at 1150 °C in a nitrogen environment. Ohmic

---

[1]) Corresponding author; Phone: +49 391 67 12740; Fax: +49 391 67 11130; e-mail: andre.krtschil@physik.uni-magdeburg.de

contacts were realized by dc sputtered aluminum followed by an annealing at 500 °C for 10 min in a nitrogen atmosphere and a cap layer of sputtered platinum. The Schottky contacts were created by dc sputtered Pt layers and show rectifying behaviour in I–V and C–V measurements.

The deep level spectra before and after RTA treatment as well as those from the non-implanted layer were analyzed by deep level transient spectroscopy (DLTS) and thermal admittance spectroscopy (TAS) in the temperature range between 80 and 500 K. DLTS was performed at 3 V reverse bias voltage with a pulse voltage of 0 V, whereas TAS was carried out under zero bias conditions. Both spectroscopic techniques detect thermally induced carrier emissions from majority traps, e.g. electron traps in our n-type samples, which are described in detail in [2]. Information on very deep states were obtained from optical admittance spectroscopy (OAS) investigations in the wavelength range from 300 to 3000 nm, where optically induced carrier emissions from the defects to the next or to the opposite band were detected resulting in peaks at the corresponding optical transition energies. For more details please refer to [2], too. Besides these electrical investigations, photoluminescence (PL) experiments at room temperature excited by the 325 nm line of a He–Cd laser giving information on defect-related luminescence properties were performed.

Results and Discussion Before the impact of RTA is analyzed, the implantation induced generation of defects is discussed briefly. Earlier investigations [3, 4] at these samples before RTA have shown, that four additional electron traps called A-D with thermal activation energies of about 140 meV, 260 meV, 700–800 meV, and 1 eV were created due to implantation in contrast to the virgin sample with an exclusive signal from trap B with a concentration below 4×10^{13} cm^{-3}. This radiation damage is independent of the implanted ion species indicating an intrinsic nature of the participating defects. The depth profiles of these electron traps were determined with bias dependent DLTS and exhibit a good correlation either with the profile of the introduced ions or with those of the generated vacancies as calculated by transport of ions in matter (TRIM) simulations. The maximum defect concentrations of traps B, C, and D are about 2×10^{14}, 2×10^{15}, and 2×10^{15} cm^{-3}, respectively [4]. The concentration of trap A was not determined since its emission maximum in DLTS is outside the temperature range we used. An estimation from the height of TAS data yields comparable concentrations for traps A and B. Using OAS, we got information on additionally generated very deep and midgap states and found a drastical increase of defect-to-band transitions in the so-called blue band (300–800 meV below gap) and in the yellow band between 1.8 and 2.5 eV including a new transition in the latter one [3, 4]. Again, these effects are not significantly influenced by the ion species.

Now the changes in this scenario after an additional RTA treatment which is intended to reduce the radiation damage are considered. In Fig. 1, TAS spectra of Si- and S-implanted layers before and after RTA are shown. Note, that the spectra of as-implanted samples reveal only small signals from traps A and B (sometimes A and B completely vanish within the background signal). In contrast, after RTA treatment the concentration of both of these traps is drastically enhanced, observable through the increased resistance peaks (Fig. 1a) and capacitance steps (Fig. 1b). As obtained from DLTS the concentration of trap B increases to 6×10^{14}–1×10^{15} cm^{-3}. Moreover, after RTA a new electron trap E with a thermal activation energy of 120–190 meV

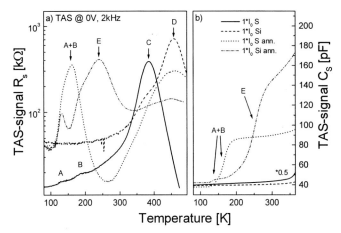

Fig. 1. TAS-spectra: a) serial resistance, b) capacitance, of S- and Si-ion implanted layers before and after thermal annealing, evidencing the drastic increase of distinct trap contributions after RTA

only appears in the Si ion implanted sample in comparable concentrations like trap B. On the other hand, the traps C and D remain widely unaffected or are only slightly enhanced without a systematic annealing effect similar to the effective carrier concentration which remains nearly the same. A similar behaviour can be observed for the midgap states in OAS as plotted in Fig. 2a where the normalized spectra of the Si ion implanted sample set is presented, including the non-implanted reference sample (#1), an as-implanted sample (#2), and layers implanted with different ion doses annealed afterwards by RTA (#3, 4, and 5). The first obvious point when observing the OAS spectra is the strong (factor 1/500) reduction of the OAS-signal in

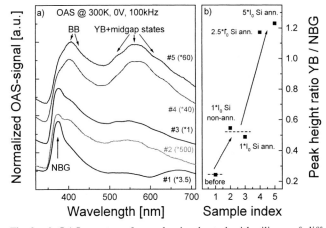

Fig. 2. a) OAS spectra of samples implanted with silicon of different ion doses (I_0 denotes an ion dose of 1×10^{14} cm^{-2}) and partly after RTA treatment. The spectra are normalized according to the legend and shifted against each other for better comparison. b) As a quantity of radiation damage, the ratio of the peak around 550 nm, originating from additionally generated midgap levels, and the NBG peak is plotted for these layers

the as-implanted sample #2 through the implantation induced defects. After annealing (#3), the signal strength increases back to the level of the unimplanted sample (#1). With increasing implanted doses (#4 and 5) the restoration of the signal level after annealing becomes worser (factor 1/40 and 1/60, respectively). As a quantity for the radiation damage responsible for the deep traps, the height of the OAS peak at 550 nm originating from implantation induced midgap states was related to that one of the near band gap (NBG) peak, where optically induced transitions between shallow donors/acceptors and the opposite band become apparent (Fig. 2b). It can be seen, that the RTA treatment of the sample implanted with the lowest dose (#3) provides only a minor reduction of these deep traps in comparison to the as-implanted sample as the signal level remains far away from the background of the non-implanted sample. With increasing implanted dose the number of midgap states increases correspondingly (#4, 5).

Summarizing we found, that after implantation with a relative high ion dose the positive impact of the used RTA treatment on electrically active deep defects is quite marginal proving its inefficiency in the present form. Moreover, this RTA procedure also contains negative aspects: the enhancement of shallower traps (A, B) and the introduction or activation of a new level (E). This relatively high amount of electrically active defects negatively affects the parameters of any device applications basing on such implanted GaN layers, e.g. the carrier mobility, the high frequency behaviour, or the luminescence properties.

In order to explain these unexpected results, we have to consider the decomposition of GaN at the annealing temperature, e.g. the loss of nitrogen from the surface, resulting in more intrinsic point defects. On the other hand, the RTA step primarily stimulates the distinct atoms to diffuse enabling a re-arrangement of already existing defect configurations or the creation of new defect structures. Both processes are hardly to separate from each other, but may contribute to the effects observed. This argumentation is also supported by room temperature PL spectra of these layers (not shown here). After implantation, the luminescence is quite poor and the near band edge luminescence completely disappears and also remains vanished after RTA indicating its inefficiency again. And in spite of RTA, the broad defect related bands around 570 nm and 700 nm are amplified by one order of magnitude. From the literature, comparable results in PL are already known and summarized in [1].

Conclusions The impact of a typical RTA procedure, often used for GaN to reduce radiation damage, on electrically active defect states was analyzed by several spectroscopy techniques. We found, that after this RTA treatment the radiation induced defects were not significantly reduced, evidencing its inefficiency for device applications in the present form. Remarkably, after RTA some shallower electron traps were drastically enhanced and newly created either due to the loss of nitrogen from the surface or due to internal re-arrangement effects of different defect configurations. These results are supported by PL investigations.

Acknowledgements The authors would like to thank M. Strassburg, Otto-von-Guericke-University Magdeburg, for the PL measurements. The gallium nitride samples were supplied by A. Lell and S. Bader, OSRAM Semiconductors, Regensburg.

References

[1] S.O. KUCHEYEV, J.S. WILLIAMS, and S.J. PEARTON, Mater. Sci. Eng. **33**, 51 (2001).
[2] P. BLOOD and J.W. ORTON, The Electrical Characterization of Semiconductors, Academic Press, London 1992.
[3] A. KRTSCHIL, A. KIELBURG, H. WITTE, A. KROST, J. CHRISTEN, A. WENZEL, and B. RAUSCHENBACH, Proc. Internat. Workshop on Nitride Semiconductors, IPAP Conf. Ser. **1**, 625 (2001).
[4] A. KRTSCHIL, A. KIELBURG, H. WITTE, A. KROST, J. CHRISTEN, A. WENZEL, and B. RAUSCHENBACH, Proc. E-MRS Spring Meeting 2001, Mater. Sci. Eng. B (2001), in print.

Annealing Behaviour of GaN after Implantation with Hafnium and Indium

K. Lorenz[1]), F. Ruske, and R. Vianden

Institut für Strahlen- und Kernphysik, University of Bonn, Nussallee 14-16, D-53115 Bonn, Germany

(Received June 26, 2001; accepted July 4, 2001)

Subject classification: 61.72.Vv; 68.55.Ln; S7.14

The annealing behaviour of GaN after implantation of ^{181}Hf and ^{111}In was studied using the perturbed angular correlation (PAC) technique. During annealing most Hf probes are built in on substitutional Ga sites and the level of lattice damage decreases with annealing temperature T_A. In the case of In the results also show a good recovery of the crystal lattice. However a large variation of the quadrupole interaction frequency is observed indicating a considerable change in the local lattice geometry. The dependency of this variation on the implantation dose is investigated. The values derived for the electric field gradients at the probe sites are compared to values determined by NMR measurements.

Introduction The integration of GaN devices into circuits requires a technology for selective area doping. Ion implantation is widely used in conventional semiconductors like Si or GaAs. However, for GaN the production and annealing of radiation damage is not yet fully understood and further investigations are necessary.

In this work the perturbed angular correlation (PAC) technique was employed to study the incorporation of the probes ^{111}In and ^{181}Hf by implantation and the annealing behaviour of the lattice. The PAC technique delivers information on the immediate lattice surroundings of an implanted probe atom and thus makes it possible to observe the reconstruction of the lattice and the interaction of the probe nucleus with defects. Hafnium is an interesting probe because its elemental properties are similar to those of rare earth atoms. Indium is especially well suited to perform annealing studies in GaN because it is isoelectronic to Ga and has similar chemical properties.

Experimental Details The radioactive probe atoms ^{181}Hf and ^{111}In were implanted at room temperature (RT) into commercial MOCVD GaN films grown on sapphire (CREE Inc., Durham, NC, USA). The implantation energy was 160 keV and typical doses of 10^{13} at/cm^2 were used. For some samples analysed with ^{111}In pre-implantations of stable In were carried out with doses of 5×10^{13} at/cm^2 and 1.2×10^{14} at/cm^2. An isochronal annealing program was carried out in a rapid thermal annealing apparatus to study the lattice recovery, using holding times of 120 s at temperatures up to 1373 K. The samples were placed between graphite strips in a N$_2$ atmosphere with their surface protected by a proximity cap. Directly after implantation and following every annealing step PAC spectra were taken at RT. The PAC technique measures the hyperfine interaction of an electric field gradient (EFG) at the site of a radioactive probe with the quadrupole moment of the intermediate state of a $\gamma\gamma$-cascade in the daughter nucleus.

---

[1]) Corresponding author; Phone: +49 228 73 3000, Fax: +49 228 73 2505, e-mail: lorenz@iskp.uni-bonn.de

Usually, the time dependent anisotropy $R(t)$ is plotted. It contains information on the perturbation of the emitted γ radiation pattern. From this the parameters of the EFG causing the perturbation can be derived. It is usually described by the quadrupole interaction frequency (QIF) v_Q and the asymmetry parameter η. The QIF is proportional to the product of the quadrupole moment Q of the intermediate state and the principal component of the EFG tensor V_{zz}. The quadrupole moment for the ^{181}Ta and ^{111}Cd cascades are known, $Q(\text{Ta}) = (+)2.36(5)\,b$ [1] and $Q(\text{Cd}) = (+)0.83(13)\,b$ [2].

Most measurements were carried out in a geometry where the c-axis of the GaN samples is aligned with the angle bisector of two detectors under 90°. Other orientations of the sample relatively to the detectors were used whenever necessary to derive the orientation of the EFG. In a hexagonal lattice like that of wurtzite GaN it can be expected that probe atoms on undisturbed lattice sites experience a unique EFG with an orientation along the c-axis. Structural defects in the lattice due to the implantation lead to additional EFGs. The PAC spectrum then gives information about the fractions f_s of the probe atoms that underlie a certain EFG, i.e. which have the same lattice environment. In the analysis of the data a Lorentzian distribution with width δ was used to characterise a distribution of interaction frequencies around a certain v_Q. A detailed description of the PAC technique and its application in solid state physics has been given by Schatz et al. [3].

Results and Discussion Typical PAC spectra of a well annealed GaN samples with the probes Hf and In are shown in Fig. 1[2]. For Hf the dominant QIF $v_{Q1}(\text{Hf}) = 338(2)$ MHz is caused by the lattice EFG and the changes observed during the annealing program show the typical behaviour of probes on regular lattice sites. With increasing T_A the fraction $f_{s1}(\text{Hf})$ of probes experiencing this EFG grows from 36% directly after implantation to 80% after the annealing step at 1373 K. The increase of this fraction with T_A and a simultaneous drop of the damping parameter δ suggest a good recovery of the crystal lattice where 80% of the probe atoms are incorporated on undisturbed lattice sites. These results are in good agreement with RBS/channeling measurements showing that after annealing a fraction of 90% of the implanted Hf atoms are located on substitutional Ga sites [4]. The discrepancy between the fractions obtained by the two techniques can be ex-

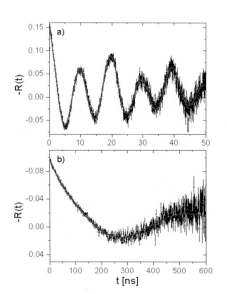

Fig. 1 (colour). Time dependent anisotropy observed for a) implanted ^{181}Hf and b) implanted ^{111}In in wurtzite GaN after annealing for 120 s at 1273 K. The observed quadrupole interaction frequencies are $v_{Q1}(\text{Hf}) = 338(2)$ MHz and $v_{Q1}(\text{In}) = 6.3(1)$ MHz

---

[2]) Colour figure is published online (www.physica-status-solidi.com).

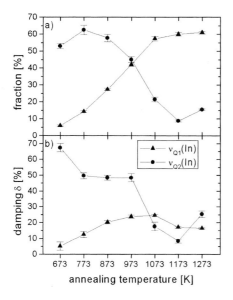

Fig. 2. Results for the sample without pre-implantation. a) Fractions f_s of ^{111}In probes showing the QIF $\nu_{Q1}(\text{In})$ and $\nu_{Q2}(\text{In})$ after various steps of an isochronal annealing program; b) width δ of the frequency distributions

plained by the fact that RBS is much less sensitive to defects near the impurity atoms. A detailed description of the PAC measurements can be found in [5]; similar results were found by Bartels et al. [6].

Also for In, emission channeling experiments yielded a substitutional fraction of 90% [7]. Figure 2 shows the variation of the fractions f_s and the damping parameter δ with T_A for the sample implanted with the probe ^{111}In with a dose of 1×10^{13} at/cm^2. To fit the PAC data three different frequencies are necessary. The first QIF, $\nu_{Q1}(\text{In})$, decreases with T_A and then stabilises at values around $\nu_{Q1}(\text{In}) = 6.3(1)$ MHz (Fig. 3). It shows all properties expected for In probes on substitutional, undisturbed lattice sites: Its fraction $f_{s1}(\text{In})$ increases with T_A to a maximal value of 60% and the damping reaches low values for the highest T_A (Fig. 2). The orientation of the EFG is parallel to the c-axis. The second frequency $\nu_{Q2}(\text{In})$, which is needed to reproduce the data, initially has a large fraction and a large distribution around $\nu_{Q2}(\text{In}) = 13(1)$ MHz. With T_A its fraction as well as its damping drops constantly. The corresponding EFG is probably caused by the superposition of the lattice EFG on substitutional sites with an EFG due to a point defect or an extended regular defect complex (e.g. a dislocation) in the next neighbourhood of the probe atom. The decrease in this fraction is due to the annealing of the surrounding defects resulting in an increase of the first fraction. A third, strongly damped frequency causes the strong drop in anisotropy in the first points of the spectra which is typical for probes in a highly damaged or amorphous environment. Similar results were reported by Burchard et al. and Ronning et al. [7, 8]. However, the large change in the QIF of

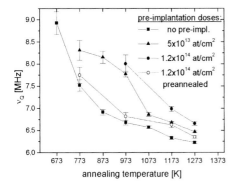

Fig. 3. Variation of $\nu_{Q1}(\text{In})$ with annealing temperature for the sample without pre-implantation and for the samples pre-implanted with different doses of stable In. One sample was pre-annealed between the pre-implantation and the implantation of ^{111}In

up to 30% with T_A was not reported up to now. The variation is especially surprising, since, in contrast, for the Hf probe the change of ν_{Q1}(Hf) with T_A is only 3% (data not shown).

In order to study the influence of the In concentration on this variation the development of ν_{Q1}(In) with T_A for samples pre-implanted with stable In was investigated (Fig. 3). As in the first sample, ν_{Q1}(In) decreases with T_A, but the absolute value for low T_A is larger for the higher implantation doses. For higher T_A, ν_{Q1}(In) approaches the value found for the low dose sample. An explanation could be the presence of strain in the lattice introduced by In implantation which is then relaxed by the annealing treatment. A slight variation in the lattice parameters due to strain is expected to have a big effect on the value of the lattice EFG. A lattice expansion due to implantation doping has also been observed by X-ray diffraction [9]. Measurements on a sample which has been annealed between the pre-implantation and the implantation of the probe atoms support this assumption, because the values for ν_{Q1}(In) lie only slightly above the values for the sample without pre-implantation and significantly lower than those of the sample pre-implanted with the same dose but without intermediate annealing. It is possible, that the unusually large change of ν_{Q1}(In) with T_A as mentioned above, is also due to a reduction of lattice strain during the annealing program. However, this would imply that the implantation of In initially causes a higher strain than Hf.

Assuming that Hf and In are both located on substitutional Ga sites the ratio between the two frequencies should be determined only by their atomic and nuclear properties. Taking into account the difference in the quadrupole moments and the different Sternheimer factors, which account for the different contribution of the electronic shell to the EFG, $((1 - \gamma_\infty) = 62$ for Hf, $(1 - \gamma_\infty) = 30$ for In), the ratio ν_{Q1}(Hf):ν_{Q1}(In) should have values around 6:1. Indeed in many materials a factor of six to ten is observed [10]. In GaN, in contrast, this ratio has values of 50:1. The lattice EFG at the Ga site can be calculated to be $V_{zz} = 0.96 \times 10^{16}$ V/cm^2 for Hf and $V_{zz} = 0.11 \times 10^{16}$ V/cm^2 for In. For Hf this agrees reasonably well with the value of $V_{zz} = 0.65 \times 10^{16}$ V/cm^2 derived for the Ga site from NMR measurements [11] wheras the value for In is six times lower.

This and the fact that the fraction of substitutional In remains significantly below the value for Hf is especially astonishing since the chemical properties of In are much closer to those of Ga than those of Hf. It is an additional indication of an unusual behaviour of implanted In in GaN as compared to Hf.

Acknowledgements This work was partially supported by BMBF (contract No. MA06.06K) and by the DFG (VI77/3-1). We gratefully acknowledge Global Lights Industries GmbH, Kamp-Lintfort, Germany for support.

References

[1] T. Butz and A. Lerf, Phys. Lett. **97A**, 217 (1983).
[2] C.M. Lederer and V.S. Shirley (Eds.), Table of Isotopes, John Wiley & Sons, New York 1978 (App. VII: Table of nuclear moments).
[3] G. Schatz, A. Weidinger, and J.A. Gardner, Nuclear Condensed Matter Physics, John Wiley & Sons, Chichester 1996.
[4] E. Alves, M. F. da Silva, J. G. Marques, J. C. Soares, and K. Freitag, Mater. Sci. Eng. B **59**, 207 (1999).

[5] K. LORENZ, R. VIANDEN, S. J. PEARTON, C. R. ABERNATHY, and J.M. ZAVADA, MRS Internet J. Nitride Semicond. Res. **5**, 5 (2000).
[6] J. BARTELS, K. FREITAG, J. G. MARQUES, J. C. SOARES, and R. VIANDEN, Hyperfine Interact. **120/121**, 397 (1999).
[7] C. RONNING, M. DALMER, M. DEICHER, M. RESTLE, M. D. BREMSER, R. F. DAVIS, and H. HOFSÄSS, Mater. Res. Soc. Symp. Proc. **468**, 407 (1997).
[8] A. BURCHARD, M. DEICHER, D. FORKEL-WIRTH, E. E. HALLER, R. MAGERLE, A. PROSPERO, and A. STÖTZLER, Mater. Sci. Forum **258–263**, 1099 (1997).
[9] C. LIU, A. WENZEL, K. VOLZ, and B. RAUSCHENBACH, Nucl. Instrum. Methods Phys. Res. B **148**, 396 (1999).
[10] R. VIANDEN, Hyperfine Interact. **35**, 1079 (1987).
[11] G. DENNINGER and D. REISER, Phys. Rev. **55**, 5073 (1997).

Magnetic Properties of Mn and Fe-Implanted p-GaN

N. Theodoropoulou (a), M. E. Overberg (b) S. N. G. Chu (c), A. F. Hebard (a), C. R. Abernathy (b), R. G. Wilson (d), J. M. Zavada (e), K. P. Lee (b), and S. J. Pearton[1]) (b)

(a) Department of Physics, University of Florida, Gainesville, FL 32611, USA

(b) Department of Materials Science and Engineering, University of Florida, Gainesville, FL 32611, USA

(c) Bell Laboratories, Lucent Technologies, Murray Hill, NJ 07974, USA

(d) Consultant, Stevenson Ranch, CA 91381, USA

(e) Army Research Office, Research Triangle Park, NC 27709, USA

(Received June 29, 2001; accepted July 3, 2001)

Subject classification: 61.72.Vv; 68.55.Ln; 75.30.Hx; 75.50.Pp; S7.14

The structural and magnetic properties of p-GaN implanted with high doses of Mn^+ or Fe^+ (0.1–5 at%) and subsequently annealed at 700–1000 °C were examined by transmission electron microscopy, selected-area diffraction patterns, X-ray diffraction and SQUID magnetometry. The implanted samples showed paramagnetic behavior on a large diamagnetic background signal for implantation doses below 3 at% Mn or Fe. At higher doses the samples showed signatures of ferromagnetism with Curie temperatures <250 K for Mn and <150 K for Fe implantation. The structural analysis of the Mn-implanted GaN showed regions consistent with the formation of $Ga_xMn_{1-x}N$ platelets occupying ~5% of the implanted volume. An estimate of ~$(5.5 \pm 1.9)\,\mu_B$ per Mn was obtained, consistent with the expected value (5.0) for a half-filled shell. The formation of secondary phases such as Mn_xGa_y or Mn_xN_y was excluded by careful diffraction analysis. The implantation process may have application in forming selected-area contact regions for spin-polarized carrier injection in device structures and in enabling a quick determination of the Curie temperatures in dilute magnetic semiconductor host materials.

Introduction Recent advances in the carrier-induced ferromagnetism in dilute magnetic compound semiconductors such as (In,Mn)As and (Ga,Mn)As has promoted interest in their potential application to new classes of devices based on spin-polarized transport or integration of magnetic, optical and electronic functions on a single chip [1–5]. The ability to control the magnetic properties through application of an electric field (i.e. field gating to manipulate the carrier density) to the material has recently been demonstrated by Ohno et al. [6]. The Curie temperatures T_C of (In,Mn)As [2] and (Ga,Mn)As [3, 7] are relatively low (~35 and 110 K, respectively) and from practical considerations it is desirable to find materials with higher values. Recent calculations based on a Zener model of ferromagnetism predict the possibility that wide bandgap systems such as (Ga,Mn)N and (Zn,Mn)O might have T_C values above room temperature [7].

Currently, little is known about the properties of GaN doped with impurities that might induce ferromagnetic behavior. Some initial reports have appeared on microcrystalline $Ga_{1-x}Mn_xN$ with Mn contents up to $x = 0.005$ which exhibited ferromagnetic

---

[1]) Corresponding author; Phone: 352/846-1086; Fax: 352/846-1182; e-mail: spear@mse.ufl.edu

behavior [8, 9] Akinaga et al. [10] reported ferromagnetic properties at <100 K in heavily Fe-doped GaN grown by low temperature (380 °C) molecular beam epitaxy. We have found apparent ferromagnetic behavior in Mn-implanted p-GaN at temperatures up to ~250 K for implanted Mn concentrations of 3–5 at% [11]. The implantation process is a simple approach for introducing magnetic ions into different host materials and could readily be used for making selected-area contact regions for injection of spin-polarized current into device structures.

In this paper, we report on the properties of p-GaN implanted with Fe or Mn at doses designed to produce concentrations of 3–5 at% at the peak of the implanted profile. Under these conditions, samples annealed at 700 °C do not show any evidence of secondary phase formation (at least to the sensitivity of transmission electron microscopy and selected area diffraction pattern analysis).

Experimental The p-GaN (3×10^{17} cm^{-3}) samples were grown by metal organic chemical vapor deposition on sapphire. The total epilayer thickness was 4 μm. Fe$^+$ or Mn$^+$ ions were implanted at an energy of 250 keV at dose of $\sim(3-5) \times 10^{16}$ cm^{-2} to produce average volume concentrations of $\sim(3-5)$ at% in the top ~2000 Å of the GaN. Amorphization of the implanted region was avoided by holding the samples at ~350 °C during the implantation [12]. Annealing was performed at 700 °C for 5 min under flowing N$_2$ gas with the samples face-down on another GaN wafer. The samples were examined by transmission electron microscopy (TEM) and selected area diffraction pattern (SADP) analysis, while the magnetic properties were measured in a Quantum Design MPMS SQUID magnetometer.

Results and Discussion Figure 1 shows the hysteresis curve at 10 K for the 5 at% Fe-implanted GaN annealed at 700 °C. From the difference in magnetization for field-cooled versus zero field-cooled samples, we could observe a ferromagnetic contribution present in the films below ~50 K (Fig. 2). The origin of the magnetic behavior in the Fe-implanted GaN is not yet clear, since the effective hole concentration is certainly

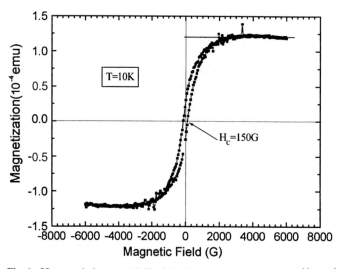

Fig. 1. Hysteresis loop at 10 K of GaN implanted with 5×10^{16} cm^{-2} Fe$^+$ and annealed at 700 °C

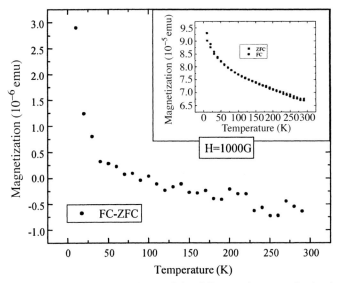

Fig. 2. Temperature dependence of the difference in magnetization between field-cooled ($B = 0.1$ T) and zero field-cooled conditions for the sample of Fig. 1. The inset shows the raw data

less than 2×10^{17} cm^{-3} present in the starting sample due to residual implant damage. Theory suggests much higher hole densities ($\geq 10^{20}$ cm^{-3}) are necessary for carrier-mediated ferromagnetism [7]. In the 3 at% Fe-implanted samples, the ferromagnetism was present at <150 K, which might be due to lower level of implant damage and higher high concentration.

The TEM cross-sectional view of Mn-implanted samples (3×10^{16} cm^{-2}) showed a buried band of dislocation loops and platelet structures. Similar results were obtained after 1000 °C anneals, with the platelet region being, on average, slightly larger than for the 700 °C annealed samples. While the end-of-range damage is a typically feature of implanted semiconductors, the platelet structures appear to be related to the chemical nature of Mn since when impurities such as Si or Au are implanted at similar doses platelets are not observed.

Figure 3 (left) shows SADP results from a GaN sample implanted with a Mn$^+$ ion dose of 5×10^{16} cm^{-2} and annealed at 1000 °C. The diffraction pattern from the GaN

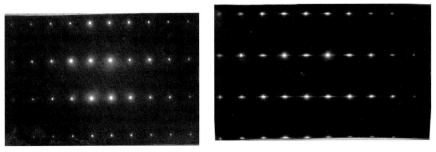

Fig. 3. SADPs from unimplanted (left) and implanted (right) regions of Mn-implanted (3×10^{16} cm^{-2}) GaN

away from the platelet region (right) shows clear spots in [2$\bar{1}$10] zone diffraction. The same analysis on a platelet region shows streaks in [2$\bar{1}$10] zone diffraction, but no extra spots due to different phases. The fact that the diffraction pattern shows only satellite spots around those of the GaN, with hexagonal symmetry, is an indication that the platelet structures are most likely $Ga_{1-x}Mn_xN$ with the same lattice structure as GaN but different lattice constant. The SADP patterns combined with energy dispersion X-ray (EDX) spectra identified these regions as $Ga_xMn_{1-x}N$ with a different (smaller) lattice constant than the host GaN. The presence of GaMnN corresponded to ferromagnetic behavior of the samples below ~250 K. The only secondary phase with hexagonal symmetry that could form in the Mn-implanted GaN would be Mn_3Ga, but this is ruled out on the grounds that it was not consistent with the EDX measurements or with the magnetization data since Mn_3Ga is ferromagnetic with a very high Curie temperature (>600 K).

Conclusions In summary, high dose implantation of Fe and Mn into p-GaN produced ferromagnetism. This behavior is consistent with previous results on epitaxial GaN(Fe). Future work should focus on the improvement in Curie temperature by increasing both Fe an Mn solubility and the hole concentration in the GaN.

Acknowledgements The work at University of Florida is partially supported by National Science Foundation, while that of RGW is partially supported by Army Research Office.

References

[1] G. A. Prinz, Science **282**, 1660 (1998).
[2] H. Ohno, Science **281**, 951 (1998).
[3] T. Dietl, A. Haury, and Y. Merle d'Aubigne, Phys. Rev. B **55**, R3347 (1997).
[4] D. D. Awschalom and R. K. Kawakami, Nature **408**, 923 (2000).
[5] B. T. Jonker, Y. D. Park, B. R. Bennet, H. D. Cheong, G. Kioseoglou, and A. Petrou, Phys. Rev. B **62**, 8120 (2000).
[6] H. Ohno, D. Chibu, F. Matsukura, T. Damiya, E. Abe, T. Dietl, Y. Ohno, and K. Ohtani, Nature **408**, 944 (2000).
[7] T. Dietl, H. Ohno, F. Matsukura, J. Cibert, and D. Ferrand, Science **287**, 1619 (2000).
[8] W. Gebicki, J. Strzeszewski, G. Kamler, T. Szyszko, and S. Podsliadlo, Appl. Phys. Lett. **76**, 3870 (2000).
[9] M. Zajac, R. Doradzinski, J. Gosk, J. Szczyko, M. Lefeld-Sosnowska, M. Kaminska, A. Twardowski, M. Palczewska, E. Grzanka, and W. Gebicki, Appl. Phys. Lett. **78**, 1276 (2001).
[10] H. Akinaga, S. Nemeth, J. de Boeck, L. Nistor, H. Bender, G. Borghs, H. Otuchi, and M. Oshima, Appl. Phys. Lett. **77**, 4377 (2000).
[11] N. Theodoropoulou, A. F. Hebard, M. E. Overberg, C. R. Abernathy, S. J. Pearton, S. N. G. Chu, and R. G. Wilson, Appl. Phys. Lett. **78**, 3475 (2001).
[12] S. O. Kucheyev, J. S. Williams, C. Jagadish, J. Zou, and G. Li, Phys. Rev. B **62**, 7510 (2000).

Outgoing Multiphonon Resonant Raman Scattering in Be- and C-Implanted GaN

W. H. Sun[1]), S. J. Chua, L. S. Wang, X. H. Zhang, and M. S. Hao

Institute of Materials Research and Engineering, 3 Research Link, Singapore 117602

(Received June 29, 2001; accepted August 4, 2001)

Subject classification: 61.72.Vv; 63.20.Ls; 63.20.Mt; 78.30.Fs; 78.55.Cr; S7.14

We have performed outgoing resonant Raman scattering and photoluminescence measurements on as-grown, Be- and C-implanted GaN in the temperature range of 77–330 K. In implanted GaN after postimplantation annealing at 1100 °C, LO multiphonons up to the seventh order were observed with the very strong 4LO and 5LO modes at $\sim$2955 and $\sim$3690 cm^{-1}, respectively, showing extraordinary resonance behavior. With increasing sample temperature, these two modes significantly decreased and increased in intensity, respectively. The phenomenon is attributed to the variation of resonant conditions due to the shift of the bandgap energy. Meanwhile, the combination of E_2(high) and quasi-LO phonons was strongly enhanced by quasi-LO phonon involving and thus the corresponding overtones can be clearly observed even up to the sixth order ($m = 6$).

Introduction Resonant Raman scattering by phonons is known for many semiconductors including III–V compounds with fundamental bandgap energy within or below the visible region. For these materials resonant Raman scattering has been shown to be an important technique to get into the basic physical properties of the materials. Some research work has been carried out on ion-implanted GaN by Raman scattering under off-resonance conditions [1, 2]. There is little work on resonant Raman scattering in ion-implanted GaN. In the present study, we have investigated the outgoing multiphonon resonant Raman scattering (OMRRS) and photoluminescence (PL) in Be and C-implanted GaN in the temperature range of 77–330 K. Temperature dependence of the OMRRS involving E_2(high) and LO phonons with m up to seven is discussed.

Experimental The MOCVD-grown GaN samples were implanted at room temperature by either a dose of 5×10^{14} cm^{-2} Be or C ions with an energy of 50 keV. After implantation, rapid thermal annealing (RTA) for 40 s at temperatures between 600 and 1100 °C was performed under an N$_2$ environment. Under optical excitation by the 325 nm line of a 30 mW He–Cd laser, OMRRS and micro-PL measurements were performed. All Raman spectra were recorded in backscattering configuration from the growth surface with light propagating parallel to the c-axis. The scattered light was not analyzed for its polarization.

Results Figure 1[2]) shows room temperature Raman spectra of Be-implanted GaN. Curves a, b, c and d show the spectra of an as-grown GaN sample, an as-implanted GaN sample, an implanted GaN sample annealed by RTA at 600 °C for 40 s and an implanted GaN sample annealed by RTA at 1100 °C for 40 s, respectively. For as-grown GaN, strong PL emission can be observed without obvious Raman bands. After

---

[1]) Corresponding author; Phone: (65)8748358, Fax: (65)8720785, e-mail: sun-wh@imre.org.sg
[2]) Colour figure is published online (www.physica-status-solidi.com).

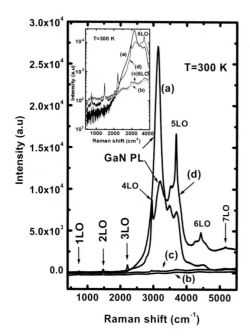

Fig. 1 (colour). Raman scattering spectra excited by 325 nm laser lines; (a) as-grown sample, (b) Be-implanted GaN sample without annealing, (c) Be-implanted GaN sample with postimplantation RTA at 600 °C, and (d) Be-implanted GaN sample with postimplantation RTA at 1100 °C. The first, second, third, fourth, fifth, sixth, seventh LO phonons are labeled as 1LO, 2LO, 3LO, 4LO, 5LO, 6LO and 7LO, respectively. The inset shows the multi-phonon resonant Raman scattering spectra in the frequency range of 500–4000 cm^{-1} with a logarithmic y-axis

Be-implantation, PL emission around 3000 cm^{-1} died out but two sharp peaks at 735.3 and 1471.7 cm^{-1} appeared in Fig. 1, curve b. After postimplantation annealing at 600 °C, five sharp peaks at around 736, 1466, 2208, 2946 and 3688 cm^{-1} appeared with very weak PL. It should be noted that these Raman peaks have almost a common energy interval of 90.5–92.5 meV, which may correspond to the energy of LO phonon in GaN [2–4]. Therefore, these modes most probably come from LO multiphonon resonant Raman scattering, being similar to the observation and assignment for A_1(LO) phonon by Behr at al. [3]. Furthermore, for the Be-implanted GaN samples after post-implantation RTA at 1100 °C, the multiphonon scattering peak pronouncedly increased in intensity with increasing order up to the fifth order, and both the fourth and fifth order modes in intensity were about two orders of magnitude greater than 1LO. As the excitation energy of 3.815 eV is above the band gap of GaN, the outgoing resonant condition is satisfied. Therefore, the fourth and fifth modes are resonantly enhanced due to their energy positions being very close to the band gap energy. On the other hand, as shown in Fig. 1, curve d, LO multiphonon modes up to the seventh order were observed, which indicate that implantation and subsequent RTA play an important role in the presence of LO multiphonon scattering. Although overtones of the LO phonon up to order $m = 4$–9 have been reported mostly for many II–VI compounds [5], the observation of LO multiphonon scattering up to the seventh order is unusual for III–V semiconductors.

In Figs. 2(I) and (II), the room temperature OMRRS spectra of the Be and C-implanted GaN with postimplantation RTA at 900 °C for 40 s are plotted, respectively. For the Be-implanted GaN, many features, such as 572.27, 740.03, 1309.58, 1475.8, 2041.56, 2214.4, ~2780, 2947.4, 3091.1, 3525.9, 3683.2, 4252.15, 4419.0, and ~5155 cm^{-1} modes, were clearly observed; similarly there existed prominent peaks at 566, 724.39, 1463.96, 1550.98, 2201.06, 2324.93, 2945.33, 3508.39, 3676.28, 4410.31 cm^{-1}, as indicated in Fig. 2(II) for the C-implanted GaN sample.

Examining the energy positions of those peaks, the phonon mode at 572 (566) cm^{-1} in Fig. 2(I,II) corresponds to E_2(high) in GaN, and 740.03(724.39), 1475.8(1463.96), 2214.4(2201.06), 2947.4(2945.3), 3682(3676.28), 4419.0(4410.31) and ~5155 cm^{-1} modes

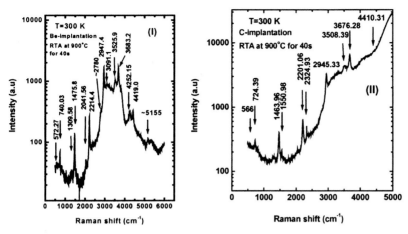

Fig. 2. OMRRS spectra of I) Be-implanted GaN and II) C-implanted GaN samples with postimplantation RTA at 900 °C

should, respectively, originate from the photon scattering by the first-, second-, third-, fourth-, fifth-, sixth- and seventh-order LO phonons. However, we find that there exist frequency differences for the same modes from the Be and C-implanted samples. Additionally, as shown in Fig. 2(I), we note that frequency of 1309.58 cm^{-1} is very close to that of combination tone of E_2 and A_1(LO) modes, and thus the 1309.58 cm^{-1} mode is tentatively assigned to the combination process of the two modes, being similar to that assignment for 1313 cm^{-1} Raman mode in GaN by Siegle et al. [4]. Interestingly, the neighbouring frequency spacing of 1309.58, 2041.40, ~2780, 3525.9 and 4252.15 cm^{-1} Raman modes is, respectively, 731.82, 738.6, 745.9 and 726.25 cm^{-1}, and is close, but not exactly equal to that of pure A_1(LO) mode. For those LO multiphonon resonant Raman modes in Fig. 1, the same phenomena also exist. As implantation generates lattice disorder and defects in GaN, even damages crystal surface to some extent, the incident light was not exactly parallel to the c-axis and the selection rule is broken. Most likely, the frequency of 738.6 and 745.9 cm^{-1} may correspond to that of a quasi-LO mode, which is a mixture of the A_1(LO) and E_1(LO) modes at zone center [6]. Therefore, the 2041.40, ~2780, 3525.9 and 4252.15 cm^{-1} modes probably arise from the third-, fourth-, fifth-, sixth-order OMRRS, respectively, involving E_2(high) and quasi-LO phonons in Be-implanted GaN.

In the OMRRS spectrum (shown in Fig. 2(II)) in our C-implanted GaN sample after postimplantation annealing at 900 °C for 40 s, similar Raman features were observed and can also be assigned.

Figure 3 shows the OMRRS spectra and PL in Be-implanted GaN after postimplantation RTA at 1100 °C with variable temperature in the range of 77–333 K. At temperature of 77 K, the intensity of the multiphonon scattering was strongly enhanced with the multiphonon order up to the fourth order and while rapidly decreased from the fifth to seventh orders. The intensity of the fourth-order scattering was approximately four orders of magnitude greater than that of 1LO mode. The position of PL peak is about 20 meV higher than the energy of the fourth-order LO scattering photon. Using the Varshni empirical equation (only considering the A valence band) [7], the interband transition energy at 77 K can be deduced, E_0(77 K) = 3.481 eV, 13 meV

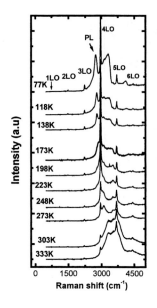

Fig. 3. OMRRS spectra of Be-implanted GaN samples with post-implantation RTA at 1100 °C in the sample temperature range of 77–333 K

larger than the PL position of 3.468 eV shown in Fig. 3. This shows that the 3.468 eV peak near band edge emission might arise from the recombination of excitons. With increasing sample temperature, the PL position shifts to lower energy, indicating the transition energy decreases due to the shift of the bandgap energy of GaN. Owing to the resonant effect, the closer the fourth-order phonon mode and the peak of PL are, the stronger this phonon mode should be. However, since the fourth-order phonon mode is superimposed the broad PL spectrum and the non-radiation recombination centers may play a crucial role, its intensity decreased as the sample temperature increased. When the sample temperature reached above 248 K, the fifth-order phonon mode exhibited strong enhancement in intensity due to the more approaching OMRRS condition. At above 273 K, the intensity of this mode exceeded that of the fourth-order mode.

Conclusions The OMRRS up to the seventh order in Be and C-implanted and subsequently annealed GaN by RTA at 1100 °C was observed. With the increase of sample temperature, the fourth and fifth multiphonon modes in intensity pronouncedly decreased and increased, respectively, due to the approaching resonance condition with detuning the band gap. We found that there were differences in the frequency spacing of the LO OMRRS modes, indicating that quasi-LO phonons involve in the OMRRS. On the other hand, the OMRRS modes up to the sixth order involving both E_2(high) and quasi-LO phonons were first observed and assigned.

References

[1] W. Limmer, W. Ritter, R. Sauer, B. Mensching, C. Liu, and B. Rauschenbach, Appl. Phys. Lett. **72**, 2589 (1998).
[2] W. H. Sun, S. T. Wang, J. C. Zhang, K. M. Chen, G. G. Qin, Y. Z. Tong, Z. J. Yang, G. Y. Zhang, Y. M. Pu, and Q. L. Zhang, J. Appl. Phys. **88**, 5662 (2000).
[3] D. Behr, R. Niebuhr, J. Wagner, K.-H. Bachem, and U. Kaufman, Appl. Phys. Lett. **70**, 363 (1997).
[4] H. Siegle, G. Kaczmarczyk, L. Filippidis, A. P. Litvinchuk, A. Hoffmann, and C. Thomsen, Phys. Rev. B **55**, 7000 (1997).
[5] Z. C. Feng, S. Perkowitz, J. M. Wrobel, and J. J. Dubowski, Phys. Rev. B **39**, 12997 (1989), and references therein.
[6] Guanghong Wei, Jian Zi, Kaiming Zhang, and Xide Xie, J. Appl. Phys. **82**, 4693 (1997).
[7] W. Shan, T. J. Schmidt, X. H. Yang, S. J. Hwuang, J. J. Song, and B. Goldenberg, Appl. Phys. Lett. **66**, 985 (1995).

Influence of Dopants on Defect Formation in GaN

Z. Liliental-Weber[1]) (a), J. Jasinski (a), M. Benamara[*]) (a), I. Grzegory (b), S. Porowski (b), D. J. H. Lampert (c), C. J. Eiting (c), and R. D. Dupuis (c)

(a) Lawrence Berkeley National Laboratory, Berkeley, m/s 62/203, CA 94720, USA

(b) High Pressure Research Center "Unipress", Polish Academy of Sciences, Warsaw, Poland

(c) Microelectronics Research Center, University of Texas at Austin, Austin TX 78712, USA

(Received July 16, 2001; accepted July 19, 2001)

Subject classification: 68.37.Lp; 68.55.Ln; S7.14

The influence of p-dopants (Mg and Be) on the structure of GaN has been studied using Transmission Electron Microscopy (TEM). Bulk GaN:Mg and GaN:Be crystals grown by a high pressure and high temperature process and GaN:Mg grown by metal-organic chemical-vapor deposition (MOCVD) have been studied. A structural dependence on growth polarity was observed in the bulk crystals. Spontaneous ordering in bulk GaN:Mg on c-plane (formation of Mg-rich planar defects with characteristics of inversion domains) was observed for growth in the N to Ga polar direction (N polarity). On the opposite side of the crystal (growth in the Ga to N polar direction) Mg-rich pyramidal defects empty inside (pinholes) were observed. Both these defects were also observed in MOCVD grown crystals. Pyramidal defects were also observed in the bulk GaN:Be crystals.

Introduction GaN can easily be grown with n-conductivity but obtaining p-doping is rather difficult. Only recently p-doping was also obtained by Be doping [1]. Mg is most commonly used as the p-dopant in GaN; however, higher hole concentrations can only be obtained after thermal annealing [2] in order to dissociate Mg–H complexes. Material made using this process has been used to fabricate light emitting diodes (LEDs) [3] and lasers [4]. Despite this success, many aspects of Mg-doping in GaN are still not fully understood. Transmission electron microscopy (TEM) studies show the formation of different types of Mg-rich defects in many bulk crystals. The type of defects formed depends strongly on the growth polarity. In the bulk GaN:Mg platelet crystals, planar defects distributed in equal distances (20 unit cells of GaN) leading to superlattice reflections in the diffraction pattern can be observed [5–7] for growth with N-polarity. Similar defects are also formed in layers grown by metal-organic chemical-vapor deposition (MOCVD) with Mg δ-doping [6, 7]. For majority of platelet crystals grown with Ga polarity, formation of pyramidal defects empty inside but rich in Mg on their walls was observed. These pyramidal defects are also typical for all MOCVD-grown crystals, and their high concentration can be observed in all crystals for a wide range of Mg concentration.

Bulk GaN:Be crystals are somehow similar to GaN:Mg. As-grown surface is rough for growth with N-polarity. However, spontaneous ordering was not observed. However, for growth with Ga polarity, pyramidal defects are also observed in some crystals.

---

[1]) Corresponding author; Phone 510-486-6276; Fax 510-486-4995; e-mail: z_liliental-weber@lbl.gov
[*]) Present address: Max Planck Institute, Halle, Germany.

Experimental Different types of GaN crystals doped by Mg and some with Be have been studied using plan-view and cross-sectioned samples. Bulk crystals were grown by the high nitrogen pressure method [8] from a solution of liquid gallium containing 0.1–0.5 at% Mg [9]. Similar procedure was applied to GaN:Be crystals. Two types of samples were grown using MOCVD where Mg was added either continuously during the growth or by the δ-doping method (one monolayer at each 100 Å of GaN). The details of this growth have been described earlier [6, 7]. Following the growth, the temperature was lowered to 850 °C, the ambient was switched to nitrogen only, and a 10 min in-situ anneal was performed to dissociate the Mg–H complexes and activate the Mg atoms [10]. The same growth temperature and post annealing was applied to the crystals for which Mg was added continuously during growth. All GaN:Mg and the GaN:Be crystals have been studied using TEM on plan-view and cross-section samples, transparent for electrons, prepared by standard methods using Topcon 002B and JEOL ARM with accelerating voltages of 200 and 800 kV, respectively.

Results A strong dependence of structure on crystal growth polarity was observed in bulk platelet crystals. For the crystal side grown with N polarity (N to Ga direction) many crystals were free of defects, but Mg-rich planar defects were observed in some crystals, despite of the fact that Mg concentrations determined by SIMS remain practically unchanged. These defects can be perfectly ordered for several micrometers from the sample surface. They can form a perfect array of planar defects separated from each other by 20 unit cells of GaN [5–7] (Fig. 1a). Perfect ordering leads to superlattice spots dividing the (0001) lattice distance into 20 equal parts (not shown here). The following TEM results, e.g. presence of split (0001) reflection, reverse contrast for multi-beam dark-field (0001) and (000$\bar{1}$) reflections, symmetrical contrast in dark field (top and bottom surface of TEM foil) indicate that these planar defects are inversion domains [5–7] formed on c-planes with a very narrow width along the c-axis. In addition, it was observed that pairs of inversion boundaries in one domain are separated by

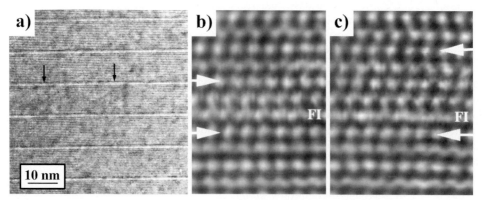

Fig. 1. a) TEM micrograph showing equi-distant arrangement of planar defects (inversion domain, ID) formed in bulk GaN crystal grown with N-polarity. The arrows indicate areas from which higher magnification micrographs are shown in b) and c), respectively. The white arrows in b) and c) indicate the position of perfect GaN matrix on both side of the planar defect. FI is a flat interface of the inversion boundary which remains on the same c-plane, but the upper interface, called corrugated interface (CI) is changing from place to place

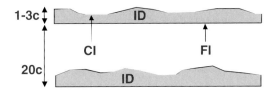

Fig. 2. Schematic showing the distribution of inversion domains (ID) in GaN grown with N-polarity. The thickness of particular domains is changing locally but the distance between equivalent domains remains $20c$ (unit cells)

not more than one to three unit cells. One of the domain boundaries called the flat interface (FI) shown in Figs. 1b and c always remains on the same plane. The upper boundary, called the corrugated interface, changes its position, so that the domain wall thickness is changing from $3/2c$ to $5/2c$ as shown by high resolution images taken in two areas from the same defects (Figs. 1b and c). The distribution of these inversion domains is shown graphically on Fig. 2. The local change of thickness within one domain gives additional evidence that the observed planar defects are no stacking faults. EDX studies gave evidence of Mg enhancement within the domains [5]. This perfect ordering was not observed in all crystals studied. In some crystals this perfect ordering was interrupted by larger areas of hexagonal GaN or this ordering can be interrupted by stacking faults.

The opposite side of the bulk crystals (grown with Ga-polarity) had a completely different defect structure as shown in Fig. 3a. They appear in $[11\bar{2}0]$ cross-section TEM

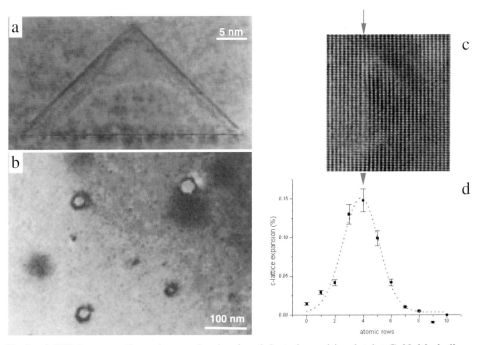

Fig. 3. a) TEM cross-section micrographs showing defects formed in platelet GaN:Mg bulk crystals on the side grown with Ga polarity observed in cross-section, b) the same defects observed in plan-view configuration, c) high resolution micrograph observed in MOCVD grown crystals showing expansion of the interplanar distances in the defect base, d) measured lattice expansion in the defect base

micrographs as triangular features with a base on the (0001) c-plane and six side facets inclined 45° to the basal plane. The dimension of the largest defects varies. The longest base length of these defects is in the range of 100 nm and the smallest of about 3–5 nm. The density of these defects is in the range of 2.5×10^9 cm^{-2}. All these triangles were oriented in a direction with the base closest and parallel to the sample surface with Ga-polarity, e.g. from the triangle tip to the base, the long bond direction along the c-axis is from Ga to N. Study of these defects in plan-view configuration proves that these defects are empty inside (Fig. 3b). EDX studies showed that Mg is segregated on the base and side-walls of these defects. High resolution TEM images from the defect base show the formation of planar defect.

GaN:Mg layers grown by MOCVD Since the majority of layers applied in devices are grown by MOCVD it was also interesting to observe the structure of p-doped layers to learn why p-doping is so difficult [2]. Two types of crystals have been studied: one where Mg was added as delta doping, to study if addition of a monolayer of Mg at equal distances will lead to the ordering observed in bulk crystals and the second where Mg was added continuously, as used for growth in devices.

Cross-section TEM studies of Mg δ-doped samples show both types of defects observed for the two opposite growth polarities in bulk GaN:Mg, e.g. planar defects observed earlier for N growth polarity and pyramidal defects observed in the bulk GaN:Mg for growth with Ga polarity [5–7]. In a crystal with Mg δ-doping, planar defects were formed followed by a high density of pyramidal defects. In this crystal a layer about 100 nm thick with planar defects (Fig. 4a), like those observed in bulk crystals grown with N-polarity, was observed at about 150 nm distance from the buffer layer. This suggest that some accumulation of Mg on the surface is needed for the first Mg-rich planar defect to appear. Further growth resulted in a random distribution of pyramidal defects with a density of ~10^{10} cm^{-2}. High-resolution TEM images from the defect base show an expansion of the interplanar distance (Fig. 3c): 14% expansion was

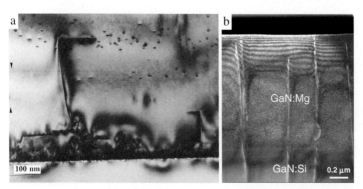

Fig. 4. a) Cross-section TEM micrograph showing a layer with δ-doping grown with GaN buffer layer on sapphire. The area between the arrows shows planar defects similar to those observed in bulk platelet crystals grown with N-polarity. These planar defects delay to appear compared to the expected location of the first monolayers of Mg. Above the top arrow, Mg monolayers were not found, but pyramidal defects (same as in bulk crystals grown with Ga polarity) in high density were observed. Some of these pyramidal defects started to appear simultaneously with the planar defects. b) A layer of GaN:Mg with continuous Mg doping. Note the high density of pyramidal defects (white dots)

Table 1
Calculated expansion of interplanar distances in the pyramidal defect base for different materials

		GaN	
		$c/2 = 2.585$ Å	$a = 3.18$ Å
MgO	$a/2 = 2.585$ Å	18%	
MgO$_2$	$a/2 = 2.422$ Å	6%	
Mg	$a = 2.95$ Å	14%	7%
Mg$_3$N$_2$		4%	

measured (Fig. 3d). This value was compared to the expansion expected if Mg, MgO, MgO$_2$, or predicted theoretically Mg$_3$N$_2$ phase would be formed on the defect base. The best agreement was obtained for Mg accumulating on the c-planes in the base of these defects (Table 1).

SIMS analysis in the layers with Mg δ-doping indicated that the Mg concentration was steadily increasing and it reached a level of 4×10^{19} cm^{-3} in the area with planar defects [7]. In the area where the hollow defects (rectangular and triangular) were formed Mg concentration dropped to 2×10^{19} cm^{-3} and stayed almost constant with a slow increase to reach a concentration of 4×10^{19} cm^{-3} at the sample surface [7].

In the samples where Mg was added continuously (with comparable Mg concentration to that in the previous samples), no planar defects were observed but a high density of pyramidal defects was present (Fig. 4b). These pyramids are also empty inside but their size was much smaller compared to those observed in bulk platelets

Table 2
SIMS analysis of bulk platelets for growth with N and Ga-polarity where either ordering (ord.) or pyramidal defects (def.) were observed and defect free samples (no ord. and no def.) compared to MOCVD samples grown with Mg delta (δ) doping and continuous (c) Mg doping. Concentration for particular elements (in cm^{-3})

sample	Mg	O	C	Si
A N-ord. Ga-def.	$2 \times 10^{20} - 6 \times 10^{19}$ 6×10^{19}	2.5×10^{19}	1.5×10^{17}	4×10^{16}
B N-par. ord. Ga-no def.	5×10^{19}	3×10^{19}	5×10^{19}	$1 \times 10^{16} - 1 \times 10^{17}$
C N-no ord. Ga-def.	$(4-7) \times 10^{19}$	2×10^{19}	3×10^{17}	$(1-3) \times 10^{17}$
D N-no ord. Ga-no def.	$(4-6) \times 10^{19}$	7×10^{19}	1.5×10^{18}	$(1-2) \times 10^{19}$
MOCVD-δ ord. + def.	$(1.5-3) \times 10^{19}$	5×10^{16}	5×10^{16}	–
MOCVD-c def.	3×10^{19}	5×10^{16}	5×10^{16}	

(8 nm in the base). At the base of these defects stacking faults were not formed as in the larger pyramids observed in the bulk samples. However, an expansion of the lattice parameter of about 14% was observed in the two basal parallel layers (Figs. 3c and d). The expansion agrees well with what would be expected if Ga atoms were substituted by Mg.

Discussion of Mg results TEM studies of bulk crystals grown under high pressure and high temperature doped with Mg and those grown by MOCVD show the formation of structural defects for a wide range of Mg concentration (Table 2). This would explain (at least to some extent) why it is so difficult to obtain high hole concentration despite nominal higher Mg concentration in the crystal. Since p-conductivity was not obtained in bulk crystals, but GaN:Mg samples became semi-insulating, some compensation of oxygen presence in these samples took place. Ordering of planar defects observed in GaN:Mg formed for growth with N-polarity appears to be similar to the polytypoids formed in AlN rich in oxygen [11]. But it appears that similar defects were formed in MOCVD grown samples, where the oxygen concentration was three orders of magnitude lower. Therefore, it is believed that mainly Mg is responsible for the formation of these defects. Our studies of bulk samples from two opposite sides, e.g. grown with N and Ga-polarity, do not indicate substantial difference in Mg incorporation for different growth polarities as predicted theoretically [12] and observed in MBE grown samples [13], where up to 30 times more Mg was incorporated in Ga-polarity, as determined by secondary ion mass spectrometry. The last authors [13] also indicated accumulation of Mg on the sample surface which would be consistent with our results.

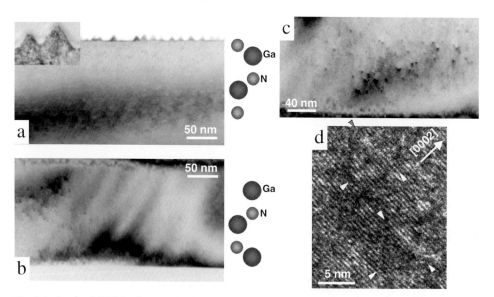

Fig. 5 (colour). a) TEM micrograph of GaN:Be platelet with corrugated surface. b) The same sample from the opposite side showing flat surface. c) Part of another platelet crystal grown with Ga polarity with pyramidal defects (small dots). d) High resolution image of two pyramids the corners of which were indicated by arrows

GaN:Be bulk samples TEM studies of Be-doped GaN platelet crystals show many similarities to the samples grown by the same method with Mg. However, ordering was not observed in these samples. For growth with N-polarity as-grown crystals appear to have rough surface (Fig. 5a[2])) and the growth with Ga polarity results in smooth surface (Fig. 5b). Only three crystals have been studied. Two of them did not show any structural defects, but pyramidal defects have been observed in one of them Figs. 5c and d. The size of these pyramids appear much smaller in comparison to those observed in Mg-doped platelet samples. The length of the base of these defects is in the range of 5–10 nm, similarly to the sizes observed in GaN:Mg grown by MOCVD. The density of these defects appears also lower in comparison to Mg-doped crystals. These samples are also semi-insulating.

Summary TEM studies of Mg-doped GaN crystals show the formation of structural defects for a wide range of Mg concentration strongly influenced by the crystal growth polarity. Similar defects are formed for bulk crystals grown under high hydrostatic pressure and for those grown by MOCVD. In some bulk crystals for the plate side grown with N-polarity equally distributed planar defects were observed (20 unit cells). Each defect appears to be an inversion domain with domain thickness of about one to three unit cells along the *c*-axis. No inversion in the GaN between these planar defects was observed.

Pyramidal defects were observed for growth with Ga polarity, also due to Mg segregation. These defects were observed in the majority of bulk GaN samples and in all samples grown by MOCVD. When these pyramids are small, a 14% lattice expansion can be detected in the bases of the pyramids consistent with the substitution of Ga by Mg atoms.

The samples grown with Be doping show similarities with Mg-doped samples in respect to their surface roughness and formation of pyramidal defects in some crystals. Planar defects were not observed in these crystals. Semi-insulating properties were measured in the crystals doped with Mg and Be.

Acknowledgement This work was supported by the U.S. Department of Energy, under the Contract No. DE-AC03-76SF00098.

References

[1] T. H. MYERS, A. J. PTAK, L. WANG, and N. C. GILES, Proc. Internat. Workshop Nitride Semiconductors, Nagoya (Japan), 24–27 Sept. 2000; IPAP, Tokyo 2000.
[2] H. AMANO, M. KITO, K. HIRAMATSU, and I. AKASAKI, Inst. Phys. Conf. Ser. **106**, 725 (1989).
[3] S. NAKAMURA, M. SENOH, and T. MUKAI, Jpn. J. Appl. Phys. **31**, L1708 (1991).
[4] S. NAKAMURA, Proc. 24th Internat. Symp. Compound Semiconductors, San Diego, CA, September 8–11, 1997.
[5] Z. LILIENTAL-WEBER, M. BENAMARA, J. WASHBURN, I. GRZEGORY, and S. POROWSKI, Phys. Rev. Lett. **83**, 2370 (1999).
[6] Z. LILIENTAL-WEBER, M. BENAMARA, W. SWIDER, J. WASHBURN, I. GRZEGORY, S. POROWSKI, D. J. H. LAMBERT, C. J. EITING, and R. D. DUPUIS, Appl. Phys. Lett. **75**, 4159 (1999).
[7] Z. LILIENTAL-WEBER, M. BENAMARA, W. SWIDER, J. WASHBURN, I. GRZEGORY, S. POROWSKI, D. J. H. LAMBERT, C. J. EITING, and R. D. DUPUIS, Physica B **273–274**, 124 (1999).

---

[2]) Colour figure is published online (www.physica-status-solidi.com).

[8] I. Grzegory, J. Jun, M. Bockowski, S. Krukowski, M. Wroblewski, B. Lucznik, and S. Porowski, J. Phys. Chem. Solids **56**, 639 (1995).
[9] S. Porowski, M. Bockowski, B. Lucznik, I. Grzegory, M. Wroblewski, H. Teisseyre, M. Leszczynski, E. Litwin-Staszewska, T .Suski, P. Trautman, K. Pakula, and J. M. Baranowski, Acta Phys. Polon. A **92**, 958 (1997).
[10] C. J. Eiting, P. A. Grudowski, J. S. Park, D. J. H. Lambert, B. S. Shelton, and R. D. Dupuis, J. Electrochem. Soc. **144**, L219 (1997).
[11] R. A. Youngman, A. D. Westwood, and M. R. McCartney, Mater. Res. Soc. Symp. Proc. **319**, 45 (1994).
[12] C. Bungaro, K. Rapcewicz, and J. Bernholc, Phys. Rev. B **59**, 9771 (1999).
[13] A. J. Ptak, T. H. Myers, L. T. Romano, C. G. Van de Walle, and J. E. Northrup, Appl. Phys. Lett. **78**, 285 (2001).

Observation of Mg-Rich Precipitates in the p-Type Doping of GaN-Based Laser Diodes

M. Hansen[1]) (a), L. F. Chen (a), J. S. Speck, and S. P. DenBaars (a)

(a) *Materials and Electrical and Computer Engineering Departments, University of California, Santa Barbara, CA, 93106-5050, USA*

(Received June 21, 2001; accepted June 28, 2001)

Subject classification: 42.55.Px; 68.37.Lp; S7.14

Uniformly distributed precipitates have been observed by TEM in the p-type layers of laser structures. The precipitate density decreases with decreasing flow of biscyclopentadienyl-magnesium (Cp$_2$Mg), which affects the hole concentrations in the p-type layers. The higher hole concentration, with the reduced precipitate density, reduces the threshold current density and improves the internal quantum efficiency because of the higher number of holes available for radiative recombination. The threshold current density is also reduced 30% from 20.8 V for lasers with a high precipitate density compared to 14.3 V for lasers with a lower precipitate density.

The successful realization of p-type doping in GaN with Mg has enabled rapid progress in the growth of high quality optoelectronic devices, such as laser diodes [1, 2]. However, the deep nature of the Mg acceptor has created stumbling blocks to obtain low resistivity p-type films, and contributes to high operating voltages of laser diodes. The high resistivity of the Mg-doped layers is responsible for a large fraction of the operating voltage, which is a key obstacle to obtain CW laser diodes.

p-type films are very heavily Mg-doped in order to provide a high hole concentration. The mobility decreases rapidly with increasing Mg concentration in GaN:Mg due to compensation effects, resulting in a nearly constant $q\,\mu\,p$ product of ~1 Ω cm for hole concentrations in the range of 10^{17}–10^{18} cm^{-3} [3]. When TEM was performed on several laser samples, uniformly distributed precipitates were revealed throughout the Mg-doped layers as shown in Fig. 1. These precipitates were approximately 5 nm in size, and presumed to be Mg$_3$N$_2$, which was previously reported [4–6]. The presence of the precipitates indicates that the p-type layers in the laser structure are doped too heavily.

The observation of precipitates indicates that the Mg solubility limit has been reached and the structure was over-doped. In the light of these results, the Mg-doping level of the laser was investigated. The details of the laser structures are described elsewhere [7]. The doping level was varied by adjusting the biscyclopentadienyl-magnesium (Cp$_2$Mg) flow, the Mg precursor, while holding the TMGa flow, and hence the growth rate, constant. The laser structures contain a 0.1 µm GaN:Mg contact layer to provide ohmic contacts because the low doping levels inherent with p-AlGaN will cause very resistive contacts. The doping in the contact layer is ramped higher throughout the layer. The GaN:Mg SCH layer, AlGaN/GaN:Mg SLS cladding, and the GaN:Mg contact layer are grown at 2 µm/h at 1010 °C with a TMGa flow of 19.2 µmol/min and NH$_3$

---

[1]) Corresponding author; Phone: (805) 893-8869; Fax: (805) 893-5263; e-mail: monica@engineering.ucsb.edu

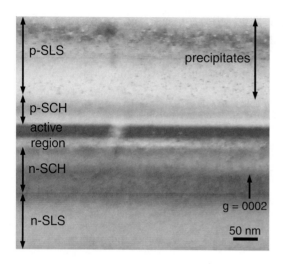

Fig. 1. Cross-sectional TEM micrograph of the laser structure showing precipitates in the p-type layers for a standard Cp_2Mg flow

flow of 26.7 mmol/min. The standard Cp_2Mg flow was 207 nmol/min, and was changed from 20 to 414 nmol/min in the p-contact layer. The reduced Mg-doping structure had a Cp_2Mg flow of 171 nmol/min, and was changed from 171 to 342 nmol/min in the p-contact layer. The growth conditions are summarized in Table 1.

The density of precipitates is reduced as the flow of Cp_2Mg is reduced, which is seen by comparing Fig. 2 with Fig. 1. Hall effect measurements were done on the laser structure to measure the mobility and hole concentration of the p-type layers (p-SCH, p-cladding and p-contact layer). The standard Cp_2Mg flow yields a mobility of 13.2 cm^2/Vs and a hole concentration of 2×10^{17} cm^{-3} in the laser structure. The reduced Cp_2Mg flow gives a mobility of 11.2 cm^2/Vs and a hole concentration of 4×10^{17} cm^{-3}.

The solubility of Mg in GaN is a factor limiting higher hole concentrations [4, 5]. The solubility is limited by the competition between incorporation of Mg on a Ga site as an acceptor and Mg forming Mg_3N_2 precipitates. As the Mg concentration in the film becomes too high, the hole concentration in the film begins to drop. The maximum hole concentration typically occurs for Mg concentrations of about mid 10^{19} to 10^{20} cm^{-3}. The reduced acceptor energy indicates that much of the Mg is not incorporating in the desired substitutional site [8]. It is likely that the decrease in the hole concentration results from the Mg forming precipitates instead of incorporating substitutionally on a Ga site. The lower density of precipitates explains the higher hole concentration for the reduced Mg doping level.

Up to this point, precipitates had not been previously observed in our bulk GaN:Mg or AlGaN:Mg films grown with a wide range of doping levels. The origin of the precipitates was investigated since the precipitates in the p-type layers had only been observed by TEM in laser structures, and not in other structures regardless of doping level. GaN:Mg films were grown on both InGaN:Si and AlGaN:Si layers with the same growth conditions and doping levels used in laser structures. The Mg-doped layer, on either InGaN or AlGaN alone, did not show precipitates. Precipitates only appear

Table 1
Growth conditions for the Mg doping level study

sample	TMGa flow (μmol/min)	Cp_2Mg flow (nmol/min)	Cp_2Mg flow in contact layer (nmol/min)
990901LA	19.2	207	207 → 414
990830LA	19.2	171	171 → 342

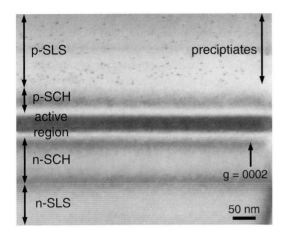

Fig. 2. Cross-sectional TEM micrograph of the laser structure showing reduced precipitate density in the p-type layers for a reduced Cp$_2$Mg flow

when both InGaN and AlGaN is grown prior to the Mg-doped layers, as in the laser structures with the n-type AlGaN/GaN cladding layer and the InGaN MQW active region. It is a combination of the growth conditions for the AlGaN and InGaN layers and the reactor environment they create that leads to precipitate formation.

There is an improvement of the threshold characteristics of the laser diode for reduced Mg doping levels. For a $3 \times 400\ \mu m^2$ device, the threshold current is reduced from 363 to 221 mA, and the threshold voltage is reduced from 20.8 to 14.3 V when the Cp$_2$Mg flow is reduced (Fig. 3). The improvement of the threshold current of the reduced Mg laser can be attributed to the increase in the internal efficiency. The internal efficiency is 13.9% for the reduced Cp$_2$Mg flow and 7.6% for the standard Cp$_2$Mg flow. The hole concentration for the reduced Cp$_2$Mg flow is a factor of two higher than the standard Cp$_2$Mg flow. The increased internal efficiency is likely due to higher hole concentration contributing more holes for radiative recombination. The lower threshold voltage is a result of the lower resistance of the diode for the reduced Mg doping level. The dc current–voltage (I–V) characteristics of the diodes were measured, and at 20 mA, the resistance was 24.5 Ω for the reduced Mg flow and 50 Ω for the standard flow, which corresponds to a 0.5 V difference. At threshold current levels, this corresponds to at least a 5 V increase. The increased diode resistance leads to higher threshold voltages in the lasers with the higher density of precipitates.

Precipitates have been observed by TEM in the p-type layers for laser structures. These precipitates, approximately 5 nm in size, have only been observed in laser structures. The origin of the precipitates was investigated and found to be caused by the reactor conditions established with growth of AlGaN and InGaN before the p-type layers, as is the case in a laser structure. The precipitate density decreases with decreasing Cp$_2$Mg flow. The density of precipitates affects the

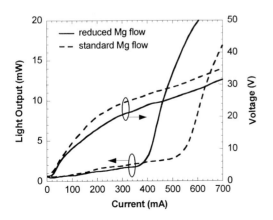

Fig. 3. L–I–V characteristics for laser structures with the standard and reduced Mg-doping level

operating characteristics of laser diodes because of the impact it has on hole concentration. The higher hole concentration with the reduced precipitate density reduces the threshold current density and improves the internal quantum efficiency because of the higher number of holes available for radiative recombination. The threshold current density is also reduced 30% from 20.8 V for lasers with a high precipitate density compared to 14.3 V for lasers with a lower precipitate density. This is likely due to the higher resistance for the lasers with a higher precipitate density.

References

[1] H. AMANO, M. KITO, K. HIRAMATSU, and I. AKASAKI, Jpn. J. Appl. Phys. **28**, L2112 (1989).
[2] S. NAKAMURA, M. SENOH, and T. MUKAI, Jpn. J. Appl. Phys. **30**, L1708 (1991).
[3] P. KOZODOY, Ph.D. Thesis, University of California at Santa Barbara, 1999.
[4] J. NEUGEBAUER and C. G. VAN DE WALLE, Mater. Res. Soc. Symp. Proc. **395**, 645 (1996).
[5] C. G. VAN DE WALLE and J. NEUGEBAUER, Mater. Res. Soc. Symp. Proc. **449**, 861 (1997).
[6] D. P. BOUR, H. F. CHUNG, W. GOTZ, L. ROMANO, B. S. KRUSOR, D. HOFSETTER, S. RUDAZ, C. P. KUO, F. A. PONCE, N. M. JOHNSON, M. G. CRAFORD, and R. D. BRINGANS, Mater. Res. Soc. Symp. Proc. **449**, 509 (1997).
[7] M. HANSEN, P. FINI, L. ZHAO, A. C. ABARE, L. A. COLDREN, J. S. SPECK, and S. P. DENBAARS, Appl. Phys. Lett. **76**, 529 (2000).
[8] J. H. EDGAR, S. STRITE, I. AKASAKI, H. AMANO, and C. WETZEL, Gallium Nitride and Related Semiconductors, EMIS Datarev. Ser. No. 23, INSPEC, IEE, London 1999.

Activation of p-Type GaN with Irradiation of the Second Harmonics of a Q-Switched Nd:YAG Laser

Yung-Chen Cheng (a), Chi-Chih Liao (a), Shih-Wei Feng (a), Chih-Chung Yang[1]) (a), Yen-Sheng Lin (b), Kung-Jeng Ma (b), and Jen-Inn Chyi (c)

(a) Department of Electrical Engineering and Graduate Institute of Electro-Optical Engineering, National Taiwan University, 1, Roosevelt Road, Sec. 4, Taipei, Taiwan

(b) Department of Mechanical Engineering, Chung Cheng Institute of Technology, Tahsi, Taoyuan, Taiwan

(c) Department of Electrical Engineering, National Central University, Chung-Li, Taiwan

(Received June 23, 2001; accepted August 1, 2001)

Subject classification: 61.80.Ba; 71.55.Eq; 78.55.Cr; S7.14

Efficient activation of Mg acceptors for obtaining a high hole concentration is a challenging topic. In this paper, we report the results of Mg acceptor activation in GaN with irradiation of the second-harmonic photons (532 nm in wavelength) of a Q-switched Nd:YAG laser. This laser was used to irradiate two Mg-doped GaN samples of different doping concentrations. With doping concentration of 1.2×10^{18} cm^{-3}, a hole concentration of 2.66×10^{17} cm^{-3} was obtained after laser-induced activation. The average temperature of samples during laser irradiation was around 30 °C. Hence, it was speculated that the irradiation process was very unlikely to be thermal annealing. PL measurements revealed that laser-induced activation could ionize not only the shallow but also the deep donors.

1. Introduction The second harmonic photons (532 nm) of a Q-switched Nd:YAG laser were used to irradiate two Mg-doped GaN samples. The MOCVD grown samples consisted of Mg-doped GaN of 1.6 µm for sample A and 0.615 µm for sample B after the growth of 0.2 µm (sample A) and 1.6 µm (sample B) nucleation layers on sapphire substrates. The growth temperature was 1020 °C. The concentrations of Mg doping were 6.3×10^{17} cm^{-3} (sample A) and 1.2×10^{18} cm^{-3} (sample B). To obtain uniform laser fluence, the laser output of 1 mm in beam radius was first expanded into a radius size of 2 cm through a defocusing lens. The central portion of the enlarged laser beam, which was believed to be quite uniform in fluence, was used for laser irradiation onto samples of 5×5 mm^2 in dimension. After laser irradiation, metal contacts were prepared, followed by thermal annealing at 750 °C for 20 s in ambient nitrogen. It was confirmed that this short-period thermal annealing process could not activate the GaN samples.

2. Experimental Results For comparison, we also conducted thermal activation with the same sample wafers. Thermal activation experiments were conducted with thermal annealing at various temperatures for 20 min (for sample A) or 30 min (for sample B). Figure 1 shows the variations of hole concentrations for samples A and B. One can see that hole concentrations of about 6×10^{16} and 2×10^{17} cm^{-3} for samples A and B,

---

[1]) Corresponding author; Phone: 886-2-23657624; Fax: 886-2-23652637; e-mail: ccy@cc.ee.ntu.edu.tw

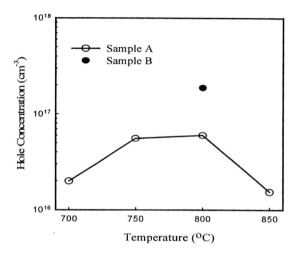

Fig. 1. Hole concentration vs. thermally activated temperature for samples A (empty circles) and B (filled circle)

respectively, were achieved with annealing at 800 °C. Because of the limited size of sample B, only one data point is available. Both of them are within reasonable ranges based on previously obtained data Refs. [1–3].

The hole concentration results of laser-induced activation with various laser fluence levels are shown in Fig. 2. Samples A and B were irradiated with 50000 and 60000 laser pulses, respectively. For sample A, the hole concentration ranges from 4.6×10^{15} to 1.25×10^{16} cm^{-3} when laser fluence increases from 0.094 to 0.3 mJ/cm^2. Also, for sample B the hole concentration ranges from 1.5×10^{16} to 9.45×10^{16} cm^{-3} when laser fluence increases from 0.154 to 0.46 mJ/cm^2. The hole concentration results of laser-induced activation with fixed laser fluence but different laser pulse numbers are shown in Fig. 3. The laser fluence levels were fixed at 0.25 mJ/cm^2 for sample A and at 0.23 mJ/cm^2 for sample B. For sample A, the case of 60000 laser pulses generated the highest hole concentration (4.56×10^{16} cm^{-3}), which is comparable to the level of thermally activated samples. For sample B, the case of 50000 laser pulses resulted in the highest hole concentration (2.66×10^{17} cm^{-3}).

3. Discussion of Activation Mechanisms

As mentioned previously, the thermal process is believed to induce the dissociation of Mg–H bonds in MOCVD-grown samples. In other words, the thermal energy ionizes shallow H donors and activates Mg acceptors. The shallow H donors and Mg acceptors correspond to the fluorescence peak near 3.2 eV [4, 5]. The deep donors formed by a complex of H and nitrogen vacancies lead to the fluorescence peak near 2.8 eV. This PL feature has been observed in the thermally activated samples. To identify the mechanisms of laser-induced activation, we conducted PL measurements of sample A under different conditions, and the results are shown in Fig. 4. Here, the

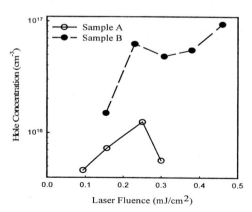

Fig. 2. Hole concentration vs. laser fluence in laser-induced activation for samples A (empty circles) and B (filled circles)

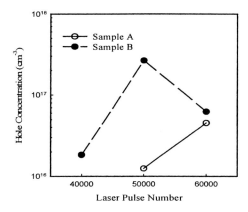

Fig. 3. Hole concentration vs. laser pulse number in laser-induced activation for samples A (empty circles) and B (filled circles)

three curves correspond to PL spectra of an as-grown sample (continuous line), a laser-activated sample (dashed line), and a thermally activated sample (dash-dotted line). The laser-activated sample was prepared under the conditions of 60000 pulses at 0.25 mJ/cm^2 fluence. The thermally activated sample was prepared by annealing at 800 °C for 20 min. The hole concentration of the former is only slightly lower than that of the latter. Although the peaks around both 2.8 and 3.2 eV are not prominent, the redshift trend of these spectra is consistent with findings previously reported [4, 5]. The as-grown sample has a peak near 3.2 eV, indicating the existence of Mg–H complexes and their dominance over the deactivation mechanism. The PL peak of the thermally activated sample shifts to around 2.9 eV, revealing that the Mg–H complexes have essentially been eliminated and the dominating deactivation mechanism is the effect of nitrogen vacancies. In the case of laser-induced activation, the PL spectrum lies in between. Such a spectral distribution may indicate that the used photons (with an energy of 2.33 eV) may partially ionize not only the shallow but also the deep donors.

To understand whether the laser-induced process was thermal, we measured the temperature of the sample during laser irradiation. The measured temperature was around 30 °C, only a few degrees above room temperature. This temperature was an average value calculated from repeated pulse irradiation. A question may be raised about the possibility that high temperatures existed for a short duration (in the microsecond range) after the irradiation of one pulse. If this is true, the continuous high temperature duration would add up to with at most several seconds in the case of 40000 pulses laser irradiation (assuming 100 μs duration high temperature for one laser pulse irradiation). Even if the transient temperature was as high as 800 °C, there was no evidence of thermal activation for such a short time period. It should be noted that the laser photon energy expended was 2.33 eV, which is lower than the absorption band of the samples, except possibly that caused by defects. Also, it is almost impossible for such photons to in-

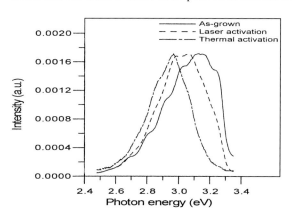

Fig. 4. PL spectra of an as-grown sample (continuous line), a laser-activated sample (dashed line), and a thermally activated sample (dash-dotted line)

duce efficient phonon generation. Hence, the process involved in laser-induced activation is very unlikely to be thermal. Note that the required energy for ionizing the shallow donors, i.e., breaking the Mg–H bonds, is only around 1.5 eV [6]. Increasing this value by 0.4 eV, the incident photons can possibly ionize the deep donors. Therefore, it is reasonable to speculate that photon-induced ionization of shallow and possibly deep donors is the mechanism of the laser-induced activation process. In other words, the relevant bonds are directly broken by the incident photons. However, such speculation requires further investigation.

4. Conclusions In summary, we have successfully activated Mg acceptors in p-type GaN samples with laser irradiation at a photon energy lower than the major absorption band. With appropriate laser fluence levels and exposure periods, the hole concentration reached the level obtained with thermal activation. Temperature measurements indicated that the laser-induced process was non-thermal. The physical mechanism of this process should be a topic for further investigation.

Acknowledgement This research was sponsored by Chung Shan Institute of Science and Technology, Taiwan, R.O.C.

References

[1] S. NAKAMURA, N. IWASA, M. SENOH, and T. MUKAI, Jpn. J. Appl. Phys. **31**, 1258 (1992).
[2] S. NAKAMURA, T. MUKAI, M. SENOH, and N. IWASA, Jpn. J. Appl. Phys. **31**, L139 (1992).
[3] S. NAKAMURA and G. FASOL, The Blue Laser Diode, Springer-Verlag, Berlin 1997.
[4] U. KAUFMANN, M. KUNZER, M. MAIER, H. OBLOH, A. RAMAKRISHNAN, B. SANTIC, and P. SCHLOTT, Appl. Phys. Lett. **72**, 1326 (1998).
[5] F. SHAHEDIPOUR and B. W. WESSELS, Appl. Phys. Lett. **76**, 3011 (2000).
[6] J. NEUGEBAUER and C. G. VAN DE WALLE, Phys. Rev. Lett. **75**, 4452 (1995).

Time- and Temperature-Resolved Photoluminescence of GaN:Mg Epitaxial Layers Grown by MOVPE

A. L. Gurskii (a), I. P. Marko (a), E. V. Lutsenko (a), V. N. Pavlovskii (a), V. Z. Zubialevich (a), G. P. Yablonskii (a), B. Schineller[1]) (b), O. Schön (b), and M. Heuken (b)

(a) Stepanov's Institute of Physics, National Acad. Sci. of Belarus, F. Skaryna Ave. 68, 220072 Minsk, Belarus

(b) AIXTRON AG, Kackertstr. 15–17, D-52072 Aachen, Germany

(Received June 21, 2001; accepted August 4, 2001)

Subject classification: 68.55.Ln; 71.55.Eq; 78.55.Cr; 78.66.Fd; S7.14

Time-integrated and time-resolved photoluminescence (PL) spectra as well as the luminescence transients of moderately doped GaN:Mg samples grown by MOVPE were studied between 80 K and 380 K at pulse excitation by a nitrogen laser beam in order to clarify the mechanism of the large blue shift of the 2.8 eV PL band above room temperature. Based on the performed study, the new band at 3.05 eV dominating in PL spectra above room temperature is attributed to the donor-to-valence band recombination. The corresponding donor ionization energy is about 290 meV. The blue shift of the spectra is therefore explained as a result of ionization of shallow acceptor states involved together with deep donors in donor–acceptor recombination forming the 2.8 eV band below room temperature.

Introduction GaN and related materials are very promising not only for optoelectronic devices but also for applications in high-power and high-temperature electronics [1]. Therefore, the investigation of the properties of this material system in the region above room temperature is of big interest. The information about the photoluminescence (PL) properties may be especially useful for obtaining information about the recombination mechanisms, and, thus, about the defect structure and the causes of conductivity compensation. Recently, we reported on the large (more than 100 meV) blue shift of the PL spectra of moderately doped GaN:Mg with increasing temperature above 300 K [2]. In this paper, we report on results of time-integrated and time-resolved PL spectroscopy of near-band-edge spectra of GaN:Mg with the aim to understand the reason of this shift and to clarify the corresponding recombination mechanisms.

Experimental The samples were grown on c-plane sapphire substrates by metalorganic vapor phase epitaxy (MOVPE) in AIXTRON planetary reactors. The growth details are described in [3]. The GaN:Mg layers demonstrated a high resistance after growth which could be converted to the p-type conductivity after the thermal annealing procedure. PL was excited by a pulsed nitrogen laser (power $P = 20$ kW, pulse duration at half-width $t = 8$ ns, repetition rate $f = 1000$ Hz) at $T = 80$–350 K. PL spectra were recorded both in time-integrated and time-resolved regime. In addition, PL transients at different wavelengths were measured with time resolution of about 3 ns.

---

[1]) Corresponding author; Phone: +49-241-8909-0; Fax: +49-241-8909-40; e-mail: bschi@aixtron.com

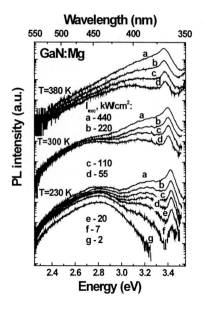

Fig. 1. Near-band-edge PL spectra of a GaN:Mg epitaxial layer recorded at $T = 230$ K, 300 K and 380 K, plotted in logarithmic scale of intensity, at different excitation conditions

Results and Discussion Figure 1 represents the near-band-edge (NBE) luminescence spectra of the GaN epitaxial layer with 0.1% Mg recorded at three temperatures: room temperature as well as one below and one above the interval where the main change of PL spectra appears. The excitation intensity was varied in order to recognize the peculiarities typical, e.g., for donor–acceptor pair (DAP) recombination. At least three bands are visible in the spectra: the band–band recombination (main sharp peak around 3.4–3.45 eV), the broad blue 2.8 eV band (dominates in the spectra at $T = 230$ K and low excitation intensity), and the 3.2 eV band having some structure like one formed by LO replicas. The relative intensity of this band increases with excitation intensity I_{exc}. At low temperatures about 80 K the band at 3.26 eV is often visible in GaN:Mg and can be attributed to the free-to-acceptor (FA) recombination involving shallow Mg acceptor state with ionization energy $E_A \approx 190$ meV. However, this band vanishes quickly with increasing temperature (Fig. 2 in [2]) and should not be observed at room temperature and above because of thermal ionization of the shallow acceptor states. Therefore we suppose that the band appearing on the same place at high temperatures is caused by another recombination mechanism.

To check this supposition, we performed a deconvolution of the NBE PL spectra by the Alentsev-Fok method [4]. The results of deconvolution are given in Fig. 2. One can see that the NBE PL spectrum at $T = 380$ K consists of at least two overlapped bands labeled Band 1 and Band 2. Band 2 at higher energy has a remarkable long-wavelength tail. It may be caused by an additional impurity re-

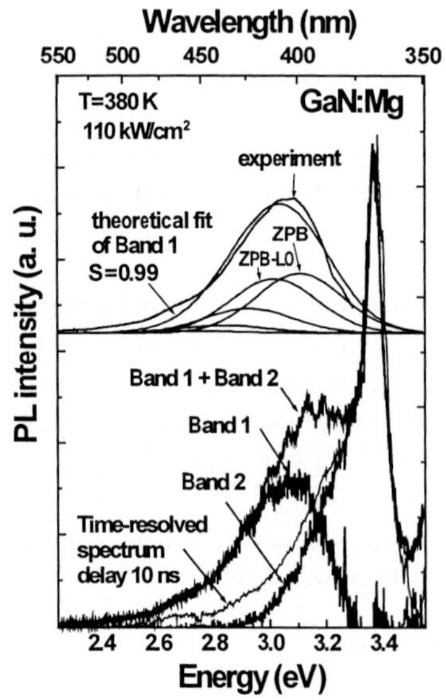

Fig. 2. The results of deconvolution of the NBE PL spectrum at $T = 380$ K into two bands (below) and the theoretical fit of Band 1 by the set containing a zero-phonon line and its LO replicas (above)

combination channel, but the present data are insufficient to separate this probable band from the main peak at 3.4 eV caused by the band–band recombination. The maximum of Band 1 is placed near 3.0 eV at this temperature. One can see also that the spectrum is still formed mainly by Band 2 after delay time of 10 ns from the pulse maximum, and the contribution of Band 1 is small which evidences on the large decay time of Band 1.

Additional information about the nature of the 3.2 eV band at higher temperatures can be extracted from PL transients at different wavelengths corresponding to the bands present in the PL spectra at certain temperatures. The transients are plotted in Fig. 3 for two temperatures: 81 K (at this temperature the FA band is well resolved) and 380 K (at this temperature the Band 1 with similar maximum position dominates in the spectrum together with band–band recombination forming Band 2). Depending on the recombination mechanism, the PL decay may be exponential, hyperbolic or multi-exponential. Therefore, the same transients are plotted in logarithmic (Figs. 3a, c) and square root scale (Figs. 3b, d), since the exponential and hyperbolic decay should give the straight line in the first and second scale, respectively.

One can see that at both $T = 81$ K and 380 K the decay of band–band recombination remains unchanged and can be well approximated by the single exponent with decay time $\tau = 15$ ns. The 2.8 eV band has at $T = 81$ K the multiexponential decay with relatively long effective decay time (hundreds of nanoseconds). The decay of the 3.2 eV band can be described neither by the exponential nor by the hyperbolic law. It evidences on the multiexponential decay which is typical for DAP recombination [5]. As seen from Figs. 3a, b, the 3.21 eV band at $T = 81$ K obeys the hyperbolic decay law rather than the exponential one. It is usual for FA recombination in the case of the weak influence of trap centers [6]. The recombination coefficient $\beta = 0.005$ was esti-

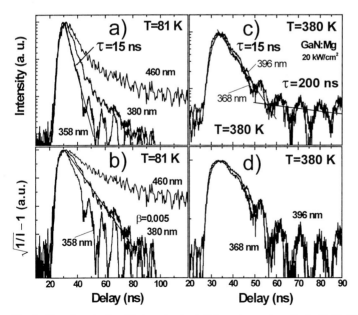

Fig. 3. Near-band-edge PL transients at different wavelengths in GaN:Mg recorded at $T = 81$ K (a, b) and $T = 380$ K (c, d) at $I_{exc} = 20$ kW/cm^2, plotted in logarithmic (a, c) and square root scale (b, d)

mated from the PL decay (Fig. 3b). An attempt to fit this part of decay at 380 nm by the exponent law yields the value of $\tau = 25$ ns.

At $T = 380$ K, the decay can be described rather by the exponential law than the hyperbolic one (Figs. 3c, d). Two decay regions are observed: the fast one with $\tau = 15$ ns (band-to-band recombination) and the slow one with τ about 200 ns. This value is far away from $\tau = 25$ ns estimated for 3.2 eV band at $T = 81$ K. Thus, time-resolved spectra give an evidence that this slow decay is related to the 3.2 eV band observed in NBE PL spectra at high temperatures (Band 1 in Fig. 2). We suppose that this band is caused by the recombination from the deep donor state to the valence band, and the reconstruction of the PL spectrum near room temperature with the quenching of the 2.8 eV band and appearance of this new band is caused by thermal ionization of shallow acceptors involved in formation of the 2.8 eV band.

In order to estimate the energy position of the deep donor E_D, we take into account the electron–phonon interaction. We performed calculations of the Huang-Rhys factor S defining the intensity distribution of LO replica in NBE PL spectra for the donor-to-valence band transitions in GaN using the same method as in [2], and obtained the values $S = 0.85$ for $E_D = 280$ meV and $S = 1.01$ for $E_D = 350$ meV. The fit of the experimental curve (Band 1) with the formula from [7] taking into account the S factor is given in the upper part of Fig. 2. From this fit, we obtain the value of $S = 0.99$ and the maximum of zero-phonon band (ZPB) at 3.09 eV. Since the band-band recombination has its maximum at 3.37 eV, we obtain from these data $E_D \approx 285$ meV, which is in fair agreement with the estimated values of S and with results and model proposed for the 2.8 eV band in [8]. Thus, our results evidence the presence of deep donor state with $E_D \approx 280-290$ meV in GaN:Mg.

Conclusion Based on investigations of the time-integrated and time-resolved NBE PL spectra of GaN:Mg under different excitation conditions, and on the study of PL transients on different wavelengths we explain the reconstruction of NBE PL spectra of GaN in the region above room temperature by a transition from the DAP recombination forming the 2.8 eV band to a transition from deep donor states to the valence band resulting from the thermal ionization of shallow acceptors.

Acknowledgements This work was supported by grant F99-117 of the Foundation of Fundamental Research of Belarus as well as by ISTC project #B-176.

References

[1] S. C. JAIN, M. WILLANDER, J. NARAYAN, and R. VAN OVERSTRAETEN, J. Appl. Phys. **87**, 966 (2000).
[2] A. L. GURSKII, M. GERMAIN, S. V. VOITIKOV, E. V. LUTSENKO, I. P. MARKO, V. N. PAVLOVSKII, B. SCHINELLER, O. SCHÖN, M. HEUKEN, E. KARTHEUSER, K. HEIME, and G. P. YABLONSKII, in: Proc. Internat. Workshop on Nitride Semiconductors, IPAP Conf. Ser. **1**, 591 (2000).
[3] C. VON EICHEL-STREIBER, O. SCHÖN, R. BECCARD, D. SCHMITZ, M. HEUKEN, and H. JURGENSEN, J. Cryst. Growth **190**, 344 (1998).
[4] M. V. FOK, Trudy FIAN SSSR **59**, 3 (1972) (in Russian).
[5] D. G. THOMAS, J. J. HOPFIELD, and W. M. AUGUSTINIAK, Phys. Rev. **140**, A202 (1965).
[6] A. M. GURVICH, Introduction to the Physical Chemistry of Crystallophosphors, Vysshaya Shkola Press, Moscow 1982 (p. 376) (in Russian).
[7] O. E. GOEDE and E. GUTSCHE, phys. stat. sol. **17**, 911 (1966).
[8] U. KAUFMANN, M. KUNZER, H. OBLOH, M. MAIER, CH. MANZ, A. RAMAKRISHNAN, and B. SANTIC, Phys. Rev. B **59**, 5561 (1999).

Spatial Fluctuations and Localisation Effects in Optical Characteristics of p-Doped GaN Films

E. M. Goldys[1]), M. Godlewski, E. Kaminska, A. Piotrowska, and K. S. A. Butcher

Division of Information and Communication Sciences, Macquarie University, Sydney 2109 NSW, Australia

(Received June 19, 2001; accepted August 4, 2001)

Subject classification: 71.35.Gg; 76.70.Hb; 78.55.Cr; S7.14

We report the observation of several intense satellite lines in the emission of p-GaN excited at 325 nm with separations suggesting a multiple LO phonon-related process. The observed phenomenon is interpreted as hot exciton luminescence enhanced by potential fluctuations and localisation. Using optically detected magnetic resonance and photoluminescence kinetics we explore the compensation processes underpinning these localisation effects.

Introduction The in-plane uniformity of GaN films and structures is an important factor in device fabrication. In this work we explore the in-plane uniformity of p-GaN films where we found new optical characteristics explained by the strong influence of localisation processes. This finding is important as localisation effects in GaN were anticipated earlier, but detailed evidence was not available. Localisation effects may help explain the relative immunity of GaN light emitters to structural defects brought about by lattice mismatched growth.

We examined the optical emission spectra and spatially resolved emission maps of p-type GaN films grown by MOCVD. This analysis uses the results of ODMR and photoluminescence kinetics, which emphasise the importance of the compensation effects in p-GaN and their contribution to potential fluctuations.

Experiment The (0001)-oriented p-type GaN films were grown on sapphire substrates with an AlN buffer by metalorganic chemical vapour deposition (MOCVD), with hole concentrations between 1 and 5×10^{17} cm^{-3}. Some films have been subjected to rapid thermal annealing at temperatures between 700 and 1150 °C for up to 5 min. As control specimens we used MOCVD GaN layers grown on sapphire with electron concentration of 1×10^{17} cm^{-3}. The photoluminescence (PL) and Raman spectra were measured in a Renishaw microRaman system with a resolution of 1.6 meV using a polarised 325 nm excitation at an excitation power density of 120 kW/cm^2. The emission spectra at single locations and emission maps, that is spatial distributions of emission from the sample, were taken at a spatial resolution typically of the order of 2×2 μm^2 and were measured at room temperature. The PL and PL kinetics were performed at 2 K using 352 nm excitation from an Ar laser, a mode-locked Ti:sapphire frequency-doubled laser at 340 nm with 2 ps pulses, or a frequency-tripled Nd:YAG laser at 355 nm, with 2 ns pulses. The ODMR studies used an X-band (9–10 GHz sys-

---

[1]) Corresponding author; Phone: +61 2 9850 8902; Fax: +61 2 9850 8115; e-mail: goldys@ics.mq.edu.au

tem) with on–off modulated power up to about 250 mW using 351 nm UV excitation at 2 K. The microwave-induced changes in emission were detected using a lock-in technique and a photomultiplier.

Results and Discussion In all samples we observed a complex broad band with a maximum at 3.15 eV (blue emission (BL)), related to donor–acceptor pair recombination involving Mg acceptors. In addition to the band, a series of intense satellite lines appear on its high energy wing. The distance of these lines from the laser energy reflects with great accuracy multiples of the LO phonon energy (91.5 meV). Up to six satellites could be observed, with the intensity as high as about one third of the emission band in some samples. In all samples the second satellite is the most intense (Fig. 1) and the third and fourth are still more intense than the first. The width of the higher satellites increases slowly, from 2.6 ± 0.2 eV (1LO) to 4 meV (2LO), up to 5–6 meV (higher satellites). The satellites were barely observable in n-type specimens, where only three very weak features could be seen.

Spatial maps of the first, second and third satellite (not shown) reveal strong intensity variations (up to 50%), rapid on the scale of the spatial resolution, but the peak energy remains constant. The variations in the intensity of the first peak are not correlated with similar variations in the second and the third peak. We also found no correlation with the spatial map of the integrated intensity of the entire blue band and of its selected parts (near 3.2 and 3.05 eV).

We verified whether similar satellites are observed at other laser excitation energies. To this aim we carried out similar measurements at 514 nm and at 800 nm (not shown). At 514 nm we observed Raman scattering at non-resonant conditions, similar to that reported earlier.

We attribute the satellite lines to a hot luminescence effect mediated by the presence of potential fluctuations, in analogy to earlier works in other materials [1]. Similar satellite lines separated by multiple LO phonon energies have been observed in other semiconductors, most notably in II–VI, but also in selected III–V compounds, such as InP, and recently in hexagonal GaN [2]. Two possible mechanisms have been used to explain similar satellite lines, namely higher-order multiphonon Raman scattering and hot exciton luminescence. Both these processes may coexist and give rise to peaks at identical energies $h\nu_{laser} - Nh\nu_{LO}$. The multiphonon Raman scattering is a coherent process, prominent in polar semiconductors. Hot exciton luminescence is an incoherent process in which hot excitons created by laser excitation lose energy in steps of $Nh\nu_{LO}$, these

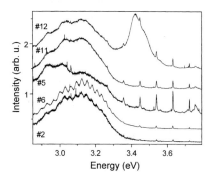

Fig. 1. Room temperature emission spectra of p-GaN films excited at 325 nm

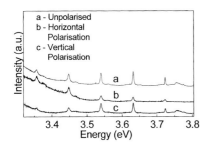

Fig. 2. Polarised spectra of the satellite lines: (a) unpolarised, (b) polarised perpendicularly to the laser polarisation, (c) the same polarisation as the laser

phonons are favoured in the cascade process as thermalisation via acoustic phonons is slower. Already the first event of the LO phonon emission allows a large fraction of the hot exciton population to acquire a near-zero momentum, and thus they are able to recombine radiatively giving rise to an emission line.

The two mechanisms may be distinguished from polarisation studies, and the presence (or lack) of coherence can be demonstrated by the polarisation properties of the peaks. To establish the likely mechanism in our samples we have measured the emission spectra in two polarisations (Fig. 2). We observed that the higher order satellites, observed at the same polarisation as the exciting laser, progressively lose the degree of polarisation, supporting the hot PL hypothesis. At the same time, the satellites observed in the alternative polarisation progressively increase in intensity. On the other hand, the intensity of a Raman scattering process in consecutive orders should behave as described in Refs. [3] and [4]. For example the second component should be proportional to C^2, where C is the electron–phonon coupling constant. As our Fig. 2 shows, the behaviour of the observed satellites is different.

We now discuss the issue of variations in the absolute intensity of the satellites with respect to other available benchmarks, such as Raman signatures of other phonons, laser line and other emission bands. We have shown that relative intensities of the satellites (Fig. 1) vary from sample to sample. Moreover, the spatial maps of the three satellites (1LO, 2LO and 3LO) taken from the same region of the sample show very rapid in-plane variations, uncorrelated with one another. These pieces of evidence reveal an important influence of disorder on the process of hot luminescence enhancement. Such influence is well documented, and for example Pelekanos et al. [1] found that localised states play an important role in the hot luminescence process. In GaN some disorder may be expected, for example it is known that GaN doped with Mg is compensated to a significant degree, a fact which can be deduced from temperature-dependent Hall measurements, and also reflected in the shift of the blue emission band with excitation power [5]. In GaN layers grown on sapphire the peak position of the BL varies with the doping level and shifts to lower energies with increasing Mg concentration [6, 7]. Such a shift was attributed to either large potential fluctuations in heavily doped samples, or to deep donors in the GaN lattice created during doping with acceptors by the self-compensation process [6, 7]. In such a compensated material large potential fluctuations due to randomly distributed charged impurities are expected. The model of strong potential fluctuations was recently supported by the results of Oh et al. [8] who observed that the BL strongly shifts to higher energies with increasing excitation density. The BL at low excitation intensities has a maximum at about 2.85 eV and progressively shifts to about 3.2 eV at high excitation conditions. The disorder is capable of

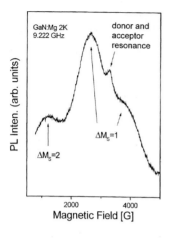

Fig. 3. Anisotropic spin triplet resonance detected in the spectral region of the YL emission in heavily Mg-compensated bulk GaN

inducing localisation effects that relax the wavevector conservation rule and allow for the participation in hot luminescence of optical phonons over much of the LO-phonon dispersion curve. Charge compensation may lead to large potential fluctuations and the creation of potential barriers for carrier/exciton migration. Fast tunnelling of electron–hole pairs from neighbouring potential minima, on a faster timescale than the slow energy relaxation process via acoustic phonons, may effectively short-circuit the latter mechanism. This tunnelling process explains why in some samples we observe satellites that lie below the GaN bandgap (sample #11). Such satellites can arise as a result of the spatially indirect process of recombination.

The possible source of these fluctuations relates to compensation effects in p-GaN. One such process is compensation of Mg by donors observed through optically detected magnetic resonance. The second is a complex multi-band character of the BL with possible involvement of several different defect centres. In Fig. 3 we show the ODMR signals observed in Mg doped GaN detected in the spectral region of the yellow emission (YL). The anisotropic ODMR spectrum is composed of a group of resonance lines around $g = 2$ and another group at about $g = 4$. Such ODMR spectra were previously observed for excitons bound at neutral complex centres and explained by the spin triplet resonance. We thus propose that the new ODMR spectrum observed by us has a similar origin and indicates recombination at a close donor–acceptor pair forming a neutral complex centre.

Recently, Mg–O complexes were found theoretically to have a fairly low formation energy [9]. The Mg–O complex consists of a single donor, which compensates a single acceptor, thus it is a close analogue of a donor–acceptor pair and a possible candidate

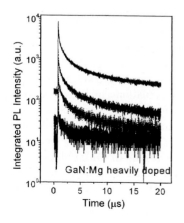

Fig. 4. The PL kinetics of the BL emission observed in heavily compensated Mg-doped sample measured at different energies within the BL: (from top to bottom) at the high energy wing, at the maximum, at the low energy wing and at the maximum of the YL

for the neutral complex centre observed in the ODMR experiments. Mg–H complexes cannot also be ruled out. Compensation is also apparent in the PL kinetics studies (Fig. 4). The PL decay spectra taken at 2 K for a heavily Mg doped crystal show non-exponential PL kinetics. The fast decay is followed by a slow decay component, with a time constant up to 500 µs. The PL decay time is photon energy dependent. Longer decay times are observed at the low energy tail of the PL emission, a characteristic property of donor–acceptor pair recombination transitions.

References

[1] N. Pelekanos, J. Ding Q. Fu, A.V. Nurmiko, S.M. Durbin, M. Kobayashi, and R.L. Gunshor, Phys. Rev. B **43**, 9354 (1991).
[2] D. Behr, J. Wagner, J. Schneider, H. Amano, and I. Akasaki, Appl. Phys. Lett. **68**, 2404 (1996).
[3] R. Zeyher, Solid State Commun. **16**, 49 (1975).
[4] Z.C. Feng, S. Perkowitz, and P. Becla, Solid State Commun. **78**, 1011 (1991).
[5] L. Eckey, U. von Gfug, J. Holst, A. Hoffmann, A. Kascher, H. Siegle, C. Thomsen, B. Schineller, K. Heime, M. Heuken, O. Schön, and R. Beccard, Appl. Phys. Lett. **84**, 5828 (1998).
[6] U. Kaufman, M. Kunzer, M. Maier, H. Obloh, A. Ramakrishnan, and B. Santic, Appl. Phys. Lett. **72**, 1326 (1998).
[7] L. Eckey, U. Gfug, J. Holst, A. Hoffmann, B. Schineller, K. Heime, M. Heuken, O. Schön, and R. Beccard, Proc. Second Internat. Conf. on Nitride Semiconductors, Tokushima, Japan, 1997 (p. 58).
[8] E. Oh, H. Park, and Y. Park, Appl. Phys. Lett. **72**, 70 (1998).
[9] C.G. Van De Walle, C. Stampfl, and J. Neugebauer, Proc. Second Internat. Conf. on Nitride Semiconductors, Tokushima, Japan, 1997 (p. 386).

Characterization of Mg-Doped GaN Micro-Crystals Grown by Direct Reaction of Gallium and Ammonia

S. H. Lee (a), K. S. Nahm[1]) (a), E.-K. Suh (b), and M. H. Hong (c)

(a) Semiconductor Physics Research Center, School of Chemical Engineering and Technology, Chonbuk National University, Chonju 561-765, Korea

(b) Semiconductor Physics Research Center, Department of Semiconductor Science and Technology, Chonbuk National University, Chonju 561-765, Korea

(c) POSCO Technical Research Lab., Kwangyang, 545-090, Korea

(Received June 23, 2001; accepted August 4, 2001)

Subject classification: 68.37.Hk; 68.37.Lp; 68.55.Ln; 78.55.Cr; 78.60.Hk; 81.10.Dn; S7.14

Mg-doped GaN micro-crystals were prepared by the direct reaction of metal gallium and ammonia using magnesium chloride ($MgCl_2$) as Mg doping source. The growth of microcrystalline hexagonal GaN was clearly found from scanning electron microscopy and TEM measurements. The grain size of Mg-doped GaN crystals was larger than that of undoped GaN. Room temperature PL and CL spectra for Mg-doped GaN micro-crystal showed the blue emission at 2.75 and 2.85 eV, respectively.

Introduction III–V nitrides (GaN, AlGaN, InGaN etc.) have already made a dramatic impact on the development of blue and green light-emitting diodes and laser diodes [1]. The preparation of GaN powder crystals is of great interest in sublimation growth of bulk GaN as well as in nano-scale optical technology [2, 3]. The application of micro-crystal GaN to a phosphor material has been an important technique to satisfy the demand for a blue phosphor applicable to vacuum fluorescent displays (VFDs) [4]. But the realization of optical devices has been hampered by the lack of a developed synthesis method for high quality and doped nano-crystalline GaN materials. In this work, we have synthesized Mg-doped GaN micro-crystal by a direct reaction of gallium and ammonia using magnesium chloride ($MgCl_2$) as Mg doping source, and characterized the structural and optical properties of the prepared crystals.

Experimental Mg-doped GaN micro-crystals were grown by the direct reaction of liquid gallium with ammonia at a temperature of 1000–1100 C and 1 atm in a quartz tubular reactor heated in electric furnace. Ga metal (99.99%, 12g) and NH_3 gas (99.99%) were used for Ga and N sources, respectively. Magnesium chloride ($MgCl_2$, 0.1 wt% of Mg relative to Ga) was used as Mg dopant source. The growth temperature was monitored with a Ru/Pt thermocouple and controlled by a temperature controller. At the growth temperature, 25 sccm NH_3 was introduced into the reactor loaded with solid reactants and the reaction started. After the growth, unreacted Ga or Mg was extracted from the product by dissolving in HCl aqueous solution for separation of GaN crystals. The purified GaN micro-crystals were rinsed with D.I. water and were dried in a vacuum oven.

Results and Discussion Figure 1 shows scanning electron microscopy (SEM) images for undoped and Mg-doped GaN micro-crystals synthesized. Dark-gray-colored micro-

---

[1]) Corresponding author; Phone: +82 63 270 2311; Fax: +82 63 270 2306; e-mail: nahmks@moak.chonbuk.ac.kr

Fig. 1. SEM images for a) undoped and b) Mg-doped GaN micro-crystals

crystals are formed with typical size of 1–5 µm and 5–10 µm for undoped and Mg-doped GaN crystals, respectively. The morphology of undoped GaN crystals is various polyhedral and rounded particles as reported in previous papers (Fig. 1a) [5]. Meanwhile, Mg-doped GaN crystals consist of various polyhedral particles, but most of them are observed to have a hexagonal shape as shown in Fig. 1b. It is interesting to note that Mg-doped GaN crystal has a vivid hexagonal shape with larger particle size. At present, we speculate that the difference between undoped and Mg-doped GaN crystals in the size and shape is attributed to the existence of magnesium in the growth of GaN microcrystal. Mg has lattice parameters ($a = 0.32$ nm and $c = 0.52$ nm) very similar to those of GaN ($a = 0.318$ nm and $c = 0.517$ nm). It is likely that magnesium inflicts some positive effects on the growing surface to result in the structural stabilization and to increase the size of GaN micro-crystals. It was reported that the existence of Mg impurity in GaN growth slightly accelerates the growth in c-direction [6].

Figure 2 shows dark-field images of transmission electron microscopy (TEM) for Mg-doped GaN micro-crystal with the corresponding selected area diffraction pattern (SADP) along the electron beam direction $\mathbf{B} = [01\bar{1}0]$ with the reflection vectors $\mathbf{g} = 0002$ (a) and $\mathbf{g} = \bar{2}110$ (b), respectively. All the GaN crystals observed in the present study by SADP are identified to have 2H hexagonal structure. Dark field micrographs show that the crystal consists of single crystal although some fringes due to difference of crystal thickness are observed.

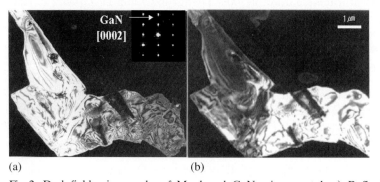

Fig. 2. Dark-field micrographs of Mg-doped GaN micro-crystal. a) Reflection vector $\mathbf{g} = 0002$; b) $\mathbf{g} = \bar{2}110$. Inset in a): Corresponding selected area diffraction pattern (SADP) along the electron beam direction $\mathbf{B} = [0110]$

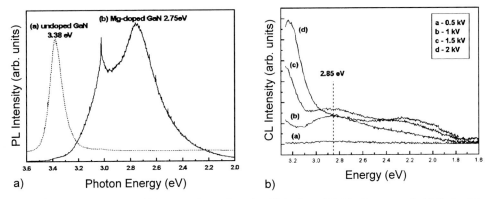

Fig. 3. a) Room temperature PL spectra for undoped and Mg-doped GaN micro-crystals. b) Cathodoluminescence spectra at room temperature for Mg-doped GaN powder at different accelerating voltages

Photoluminescence (PL) measurements were carried out to characterize optical properties of Mg-doped GaN micro-crystals using a He–Cd laser with 325 nm wavelength and 15 mW power. Figure 3a presents PL spectra measured at room temperature for undoped and Mg-doped GaN micro-crystals. Undoped GaN shows a strong band edge emission at an energy position of 3.38 eV with a negligible yellow emission in deep level, whereas Mg-doped GaN emits a broad peak centered at 2.75 eV, corresponding to band edge emission of heavily doped MOCVD material. A sharp peak at 3.02 eV is not induced by the sample but by the measurement system.

Figure 3b shows cathodoluminescence spectra measured at different accelerating voltages (0.5–2 kV) for Mg-doped GaN micro-crystals. It is seen that blue emission (2.85 eV) begins to be detected at ≈0.5 kV and the emission shifts toward green color as the accelerator voltage increases. At accelerating voltages above 1 kV, GaN host wide band peak starts to appear at ≈3.25 eV and grows in intensity with increasing accelerating voltage. These lines are commonly observed in room temperature PL and CL measurements of Mg-doped GaN. The ≈3.25 eV line is widely attributed to a free electron–Mg acceptor transition and the broad emission at 2.85 eV could be assigned to a deep Mg-related complex [7, 8].

Acknowledgements This work was supported by KRF Grant (KRF-99-005-D00036) through the Semiconductor Physics Research Center at Chonbuk National University.

References

[1] S. Nakamura, T. Mukai, and M. Senoh, Appl. Phys. Lett. **64**, 1687 (1994).
[2] S. Kurai, T. Abe, Y. Naoi, and S. Sakai, Jpn. J. Appl. Phys. **35**, 1637 (1996).
[3] K. E. Gonsalves, S. D. Rangarajan, G. Carlson, J. Kumar, K. Yang, M. Benaissa, and M. Jose-Yacanan, Appl. Phys. Lett. **71**, 2175 (1997).
[4] F. Kataoka, Y. Satoh, Y. Suda, K. Honda, and H. Toki, Physics and Chemistry of Luminescent Materials, ECS, 1999 (p. 17).
[5] C. M. Balkas and R. F. Davis, J. Am. Ceram. Soc. **79**, 2309 (1996).
[6] S. Porowski, J. Cryst. Growth **189/190**, 153 (1998).
[7] M. Smith, G. D. Chen, J. Y. Lin, H. X. Jiang, A. Salvador, B. N. Sverdlov, A. Botchkarev, H. Morkoc, and B. Goldenberg, Appl. Phys. Lett. **68**, 1883 (1996).
[8] M. Gross, G. Henn, J. Ziegler, P. Allenspacher, C. Cychy, and H. Schröder, Mater. Sci. Eng. B **59**, 94 (1999).

Activation of Mg Acceptor in GaN:Mg with Pulsed KrF (248 nm) Excimer Laser Irradiation

Dong-Joon Kim, Hyun-Min Kim, Myung-Geun Han, Yong-Tae Moon, Seonghoon Lee, and Seong-Ju Park[1])

Department of Materials Science and Engineering and Center for Optoelectronic Materials Research, Kwangju Institute of Science and Technology, Kwangju 500-712, Korea

(Received June 24, 2001; accepted August 4, 2001)

Subject classification: 61.10.Eq; 61.80.Ba; 68.35.Bs; 73.61.Ey; 81.15.Gh; S7.14

We report on the activation of Mg acceptors in Mg-doped GaN films, grown by metalorganic chemical vapor deposition, via the use of a pulsed KrF (248 nm) excimer laser irradiation. The as-grown GaN:Mg, which was irradiated by the KrF excimer laser at a laser energy density of 590 mJ/cm^2 in a N_2 ambient showed a hole concentration of 4.42×10^{17} cm^{-3}. Furthermore the hole concentration in GaN:Mg, which was activated by a conventional rapid thermal annealing, was increased from 4.3×10^{17} to 9.42×10^{17} cm^{-3} as the result of subsequent laser irradiation. These results suggest that a pulsed KrF excimer laser irradiation can dramatically enhance the p-type conductivity of GaN:Mg by efficiently dissociating the Mg–H complexes.

Introduction Preparation of highly p-type conductive Mg-doped GaN films, grown by metalorganic chemical vapor deposition (MOCVD), is a major issue in the fabrication of high brightness light emitting diodes and high power short wavelength laser diodes [1]. However, the as-grown Mg-doped GaN films are highly resistive due to the formation of electrically inactive Mg–H complexes in the case of the MOCVD growth of GaN:Mg [2]. In general, an additional post-growth treatment is required to activate the Mg acceptors, in order to achieve p-type conductivity. It is well known that GaN:Mg, annealed in a N_2 ambient, shows p-type conductivity as the result of the dissociation of Mg–H complexes [3]. In this work, we report on an investigation of the electrical activation of Mg acceptors in Mg-doped GaN films with a pulsed KrF (248 nm) excimer laser irradiation. For the GaN:Mg, which had undergone rapid thermal annealing (RTA) and subsequent laser irradiation, the hole concentration was increased from 4.3×10^{17} to 9.42×10^{17} cm^{-3}. This high hole concentration can lead to low Ohmic contact resistance, giving rise to an improved performance and reliability of optoelectronic devices by reducing the heat generation during operation. In addition, the laser irradiation method has several potential advantages such as very high speed and selective area processes for Mg activation.

Experimental The Mg-doped GaN thin films were grown on c-face (0001) sapphire substrates by low-pressure MOCVD in a vertical rotating disc reactor (Emcore D-125^{TM}). The growth structure consisted of p-GaN (1 μm)/GaN nucleation layer (30 nm)/sapphire [4]. As-grown Mg-doped GaN thin films showed semi-insulating characteristics. In order to obtain p-type conductivity, a pulsed KrF (248 nm) excimer laser was employed to irradiate the semi-insulating GaN:Mg with various energy densities. In addition,

---

[1]) Corresponding author; Phone: +82-62-970-2309; Fax: +82-62-970-2304; e-mail: sjpark@kjist.ac.kr

Table 1

Hole concentrations of GaN:Mg thin films irradiated by KrF (248 nm) excimer laser under nitrogen and oxygen ambient gases. The repetition rate and the number of laser pulses was 5 Hz and 600, respectively.

laser energy density	590 mJ/cm^2	525 mJ/cm^2	455 mJ/cm^2
nitrogen (500 Torr)	4.42×10^{17} cm^{-3}	7.35×10^{16} cm^{-3}	5.52×10^{16} cm^{-3}
oxygen (500 Torr)	highly resistive	highly resistive	highly resistive

GaN:Mg, which had been activated by a conventional RTA process, was also treated by laser irradiation to further improve the hole concentration of GaN:Mg. The electrical activation of GaN:Mg was ascertained by using a Hall effect measurement with a van der Pauw geometry. The surface roughness and structural properties of the resulting GaN:Mg films were also characterized by means of atomic force microscopy (AFM) and X-ray diffraction measurements, respectively, to assess the laser induced damages to the films.

Results and Discussion The hole concentrations of the GaN:Mg thin films, which were irradiated by various energy densities using a pulsed KrF (248 nm) excimer laser under two different ambient gases, are tabulated in Table 1. When the GaN:Mg samples were irradiated under N$_2$, all samples showed p-type conductivity and, in particular, the sample irradiated by the laser under N$_2$ at a laser energy density of 590 mJ/cm^2 showed a hole concentration of 4.42×10^{17} cm^{-3}. This value is comparable to that of GaN:Mg activated by a conventional RTA process. The calculated dissociation energy barrier for the Mg–H complex is reported to be about 1.5 eV [5].

Therefore, a KrF excimer laser light with a photon energy of about 5 eV appears to activate the Mg acceptors in GaN:Mg thin films by dissociating the Mg–H complexes, giving rise to p-type conductivity. However, as shown in Table 1, the samples irradiated under an O$_2$ ambient showed no improvement in electrical properties, which is probably due to the surface oxidation of GaN:Mg by oxygen, during the laser irradiation.

To further investigate the effect of the laser irradiation on the electrical activation of Mg acceptors, a KrF excimer laser irradiated the GaN:Mg which had been activated by an RTA process at 950 °C for 1 min. Before the laser irradiation, the GaN:Mg samples were coated with SiO$_2$, which was deposited by plasma enhanced CVD, to protect the sample surface during laser irradiation and enhance the activation efficiency [6]. The SiO$_2$ layer was completely etched out, in order to conduct a Hall measurement after the laser irradiation was performed. The

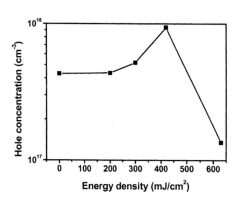

Fig. 1. Hole concentrations of GaN:Mg films irradiated by KrF (248 nm) excimer laser at various energy densities. The repetition rate and the number of laser pulses was 5 Hz and 100, respectively

Table 2

AFM rms roughness (10 × 10 μm^2) of GaN:Mg thin films irradiated by KrF (248 nm) excimer laser at various energy densities. The repetition rate and the number of laser pulses was 5 Hz and 100, respectively

	as-grown	200 mJ/cm^2	300 mJ/cm^2	420 mJ/cm^2	620 mJ/cm^2
AFM rms roughness (nm)	0.85	0.89	0.9	0.95	> 100

GaN:Mg sample which was activated by an RTA showed a hole concentration of 4.3×10^{17} cm^{-3}. Figure 1 shows that the hole concentration of the GaN:Mg sample was increased up to 9.42×10^{17} cm^{-3} when the excimer laser irradiated the GaN:Mg, which had been activated by an RTA, at a laser energy density of 420 mJ/cm^2. This result is probably due to the effective activation of Mg acceptors by photon-assisted annealing through SiO$_2$ layer which suppresses the out-diffusion of Mg [6]. It is also possible to consider that the laser irradiation and thermal annealing may activate different types of Mg–H complexes. However, a further increase in the laser energy density resulted in a decrease in hole concentration. In order to understand this result, the surface morphologies of GaN:Mg samples irradiated with the KrF excimer laser were examined. As shown in Table 2, the AFM surface root-mean-square (rms) roughness, which was measured after removing SiO$_2$ layer, was similar to that of the as-grown state of GaN:Mg in a range of laser energy density from 200 to 420 mJ/cm^2. However, the sample which had been treated by the excimer laser at a laser energy density of 620 mJ/cm^2, showed a large rms roughness value of over 100 nm. This can be attributed to the laser ablation of GaN:Mg. It would be expected that nitrogen would be preferentially desorbed from the surface of GaN:Mg at an early stage of ablation. This nitrogen loss at the surface may result in the decrease in hole concentration. Ahn et al. [4] reported that the hole concentration of GaN:Mg was reduced by increasing the RTA time to over 900 °C due to nitrogen loss at the surface and also showed that the AFM surface roughness was significantly increased. Figure 2 represents the full width at half maximum (FWHM) variations for (002) planes of GaN:Mg thin films with increasing laser energy density. It also shows that the FWHM of GaN:Mg irradiated at a laser energy density of 620 mJ/cm^2 is exceptionally large, indicating that the structural property of the sample has drastically been deteriorated. From these results, we conclude that the laser activation of GaN:Mg with an optimum energy density is very effective in further increasing the hole concentration of GaN:Mg which has previously been activated by a conventional RTA process.

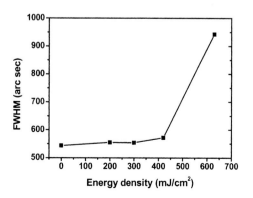

Fig. 2. FWHMs of X-ray rocking curve for the (002) planes of GaN:Mg films irradiated by KrF (248 nm) excimer laser at various energy densities. The repetition rate and the number of laser pulses was 5 Hz and 100, respectively

Conclusions The electrical activation of GaN:Mg irradiated by a pulsed KrF (248 nm) excimer laser has been investigated. A hole concentration of 4.42×10^{17} cm^{-3} was routinely obtained in an as-grown sample, which was irradiated by an excimer laser at a laser energy density of 590 mJ/cm^2 under an N$_2$ environment. A subsequent excimer laser irradiation of a sample following a conventional RTA process further increased the hole concentration of the GaN:Mg from 4.3×10^{17} to 9.42×10^{17} cm^{-3}. These results indicate that the short excimer laser irradiation is very effective in the dissociation of Mg–H complexes of GaN:Mg, and results in a high p-type conductivity.

Acknowledgement This work was partially supported by the Brain Korea 21 program and the Ministry of Commerce, Industry and Energy in Korea.

References

[1] I. AKASAKI and H. AMANANO, J. Electrochem. Soc. **141**, 2266 (1994).
[2] S. M. MYERS, A. F. WRIGHT, G. A. PETERSEN, C. H. SEAGER, W. R. WAMPLER, M. H. CRAWFORD, and J. HAN, J. Appl. Phys. **88**, 4676 (2000).
[3] W. GÖTZ, N. M. JOHNSON, J. WALKER, D. P. BOUR, and R. A. STREET, Appl. Phys. Lett. **68**, 667 (1996).
[4] K. S. AHN, D. J. KIM, Y. T. MOON, H. G. KIM, and S. J. PARK, J. Vac. Sci. Technol. B **19**, 215 (2001).
[5] J. NEUGEBAUER and C. G. VAN DE WALLE, Phys. Rev. Lett. **75**, 4452 (1995).
[6] C. R. LEE, J. Y. LEEM, and B. G. AHN, J. Cryst. Growth **216**, 62 (2000).

Analysis of Time-Resolved Donor–Acceptor-Pair Recombination in MBE and MOVPE Grown GaN:Mg

S. Strauf[1]), S. M. Ulrich, P. Michler, J. Gutowski, T. Böttcher, S. Figge, S. Einfeldt, and D. Hommel

Institut für Festkörperphysik, Universität Bremen, D-28334 Bremen, Germany

(Received June 29, 2001; accepted July 14, 2001)

Subject classification: 68.55.Ln; 71.35.Cc; 71.55.Eq; 78.55.Cr; 81.15.Hi; 81.15.Kk; S7.14

We have investigated the dynamics of the donor–acceptor-pair (DAP) recombination in Mg-doped GaN layers grown by MOVPE as well as MBE. The observed nonexponential decay curves are perfectly described if all parallel decay channels for each donor impurity surrounded by the acceptor impurities are included. Best fits have been obtained with a donor binding energy of 32 ± 2 meV. Additionally, under resonant excitation of the donor-bound-exciton complex the excited state of the donor could be clearly identified. From these data a donor binding energy of 29.9 ± 1.0 meV has been estimated, in good agreement with the value obtained from the DAP decay. We emphasize that the analysis of the DAP decay yields an accurate estimate for the neutral acceptor concentration in GaN:Mg without any need for further electrical measurements.

Introduction A pronounced nonexponential decay of the donor-acceptor-pair (DAP) recombination has often been observed in semiconductors [1–5]. An ample theoretical description of the DAP decay has already been given by Thomas et al. [1] and recently successfully applied by us for the case of ZnSe:N [2] and by Hofmann et al. for the yellow luminescence band in GaN [3]. Concerning the Mg-related DAP emission in GaN:Mg some work claims a biexponential ansatz to extract a kind of lifetime [4, 5]. However, considerable information was then ignored which is hidden in the luminescence of the distant pairs where the decay law is nonexponential. The purpose of the present paper is to present a detailed analysis of the dynamics of the Mg-related DAP recombination band in GaN:Mg.

Sample Preparation and Experimental Setup The MBE samples were grown in an EPI 930 MBE system equipped with an EPI Unibulb nitrogen plasma source. Layers with thickness of about 1 µm were deposited on sapphire substrates after growth of a 40 nm undoped GaN layer which was intended to suppress any modification of the nucleation by the dopant. Doping was performed using a standard effusion cell for Mg. The MOVPE samples were grown in a vertical type Thomas Swan reactor equipped with a close spaced showerhead using TMGa and NH_3 as precursors. About 0.5 µm thick GaN:Mg layers doped with bis-cyclopentadienyl-Mg as precursor were grown on undoped and fully coalesced GaN layers. An argon laser operated at 351 nm served as excitation source for time-resolved measurements, being pulsed by an external acousto-optic deflector which supplies 50 ns pulses at a repetition rate of 4 kHz. The DAP spectra were recorded with a gated optical-multichannel-analyzer system whose time window was set to 50 ns. For resonant excitation the frequency-doubled emission of a

---

[1]) Corresponding author; Phone: +49 421 218 4430; Fax: +49 421 218 7318; e-mail: strauf@physik.uni-bremen.de

synchronously pumped dye-laser system equipped with Pyridin 1 has been used, which supplies 10 ps pulses at 82 MHz tunable in the spectral range of 340–380 nm. All samples were mounted on the cold-finger of a He cryostat at 4 K.

Results and Discussion The intensity $J_E(t)$ of emitted DAP photons with energy E at a time t after the exciting pulse is given by [1, 2]

$$J_E(t) = W(r) \exp(-W(r)t) \langle Q(t) \rangle, \qquad (1)$$

in which

$$\langle Q(t) \rangle = \exp\left[4\pi N_A \int_0^\infty (\exp(-W(r)t) - 1) r^2 \, dr\right] \qquad (2)$$

denotes the ensemble average of the donor occupation probability at time t taking into account all parallel possible decay channels of each donor with all N_A sourrounding acceptors. The pair separation r can be calculated from the recombination energy $E = E_g - E_A - E_D + E_C$ where E_g is the bandgap, E_D and E_A are the donor and acceptor binding energies, respectively, and the Coulomb term E_C is given by

$$E_C = \frac{e^2}{4\pi\epsilon_0 \epsilon(0) r}, \qquad (3)$$

$\epsilon(0)$ is the static dielectric constant. The recombination rate reads $W(r) = W_0 |I(r)|^2$, where $I(r)$ is the overlap of the donor and acceptor wave functions at pair separation r and W_0 is the recombination constant. To account for the chemical shift of impurity binding energies we used the so-called scaled effective-mass theory to calculate the relation between the binding energies and the effective Bohr radii as described in [2]. The dynamics of the DAP decay according to Eq. (1) are fully determined by the parameter set E_g, W_0, N_A, E_D, E_A.

The time-integrated photoluminescence (PL) spectra of moderately doped GaN:Mg layers are dominated by a pronounced DAP transition at about 3.26 eV with clearly structured LO-phonon replicas at an energy separation of 92 meV (not shown). With increasing excitation density the DAP zero-phonon-line maximum shifts by about 20 meV to higher energies because of the increasing dominance of close pairs (compare, e.g., Fig. 2 in [6]). Fig. 1a shows the DAP recombination at different time delays

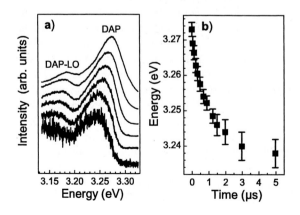

Fig. 1. a) Time-resolved photoluminescence of an MBE grown GaN:Mg layer at different delays (from top to bottom: 0, 125, 325, 675, 1225, 2975 ns) after pulsed excitation at 351 nm. Spectra are normalized and shifted vertically for better clarity. b) Energy position of DAP zero-phonon-line maximum as a function of delay time. $T = 4$ K

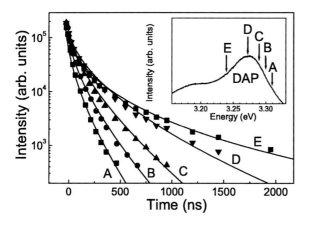

Fig. 2. Decay of DAP photoluminescence intensity (symbols) at various energy positions A–E as can be seen from the inset. The solid lines are theoretical curves according to Eq. (1), see text. $T = 4$ K

after pulsed excitation exemplarily for an MBE grown p-GaN:Mg layer with a free-hole concentration of $p = 1 \times 10^{17}$ cm^{-3} (at 300 K). With increasing time the maximum of the DAP zero-phonon line shifts by about 30 meV to lower energies as depicted in Fig. 1b. This behavior is unambiguously reflecting the nature of a DAP recombination process due to the higher recombination rate of closer pairs provided by their stronger wavefunction overlap. Figure 2 shows the corresponding DAP decay curves (symbols) at various energy positions referred to as A–E as can be seen from the inset. The solid lines represent the theoretical curves according to Eq. (1). With a fixed set of five parameters the decay characteristics of all energies between 3.233 and 3.306 eV can perfectly be reproduced. The bandgap for this sample has been fixed to 3.504 eV as measured by photoreflectance [6]. E_A is fixed to 245 meV as estimated from the corresponding free-to-bound transition of the Mg acceptor at higher temperatures already shown in [7]. The nearly exponential decay curves of the close pairs at energy positions A and B are not very sensitive to N_A but are strongly influenced by variations of E_D and W_0. At these energies best fits to the experimental data have been found for $E_D = 32 \pm 2$ meV and $W_0 = 1.6 \pm 0.2 \times 10^8$ s^{-1}. In contrast, the pronounced nonexponential decay curves of the distant pairs at energy positions D and E are highly sensitive to N_A where we found an optimum value of $N_A = 3.2 \pm 0.3 \times 10^{18}$ cm^{-3}. Compared to the overall concentration of Mg impurities of 1.6×10^{20} cm^{-3} in this layer as determined by secondary-ion-mass spectroscopy, it is found that only about 2% of the Mg atoms are incorporated on lattice site forming shallow Mg acceptors.

We investigated further a moderately doped MOVPE grown GaN:Mg layer which shows very similar data as those depicted in Figs. 1 and 2 for the MBE sample. The bandgap for this sample is 3.512 eV as estimated from reflection measurements, corresponding to a higher compressive biaxial strain component compared to the MBE sample. The DAP decay curves could be fitted with the same values for W_0, E_D and E_A as given above. A lower value of $N_A = 1.9 \pm 0.3 \times 10^{18}$ cm^{-3} has been found when compared to the MBE sample, in agreement with the vanishing free-hole concentration at 300 K in this sample. Therefore, the dynamics of the shallow DAP recombination in GaN:Mg is independent of the actual strain situation, the specific growth method (MBE or MOVPE) and the details of the doping.

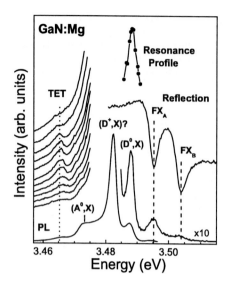

Fig. 3. PL and reflection spectra in the exciton region of an MOVPE grown GaN:Mg layer. TET spectra are shown on the left for increasing excitation energies (from top to bottom) across the (D^0, X) transition. The corresponding resonance profile of the TET intensity is shown on the top. $T = 4$ K

Additionally, E_D has been deduced from the two-electron transition (TET) spectra generated by the recombination of the neutral donor-bound-exciton complex (D^0, X) leaving the donor electron in one of its excited states. Figure 3 shows the PL and the reflection spectra of a lightly doped MOVPE grown GaN:Mg layer in the exciton region. Pronounced free A- and B-band exciton transitions (FX_A and FX_B) and the (D^0, X) recombination located 7.3 ± 0.5 meV below FX_A are observed. Two additional transitions with localisation energies of 12.8 ± 0.5 and 21 ± 2 meV have been found. While the latter is attributed to the Mg-related acceptor-bound-exciton (A^0, X) complex, the former is probably caused by the ionized donor-bound-exciton (D^+, X) transition which will be discussed elsewhere. Under resonant excitation of the (D^0, X) transition the first excited state of the donor (2S) could be clearly identified in the TET spectra as shown on the left in Fig. 3. The maximum of the TET intensity is found to correspond to the central peak position of the (D^0, X) transition as shown by the resonance profile of the TET intensity on the top in Fig. 3. From the 1S–2S energy separation of 22.4 ± 1.0 meV an effective mass donor binding energy of 29.9 ± 1.0 meV has been estimated, in good agreement with the value obtained from the DAP decay.

Conclusions In summary, the decay characteristics of the Mg-related DAP recombination in GaN:Mg can be perfectly reproduced if all parallel decay channels are included. Values of $W_0 = 1.6 \pm 0.2 \times 10^8$ s^{-1}, $E_A = 245 \pm 10$ meV and $E_D = 32 \pm 2$ meV have been found being independent of the actual strain situation, the specific growth method (MBE or MOVPE) and the details of the doping. Additionally, a donor binding energy of $E_D = 29.9 \pm 1.0$ meV has been deduced from two-electron transition spectra in good agreement with the value obtained from the DAP decay. We emphasize that the acceptor concentration can be accurately estimated for each sample from the analysis of the DAP decay without any need for further electrical measurements.

Acknowledgements The authors wish to thank D.M. Hofmann and B.K. Meyer for providing SIMS measurements and P. Bäume for helpful discussions. This work was partially supported by the Deutsche Forschungsgemeinschaft.

References

[1] D.G. THOMAS, J.J. HOPFIELD, and W.M. AUGUSTYANIAK, Phys. Rev. B **140**, 202 (1965).
[2] P. BÄUME, S. STRAUF, J. GUTOWSKI, M. BEHRINGER, and D. HOMMEL, J. Cryst. Growth **184/185**, 531 (1998).

[3] D.M. Hofmann, D. Kovalev, G. Steude, B.K. Meyer, A. Hoffmann, L. Eckey, R. Heitz, T. Detchprom, H. Amano, and I. Akasaki, Phys. Rev. B **52**, 16702 (1995).
[4] M. Smith, G.D. Chen, J.Y. Lin, H.X. Jiang, A. Salvador, B.N. Sverdlov, A. Botchkarev, H. Morkoc, and B. Goldenberg, Appl. Phys. Lett. **68**, 1883 (1996).
[5] H. Teisseyre, B. Kozankiewicz, M. Leszczynski, I. Grzegory, T. Suski, M. Bockowski, S. Porowski, K. Pakula, P.M. Mensz, and I.B. Bhat, phys. stat. sol. (b) **198**, 235 (1996).
[6] S. Strauf, P. Michler, J. Gutowski, U. Birkle, M. Fehrer, S. Einfeldt, and D. Hommel, phys. stat. sol (b) **216**, 557 (1999).
[7] S. Strauf, S.M. Ulrich, P. Michler, J. Gutowski, V. Kirchner, S. Figge, S. Einfeldt, and D. Hommel, in: Proc. Internat. Workshop on Nitride Semicond., IPAP Conf. Ser. **1**, 721 (2000).

Investigation of Defect Levels in Mg-Doped GaN Schottky Structures by Thermal Admittance Spectroscopy

N. D. Nguyen[1]) (a), M. Germain (a), M. Schmeits (a), B. Schineller (b), and M. Heuken (b, c)

(a) Institut de Physique, Université de Liège, B5, B-4000 Liège, Belgium

(b) AIXTRON AG, Kackerstr. 15-17, D-52072 Aachen, Germany

(c) Institut für Halbleitertechnik, RWTH Aachen, Templergraben 55, D-52056 Aachen, Germany

(Received June 21, 2001; accepted August 4, 2001)

Subject classification: 71.55.Eq; 73.30.+y; 73.40.–c; 73.61.Ey; 84.37.+q; S7.14

Schottky structures based on Mg-doped GaN layers grown by metalorganic chemical vapor deposition (MOCVD) on sapphire substrate are studied by thermal admittance spectroscopy from 90 K to room temperature. Evidence of two impurity levels results from the analysis of the observed peaks in the conductance curves, whose positions and strengths are temperature dependent. The experimental results are analyzed within a detailed theoretical study of the steady-state and small-signal electrical characteristics of the structure. Numerical simulations are based on the solution of the basic semiconductor equations for the structure consisting of two Schottky diodes connected back-to-back by a conduction channel formed by the GaN layer.

Introduction Successful Mg-doping of GaN has led to a breakthrough in the use of this wide band-gap compound semiconductor both for optoelectronics and for high-power, high-frequency electronic devices. As Schottky-type junctions are the building blocks of these devices, we have performed a thorough analysis of the conduction mechanism in GaN:Mg double Schottky structures. We use numerical simulation to fully explain the experimental results obtained by thermal admittance spectroscopy. It is shown that, due to the simultaneous role of Mg both as a dopant and as a deep impurity, the classical analysis of the Arrhenius plot underestimates the defect ionization energy. Moreover, we show the existence of a second shallow acceptor state with an activation energy of several tens of meV.

Experimental Results Epitaxial growth by metalorganic chemical vapor deposition (MOCVD) of the GaN layers studied in this work has been detailed elsewhere [1]. The structures consist of a nucleation layer grown directly on top of the sapphire substrate, followed by a 1 μm thick undoped buffer layer and a 2 μm thick Mg-doped layer. Coplanar metallic contacts (Ni/Au) of 0.5 mm diameter and separated by 2 mm, were evaporated on top of the layers.

In Fig. 1, frequency responses of the conductance G divided by $\omega = 2\pi f$, where f is the measurement frequency, of a typical sample, are shown for temperatures ranging from room temperature (RT) down to 90 K. Between RT and 200 K, the G/ω curves

---

[1]) Corresponding author; Tel.: +32-4-3663722; Fax: +32-4-3662990; e-mail: nd.nguyen@ulg.ac.be

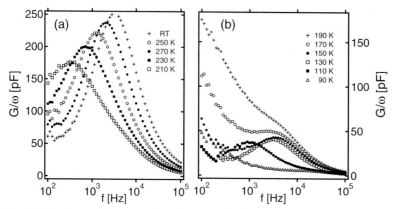

Fig. 1. Experimental curves of conductance G divided by $\omega = 2\pi f$ as function of frequency f for temperatures ranging from a) 293 (RT) to 210 K and b) 190 to 90 K (curves from top to bottom)

show a large peak at the cutoff frequencies f_c in C–f curves (not shown here). The amplitude of this peak decreases as the temperature is lowered. At around 200 K, a shoulder emerges from the high-frequency side of the mentioned peak and is resolved into another peak for temperatures below 170 K. Its position is also temperature dependent.

Numerical Simulation It is known that Mg gives rise to an acceptor-type energy level in the GaN bandgap. Values ranging from 100 to 250 meV above the valence band edge E_V are reported [2–6]. We denote by E_{tA} and N_{tA} the energy position and the total concentration of the level corresponding to the acceptor Mg impurity. In the same way, we introduce E_{tB} and N_{tB}, the energy position and the total concentration of the acceptor level whose existence is suggested by the experimental characteristics at low temperature. To investigate their role in the electrical characteristics of the Schottky diodes, explicit inclusion of the related active population in the semiconductor equations is necessary [7].

The classical basic semiconductor equations are solved numerically in order to obtain as function of the one-dimensional x-coordinate, the hole concentration p, and the occupied level concentrations n_{tA} and n_{tB} which are, with the electrical potential, the primary unknowns involved in the as-formulated system of equations. After solving the latter for the steady-state conditions, the small-signal amplitudes resulting from the application of an ac voltage of frequency $f = \omega/2\pi$ are calculated. A fixed Schottky barrier height $q\phi_b$, which is chosen identical for both contacts, is imposed in the expression of the boundary conditions. As a result of the ac calculation, the total current density, composed of the hole current density and the displacement current, is obtained. In a final step, we determine the conductance G and the capacitance C from the complex admittance $Y = G + i\omega C$ [7]. Further physical parameters are necessary to effectively reproduce the experimental admittance curves. In addition to $q\phi_b$, E_{tA} and E_{tB}, N_{tA} and N_{tB}, which have already been introduced, one needs the values of the defect thermal capture cross sections, σ_A and σ_B. The hole effective mass is taken as $m_h = 0.8 m_0$. The value of the hole mobility μ_h is fixed at 10 cm^2/Vs above $T_0 = 150$ K and given a $(T/T_0)^\alpha$ dependence below that temperature, with α being a free parameter. This T-dependence reproduces the decrease of μ_h with temperature which has been observed experimentally [4]. In order to reproduce with a one-dimensional formalism the electrical characteristics of a structure which is three-dimensional by nature, we have introduced a geometrical parameter R_μ multiplying the carrier mobilities and whose value is of the order of the ratio between the GaN layer thickness, which is in the micrometer range, and the contact diameter, which is in the millimeter range.

Discussion With the numerical procedure explained above, the electrical characteristics can be calculated at various temperatures, starting from an initial guess for the free parameters. A least-square fit procedure, based on the comparison of the electrical characteristics obtained from the numerical calculation to the experimental data, is used to determine the best values of the parameters. This optimization procedure yields the following values: $(E_{tA} - E_V) = 210$ meV, $(E_{tB} - E_V) = 30$ meV, $N_{tA} = 1 \times 10^{19}$ cm^{-3}, $N_{tB} = 2 \times 10^{16}$ cm^{-3}, $\sigma_A = 2 \times 10^{-19}$ cm^2, $\sigma_B = 2 \times 10^{-19}$ cm^2, $q\phi_b = 1.05$ eV, $R_\mu = 0.007$, and $\alpha = 3$. The resulting G/ω curves for six temperatures are shown in Fig. 2a. The as-obtained values of the parameters are comparable to those found in the literature for the Schottky barrier height [8, 9] in Au–Ni/p-GaN contacts, the activation energy of the Mg-related acceptor level determined by admittance spectroscopy [2, 3] or by Hall measurements [3, 4], and the thermal capture cross section of the defect A level [3]. The small values obtained for the activation energy and the total concentration of the defect B level, combined with the fact that its effect only appears below 150 K with a weaker amplitude than for level A, may greatly reduce the possibility to easily detect such a level. Nevertheless, more than one acceptor level has been observed in experimental works [2].

In Figs. 2b to g, we show the calculated capacitance and conductance (G/ω) curves of two different structure types between 350 and 100 K. One (labeled LD, for "long

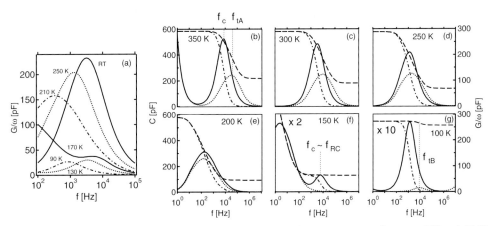

Fig. 2. a) Theoretical G/ω curves as function of frequency for temperatures between RT and 90 K. b) to g) Theoretical capacitance C (dashed lines) and conductance G/ω (dotted lines) curves for the short diode (SD) case, and calculated capacitance C (dash-dotted lines) and conductance G/ω (full lines) curves for the long diode (LD) case. Results are shown for temperatures between 350 and 100 K. The curves are magnified for temperatures of 150 K ($\times 2$) and 100 K ($\times 10$). Positions of the cutoff frequencies f_c and of the defect transition frequencies f_{tA} and f_{tB} are shown for some G/ω curves

diode") is the complete double Schottky diode structure which has been studied up to now. The other one is essentially the same structure but with a total length of 1 µm and $R_\mu = 1$ (labeled SD, for "short diode"). In this case, series resistance effects, which are explained below, are expected to play no role. The electrical characteristics for temperatures above 200 K are first discussed. In Figs. 2b, c, d and e, the cutoff frequency in the capacitance curve and the peak in the G/ω curve for the SD case are the signature of the Mg-related defect whose transition frequency is denoted by f_{tA}. Starting with this observation, we found that for $T > 240$ K, the cutoff frequency f_c in the LD case is mainly due to an electrical cutoff mechanism, i.e. $f_c \approx f_{RC}$, with $f_{RC} = 1/(2\pi R_S C)$, where R_S is the series resistance of the structure and C the low-frequency value of the capacitance, while for $T < 240$ K, f_c can be identified with f_{tA}. The cutoff frequency for the LD case is therefore the feature resulting from the combination of two mechanisms which are both present in the 200–350 K temperature range. An Arrhenius plot based on the temperature dependence of the peak position between 200 and 300 K yields activation energies of 176 and 107 meV, respectively, for the SD case and the LD case, which are both distinct from the input value of 210 meV. This is partly explained by the fact that the Mg impurity plays a double role: it acts as a dopant and as a deep impurity. The assumptions leading to the identification of the defect transition frequency with the thermal emission rate are not fulfilled [7, 10], even when resistance effects do not take place, i.e. in the SD case.

Below 200 K, the main peak in G/ω moves below the 100 Hz limit of the measurement equipment. A shoulder appears in the 10^3–10^4 Hz frequency range and leads to a second resolved peak around 4×10^4 Hz for temperatures around 150 K (Fig. 2f). The position of the latter is once more that of a $R_S C$ electrical cutoff whose evolution with temperature is mainly due to the variation of the series resistance R_S. The related cutoff frequency $1/(2\pi R_S C)$ is proportional to the conductivity of the layer and therefore proportional to p and μ_h. The calculated bulk concentration of holes p between 130 and 160 K varies only weakly with temperature, this explains why the displacement of this peak is small in this temperature range. Below 130 K, the displacement of the cutoff frequency is explained by the decrease of p due to the variation of the bulk value of $(E_F - E_{tB})$, where E_F is the Fermi energy, and by the reduction of μ_h due to various scattering mechanisms. In Fig. 2g, the transition frequency f_{tB} of defect B appears at 10^4 Hz in the admittance curves of the SD case at 100 K.

Conclusion This work shows that only a complete numerical study allows a correct interpretation of the admittance curves. In particular, it is shown that deducing activation energies from an Arrhenius plot of the cutoff frequencies would lead, in the case of the system we have studied, to an underestimation of the activation energies.

Acknowledgements Financial support by the Belgian Fonds National de la Recherche Scientifique (Contract No. 9.4565.96F) and by an INTAS grant No. N97-0995 are gratefully acknowledged.

References

[1] N. D. NGUYEN, M. GERMAIN, M. SCHMEITS, R. EVRARD, B. SCHINELLER, and M. HEUKEN, J. Cryst. Growth **230**, 600 (2001).
[2] J. W. HUANG, T. F. KUECH, H. LU, and I. BHAT, Appl. Phys. Lett. **68**, 2392 (1996).

[3] D. J. Kim, D. Y. Ryu, N. A. Bojarczuk, J. Karasinski, S. Guha, S. H. Lee, and J. H. Lee, J. Appl. Phys. **88**, 2564 (2000).
[4] W. Götz, R. S. Kern, C. H. Chen, H. Liu, D. A. Steigerwald, and R. M. Fletcher, Mater. Sci. Eng. B **59**, 211 (1999).
[5] U. Kaufmann, M. Kunzer, M. Maier, H. Obloh, A. Ramakrishnan, B. Santic, and P. Schlotter, Appl. Phys. Lett. **72**, 1326 (1998).
[6] A. K. Viswanath, E. Shin, J. I. Lee, S. Yu, D. Kim, B. Kim, Y. Choi, and C. H. Hong, J. Appl. Phys. **83**, 2272 (1998).
[7] M. Schmeits, N. D. Nguyen, and M. Germain, J. Appl. Phys. **89**, 1890 (2001).
[8] E. Monroy, F. Calle, J. L. Pau, F. J. Sanchez, E. Munoz, F. Omnes, B. Beaumont, and P. Gibart, J. Appl. Phys. **88**, 2081 (2000).
[9] Z. Z. Bandic, P. M. Bridger, E. C. Piquette, and T. C. McGill, Appl. Phys. Lett. **73**, 3276 (1998).
[10] P. Kozodoy, S. P. DenBaars, and U. K. Mishra, J. Appl. Phys. **87**, 770 (2000).

Low-Temperature Activation of Mg-Doped GaN with Pd Thin Films

I. Waki[1]) (a), H. Fujioka (a), M. Oshima (a), H. Miki (b), and M. Okuyama (b)

(a) Department of Applied Chemistry, The University of Tokyo, 7-3-1 Hongo, Bunkyo-ku, Tokyo, 113-8656, Japan

(b) Chichibu Research Laboratory, Central Research Laboratory, Showa Denko K.K., 1505 Shimokagemori, Chichibu-shi, Saitama, 369-1871, Japan

(Received June 25, 2001; accepted July 8, 2001)

Subject classification: 72.80.Ey; 73.61.Ey; S1; S7.14

The activation of metalorganic chemical vapor deposition (MOCVD)-grown Mg-doped GaN by N_2 annealing with thin Pd films has been investigated. p-type GaN with a hole concentration of 7×10^{16} cm^{-3} has been obtained at an annealing temperature as low as 200 °C using this technique. Thermal desorption spectroscopy (TDS) measurements have revealed that hydrogen is effectively removed from the Mg-doped GaN layer by the use of the Pd film.

Introduction Mg has been commonly used as an acceptor impurity in GaN-based materials. However, the carrier concentration achieved in p-type GaN is not high enough because of hydrogen passivation and compensation by residual donors [1, 2]. SIMS analysis has revealed that hydrogen atoms tend to be incorporated into GaN in the same order of magnitude with Mg atoms, suggesting a formation of Mg–H complex. Since as-grown Mg-doped GaN usually shows semi-insulating behavior, post-growth treatment is required to convert it into p-type GaN. Thermal annealing at temperatures above 700 °C in N_2 is most widely used for this purpose [3] because hydrogen atoms can be thermally removed from GaN at these temperatures.

Recently, we have shown that the p-type conduction of the film can be achieved by annealing at 200 °C with the use of a thin Ni film deposited on the surface of GaN [4]. It has been revealed that the Ni film acts as a catalyst to remove hydrogen from the surface of GaN. However, annealing with the Ni film at 800 °C also caused reduction in hole concentration. This phenomenon is due to the formation of nitrogen vacancies because the nitrogen desorption is enhanced by the Ni at temperatures above 460 °C as shown by our TDS study [5]. Therefore, it is important to find a suitable catalyst, which enhances only hydrogen desorption.

Experiments The samples used in this study were grown by metalorganic chemical vapor deposition (MOCVD). A Mg-doped GaN (0.85 μm)/undoped GaN (1.8 μm) structure was grown on a sapphire (0001) substrate using the low-temperature buffer layer technique. A Pd film was entirely deposited on the sample surface in a thermal evaporation chamber. The samples without Pd were also prepared by annealing in N_2 as references. The samples were then introduced into a quartz tube furnace for activation annealing at temperatures ranging from 200 to 800 °C in N_2 for 10 min. To remove

---

[1]) Corresponding author; Tel.: +81-3-5841-6753; Fax: +81-3-5841-6027; e-mail: waki@hotaka.t.u-tokyo.ac.jp

the Pd films on the surfaces, the samples were cleaned in boiling aquaregia for 15 min at room temperature. For the Hall effect measurements, Ni/Au electrodes were deposited for ohmic contacts. TDS measurements were performed to investigate hydrogen desorption characteristics in an ultra-high vacuum condition using EMD-WA1000S (ESCO Ltd.). After the baking of a sample stage in the chamber, the sample was set on the stage, and heated at a rate of 1.0 K/s by an infrared heating unit.

Results and Discussion Figure 1 shows resistivities of the samples activated with and without the Pd films as a function of annealing temperature. The resistivities dramatically decreased particularly at temperatures below 600 °C by the use of the Pd films. P-type GaN with a hole concentration of 7×10^{16} cm^{-3} and a resistivity of 9.1 Ω cm was obtained even at an annealing temperature as low as 200 °C for the sample activated with the Pd film. This hole concentration is higher than that of the sample activated with a Ni film at the same annealing temperature reported in our previous paper [4]. In addition, a minimum resistivity (3.6 Ω cm) with a hole concentration of 2×10^{17} cm^{-3} was achieved at the annealing temperature of 400 °C. On the other hand for the samples without the Pd films, p-type GaN was obtained by annealing at temperatures above 600 °C, and the minimum resistivity (3.1 Ω cm) with a hole concentration of 2.5×10^{17} cm^{-3} was achieved at the annealing temperature of 800 °C. The resistivity of the sample activated with Pd at 400 °C is five orders of magnitude lower than that of the sample activated without Pd at this annealing temperature. The enhancement of the acceptor activation is probably due to the catalytic effect of Pd for hydrogen desorption. Although annealing at 800 °C with the Ni film causes increase in the resistivity due to the nitrogen desorption [4, 5], we did not observe the increase for the sample activated with the Pd film. These results indicate that the nitrogen desorption is not enhanced by the Pd film at this temperature range. It should be noted that the resistivities of the samples activated with the Pd films did not change in the annealing tem-

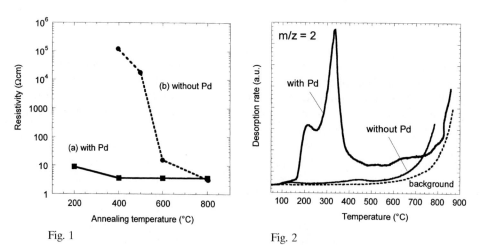

Fig. 1

Fig. 2

Fig. 1. Resistivity of the samples activated (a) with Pd films and (b) without Pd films in N$_2$ for 10 min. Hall effect measurements were carried out at room temperature

Fig. 2. TDS spectra of $m/z = 2$ for the samples with and without the Pd film. A background is also shown by the dotted line. The heating rate was kept at 1.0 K/s

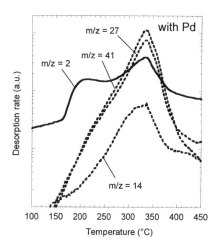

Fig. 3. TDS spectra of $m/z = 2$, 14, 27 and 41 for the sample with the Pd film. The heating rate was kept at 1.0 K/s

perature range from 400 to 800 °C. This is probably because the great majority of the hydrogen atoms which passivate Mg acceptors were removed from GaN by the annealing at 400 °C with the Pd film. Although the maximum hole concentration for both samples achieved in this study was somewhat lower compared with that achieved by the use of Ni in our recent study [4], this phenomenon is probably due to the wafer-to-wafer distribution of the crystal quality.

Figure 2 shows the TDS spectra for a mass to charge ratio (m/z) of 2 at the heating rate of 1.0 K/s. The desorption rate for $m/z = 2$ was considerably increased by the presence of the Pd film as can be judged from the two peaks around 210 °C and 335 °C. On the contrary, no peak was observed for the sample without the Pd film around this temperature and the hydrogen desorption rate gradually increased at temperatures above 600 °C. These results can well explain the difference in the shape of the two curves in Fig. 1. It also should be noted that some portion of hydrogen in Fig. 2 can be attributed to hydrocarbons on the sample surface. Hence, it is necessary to compare the spectra for $m/z = 2$ with the other spectra for further understanding.

Figure 3 shows the TDS spectra ($m/z = 2$, 14, 27 and 41) for the sample with the Pd film at heating rate of 1.0 K/s. The desorption rate is shown in a logarithmic scale in this figure. The m/z of 2, 14, 27 and 41 can be assigned to H_2^+, N^+ (or CH_2^+), $C_2H_3^+$ and $C_3H_5^+$, respectively. The peak observed around 210 °C for $m/z = 2$ was assigned to hydrogen from GaN because the peak shape was independent of those for $m/z = 14$, 27 or 41 around this temperature. Whereas, the other peak observed around 335 °C showed a similar temperature dependence to those for $m/z = 14$, 27 and 41 in this temperature range. We, therefore, conclude that this second hydrogen desorption peak includes the hydrogen partly from hydrocarbons which remained on the surface of the sample.

Conclusions P-type GaN has been obtained at an annealing temperature as low as 200 °C with the use of thin Pd catalytic films, and a minimum resistivity with a hole concentration of 2×10^{17} cm^{-3} was achieved at an annealing temperature of 400 °C. TDS measurements have revealed that hydrogen was effectively removed from the Mg-doped GaN layer by the use of Pd.

References

[1] Y. OHBA and A. HATANO, Jpn. J. Appl. Phys. **33**, L1367 (1994).
[2] P. KOZODOY, H. XING, S. P. DENBAARS, U. K. MISHRA, A. SAXLER, R. PERRIN, S. ELHAMRI, and W. C. MITCHEL, J. Appl. Phys. **87**, 1832 (2000).
[3] S. NAKAMURA, T. MUKAI, M. SENOH, and N. IWASA, Jpn. J. Appl. Phys. **31**, L139 (1992).
[4] I. WAKI, H. FUJIOKA, M. OSHIMA, H. MIKI, and A. FUKIZAWA, Appl. Phys. Lett. **78**, 2899 (2001).
[5] I. WAKI, H. FUJIOKA, M. OSHIMA, H. MIKI, and M. OKUYAMA, J. Appl. Phys., accepted.

Threading Dislocations and Optical Properties of GaN and GaInN

T. Miyajima[1]) (a), T. Hino (b), S. Tomiya (c), K. Yanashima (a), H. Nakajima (a),
T. Araki (d), Y. Nanishi (d), A. Satake (e), Y. Masumoto (e), K. Akimoto (f),
T. Kobayashi (a), and M. Ikeda (b)

(a) Core Technology & Network Company, Sony Corporation, 4-14-1 Asahi-cho, Atsugi, Kanagawa 243-0014, Japan

(b) Sony Shiroishi Semiconductor Inc., 3-53-2 Shiratori, Shiroishi, Miyagi 989-0734, Japan

(c) Environment & Analysis Tech. Dept., Sony Corporation, 2-1-1 Shin-sakuragaoka, Yokohama, Kanagawa 240-0036, Japan

(d) Dept. of Photonics, Ritsumeikan University, Kusatsu, Shiga 525-8577, Japan

(e) Institute of Physics, University of Tsukuba, Tsukuba, Ibaraki 305-8571, Japan

(f) Institute of Applied Physics, University of Tsukuba, Tsukuba, Ibaraki 305-8573, Japan

(Received June 29, 2001; accepted August 4, 2001)

Subject classification: 61.72.Ff; 68.37.Ps; 78.47.+p; 78.55.Cr; 78.60.Hk; 81.15.Gh; S7.14

We categorized threading dislocations in GaN and GaInN multiple quantum wells and epitaxially lateral overgrown GaN into three types of line defects (edge, screw and mixed dislocations), and investigated the optical properties. It was confirmed by cathodoluminescence measurements that not only screw and mixed dislocations but also edge dislocations act as non-radiative centers in GaN. Epitaxial lateral overgrowth (ELO) technique can reduce the densities of all line-defects in a several μm wide wing region. Growth steps in the wing region were disturbed by the defects which were left in a seed region, and a complicated structure was formed at the surface of GaN and GaInN layers grown on ELO-GaN at low temperature. We believe that this surface structure formed by high supersaturation is a cause of In compositional spatial fluctuation and phase separation of GaInN alloy.

1. Introduction Although many threading dislocations of 10^8–10^{10} cm^{-2} exist in a GaN epitaxial layer grown on sapphire substrate, GaN-based light-emitting diodes (LEDs) have a high external quantum efficiency [1] and it has been suggested that the dislocations do not act as non-radiative centers. However, it was recently demonstrated that the reduction of dislocations can improve the lifetime of GaN-based laser diodes [2, 3]. To understand this apparent contradiction, it is necessary to know not only the radiative emission mechanism of the GaInN active layer with threading dislocations [4–7], but also the optical properties of the threading dislocations themselves.

It has been reported by Rosner et al. [8] that threading dislocations act as non-radiative centers in the GaN layer. Furthermore, Sugahara et al. [9] have confirmed this result using TEM and cathodoluminescence (CL). We have developed an etch-pit observation method to categorize the threading dislocations into three types of line defects – edge, screw and mixed dislocations –, and discussed the optical properties of each line defects [10, 11]. More recently, several researchers distinguished the line defects and discussed their properties [12, 13].

---

[1]) Corresponding author; Phone: +81 46 230 5089; Fax: +81 46 230 5775;
e-mail: takao.miyajima@jp.sony.com

Recently, we reduced threading dislocation density of GaN to 10^6 cm^{-2} using epitaxial lateral overgrowth (ELO) [14–16] without a SiO$_2$ mask, which was developed by Zheleva et al. [17], and demonstrated a practical lifetime of more than 1000 hours for a GaN-based laser diode with an output power of 30 mW [18]. This ELO without a SiO$_2$ mask has the advantage of a smaller c-axis tilting [19, 20] in the wing region.

In this paper, we discuss the role of the threading dislocations in epitaxial lateral overgrown GaN (ELO-GaN) and GaInN multiple quantum wells on ELO-GaN after reviewing our result about the threading dislocations in GaN.

2. Experimental The samples were grown on c-face (0001) sapphire substrates using metal-organic chemical vapor deposition (MOCVD) with low-temperature buffer layer. The growth condition is described elsewhere [21]. The sample of GaN:Si consists of 1.5 μm thick Si-doped GaN on 1.0 μm thick undoped GaN. All samples of GaN:Si have almost the same free electron concentration and mobility, which were typically 2.0×10^{18} cm^{-3} and 265 cm^2/Vs, respectively. The sample of ELO-GaN consists of 2.5 μm thick Si-doped GaN on a seed layer of 2 μm thick Si-doped GaN. The growth of ELO-GaN is described elsewhere [22]. The sample of GaInN multiple quantum wells (MQWs) consists of three pairs of 3.5 nm thick Ga$_{0.90}$In$_{0.10}$N:Si well layer and a 3.5 nm thick Ga$_{0.98}$In$_{0.02}$N:Si barrier layer, and 50 nm thick GaN:Si capping layer on ELO-GaN. The Si concentration in ELO-GaN and GaInN MQWs was 1×10^{18} and 1×10^{19} cm^{-3}, respectively.

For etch-pit observation, GaN surfaces were etched in a horizontal reactor by HCl gas at the temperature of 600 °C during 30 min [10]. The HCl gas was diluted by N$_2$ gas, and the flow rate of [HCl]/[N$_2$] was 0.2. The etching rate was approximately 2 nm/min in the flat region. The number and shape of etch pits were observed using SEM.

Photoluminescence (PL) measurements were performed using a 25 mW He–Cd laser emitting at 325 nm at room temperature. The excitation density was 0.14 W/cm^2. Cathodolumenescence (CL) measurements were performed at room temperature using an electron acceleration energy of 5 keV.

3. Results and Discussion
3.1 Categorization of threading dislocations in GaN
Three types of etch pits, labeled α, β and γ, were observed in an etched GaN surface as shown in the SEM image of Fig. 1. The α-, β- and γ-type etch pits have a well-defined hexagonal shape with a large core, an ambiguous hexagonal shape with a small core, and a polygonal shape without a core, respectively. Figure 2 shows the depth profiles of the etch pits observed using

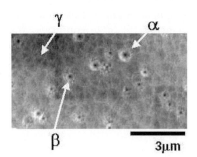

Fig. 1. Typical SEM image of etched GaN surface

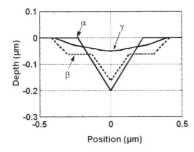

Fig. 2. Schematic depth profiles of α-, β- and γ-type etch pits

AFM. The α-type etch pit has a triangular shape, which has a {1$\bar{1}$02} facet. The γ-type etch pits has an arc shape. The β-type etch pit has a combination of triangular and trapezoidal shape, which seems the results of a combination of α- and γ-type.

Figure 3 shows the cross sectional TEM image of the three types of etch pits. The α-type etch pits originate mostly from a screw dislocation with Burgers vector of $\mathbf{b} = \langle 0001 \rangle$, and slightly nanopipe. The β- and γ-type etch pits originate from a mixed dislocation with $\mathbf{b} = 1/3\langle 11\bar{2}3 \rangle$ and an edge dislocation with $\mathbf{b} = 1/3\langle 11\bar{2}0 \rangle$, respectively. The Burgers vector $\mathbf{b}$ was determined using the relation $\mathbf{g} \cdot \mathbf{b} = 0$, where $\mathbf{g}$ is the diffraction vector.

This etch-pit observation was used for categorizing threading dislocations into the line defects and estimating the density of each line defect in this study.

3.2 Optical properties of edge, screw and mixed dislocations in GaN:Si

Band-edge emission was observed at an energy of 3.42 eV in the room-temperature PL spectra of Si-doped GaN. The Si concentration was maintained to be 2.0×10^{18} cm^{-3}. The emission should be ascribed to donor-bound excitons. A deep emission was also observed around 2.3 eV, but the intensity was 1/10 to 1/100 lower than that of band-edge emission. The PL intensity of band-edge emission is strongly related to the density of screw and mixed dislocations, as shown in Fig. 4, although the edge dislocation was majority in the total density of the three line defects and the density was relatively constant. The density of the edge dislocations was $2-3 \times 10^8$ cm^{-2}. This dependency indicates, therefore, that the screw and mixed dislocations should act as strong non-radiative centers in the GaN layer. This result is supported by Northrup [23] who theoretically predicted that a screw dislocation with Ga-filled core would be a strong non-radiative center in GaN.

At the time, we believe that the edge dislocations of other line defects might not act as non-radiative centers, as reported by Elsner et al. [24] who have theoretically suggested that edge and open-core screw dislocations are electrically inactive. However, high spatial resolution CL measurements show that edge dislocations also act as non-

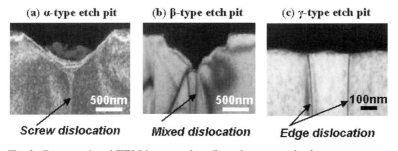

Fig. 3. Cross sectional TEM images of α-, β- and γ-type etch pits

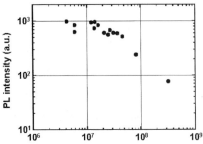

Fig. 4. Room-temperature PL intensity dependence on the total density of screw and mixed dislocations for Si-doped GaN. Si concentration was maintained to be 2.0×10^{18} cm^{-3}. The dislocation densities were estimated using etch-pit observation

radiative center. Figure 5 shows a SEM image of etched GaN:Si surface and a monochromated CL image of as-grown GaN:Si surface. These images were taken using two 5×5 mm^2 samples which were cut from the next places of the same wafer. The CL image was taken at a wavelength of 362 nm which corresponds to the band-edge emission of GaN:Si. Using etch-pit observation, the density of edge, screw and mixed dislocations was estimated to be 1.0×10^9, 1.1×10^7 and 2.1×10^7 cm^{-2}, respectively. The density of dark spots in the CL images was around 5×10^8 cm^{-2}, which is close to the density of edge dislocations. This result shows that edge dislocations act as non-radiative centers in GaN.

3.3 Optical parameters of GaN:Si

Figure 6 shows the room temperature PL decay of band-edge emission for samples A and B of GaN:Si. The density of edge, screw and mixed dislocations for sample A was 2.2×10^8, 7.4×10^5 and 3.0×10^6 cm^{-2}, respectively. The density of edge, screw and mixed dislocations for sample B was 2.6×10^8, 2.4×10^8 and 8.2×10^7 cm^{-2}, respectively. The set-up of time-resolved PL measurements is described elsewhere [11]. The PL lifetime (τ_{PL}) of samples A and B was estimated to be τ_{PL} = 250 and 68 ps, respectively. We believe that the PL lifetime of sample B is shortened by the increase of threading dislocations, and that even the PL lifetime (τ_{PL}) of sample A could be determined by the non-radiative lifetime.

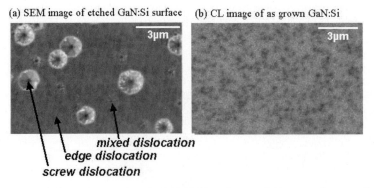

Fig. 5. a) SEM image of etched GaN:Si surface and b) monochromated CL image of as-grown GaN:Si surface at a wavelength of 362 nm. These images were taken using two 5×5 mm^2 samples which were cut from next places of the same wafer

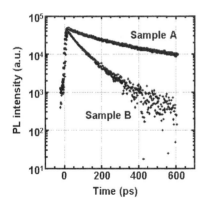

Fig. 6. Time-resolved PL spectra at room temperature for samples A and B of GaN:Si. The density of edge, screw and mixed dislocations for sample A was 2.2×10^8, 7.4×10^5 and 3.0×10^6 cm^{-2}, respectively. The density of edge, screw and mixed dislocations for sample B was 2.6×10^8, 2.4×10^8 and 8.2×10^7 cm^{-2}, respectively. The excitation energy was 3.594 eV and the detection energy 3.415 eV

The radiative efficiency (η) at room temperature was estimated to be $\eta = 0.18$ [11] using photo-calorimetric spectroscopy (PAC), where a re-absorption of photons in the sample was ignored. This estimation was confirmed by the ratio of PL integral intensities at 300 K (I_{300}) and at 5 K (I_5) of $I_{300}/I_5 = 0.15$, which is the radiative efficiency at room temperature if η is assumed to be 1.0 at $T = 5$ K. Then, radiative (τ_r) and non-radiative lifetime (τ_{nr}) of sample A were calculated to be $\tau_r = 1.4$ ns and $\tau_{nr} = 300$ ps using the two equations $1/\tau_{PL} = 1/\tau_r + 1/\tau_{nr}$ and $\eta = (1/\tau_r)/(1/\tau_r + 1/\tau_{nr})$. Because the difference between radiative lifetime and non-radiative lifetime was not large, the PL lifetime did not have a strong dependence on threading dislocation density.

Using the radiative lifetime of $\tau_r = 1.4$ ns, the carrier concentration of $n = 1.9 \times 10^{18}$ cm^{-3} and the equation $\tau_r = 1/Bn$, the radiative recombination coefficient (B) was calculated to be 3.8×10^{-10} cm^3/s. This is close to 7×10^{-10} cm^3/s which was theoretically calculated by Brandt et al. [25].

3.4 Threading dislocations in epitaxial lateral overgrown GaN
Etch-pit observation was performed for ELO-GaN as shown in Fig. 7. In the wing region which corresponds to the laterally overgrown GaN, all three types of line defects were reduced. The edge-dislocation density was 1×10^6 cm^{-2} and each of screw and mixed dislocation density was less than 10^5 cm^{-2}. The same density of threading dislocations was observed in the seed region as in the GaN layer grown simply on the sapphire substrate. Edge and

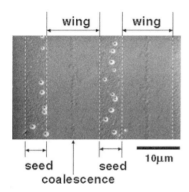

Fig. 7. SEM image of etched ELO-GaN surface

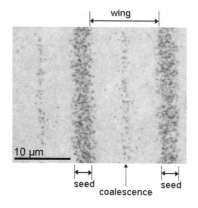

Fig. 8. CL image of as-grown ELO-GaN surface

mixed dislocations of around a 10^8 cm^{-2} density were also observed in the coalescence region. Similar results were obtained in plan-view TEM observations.

A room-temperature CL image of ELO-GaN is shown in Fig. 8. In the seed region, there are many dark spots which originated from the threading dislocations which penetrated from the seed GaN layer and were generated in the coalescence region. The density was several times 10^8 cm^{-2}. On the other hand, there are a few dark spots in the wing region. The density was about 1×10^6 cm^{-2}. In practice, when the laser stripe forms in the wing region, a long practical lifetime of over 1000 hours for a GaN-based laser diode has been demonstrated even with a high output power of 30 mW at a high temperature atmosphere of 50 °C [18]. The threading-dislocation density in the wing region is still high as compared with that in a GaAs- and InP-based laser diodes. We believe that the rate of proliferation of non-radiative centers is very low in GaN-based laser diodes, possibly due to the innate character of the GaN-based semiconductor.

3.5 Threading dislocations in GaInN multiple quantum wells Figure 9 shows a CL image of GaInN MQWs grown simply on a sapphire substrate. Two types of dark spots were clearly observed in it. The small dark spots with a diameter of 50 nm could be ascribed to edge dislocations because the density of 1×10^9 cm^{-2} is close to the edge-dislocation density of 3.1×10^8 cm^{-2}. The large dark spots with a diameter of 110 nm could be ascribed to screw or mixed dislocations because the density of 2.3×10^7 cm^{-2} is close to the screw-dislocation density of 1.3×10^7 cm^{-2} and the mixed-dislocation density of 2.2×10^7 cm^{-2}. We believe that the diameters of dark spots do not show minority carrier diffusion length of defect but show the growth pits originated by threading dislocations because the growth pits (holes) were easily formed from the threading dislocations in GaInN layer or GaN layer grown at low temperature [12, 26, 27]. The density of growth pits can be reduced using ELO-GaN which has the wing region with low dislocation density of 10^6 cm^{-2}. However we observed another problem

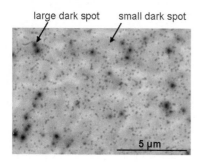

Fig. 9. CL surface image of GaInN MQWs grown on a sapphire substrate

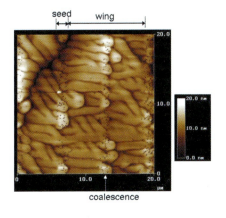

Fig. 10. AFM surface image of GaInN/GaN MQWs grown on ELO-GaN

related to the dislocation in the seed region of ELO-GaN, as shown in the next section.

In the CL image, an inhomogeneity of CL intensity is observed and is not related to the position of dark spots. We believe that this inhomogeneity could be related to the In compositional fluctuation, and one of the causes is the growth steps on the surface as described later.

3.6 Threading dislocations in GaInN multiple quantum wells on ELO-GaN Figure 10[2]) shows an AFM image of the surface of GaInN MQWs grown on ELO-GaN. A few pits were observed in the wing region, but many pits of 10^9 cm^{-2} were observed in the seed and the coalescence region. As can be seen in the figure, growth steps were disturbed by threading dislocations in the seed and coalescence region, and formed a complicated surface structure. Similar phenomena have been observed by Hansen et al. [28]. Generally when the growth temperature decreases, the degree of supersaturation increases, the spiral growth easily occurs, and the curvature radius decreases. We believe that this growth situation takes place during the low-temperature growth of GaN and GaInN, so that a complicated surface structure is formed on the surface. There could be screw and mixed dislocations with large Burgers vector along c-axis at the center of the spiral growth. The surface structure can be one of the causes of In fluctuation and phase separation in the GaInN layer.

4. Conclusion Mixed and screw dislocations, which have a large Burgers vector along c-axis of GaN, strongly affect not only the optical properties of GaN but also the surface structure of GaN and GaInN although the density of mixed and screw dislocations is about 1/10 times less than that of edge dislocations. ELO can reduce the overall line defect density to less than 10^6 cm^{-2} in the wing region. However, threading dislocations, especially mixed and screw dislocations in the seed region, disturb the growth steps of GaN and GaInN grown at low temperature. We believe that the disturbed growth steps, which are formed for high supersaturation, are one of the causes of In fluctuation and phase separation of GaInN alloy.

Acknowledgements The authors would like to thank Prof. T. Nishinaga for his advice of growth mechanism, T. Asano for growing samples, and T. Yamaguchi and S. Kijima for fabricating ELO-GaN, T. Asatsuma for his help of AFM measurements, and K. Funato for fruitful discussion. They also would like to thank Dr. O. Kumagai for the encouragement during this research.

References

[1] S.D. LESTER, F.A. PONCE, M.G. CRAFORD, and D.A. STEIGERWALD, Appl. Phys. Lett. **66**, 1249 (1995).
[2] S. NAKAMURA, M. SENOH, S. NAGAHAMA, N. IWASA, T. YAMADA, T. MATSUSHITA, H. KIYOKU, Y. SUGIMOTO, T. KOZAKI, H. UMEMOTO, M. SANO, and K. CHOCHO, 2nd Internat. Conf. on Nitride Semicond., Tokushima (Japan) 1997 (p. 444).
[3] S. UCHIDA, S. KIJIMA, T. TOJYO, S. ANSAI, M. TAKEYA, T. HINO, K. SHIBUYA, S. IKEDA, T. ASANO, K. YANASHIMA, S. HASHIMOTO, T. ASATSUMA, M. OZAWA, T. KOBAYASHI, Y.YABUKI, T. AOKI, and M. IKEDA, Proc. SPIE **3947**, 156 (2000).
[4] S. CHICHIBU, T. AZAHATA, T. SOTA and S. NAKAMURA, Appl. Phys. Lett. **69**, 4188 (1996).
[5] Y. NARUKAWA, Y. KAWAKAMI, SZ. FUJITA, SG. FUJITA, and S. NAKAMURA, Phys. Rev. B **55**, R1938 (1997).
[6] A. SATAKE, Y. MASUMOTO, T. MIYAJIMA, T. ASATSUMA, F. NAKAMURA, and M. IKEDA, Phys. Rev. B **57**, R2041 (1998).
[7] A. YAMAGUCHI, Y. MOCHIZUKI, and M. MIZUTA, Ext. Abstr. SSDM, Tokyo 1999 (pp. 64).
[8] S.J. ROSNER, E.C. CARR, M.J. LUDOWISE, G. GIROLAMI, and H.I. ERIKSON, Appl. Phys. Lett. **70**, 420 (1997).
[9] T. SUGAHARA, H. SATO, M. HAO, Y. NAOI, S. KURAI, S. TOTTORI, K. YAMASHITA, K. NISHINO, L.T. ROMANO, and S. SAKAI, Jpn. J. Appl. Phys. **37**, L398 (1998).
[10] T. HINO, S. TOMIYA, T. MIYAJIMA, K. YANASHIMA, S. HASHIMOTO, and M. IKEDA, Appl. Phys. Lett. **76**, 3421 (2000).
[11] T. MIYAJIMA, T. HINO, S. TOMIYA, A. SATAKE, E. TOKUNAGA, Y. MASUMOTO, T. MARUYAMA, M. IKEYA, S. MORISHIMA, K. AKIMOTO, K. YANASHIMA, S. HASHIMOTO, T. KOBAYASHI, and M. IKEDA, Proc. Internat. Workshop on Nitride Semiconductors, IPAP Conf. Ser. **1**, 536 (2000).
[12] D. CHERNS, S.J. HENLEY, and F.A. PONCE, Appl. Phys. Lett. **78**, 2691 (2001).
[13] J.W.P. HSU, M.J. MANFRA, D.V. LANG, S. RICHTER, S.N.G. CHU, A.M. SERGENT, R.N. KLEIMAN, L.N. PFEIFFER, and R.J. MOLNAR, Appl. Phys. Lett. **78**, 1685 (2001).
[14] T. NISHINAGA, T. NAKANO, and S. ZHANG, Jpn. J. Appl. Phys. **27**, L964 (1988).
[15] A. USUI, H. SUNAKAWA, A. SAKAI, and A. YAMAGUCHI, Jpn. J. Appl. Phys. **36**, L899 (1997).
[16] O.H. NAM, M.D. BREMSER, T. ZHELEVA, and R.F. DAVIS, Appl. Phys. Lett. **71**, 2638 (1997).
[17] T.S. ZHELEVA, S.A. SMITH, D.B. THOMSON, T. GEHRKE, K.J. LINTHICUM, P. RAJAGOPAL, E. CARLSON, W.M. ASHMAWI, and R.F. DAVIS, MRS Internet J. Nitride Semicond. Res. **4S1**, G3.38 (1999).
[18] M. IKEDA and S. UCHIDA, Proc. of China–Japan Workshop on Nitride Semiconductor Materials and Devices, Shanghai (China) 2001 (p. 100).
[19] A. SAKAI, H. SUNAKAWA, and A. USUI, Appl. Phys. Lett. **73**, 481 (1998).
[20] S. TOMIYA, T.HINO, S. KIJIMA, T. ASANO, H. NAKAJIMA, K. FUNATO, T. ASATSUMA, T. MIYAJIMA, K. KOBAYASHI, and M. IKEDA, Proc. Internat. Workshop on Nitride Semiconductors, IPAP Conf. Ser. **1**, 284 (2000).
[21] T. ASANO, K. YANASHIMA, T. ASATSUMA, T. HINO, T. YAMAGUCHI, S. TOMIYA, K. FUNATO, T. KOBAYASHI, and M. IKEDA, phys. stat. sol. (a) **176**, 23 (1999).
[22] M. TAKEYA, K. YANASHIMA, T .ASANO, T. HINO, S. IKEDA, K. SHIBUYA, S. KIJIMA, T. OJYO, S. ANSAI, S. UCHIDA, Y. YABUKI, T. AOKI, T. ASATSUMA, M. OZAWA, T. KOBAYASHI, E. MORITA, and M. IKEDA, J. Cryst. Growth. **221**, 646 (2000).
[23] J.E. NORTHRUP, Appl. Phys. Lett. **78**, 2288 (2001).
[24] J. ELSNER, R. JONES, P. K. SITCH, V. D. POREZAG, M. ELSNER, TH. FRAUENHEIM, M. I. HEGGIE, S. ÖBERG, and P.R. BRIDDON, Phys. Rev. Lett. **79**, 3672 (1997).
[25] O. BRANDT, H.-J. WÜNSCHE, H. YANG, J. MÜLLHÄUSER, R. KLANN, and K. PLOOG, J. Cryst. Growth **189/190**, 790 (1998).
[26] C.J. SUN, M.Z. ANWAR, Q. CHEN, J.W. YANG, M.A. KAHN, M.S. SHUR, A.D. BYKHOVSKI, Z. LILIENTHAL-WEBER, C. KISIELOWSKI, M. SMITH, J.Y. LIN, and H.X. XIANG, Appl. Phys. Lett. **70**, 2978 (1997).
[27] Y. KAWAGUCHI, M. SHIMIZU, M. YAMAGUCHI, K. HIRAMATSU, N. SAWAKI, W. TAKI, H. TSUDA, N. KUWANO, K. OKI, T. ZHELEVA, and R. F. DAVIS, Proc. 2nd Internat. Conf. Nitride Semiconductors, Tokushima (Japan), Oct. 27–31, 1997 (p. 22).
[28] M. HANSEN, P. FINI, B. HEYING, J.S. SPECK, and S.P. DENBAARS, Proc. Internat. Workshop on Nitride Semiconductors, IPAP Conf. Ser. **1**, 316 (2000).

Mosaicity of GaN Epitaxial Layers: Simulation and Experiment

R. Chierchia[1]), T. Böttcher, S. Figge, M. Diesselberg, H. Heinke and D. Hommel

University of Bremen, Institute of Solid State Physics, P.O.Box 330440, 28334 Bremen, Germany

(Received June 22, 2001; accepted July 4, 2001)

Subject classification: 61.10.Nz; 61.72.Dd; 68.55.Jk; S7.14

High-resolution X-ray diffraction has been used to analyze GaN epilayers with varying coalescence thickness which were grown by MOVPE on (0001) oriented sapphire. The decrease of the density of edge type threading dislocations with increasing coalescence thickness causes a marked difference in the mosaicity of the samples. As the defects form along the grain boundaries, this corresponds to an increase in lateral coherence length with increasing coalescence thickness. The lateral coherence length has been obtained from simulations of reciprocal lattice points of off-axis Bragg reflections, measured in asymmetric diffraction geometry.

Introduction The grain coalescence process during metal organic vapour phase epitaxy (MOVPE) of GaN on sapphire strongly influences the density of edge type threading dislocations (named edge dislocation density in the following). GaN is commonly grown utilizing a low temperature nucleation layer. During the successive high temperature growth three-dimensional islands form which coalesce towards a smooth film. This causes a reduction of the edge dislocation density with increasing coalescence thickness as the edge dislocations form along the crystallite boundaries to compensate for the in-plane rotation (twist) of the islands [1–3]. The mosaic structure of the GaN layer is determined by a set of four parameters wich are illustrated in Fig. 1.

The angular mismatch of the crystallites is given by the tilt and the twist, describing the angular distribution of the crystallographic orientation of the mosaic blocks perpendicular to and within the growth plane. The lateral and vertical dimensions of the crystallites are given by the lateral and vertical coherence length. All parameters can be accessed by X-ray diffraction. This work investigates the effect of tilt and lateral coherence length of the crystallites on the distribution of scattered intensity in reciprocal space. This will be discussed for the example of epitaxial layers grown with different coalescence thickness.

Experiment The GaN layers were grown on (0001) oriented sapphire by MOVPE at a growth temperature of 1050 °C. Through changes in growth conditions the coalescence thickness was varied between 250 and 4200 nm [2, 4] corresponding to smaller and larger in-plane average grain diameters, respectively. The total thickness of the samples was between 2000 and 5000 nm. The X-ray measurements were performed using high-resolution X-ray diffractometers (Philips MRD and X'Pert MRD) equipped with a Cu

---
[1]) Corresponding author; Phone: +49 421 218 3380; Fax: +49 421 218 4581; e-mail: chierchi@physik.uni-bremen.de

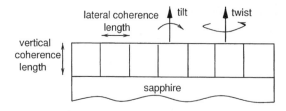

Fig. 1. Illustration of the characteristic parameters of a mosaic layer

sealed anode and a four-crystal monochromator in Ge(220) configuration. All data were recorded by triple axis measurements using a two or three reflection analyzer. In particular, rocking curves in symmetric diffraction geometry (symmetric reflections) and reciprocal space maps (RSMs) in asymmetric diffraction geometry (asymmetric reflections) were recorded.

Impact of Mosaicity on Reciprocal Lattice Points Tilt and lateral coherence length can be accessed by measurements of symmetric or asymmetric reflections. In symmetric diffraction geometry, the results are obtained from a Williamson-Hall plot [5] of the (002), (004), (006) diffraction peak widths, whereas for asymmetric diffraction geometry reciprocal space maps have to be recorded. For highly imperfect epitaxial layers, one commonly observes broadened reciprocal lattice points (RLPs), i.e. distributions of scattered intensity. These RLPs often exhibit elliptical shape. The orientation of the main axis of these ellipses depends on the characteristics of the mosaic structure of the layer. The main axis will be oriented parallel to q_x (if z is oriented along the surface normal of the sample), if the finite lateral size of the crystallites dominates the mosaic structure. In contrast, the RLPs will be dominantly broadened perpendicular to the diffraction vector q, if the tilt of the crystallites determines the mosaicity. All cases in between can be described by an inclination angle α between the main axis of the RLP ellipses and the q_x-axis which varies between 0 (finite size) and ϕ (tilt), with ϕ being the inclination angle of the reflecting lattice planes with respect to the sample surface.

RLPs of mosaic layers can be simulated using a model developed by Holy et al. [6], which is based on the kinematical diffraction theory. Examples of simulated RSMs for the (105) reflection of GaN obtained from the model are presented in Fig. 2.

The input parameters tilt and lateral coherence length have been varied for the simulations. The extreme case of dominant lateral coherence length is shown in Fig. 2a.

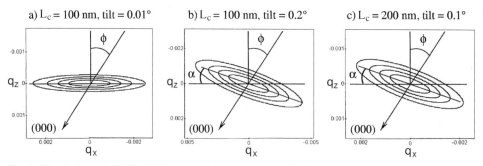

Fig. 2. Simulation of (105) RSMs according to a model developed by V. Holy [6]. a) Dominant lateral coherence length; b), c) dominant tilt. The input parameters are given on top of the figure

There the RLP is strictly aligned along the q_x-direction pointing to the dominant effect of the lateral coherence length. In contrast, the RLP-ellipse is clearly rotated in Fig. 2b which is attributed to a stronger influence of the tilt. However, the orientation of the RLP is sensitive to the product of tilt and lateral coherence length. This is obvious by a comparison of Fig. 2b and c, which are simulated for a lateral coherence length of 100 nm and 200 nm while assuming a tilt of 0.2° and 0.1°, respectively. Hence, an independent evaluation of both parameters from RSMs of asymmetric reflections only is not possible. For this reason, the tilt determined by measurements in symmetric diffraction geometry was assumed for the evaluation of RSMs of asymmetric reflections in the present study.

Experimental Results and Discussion Williamson-Hall plots of the rocking curve widths of symmetric reflections enable to determine the tilt and lateral coherence length. As can be seen in Table 1, the tilt found for all layers is rather small and the lateral coherence length is around 1000 nm. As there is no obvious dependence on the coalescence thickness, the tilt and the lateral coherence length as determined from symmetric reflections are probably determined during nucleation.

This observation can be understood by the specific sensitivity of different reflections to various types of dislocations [5, 7]. Symmetric reflections are sensitive to the screw-type threading dislocations only. Thus the determined lateral coherence length L_{c1} characterizes the mosaic structure correlated with the screw dislocations. In order to obtain information on the mosaicity defined by edge dislocations, asymmetric RLPs have to be analyzed.

Therefore reciprocal space maps of three different asymmetric reflections ((105), $\phi = 20.6°$; (114), $\phi = 39.1°$; (213), $\phi = 58.9°$) were performed for each of the samples studied, and the orientation of the RLP-ellipses was determined. The corresponding data for the angle α is shown in Fig. 3. The inclination angle α of the RLP is plotted versus the inclination of the reflecting lattice planes ϕ. A clear trend is observed, as an increase of the coalescence thickness causes an increase of the inclination of the RLPs. Thus, the lateral coherence length rises and the tilt dominates the broadening of the RLPs at larger coalescence thickness.

The experimental data were compared with theoretical ones, resulting from simulations of reciprocal space maps for the reflections mentioned. For the simulation, the tilt was held constant at the values given in Table 1 and the lateral coherence length was varied to obtain best agreement in the α-versus-ϕ curves. The results of that procedure are shown in Fig. 3 and Table 1.

As can be seen in Fig. 3, the simulated data are in rather good agreement with the measured values. Deviations exist for larger lattice plane inclinations, which become

Table 1

Tilt and lateral coherence lengths as determined from measurements in symmetric and asymmetric diffraction geometry for GaN layers with different coalescence thicknesses

coalescence thickness (nm)	tilt (°)	L_{c1} symm. (nm)	L_{c2} asymm. (nm)
250	0.063	590	190
350	0.081	880	200
1000	0.056	1550	300
4200	0.047	950	540

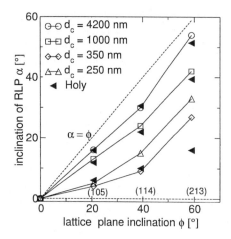

Fig. 3. Experimental (open symbols) and simulated data (close symbols) for the orientation of RLPs for different asymmetric reflections determined for GaN layers with different coalescence thickness d_c. The angles α and ϕ are defined in Fig. 2

more clear for lower α values. The resulting lateral coherence length increases with increasing coalescence thickness (see Table 1). This supports the model that the edge dislocations originate from the island coalescence process.

Summary In conclusion, the mosaicity of GaN epitaxial layers characterized by different coalescence thicknesses has been analyzed by X-ray diffraction. In particular, a lateral coherence length has been obtained both from simulations of reciprocal lattice points of asymmetric reflections and from Williamson-Hall plots of the rocking curve widths of symmetric reflections. Only the lateral coherence length determined from the asymmetric reflections increases with increasing coalescence thickness. This is in agreement with a decrease of the edge dislocation density, and the assumption that the defects form along the grain boundaries.

Acknowledgement The authors thank V. Holy for providing a program used as a basis for the simulations.

References

[1] F.A. PONCE, MRS Bull. **22**, 51 (1997).
[2] T. BÖTTCHER, S. EINFELDT, S. FIGGE, R. CHIERCHIA, H. HEINKE, D. HOMMEL, and J.S. SPECK, Appl. Phys. Lett. **78**, 1976 (2001).
[3] X.H. WU, P. FINI, E.J. TARSA, B. HEYING, S. KELLER, U.K. MISHRA, S.P. DENBAARS, and J.S. SPECK, J. Cryst. Growth **189/190**, 231 (1998).
[4] S. FIGGE, T. BÖTTCHER, S. EINFELDT, and D. HOMMEL J. Cryst. Growth **221**, 262 (2000).
[5] T. METZGER, R. HÖPPLER, E. BORN, O. AMBACHER, M. STUTZMANN, R. STÖMMER, M. SCHUSTER, H. GÖBEL, S. CHRISTIANSEN, M. ALBRECHT, and H.P. STRUNK, Philos. Mag. A **77**, 1013 (1998).
[6] V. HOLY, J. KUBENA, E. ABRAMOF, K. LISCHKA, A. PESEK, and E. KOPPENSTEINER, J. Appl. Phys. **74**, 1736 (1993).
[7] B. HEYING, X.H. WU, S. KELLER, Y. LI, D. KAPOLNEK, B.P. KELLER, S.P. DENBAARS, and J.S. SPECK, Appl. Phys. Lett. **68**, 643 (1995).

Correlation of Defects and Local Bandgap Variations in GaInN/GaN/AlGaN LEDs

F. Hitzel (a), A. Hangleiter (a), S. Bader (b), H.-J. Lugauer (b), and V. Härle (b)

(a) Institute of Technical Physics, TU Braunschweig, Germany

(b) Osram Opto Semiconductors, Regensburg, Germany

(Received June 21, 2001; accepted August 14, 2001)

Subject classification: 68.55.Ln; 71.55.Eq; 78.55.Cr; 78.60.Fi; 85.60.Jb; S7.14

Currently even the best available GaN device structures have an extremely high defect density. Their surprisingly small influence on the carrier recombination rates and brightness of LEDs is still not really understood. We use a scanning nearfield optical microscope (SNOM) to obtain local photoluminescence and electroluminescence spectra of GaN/GaInN/AlGaN LED structures with a lateral resolution better than 200 nm. The optical information is compared with defect positions made visible for atomic force microscopy by wet chemical etching. Peak wavelength maps obtained from the SNOM data show spatial regions of different emission wavelength with typical dimensions of 200–500 nm, associated with the defect structure. We observed no evidence for small scale fluctuations of emission wavelength as expected in case of phase separation of InGaN.

GaN based III–V semiconductors are widely used for making short wavelength LEDs and lasers [1, 2]. Because of the large lattice mismatch between substrate (typically sapphire or SiC) and GaN these structures contain a large number of misfit dislocations.

Despite their very high defect density in the range of 10^8–10^{10} cm^{-2} LED devices made from those materials have a very high brightness. The reason why the high defect densities apparently do not degrade luminescence efficiency significantly, i.e., why they do not behave as nonradiative recombination centres as these defects do in other materials, is not really understood [3]. It is speculated that carrier localization caused by random potential fluctuations due to phase separation of GaInN into In-rich and In-poor regions leads to reduced nonradiative recombination [4, 5].

Even though the dislocation density is highest close to the GaN/substrate interface, a large number of threading dislocations are starting off the nucleation layer and propagate straight through the whole structure to the surface. This allows to use atomic force microscopy (AFM), decoration of dislocations by wet or dry etching, or plan view transmission electron microscopy (TEM) from the top surface to visualize the positions of defects in lower-lying layers.

Our aim was to study the correlation between defect positions and local emission wavelength in GaInN quantum well LED structures. This should allow us to find possible localization of carriers away from the nonradiative defects.

Our samples were GaInN/GaN/AlGaN quantum well LED structures grown on SiC or sapphire. We used a wet chemical etching procedure to make defects visible at the surface for AFM scans. For samples grown on SiC substrates, best results were achieved by etching with pure KOH at 170 °C for 3 h or with pure H_3PO_4 at 185 °C by etching for 2 h. The KOH seemed to flatten the surface of the structure, too, while H_3PO_4 seemed to etch only at defect positions. For samples grown on sapphire the same process is used, but we

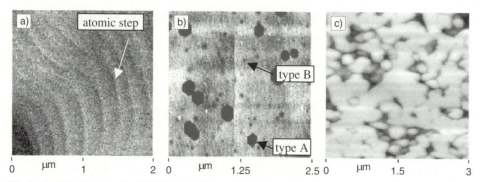

Fig. 1. AFM micrographs. a) Typical surface of a GaN sample after epitaxy. b) Etched GaN sample grown on SiC, there exist two different types of defects (bigger type A and smaller type B). c) Etched GaN sample grown on sapphire. Both etched samples were treated with 150°C phosphoric acid. Etch rates with samples grown on sapphire have been more than ten times higher than with samples grown on SiC

achieved etch rates that are ten times higher than with GaN samples grown on SiC substrates. Of course all etching depths are smaller than the distances between surface and quantum well. After etching we used a scanning electron microscope (SEM) and an atomic force microscope (AFM) for determination of etch pit densities.

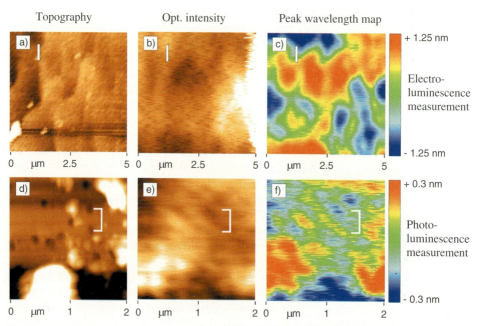

Fig. 2. SNOM images of two different samples. a)–c) come from one scan of a GaN-LED emitting at 500 nm, d)–f) come from an etched GaN-LED structure emitting at 405 nm. Both LED structures were grown on SiC. Pictures a) and d) are topography images, b) and e) are optical emission intensity images (spectrally integrated), c) and f) are wavelength maps of the maximum emission wavelength. Pictures b) and c) show the electroluminescence signal, e) and f) show the photoluminescence signal. The markers in the images help finding the same place in all images of one scan

For the measurements of local photoluminescence we used a scanning nearfield optical microscope (SNOM) which was operated in illumination-collection mode. For the scanning we used a graded index multimode UV fibre. The tip was etched using a tube etching process [6] (tip radius <100 nm) and was left uncoated. With this equipment we achieved a spatial resolution mostly better than 100 nm. For the photoluminescence measurements we used an argon ion laser operated at 379.5 nm as excitation source. The detection system was an imaging monochromator with a CCD detector.

In taking a complete optical spectrum at each point of the surface we were able to collect about 12000–20000 optical spectra within about 10–15 min. The high scanning speed reduced the influence of piezo motor drift of the scanning unit and minimized the influence of low frequency vibrations. The data can be converted to wavelength image maps, as we call it here, by taking the wavelength of maximum emission intensity at each point of the surface. The wavelength maps have the big advantage that they are independent of laser and photoluminescence light intensity variations and therefore the problem of correlation between tip–surface distance and optical intensity is overcome.

As the fibre–sample distance in the near-field microscope is controlled like in a normal AFM, each SNOM scan provides topography information at the same time. With the SNOM we took photoluminescence and electroluminescence measurements on unetched pieces of the same structures as mentioned above. Additionally we performed photoluminescence SNOM scans on the etched samples. All SNOM images were taken at room temperature.

When an AFM scan is taken immediately after growth, as shown in Fig. 1a, our sample surfaces exhibit a nice terrace structure, indicative of step-flow growth. After H_3PO_4 etching the samples grown on SiC show two types of defects (Fig. 1b), which differ in the size of the etched pit. The density of the bigger type A was about 2×10^8 cm^{-2}, the density of the smaller etched pits (type B) was about 1×10^9 cm^{-2}. These values are well in the range of commonly reported GaN defect densities. Therefore, the observed etch pits are likely at the positions of defects. A similar hierarchy of defects has already been observed in [7]. In contrast, samples grown on sapphire show a columnar structure after etching. Etching rates are more than ten times higher than for samples grown on SiC.

In Fig. 2 we present two series of images derived from two SNOM scans of two different samples. The images a)–c) show the results of a single SNOM scan of an GaN LED emitting at 500 nm. The optical images show the electroluminescence signal of the sample. In the topography image a) some features can be seen that are supposed to be correlated to lattice defects; because the sample is not etched the defects cannot be seen directly in the topography image. The steps in the topography image have a height of 1–2 nm. The wavelength map c) was created by taking the wavelength of the emission maximum and subtracting the average wavelength. It shows that there exist regions of comparable emission wavelength that are directly correlated to the defect structures visible in the topography image. The white marker indicates a change in the topography image, at the same point there are changes in the optical images, too. In image b) the optical intensity changes by about 30%.

Images d)–f) show a scan of an etched GaN-LED structure, emitting at 405 nm. e) and f) show the photoluminescence signal of the sample. In the topography images the bigger type of defects is visible, the resolution of the AFM information obtained from the SNOM is too limited to see the smaller type of defects. In the wavelength map there are similar structures like in the electroluminescence scan. Here the sizes of regions of comparable emission wavelength are of the order of defect distances as well. Of course, there is no influence of the etching procedure on SNOM results, since SNOM scans of the same structure before etching showed similar optical images.

The emission wavelength fluctuations we saw are very reproducible. Our SNOM measurements indicate a direct correlation between defects and changes in emission wavelength. Let us now discuss the results in some detail. Obviously, the unexpectedly large quantum efficiency of such LED structures means that carriers are kept away from defects by some mechanism. Effects that can cause changes in emission wavelength are local variations in In composition in the quantum well or changes in well width and strain. The emission wavelength variation can be the result of one or of all of them.

In the case of variation of In composition there would be an influence of the defects on the In composition during growth. If this effect would be responsible for keeping away carriers from defects the In concentration should be lower in the vicinity of the defects. If the variation in well thickness would be the dominant effect, the quantum well thickness should be smaller around the defects for keeping carriers away. Concerning strain we can say that less strain means less piezoelectric field and therefore a larger bandgap [8, 9]. So strain should be somewhat relaxed at defect position for keeping away carriers from defects.

Out of our measurements we cannot determine which of these effects are most responsible for the emission wavelength variations. But it can be seen that emission wavelength is correlated with defect positions and is not caused by random In compositional variations which are independent of defects.

In summary, we performed scanning nearfield optical microscope (SNOM) measurements on GaN/InGaN/AlGaN LED structures and compared them to the defect densities and defect positions in the samples. We observe variations of emission wavelength, regions with similar emission wavelengths having sizes in the range of defect distances. A correlation between defects and emission wavelength can be directly seen. Variations in emission wavelength can be caused by defect correlated In composition changes, changes of quantum well width or relaxed strain at defect positions.

References

[1] S. Nakamura, M. Senoh, N. Iwasa, and S. Nagahama, Appl. Phys. Lett. **67**, 1868 (1995).
[2] S. Nakamura, IEEE J. Sel. Top. Quantum Electron. **3**, 435 (1997).
[3] S. D. Lester, F. A. Ponce, M. G. Craford, and D. A. Steigerwald, Appl. Phys. Lett. **66**, 1249 (1995).
[4] X. Zhang, D. H. Rich, J. T. Kobayashi, N. P. Kobayashi, and P. D. Dapkus, Appl. Phys. Lett. **73**, 1432 (1998).
[5] T. Someya and Y. Arakawa, Jpn. J. Appl. Phys. **38**, L1216 (1999).
[6] R. Stöckle, C. Fokas, V. Deckert, and R. Zenobi, Appl. Phys. Lett. **75**, 160 (1999).
[7] P. Visconti, K. M. Jones, M. A. Reshchikov, R. Cingolani, and H. Morkoç, Appl. Phys. Lett. **77**, 3532 (2000).
[8] J. S. Im, H. Kollmer, J. Off, A. Sohmer, F. Scholz, and A. Hangleiter, Phys. Rev. B **57**, R9435 (1998).
[9] H. Kollmer, Jin Seo Im, S. Heppel, J. Off, F. Scholz, and A. Hangleiter, Appl. Phys. Lett. **74**, 82 (1999).

Energetic Calculation of Coincidence Grain Boundaries with a Modified Stillinger-Weber Potential

J. Chen (a), G. Nouet[1]) (b), and P. Ruterana (b)

(a) Laboratoire Universitaire de Recherche Scientifique d'Alençon,
Institut Universitaire de Technologie, F-61250 Damilly, France

b) Equipe Structure et Comportement Thermomécanique,
Laboratoire de Cristallographie et Sciences des Matériaux, UMR6508, CNRS,
Institut des Sciences de la Matière et du Rayonnement, 6 Bld Maréchal Juin,
F-14050 Caen cedex, France

(Received June 21, 2001; accepted July 7, 2001)

Subject classification: 61.72.Mm; 71.15.Nc; S7.14

Potential energy of two coincidence grain boundaries, $\Sigma = 7$ and 19, was calculated with a Stillinger-Weber potential previously adapted to wurtzite GaN in order to take into account the Ga–Ga and N–N wrong bonds. Their atomic structures, determined before by high resolution electron microscopy, are based on 5/7 and 8 atom ring structural units. By using periodical boundary conditions, it is shown that the high angle grain boundary $\Sigma = 7$ ($\Theta = 21°79/[0001]$) has always a smaller energy than $\Sigma = 19$ ($\Theta = 13°17/[0001]$).

Introduction The large number of threading dislocations in gallium nitride is a well-known problem, their density can be as high as 10^{10} cm^{-2}. The properties of the materials depend directly on the nature and the number of the defects. It is then important to determine theoretically the atomic structure of the defects. The atomic structures of the **a** edge threading dislocation were analysed by High Resolution Electron Microscopy (HREM) [1, 2] and by Z-contrast microscopy [3]. The 8 atom ring core of this dislocation was previously investigated within the local-density approximation by Elsner et al. [4]. Wright and Grossner [5] analysed intermediate structures between the 8 and 5/7 atom ring cores by considering open-core structure similar to the 5/7 atom ring core identified by HREM [1, 2]. Three types of cores were identified by HREM [1, 2], they correspond to cores with 4, 8 or 5/7 atom rings. The Stillinger-Weber (SW) empirical potential was modified to take into account the Ga–Ga and N–N bonds, which form in inversion domain boundaries (IDBs) as well as in the 5/7 atom ring configuration [6]. This modified potential was then used to calculate the potential energy of the three atomic configurations [7]. Coincidence grain boundaries were identified in GaN and it was shown that their atomic structures were based on various combinations of the different atomic configurations of the **a** edge dislocation [2]. For special misorientation, corresponding to coincidence grain boundaries, low-energy cusps are expected to form [8]. Such empirical potentials are still useful in order to have a physical insight on the structure of large systems. In the following we discuss the results obtained for two grain boundaries $\Sigma = 7$ and 19.

Atomic Models and Boundary Conditions By HREM, grain boundaries in wurtzite GaN were identified with rotation around the [0001] common axis and with rotation

---

[1]) Corresponding author; Phone: +33 231 45 26 47; Fax: +33 231 45 26 73; e-mail: nouet@ismra.fr

angles very close to the theoretical values: 13.17° and 21.79°. In the coincidence site lattice formalism, they can be characterised in the hexagonal system by the coincidence indices, Σ = 19 and 7, respectively. The boundary planes are symmetric and have a tilt character. The atomic structures of Σ = 7 and 19 were constructed theoretically on the basis of the experimental results. It was shown that they consist of regularly 5/7 and/or 8 atom rings corresponding to the cores of the **a** edge dislocation. For the two orientations the periods contain only one **a** dislocation; they differ by the number of bulk 6 atom rings. Some of these rings are bordering the boundary plane: A, and the others are inside: B. Thus, the periods are: .B A B 8 A. and .B A 5/7 A. for Σ = 19, .B 8 A. and 5/7 A. for Σ = 7. It is worth noting that these atomic structures described by rings or structural units are rather similar to those found in silicon for which only the 5/7 atom ring is used [8]. However, their geometrical description (Σ, angle and rotation axis) is different due to the crystal system.

For the calculations, the periodical boundary condition was used. The relaxed atom box contained two grain boundaries with opposite orientations. The distance between the two grain boundaries was chosen large enough to minimise the interference effect. The number of atoms is 400 for Σ = 7 and 1100 for Σ = 19.

Stillinger-Weber Potential In the standard Stillinger-Weber potential (SW) [9], there are two-body and three-body terms which, respectively, correspond to a pair and pseudo-many-body interaction potential. The three-body term gives the angular distortion which is important in the description of the deformed structure. This potential is unable to take into account the chemical nature of the bonds. We have modified its parameters on the basis of results obtained previously for other III–V compounds and used it in the energy investigation of the planar defects, agreement with ab initio calculations was attained for the different structures of IDBs [6, 10]. For the relaxation of the grain boundaries, the cut-off radius of the potential was chosen: $1.2r_0$ corresponding to the nearest-neighbours with r_0 = 1.95, 2.93, and 1.11 Å for the Ga–N, Ga–Ga and N–N interactions, respectively [11].

Result and Discussion The relaxation with the SW empirical potential for the two boundaries show that Σ = 7 whose period is formed by one **a** dislocation 5/7 or 8 atom ring has always a smaller energy than that of the Σ = 19 whose period contains extra sixfold perfect crystal cells (Table 1). This is in agreement with the usual energy cups already demonstrated for the cubic materials [8].

The excess energy is seen to be located on the highly deformed area: the wrong bonds inside the 5/7 atom ring or the dangling bond inside the 8 atom ring (large dots in Fig. 1). The changes in bond lengths and angles are shown in Tables 2 and 3 for the two boundary atomic configurations. We notice that the deformation is, like the energy variation, located mostly at the wrong bond atoms inside the 5/7 atom rings. This holds

Table 1
Potential energy (in mJ/m^2)

7 (5/7)	782	7 (8)	1014
19 (5/7)	916	19 (8)	1255

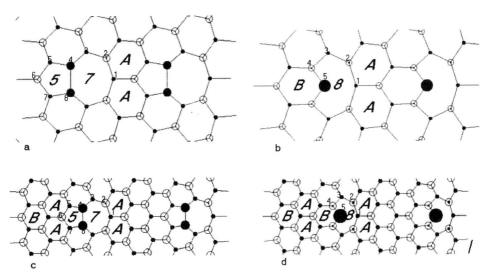

Fig. 1. $\Sigma = 7$: a) 5/7 and b) 8 atomic configurations; $\Sigma = 19$: c) 5/7 and d) 8 atomic configurations

Table 2

atom No.	$\Sigma = 7$, 8 atomic configuration						$\Sigma = 19$, 8 atomic configuration					
	length			angle			length			angle		
	min	max	mean	min	max	mean	min	max	mean	min	max	mean
1 Ga	1.982	2.056	2.019	99.86	130.63	108.88	1.992	2.070	2.031	101.29	129.27	108.91
2 N	1.976	2.056	2.005	93.64	120.02	109.05	1.978	2.070	2.013	89.67	121.79	108.97
3 Ga	1.934	1.994	1.956	99.07	123.26	109.20	1.922	2.006	1.958	96.20	127.90	109.02
4 N	1.922	1.949	1.935	87.28	121.30	109.15	1.922	1.938	1.928	88.44	118.97	109.18
5 Ga	**1.922**	**1.942**	**1.929**	**97.44**	**110.06**	**105.85**	**1.922**	**1.942**	**1.929**	**95.49**	**109.88**	**105.09**

Table 3

at. No.	$\Sigma = 7$ 5/7 atomic configuration						$\Sigma = 19$ 5/7 atomic configuration					
	length			angle			length			angle		
	min	max	mean	min	max	mean	min	max	mean	min	max	mean
1 Ga	1.952	1.996	1.975	100.72	126.14	109.12	1.976	2.022	1.999	103.23	124.16	109.26
2 N	1.961	1.996	1.975	104.94	112.65	109.41	1.969	2.022	1.989	99.65	116.44	109.40
3 Ga	1.952	1.978	1.966	106.71	112.01	109.47	1.953	1.989	1.970	106.58	116.18	109.47
4 N	**1.622**	**1.976**	**1.883**	**100.11**	**135.11**	**108.74**	**1.620**	**1.972**	**1.878**	**100.15**	**136.44**	**108.73**
5 Ga	1.935	1.959	1.943	102.80	117.47	109.44	1.920	1.956	1.935	101.70	121.43	109.36
6 N	1.938	1.952	1.944	92.18	116.81	109.18	1.919	1.933	1.924	99.40	115.42	109.27
7 Ga	1.935	1.959	1.943	102.80	117.47	109.44	1.919	1.956	1.935	101.56	121.56	109.36
8 N	**1.622**	**1.976**	**1.883**	**100.11**	**135.11**	**108.74**	**1.620**	**1.972**	**1.878**	**100.16**	**136.24**	**108.73**
11 N	1.953	1.998	1.976	100.46	126.79	109.17	1.976	2.025	2.000	102.64	124.76	109.02
12 Ga	1.959	1.998	1.974	100.98	113.94	108.98	1.969	2.024	1.989	96.16	116.81	108.94
13 N	1.935	1.962	1.952	98.98	114.98	108.14	1.929	1.976	1.957	96.12	119.29	108.01
14 Ga	**1.935**	**2.315**	**2.040**	**79.89**	**126.57**	**106.78**	**1.929**	**2.315**	**2.037**	**79.85**	**127.85**	**106.84**
15 N	1.917	1.942	1.933	105.38	112.33	108.66	1.910	1.934	1.923	105.32	112.57	108.68
16 Ga	1.938	1.953	1.944	96.54	114.96	109.29	1.918	1.933	1.924	97.79	113.55	109.39
17 N	1.917	1.942	1.933	105.38	112.33	108.66	1.910	1.934	1.923	105.18	112.57	108.68
18 Ga	**1.935**	**2.315**	**2.040**	**79.89**	**126.57**	**106.78**	**1.929**	**2.315**	**2.037**	**79.84**	**127.68**	**106.84**

for the bond lengths and the bond angles. The N–N bond is quite short whereas the Ga–Ga one is larger than the GaN bond. Inside the 8 atom ring cores the variations in bond length are rather small. As can be noticed the changes are larger in bond angles with averages reaching 105.09° for the dangling bond atom. This probably explains why the SW potential gives a higher energy for the 8 atom configuration.

Conclusion Using SW potential in wurtzite GaN, it is shown that the $\Sigma = 7$ and 19 boundary energies agree with the hierarchy predicted by the coincidence theory in cubic materials. However, the calculated differences in energy are rather small and this may explain why the 8 and 5/7 atom configurations were found to coexist experimentally inside these boundaries.

Acknowledgements This work was supported by the EU under contract No. HPRN-CT-1999-00040.

References

[1] P. RUTERANA, V. POTIN, and G. NOUET, Mater. Res. Soc. Symp. Proc. **482**, 359 (1998).
[2] V. POTIN, P. RUTERANA, G. NOUET, R. C. POND, and H. MORKOÇ, Phys. Rev. B **61**, 5587 (2000).
[3] Y. XIN, S. J. PENNYCOOK, N. D. BROWNING, P. D. NELLIST, S. SIVANATHAN, F. OMNÈS, B. BEAUMONT, J. P. FAURIE, and P. GIBART, Appl. Phys. Lett. **72**, 2680 (1998).
[4] J. ELSNER, R. JONES, P. K. SITCH, V. D. POREZAG, M. ELSTNER, T. FRAUENHEIM, M. I. HEGGIE, S. OBERG, and P. R. BRIDDON, Phys. Rev.Lett. **79**, 3672 (1997).
[5] A. F. WRIGHT and U. GROSSNER, Appl. Phys. Lett. **73**, 2751 (1998).
[6] N. AICHOUNE, P. RUTERANA, A. HAIRIE, G. NOUET, and E. PAUMIER, Comp. Mater. Sci. **17**, 380 (2000).
[7] J. CHEN, P. RUTERANA, and G. NOUET, Mater. Sci. Eng. B **82**, 117 (2001).
[8] M. KOHYAMA, R. YAMAMOTO, and M. DOYOMA, phys. stat. sol. (b) **137**, 11 (1986).
[9] F. H. STILLINGER and T. A. WEBER, Phys. Rev. B **31**, 5262 (1985).
[10] J. E. NORTHRUP, J. NEUGEBAUER, and L. T. ROMANO, Phys. Rev. Lett. **77**, 103 (1996).
[11] D. E. BOUCHER, G. G. DELEO, and W. B. FOWLER, Phys. Rev. B **59**, 10064 (1999).

Structure Characterization of (Al,Ga)N Epitaxial Layers by Means of X-Ray Diffractometry

J. Kozłowski[1]), R. Paszkiewicz, and M. Tłaczała

Institute of Microsystems Technology, Wrocław University of Technology, Janiszewskiego 11/17, PL-50-372 Wrocław, Poland

(Received July 10, 2001; accepted August 3, 2001)

Subject classification: 61.10.Kw; 68.55.Jk; S7.14

The structural properties of (Al,Ga)N epitaxial layers were studied basing on X-ray diffractometric measurements. The examined layers were deposited by MOVPE on c-plane sapphire substrate. The measurements were performed on MRD-Philips diffractometer. Particularly, the density of lateral correlation length (coherence length) distribution, the most probable shape of the (Al,Ga)N/GaN blocks and twist as well as tilt mosaicities were described. The distribution of the crystallite block sizes was calculated from X-ray peak profile analysis. Twist and tilt mosaicities were examined using rocking curve mode and specially chosen configuration, where an edge of the sample was illuminated.

Introduction (Al,Ga)N structures belong to the most popular wide-bandgap semiconductors. The large lattice mismatch existing between epitaxial layer and supplied sapphire substrate (16%) gives a structure with great amount of defects. The typical layer consists of columnar crystallites with diameters equal to few hundred nanometers and the density of dislocations changes from 10^8 cm^{-2} inside the grains to 10^{10} cm^{-2} in the grain boundaries. This paper presents a new method which allows determining the distribution of lateral correlation length on the basis of peak profile analysis. The necessary peaks were obtained from high-resolution X-ray diffractometry measurements. Twist and tilt mosaicities were determined taking into account the rocking curves recorded from the planes perpendicular to the sample surface, i.e. (11.0) and (01.0).

Sample Preparation The GaN layers and $Al_xGa_{1-x}N$/GaN heterostructures were grown on α-Al_2O_3 (00.1) substrate under atmospheric pressure, single wafer, vertical flow MOVPE system redesigned for nitrides deposition [1]. Two kinds of buffers were used: low temperature GaN layer (sample s1) and AlN layer (samples s2, s3). Prior to high temperature GaN layer growth the low temperature buffer layers were annealed in NH_3/H_2 ambient for 10 min at 1040 °C. Thus, the nominally 2 μm thick high temperature GaN layer was grown with a 6600 V/III molar ratio. In the case of sample s3 GaN layer was followed by a 20 nm $Al_xGa_{1-x}N$ layer ($x = 0.17$).

X-Ray Measurements Diffractometric measurements obtained from the planes perpendicular to the sample surface are presented. A schematic view of the X-ray measurements is shown in Fig. 1. An alignment of the sample was performed by its two-dimensional movement (y-axis and Ψ-scan). Using this set-up and parallel beam optics,

---

[1]) Corresponding author; Phone: +48(71)320 2234; Fax: + 48(71) 328 3504; e-mail: janx@wtm.ite.pwr.wroc.pl

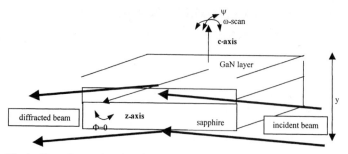

Fig. 1. Experimental set-up for X-ray edge scan measurements

it is possible to obtain the information about lattice parameter a of the whole structure, which is independent of the parameter c.

High-resolution optics enables one to measure rocking curves allowing to evaluate the twist ($\Phi = 0°$) and tilt ($\Phi = 90°$) mosaicities. The (11.0) and (01.0) planes of GaN layers were taken into account for these purposes.

The typical (Al,Ga)N layer deposited on sapphire substrate consists of column-like crystallites of height equal to the layer thickness with a hexagonal lattice. In order to determine the distribution of the lateral correlation length and lattice strains of second order the method was developed in the frame of the kinematical theory of X-ray diffraction basing on the three linear integral Fredholm equations of first kind describing X-ray diffraction on a set of deformed crystallites. The solution of the integral describing the line profile was obtained using Tikhonov's regularisation idea [2, 3]. The crystallites were modelled in three different forms: as a prism of rhombic base of edges parallel to the main crystalline axes, as a prism of hexagonal base and as cylinder of axis parallel to the [001] axis. The most probable shape of the crystallite block was interpreted as a minimum absolute mean-square error between the measured and calculated line profiles for the assumed crystallite shapes. Three profiles of peaks, recorded in the $2\theta/\omega$ high-resolution mode, were taken into account: (00.2), (01.5) and (11.4).

Results Figure 2 shows the diffractograms obtained from both perpendicular edges of sample s3. The profile as well as the 2θ positions of the (Al,Ga)N/GaN peaks could

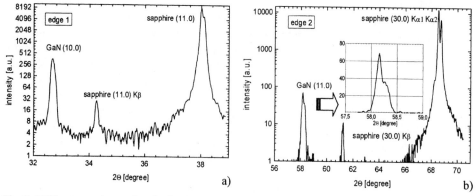

Fig. 2. Diffractometric graphs obtained using parallel beam optics (sample s3)

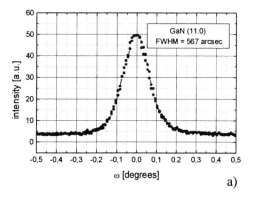 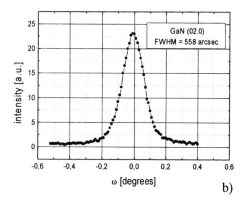

Fig. 3. Rocking curves obtained using high-resolution optics (sample s2); a) (11.0) peak, b) (02.0) peak

give the valuable information about the value and gradient of lattice parameter a. Distortion of the GaN unit cell can be determined from the a and c parameters calculated independently. Using high-resolution mode the rocking curves for these planes can be measured. Figure 3 presents the rocking curves obtained for sample s2. FWHM mainly contains the information about the twist mosaicity. It should be noticed that the perpendicular planes to the sample surface could be used in order to obtain also information about tilt mosaicity. This can be reached, when the sample surface is perpendicular to the plane determined by the incident and diffracted beams ($\Phi = 90°$). The basic rocking curve data obtained for the sample s2 are presented in Table 1.

Figure 4 presents the density of the lateral correlation length distribution calculated for samples s1 and s2. The prism of rhombic base is the most probably shape of grains in both cases.

Two various fractions of lateral correlation lengths are visible for the sample with GaN buffer layer (s1), whereas sample s2 shows only one range of this parameter from 10 to 750 nm and the mean value is equal to 250 nm.

Summary X-ray scans realized from the edge of the sample can give the following structural information: lattice parameter a value and variation in (Al,Ga)N/GaN layers (broadening, shape and 2θ position of (11.0), (02.0) peak analysis), twist and tilt mosaicity (FWHM of rocking curves recorded using (11.0) (02.0) reflections). Modelling of

Table 1
Rocking curve data (sample s2)

structure		peak	$\omega/2\theta$	FWHM [arcsec]	intensity [cps]
sapphire		(30.0)	34.1/68.2	50	11 802
		(22.0)	10.35/80.69	42	4 942
GaN	$\Phi = 0°$	(11.0) twist	28.92/57.84	567	43
	$\Phi = 0°$	(02.0) twist	3.94/67.88	558	23
	$\Phi = 90°$	(11.0) tilt	28.92/57.84	178	1 189
	$\Phi = 90°$	(02.0) tilt	3.94/67.88	256	22

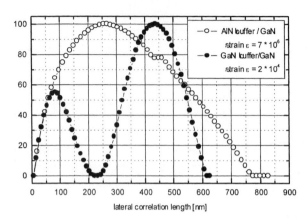

Fig. 4. Density of lateral correlation length distribution for sample s1 (full circles) and sample s2 (empty circles)

the diffracted line profiles allows to obtain the density of lateral correlation length and strain distribution, which may be interpreted as probability of finding crystallite blocks of assumed size, shape and strain in the analysed samples.

Acknowledgements The work was supported by the Polish Scientific Research Committee, Grants: 7T08A003 18, 8T11B006 18, 8T11B057 18, Polish Science Foundation Projects (TECHNO 106/2000, TECHNE 6/2001) and Wroclaw University Advanced Materials and Nanotechnology Centre Project.

References

[1] R. PASZKIEWICZ, R. KORBUTOWICZ, M. PANEK, B. PASZKIEWICZ, M. TŁACZAŁA, and S. V. NOVIKOV, Heterostructure Epitaxy and Devices – Head' 97, Vol. 48, Eds. P. KORDOS and J. NOVAK, Kluver Academic Publ., Dordrecht 1998 (p. 98).
[2] M. A. KOJDECKI, Int. J. Appl. Electromagn. Mater. **2**, 147 (1991).
[3] C. W. GROETSCH, The Theory of Tikhonov Regularization for Fredholm Equations of the First Kind, Pitman Publ. Ltd., London 1984.

Optical Micro-Characterization of Complex GaN Structures

J. Christen[1]) and T. Riemann

Institute of Experimental Physics, Otto-von-Guericke-University, P.O. Box 4120, D-39016 Magdeburg, Germany

(Received August 31, 2001; accepted September 3, 2001)

Subject classification: 61.72.Ss; 68.55.Ln; 78.30.Fs; 78.60.Hk; S7.14

Various techniques of pattern-controlled epitaxy like epitaxial lateral overgrowth (ELO) and lateral seeding epitaxy have successfully been used to overcome the lattice mismatch problem, however, resulting in a complex system of self-organized growth domains. For a detailed understanding a correlation of the structural, electronic, and optical properties on a micro-scale is mandatory. Scanning cathodoluminescence (CL) microscopy provides a powerful tool compiling low temperatures, spatial resolution $\Delta x < 45$ nm and high spectral resolution. ELO GaN on stripe patterned SiO_2 and W masks were characterized by CL microscopy, directly visualizing the formation of different growth domains: 1. the coherently grown regions between the stripes; 2. the ELO areas coalescing on the center of the masks and showing a blue shifted and strongly broadened CL where μ-Raman evidences the reduction of local strain and probes the free carrier concentration which is dramatically changed over the masks; 3. the coalescence regions with poor epitaxial perfection including voids and strong incorporation of impurities; 4. the transition regions between 1. and 2. at the very edges of the masks characterized by high defect density and strongly blue shifted CL. Plan view CL directly images the local areas of improved crystal perfection only involving part of the total ELO regions. The formation of specific micro-domains in 5 μm thick crack-free $Al_{0.19}Ga_{0.81}N$, exhibiting different Al concentrations, during the initial stage of MOVPE grown on patterned GaN substrates and the transition to finally homogeneous growth is directly visualized by cross-sectional CL.

Introduction For a detailed understanding of complex semiconductor heterostructures and the physics of devices based on them a systematic determination and correlation of the structural, chemical, electronic, and optical properties on a micro- or nanoscale is essential. Luminescence techniques belong to the most sensitive, non-destructive methods of analyzing both, intrinsic and extrinsic semiconductor properties. Combining the luminescence spectroscopy with the high spatial resolution of a scanning electron microscope (SEM), realized by the technique of cathodoluminescence (CL) microscopy, provides a powerful tool for the nano-characterization of semiconductors, their heterostructures as well as their interfaces [1].

Employing the technique of Epitaxial Lateral Overgrowth (ELO) to the group-III nitrides has been proven successful in significantly reducing the concentration of threading dislocations emanating from the underlying buffer layer. The ELO approach consists of masking parts of the defective crystalline substrate GaN "seed" layer with an amorphous layer preventing the dislocations from propagating into the overlayer during subsequent re-growth. While impurities are unintentional incorporated in the initial stages of ELOG, the biaxial strain and defect concentration are reduced on the top of the mask [2–5].

The maskless heteroepitaxy on pre-patterned substrates has been proven successful in achieving crack free AlGaN on trench-patterned GaN/sapphire substrates [6, 7].

---
[1]) Phone: +49 391 67 11259; Fax: +49 391 67 11130;
e-mail: juergen.christen@physik.uni-magdeburg.de

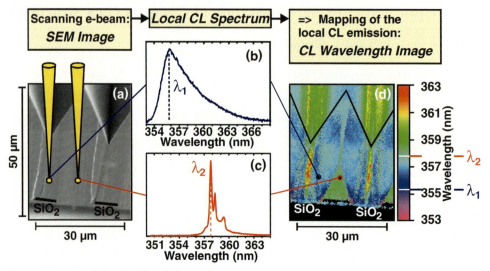

Fig. 1. Principle of the CLWI technique

The Technique of Cathodoluminescence Microscopy In contrast to photoluminescence (PL) for cathodoluminescence (CL) the sample under investigation is not excited by a laser but by the highly focused electron beam of a SEM. Our CL system combines low temperatures (5 K < T < 300 K), an overall spatial resolution of $\Delta x < 45$ nm (under optimum conditions) with high spectral resolution. However, the unique feature of our CL system is the imaging of complete CL spectra (Fig. 1[2]): While the focused e-beam is digitally scanned over typically 256 × 200 pixels (Fig. 1a) a complete CL spectrum $I(\lambda)_{x,y}$ is recorded at each pixel (x, y) and stored (e.g. Figs. 1b and c). A three-dimensional data set $I(x, y, \lambda)$ is obtained and consecutively evaluated. All type of data cross sections through this $I(x, y, \lambda)$ tensor can be generated, e.g. sets of monochromatic CL images $I(x, y, \lambda_1)$, $I(x, y, \lambda_2)$,...; local CL spot spectra $I(x_1, y_1, \lambda)$, $I(x_2, y_2, \lambda)$, ...; CL spectrum linescans $I(s, \lambda)$ (s = arbitrary linescan); as well as CL wavelength images (CLWI) $\lambda(x, y)$, i.e. mappings of the local emission peak wavelength (Fig. 1d) [1].

Epitaxial Lateral Overgrowth GaN ELOG The *SiO$_2$-masked* ELOG samples consist of a 3 µm thick GaN epilayer grown by metal organic vapor phase epitaxy (MOVPE) on (0001) sapphire subsequently structured with SiO$_2$ stripes parallel to $\langle 11\bar{2}0 \rangle$. The selective lateral overgrowth was achieved with 50 µm thick hydride vapor phase epitaxy (HVPE) GaN [2, 3]. Its SEM cross section is depicted in Fig. 1a showing the facetted surface achieved for this mask orientation (SiO$_2$ masks in $\langle 1100 \rangle$ direction result in a planar ELOG surface). The CLWI in Fig. 1d is a mapping of the local emission wavelength and directly visualizes three different growth regions: the GaN buffer layer, the overgrown region above the SiO$_2$ and the area of coherent growth between the SiO$_2$ pattern. In contrast to the spectral position of (D^0, X) in completely relaxed GaN (357.2 nm) the buffer layer shows a blue-shifted (D^0, X) emission at 356.4 nm according to a compressive biaxial stress of 0.8 GPa [8]. In the coherently grown region a monochromatic triangle (green color coded in Fig. 1d) of almost homogeneous emission at

358 nm is visible evolving in the center between the SiO_2 stripes (Fig. 1c). The overgrowth region is dominated by a blue-shifted emission around 356 nm and is very inhomogeneous showing stripe-like patterns in c-direction. Strongly red shifted, extrinsic CL (362 nm) dominates the very center of the overgrowth region, i.e. the ELO coalescence area. At the outer edges of the SiO_2 stripes a strong blue-shift (354 nm) is obtained (Fig. 1b) [2, 3].

ELOG samples using *tungsten mask* were investigated in comparison. In Fig. 2a a cross-sectional CL wavelength image of an ELOG sample with the tungsten stripes orientated along $\langle 1\bar{1}00 \rangle$ is depicted, showing strongly varying local CL emission wavelength in the initial ELO domains. As in the case of the SiO_2 masked samples, the strongest fluctuations of the emission wavelength occur directly above the tungsten mask. In contrast to the spectral position of (D^0, X) in completely relaxed GaN (357.2 nm) this overgrown region above the W stripes is dominated by a weak, blue-shifted and broad emission around 355 nm. In cross-section this area of blue-shifted emission always exhibits a typical form (rabbit ear structure), which obviously differs from the specific structures of the overgrown region found for the SiO_2 masked samples. Its size is self-limited by the coalescence of the laterally advancing facets above the W masks. In the coherently grown region evolving in the center between the W stripes only sharp excitonic lines are visible. After 15 μm distance to the substrate facet growth stops and only perfect $\langle 0001 \rangle$ growth occurs both between and above the tungsten stripes. Accordingly, at the sample surface no influence of the mask position is detected in the luminescence, showing the main excitonic line at 357 nm [4, 5]. This is directly visualized in Fig. 2 compiling sets of spatially resolved local CL spectra obtained from the various different ELOG growth domains. Strongly blue-shifted and broadened (e, h)-plasma recombination is exclusively found in the "rabbit ear" regions at the edges on top of the tungsten masks [9]. A set of monochromatic CL images re-

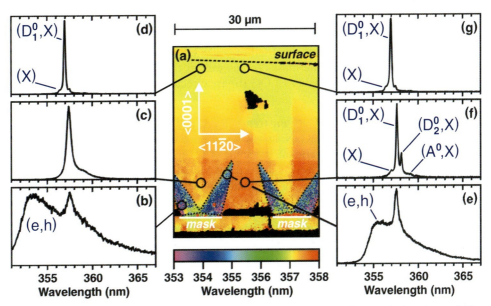

Fig. 2. a) Cross sectional CL wavelength image of the different ELOG growth domains and b) to e) local spot mode CL spectra

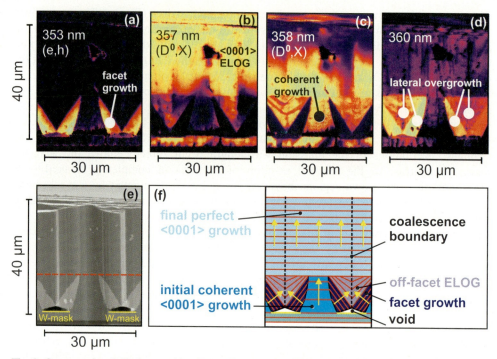

Fig. 3. Cross sectional CL images identifying the different ELOG growth domains

corded at the wavelength of (e, h)-plasma recombination (a), (D^0, X) main residual donor (b), (D^0, X) second donor (c), as well as (A^0, X) (d) is plotted in Fig. 3 together with the masked SEM image (e), correlating with the different micro-domains depicted in the ELOG growth model (f). The individual self-organized domains of initial facet growth (a) and coherent growth between the masks (c), as well as the subsequent off-facet growth (d) and the final stage of high quality ELOG (b) are directly imaged by CL [10].

Lateral Seeding Epitaxy of AlGaN on trench patterned GaN The formation of specific micro-domains in thick AlGaN layers, exhibiting different Al concentrations, is directly visualized by cross-sectional scanning CL microscopy (Fig. 4). Crack-free, 5 μm thick layers ([Al] = 0.19) were achieved by MOVPE growth on patterned GaN/LT-AlN/sapphire templates. Prior to the final AlGaN deposition, a periodic grid of trenches along $\langle 1\bar{1}00 \rangle$ (5 μm wide/1 μm deep) was fabricated in the 3 μm GaN layer and subsequently overgrown with a LT-AlN interlayer [7]. Three clearly separated spectral components appearing in laterally integrated CL spectra (b) can by unambiguously assigned to three different species of micro-domains. This fact is illustrated in the CL wavelength image (c) showing the identical area as the SEM image (a). The *onset of AlGaN growth on the terrace regions (A)* gives rise to an emission at 305 nm as seen in the CL intensity image (d). In direct contrast, *initial AlGaN growth above the trenches (B)* leads to an emission band centered around 335 nm (f). After 3 μm AlGaN deposition an emission wavelength around 315 nm is observed indepen-

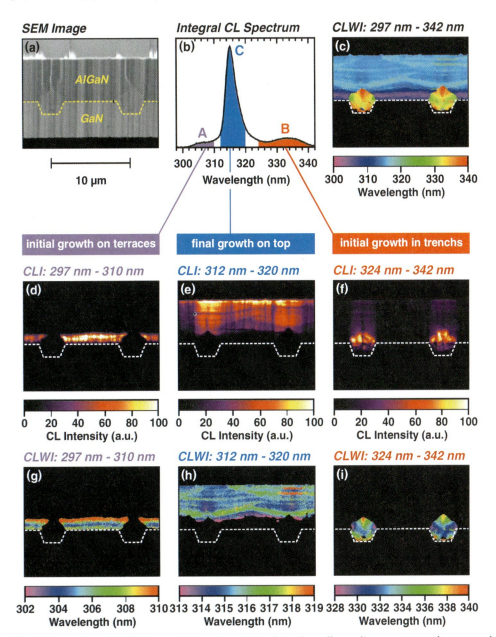

Fig. 4. Formation of AlGaN growth domains during lateral seeding epitaxy over trench-patterned GaN/sapphire templates

dent of the trench position in the *final domain C* (e). Above the trenches we find a drastic increase of quantum efficiency, indicating an improvement of AlGaN quality.

All three types of micro-domains *A*, *B*, and *C* additionally show specific modulations of their average [Al] as illustrated in the expanded CL wavelength images (g) to (i).

The optical data can be perfectly correlated with the microscopic structural properties, e.g. the local [Al], by comparing the CL results with micro-Raman spectroscopy mappings [11].

Conclusion Epitaxial lateral overgrowth GaN structures with SiO_2 and W mask orientated along $\langle 11\bar{2}0 \rangle$ and $\langle 1\bar{1}00 \rangle$ directions were characterized and regions of different growth regimes were identified and directly imaged by spectrally resolved scanning cathodoluminescence microscopy. In all ELO structures the coherently grown region shows perfect excitonic CL, i.e. high crystallographic quality. In the ELO sample with SiO_2 stripes in $\langle 11\bar{2}0 \rangle$ direction the coherently grown area forms a sharply defined triangle in the middle of the structure. This pattern orientation shows a strong $\langle 1\bar{1}01 \rangle$ facetting in the surfaces morphology. In contrast planar surface morphology is achieved for the tungsten-masked samples. The region of initial overgrowth is always dominated by a strongly blue shifted, weak and broad CL (e, h)-plasma emission evidencing the high local free carrier concentration and high local defect density. Continuous ELO above the self-limited micro-domains finally results in homogeneous ELOG layers of perfect luminescence quality.

The formation of Al-accumulated and Al-depleted micro-domains in 5 μm thick crack-free AlGaN layers MOVPE grown on patterned GaN/LT-AlN/Sapphire substrates is directly visualized by cross-sectional scanning CL microscopy.

Acknowledgements We are grateful to K. Hiramatsu (Mie University, Tsu), T. Shibata and N. Sawaki (Nagoya University), as well as H. Amano and I. Akasaki (Meijo University, Nagoya) for providing the excellent samples. The authors like to express their special thanks to F. Bertram (Otto-von-Guericke-University, Magdeburg). This work is supported by the Deutsche Forschungsgemeinschaft (CH 87/4-2, Tho662/4-2).

References

[1] J. Christen, M. Grundmann, and D. Bimberg, J. Vac. Sci. Technol. B **9**, 2358 (1991).
[2] A. Kaschner, A. Hoffmann, C. Thomsen, F. Bertram, T. Riemann, J. Christen, K. Hiramatsu, T. Shibata, and N. Sawaki, Appl. Phys. Lett. **74**, 3320 (1999).
[3] F. Bertram, T. Riemann, J. Christen, A. Kaschner, A. Hoffmann, C. Thomsen, K. Hiramatsu, T. Shibata, and N. Sawaki, App. Phys. Lett. **74**, 359 (1998).
[4] Y. Kawaguchi, S. Nambu, H. Sone, T. Shibata, H. Matsushima, M. Yamaguchi, H. Miyake, H. Hiramatsu, and N. Sawaki, Jpn. J. Appl. Phys. **37**, L845 (1998).
[5] H. Sone, S. Nambu, Y. Kawaguchi, M. Yamaguchi, H. Miyake, K. Hiramatsu, Y. Iyechika, T. Maeda, and N. Sawaki, Jpn. J. Appl. Phys. **38**, L356 (1999).
[6] T. Detchprohm, M. Yano, S. Sano, R. Nakamura, S. Mochiduki, T. Nakamura, H. Amano, and I. Akasaki, Jpn. J. Appl. Phys. **40**, L16 (2001).
[7] M. Iwaya, R. Nakamura, S. Terao, T. Ukai, S. Kamiyama, H. Amano, and I. Akasaki, Proc. Internat. Workshop on Nitride Semiconductors (IWN 2000), Nagoya (Japan); IPAP Conf. Ser. **1**, 833 (2000).
[8] H. Siegle, A. Hoffmann, L. Eckey, C. Thomsen, J. Christen, F. Bertram, M. Schmidt, D. Rudloff, and K. Hiramatsu, Appl. Phys. Lett. **71**, 2490 (1997).
[9] A. Kaschner, A. Hoffmann, C. Thomsen, F. Bertram, T. Riemann, J. Christen, K. Hiramatsu, H. Sone, and N. Sawaki, Appl. Phys. Lett. **76**, 3418 (2000).
[10] K. Hiramatsu, A. Motogaito, H. Miyake, Y. Honda, Y. Iyechika, T. Maeda, F. Bertram, J. Christen, and A. Hoffmann, IEICE Trans. Electron. E **83**, 620 (2000).
[11] T. Riemann, J. Christen, A. Kaschner, A. Laades, A. Hoffmann, C. Thomsen, M. Iwaya, H. Amano, and I. Akasaki, unpublished.

Optical Properties of III-Nitride Ternary Compounds

A. Baldanzi (a), E. Bellotti[1]) (b), and M. Goano (a)

(a) INFM and Dipartimento di Elettronica, Politecnico di Torino,
Corso Duca degli Abruzzi 24, I-10129 Torino, Italy

(b) Electrical and Computer Engineering Department, Boston University,
8 Saint Mary's Street, Boston, MA 02215-2421, USA

(Received July 4, 2001; accepted August 16, 2001)

Subject classification: 78.20.Ci; 78.66.Fd; S7.14

As a basic step in the evaluation of optical properties of the III-nitrides, the dielectric function of AlGaN and InGaN has been computed using the information from the full-band electronic structure in the framework of the random phase approximation. Both real and imaginary parts have been evaluated, and the refractive index has been compared with available parametrizations. Sellmeier coefficients for the refractive index at energies below the band gap have been derived as functions of the alloy composition.

Introduction Because of the increasing importance of the III-nitride ternary alloys for applications in optoelectronic devices, their optical properties have to be determined as functions of the mole fraction. Optical data from several, though not univocal, experimental measurements are available for GaN (see e.g. [1–5]) and for AlGaN films [6–12], while little information is available for InGaN. (Other useful references are [13–20].) The limitations of the various measurement techniques, the differences in substrates and in growth technologies, and the uncertainties in the quality, uniformity, thickness, defect density, and composition of the alloy samples, make theoretical calculations of the optical properties still useful in order to provide complete and coherent information for device design and optimization. In the following sections, the complex dielectric function and the refractive index of AlGaN and InGaN, calculated from state-of-the-art empirical pseudopotential (EPM) band structures, will be presented and compared with available parametrizations.

Theory and Implementation Our computations are based on the theory of the dielectric screening associated with electronic interband transitions in crystals as formulated by Ehrenreich and Cohen [21] in the mean field approximation. The real part of the frequency and wave-vector dependent dielectric function can be expressed as

$$\epsilon_1(\mathbf{q}, \omega) = \epsilon_0 + \frac{e^2}{\Omega q^2} \sum_{\mathbf{k},c,v} |\langle \mathbf{k}, c \mid \mathbf{k}+\mathbf{q}, v \rangle|^2 \{[E_c(\mathbf{k}) - E_v(\mathbf{k}+\mathbf{q}) - \hbar\omega]^{-1} + [E_c(\mathbf{k}) - E_v(\mathbf{k}+\mathbf{q}) + \hbar\omega]^{-1}\}, \quad (1)$$

where $\mathbf{k}$ is summed over the First Brillouin Zone (FBZ), v and c represent the indices of occupied valence and empty conduction bands, and Ω is the atomic volume of the lattice. Energy bands and overlap integrals are, together with the crystal structure, the

---

[1]) Corresponding author; Phone: +1 617 358-1576; Fax: +1 617 353-6440; e-mail: bellotti@bu.edu

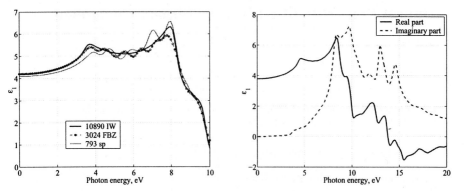

Fig. 1. Left: convergence study on the real part of GaN dielectric function, computed through a sum over 793 special points, 3024 evenly spaced points in the FBZ [22], and 10890 evenly spaced points in the IW. Right: calculated real and imaginary parts of the dielectric function for $Al_{0.4}Ga_{0.6}N$

inputs of the computational process. In the long wavelength limit a $\mathbf{k} \cdot \mathbf{p}$ expansion of the overlap integral is used [22],

$$\lim_{\mathbf{q} \to 0} \epsilon_1(\mathbf{q}, \omega) = \epsilon_0 + \frac{2e^2\hbar^2}{m_0^2 \Omega} \sum_{\mathbf{k},c,v} \frac{|\hat{\mathbf{q}} \cdot \mathbf{p_{cv}}|^2}{[E_c(\mathbf{k}) - E_v(\mathbf{k})] \{[E_c(\mathbf{k}) - E_v(\mathbf{k})]^2 - (\hbar\omega)^2\}}, \quad (2)$$

where $\hat{\mathbf{q}}$ is the unit vector in the $\mathbf{q}$ direction, and the overlap integral has been replaced by the matrix element $\mathbf{p_{cv}} = \langle \mathbf{k}, c | \mathbf{P} | \mathbf{k}, v \rangle$, where $\mathbf{P}$ is the momentum operator. The integration over the FBZ can be performed following [23] or resorting to a special point technique [24].

The energy bands and wavefunctions employed in the present work for the approximation of (1) and (2) are obtained from recent nonlocal EPM results [25, 26], where the band structure of the ternary alloys was derived through the virtual crystal approximation. Summations involve eight valence bands and twelve conduction bands; no appreciable change in the results may be observed if the number of conduction bands is increased. A total of 197 plane waves is used for expanding the pseudowavefunctions in

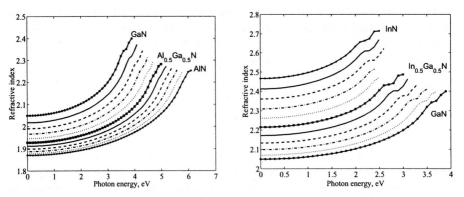

Fig. 2. Calculated refractive indices of AlGaN and InGaN for different values of the molar fraction

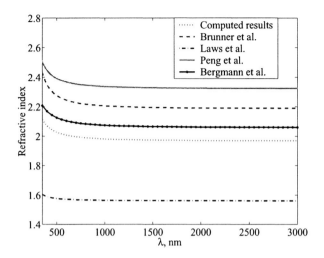

Fig. 3. Comparison between the present computation for the refractive index of $Al_{0.4}Ga_{0.6}N$ and fitting expressions from experimental values

the computation of overlap integrals [25]. Apart from the greater accuracy of the band structure, the improvements of the present computation over the approach proposed in [22] include the use of a much finer mesh in the reciprocal space. Instead of picking 3024 evenly spaced points in the entire FBZ, we divide the Irreducible Wedge (IW, which amounts to 1/24 of the FBZ) into 10890 cubes. As shown in Fig. 1, the impact is small for photon energies lower than the band gap, but significant improvements may be noticed for shorter wavelengths. Moreover, the imaginary part of the dielectric function is now evaluated through the Kramers-Kronig relation, implemented with a robust public domain QUADPACK routine [27], thus removing the uncertainties associated with the approximation of δ-functions.

Results The calculated refractive index for AlGaN and InGaN along (001) is presented in Fig. 2, and a comparison with fitting expressions from measurements [6, 8, 10, 20] is provided for $Al_{0.4}Ga_{0.6}N$ in Fig. 3. Lack of experimental refractive index data pre-

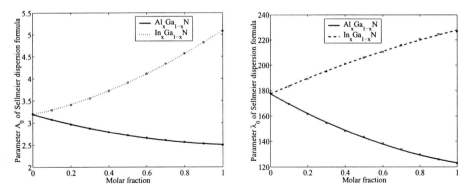

Fig. 4. Parameters A_0 and λ_0 of the Sellmeier formula corresponding to the present computation of AlGaN and InGaN refractive indices. The best-fit expressions are
$A_0(AlGaN) = 3.19311 - 1.21906x + 0.532694x^2$, $A_0(InGaN) = 3.18673 + 0.961966x + 0.948393x^2$,
$\lambda_0(AlGaN) = 177.545 - 83.5651x + 29.2846x^2$, $\lambda_0(InGaN) = 177.3735 + 64.14402x - 13.73745x^2$

vents a similar comparison for InGaN. The refractive index below the energy gap can be suitably described by the first-order Sellmeier dispersion formula

$$n(\lambda)^2 = 1 + \frac{A_0 \lambda^2}{\lambda^2 - \lambda_0^2} \qquad (3)$$

and the values of the fitting parameters A_0 and λ_0 corresponding to our refractive index computations are presented in Fig. 4 as functions of the alloy molar fraction.

References

[1] P. PERLIN, I. GORCZYCA, N. E. CHRISTENSEN, I. GRZEGORY, H. TEISSEYRE, and T. SUSKI, Phys. Rev. B **45**, 13307 (1992).
[2] S. LOGOTHETIDIS, J. PETALAS, M. CARDONA, and T. D. MOUSTAKAS, Phys. Rev. B **50**, 18017 (1994).
[3] M. A. VIDAL, G. RAMÍREZ-FLORES, H. NAVARRO-CONTRERAS, A. LASTRAS-MARTÍNEZ, R. C. POWELL, and J. E. GREENE, Appl. Phys. Lett. **68**, 441 (1996).
[4] T. KAWASHIMA, H. YOSHIKAWA, S. ADACHI, S. FUKE, and K. OHTSUKA, J. Appl. Phys. **82**, 3528 (1997).
[5] S. PETERS, T. SCHMIDTLING, T. TREPK, U. W. POHL, J.-T. ZETTLER, and W. RICHTER, J. Appl. Phys. **88**, 4085 (2000).
[6] T. PENG and J. PIPREK, Electron. Lett. **32**, 2285 (1996).
[7] O. AMBACHER, M. ARZBERGER, D. BRUNNER, H. ANGERER, F. FREUDENBERG, N. ESSER, T. WETHKAMP, K. WILMERS, W. RICHTER, and M. STUTZMANN, MRS Internet J. Nitride Semicond. Res. **2**, No. 22 (1997).
[8] D. BRUNNER, H. ANGERER, E. BUSTARRET, F. FREUDENBERG, R. HÖPLER, R. DIMITROV, O. AMBACHER, and M. STUTZMANN, J. Appl. Phys. **82**, 5090 (1997).
[9] G. YU, H. ISHIKAWA, M. UMENO, T. EGAWA, J. WATANABE, T. SOGA, and T. JIMBO, Appl. Phys. Lett. **73**, 1472 (1998).
[10] M. J. BERGMANN, Ü. ÖZGÜR, H. C. CASEY, JR., H. O. EVERITT, and J. F. MUTH, Appl. Phys. Lett. **75**, 67 (1999).
[11] U. TISCH, B. MEYLER, O. KATZ, E. FINKMAN, and J. SALZMAN, J. Appl. Phys. **89**, 2676 (2001).
[12] J. WAGNER, H. OBLOH, M. KUNZER, M. MAIER, K. KÖHLER, and B. JOHS, J. Appl. Phys. **89**, 2779 (2001).
[13] J. L. P. HUGHES, Y. WANG, and J. E. SIPE, Phys. Rev. B **55**, 13630 (1997).
[14] L. X. BENEDICT and E. L. SHIRLEY, Phys. Rev. B **59**, 5441 (1999).
[15] X. TANG, Y. YUAN, K. WONGCHOTIGUL, and M. G. SPENCER, Appl. Phys. Lett. **70**, 3206 (1997).
[16] A. B. DJURIŠIĆ and E. H. LI, Appl. Phys. Lett. **73**, 868 (1998).
[17] A. B. DJURIŠIĆ and E. H. LI, J. Appl. Phys. **85**, 2848 (1999).
[18] A. B. DJURIŠIĆ and E. H. LI, phys. stat. sol. (b) **216**, 199 (1999).
[19] A. B. DJURIŠIĆ and E. H. LI, J. Appl. Phys. **89**, 273 (2001).
[20] G. M. LAWS, E. C. LARKINS, I. HARRISON, C. MOLLOY, and D. SOMERFORD, J. Appl. Phys. **89**, 1108 (2001).
[21] H. EHRENREICH and M. H. COHEN, Phys. Rev. **115**, 786 (1959).
[22] R. WANG, P. P. RUDEN, J. KOLNÍK, İ. H. OĞUZMAN, and K. F. BRENNAN, Mater. Res. Soc. Symp. Proc. **395**, 601 (1996).
[23] J. P. WALTER and M. L. COHEN, Phys. Rev. B **5**, 3101 (1972).
[24] D. J. CHADI and M. L. COHEN, Phys. Rev. B **8**, 5747 (1973).
[25] M. GOANO, E. BELLOTTI, E. GHILLINO, G. GHIONE, and K. F. BRENNAN, J. Appl. Phys. **88**, 6467 (2000).
[26] M. GOANO, E. BELLOTTI, E. GHILLINO, C. GARETTO, G. GHIONE, and K. F. BRENNAN, J. Appl. Phys. **88**, 6476 (2000).
[27] R. PIESSENS, E. DEDONCKER KAPENGA, C. W. ÜBERHUBER, and D. K. KAHANER, QUADPACK: A Subroutine Package for Automatic Integration, Springer-Verlag, Berlin 1983.

Investigation of Refractive Index and Optical Propagation Loss in Gallium Nitride Based Waveguides

E. Dogheche[1]) (a), P. Ruterana (b), G. Nouet (b), F. Omnes (b), and P. Gibart (c)

(a) Université de Valenciennes, Dépt Matériaux pour Intégration en Microélectronique et Microsystèmes (MIMM), Le Mont-Houy, F-59309 Valenciennes Cedex, France

(b) ESCTM, CRISMAT UMR 6508, CNRS, Institut des Sciences de la Matière et du Rayonnement, 6 Boulevard du Maréchal Juin, F-14050 Caen Cedex, France

(c) Centre de Recherche sur les Hétéroépitaxies et ses Applications (CRHEA), CNRS rue Bernard Gregory Sophia-Antipolis, F-06560 Valbonne, France

(Received June 28, 2001; accepted August 4, 2001)

Subject classification: 68.65.Ac; 78.20.Ci; 78.66.Fd; S7.14

We have investigated the optical properties of $Al_xGa_{1-x}N$ films by using the prism coupling technique. The refractive indices of AlGaN/AlN and AlGaN/GaN heterostructures were accurately measured as a function of the aluminum molar fraction x at 632.8 nm. Optical losses, which are very sensitive to defects content, were evaluated around 1.2 dB/cm in AlGaN/GaN and 1.8 dB/cm in AlGaN/AlN. The origin of these losses was related to the crystalline quality of epitaxial AlGaN layers and the nature of buffer layers.

Introduction The material properties of hexagonal III-nitride semiconductors grown on substrates like c-oriented sapphire or SiC make them of high interest for optoelectronic and electronic device applications. A new generation of light emitting devices and photodetectors is nowadays emerging using these materials. In order to improve the device performances, an understanding of materials optical properties is therefore important for future design. In this paper, we will concentrate on the use of III-nitride materials for visible waveguide device applications. This paper describes the optical performances of $Al_xGa_{1-x}N$ ($x = 0-35\%$) waveguides grown on c-oriented sapphire by metalorganic vapor phase epitaxy (MOVPE). Conventional low temperature grown AlN or GaN layers were used in all applications. The process conditions for growing epitaxial layers are reported by Omnes et al. [1].

Experimental In this work, the optical properties, such the refractive index and the optical loss, have been examined by using the prism coupling described in details by Tien et al. [2]. This technique is ideal to investigate the main optical characteristics of AlGaN/AlN and AlGaN/GaN heterostructures without destruction of samples. In particular, the film thickness and the optical constants can be determined accurately at any point of sample surface from optical information. In this experiment, a right angle rutile prism is used to couple a laser beam (that emits at 632.8 nm) to the guided modes of the structure. The latter is pressed against the base of the prism. By measuring the reflected intensity versus the angle of incidence θ, we draw the guided mode spectrum

---

[1]) Corresponding author; Tel.: (33) 3 27 53 16 80, Fax: (33) 3 27 64 66 54, e-mail: elhadj.dogheche@univ-valenciennes.fr

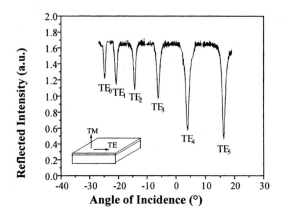

Fig. 1. Recorded transverse electric coupling response in $Al_xGa_{1-x}N$ heterostructure grown on Al_2O_3 substrate

of the samples. The reflectivity dips in the spectrum are correlated to the incident angles that serve to determine the effective mode indices.

Results As an example, we report in Fig. 1 the TE mode spectrum of AlGaN films. Six sharp reflectivity dips are observed at certain angles θ and correspond to the excitation of guided modes. These can be identified as the modes TE_0–TE_5. Six modes have been excited with the TM polarized light. From angular positions of the guided modes, we computed the corresponding effective refractive indices. Using the procedure described by Ulrich and Torge [3], the ordinary refractive indices have been determined: $n_{TE} = 2.3097 \pm 0.004$ and $n_{TE} = 2.3057 \pm 0.004$ for $Al_xGa_{1-x}N/GaN$ and $Al_xGa_{1-x}N/AlN$ heterostructures ($x = 0.17$), respectively. The extraordinary refractive index has been obtained from the TM mode data: $n_{TM} = 2.3411 \pm 0.006$ for $Al_xGa_{1-x}N/GaN$ and $n_{TM} = 2.3423 \pm 0.006$ for $Al_xGa_{1-x}N/AlN$ ($x = 0.17$). Furthermore, we have investigated the relationship between the refractive indices of $Al_xGa_{1-x}N$ alloys and the aluminum content x at 632.8 nm, as plotted in Fig. 2. These values agree with those reported by Yu et al. [4] for gallium nitride thin films using the spectroscopic ellipsometry technique.

The waveguide mode distribution plotted in Fig. 1 clearly shows optical couplings with a narrow width ($< 0.5°$). We demonstrate here the possibility to guide optical waves in AlGaN/AlN and AlGaN/GaN heterostructures. In this technique, the optical attenuation was evaluated from the measurements of the scattered intensity as a function of the propagation along the guide [5]. We considered that the scattered light intensity is proportional to the total light intensity inside the guide as assumed for

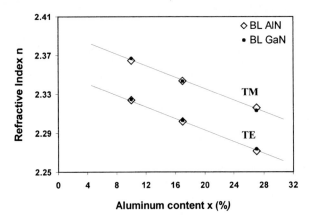

Fig. 2. Dependence of $Al_xGa_{1-x}N/GaN$ and $Al_xGa_{1-x}N/AlN$ refractive indices (n_{TE} and n_{TM}) with the aluminum content x at 632.8 nm

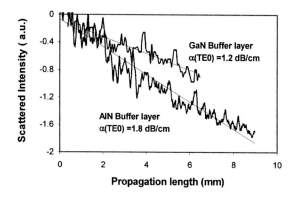

Fig. 3. Scattered intensity measured from $Al_xGa_{1-x}N$ heterostructures using prism coupling technique, showing the influence of buffer layers on optical losses in $Al_xGa_{1-x}N/GaN$ and Al_xGa_{1-x}/AlN heterostructures

homogeneous film waveguides. Data were extracted from digitally recorded pictures. We consider the loss of the mode as the decay of the streak intensity with distance from the prism. The influence of GaN or AlN buffer layers on propagation losses was found to be important as observed in Fig. 3. We have measured losses α_{AlN} of 1.2 dB/cm for AlGaN/GaN and α_{GaN} of 1.8 ± 0.1 dB/cm for AlGaN/AlN for the mode TE_0. This result shows that losses are stronger increased with AlN buffer layer than using GaN buffer layer. As assumed in the field of guided-wave optics, the main loss mechanisms in semiconductor materials are scattering and absorption in the volume of the layer and in the transition layers at boundaries. The interface defects and surface roughness can be dominant contributors to scattering loss.

We have used transmission electron microscopy (TEM) analysis to investigate the growth properties of AlGaN/AlN and AlGaN/GaN heterostructures, in particular at the film–substrate interface. In both AlN or GaN based heterostructures, results have shown a clear evidence for the formation of an interface layer, extending in a relative small thickness as reported in our earlier work [6]. The presence of defects, i.e. threading, misfit dislocations and loops, near the buffer layer may explain the difference in optical loss values.

Conclusion In summary, we have investigated the optical properties of $Al_xGa_{1-x}N$ films by prism coupling technique. The refractive indices of AlGaN/AlN and AlGaN/GaN heterostructures were accurately measured as a function of the aluminum molar fraction x. Optical losses, which are very sensitive to defects, have been evaluated around 1.2 dB/cm in AlGaN/GaN and 1.8 dB/cm in AlGaN/AlN. We have observed a correlation between optical properties and the interface properties: the origin of the losses is related to the quality of the epitaxial AlGaN layer and the nature of the buffer layer.

Acknowledgement Four of the authors (P.R., P.G., G.N., F.O.) acknowledge the support of the EU under contract number HPRN-CT-1999-00040

References

[1] F. OMNES, N. MARENCO, S. HAFFOUZ, H. LARECHE, PH. DE MIERRY, and B. BEAUMONT, Mater. Sci. Eng. B **59**, 401 (1999).
[2] P.K. TIEN, R. ULRICH, and J.R. MARTIN, Appl. Phys. Lett. **14**, 291 (1969).

[3] R. ULRICH and R. TORGE, Appl. Opt. **12**, 2901 (1973).
[4] G. YU, H. ISHIKAWA, M. UMENO, T. EGAWA, J. WATANABE, T. JIMBO, and T. SOGA, Appl. Phys. Lett. **72**, 2202 (1998).
[5] Y. OKAMURA, S. YOSHINAKA, and S. YAMAMOTO, Appl. Opt. **22**, 3892 (1983).
[6] E. DOGHECHE, B. BELGACEM, D. RÉMIENS, P. RUTERANA, and F. OMNES, Appl. Phys. Lett. **75**, 3324 (1999).

Extremely Slow Relaxation Process of a Yellow-Luminescence-Related State in GaN Revealed by Two-Wavelength Excited Photoluminescence

J. M. Zanardi Ocampo (a), N. Kamata[1]) (a), W. Okamoto (a), K. Yamada (a), K. Hoshino (b), T. Someya (b), and Y. Arakawa (b)

(a) Department of Functional Materials Science, Saitama University, 255 Shimo-Ohkubo, Saitama-shi, Saitama 338-8570, Japan

(b) Research Center for Advanced Science and Technology, University of Tokyo, 4-6-1 Komaba, Meguro-ku, Tokyo 153-8505, Japan

(Received June 21, 2001; accepted August 4, 2001)

Subject classification: 71.55.Eq; 73.20.Hb; 78.55.Cr; S7.14

In addition to detection of nonradiative recombination centers by a below-gap excitation (BGE) technique at steady state, we have studied the time-response of both donor–acceptor pair luminescence and yellow luminescence (YL) in GaN after switching off the BGE light. An extremely slow recovery of YL with a time constant of up to 28 s was observed, which is interpreted as a relaxation process of electrons from the state chosen by a BGE energy of 1.17 eV. The dependence of recovery time on the BGE energy and on the above-gap and below-gap excitation power densities became clear for the first time; it reflects the carrier recombination dynamics including YL among below-gap states.

1. Introduction In the last decade, developments in crystal growth technique made possible a drastic improvement of optical and electronic properties of GaN-based semiconductors [1]. Their application to light emitters as well as high power electronic devices [2] has become a matter of worldwide interest. For the improvement of light emission efficiency, it is indispensable to characterize below-gap states that act as nonradiative recombination (NRR) centers. A pair of below-gap states results in yellow luminescence (YL) [3, 4].

By a two-wavelength excited photoluminescence (TWEPL) technique [5], which utilizes below-gap excitation (BGE) light in addition to that of above-gap excitation (AGE), we reported on detection of below gap states in GaN [6–8]. Here we announce results on time-resolved behaviour of TWEPL in undoped GaN, in which temporal response of photoluminescence (PL) was measured after turning off the BGE light. An extremely slow recovery of YL provides an important clue about the recombination dynamics via below-gap states.

2. Experimental
2.1 Sample Over a (0001) sapphire substrate, a 25 nm thick low temperature GaN buffer layer and a 2.1 μm thick unintentionally-doped GaN layer were grown sequentially by MOCVD technique. The residual concentration of electrons was 4.1×10^{16} cm^{-3}.

---

[1]) Corresponding author; Phone: +81 48 858 3529, Fax: +81 48 858 9131,
e-mail: kamata@fms.saitama-u.ac.jp

2.2 TWEPL measurement Conventional PL, whose intensity is denoted as I_A, was obtained by irradiating the sample with an AGE provided by a combination of a D_2 lamp and interference filter, giving an excitation energy E_A = 4.12 eV with a variable excitation power density P_A. A second light source used as a BGE ($E_B < E_g$) was obtained from the 1.064 μm line (E_B = 1.17 eV) of a Nd:YAG laser (power density P_B) and focused in superposition to the AGE at the same spot on the sample. The PL intensity under this double excitation is denoted as I_{A+B} and the ratio between PL intensities at both excitation conditions is expressed as: $\Delta PL = I_{A+B}/I_A$. A decrease in PL after the addition of BGE ($\Delta PL < 1$) is interpreted by interstate excitation between two below-gap states promoted by BGE [5, 6].

During measurements, the sample is immersed in liquid nitrogen so as to avoid thermal effects. The intensity of PL at a desired wavelength is detected by a photomultiplier through a monochromator and processed with a single-photon-counting technique. AGE was provided continuously whilst the BGE was chopped so as to obtain data from both single and doubly excited regimes and the transition between them. The time resolution of the system in this work was 1 ms.

3. Results and Discussion

3.1 A slow recovery process of the quenched PL The time response of YL (2.26 eV) after turning on and off the BGE light is shown in Fig. 1. A steep decrease in YL intensity was observed as soon as the BGE was superposed on the AGE (BGE on), while it gradually recovered to its original value after cutting off the BGE. We here introduce the recovery time τ, which is calculated as the best fitted result of the equation: $I(t) = (I_A - I_{A+B})\{1 - \exp(-t/\tau)\} + I_{A+B}$.

The inset in Fig. 1 shows time integrated conventional PL spectrum (solid line) and TWEPL spectrum (dashed line) at the excitation conditions P_A = 363 nW/mm² and P_B = 5.17 mW/mm². Two main peaks, a dominating YL centered at 2.26 eV and a donor–acceptor pair (DAP) transition at 3.21 eV, are clearly observed along with a decrease in PL intensity for both of them due to the BGE effect [5]. In the following, we will introduce experimental results of ΔPL and τ measured for these two characteristic peaks.

3.2 The quenched PL and the recovery time τ The empirical dependence of ΔPL and τ as a function of P_A was studied for a constant P_B of 5.18 mW/mm² and 9.55 mW/mm² for YL and DAP luminescence, respectively. The behaviour of PL quenching for both YL and DAP peaks, which is accentuated from a value of roughly ΔPL = 0.6 to nearly

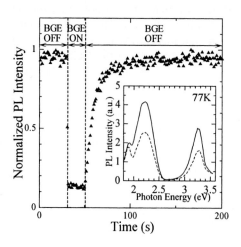

Fig. 1. Temporal YL response. Inset: conventional PL spectrum (solid line) and TWEPL spectrum (dashed line)

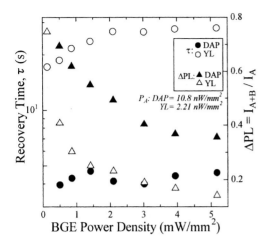

Fig. 2. Measured τ and ΔPL as a function of BGE power density

ΔPL = 0.15 as P_A decreases from 363 nW/mm² to 10 nW/mm², is in concordance with theoretically expected results for TWEPL [5–8]. On the other hand, for YL the characteristic τ increases monotonically from an initial value of 27 ms up to 17.8 s as P_A ranges from 347 nW/mm² down to 2 nW/mm². The DAP transition behaves similarly but τ varies between an initial 40 ms up to 9.98 s. The increase of τ as reducing P_A is explained by a correspondingly decreasing number of photoexcited electrons and holes in conduction and valence bands, respectively, which are necessary to promote recombination processes.

Experimental results of ΔPL and τ as a function of P_B are plotted in Fig. 2, where P_A was kept constant at 2.21 nW/mm² and 10.8 nW/mm² for YL and DAP luminescence respectively. A decreasing ΔPL as P_B is augmented, obtained for both the YL and DAP transition, is in concordance with theory since a higher number of BGE photons will produce an increase of the NRR pathways connected to the interstate excitation process [5–8].

Conversely, for the measurements of YL, as P_B was intensified from 0.14 to 2.09 mW/mm² the value of τ progressively dilated from 17.2 to 26.8 s and from that point on, it showed an apparent saturation tendency with a maximum value of τ = 28.0 s at a P_B = 5.18 mW/mm². The recovery time of the DAP transition, instead, showed a roughly constant value between 3.95 s and 4.58 s along the whole range of P_B.

An increasing value of τ as the BGE power density is augmented is a reflection of the longer time it takes for electrons trapped in the upper below-gap state by BGE to recombine with a fixed number of holes in the valence band (created by a constant AGE flux), and/or for a fixed number of conduction photoelectrons to fill the basal below-gap state, which was increasingly depleted of electrons every time a more intense BGE was applied. The critical excitation density P_B = 2.09 mW/mm² sets down the beginning of coincidental saturation tendencies of both τ and ΔPL, which might be related to a trap filling effect. This is a promising phenomenon in regard of a quantitative characterization of the trap parameters involved in the process, i.e. trap density and electron and hole capture rates [5].

3.3 The recovery time τ for different below-gap states

In order to investigate the dynamics of not only one combination of below-gap states (as selected by a fixed E_B) but further extend the study to different traps, we substituted the BGE light source by a combination of Xe lamp and a monochromator. Figure 3 depicts how the recovery time of the YL suffers a dilation and subsequent shortening as the BGE photon energy increases: 2.22, 3.14 and 2.78 s for an E_B of 1.6, 2.0 and 2.6 eV, respectively. The curve

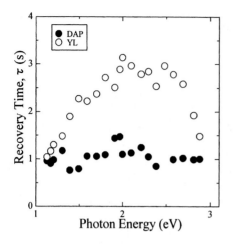

Fig. 3. BGE energy dependent τ values of YL and DAP luminescence

of recovery time is complementary (opposite curvature) to that obtained for ΔPL, which is also remarkably steep in the range 1.6 eV $\leq E_B \leq$ 2.6 eV. We have previously ascribed that kind of results to a collection of states spreading just below the conduction band, which are connected with YL in GaN [8].

An enhanced carrier capture process linked with a higher density of traps or electron/hole capture rates is reflected as a deeper decrease in PL and an elongation of the carrier detrapping time indicated by τ. That credits for the analogy between ΔPL and τ for different below-gap states selected by a tuneable BGE energy.

4. Conclusions We have studied the time-response of both YL and DAP luminescence in undoped GaN grown by MOCVD technique after switching off the BGE light. An extremely slow recovery of YL with a time constant of up to 28 s was observed, which is interpreted as a relaxation process of electrons from the state chosen by a BGE energy of 1.17 eV. The dependence of recovery time on the BGE energy, AGE and BGE power densities became clear for the first time, which reflects the carrier recombination dynamics including YL among below-gap states.

Acknowledgements The authors would like to thank Mr. M. Hirasawa for assisting experiments, Shimadzu Foundation and Japanese Society for Promotion of Science for supporting this work by Shimadzu Scientific Grant and Grant-in-Aid for Scientific Research (C), respectively.

References

[1] S. C. JAIN, M. WILLANDER, J. NARAYAN, and R. VAN OVERSTRAETEN, J. Appl. Phys. **87**, 965 (2000).
[2] H. MORKOÇ, S. STRITE, G. B. GAO, M. E. LIN, B. SVERDLOV, and M. BURNS, J. Appl. Phys. **76**, 1363 (1994).
[3] T. L. TANSLEY, E. M. GOLDYS, M. GOLDLEWSKY, B. ZHOU, and H. Y. ZUO, GaN and Related Materials I, Ed. S. J. PEARTON, Gordon and Breach, New York 1997 (p. 274).
[4] E. CALLEJA, F. J. SÁNCHEZ, D. BASAK, M. A. SÁNCHEZ-GARCÍA, E. MUÑOZ, I. IZPURA, F. CALLE, J. M. G. TIJERO, J. L. SÁNCHEZ-ROJAS, B. BEAUMONT, P. LORENZINI, and P. GIBART, Phys. Rev. B **55**, 4689 (1997).
[5] E. KANOH, K. HOSHINO, N. KAMATA, K. YAMADA, M. NISHIOKA, and Y. ARAKAWA, J. Lumin. **63**, 235 (1995).
[6] J. M. ZANARDI OCAMPO, N. KAMATA, K. HOSHINO, K. ENDOH, K. YAMADA, M. NISHIOKA, T. SOMEYA, Y. ARAKAWA et al., J. Lumin. **87–89**, 363 (2000).
[7] N. KAMATA, M. HIRASAWA, J. M. ZANARDI OCAMPO, K. HOSHINO, K. YAMADA, T. SOMEYA, and Y. ARAKAWA, in: Proc. Internat. Conf. on Physics of Semiconductors, Osaka (Japan), September 17–22, 2000; Springer Proc. in Physics, Vol. 87, Springer-Verlag, Berlin 2001 (p. 1521).
[8] J. M. ZANARDI OCAMPO, N. KAMATA, M. HIRASAWA, K. YAMADA, K. HOSHINO, T. SOMEYA, and Y. ARAKAWA, in: Proc. Internat. Workshop on Nitride Semicond., Nagoya (Japan), September 24–27, 2000; IPAP Conf. Ser. **1**, 544 (2000).

Infrared Ellipsometry – a Novel Tool for Characterization of Group-III Nitride Heterostructures for Optoelectronic Device Applications

M. Schubert[1]) (a, e), A. Kasic (a), S. Einfeldt (b), D. Hommel (b), U. Köhler (c), D. J. As (c), J. Off (d), B. Kuhn (d), F. Scholz (d), and J. A. Woollam (e)

(a) Universität Leipzig, Fakultät für Physik und Geowissenschaften, Arbeitsgruppen Festkörperoptik und Halbleiterphysik, D-04103 Leipzig, Germany

(b) University Bremen, Institut für Festkörperphysik, Bereich Halbleiterepitaxie, D-28359 Bremen, Germany

(c) Universität Paderborn, Fachbereich Physik–Optoelektronik, D-33098 Paderborn, Germany

(d) Universität Stuttgart, 4. Physikalisches Institut, Kristalllabor, D-70550 Stuttgart, Germany

(e) University of Nebraska-Lincoln, Center for Microelectronic and Optical Materials Research, Department of Electrical Engineering, NE 68588-0511, USA

(Received July 10, 2001; accepted August 4, 2001)

Subject classification: 78.20.Ci; 78.30.Fs; 78.66.Fd; 78.67.De; S7.14

We demonstrate the application of spectroscopic infrared ellipsometry to determine nondestructively the free-carrier distribution in group-III nitride heterostructures, such as for optoelectronic and electronic device applications. Results are shown for a blue-light emitting diode structure based on wurtzite III–N materials grown on (0001) sapphire by metal-organic vapor phase epitaxy.

Success in growth of wide-band-gap group-III nitride alloys (Al, Ga, In)N over the past several years led to fast development of nitride-based semiconductor device research. In particular, the achievement of p- and n-type doping of respective device constituents resulted in production of short-wavelength light-emitting diodes (LEDs), laser diodes, and high temperature, high power, high frequency electronic devices. Control and characterization of doping profiles in complex heterostructures represents a challenge. Nondestructive and noninvasive tools are beneficial for both research and production environments.

Polar lattice modes and absorption by free carriers (FC) dominate the infrared (ir) dielectric function (DF) of III–N compounds. Precise knowledge of the ir-DF provides fundamental material properties such as lattice phonon frequencies and broadening parameters, as well as mobility, concentration, and effective mass parameters of p- and n-type FCs. Ellipsometry for ir wavelengths can precisely determine the spectral dependence and anisotropy of a thin-film DF without the need for numerical Kramers-Kronig inversion, or reflectivity standards [1]. Advanced spectroscopic ellipsometry (SE) approaches allow differentiating between the optical responses of multiple layers within complex heterostructures [2]. SE approaches were recently progressed for determina-

---

[1]) Corresponding author; Phone: +49(341) 9732683; Fax: +49(341) 9732699;
e-mail: mschub@physik.uni-leipzig.de

tion of phonon and FC properties of III–N compounds in ordinary heterostructures [3–5]. Single-layer n- or p-type GaN samples were studied by ir-SE and Hall-effect measurements, and optically and electrically determined FC concentrations were found highly consistent, respectively [3]. In this work we use ir-SE for measurement of p- and n-type FC properties within an LED structure.

The SE parameters Ψ and Δ are defined by the complex ratio ϱ of the p- and s-polarized reflectance coefficients r_p and r_s, respectively [2]

$$\varrho \equiv \frac{r_p}{r_s} = \tan \Psi \exp i\Delta, \tag{1}$$

and depend on the angle of incidence Φ_a, the thickness d of each layer, the anisotropic dielectric functions ε_j of the substrate and all layers of the sample [2, 4]. Model dielectric function (MDF) expressions are used for model description of the ir-DF for the III–N compounds [3, 4]. The MDF contains the transverse (TO) and longitudinal optical (LO) phonon mode, the FC concentration N, effective mass m^*, and mobility μ parameters. Figure 1[2]) depicts an excerpt from our ir-SE database of wurtzite and cubic ternary III–N alloy phonon modes. These frequencies depend on both composition and layer strain. Both effects must be studied and quantified prior to model analysis of complex heterostructures. Such heterostructures are analyzed by adjusting relevant ir-DF parameters of the individual constituents (layers) during the ir-SE best-fit regression calculation, and by comparing the results with those obtained from reference samples.

The LED structure studied here consists of wurtzite III–N materials, which were grown by MOCVD on c-plane sapphire and annealed subsequently under N_2-atmos-

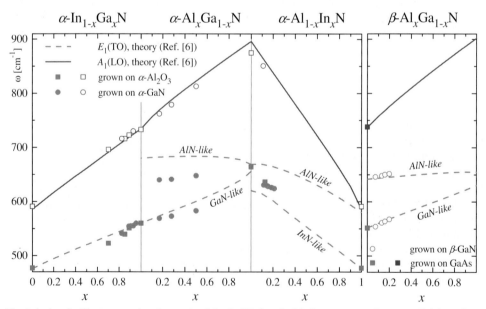

Fig. 1 (colour). Phonon modes determined by ir-SE (symbols) from ternary hexagonal (α) and cubic (β) III–N alloys grown on various substrates. For further details see Refs. [3, 4]. Lines reprint theoretical results given in Ref. [6]

---

[2]) Colour figures are published online (www.physica-status-solidi.com), where indicated.

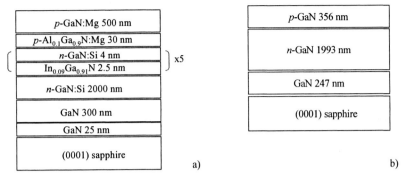

Fig. 2. a) Cross-section of the LED structure studied in this work. The thicknesses refer to the intended growth rates and durations. b) ir-SE model used during the best-fit calculations shown in Fig. 3. The thicknesses refer to the best-fit results

phere for Mg donor activation. The growth sequence of the LED structure is depicted in Fig. 2a. A thick Si-doped n-type GaN layer is deposited onto the undoped buffer/nucleation layer sequence. The active region consists of a Si doped $In_{0.09}Ga_{0.91}N$/GaN multiple quantum well (MQW), which is capped by a 30 nm thick Mg-doped $Al_{0.1}Ga_{0.9}N$ electron-blocking layer. A high Mg doped GaN layer is used for the top contact. The LED structure provides emission at $\lambda \approx 395$ nm under electrical excitation. SE parameters were measured at 300 K for wavenumbers from 333 to 5000 cm^{-1} with 2 cm^{-1} resolution, and at 40°, 55°, and 70° angle of incidence.

Figure 3 shows experimental and best-fit calculated Ψ spectra for our LED structure. The sapphire lattice modes dominate the spectra for wavenumbers $\omega < 1200$ cm^{-1}. Above $\omega \sim 1200$ cm^{-1} optical thickness interference oscillations within the epilayers occur. The GaN and Al_2O_3 ir-DFs for polarizations parallel ($\varepsilon_\parallel$) and perpendicular ($\varepsilon_\perp$) to the material's c-axis were parameterized as described previously [3, 4]. The ir-SE model does not account for the InGaN/GaN MQW, and also not for the AlGaN electron-blocking layer. The ir-SE spectra did not provide sufficient sensitivity to AlGaN- or InGaN-related phonon absorptions, which are here subsumed by the strong GaN-related phonon and FC absorption effects. The thickness of both components contributes ≈3% to the overall device thickness only. The GaN $E_1(TO)$ mode resonance absorption is indicated by vertical arrows in Fig. 3. The ir-SE data analysis distinguishes between three different GaN layers, which reveal common lattice mode parameters but different FC properties. The GaN lattice mode parameters obtained from the best-fit calculations are $E_1(TO) = 560 \pm 0.2$ cm^{-1}, $E_1(LO) = 742 \pm 0.3$ cm^{-1}, $A_1(LO) = 735.8 \pm 0.3$ cm^{-1}, $\gamma[E_1(TO)] = \gamma[A_1(TO)] = 4 \pm 1$ cm^{-1}, and $\varepsilon_\infty = 5.17 \pm 0.1$ (isotropically averaged between $\varepsilon_{\infty\parallel}$, $\varepsilon_{\infty\perp}$) in good agreement with previous results for GaN. The $A_1(TO)$ mode is symmetry forbidden and not significant for the ir-SE spectra, but needed for the model calculation, and was set to 536 cm^{-1} [3, 4]. The nucleation/undoped buffer layer sequence is modeled as a single layer without FC contribution. This layer causes the strong resonance in Ψ due to the GaN $A_1(LO)$ mode (Fig. 3). The n-type layer is modeled assuming a FC absorption (Drude) term with $m_e^* = 0.24 m_e$. The concentration $N_e = 2.2 \pm 0.2 \times 10^{18}$ cm^{-3} and mobility parameter $\mu_e = 110 \pm 20$ cm^2/(Vs) follow from the best-fit calculation for the n-type layer. The high-frequency coupled LO-plasmon mode (LPP$^+$) in this layer amounts to ≈792 cm^{-1}

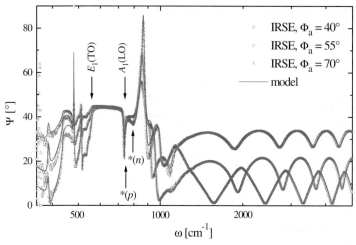

Fig. 3 (colour). Experimental (symbols) and best-fit calculation (solid lines) of ir-SE Ψ data for different angles of incidence Φ_a from the LED structure shown in Fig. 2. The best-fit calculation spectra were obtained with the model depicted in Fig. 2b

(denoted by "*(n)" in Fig. 3). For the p-type layer we assumed $m_h^* = 0.8\,m_e$ and obtained $N_h = 6 \pm 2 \times 10^{16}$ cm^{-3} and $\mu_h \approx 30$ cm^2/(Vs). The LPP$^+$ mode (denoted by "*(p)" in Fig. 3) amounts to ≈ 733.5 cm^{-1}, and is just distinguishable from the uncoupled A_1(LO) mode. The n- and p-type FC concentrations are in good agreement with those intended by choosing the doping rate during the MOCVD growth. According to the best-fit thickness parameters, the p–n junction is located ≈ 356 nm below the surface of the LED structure. However, non-uniform carrier distribution due to surface or interface states, or spontaneous and strain-induced polarization fields may cause inhomogeneous carrier depth profiles [7], which were here assumed to be constant between, and abrupt across interfaces.

Acknowledgements This work was done under DFG grants Rh 28/3-1,2, and under NSF Contract DMI-9901510. M.S. and A.K. thank the CMOMR for further support.

References

[1] D. E. Aspnes, in: Handbook of Optical Constants of Solids, Academic Press, New York 1998.
[2] G. E. Jellison, Thin Solid Films **313/314**, 33 (1998), and references therein.
[3] A. Kasic et al., Phys. Rev. B **62**, 7365 (2000);
   J. Appl. Phys. **89**, 3720 (2001);
   Appl. Phys. Lett. **78**, 1526 (2001);
   Mater. Sci. Eng. B **82**, 74 (2001).
[4] M. Schubert et al., Phys. Rev. B **61**, 8187 (2000);
   MRS Internet J. Nitride Semicond. Res. **4**, 11 (1999);
   MRS Internet J. Nitride Semicond. Res. **5**, W11.39 (2000);
   phys. stat. sol. (b) **216**, 655 (1999);
   Mater. Sci. Eng. B **82**, 178 (2001).
[5] G. Leibiger et al., J. Appl. Phys. **89**, 4927 (2001);
   J. Appl. Phys. **89**, 294 (2001).
[6] H. Grille, C. Schnittler, and F. Bechstedt, Phys. Rev. B **61**, 6091 (2000).
[7] R. Dimitrov et al., J. Appl. Phys. **87**, 3375 (2000).

Electronic Defect States Observed by Cathodoluminescence Spectroscopy at GaN/Sapphire Interfaces

X. L. Sun[1]) (a), S. H. Goss (a), L. J. Brillson (a, b, c), D. C. Look (d), and R. J. Molnar (e)

(a) Department of Electrical Engineering, The Ohio State University, Columbus, OH, USA

(b) Center for Materials Research, The Ohio State University, Columbus, OH, USA

(c) Department of Physics, The Ohio State University, Columbus, OH, USA

(d) Semiconductor Research Center, Wright State University, Dayton, OH, USA

(e) Massachusetts Institute of Technology, Lincoln Labs, Lexington, MA, USA

(Received June 22, 2001; accepted August 4, 2001)

Subject classification: 68.35.Dv; 68.37.Hk; 78.60.Hk; 81.05.Ea; S7.14; S10.1

Cathodoluminescence (CL) imaging and temperature-dependent cathodoluminescence spectroscopy (CLS) have been used to probe the spatial distribution and energies of electronic defects near GaN/Al_2O_3 interfaces grown by hydride vapor phase epitaxy (HVPE). Cross sectional secondary electron microscopy imaging, CLS and CL imaging show systematic variations in defect emissions with a wide range of HVPE GaN/sapphire electronic properties. Highly degenerate interface regions give rise to above bandgap emissions due to band filling and free electron recombination. Besides the common donor and acceptor bound exciton, CLS and CL images also reveal emissions due to excitons bound to stacking faults and cubic phase GaN.

Introduction The group-III nitrides are attractive materials for optoelectronic devices such as GaN-based laser, light-emitting diodes and ultraviolet photodetectors, and high temperature, high power electronic devices [1, 2]. The present or future applications of GaN based devices present an interesting challenge for the further development and the physical understanding of the materials involved. However, numerous electronically active sites within this material are relatively unexplored, for instance, the nature of the residual acceptor and donor [3] and the identification of various emissions related to defects [4].

The localized states at the interface of GaN/sapphire are of significant concern for achieving high electronic quality. In particular, degenerate doping and high conductance [5, 6] can occur near this interface with sapphire, the most common growth substrate, degrading control of transport in the overall epilayer. Hence, impurity diffusion, interface reactions and related defect formation are important to understand and control. In order to determine the physical nature of these dopants and defects, we carried out cross sectional secondary electron (SE) imaging, temperature-dependent (10–300 K) cathodoluminescence spectroscopy (CLS), and CL imaging of these interfaces. Depth-dependent CLS and CL images provide information for detecting the location and physical origin of defects and impurity doping at GaN/Al_2O_3 interfaces.

Experiments The GaN epilayer was deposited on c-plane sapphire by HVPE. A ZnO layer was sputter deposited to prepare the sapphire substrate for GaN growth and was

---

[1]) Corresponding author; Phone: (614) 2922463; Fax: (614) 6884688; e-mail: sun.131@osu.edu

thermally desorbed before epitaxial growth [7]. The GaN epilayer is 5 µm thick with a sheet interface concentration of 6×10^{14} cm^{-2}. A modified JEOL 7800F SEM Auger microprobe (base pressure 8×10^{-11} Torr) fitted with an Oxford Scientific monochromator with a resolution of 0.5 nm, liquid He cold stage, and visible-UV sensitive photomultiplier tube provided CL spectra and images. The incident electron beam energy was 5 keV and current was 1 nA over an area less than 0.5×0.5 µm^2 raster square area. Excitation occurred ≈160 nm below the surface to minimize any surface artifacts. Depth-dependent CLS were taken as a function of the distance (d_{int}) from the GaN/Al$_2$O$_3$ interface.

Results and Discussions Three spectra taken at different distances from the GaN/Al$_2$O$_3$ interface are shown in Fig. 1. The spectra taken in the interface region ($d_{int} = 0.3$ µm) have an extra peak above the band gap (at 3.517 eV). As the distance to the interface increases, the intensity and energy decrease ($d_{int} = 1$ µm) and then disappear near the surface ($d_{int} = 4$ µm). A possible explanation for this broadening to higher energies is band filling at these degenerate-doping levels and the free-electron recombination across the band gap (FERB) as reported by Arnaudov et al. [8]. The spectra taken at $d_{int} = 4$ µm region show strong neutral donor bound exciton (D^{0}X) emission at 3.483 eV. The full width at half maximum (FWHM) of the D^{0}X line is 10 meV, which shows the good crystalline quality of this sample. This line may also contain contributions from free excitons which should have energies only about 6 meV above those of the donor-bound excitons. A peak at 3.464 eV is assigned to an exciton bound to an acceptor [9]. An emission at 3.503 eV is also observed which is probably an excited state of the free exciton, which, should be about 3/4 E_{ex} higher [10].

The CL spectra at different depths also indicate that the quality of the GaN epitaxial layer improves when the film gets thicker. However, as d_{int} increases away from the interface, three additional optical transitions at 3.41, 3.30, 3.28 and 3.19 eV appear; the peaks at 3.28 and 3.19 eV are the characteristic neutral donor–acceptor pair recombination (D^{0}A^0) and its LO phonon replica (D^{0}A^0–LO) in the GaN bulk.

In order to understand the spatial dependence and physical nature of these peaks, we further obtained monochromatic CL images in the whole cross section of this sample at 10 K. Figure 2 shows the SEM image and the monochromatic spatial maps of some different peaks as indicated in Fig. 1. In the SEM image as shown in Fig. 2a, the small triangle area at the lower right corresponds to the sapphire substrate, whereas the triangular region at the upper left corresponds to the free GaN surface. The CL images in Fig. 2 correspond to the same region as the SEM image. The emissions shown in Figs. 2b–d are 3.503, 3.483 and 3.30 eV, respectively. The arrow in each image indicates the interface. The en-

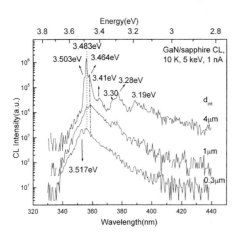

Fig. 1. CL spectra of GaN/sapphire sample taken at $d_{int} = 0.3$ µm, 1 µm and 4 µm as indicated in the figure

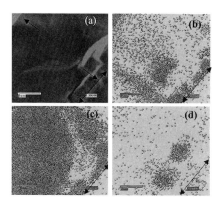

Fig. 2. a) SEM image and b)–d) CL images shown for emissions of b) 3.503 eV, c) 3.483 eV and d) 3.30 eV, taken at the cross section of GaN/sapphire

hanced contrast and brightness of these images serve to indicate the brighter emitting areas.

Additional clues to the origins of the emissions as shown in Figs. 1 and 2 are provided by temperature dependent (10–300 K) experiments. Figure 3 shows the temperature-dependent spectra taken from the bulk epilayer ($d_{int} = 4$ μm). Figure 4a gives the temperature dependence of optical transition energies corresponding to Fig. 3. The energy of all four emissions shown decreases with increasing temperature. Figure 4b is an Arrhenius plot of the integrated intensities of the D^0X, the 3.41 eV and the 3.30 eV peak.

The data presented in Figs. 1–4 provide evidence for several near-interface phenomena. As shown in Figs. 2b and c, the CL images of the 3.503 eV transition and the 3.483 eV D^0X were observed in the same regions, while the intensity of the 3.503 eV transition is much weaker than that of the D^0X. As Fig. 4b shows, the peak at 3.483 eV quenches first with an energy of 4 meV and then quenches with an energy of 12 meV, presumably due to two competitive recombination channels. The small energy quenching of the D^0X corresponds to thermal detrapping towards the free exciton band (≈ 6 meV) while the second nonradiative path could be the ionization of the neutral donor involved in the excitonic complex [11].

As temperature increases, the 3.41 eV peak in Fig. 3 becomes pronounced and quenches very slowly. This peak displays quenching behavior that suggests stacking faults [12]. The low activation energy of quenching is 5–7 meV as shown in Fig. 4b and corresponds to the escape of the electron. The 3.30 eV peak is mostly distinct at $d_{int} = 1$–1.5 μm as shown in Fig. 2d. This may be indicative of another phase besides wurtzite GaN. The 3.30 eV peak quenches with an energy of 10 meV, which is not like the phonon replica of the 3.41 eV peak. Therefore, we believe that the 3.30 eV peak is related to a second phase. From its energy, it could be due to either ZnO or cubic phase GaN domains that often coexist with hexagonal phase. Secondary mass ion spectroscopy (SIMS) shows no Zn in the samples. Therefore, this peak is more likely to be a cubic domain, since stacking faults are always related to the formation of cubic phase GaN.

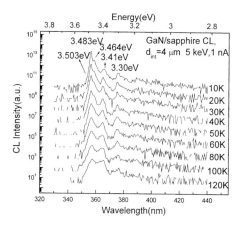

Fig. 3. T-dependence of GaN CL spectra

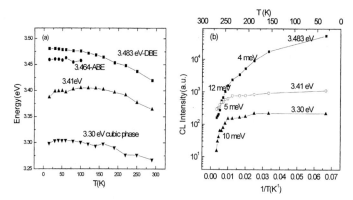

Fig. 4. T-dependence of a) the energies and b) the intensities of the transitions at 3.483, 3.41 and 3.30 eV as shown in Fig. 3

Conclusions CL imaging and temperature-dependent CLS show systematic variations in defect emissions with a wide range of HVPE GaN/sapphire electronic properties. We report large differences between interface and bulk in terms of band edge and deep level emission. These emissions correspond to regions of high carrier concentration, impurity doping, and new phase formation. Highly degenerate interface regions give rise to above bandgap emissions due to band filling and FERB. Combined with the cross sectional SEM and CL images of the defects, the spatial data provide new information on the nature of impurity diffusion and doping at the GaN/Al$_2$O$_3$ interface.

Acknowledgements This work is supported by the Office of Naval Research, the National Science Foundation, the Air Force Office of Scientific Research, and the Department of Energy (depth-dependent measurements). The Lincoln Laboratory portion of this work was sponsored by the Office of Naval Research under Air Force Contract No. F19628-00-C-0002.

References

[1] S. NAKAMURA and G. FASOL, The Blue Laser Diode, Springer, Berlin 1997 (pp. 1–10).
[2] M. RAZEGHI and A. ROGALSKI, J. Appl. Phys. **79**, 7433 (1997).
[3] S. FISHER, C. WETZEL, E. E. HALLER, and B. K. MEYER, Appl. Phys. Lett. **67**, 1298 (1995).
[4] B. MONEMAR, J. Cryst. Growth **189/190**, 1 (1998).
[5] W. GÖTZ, J. WALKER, L.T. ROMANO, N.M. JOHNSON, and R.J. MOLNAR, Mater. Res. Soc. Symp. Proc. **449**, 525 (1997).
[6] D.C. LOOK and R.J. MOLNAR, Appl. Phys. Lett. **70**, 3377 (1997).
[7] L.T. ROMANO, B. S. KRUSOR, and R.J. MOLNAR, Appl. Phys. Lett. **71**, 2283 (1997).
[8] B. ARNAUDOV, T. PASKOVA, E. M. GOLDYS, R. YAKIMOVA, S. EVTIMOVA, I. G. IVANOV, A. HENRY, and B. MONEMAR, J. Appl. Phys. **85**, 7888 (1999).
[9] T. PASKOVA, E. M. GOLDYS, R. YAKIMOVA, E. B. SVEDBERG, A. HENRY, and B. MONEMAR, J. Cryst. Growth **208**, 18 (2000).
[10] J. SKROMME, J. JAYAPALAN, R. P. VAUDO, and V. M. PHANSE, Appl. Phys. Lett. **74**, 2358 (1999).
[11] B. MEYER, Mater. Res. Soc. Symp. Proc. **449**, 497 (1997).
[12] M. ALBRECHT, S. CHRISTIANSEN, G. SALVIATI, C. ZANOTTI-FREGONARA, Y. T. REBANE, Y. G. SHRETER, M. MAYER, A. PELZMANN, M. KAMP, K.J. EBELING, M. D. BREMSER, R. F. DAVIS, and H. P. STRUNK, Mater. Res. Soc. Symp. Proc. **468**, 293 (1997).

Scanning Tunneling Luminescence Studies of Nitride Semiconductor Thin Films under Ambient Conditions

S. K. Manson-Smith, C. Trager-Cowan[1]), and K. P. O'Donnell

Department of Physics and Applied Physics, University of Strathclyde, Glasgow G4 0NG, Scotland, UK

(Received June 26, 2001; accepted August 14, 2001)

Subject classification: 68.37.Ef; 78.55.Cr; 78.67.De; 85.60.Jb; S7.14

We have investigated the properties of a commercial light-emitting diode (LED) structure containing an InGaN single quantum well (SQW) by scanning tunneling luminescence (STL). Data was acquired under ambient conditions, i.e., in air and at room temperature, using our unique STL microscope with a novel light collection geometry. Scanning tunneling microscopy (STM) images revealed the presence of hexagonal pits in the structure, with STL images showing strong luminescence from these pits. The variation of STL intensity with bias voltage shows the STL threshold at -2.1 V is numerically similar to the peak position of the SQW luminescence band. A slight shoulder at -2.8 V corresponds to the plateau of the delocalised absorption profile, observed in macroscopic measurements. The peak observed at -3.2 V is close to the observed GaN band edge emission.

Introduction Nitride thin films are currently arousing considerable excitement because of their suitability for UV and visible light emitting diodes. Scanning tunneling luminescence (STL) is an exciting technique with which to study nitride thin films because it is possible to interrogate material properties with sub-nanometer resolution [1]. STL microscopy combines the high resolution topographic imaging capabilities of the scanning tunneling microscope (STM) with the analytical power of luminescence spectroscopy. The STM tunneling current acts as a highly localized source of electrons (or holes) that generate luminescence in certain materials. If the luminescence induced by the current-injecting probe is detected concurrently with the height variation of the probe as it is raster-scanned over the surface of a sample, the sample's physical morphology and luminescence properties may be simultaneously mapped and compared. Spectroscopic resolution can be obtained by using filters. STL-induced spectra can be obtained by keeping the position of the STM tip fixed, and measuring the spectrum of the collected luminescence. Additionally, the variation of luminescence intensity with tunnel current and with bias voltage can also provide information on recombination processes and material properties [1, 2].

The use of STL to characterize nitride semiconductors is still very new. To date, to our knowledge [3–7], all STL measurements obtained from nitrides have been carried out under UHV conditions and in some cases the samples have been cooled. Light collection has been achieved using a lens sited above the tunneling region. Spectrally filtered STL images have been obtained from GaN epilayers [5, 6] and InGaN MQW samples [7] with STL images from the MQW samples showing luminescence variations on the 30–100 nm scale [7]. STL spectra published to date have been acquired at voltage and current values where the dominant mechanism of light emission is likely to be due to the impact ionization of electron–hole pairs by electrons, i.e., CL as opposed to tunnel-

---

[1]) Corresponding author; Phone: +44 141 548 3465; Fax: +44 141 552 2891; e-mail: cacs19@strath.ac.uk

ing luminescence [4]. The variation of luminescence intensity with bias voltage has been measured for GaN epilayers and InGaN MQW samples. For GaN epilayers the onset of luminescence has been observed at bias voltages corresponding to energies below the GaN bandgap, this may be due to deep-level trap states in the bandgap [3] or to surface related phenomena [5]. For the InGaN MQW samples, the onset of luminescence was observed to be closer to the GaN bandgap than the emission peak of the InGaN MQW [7].

In this paper, we describe our unique STL microscope with a novel light collection geometry. All measurements are carried out under ambient conditions, i.e., in air and at room temperature. We have obtained STL images and the variation of luminescence intensity with bias voltage from an InGaN single quantum well (SQW) LED structure. Our novel instrumentation and results are described below.

Experimental The sample under investigation is a commercial light-emitting diode (LED) structure containing an InGaN single quantum well (SQW), supplied unmounted in wafer form by Nichia Chemical Industries Ltd., Japan. The sample was grown by two-flow metalorganic chemical vapour deposition (MOCVD), and consists of a 30 nm GaN buffer layer grown at a low temperature (550 °C), a 0.7 µm layer of undoped GaN, a 3.3 µm layer of n-type GaN:Si, a 40 nm layer of undoped GaN, a 2.5 nm active layer of undoped InGaN, a 30 nm layer of p-type AlGaN:Mg and a 0.2 µm layer of p-type GaN:Mg [8].

The experiments described in this paper were carried out using a unique STL microscope that we describe below. A schematic diagram of the instrument is shown in Fig. 1.

The microscope operates in air with the samples at room temperature. The tips used are cut 0.5 mm diameter Pt(80%)Ir(20%) wire. A tiny inertial stick-slip motor is used for coarse approach of tip to sample. Its small size and light weight reduces susceptibility to vibration.

The microscope uses a novel light collection setup. As the samples we are studying are grown on transparent substrates, we may collect the STL from the opposite side of the sample with respect to the tunneling tip (see Fig. 2). Our PMT sits directly underneath (3 mm away) from the sample. This removes the need for light collection optics

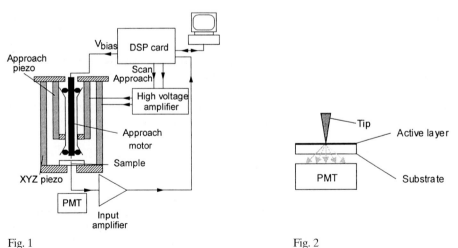

Fig. 1   Fig. 2

Fig. 1. STL microscope schematic

Fig. 2. Light generated at the tunneling junction is detected through the transparent substrate

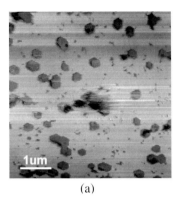 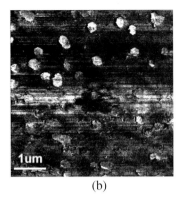

(a) (b)

Fig. 3. a) STM and b) STL images of LED InGaN SQW structure

and the associated losses and provides a collection solid angle of nearly $2\pi sr$. A low background (9 counts per second (cps)) bialkali tube is used which does not require cooling, simplifying design. Filters can be placed between the PMT and sample to allow monochromatic images to be obtained. The use of a tunable filter will allow STL spectra to be acquired.

Finally, a digital (DSP-based) controller provides an easy-to-use, easy-to-modify control system.

Results and Discussion STM (topographic) and panchromatic STL (luminescence) images are shown in Figs. 3a and b, respectively. Images were acquired simultaneously, at a tip bias voltage of -3.3 V and a constant tunneling current of 500 pA with the tip to sample distance kept approximately constant. As the top of the LED InGaN SQW structure is p-type, for tunneling to occur, the tip bias voltage needs to be negative with electrons injected into the structure [9]. No images are obtained for a positive tip bias voltage in the range of voltages available ($\leq +4$ V). We estimate that for these samples, we obtain a spatial resolution of order 100 nm.

The STM image of Fig. 3a reveals the presence of hexagonal pits in the structure. The STL image (Fig. 3b) shows strong luminescence from these pits. The pits are typically 20 nm deep with the luminescence increasing from around 300 cps outside the pits to around 20 kcps at the bottom of the pit. Figure 4 shows schematically why the presence of pits in the structure should result in such an increase in brightnesss of the STL. When the tip is in a pit, a significant number of tunneling electrons reach and thus recombine in the SQW, generating bright luminescence compared to the luminescence generated in the GaN/AlGaN capping layer.

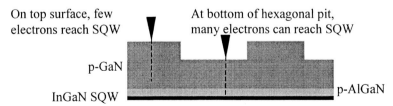

Fig. 4. Schematic diagram illustrating why STL can be observed from hexagonal pits (note not drawn to scale)

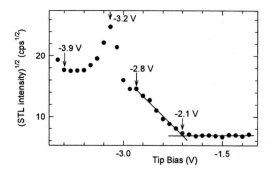

Fig. 5. (STL intensity)$^{1/2}$ versus tip bias voltage

As illustrated in Fig. 4, in order to get to the SQW, even when the tip is in a pit, the injected electrons need to pass through the GaN/AlGaN capping layer. We surmise that the tunneling electrons reach the InGaN SQW due to a combination of band bending at the sample surface due to Fermi level pinning (the tunneling electrons have to overcome a depletion layer at the sample surface before recombining [9] and carrier diffusion (the diffusion length is typically 100 nm in nitride materials). That is injected electrons cross the depletion layer in the GaN/AlGaN cap and then diffuse to the InGaN SQW before recombining. The presence of the pits reduces the distance travelled by the injected electrons sufficiently to result in recombination in the SQW, thus producing observable luminescence and thus contrast in the STL image.

Figure 5 shows (STL intensity)$^{1/2}$ versus tip bias voltage, such a plot reveals the parabolic dependence of luminescence intensity with bias voltage [3, 9]. The STL threshold at -2.1 V is numerically similar to the peak position of the SQW luminescence band [10]. The slight shoulder at -2.8 V corresponds to the plateau of the delocalised absorption profile, observed in macroscopic measurements [10]. The peak observed at -3.2 V is close to the observed GaN band edge emission, finally the increase of luminescence observed at -3.9 V corresponds to the expected emission energy of the $Al_{0.2}Ga_{0.8}N$ layer. Following the work of Tsuruoka et al. [11] and Garni et al. [3], we tentatively suggest that our observed variations in STL bias dependence are due to changes in the density of states in our sample.

Acknowledgement Thank you to Professor Shuji Nakamura and Nichia Chemical Industries, Japan for the sample. We are grateful to EPSRC for funding for this work.

References

[1] S. F. ALVARADO, PH. RENAUD, D. L. ABRAHAM, CH. SCHÖNENBERGER, D. J. ARENT, and H. P. MEIER, J. Vac. Sci. Technol. B **9**, 409 (119).
[2] R. BERNDT and J. K. GIMZEWSKI, Phys. Rev. B **45**, 14095 (1992).
[3] B. GARNI, J. MA, N. PERKINS, J. LIU, T. F. KUECH, and M. G. LAGALLY, Appl. Phys. Lett. **68**, 1380 (1996).
[4] M. ORTSIEFER, A. LIEBHEIT, M. SCHWARTZKOPFF, P. RADOJKOVIC, T. GABRIEL, and E. HARTMANN, Appl. Phys. A **66**, S371 (1998).
[5] S. EVOY, C. K. HARNETT, H. G. CRAIGHEAD, T. J. EUSTIS, W. A. DAVIS, M. J. MURPHY, W. J. SCHAFF, and L. F. EASTMAN, J. Vac. Sci. Technol. B **16**, 1943 (1998).
[6] S. EVOY, H. G. CRAIGHEAD, S. KELLER, U. K. MISHRA, and S. P. DENBAARS, J. Vac. Sci. Technol. B **17**, 29 (1999).
[7] S. EVOY, C. K. HARNETT, H. G. CRAIGHEAD, S. KELLER, U. K. MISHRA, and S. P. DENBAARS, Appl. Phys. Lett. **74**, 1457 (1999).
[8] T. MUKAI, H. NARIMATSU, and S. NAKAMURA, Jpn. J. Appl. Phys. **37**, L479 (1998).
[9] P. RENAUD and S. F. ALVARADO, Phys. Rev. B **44**, 6340 (1991).
[10] S. K. MANSON-SMITH, C. TRAGER-COWAN, R. W. MARTIN, and K. P. O'DONNELL, paper in preparation.
[11] T. TSURUOKA, Y. OHIZUMO, S. USHIODA, Y. OHNO, and H. OHNO, Appl. Phys. Lett. **73**, 1544 (1998).

Raman Scattering and Photoluminescence of Mg-Implanted GaN Films

Lianshan Wang[1]), Soo Jin Chua, and Wenhong Sun

Optoelectronics & Photonics, Institute of Materials Research and Engineering, 3 Research Link, 117602, Singapore

(Received July 1, 2001; accepted August 4, 2001)

Subject classification: 61.72.Vv; 68.55.Ln; 78.30.Fs; 78.55.Cr; S7.14

As-grown highly resistive (>10^7 Ω cm) GaN samples were exposed to implantation of ^{24}Mg$^+$ ions with the energy of 120 keV and dose of 2×10^{14} cm^{-2}. Isochronal rapid thermal annealing (RTA) was carried out for the Mg-implanted samples at 850, 950 and 1050 °C, respectively. Under $z(x, x)z'$ backscattering configuration, A$_1$(LO) and E$_2$(high) Raman modes of 735 cm^{-1} and 569 cm^{-1}, respectively, were observed for the as-grown and Mg-implanted samples, but a Raman mode of 662 cm^{-1} appeared only in the implanted samples. We tentatively assigned the mode to a Mg acceptor-related vibration mode. Room temperature photoluminescence revealed that the as-grown films displayed a strong yellow luminescence of 560 nm (2.21 eV), while Mg-implanted samples only exhibited a broad blue-violet luminescence peaking at 400 nm (≈3.11 eV) which decreased with increasing RTA temperature. Our observation supported the Ga vacancy model responsible for yellow luminescence.

Introduction The family of III–V nitrides has recently attracted much attention because of their wide spectrum suitable for optoelectronic and electronic devices [1, 2]. For most of the GaN materials used so far for device fabrication, dopants are introduced into GaN films during growth [1]. Ion implantation is the selective doping of some certain area on the samples by the introduction of elements. However, damage of the crystal lattice induced by implantation needs to be removed by subsequent thermal annealing [3]. In this paper, we study Mg-implanted GaN films by the combination of Raman scattering and PL spectroscopy. A Raman mode of 662 cm^{-1} will be assigned, and the dependence of a photoluminescence peak intensity at 3.11 eV on the rapid thermal annealing temperature will be revealed.

Experimental As-grown highly resistive (>10^7 Ω cm) GaN films were grown on (0001) sapphire by MOCVD to a thickness of 1.2 μm. Samples cut off from the wafer were homogeneously implanted by ^{24}Mg$^+$ with an energy of 120 keV and the dose of 2×10^{14} cm^{-2}, at room temperature. After implantation, a sample was cut into four pieces and isochronal rapid thermal annealing was performed in flowing N$_2$ for 20 s at 850, 950 and 1050 °C, respectively, for three of them.

Raman measurement was carried out at room temperature using the 514.5 nm line of an Ar$^+$ laser as excitation source. The scattering light was detected in backscattering geometry of $z(x, x)z'$ whose z-direction was parallel to the c-axis of the GaN films. PL was excited with the 325 nm line, 20 mW, of a He–Cd laser.

Result and Discussion Wurtzite GaN has space group C6$_v^4$ (C63mc) with all atoms occupying C3v sites. According to the group theory, Raman-active phonons were pre-

---

[1]) Corresponding author; Phone: +65 874 8355, Fax: +65 872 0785, e-mail: ls-wang@imre.org.sg

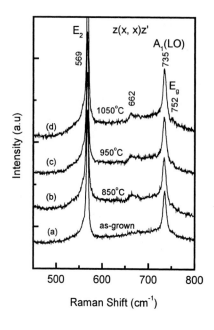

Fig. 1. Raman spectra of the as-grown and post-implantation RTA GaN films. The 569 cm^{-1} and 735 cm^{-1} peaks are assigned to E_2 and A_1(LO) modes, respectively, and a new mode at 662 cm^{-1} is observed. The sapphire line E_g at 752 cm^{-1} is overlapped on the A_1(LO) mode as a shoulder

dicted for the wurtzite structure as the following [4]: an A_1 branch, an E_1 branch and two E_2 branches. In the backscattering geometry of $z(x, x)z'$, the TO branch of the A_1 mode and the TO and LO branches of the E_1 mode are forbidden. The LO branch of A_1 mode and the E_2 modes are allowed.

A typical Raman scattering spectrum for the as-grown and post-implantation annealed GaN films is given in Fig. 1. E_2 and A_1(LO) modes are observed in all samples with E_2 peak at 569 cm^{-1} and A_1(LO) peak at 735 cm^{-1}. The sapphire E_g line at 752 cm^{-1} is also observed as a weaker shoulder overlapping on the spectum. This assignment is in line with that of Azuhata et al. [5]. A new Raman mode at 662 cm^{-1} is evident for all post-implantation thermal-annealed samples, and the peaks become sharper with increasing RTA temperatures, while the as-grown sample does not show this mode.

The room temperature PL spectrum is shown in Fig. 2 for the as-grown GaN film. The near band edge is close to 3.4 eV with line width of 64 meV which is equal to $2.5\,kT$ at 300 K. The second luminescence line (yellow luminescence) is centered at 2.21 eV ($\approx$560 nm).

PL spectra of the post-implantation annealed films are presented in Fig. 3. No signal is observed for the as-implanted GaN sample, indicative of the deterioration of crystal quality due to the implantation damage. The broad blue-violet emission with main peaks at 3.11 eV occurs for all post-implantation annealed samples, which is generally considered as the emission with respect to the shallow Mg acceptor, but the 2.06 eV emission is absent [6]. With increasing RTA temperature, the intensity of the main peaks

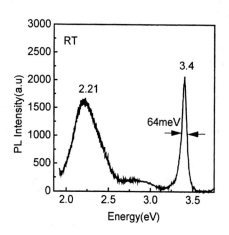

Fig. 2. RT PL spectrum of the as-grown GaN film. The near band edge is close to 3.4 eV with line width of 64 meV, and the yellow luminescence at 2.21 eV is evident

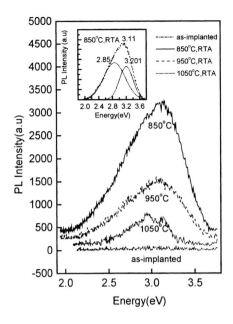

Fig. 3. RT PL spectra for Mg-implanted GaN films. The envelope of blue-violet emission with the main peak at 3.11 eV is revealed. Fitting of the main peak with two Gassian-like peaks, 3.201 eV and 2.85 eV, is shown in the inset for the 850 °C RTA sample

reduces. It is intriguing that the intensity of the 950 °C RTA sample is about half that of the 850 °C RTA sample. Likewise, the intensity of the 1050 °C RTA sample is only close to 50% intensity of the 950 °C RTA sample. The cause of the dependence of the intensities of blue-violet emission on RTA temperature is currently unknown. However, the intensity of the 3.11 eV emission has an optimal annealing temperature (850 °C) for our samples. It is probable that the effective number of blue-violet emission centers drops with raising temperature due to the compensation of N vacancy. We made a fit for the broad PL peak of the 850 °C RTA sample with two Gaussian peaks as shown in the inset of Fig. 3. The fitting demonstrates the two energy levels related to Mg impurity, one at 3.201 eV, the other at 2.85 eV, which have been reported in the literature [6, 7]. The transition of 3.201 eV is due to the recombination from the conductance band to the shallow acceptor, while the 2.85 eV peak is attributed to the emission from the conductance band to a deep Mg-related level, or a complex [7].

Furthermore, in our implanted samples no yellow luminescence appears. We guess that the most probable candidate responsible for yellow luminescence is Ga vacancy [8], because Mg atoms should occupy the interstitial sites after implantation and before rapid thermal annealing. After annealing, the interstitial Mg atoms moved to Ga vacancy, thus leading to the absence of yellow luminescence. From the line width and intensity of E_2 and $A_1(LO)$ mode, we can see that the lattice damage induced by the implantation process has been removed after RTA at 850 °C. Albeit, at further enhancing temperature, the number of Mg atoms at Ga vacancy further increased until all Mg atoms were placed on the lattice sites, and not in the interstitial. On the other hand, using too high a temperature possibly gave rise to the increase of N vacancies.

Kuball et al. [9] investigated the thermal stability of Mg-doped GaN by Raman scattering and found that the line width of $A_1(LO)$ mode was broadened due to the thermal degradation, when the sample temperature was raised between 900 and 1000 °C. New modes at 633 and 652 cm^{-1} were observed at annealing temperatures between 1000 and 1100 °C. It was assumed that the emergence of the new discrete Raman lines in Raman spectra resulted from macroscopic disorder [9]. Conversely, our samples seem not to have any thermal deterioration or macroscopic disorder judging from the intensity and the line width of E_2 and $A_1(LO)$ modes. In addition, the annealing time in our samples was much shorter than that used in their films. From its intensity and its emergence only in Mg-implanted samples, we tentatively assign the 662 cm^{-1} mode to the Mg-related vibration mode, very possibly from the complex of a Mg atom at Ga site and a neighbor N vacancy. Our conjecture is also supported by the intensity change of the blue-violet emission.

Summary Mg-implanted GaN films were investigated by means of Raman scattering and PL spectroscopy. Our results showed that an envelope of blue-violet emission was observed and its intensity was lowered with RTA temperature. In Mg-implanted samples, yellow luminescence was not observed although it existed in the as-grown sample. This result supports the Ga vacancy model as the origin of yellow luminescence. The Raman mode of 662 cm^{-1} was tentatively assigned to the Mg-related vibration mode.

References

[1] S. NAKAMURA, T. MUKAI, and M. SENOH, Appl. Phys. Lett. **64**, 1687 (1994).
[2] S. NAKAMURA, M. SENOH, S. NAGAHAMA, N. IWASA, T. YAMADA, T. MATSUSHITA, H. KIYOKU, Y. SUGIMOTO, T. KOZAKI, H. UMENOTO, M. SANO, and K. CHOCHO, Appl. Phys. Lett. **72**, 211 (1998).
[3] J. C. ZOLPER, J. Cryst. Growth **178**, 157 (1997)
[4] C. A. ARGUELLO, D. L. ROUSSEAU, and S. P. S. PORTO, Phys. Rev. **181**, 1351 (1969).
[5] T. AZUHATA, T. SOTA, K. SUSZUKI, and S. NAKAMURA, J. Phys.: Condens. Matter **7**, 129 (1995).
[6] J. I. PANKOVE, and J. A. HUTCHBY, J. Appl. Phys. **47**, 5387 (1994).
[7] U. KAUFMANN, M. KUZER, M. MAIER, H. OBLOH, A. RAMAKRISHNAN, B. SANTIC, and P. SCHLOTTER, Appl. Phys. Lett. **72**, 1326 (1994).
[8] J. NEUGEBAUER, and C. G. VAN DE WALLE, Appl. Phys. Lett. **69**, 503 (1996).
[9] M. KUBALL, F. DEMANGEOT, J. FRANDON, M. A. RENUCCI, J. MASSIES, N. GRANDJEAN, R. L. AULOMBARD, and O. BRIOT, Appl. Phys. Lett. **73**, 960 (1998).

Effects of Indium Segregation and Well-Width Fluctuations on Optical Properties of InGaN/GaN Quantum Wells

A. Soltani Vala, M. J. Godfrey[1]), and P. Dawson

Department of Physics, University of Manchester Institute of Science and Technology, P.O. Box 88, Manchester M60 1QD, UK

(Received June 26, 2001; accepted August 4, 2001)

Subject classification: 73.21.Fg; 77.65.Ly; 78.47.+p; 78.55.Cr; S7.14

We report the results of calculations for the energies of confined electrons and holes and their wavefunction overlap in $In_xGa_{1-x}N$/GaN quantum wells (QWs) with an indium concentration of $x = 15\%$ in the well material. It is known that the observed increase in the photoluminescence lifetime with increasing well width can be explained qualitatively by the reduction in overlap of the electron and hole wave functions, which is caused by the piezoelectric field in the strained QW material. We show that the energy dependence of the lifetime measured across the emission line can be explained in a similar way, as the result of ± 1-monolayer variations in the QW width. We also calculate the energies and electron–hole wave-function overlap for carriers trapped within indium-rich regions of the QW, taking into account the relaxation of the strain field in and around the indium fluctuation. Our results indicate that well-width fluctuations lead to a stronger energy dependence of the lifetime: the magnitude of the effect is the same order as in experiment, and shows a similar increase with increasing well width.

Introduction The measured photon energy of the photoluminescence (PL) from InGaN quantum wells (QWs) shows a consistent red-shift, compared with the band gap of bulk InGaN alloy material of the same composition. In addition, the PL decay time depends on the photon energy, increasing as the energy of the emission is scanned from the high to the low energy side of the emission line.

Various models have been put forward in partial explanation of these results. Firstly, it has been postulated [1] that there exist localized states whose energies extend into the band gap of the InGaN QW with a density that decreases exponentially away from the band edges. In this *band tail model* it can be shown that the decay time increases towards lower transition energies, but the model offers no explanation for the strong dependence on well width of the decay time. Moreover, if the timescale of the PL decay is to be explained by the relatively slow relaxation of carriers through these localized states, then, by a similar argument, the onset of the PL should be expected to be delayed on a similar timescale. No such effect on the rise time is observed.

A second model proposed does not require the existence of localized states, but instead relies on the quantum confined Stark effect (QCSE) in the InGaN wells [2]. Strong spontaneous and piezoelectric polarization effects [3] are expected, the latter arising from the large difference between the in-plane lattice parameters of InN and GaN. The electron and hole wave functions are confined near the edges of the QW and this accounts for both the large downward shifts in emission energy (compared with the

---
[1]) Corresponding author; Phone: +44 161 200 3182; Fax: +44 161 200 3941; e-mail: m.j.godfrey@umist.ac.uk

InGaN band edge) and for the well-width dependence of the radiative lifetime, owing to reduced overlap of the electron and hole wave functions.

Thirdly, the effects of indium segregation are often invoked, to the extent that quantum-dot-like structures of different sizes and composition will have different transition energies and electron–hole overlap integrals. This model has been reviewed by Chichibu et al. [4].

We report the results of calculations that indicate the relative importance of the second and third of the above mechanisms that may lead to an energy dependence of the lifetime. The first, and simpler, calculation is based on the QCSE, in which the piezoelectric field is the dominant effect. We show that the dependence on energy of the PL decay time is largely explained as an effect of *well-width fluctuations*: carriers in wider quantum wells have smaller transition energies and a reduced wave-function overlap, which significantly increases the lifetime. In our second calculation, we consider the *additional* effect of an indium-rich region within a QW of fixed width: a Green's-function method is used to solve the equations of elasticity for this inhomogeneous problem, and so find the piezoelectric potential associated with the indium-rich region. We find that the effect on the transition energy and wave-function overlap is relatively small, unless the indium-rich region has a very large extent in the plane of the QW.

Simple QCSE Model Owing to the large lattice mismatch and the substantial value of the piezoelectric constant, there is a very large piezoelectric field in the active region of InGaN/GaN QWs. The strain-induced piezoelectric field along the [0001] axis is given by

$$E_{\text{piez}} = \frac{2}{\varepsilon_r} \left(\frac{c_{13}}{c_{33}} e_{33} - e_{31} \right) \varepsilon_{xx},$$

where c_{ij}, e_{ij}, and ε_{xx} are, respectively, the elastic stiffness constants, the piezoelectric constants, and the in-plane strain. We use a simple one-band effective mass theory to calculate the transition energy and wave-function overlap integral. Due to lack of certainty in the effective mass (and other) parameters of nitride semiconductors, we believe that this simple model with a only a small number of parameters is an appropriate starting point. We use a finite difference method to solve the effective-mass Schrödinger equations for an electron and a hole, and so find the quantization energies, the wave functions, and the overlap integral of the electron and hole wave functions. Effective-mass parameters are taken from Ref. [5] and elastic and piezoelectric constants from Refs. [6, 7].

In Fig. 1 we have plotted, as a function of well width, the squared overlap integral of the electron and hole wave functions. The result

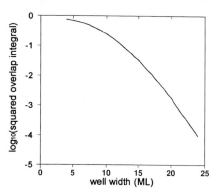

Fig. 1. Squared overlap integral of electron and hole wave functions versus well width

shows that, in the In$_x$Ga$_{1-x}$N/GaN system, the squared overlap integral, which determines the optical transition rate, decreases rapidly with increasing QW width. We particularly wish to emphasize that the overlap changes significantly even for *small* fluctuations in the well width: for a ±1-monolayer variation in the well width, the overlap changes by factors of 1.9 (4.0) for a 25 Å (50 Å) QW, but the corresponding change in the transition energy is approximately 60 meV in each case. These results are in semi-quantitative agreement with the observed variation of the PL decay time across the breadth of the emission line, if this breadth (typically 50 meV) is taken to be the full width at half maximum of the PL intensity [8]. As found in experiment, the PL decay time shows a greater variation across the emission line in samples with wide QWs than in those with narrow QWs, but the linewidth itself is relatively constant.

Effect of an Indium Fluctuation We consider the additional strain and resulting piezoelectric potential due to a fluctuation $X(\mathbf{r})$ in the molar concentration of indium within a QW. By use of the Green's tensor G_{in} of a homogeneous medium, which satisfies

$$\lambda_{lm,ij} \frac{\partial^2 G_{in}(\mathbf{r})}{\partial x_m \, \partial x_j} = -\delta(\mathbf{r}) \, \delta_{ln} \,, \tag{1}$$

it is a straightforward matter to show that, to first order in $X(\mathbf{r})$, the additional material displacement $\mathbf{U}'(\mathbf{r})$ due to the composition fluctuation is

$$U'_i(\mathbf{r}) = -B \int G_{in}(\mathbf{r} - \mathbf{r}') \frac{\partial X(\mathbf{r}')}{\partial x'_n} \, dV' \,, \quad \text{where} \quad B = 3\kappa \Delta a / a_{av} \,. \tag{2}$$

Here, Δa is the difference between the in-plane lattice parameters of InN and GaN; a_{av} is the lattice parameter of homogeneous bulk material with the same composition as in the QW; and κ is the bulk modulus. For simplicity, we use the isotropic approximation to the elastic properties throughout and, as implied by the use of Eq. (1), we neglect any dependence of the elastic constants on alloy composition.

Equation (2) leads to the following expression for the Fourier transform $\tilde{U}'_{ij}$ of the additional strain U'_{ij}, in terms of the Fourier transforms of G_{in} and $X(\mathbf{r})$,

$$\tilde{U}'_{ij}(\mathbf{k}) = B(\delta_{ij} - 12\kappa\pi^3 \{k_i \tilde{G}_{jn}(\mathbf{k}) \, k_n + k_j \tilde{G}_{in}(\mathbf{k}) \, k_n\}) \, \tilde{X}(\mathbf{k})/3\kappa \,. \tag{3}$$

(In Refs. [9, 10], a similar expression has been derived for the strain field associated with a quantum dot of uniform composition, where, in that application, $X(\mathbf{r})$ would be the *characteristic function* of the dot: unity for $\mathbf{r}$ inside the quantum dot and zero outside.) Equation (3) may be re-expressed, using the Lamé coefficients λ and μ, as

$$\tilde{U}'_{ij}(\mathbf{k}) = \varepsilon_0 \tilde{X}(\mathbf{k}) \, (\delta_{ij} - \beta k_i k_j / k^2) \,, \quad \text{where} \quad \varepsilon_0 = \frac{\Delta a}{a_{av}} \quad \text{and} \quad \beta = \frac{(3\lambda + 2\mu)}{(\lambda + 2\mu)} \,.$$

The piezoelectric potential associated with the additional strain field can be calculated by solving the Poisson equation: the additional charge density ϱ is related to U'_{kl} through the expressions $\varrho = -\nabla \cdot \mathbf{P}$ and $P_i = e_{i,kl} U'_{kl}$, where $\mathbf{P}$ and $e_{i,kl}$ are the piezoelectric polarization and the piezoelectric tensor respectively. As we can calculate the Fourier transform of the strain tensor, it is simplest to solve the Poisson equation in Fourier space and obtain the inhomogeneous piezoelectric potential by inverse Fourier

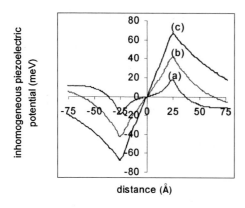

Fig. 2. Calculated additional piezoelectric potential through the centre of the indium fluctuation, along the direction normal to the QW. Curves a, b, and c correspond to indium fluctuations of radius r_0 = 50, 100, and 200 Å, respectively

transformation. For ease of calculation we have used the following form for $X(\mathbf{r})$:

$$X(\mathbf{r}) = X_0 \exp\{-(x^2+y^2)/r_0^2\}\,\theta(z)\,\theta(L_\mathrm{w}-z),$$

which is non-zero only within the QW, $0 < z < L_\mathrm{w}$, and, in the plane of the well, decreases exponentially on a lengthscale r_0.

We have solved the resulting axially-symmetric Schrödinger equation to obtain the squared electron–hole overlap integral and the transition energy for a mean indium concentration $x_{av} = 15\%$ and a (large) fluctuation of magnitude $X_0 = 5\%$, for values of r_0 in the range 50–200 Å. In Fig. 2 we plot the *additional* piezoelectric contribution to the potential in a 50 Å QW. Over the stated range for r_0 we find that the squared overlap integral decreases by only a factor of two, while the transition energy decreases by 109 meV. This result shows that the energy dependence of the PL decay time across the emission line could be only *partly* explained even by the presence of *large* indium fluctuations with a wide range of sizes. We emphasize that the resulting energy dependence of the lifetime is therefore predicted to be weak, compared with the effect of 1 ML well-width variations considered above.

References

[1] A. SATAKE and Y. MASUMOTO, Phys. Rev. B. **57**, 2041 (1998).
[2] A. HANGLEITER, J. S. IM, H. KOLLMER, S. HEPPEL, J. OFF, and F. SCHOLZ, MRS J. Nitride Semicond. Res. **3**, 15 (1998).
[3] F. BERNARDINI, V. FIORENTINI, and D. VANDERBILT, Phys. Rev. B. **56**, 10024 (1997).
[4] S. CHICHIBU, T. SOTA, K. WADA, and S. NAKAMURA, J. Vac. Sci. Technol. B **16**, 2204 (1998).
[5] M. SUZUKI and T. UENOYAMA, Phys. Rev. B **52**, 8132 (1995).
[6] A. F. WRIGHT, J. Appl. Phys. **82**, 2833 (1997).
[7] A. D. BYKHOVSKI, V. V. KAMINSKI, M. S. SUR, Q. C. CHEN, and M. A. KHAN, Appl. Phys. Lett. **68**, 818 (1996).
[8] J. A. DAVIDSON, P. DAWSON, T. WANG, T. SUGAHARA, J. W. ORTON, and S. SAKAI, Semicond. Sci. Technol. **15**, 497 (2000).
[9] A. D. ANDREEV, J. R. DOWNES, D. A. FAUX, and E. P. O'REILLY, J. Appl. Phys. **86**, 297 (1999).
[10] A. D. ANDREEV and E. P. O'REILLY, Thin Solid Films **364**, 291 (2000).

First-Principles Calculations of Optical Properties of AlN, GaN, and InN Compounds under Hydrostatic Pressure

B. Abbar (a), B. Bouhafs (a), H. Aourag (a), G. Nouet[1]) (b), and P. Ruterana (b)

(a) Department of Physics, CMSL, University Sidi Bel-Abbes, Sidi-Bel Abbes 22000, Algeria

(b) ESCTM-CRISMAT, UMR 6508 CNRS, ISMRA, 6, Bld Maréchal Juin, F-14050 Caen cedex, France

(Received June 25, 2001; accepted July 30, 2001)

Subject classification: 78.20.Ci; 78.66.Fd; S7.14

We present first-principles full-potential linearized augmented plane wave calculations of the effect of hydrostatic pressure on the optical properties of wurtzite GaN, InN and AlN compounds. The refractive index and its variation with hydrostatic pressure are well described. The accurate calculation of linear optical function (refraction index and its pressure derivative, and both imaginary and real parts of dielectric function) is performed in the photon energy range up to 30 eV. The predicted optical constant agrees well with the available experimental data.

Introduction The aim of this work is to examine the optical properties of III–V materials GaN, InN and AlN, with emphasis on their dependence on hydrostatic pressure. The wurtzite III–V nitrides are wide band gap materials with direct optical transition, characterised by high ionicity, high bond strength, and good thermal conductivity. In the following, we study the optical properties of III–V nitrides, like dielectric function (real and imaginary parts) and refraction index under pressure. The calculation was made using the first-principles full-potential linearized augmented plane wave (FP-LAPW) method in conjunction with local density approximation (LDA) to the density functional theory (DFT).

Calculations Scalar-relativist calculations have been performed using the WIEN97 code [1] that is the implementation of the method. For the exchange and correlation potential, we use the local density approximation with a reparametrization of Ceperly-Alder data [2]. Basis functions, electron densities and potentials were expanded inside the muffin-tin spheres in combination with spherical harmonic functions with a cut-off $l_{max} = 12$, and in Fourier series in the interstitial region. Here, 2891 plane waves were used for the expansion of the charge density and potential. We use a parameter $R_{MT} \cdot K_{max} = 10$, which determines matrix size (convergence), where K_{max} is the plane wave cut-off and R_{MT} is the smallest of all atomic sphere radii. We chose the muffin-tin radii of Ga, N, In and Al to be 1.95, 1.65, 2.05, and 1.85 a.u., respectively. The value of sphere radius and K_{max} were kept constant over all the range of lattice constants considered. To correct the LDA error in the band gaps, a constant potential was applied to the conduction band states (using the scissors op-

---

[1]) Corresponding author; Phone 33 231 45 26 47; Fax: 33 231 45 26 73; e-mail nouet@ismra.fr

Table 1

The calculated dielectric constants $\varepsilon_{xyz}(0)$, and pressure coefficients of refractive index in 10^{-2} GPa^{-1} for GaN, InN and AlN

	ε_{xyz}	ε_{xy} ($\varepsilon_\perp$)	ε_z ($\varepsilon_\parallel$)	$1/n$ (dn/dp)$^-$
GaN	4.21[a]	4.17[a]	4.30[a]	−0.45[a]
	4.68[b]	4.71[b]	4.62[b]	−0.19[b]
	5.20[c]	−	−	−
InN	8.26[a]	8.34[a]	8.10[a]	−3.40[a]
	7.16[b]	7.27[b]	6.94[b]	−0.43[b]
	8.40[c]	−	−	−
AlN	4.43[a]	4.40[a]	4.49[a]	−0.21[a]
	3.86[b]	3.91[b]	3.77[b]	−0.18[b]
	4.68[c]	−	−	−

[a] Present work, [b] other result [4], [c] experiment [5]

erator which rigidly shifts the conduction band states) in order to match the calculated band gaps with the experimental data [3].

Results and Discussion The dielectric functions of GaN, InN and AlN in the wurtzite structure are resolved into two components $\varepsilon_{xy}(\omega)$, average of the spectra for the polarisation along x and y-directions and $\varepsilon_z(\omega)$, the polarisation vector parallel to the z-direction, the averaged $\varepsilon_{xyz}(\omega) = (\varepsilon_x(\omega) + \varepsilon_y(\omega) + \varepsilon_z(\omega))/3$ is also computed (see Table 1).

Figure 1 shows the variation of the imaginary part of the electronic dielectric function ($\varepsilon_{xyz}(\omega)$) at normal and under hydrostatic pressure for the III-nitrides GaN, InN

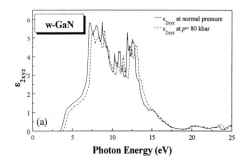
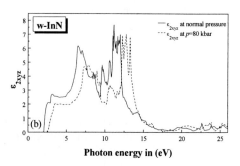
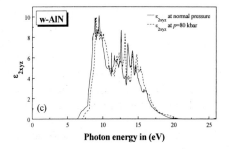

Fig. 1. Imaginary part of the dielectric function averaged over the three Cartesian directions of a) GaN, b) InN and c) AlN

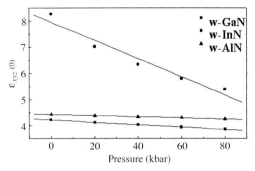
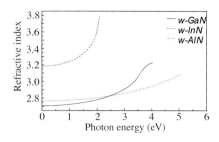

Fig. 2. Pressure dependence of $\varepsilon(0)$ of GaN, InN and AlN

Fig. 3. Refractive index of III-nitrides in the wurtzite structure

and AlN, for radiation up to 30 eV. The calculated results are shifted rigidly upwards by 1.60 eV in GaN, 1.40 eV in InN, and 2.00 eV in AlN. Some features are in common. There are three groups of peaks, the first group is in (3.40–9.00 eV) photon energy range, they are mainly due to transitions in vicinity of M. This is usually associated with E_1 transition. However, the L_{3v}–L_{1c} transition occurs at 7.19 eV. Similarly, the main peak in the spectra of InN and AlN is situated at 6.51 and 8.85 eV, respectively. The second group of peaks (11.30–12.93 eV in GaN, 9.34–10.27 eV in InN and 11.30–12.42 in AlN) comes from the transition at P and L' in GaN, and Σ, P and L in AlN. The last group of peaks (up to 13 eV) is connected mainly to transitions at Γ, M (in GaN, InN) and T, M (in AlN).

In Table 1 the dielectric constants $\varepsilon_{xyz}(0)$ and the pressure coefficients of refractive indices $1/n\,(\mathrm{d}n/\mathrm{d}p)$, are given in comparison with the results of experiments. Figure 2 shows the pressure dependence of $\varepsilon_{xyz}(0)$ for all compounds considered. We have a negative value of pressure coefficients of refractive index, in good agreement with previous calculations [4] and experiments [5].

The refraction index, shown in Fig. 3, was computed using both real and imaginary parts of the dielectric function. The values obtained were found about 3% lower than those reported in [6]. We have fitted our calculated refractive indices using the empirical formula of Peng and Piprek [7], given by the following equation:

$$n(E) = \left[a \left(\frac{E}{E_g} \right)^2 \left(2 - \left(1 + \frac{E}{E_g}\right)^{0.5} - \left(1 - \frac{E}{E_g}\right)^{0.5} \right) + b \right]^{0.5},$$

where the values of the direct energy gaps (E_g) are obtained from our optical spectra, and E is the photon energy. The parameters a and b are obtained from the fit of the calculated refractive index spectra for the three binary compounds (see Fig. 3). All parameters are listed in Table 2.

Table 2
Parameters of GaN, InN and AlN

compounds	E_g (eV)	a	b
GaN	3.50	3.92	7.74
InN	1.89	5.87	10.33
AlN	6.20	13.70	7.81

Conclusion The optical properties of wurtzite nitrides have been investigated at normal and under hydrostatic pressure. The real and imaginary parts of the dielectric function were calculated for polarisation in the xy plane, along the z-axis and the averaged values of both parts for three compounds of interest. It is shown that the refraction index decreases under pressure and we derived also the refractive index.

Acknowledgement This work is carried out with the support of the Algerian-French Ministries of Foreign Affair under project CMEP 01 MDU 516.

References

[1] P. BLAHA, K. SCHWARZ, and J. LUITZ, WIEN97, A Full Potential Linearized Augmented Plane Wave Package for Calculating Crystal Properties, K-H. SCHWARZ, Techn. University Wien (Austria), 1997.
[2] J. P. PERDEW and Y. WANG, Phys. Rev. B **45**, 13244 (1992).
[3] G. MIMTS and B. SCHLICHT, in: Narrow Gap Semiconductors, Springer-Verlag, Berlin/New York 1985.
[4] N. E. CHRISTENSEN and I. GORCZYCA, Phys. Rev. B **50**, 4397 (1994).
[5] J. MISEK and F. SROBAR, Elektrotech. Cas. **30**, 690 (1979).
[6] M. ALOUANI, L. BEREY, and N. E. CHRISTENSEN, Phys. Rev. B **37**, 1167 (1988).
[7] T. PENG and J. PIPREK, Electron. Lett. **32**, 24 (1996).

Near K-Edge Absorption Spectra of III–V Nitrides

K. Fukui[1]) (a), R. Hirai (b), A. Yamamoto (b), H. Hirayama (c), Y. Aoyagi (c), S. Yamaguchi (d), H. Amano (d), I. Akasaki (e), and S. Tanaka (f)

(a) Research Center for Development of Far-Infrared Region, Fukui University, Fukui 910-8507, Japan

(b) Department of Electrical and Electronic Engineering, Fukui University, Fukui 910-8507, Japan

(c) Institute of Physical and Chemical Research (RIKEN), Wako, Saitama 351-0198, Japan

(d) Department of Materials Science and Engineering, Meijo University, Nagoya 468-8502, Japan

(e) High-Tec Research Center, Meijo University, Nagoya 468-8502, Japan

(f) Research Institute for Electronic Science, Hokkaido University, Sapporo 060-0812, Japan

(Received June 22, 2001; accepted August 4, 2001)

Subject classification: 71.20.Nr; 78.20.Ci; 78.70.Dm; S7.14

Nitrogen and aluminum near K-edge absorption measurements of wurtzite AlN, GaN and InN, and their ternary compounds (AlGaN, InGaN and InAlN) at various molar fractions have been performed using synchrotron radiation. Using the linear polarization of synchrotron radiation, absorption measurements with different incident light angles were also performed. The spectral distribution of the nitrogen K absorption spectra clearly depends on both the incident light angles and the molar fractions of the samples. That of the aluminum K absorption spectra also shows the clear angle dependence, but it does not show the drastic molar dependence. Spectral shape comparisons among the various molar fractions, different incident angles and between the two ion sites are discussed. The numerical component analysis of the K absorption spectra is also presented.

Introduction Since the core levels are strictly localized in space, inner core excitation can select the specific ion site in the materials and give us site-specific information. This means that it is a useful method to investigate binary or ternary compounds. The spectrum near the K-core absorption edge, which excites the 1s electrons to the conduction band, represents the unoccupied p partial density of states (p-PDOS) according to the selection rule. Such kind of works for III–V nitride semiconductors has been reported by several studies in Refs. [1, 2] and references therein. They indicate the good overall agreement between the experimental and calculated results. Furthermore, since the synchrotron light source has linear polarization, the unoccupied p-PDOS is resolved into the in-plane and the out-of-plane states.

In our previous work [1], we reported the soft X-ray absorption spectra around nitrogen K (N-K) and aluminum K edge (Al-K) of the wurtzite AlN, GaN and their ternary compounds $Al_xGa_{1-x}N$. All samples clearly showed the incident light angle dependence of the absorption spectra, which directly reflected the anisotropy of the unoccupied

---

[1]) Corresponding author; Phone: +81 776 27 8562; Fax: +81 776 27 8749; e-mail: fukui@fuee.fukui-u.ac.jp

p-PDOS. Then we presented a numerical component analysis to separate the experimental K-absorption spectrum (KAS) into three partial spectra which correspond to in-plane, out-of-plane and angular independent components. The comparison of the spectral feature among the different molar fraction (x) samples is one of the standards to investigate the peak identifications in the spectra. Therefore, the combination of those two techniques with the ion-site specific nature of the core absorption measurement becomes effective to investigate the structures of the unoccupied DOS. In this paper, to investigate the structures of the unoccupied p-PDOS of the III–V nitrides, both the N-KAS and Al-KAS of the whole III–V nitrides (AlN, GaN and InN) and their compounds ($Al_xGa_{1-x}N$, $In_xGa_{1-x}N$ and $In_xAl_{1-x}N$) were carried out.

Experimental Procedure All samples are wurtzite thin films that have their c-axis perpendicular to the film surface. They were grown by the MOCVD method. All the $Al_xGa_{1-x}N$ ($0.05 \leq x \leq 1.0$) and $In_xGa_{1-x}N$ ($x \leq 0.15$) samples were fabricated on SiC substrates at RIKEN, and GaN, $In_xAl_{1-x}N$ and $In_xGa_{1-x}N$ ($0.02 \leq x \leq 1.0$) were grown on α-Al_2O_3 substrates at Nichia Chemical Industries Ltd., Meijo University and Fukui University, respectively. The sample thickness is between 0.1 and 4 μm. The molar fraction x was determined by the lattice constant or luminescence peak of each sample.

The N-K and Al-K absorption measurements were performed at the beamlines BL4B and BL8B1, BL1A and BL7A of UVSOR (synchrotron radiation light source), Institute for Molecular Science. All the experiments were performed at room temperature using the total photoelectron yield method that was measuring the sample drain current in the vacuum chamber. The incidence angle θ is defined as the angle between the incident light and the normal axis of the sample surface. The configuration between the electric field of the incidence light **E** and the rotation axis is p-polarization. To avoid a charge-up of the sample surface for the threshold energy determination experiment at $\theta = 0$, an Au mesh and a Cu box were used to cover the sample surface.

Results and Discussion Figure 1a shows the N-KAS of the $Al_xGa_{1-x}N$, $In_xGa_{1-x}N$ and $In_xAl_{1-x}N$ and Fig. 1b shows the Al-KAS of $Al_xGa_{1-x}N$ and $In_xAl_{1-x}N$. The param-

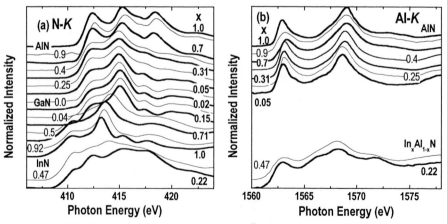

Fig. 1. Normal incidence a) nitrogen K absorption spectra of $Al_xGa_{1-x}N$, $In_xGa_{1-x}N$ and $In_xAl_{1-x}N$ (from top to bottom) and b) aluminum K absorption spectra of $Al_xGa_{1-x}N$ and $In_xAl_{1-x}N$

eter x of each spectrum corresponds to the molar fraction of Al or In of the respective ternary compound sample. The intensity is normalized at the maximum peak. The spectrum distribution of both Figs. 1a and b are in agreement with the other works [1, 2]. Many peaks and shoulders are obviously seen in each spectrum of Figs. 1a and b except the case of $In_xAl_{1-x}N$ samples that show a relatively structure-less feature. This may correspond to the X-ray diffraction results which suggest that c-axis fluctuations of $x = 0.22$ and 0.47 samples are about three and ten times larger than those of the typical $Al_xGa_{1-x}N$ samples, respectively. The N-KAS, drastically but continuously, changes its feature with x, however Al-KAS basically does not change its feature with x. As pointed out in our previous work for AlGaN [1], this difference is mainly due to the environmental difference between cation and anion. Since the substituted ions are always cation, the nearest neighbour ions of the N ion have probability of substitution. On the other hand, those of the Al ion are always N ions. The continuous change of the spectral distribution with x in the N-KAS suggests that the unoccupied p-PDOS feels the long-range cation substitution. The insensitiveness to x of the Al-KAS suggests that the Al 1s electron transition is almost intra-atomic. These results represent that the unoccupied p-PDOS mainly consists of cation p states. The similar explanation can be applied to the threshold energy shift. The threshold energy of the N-K absorption systematically shifts to the lower energy side from AlN to InN. An obvious shift can be seen in the InGaN region. However, the threshold energy of the Al-K absorption has no shift beside the spectral broadening in the InAlN region.

Both the N-KAS and Al-KAS of all samples except $In_{0.47}Al_{0.53}N$ change their spectral distributions with θ. The intensities of the peaks in the $In_{0.22}Al_{0.78}N$ KAS indicate a weak θ dependence relative to those in other samples. However, it is noted that this dependence of the intensity of each peak on θ in Figs. 1a and b is completely inversion related between $In_{0.22}Al_{0.78}N$ and the other AlGaN (and InGaN) samples. For example, the intensity of the peak around 1569 eV of AlN (maximum peak) increases its intensity with decreasing θ, however that of $In_{0.22}Al_{0.78}N$ increases with increasing θ. The similar inverse relation is observed in the N-KAS. There are not enough results to recognize this phenomenon, but this result suggests that the p-PDOS in the conduction band switches its anisotropy twice between AlN and InN. It may also suggest that the weak or no anisotropy of the KAS of InAlN is due to not only the crystallinity but also this complicated band configuration.

The partial spectra decomposed numerically from N-KAS and Al-KAS of AlN are seen in Figs. 2a and b, respectively. Details of the numerical decomposition for $Al_{0.4}Ga_{0.6}N$ are described in the previous work [1]. The spectra labelled XY, Z, NA and T correspond to in-plane, out-of-plane, angular independent components of the unoccupied p-PDOS at N- or Al-sites and the calculated KAS at $\theta = 35.26°$ using XY, Z and NA spectra. Similar spectral distributions of those partial spectra are seen in both the N-KAS and Al-KAS of all samples except InAlN. In case of $In_{0.22}Al_{0.78}N$, XY and Z partial spectra are weak and exchange their spectral distributions with each other. There are two in-plane subbands and two out-of-plane subbands at N-site and they correspond to the peaks in Fig. 1a. Then, the maximum peak in each N-KAS has the similar origin and consists of in-plane p bonding. Both the lower and higher side peaks relative to the maximum peak consist of out-of-plane p bonding. In the $Al_xGa_{1-x}N$ region, the maximum peak becomes the dominant peak with decreasing x. However,

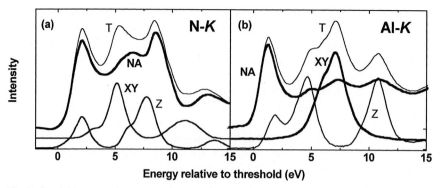

Fig. 2. Partial spectra decomposed numerically from a) nitrogen K and b) aluminum K absorption spectra of AlN

decomposition reveals that the maximum peak increase is mainly due to the increasing NA component with decreasing x. In the lower x In$_x$Ga$_{1-x}$N region ($x < 0.71$), these partial spectra resemble those of GaN. In the higher x In$_x$Ga$_{1-x}$N region, the lowest peak disappears in the Z partial spectrum. The threshold energy shift between normal and parallel incidence N-KAS is reported [2], but we cannot observe it within the limitation of $\theta \leq 60°$. If one defines the intensity ratio XY (or Z) to NA spectra as an index of the p bonding anisotropy, the anisotropy becomes weak with decreasing x in Al$_x$Ga$_{1-x}$N region and almost constant in lower x In$_x$Ga$_{1-x}$N region. There are three out-of-plane subbands and one in-plane subband at Al-site and they correspond to the peaks and shoulder in Fig. 1b. The maximum peak also consists of out-of-plane bonding. As pointed out in the previous work [1], except for the lowest peak in Z spectrum, a peak in the XY spectrum at N-site correspond to a Z peak at Al-site and *vice versa*. This is because the different bonds are formed in-plane and out-of-plane [2].

Conclusions The N-K and Al-K absorption measurements of wurtzite AlN, GaN and InN, and their ternary compounds at various molar fractions have been performed using synchrotron radiation. The KAS is almost directly compared with the unoccupied p-PDOS. The spectral distribution of N-KAS, drastically but continuously, changes with x among AlN, GaN, and InN. On the other hand, that of Al-KAS has no drastic x dependence. The threshold energies of N-KAS and Al-KAS show x dependence and x independence, respectively. They are explained as the enviroment difference between cation and anion sites.

Since all samples except In$_{0.47}$Al$_{0.53}$N show anisotropy of the unoccupied p-PDOS, the spectral distributions of different θ are different with each other. The intensity dependence of a peak in the N-KAS and Al-KAS on θ is essentially similar for any x of all the samples except InAlN, but the inverse dependence is observed in In$_{0.22}$Al$_{0.78}$N. According to the numerical decomposition of the KAS, the maximum peak in each N-KAS and Al-KAS has similar origin, respectively. These peaks consist of in-plane p bonding.

Acknowledgements This work was partly supported by the Joint Studies Program (1996-2001) of the Institute for Molecular Science. K.F. thanks Prof. S. Nakamura of Nichia Chemical Ind. Ltd. (USB at present) for providing GaN samples.

References

[1] K. Fukui, R. Hirai, A. Yamamoto, S. Naoe, and S. Tanaka, Jpn. J. Appl. Phys. **38**, Suppl. 38-1, 538 (1999), and references therein.
[2] K. Lawniczak-Jablonska, T. Suski, I. Gorczyca, N. E. Christensen, K. E. Attenkofer, R. C. C. Perera, E. M. Gullikson, J. H. Underwood, D. L. Ederer, and Z. Liliental-Weber, Phys. Rev. B **61**, 16623 (2000).

Internal Structure of Free Excitons in GaN

P. P. Paskov[1]), T. Paskova, P. O. Holtz, and B. Monemar

Department of Physics and Measurement Technology, Linköping University, S-581 83 Linköping, Sweden

(Received June 20, 2001; accepted June 28, 2001)

Subject classification: 71.35.Cc; 71.70.Gm; 78.55.Cr; S7.14

Polarized photoluminescence is used to study the fine structure of free excitons in thick GaN layers grown on differently oriented sapphire substrates. The singlet–triplet splitting of the A exciton is measured and the exchange interaction constant in GaN is determined. For the samples grown on the *a*-plane sapphire, splitting of the A and B excitons induced by the uniaxial in-plane stress is also observed.

1. Introduction In wurtzite GaN, the valence band is split into three bands due to the crystal–field and spin–orbit interactions. Consequently, three excitons, commonly denoted A, B and C, are observed in the near bandgap optical spectra of high quality samples [1]. However, each of these excitons is expected to have a substructure resulting from the electron–hole exchange interaction. The exchange interaction is known to lead to a splitting of the spin-singlet states from the spin-triplet states and also to the longitudinal-transverse splitting of the spin-singlet states [2]. Due to the insufficient experimental data available the fine structure of excitons in GaN is still subject of discussion. The two values reported so far for the singlet–triplet splitting of the A exciton are quite controversial. Eckey et al. [3] measured a value of (0.120 ± 0.1) meV in heteroepitaxial GaN sample, while recently Reynolds et al. [4] obtained a singlet–triplet splitting of 2.9 meV in a free-standing GaN layer. In this paper, we present a photoluminescence (PL) study of the free excitons in GaN layers grown by hydride vapour phase epitaxy (HVPE). The polarization measurements reveal the exciton states of different symmetry and allow us to inspect the exchange interaction in GaN.

2. Experimental For this study, two 80 μm thick GaN layers simultaneously grown in a single growth run on *c*-plane and *a*-plane sapphire were used. The PL measurements were performed with cw optical excitation ($\lambda_{\mathrm{exc}} = 266$ nm) at $T = 4$ K. In the case of $\mathbf{k} \perp \mathbf{c}$ geometry, the laser beam was focused on the side facet of the sample by a microscopic objective and the spectra are taken from a spot about 10 μm below the top surface. The luminescence signal was detected by a cooled charge coupled detector (CCD). The spectral resolution was 0.2 meV in the region around 350 nm.

3. Results and Discussion Figure 1 shows the PL spectra of the sample grown on *c*-plane sapphire for different polarization configurations. The spectra reveal emission peaks of the acceptor-bound excitons (A°X), donor-bound excitons (D°X) and free-excitons (FX$_\mathrm{A}$ and FX$_\mathrm{B}$). The assignment of the exciton PL peaks is confirmed by the reflectance spectrum of the same sample [5]. When the wave vector is perpendicular to

---

[1]) Corresponding author; Phone: +46 13 282688; Fax: +46 13 142337; e-mail: plapa@ifm.liu.se

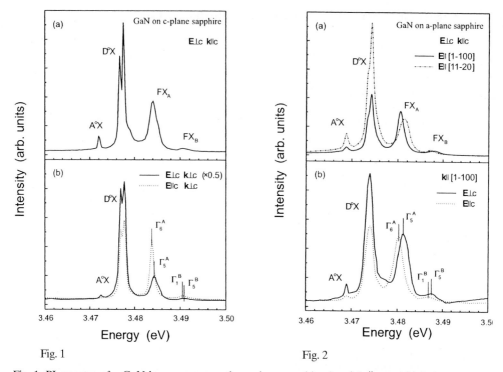

Fig. 1. PL spectra of a GaN layer grown on the c-plane sapphire for a) $\mathbf{k} \parallel \mathbf{c}$ and b) $\mathbf{k} \perp \mathbf{c}$

Fig. 2. PL spectra of the sample grown on a-plane sapphire for a) $\mathbf{k} \parallel \mathbf{c}$ and b) and $\mathbf{k} \perp \mathbf{c}$

the c-axis, a clear difference in the energy position of the free exciton peak for $\mathbf{E} \perp \mathbf{c}$ and $\mathbf{E} \parallel \mathbf{c}$ polarizations is observed (Fig. 1b). The A exciton in GaN comes from the conduction band having Γ_9 symmetry and the topmost valence band with Γ_9 symmetry and transforms according to $\Gamma_5 + \Gamma_6$. The B exciton arises from the lower valence band (with Γ_7 symmetry) and transforms as $\Gamma_1 + \Gamma_2 + \Gamma_5$. The twofold degenerate spin-singlet state Γ_5 is optically allowed for $\mathbf{E} \perp \mathbf{c}$ polarization, while the Γ_1 state is allowed for $\mathbf{E} \parallel \mathbf{c}$ polarization. The spin-triplet states Γ_6 and Γ_2 are forbidden for both polarizations. According to the optical selection rules, we assign the two free-exciton peaks for $\mathbf{E} \perp \mathbf{c}$ as the transverse Γ_5 states of the A and B excitons. In the case of $\mathbf{E} \parallel \mathbf{c}$ polarization, the observed peaks are interpreted as the Γ_6 state of the A exciton and the Γ_1 state of the B exciton. The appearance of the dipole-forbidden Γ_6 state in the spectra can be explained as being due to the mixing with the Γ_1 state of the B exciton at a finite k-vector [2]. The energy separation between the peaks assigned as Γ_5^A and Γ_6^A, which is a measure of the singlet–triplet splitting of the A exciton, is found to be (0.53 ± 0.02) meV. From the spectra in Fig. 1b, we also deduce the energy splitting of (0.240 ± 0.08) meV between the two optically-active states of the B exciton, Γ_5^B and Γ_1^B.

For the sample grown on a-plane sapphire, we found that the PL spectrum is polarization dependent even at $\mathbf{k} \parallel \mathbf{c}$. As it is seen in Fig. 2a, both the energy position and the relative intensities of the A and B excitons depend on the orientation of the polarization in the (0001) plane. This observation is attributed to the distortion of the wurtzite symmetry when the GaN is grown on the *a*-plane sapphire. Such a layer exhibits an anisotropic in-plane compression due to the anisotropy in the thermal expansion coefficient of the sapphire. The resultant uniaxial stress combined with the exchange interaction removes the twofold degeneracy of the Γ_5 state and two transverse excitons with a dipole momentum polarized parallel and perpendicular to the direction of the stress can be observed [6]. In this sample, we obtain (0.77 ± 0.02) meV and (0.69 ± 0.04) meV for the splitting of Γ_5 state of the A and B excitons, respectively. Accordingly, the energy position of the Γ_5^A and Γ_5^B states for $\mathbf{k} \perp \mathbf{c}$, $\mathbf{E} \perp \mathbf{c}$ depends on the angle between the wave vector and the direction of the uniaxial stress. The singlet–triplet splitting now is (0.94 ± 0.03) meV (Fig. 2b) indicating that the high energy component of the Γ_5^A state is detected. We should note that the Γ_6^A state gains oscillator strength because it becomes allowed with a dipole moment parallel to *c*-axis. The Γ_5^B–Γ_1^B splitting is found to be (0.69 ± 0.1) meV. Although the low energy component of the Γ_5^B state should be allowed in this configuration, the uniaxial stress shifts down the Γ_1^B state and a larger splitting than that in the layer on *c*-plane sapphire is observed.

The internal structure of the excitons in GaN is described in the framework of the effective Hamiltonian formalism [7, 8]. The exciton Hamiltonian can be written as $H = H_0 + H_{\text{def}} + H_{\text{exch}}$, where H_0 represents the crystal field and the valence band spin–orbit interaction and H_{def} describes the strain-related effects on the evolution of the band extrema. The last term accounts for the analytical part of the electron–hole exchange interaction and is given by $H_{\text{exch}} = 1/2\gamma(1-\sigma_e\sigma_h)$, where γ is the isotropic exchange interaction constant between the hole and electron spins and σ_e and σ_h denote the vector operators whose components are the Pauli spin matrices of the electron and the hole, respectively [9]. The diagonalization of the Hamiltonian is performed for the crystal-field and the spin–orbit parameters $\Delta_1 = 10$ meV, $\Delta_2 = 5.5$ meV, and $\Delta_3 = 6.0$ meV, respectively [10]. Since we are interested in the splitting between exciton states and not in their absolute energies, only one deformation potential is needed, namely $C_3 = -2C_4 = 7.2$ eV [10]. First we consider a biaxially compressed GaN, i.e. a layer grown on *c*-plane sapphire. From the empirical relation between the strain and the energy position of the A exciton [11], the in-plane stress in the sample studied here is estimated to be $\sigma_{xx} = (2.9 \pm 0.3)$ kbar. For this stress level, we calculate the energy separations Γ_5^A–Γ_6^A and Γ_5^B–Γ_1^B, adjusting the exchange constant γ until the calculated values match the measured ones. The best fit is obtained for $\gamma = (0.58 \pm 0.05)$ meV. This value is in a nice agreement with $\gamma = (0.6 \pm 0.1)$ meV estimated from the polarized reflectance measurements for $\mathbf{k} \parallel \mathbf{c}$ configuration in a thin GaN layer grown on *a*-plane sapphire [6]. The calculated

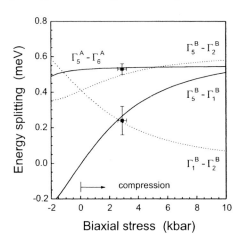

Fig. 3. Stress dependence of the energy splitting of the A and B excitons calculated with $\gamma = 0.58$ meV. The squares represent the experimentally obtained values

stress dependence of the splitting of the A and B excitons is given in Fig. 3. The results show that the singlet–triplet splitting Γ_5^A–Γ_6^A is only weakly dependent on the stress. On the other hand, the splitting between the states of the B exciton is found to be very sensitive to the stress.

In the case of the GaN grown on the *a*-plane sapphire, an additional off-diagonal term accounting for the uniaxial stress has to be included in the exciton Hamiltonian. This term has a form $\delta_3 = C_5(S_{11} - S_{12})(\sigma_{xx} - \sigma_{yy})$, where C_5 is the shear deformation potential, S_{11} and S_{12} are the compliance constants, and $\sigma_{xx} - \sigma_{yy}$ is the uniaxial stress component. Combining the deduced exchange interaction constant with the measured splitting of the Γ_5^A and Γ_5^B states, we estimate $\delta_3 = -(3.1 \pm 0.2)$ meV in our sample. Using $C_5 = -2.4$ eV [12] and the stiffness coefficients from Ref. [13], the uniaxial component of the stress is estimated to be (3.0 ± 0.2) kbar. This is almost equal to the biaxial stress $\sigma_{xx} + \sigma_{yy} = (3.5 \pm 0.2)$ kbar determined from the peak position of the A exciton leading to a conclusion that the in-plane stress in the sample is predominantly uniaxial.

4. Conclusion A comparative study of the polarized PL in GaN layers grown on *c*-plane and *a*-plane sapphire has been performed. Due to the well resolved emission peaks we were able to examine the effect of both the electron–hole exchange interaction and the uniaxial in-plane stress on the exciton structure.

Acknowledgement This work was partly supported by the European Community via the CLERMONT project PHRN-CT-1999-00132.

References

[1] B. Monemar, in Gallium Nitride I, Eds. J. I. Pankove and T. D. Moustakas, Semicond. Semimet. **50**, 305 (1998).
[2] K. Cho, Phys. Rev. B **14**, 4463 (1976).
[3] L. Eckey, A. Hoffmann, P. Thurian, I. Broser, B. K. Meyer, and K. Hiramatsu, Mater. Res. Soc. Symp. Proc. **482**, 555 (1998).
[4] D. C. Reynolds, D. C. Look, B. Jogai, A. W. Saxler, S. S. Park, and J. Y. Hahn, Appl. Phys. Lett. **77**, 2879 (2000).
[5] G. Pozina, J. P. Bergman, T. Paskova, and B. Monemar, Appl. Phys. Lett. **75**, 4124 (1999).
[6] M. Julier, J. Campo, B. Gil, J. P. Lascaray, and S. Nakamura, Phys. Rev. B **57**, R6791 (1997).
[7] D. W. Langer, R. N. Euwema, K. Era, and T. Koda, Phys. Rev. B **2**, 4005 (1970).
[8] B. Gil and O. Briot, Phys. Rev. B **55**, 2530 (1997).
[9] N. T. Thang and G. Fishman, Phys. Rev. B **31**, 2404 (1985).
[10] A. A. Yamaguchi, Y. Mochizuki, H. Sunakawa, and A. Usui, J. Appl. Phys. **83**, 4542 (1998).
[11] A. Shikanai, T. Azuhata, T. Sota, S. Chichibu, A. Kuramata, K. Horino, and S. Nakamura, J. Appl. Phys. **81**, 417 (1997).
[12] A. Alemu, B. Gil, M. Julier, and S. Nakamura, Phys. Rev. B **57**, 3761 (1998).
[13] M. Yamaguchi, T. Yagi, T. Azuhata, T. Sota, K. Suzuki, S. Chichibu, and S. Nakamura, J. Phys.: Condens. Matter **9**, 241 (1997).

Experimental and Theoretical Tools for the Study of Exciton Properties versus Disorder in Nitride-Based Quantum Structures

B. Gil (a), M. Zamfirescu (b), P. Bigenwald (c), G. Malpuech[1]) (b), and A. Kavokin (b)

(a) GES, CNRS, Université Montpellier II, Case courrier 074, F-34095 Montpellier Cedex 5, France

(b) LASMEA, CNRS, Université Blaise Pascal-Clermont-Ferrand II, F-63177 Aubière Cedex, France

(c) Université d'Avignon, 33 Rue Louis Pasteur, F-84000 Avignon, France

(Received June 26, 2001; accepted August 4, 2001)

Subject classification: 71.35.Cc; 77.65.Ly; 78.47.+p; 78.67.De; S7.14

We show that resonant Rayleigh scattering measurements combined with conventional reflectivity measurements can easily allow the determination of the most relevant parameters of nitride-based quantum well excitons. In particular, comparing the Fourier-transformed reflection spectra with calculated time-resolved reflectivities, we have found the exciton oscillator strength to decrease dramatically with increase of the QW width in the GaN/Al$_{0.07}$Ga$_{0.93}$N system. The collapse of the oscillator strength is a manifestation of the polarization field effect, as confirmed by our variational calculation. We find that only excitons in very thin quantum wells have an oscillator strength exceeding that of the exciton in bulk GaN.

1. Introduction There was a series of experimental indications that the exciton oscillator strength in GaN is extremely high [1–4]. This makes GaN and related heterostructures quite promising for study of light-exciton coupling phenomena (like the polariton effect [5], motional narrowing [6], resonant Rayleigh scattering [7], etc.). However, quite often, GaN-based quantum wells (QWs) exhibit relatively weak exciton resonances, while usually the quantum confinement leads to a significant increase of the oscillator strength. In order to interpret this important phenomenon one should be able, first, to measure with a good accuracy the exciton oscillator strength as a function of the QW parameters, and, second, to be able to calculate it taking into account not only the QW potential, but also the potential gradient induced by polarization fields in GaN. These latter, composed by piezoelectric and spontaneous polarization fields have been shown recently to have a crucial effect on the electronic structure of GaN-based heterostructures [8].

In previous papers [9, 10], we have presented variational approach that allowed to calculate the main excitonic parameters in GaN/AlGaN QWs taking into account the polarization fields and any eventual screening of them by a photoinduced free electron–hole plasma.

In this paper, we present experimental and theoretical tools for the study of exciton properties versus disorder in nitride-based quantum structures. We show that resonant Rayleigh scattering (RRS) measurements combined with conventional reflectivity measurements can easily allow the determination of the most essential parameters of quantum well (QW) excitons if they are interpreted in the framework of a Fourier spectroscopy analysis [11, 12]. These parameters are, the radiative lifetime Γ_0, the inhomogeneous broadening Δ, or the number of localized states N generated by the disorder potential in the plane of the quantum wells.

2. RRS Measurements as a Tool to Characterize QW Excitons

The RRS of light by quantum well excitons is due to the breakdown by the disorder of the translation invariance in the plane of the QWs. In other words, the disorder potential induces the localization of excitons in certain "quantum dot-like" states, and these localized excitons coherently scatter the light in all directions. The time-resolved RRS spectra are a direct probe of the statistical properties of the disorder [7]. The low intensity of the light scattered in a given direction makes the RRS experiment quite non-trivial to perform. In nitrides this problem is less important than in arsenides because of the strong disorder. Basically, we show here theoretically that the scattering experiments would be certainly the most adapted tool to study the exciton properties in nitride-based QWs. The detailed theoretical formalism allowing to describe the RRS and the reflection of light by disordered QWs are described elsewhere [7, 13], here we present the results. The ratio of the reflected intensity to the average scattered intensity is thus

$$I_{\text{ref}}/I_{\text{scat}} \approx N(\Gamma_0/\Delta)^2 . \tag{1}$$

Figure 1 shows the integrated intensity as a function of the scattering angle calculated from [7, 15]. The intensity for a scattering angle equal to zero corresponds to the reflected intensity. The average intensity obtained for the other angles is referred to as the scattered intensity. The calculation has been performed for $N = 10^6$, $\Gamma_0 = 0.5$ meV and $\Delta = 5$ meV and for the background optical index $n_b = \sqrt{\varepsilon_b} = 2.6$. The ratio between reflection and scattering is 10^4, in perfect agreement with the estimation obtained from formula (1).

It should be noted, however, that to obtain a precise information on the statistics of excitons in a QW, time-resolved RRS measurements are needed [13].

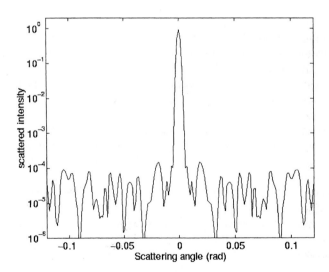

Fig. 1. Integrated scattered intensity as a function of the scattering angle of a GaN/AlGaN single quantum well calculated with (1). The parameters used are $N = 10^6$, $\Gamma_r = 0.5$ meV and $\Delta = 5$ meV and $n_b = \sqrt{\varepsilon_b} = 2.6$

3. Analysis of the Reflection Spectra of GaN/AlGaN QWs

In this section, we present a method allowing to analyze the exciton reflection spectra taking into account the exciton inhomogeneous broadening, and separating it numerically from the radiative broadening. The use of this technique, combined with the results observerved from Section 2, could in principle allow to obtain a quite complete set of parameters describing QW excitons.

Our method consists in the numerical Fourier transform of the experimental reflection spectrum of our sample and its comparison with the calculated time-resolved reflection coefficient. This technique has been applied to a series of reflection spectra taken by us from a series of high-quality MBE-grown GaN/Al$_{0.07}$Ga$_{0.93}$N QW samples containing QWs of different thicknesses.

The comparison between experimental and theoretical curves allows us to extract Γ_0 and Δ (see Figs. 2a and b). Having both cw and time-resolved reflection spectra one can extract the excitonic parameters with quite a good accuracy (see error bars in Fig. 2).

The results are summarized in Fig. 2 that shows the extracted points in comparison with theoretical curves calculated by the variational method [9, 10] for different concentrations of Al in the barriers. One can observe an extremely strong dependence of radiative broadening on the thickness of the QW. The exciton oscillator strength decreases by a factor of 30 between 4 and 12 ML thick QWs, while the same variation of thickness of a GaAs/AlGaAs QW would cause the oscillator strength decrease by only a factor of less than two. This drastic dependence of the oscillator strength on the QW width is an important peculiarity of GaN/AlGaN QWs, that should be carefully taken into account when using these wells in light-emitting optical devices.

This dependence is mostly due to the strong polarization fields that are present in the nitride-based QWs. The fields spatially separate the electron and hole mass centers by a distance close to the well thickness. This considerably suppresses the overlap of electron and hole wave functions leading to a collapse of the

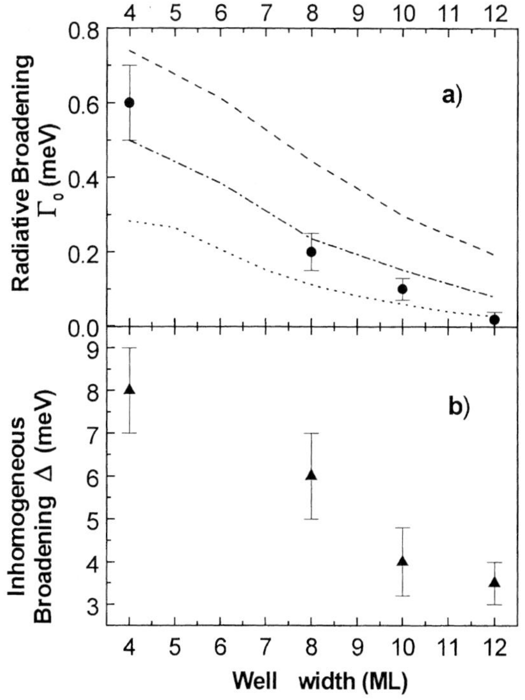

Fig. 2. Well width dependence of the a) exciton radiative damping and b) inhomogeneous broadening obtained from the experimental reflectivity spectra of GaN/Al$_{0.07}$Ga$_{0.93}$N QWs. The dashed, dash-dotted and dotted curves show the exciton radiative damping calculated variationally with the polarization field values of 500, 700 and 1000 kV/cm, respectively

oscillator strength. This tendency has been predicted theoretically by variational calculations of the oscillator strength. We present here the first direct experimental evidence of this dependence.

The comparison of our data with the result of a variationnal calculation shows that the value of the polarization field in our QWs was about 750 kV/cm. Note that this value is sensitive to the aluminium concentration in the barriers and the barrier thickness, 10 nm here. The value of 750 kV/cm seems in average be the closest of the real values of the field, even if it is different in all the wells because of the different values of strain and of the ratio of the well thickness over barrier thickness.

Figure 2b shows the evolution of the exciton inhomogeneous broadening Δ with QW width. It is quite large compared to the typical GaAs-based structures and strongly increases with the decrease of well thickness. However, it is worth to note that the ratio Γ_0/Δ in the thin GaN/AlGaN wells is higher that in most of the conventional III–V QWs (0.1 for the 8 MLs well against 0.03 in a 10 nm thick GaAs/AlGaAs QW [5]). The high value of this ratio gives much hope for the observation of exciton–polariton propagation effects in nitride-based heterostructures [6, 7, 13, 14].

4. Conclusion We have shown theoretically that studying the cw RRS spectra one can obtain complex information on the exciton statistics. We have then presented the first direct experimental study of the QW width effect on the radiative broadening and inhomogeneous broadening of excitons in $GaN/Al_{0.07}Ga_{0.93}N$ QWs. The major limitation of the accuracy of the experimental determination of these characteristics due to extremely broad exciton lines in the spectra has been eliminated by use of the original technique of numerical analysis of the spectra. The values of the radiative broadening we have obtained are in good agreement with the results of a variational calculation assuming a polarization field within the wells of 750 MV/cm. The collapse of the radiative broadening with the increase of the well width, and its high value for thin QWs are evidenced. The significant value of the radiative broadening compared to the inhomogeneous broadening in thin GaN/AlGaN QWs make them extremely promising for the observation of fine exciton–polariton effects. The stability of the excitons in nitrides at room temperature and their strong coupling with light give hope for the realization of new classes of optoelectronic devices based on microcavity polaritons.

References

[1] A. ALEMU et al., Phys. Rev. B **57**, 3761 (1998).
[2] P. LEFEBVRE et al., Phys. Rev. B **57**, R9447 (1998).
[3] M. LEROUX et al., Phys. Rev. B **60**, 1496 (1999).
[4] B. GIL et al., Phys. Rev. B **59**, 10246 (1999).
[5] L. C. ANDREANI, in: Confined Electrons and Photons, Eds. E. BURSTEIN and C. WEISBUCH, Plenum Press, New York 1995 (p. 57).
[6] G. MALPUECH and A. KAVOKIN, App. Phys. Lett. **76**, 3049 (2000).
[7] G. MALPUECH and A. KAVOKIN, phys. stat. sol. (a) **183**, 75 (2001).
[8] F. BERNARDINI and V. FIORENTINI, Phys. Rev. B **57**, R6427 (1998).
[9] P. BIGENWALD et al., Phys. Rev. B **61**, 15621 (2000).
[10] P. BIGENWALD et al., Phys. Rev. B **63**, 5315 (2001).
[11] G. MALPUECH et al., Solid State Comm. **113**, 185 (2000).
[12] M. ZAMFIRESCU et al., Phys. Rev. B (Rap. Comm.) **64**, 121304 (R), (2001).
[13] G. MALPUECH and A. KAVOKIN, J. Phys.: Condens. Matter **13**, 7075 (2001).
[14] Y. YAMAMOTO, Nature **405**, 629 (2000).

Comparison of Exciton–Biexciton with Bound Exciton–Biexciton Dynamics in GaN: Quantum Beats and Temperature Dependence of the Acoustic-Phonon Interaction

K. Kyhm[1]) (a), R. A. Taylor (a), J. F. Ryan (a), T. Aoki (b),
M. Kuwata-Gonokami (b), B. Beaumont (c), and P. Gibart (c)

(a) Clarendon Laboratory, University of Oxford, Parks Road, Oxford OX1 3PU, U.K.

(b) Department of Applied Physics, Tokyo University, 7-3-1 Hongo, Bunkyo-ku, Tokyo 113, Japan

(c) CNRS-Centre de Recherche sur l'Hétéro-Epitaxie et ses Applications, Rue B. Gregory, Parc Sophia-Antipolis, F-06560 Valbonne, France

(Received June 21, 2001; accepted August 4, 2001)

Subject classification: 71.35.–y; 71.35.Gg; 78.47.+p; 78.55.Cr; S7.14

The polarization dependence of biexcitonic signals and quantum beats between A-excitons (X_A) and A-biexcitons ($X_A X_A$) in a high-quality GaN epilayer is measured by spectrally-resolved and time-integrated four-wave mixing. With cross-linear polarised light, mixed beats with two periods are observed: the first beating period corresponds to the energy splitting between X_A and $X_A X_A$, and agrees well with the calculated $X_A X_A$ binding energy; while the second beating period corresponds to that between X_A and donor bound excitons (D^0X). The temperature-dependent homogeneous linewidth shows that the D^0X has a larger acoustic phonon coupling coefficient than the $X_A X_A$. We also measured the polarization dependent B-biexciton ($X_B X_B$) signal. The effective masses for the A- and B-hole were deduced from the binding energy.

Introduction GaN is strongly excitonic, with an exciton binding energy of $E_X \sim 25$ meV. The band-edge optical properties are dominated by the pronounced excitonic resonances arising from the three closely-spaced valence bands present in the GaN wurzite structure. Although the existence of biexcitons in GaN has been reported [1–3], little detail is known about their behaviour and dynamics. Because the energy difference between neutral donor-bound excitons (D^0X) and A-biexcitons ($X_A X_A$) is very small, the spectral broadening present in photoluminescence measurements makes it difficult to resolve the two effects. However, the beating period measured by transient degenerate four-wave mixing (DFWM) yields an accurate splitting of the resonances directly, hence giving the binding energy of D^0X and $X_A X_A$. In this paper we report the first observation of of D^0X–X_A and $X_A X_A$–X_A quantum beats in GaN. By measuring of the dephasing of the beats, the temperature dependence of the interaction with acoustic phonons is determined. We found that the D^0X has a larger acoustic phonon coupling coefficient than the $X_A X_A$ as the D^0X is more localised. Additionally, we have also measured the B-biexciton ($X_B X_B$) signal in GaN for the first time.

---

[1]) Corresponding author; Phone: +44 1865 272 393; Fax: +44 1865 271 553; e-mail: k.kyhm1@physics.ox.ac.uk

Experiments Spectrally-resolved (SR) DFWM and time-integrated (TI) DFWM measurements were made as a function of delay time and temperature using cross-linear and co-circular polarised light. A Kerr lens mode-locked Ti:sapphire laser with 120 fs pulse width at a repetition rate of 76 MHz was frequency doubled in a β-barium-borate crystal. The second harmonic pulse width was estimated to be 167 fs by the Gaussian transform of its spectral width of 7.90 meV. A 50 μm spot diameter pump beam was focused normally on to a sample mounted in a closed-cycle helium cryostat kept at 15 K. We used a two-pulse self-diffraction configuration in reflection geometry [4]. The GaN sample was a nominally undoped epilayer grown by lateral epitaxial overgrowth, which is a two-step process involving both undoped and Mg-doped GaN deposition by metal organic vapor phase epitaxy (MOVPE) [5].

Results and Discussion Figures 1a and b show the time-integrated photoluminescence spectrum and the SR-DFWM signal, respectively, measured at zero time delay with cross-linear and co-circular polarised light. It has already been shown that the dominant luminescence peak can be attributed not only to neutral donor-bound exciton (D^0X) emission but also to A-biexciton ($X_A X_A$) emission in GaN [1], and the DFWM signal of biexcitons must disappear for excitation with co-circular polarised light because of the polarization selection rules [6]. Indeed, the spectrum shown by the dash-dotted line in Fig. 1b with co-circular polarised light does not show a biexcitonic signal, although a signal from D^0X is present. However, a strong biexcitonic signal appears when cross-linear polarization is used, represented by the solid line in Fig. 1b. Both DFWM signals of D^+X and D^0X are very weak despite of their strength in the photoluminescence

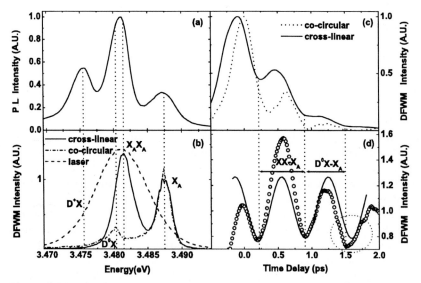

Fig. 1. a) Photoluminescence spectrum measured at 15 K in GaN. b) SR-DFWM spectrum measured at zero delay with cross-linear (solid curve) and co-circular (dash-dotted curve) polarised light. c) TI-DFWM signal as a function of delay time with cross-linear and co-circular polarised light. d) The residual modulation signal (open circles) is taken by removing the exponential decay term from b), a fitting function (solid curve) having a period of $X_A X_A - X_A$ beats is compared with the data (open circles)

spectrum. An $X_A X_A$ biexciton binding energy of (5.83 ± 0.10) meV is measured. This value is comparable to recently reported $X_A X_A$ binding energies of 5.7 meV [2, 3] for wurtzite GaN, and quite similar to a theoretical binding energy of 5.798 meV, which is calculated using Feynman's path integral method combined with Monte-Carlo techniques [7]. Figure 1c shows the TI-DFWM signal as a function of time delay at the $X_A X_A$ energy with cross-linear and co-circular polarised light. Both signal intensities are normalised for convenience, and the original signal intensity ratio is $I^{cross}/I^{co} = 6.2$. With co-circular polarised light, the decay of this signal for $T_L = 15$ K shows an oscillation with a period of (579 ± 2) fs. This beat period corresponds to an energy splitting between $D^0 X$ and X_A of $\Delta E = h/\tau = (7.145 \pm 0.098)$ meV, the $D^0 X$ binding energy. However, with cross-linear polarised light, the DWFM signal is modulated by the interference of two separate beat periods ($X_A X_A$–X_A and $D^0 X$–X_A). The B-exciton signal is eliminated by careful tuning of the laser. To measure the beating periods more precisely, the decay signal was removed by dividing the data by an exponential decay term (Fig. 1d), and an oscillating function, having the period of the $X_A X_A$–X_A beating, was compared with the measured data.

As is obvious from the figure, the two different beat periods are clearly mixed. The first beat period of (709 ± 2) fs corresponds well with the energy splitting between X_A and $X_A X_A$ ((5.832 ± 0.098) meV) while a second beat period of (579 ± 2) fs corresponds to the $D^0 X$ binding energy ((7.145 ± 0.098) meV). Bearing in mind the fact that the $X_A X_A$ signal is much larger than that of the $D^0 X$ in the SR-DFWM spectrum, the oscillation intensity of the $X_A X_A$ should be dominant during the early part of the beating signal. The $D^0 X$ beating signal would be masked. However, the $X_A X_A$ linewidth is wider than that of the $D^0 X$, implying that the $X_A X_A$ signal decays faster than the $D^0 X$. As a result, the $D^0 X$ beating signal becomes more evident as the time delay between pump and probe is increased. At the same time, the $X_A X_A$ signal weakens.

Since beating signals appear at positive delay times and have large linewidths in the SR-DFWM spectrum, these transitions are inhomogeneously broadened. In this case, the decay time of the DFWM signal T_d is equal to one quarter of the exciton dephasing time, $T_d = T_2/4$ [8]. Figure 2a depicts the temperature dependence of the homogeneous linewidth of $X_A X_A$ and $D^0 X$. The solid line depicts a fit using the equation $\Gamma^{hom} = \Gamma^{hom}(0) + \alpha T + \beta e^{E_a/k_B T}$ [9], where αT relates to acoustic phonon scattering, and the last term represents the dephasing due to thermal dissociation with an activation energy E_a. These activation energies ($E_a^{xx} = 7.9$ meV, $E_a^{D^0 X} = 10.4$ meV) correspond approximately to the respective binding energies. It has been reported that biexciton–phonon scattering is twice as fast as exciton–phonon scattering due to a larger deformation potential [10]. We also found that $D^0 X$ has a larger acoustic-phonon coefficient ($\gamma = 2.62$ μeV/K) than $X_A X_A$ ($\gamma = 0.86$ μeV/K). This implies that $D^0 X$ may have a larger deformation potential than $X_A X_A$ because the $D^0 X$ is more localised in the lattice. In a further measurement, we tuned the laser to an excitation energy between that of the X_A and X_B. The DFWM signal obtained is shown in Fig. 2b. With cross-linear polarised light, the solid line exhibits a shoulder between the X_A and X_B signal. Since this signal disappears with co-circular polarised light, this polarization dependence supports the assertion that the shoulder arises from the B-biexciton ($X_B X_B$) since this would not disappear when excited by co-circular polarised light if it was a composite biexciton such as $X_A X_B$ from polarization selection rules. A binding energy for the

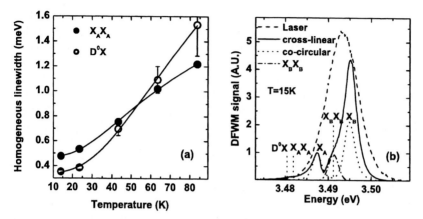

Fig. 2. a) Lattice temperature dependence of the homogeneous linewidth of $X_A X_A$ and $D^0 X$ in GaN, deduced from TI-DFWM decay rates assuming that both are inhomogeneously broadened. b) SR-DFWM spectrum measured at zero delay with cross-linear (solid curve) and co-circular (dotted curve) polarised light when the laser is tuned between the X_A energy and that of X_B. The $X_B X_B$ signal (dash-dotted curve) is extracted by using a Gaussian fit to the lineshape

$X_B X_B$ of $((3.80 \pm 0.01)\,\mathrm{meV})$ is obtained by fitting the DWFM lineshape using a series of Gaussians. Using the same method for calculating the biexciton binding energy discussed above, we can deduce a hole effective mass from the measured binding energy of the biexciton by using the electron effective mass in GaN ($0.22 m_0$). The resultant effective mass for the B-hole is slightly lighter ($1.03 m_0$) than that for the A-hole ($1.30 m_0$).

Conclusions In conclusion, we have performed femtosecond SR-DFWM and TI-DFWM spectroscopy on a high-quality GaN epilayer. With co-circular polarised light, we measured only the $D^0 X$–X_A quantum beats. However, with cross-linear polarised light, we found two beat periods ($D^0 X$–X_A and $X_A X_A$–X_A), which are mixed with different dephasing rates, and that each beat period corresponds exactly to the binding energy. We found that the strength of acoustic-phonon interaction with the $D^0 X$ is stronger than that with the $X_A X_A$ by analyzing the temperature dependent homogeneous linewidth. Finally, we have confirmed the existence of the B-biexciton, and the binding energies of the $X_A X_A$ and $X_B X_B$ have been measured. We also obtained effective masses of the A- and B-hole, and found that the B-hole effective mass is slightly lighter than the A-hole effective mass.

Acknowledgements We are grateful to Mr. J. Park for his computational support. K. Kyhm was supported by the Oriental Institute at The University of Oxford with the Sasakawa Fund.

References

[1] K. OKADA, Y. YAMADA, T. TAGUCHI, F. SASAKI, S. KOBAYASHI, T. TANI, S. NAKAMURA, and G. SHINOMIYA, Jpn. J. Appl. Phys. **35**, L787 (1996).
[2] Y. KAWAKAMI, Z. PENG, Y. NARUKAWA, SHIZUO FUJITA, SHIGEO FUJITA, and S. NAKAMURA, Appl. Phys. Lett. **69**, 1414 (1996).

[3] R. Zimmermann, A. Euteneuer, J. Mobius, D. Weber, M. R. Hofmann, and W. W. Rühle, Phys. Rev. B **56**, R12722 (1997).
[4] T. Yajima, and Y. Taira, J. Phys. Soc. Jpn. **47**, 1620 (1979).
[5] B. Beaumont, M. Valle, G. Nataf, A. Bouille, J. C. Guillaume, P. Vennegues, S. Haffouz, and P. Gibart, MRS Internet J. Nitride Semicond. Res. **3**, 20 (1998).
[6] T. F. Albrecht, K. Bott, T. Meier, A. Schulze, M. Koch, S. T. Cundiff, J. Feldmann, W. Stolz, P. Thomas, S. W. Koch, and E. O. Göbel, Phys. Rev. B **54**, 4436 (1996).
[7] W. Huang, phys. stat. sol. (b) **60**, 309 (1973).
[8] J. Shah, Ultrafast Spectroscopy of Semiconductors and semiconductor Nanostructures, Springer-Verlag, Berlin 1996.
[9] C. Dornfeld and J. M. Hvam, IEEE J. Quantum Electron. **25**, 904 (1989).
[10] W. Langbein and J. M. Hvam, Phys. Rev. B **61**, 1692 (2000).

Micro-Photoluminescence Spectroscopy of Exciton–Polaritons in GaN with the Wave Vector k Normal to the c-Axis

T. V. Shubina (a), T. Paskova (b), A. A. Toropov (a), A. V. Lebedev (a), S. V. Ivanov (a), and B. Monemar (b)

(a) Centre of Nanoheterostructure Physics, Ioffe Institute, Russian Academy of Sciences, Politekhnicheskaya 26, 194021 St. Petersburg, Russia

(b) Department of Physics and Measurement Technology, Linköping University, S-581 83 Linköping, Sweden

(Received July 2, 2001; accepted August 4, 2001)

Subject classification: 71.35.Cc; 71.36.+c; 78.55.Cr; S7.14

We report on polarized micro-photoluminescence (μ-PL) and micro-reflectance (μ-R) studies of GaN layers grown by HVPE. A strong π-polarized component in the vicinity of A exciton is observed in the μ-PL and attributed as a mixture of a bound B exciton, dominating at low temperature, and scattered A exciton–polariton states prevailing at higher temperatures. Temperature variation of exciton energies in the μ-R spectra reveals strain-induced difference between the top surface and the cleaved edges.

Introduction Optical properties of excitons in high-quality GaN can be properly described only taking into account the coupling between photons and excitons [1]. In this case a two-branch dispersion curve in k-space, consisting of an upper polariton branch (UPB) and a lower polariton branch (LPB), is expected for each allowed excitonic state. In hexagonal GaN the basic excitonic states for A band are an allowed transverse Γ_5 and forbidden Γ_6 as well as longitudinal Γ_5 states. Both B and C excitons split into three levels: Γ_5, Γ_1 and Γ_2. Optically active are the Γ_1 states in the $\mathbf{E} \parallel \mathbf{c}$ polarization and the Γ_5 states for $\mathbf{E} \perp \mathbf{c}$ [2].

Since the polariton branches originate from the appropriate excitonic levels, the light propagating either along or normal to the c-axis has to excite different polaritonic modes. Simple consideration shows that the pattern of polariton branches at $\mathbf{k} \perp \mathbf{c}$ is obviously more complicated than that for $\mathbf{k} \parallel \mathbf{c}$. For instance, for $\mathbf{k} \parallel \mathbf{c}$ in the vicinity of the A band branches (UPB$_A(\Gamma_5)$ and LPB$_A(\Gamma_5)$) there is only a similar polarized branch of LPB$_B(\Gamma_5)$, while for $\mathbf{k} \perp \mathbf{c}$, in addition to the σ-polarized LPB$_B(\Gamma_5)$, there are a π-polarized LPB$_B(\Gamma_1)$ and a longitudinal polariton (Γ_{5L}). Moreover, small energy gaps between the three top valence bands in GaN result in an appreciable overlapping and integration of polariton branches originating from different excitonic states, but with the same orientation of the electric dipole [3]. This integrated pattern of polaritons differs at $\mathbf{k} \parallel \mathbf{c}$ and $\mathbf{k} \perp \mathbf{c}$. In the latter case, differently polarized polaritons coexist at the same energies and sufficiently close k-values. In this paper, we demonstrate that polarization measurements performed with high spatial resolution and high level of excitation characteristic for micro-photoluminescence (μ-PL) make possible the observation of polariton effects related to the phenomenon.

Experimental The study was performed using ~25 μm thick samples grown on c-plane sapphire by HVPE. Structural characterization by X-ray diffractometry has shown a record small value of column tilt (~62 arcsec), providing a minimum value of depolarization due to the c-axis inclination [4]. Measurements of μ-PL and micro-reflectance (μ-R) were carried out in a He continuous flow cryostat in the 4–300 K temperature range under cw excitation by a 266 nm laser line and radiation of a tungsten lamp, respectively. The detailed description of the measurements have been presented elsewhere [5]. During the measurements special measures (marking) were taken to maintain the focusing on the same point.

Results and Discussion μ-R spectra (Fig. 1a) generally satisfy the selection rules and the expected intensity relationship of the exciton transitions [6]. A comparison of the μ-R and μ-PL spectra permits us to assign the excitonic features (Fig. 1b).

Temperature variations of the exciton energies in α- and σ-polarizations are expected to be the same. However, it was found that they are different in the reflectance spectra, being identical in those of μ-PL (Fig. 1c). Thus, we conclude that the edge μ-R probes an outer cleaved region influenced by a strong anisotropic strain, while the μ-PL originates from deeper internal regions. This is indicative of the exciton–polariton emission and is consistent with the high excitation density characteristic of the μ-PL, when different centers capturing excitons are saturated and, hence, the effective exciton–polariton diffusion length significantly increases [8].

During the μ-PL measurements we have observed unexpected enhancement of π-polarized component (X(π)) in the vicinity of A exciton (Fig. 1b), which is almost twice stronger than $FX_A(\sigma)$. This obviously contradicts to the selection rules. To elucidate the origin of that we performed power- and temperature-dependent measurements of the polarized μ-PL. Figure 2 presents results of a deconvolution of the respective spectra using Lorentzian shape contours. At 5 K the power measurements (Fig. 2a) demon-

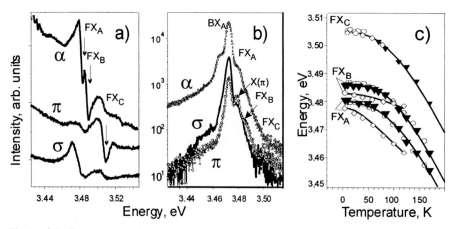

Fig. 1. a) Reflectance and b) μ-PL spectra measured at 5 K for different polarizations: α ($\mathbf{k} \parallel \mathbf{c}$, $\mathbf{E} \perp \mathbf{c}$), σ ($\mathbf{k} \perp \mathbf{c}$, $\mathbf{E} \perp \mathbf{c}$) and π ($\mathbf{k} \perp \mathbf{c}$, $\mathbf{E} \parallel \mathbf{c}$). c) Temperature dependences of exciton energies obtained from μ-R (α, open circles; σ, open diamonds) and μ-PL (solid triangles) demonstrate different curvature. (For the sake of clarity, σ-polarized μ-PL data are presented only.) The solid lines are the results of fitting done using Pässler's model [7]. FX_A, FX_B, and FX_C denote the transitions of free A, B and C excitons; BX_A is the bound A exciton, X(π) is enhanced π-polarized peak near FX_A

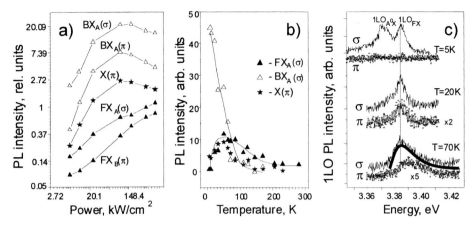

Fig. 2. Integrated intensity of polarized μ-PL peaks versus a) excitation power and b) temperature. Solid lines are only guides to the eyes. c) Spectra of 1LO phonon-assisted emission measured in π (open diamonds) and σ (thin lines) polarizations at different temperatures. The solid line represents the fitting (see text)

strate appreciable saturation of both the bound A exciton and X(π) line with the increasing power, while the α and σ-polarized components of the A and B free excitons still rise. Such a saturation evidences that significant part of the X(π) at low temperature is related to the bound B exciton. It is possible, because the separation between the free and donor bound excitons in GaN (6.4–6.8 meV) is the same, within the measurement accuracy, as the splitting between A and B excitons in our samples (6.5–6.7 meV). Therefore, we had been expecting a fast dropping of the line with increasing temperature. However, it is not observed. Moreover, the temperature-dependent measurements of the intensity provide a controversial result, namely, the X(π) line firstly rises with increasing temperature, then decreases, like FX$_A$(σ), and their intensities are almost equal at about 50 K (see Fig. 2b). At the same time both σ and π-polarized components of the bound A exciton (BX$_A$) are quickly and monotonically weakened.

It is worth to note that appearance and a certain increase of the forbidden Γ$_6$ can be observed in the same spectral range at high enough magnetic fields [9, 10] due to mixing of the forbidden Γ$_6$ and allowed Γ$_1$ states. Such magneto-optical measurement [9] has allowed the estimation of an exchange interaction constant as 0.91 meV. The splitting between X(π) and FX$_A$(σ), observed in our measurements, is small and cannot be resolved within the experimental accuracy (∼0.5 meV). The forbidden Γ$_6$ and longitudinal Γ$_{5L}$ can be also weakly observed at $\mathbf{k} \perp \mathbf{c}$ without magnetic field due to finiteness of the exciton wave vector [11]. However, this mechanism can hardly provide so strong enhancement of the π-polarized component. That enforces us to search alternative polariton-related effects. We have performed an analysis of the temperature variation of shape and intensity of a longitudinal optical (LO) phonon-assisted recombination, which is considered as a measure of a density of exciton–polariton states [12]. It was found that both 1LO and 2LO replica intensity dependencies on energy E are well described by the Maxwell-like distribution $I \sim E^\eta \exp(-E/k_B T)$ with $\eta = 3/2$ and $1/2$, respectively [13]. At 5 K the π-polarized LO replicas are negligibly small as compared to the σ-polarized ones. This is well consistent with the bound B exciton dominance at this temperature, since its LO phonon replica is hardly observed due to a weak LO phonon coupling strength, in contrast to the well visible LO replica of acceptor-bound exciton A^0X_A possessing a heavier hole mass [14]. The π-polarized component of the 1LO replica increases noticeably up to 20 K (Fig. 2c). Starting from 30–40 K maximum of the π-polarized 1LO replica begins to shift towards the energy of the B exciton following the thermally activated occupation of the B exciton band.

Thus, a density of π-polarized exciton–polariton states really exists in the 10–60 K temperature range in the vicinity of the A exciton. We believe that the main factor stimulating their appearance is the complex structure of the polariton branches at $\mathbf{k} \perp \mathbf{c}$, where bottleneck regions of the σ-polarized branches (Γ_5) are closely matched with the π-polarized $LPB_B(\Gamma_1)$. Due to the low group velocity v the σ-polarized polaritons undergo numerous scattering acts by longitudinal acoustic (LA) phonons with partial depolarization. The probability of the scattering $W^{LA} \sim k(k_B T)$ increases with temperature up to chaotic thermal vibrations [15]. On the contrary, the $LPB_B(\Gamma_1)$ branch, possessing high v, provides an effective channel for different scattered contributions to emit from the crystal. It is worth to note that the forbidden exciton Γ_6 with its long lifetime can also be scattered to the π-polarized polariton branch. Contribution from the mixed states involving the longitudinal polariton Γ_{5L} is possible as well, even at the normal excitation used, as a result of the slight tilt of the c-axis in a crystal.

In summary, polarized μ-measurements demonstrate a strong enhancement of π-polarized PL line in the vicinity of A exciton, related partly to effective scattering between differently polarized polariton branches at $\mathbf{k} \perp \mathbf{c}$. The bound B exciton contributes to the line mostly at low temperatures. The dissimilarity in optical properties of internal and cleaved edge regions is demonstrated, which may be important for operating of edge-emitting devices.

Acknowledgements This work was partly supported by the Program of the Ministry of Sciences of RF "Physics of Solid State Nanostructures" and RFBR Grants (No. 99-02-17103 and No. 00-02-17022).

References

[1] J. J. HOPFIELD, Phys. Rev. **112**, 1555 (1958).
[2] E. L. IVCHENKO and G. E. PIKUS, in: Superlattices and Other Heterostructures, Springer Series Solid-State Sciences, Springer-Verlag, Berlin 1997.
[3] B. GIL, S. CLUR, and O. BRIOT, Solid State Commun. **104**, 267 (1997).
[4] V. RATNIKOV, R. KYUTT, T. SHUBINA, T. PASKOVA, E. VALCHEVA, and B. MONEMAR, J. Appl. Phys. **88**, 6252 (2000).
[5] T. V. SHUBINA, A. A. TOROPOV, V. V. RATNIKOV, R. N. KYUTT, S. V. IVANOV, T. PASKOVA, E. VALCHEVA, and B. MONEMAR, IWN2000 IPAP Conf. Ser. **1**, 595 (2001).
[6] R. DINGLE, D. D. SELL, S. E. STOKOWSKI, and M. ILEGEMS, Phys. Rev. B **4**, 1211 (1971).
[7] R. PÄSSLER, J. Appl. Phys. **83**, 3356 (1998).
[8] V. V. KRIVOLAPCHUK, S. A. PERMAGOROV, and V. V. TRAVNIKOV, Sov. Phys. – Solid State **23**, 343 (1981).
[9] R. STEPNIEWSKI, M. POTEMSKI, A. WYSMOLEK, K. PAKULA, J. M. BARANOWSKI, J. LUSAKOWSKI, I. GRZEGORY, S. POROWSKI, G. MARTINEZ, and P. WIDER, Phys. Rev. B **60**, 4438 (1999).
[10] L. ECKEY, A. HOFFMANN, P. THURIAN, I. BROSER, B. K. MEYER, and K. HIRAMATSU, Mater. Res. Soc. Symp. Proc. **482**, 555 (1998).
[11] D. C. REYNOLDS, D. C. LOOK, B. JOGAI, A. W. SAXLER, S. S. PARK, and J. Y. HAHN, Appl. Phys. Lett. **77**, 2879 (2000).
[12] C. WEISBUCH and R. G. ULBRICH, J. Lumin. **18/19**, 27 (1979).
[13] E. F. GROSS, S. A. PERMAGOROV, and B. S. RAZBIRIN, Usp. Fiz. Nauk **14**, 104 (1971).
[14] J. JAYAPALAN, B. J. SKROMME, R. P. VAUDO, and V. M. PHANSE, Appl. Phys. Lett. **73**, 1188 (1998).
[15] W. C. TAIT and L. WEINER, Phys. Rev. **166**, 769 (1968).

Radiative and Nonradiative Exciton Lifetimes in GaN Grown by Molecular Beam Epitaxy

G. Pozina[1]) (a), J. P. Bergman (a), B. Monemar (a), B. Heying (b), and J. S. Speck (b)

(a) Department of Physics and Measurement Technology, Linköping University, S-581 83 Linköping, Sweden

(b) Materials and ECE Department, University of California, Santa Barbara, CA 93106, USA

(Received July 8, 2001; accepted August 4, 2001)

Subject classification: 71.35.Cc; 78.47.+p; 78.55.Cr; S7.14

GaN epilayers grown by molecular-beam epitaxy (MBE) have been studied by temperature dependent time-resolved photoluminescence (PL). The PL decay times for free excitons and donor-bound excitons as well as the quantum efficiency have been measured for different temperatures. Radiative and nonradiative lifetimes have been evaluated from experimental values for the quantum efficiency and the PL decay time, assuming fully radiative processes at 2 K. The so obtained temperature dependence for the radiative lifetime cannot be described by a simple $T^{3/2}$ law for the whole temperature range. The temperature behavior of radiative lifetimes suggests the presence of a strong nonradiative recombination channel even at very low temperatures.

Introduction Successful applications of GaN for ultraviolet (UV) light-emitting diodes and UV laser diodes [1, 2] make this material attractive for research in order to further improve the material quality and develop a more complete understanding of optical and structural properties. Radiative processes play an important role for III-nitride optoelectronic device performance. The understanding of the physics of recombination mechanisms in nitrides can be of significant help for further development of the device efficiency. Also, verification of the theoretical models of the radiative recombination mechanism via experimental studies of good quality materials is necessary. However, at present the discrepancy between the theory of radiative lifetimes in GaN [3] and experimental studies [4] is still significant. In this work we present results of a photoluminescence (PL) study of high quality GaN epilayers grown by molecular-beam epitaxy (MBE). The studies were complemented by temperature dependent time-resolved photoluminescence and cathodoluminescence measurements.

Experimental The investigated samples consist of a 350 nm thick GaN film grown by MBE on a 2 μm thick GaN template layer grown by MOCVD. One piece of such a MOCVD-grown GaN template was used as a reference sample. Despite the rather high threading dislocation density in the range of 10^8–10^9 cm^{-2} these MBE-grown GaN layers demonstrate an electron mobility of about 1200 cm^2/Vs at room temperature (RT), which is one of the best values reported for GaN at present. For optical measurements the samples were placed in a variable temperature cryostat. The third harmonics (λ_{exc} = 266 nm) from a Ti:sapphire laser has been used as a fs pulsed excitation source.

---

[1]) Corresponding author; Phone: +46 13 28 4479; Fax: +46 13 14 2337; e-mail: galia@ifm.liu.se

The average power in the beam was 300 W/cm². The time-resolved PL measurements employed a Hamamatsu syncroscan streak camera set up with a time resolution less than 10 ps.

Results and Discussion The low temperature PL spectrum measured for the MBE-grown GaN layer (see Fig. 1a) is dominated by two emissions related to the free A exciton (FE) recombination and to the neutral-donor bound exciton (D_0X) transition. The peak energy of the FE is 3.4917 eV at 2 K (according to our calibration) for both MBE-grown GaN and for the reference GaN, while the position of D_0X is 3.4847 eV in the MBE-GaN and 3.4858 eV in the MOCVD-GaN. (We do not show here the experimental data for the reference sample.) Since the spectral position of the FE is the same in both cases, i.e. the samples are under the same strain, the energy shift for D_0X might indicate on different origin of the donor-bound excitons in the MBE-grown GaN and in the MOCVD-grown template. In addition, the line width becomes narrower, i.e. in the MBE GaN sample the full width at half maximum (FWHM) is 3.2 and 2.1 eV for FE and for D_0X, respectively, while in the GaN-reference sample the FWHM for FE and for D_0X is 4.5 and 2.5 eV, respectively. All these characteristics suggest a better quality of the MBE-grown GaN layer compared to the MOCVD-grown GaN template. Fig. 1b shows the low temperature PL decay curves measured at the peak position of the FE and of D_0X for the MBE-grown GaN. The rise time of the PL intensity τ_{rs} is longer than the temporal resolution limit in the case of D_0X for GaN, which is most likely related to the capture processes of FEs.

The PL decay times (τ) for the FE and D_0X as well as their PL intensities have been measured as a function of temperature. We have observed that τ demonstrates a pronounced non-monotonic temperature dependence both for the FE and D_0X, see Fig. 2a, if we evaluate the PL decay time using a simple exponential decay for both the FE and D_0X. In this case the FE shows a rather fast decay at 2 K with a typical recombination time about 60 ps. The PL decay time is longer for D_0X, about 70 ps. The increase of the PL decay time with temperature of the FE could be understood as the expected behavior of the radiative lifetime. However, it is difficult to explain a similar increase of the PL lifetime for the donor-bound exciton, if we consider the decay processes for D_0X in the studied sample as independent from the FE decay. A constant decay time as a function of temperature is expected for a bound exciton. As we mentioned above,

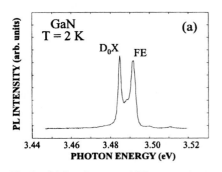
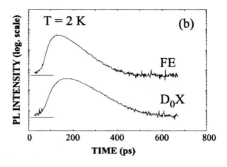

Fig. 1. a) Time integrated PL spectrum measured at $T = 2$ K for the MBE-grown GaN epilayer, b) PL decay curves detected at $T = 2$ K at the peak energy of the free excitons and the donor bound excitons for the MBE-grown GaN layer (upper) and the GaN template

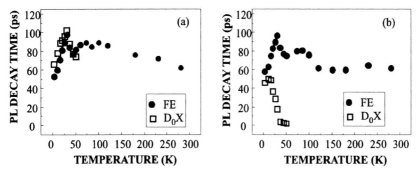

Fig. 2. a) PL decay times determined in the exponential approximation for FE (circles) and for D_0X (squares) are shown versus temperature for the MBE-grown GaN epilayer. b) The deconvoluted values of the lifetimes for the FE and for D_0X

the D_0X decay curve has a rather long rise time. Thus, we incorporate in the modeling that the donor-bound excitons are created by capture of the free excitons. Consequently, the observed D_0X decay curves are a combination of the radiative recombination from the D_0X and the feeding (capture) from the FE states. To estimate the D_0X decay time according to this model we deconvolute the measured PL decay curve for D_0X with the experimental FE decay function. The FE lifetimes were obtained by deconvolution of the experimental decay curves for FE with the instrumental response function. The deconvoluted values are shown in Fig. 2b. Now the lifetime for D_0X is practically constant up to 30 K and decreases rapidly at higher temperatures, which reflects a dissociation of the shallow bound excitons at elevated temperatures, possibly combined with transfer of D_0X to other defects before recombination [5]. The FE PL lifetime values in GaN were almost constant about 70–80 ps in the broad temperature range 100–300 K under our excitation conditions.

Figure 3 shows the temperature dependences of the radiative (squares) and nonradiative (circles) lifetimes determined using the following relations [4]:

$$\eta(T) = \frac{\tau_{\mathrm{PL}}(T)}{\tau_{\mathrm{r}}(T)}, \tag{1}$$

$$\frac{1}{\tau_{\mathrm{PL}}(T)} = \frac{1}{\tau_{\mathrm{r}}(T)} + \frac{1}{\tau_{\mathrm{nr}}(T)}. \tag{2}$$

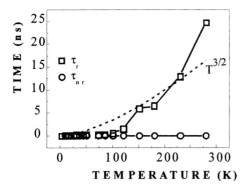

Fig. 3. Radiative (squares) and nonradiative (circles) lifetimes calculated in accordance with Eqs. (1) and (2) for the FE exciton emission are shown as a function of temperature. Solid lines are guides for the eye. The dashed line represents a $T^{3/2}$ law fit of the radiative lifetime values at low temperatures

The integrated PL intensity is normalized to its value at $T = 2$ K and it can be used as a measure for the quantum efficiency η assuming that the nonradiative channels are non-active at the lowest temperatures. The radiative lifetime of the FE estimated this way shows a sharp increase at temperatures above 100 K, giving a value of 25 ns at room temperature, which is approximately five times higher than the predicted theoretical value for the FE in GaN [3]. Our results are qualitatively similar to the data obtained by Im et al. [4] for MOCVD grown GaN layers, however, the samples investigated in [4] were of lower quality with PL lifetimes shorter than 25 ps at low temperature. We show by the dashed line in Fig. 3 the dependence $\tau_r \sim T^{3/2}$, which is expected to govern the radiative lifetime in direct gap bulk semiconductors [6]. As seen from Fig. 3 the discrepancy between theory and experimentally determined values is significant. It could be due to nonradiative channels, which exist already at very low temperatures. An indication of this is that our previous data for very thick high quality HVPE grown GaN layers show an FE decay time of about 200 ps at 2 K [7]. Indeed, increase of the excitation power within our experimental limit results in an increase of the PL decay times at 2 K by more than 20% both for the FE and for D_0X, which indicates on a presence of a strong nonradiative recombination channel even at low temperatures. Then the assumption that $\eta = 1$ at 2 K is an overestimation and gives rise to too high estimated values for the FE radiative lifetimes at high temperatures. There apparently exists a nonradiative shunt (possibly related to the dislocations) even at 2 K, and this shunt channel needs to be modeled carefully in order to estimate the true radiative lifetimes and the temperature dependence thereof.

Conclusion We have studied high quality GaN epilayers overgrown by MBE on top of a template GaN layer grown by MOCVD on sapphire. The electron mobility was about 1200 cm^2/Vs. The PL decay time is determined to 60 ps at 2 K. The radiative and nonradiative lifetimes have been extracted up to 300 K from the experimentally measured quantum efficiency and PL decay times, assuming that the 2 K data are dominated by radiative processes. Extrapolation to RT then gives a value of 25 ns for the radiative lifetime, which is much higher than predicted by theory for free excitons. The temperature dependence of radiative lifetimes suggests the existence of a rather strong nonradiative recombination channel even at low temperatures, possibly related to the dislocations.

References

[1] S. Nakamura, M. Senoh, S. Nagahama, N. Iwasa, T. Yamada, T. Matsushita, Y. Sugimoto, and H. Kiyoku, Appl. Phys. Lett. **70**, 1417 (1997).
[2] S. Nakamura, M. Senoh, S. Nagahama, N. Iwasa, T. Yamada, T. Matsushita, Y. Sugimoto, and H. Kiyoku, Jpn. J. Appl. Phys. **36**, L1059 (1997).
[3] A. Dmitriev and A. Oruzheinikov, J. Appl. Phys. **86**, 3241 (1999).
[4] J. S. Im, A. Moritz, F. Steuber, V. Härle, F. Scholz, and A. Hangleiter, Appl. Phys. Lett. **70**, 631 (1997).
[5] G. Pozina, J. P. Bergman, T. Paskova, and B. Monemar, Appl. Phys. Lett. **75**, 4124 (1999).
[6] J. Feldmann, G. Peter, E. O. Göbel, P. Dawson, K. Moore, C. Foxon, and J. Elliott, Phys. Rev. Lett. **59**, 2337 (1987).
[7] B. Monemar, J. P. Bergman, H. Amano, A. Akasaki, T. Detchprohm, K. Hiramatsu, and N. Sawaki, Proc. Internat. Symp Blue Lasers and LEDs, Chiba (Japan), March 5–7, 1996, Ohmsha Ltd., Tokyo 1996 (p. 135).

The 3.466 eV Bound Exciton in GaN

B. Monemar[1]), W. M. Chen, P. P. Paskov, T. Paskova, G. Pozina, and J. P. Bergman

Department of Physics and Measurement Technology, Linköping University, S-581 83 Linköping, Sweden

(Received June 21, 2001; accepted July 30, 2001)

Subject classification: 71.35.Cc; 71.55.Eq; 78.55.Cr; S7.14

We discuss the available optical data for the 3.466 eV bound exciton in GaN, which has been a controversial issue in the recent literature. We conclude that the experimental results are only consistent with the identification as an exciton bound at a neutral acceptor with a spin-like bound hole. The chemical identity is still not clear.

Introduction The identification of shallow donors and acceptors in GaN via their bound exciton signatures is still not clear. The photoluminescence features around 3.471 eV in unstrained GaN are due to donor bound excitons [1], as recognized also in early work [2, 3]. At lower energy, i.e. at about 3.466 eV in unstrained GaN, there is another very common feature in PL spectra, with an unknown origin. In literature it has been suggested to be due to excitons bound at a neutral deep donor [4], a neutral acceptor [5], or a charged donor [6, 7]. In this work we attempt to clarify the origin of this PL peak.

Experimental Results and Discussion Figure 1 shows schematically the electronic configuration of BEs in GaN, for three different cases: neutral donors, neutral acceptors and charged donors. The PL transitions occur between the excited state, i.e. the donor BE (DBE) state, and the neutral donor ground state for the case of donors (Fig. 1). A basic assumption in the description of the bound exciton states for neutral donors and acceptors is a dominant coupling of the like particles in the BE states, as indicated in Fig. 1 [8–11]. For a shallow neutral donor the two electrons in the BE state are as-

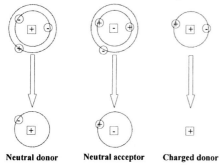

Fig. 1. Sketch of the electronic structure of bound excitons. The bound exciton states as well as the corresponding defect ground states are shown, for neutral donors, neutral acceptors and charged donors

---

[1]) Corresponding author; Phone: +46 13 281765; Fax: +46 13 142337; e-mail: bom@ifm.liu.se

sumed to pair off into a two-electron state with zero spin. The additional bound hole is then assumed to be weakly bound in the net hole-attractive Coulomb potential set up by this bound two-electron aggregate (Fig. 1).

Neutral shallow acceptor bound excitons (ABEs) are by the same arguments expected to have a two-hole state derived from two Γ_9 holes from the topmost valence band. Only one such state with $J = 0$ is allowed by the Pauli principle in wurtzite semiconductors. For deeper acceptors with a spin-like hole a similar $J = 0$ two-hole state is expected. The additional electron in the ABE state then contributes to its unpaired spin, so that the ABE state has $J = 1/2$. The ABE state is then expected to have a nearly isotropic g-tensor, ideally reflecting the shallow donor g-value in GaN, $g = 1.95$ [12].

A typical PL spectrum of nominally undoped GaN is shown in Fig. 2. Apart from the prominent split 3.477 eV BE, which is agreed to be related to neutral donors (presumably oxygen and silicon), a sharp line is seen at 3.472 eV. This is here upshifted by about 6 meV due to the residual biaxial strain in this sample, i.e. it corresponds to the 3.466 eV BE in unstrained samples [13, 14]. There are several arguments that this should be an ABE for a neutral acceptor, one argument involves the magneto-optical properties which will be discussed below. Further this BE has been found to be dominant in slightly Mg-doped GaN samples [5, 15, 16], it has a low energy acoustic phonon wing which is very characteristic of ABEs [8]. Also it has a rather strong LO phonon coupling, much stronger than for the DBE, a property which has also been found in previous studies of shallow ABEs in CdS [8]. Alternative interpretations of this BE line as being related to neutral [4] or charged donor bound excitons [6, 7, 17, 18] will be discussed below.

The chemical identity of this 3.466 eV ABE is still not confirmed. As stated above many workers have tentatively associated this ABE with Mg_{Ga} [5, 15, 16]. A difficulty is that most experiments with doped crystals up to now have been performed on heteroepitaxial material, where the ABE peak position strongly depends on the strain. While the most shallow BEs (such as the DBEs) tend to have a nearly constant distance to the FE_A position in strained layers, the ABEs, being deeper, do not follow such a simple behavior. For example in Mg-doped samples with an A exciton position at 3.499 eV at 2 K the Mg ABE is observed at 3.480 eV [19], a binding energy of 19 meV. In unstrained samples the same distance is observed to be 11–12 meV [13, 14]. The fact that this 3.466 eV ABE line is observed in almost all PL spectra of GaN in

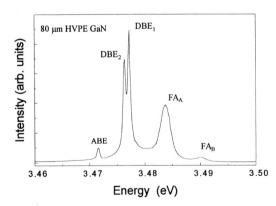

Fig 2. Photoluminescence spectrum of a thick nominally undoped HVPE grown GaN layer, showing the two dominant donor bound excitons, and the most common acceptor bound exciton at lower energy. Note that the spectra are upshifted in energy by about 6 meV compared to unstrained GaN

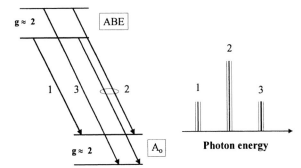

Fig. 3. Sketch of the expected photoluminescence lines of a spin-like acceptor in a magnetic field. In this case both the ABE state and the acceptor ground state will have an approxima tely isotropic splitting corresponding to $g \approx 2$, which leads to essentially a three line structure of the PL spectrum (right)

literature suggests great caution should be exercised in assigning this BE to a particular impurity. Recent SIMS data on our HVPE GaN have confirmed that Mg is often present at concentrations in the 10^{15}–10^{16} cm^{-3} range. Other possibilities cannot be excluded however.

The recent magneto-optical data for homoepitaxial GaN seem to confirm the identification of the 3.466 eV BE as an ABE [20]. The ABE state has an effective spin of the electron (Fig. 1), while the acceptor ground state has the effective hole spin. From independent magnetic resonance studies of the Mg acceptor it has been established that the Mg acceptor hole is essentially spin-like, with a g-factor close to two and nearly isotropic [21, 22]. This may be expected for a bound hole state with a sufficiently localized wave function, provided a small local lattice distortion is present [10], and may be a common property for acceptors in GaN. The magnetic field splitting of the ABE PL line should therefore show essentially an isotropic splitting pattern into three lines, as schematically indicated in Fig. 3. The experimental data from Ref. 20 may be interpreted along these lines. For the case of the magnetic field **B** parallel to the c-axis no splitting is resolved in the data presented, but for **B** ⊥ **c** a splitting into three lines is clearly seen, in agreement with the expected pattern in Fig. 3. (The discussion in Ref. [20] is slightly different, since the authors assume a Γ_9 hole for the acceptor.)

The alternative explanation of a deep neutral donor BE [4] seems not to be correct, since in this case a splitting into a two line patterns is expected, and also recently observed for the neutral donor BEs [23]. This disagrees with the published magnetooptical data for the 3.466 eV BE [20].

For the other alternative explanation with a charged donor BE [6, 7, 17, 18] there is no particle in the final state of the PL transition (Fig. 1). Therefore, the magnetic splitting should be similar to the case of the free excitons in GaN [24]. For instance, in the case **B** ⊥ **c** a splitting into a doublet would be expected, which is not observed in Ref. [20]. In addition it has to be pointed out that charged donor BEs, which are often observed in direct bandgap semiconductors [8], were typically much weaker than the neutral donor BEs. This is due to the fact that electron capture to a charged donor is very fast at the beginning of the photoexcitation, so that in a PL experiment the overwhelming majority of donors are expected to be neutral.

Conclusion In conclusion, available experimental data for the 3.466 eV BE provide strong evidence of the neutral acceptor origin of this PL line. The chemical identification is much less clear, since there may be many acceptors with spin-like bound holes in GaN.

References

[1] B. Monemar, J. Phys.: Condens. Matter **13**, 7001 (2001).
[2] R. Dingle, D. D. Sell, S. E. Stokowski, and M. Ilegems, Phys. Rev. B **4**, 1211 (1971).
[3] O. Lagerstedt and B. Monemar, J. Appl. Phys. **45**, 2266 (1974).
[4] B. K. Meyer, Mater. Res. Soc. Symp. Proc. **449**, 497 (1997).
[5] B. J. Skromme and G. L. Martinez, Mater. Res. Soc. Symp. **595**, W9.8 (1999).
[6] D. C. Reynolds, D. C. Look, B. Jogai, V. M. Phanse, and R. P. Vaudo, Solid State Commun. **103**, 533 (1997).
[7] B. Santic, C. Merz, U. Kaufmann, R. Niebuhr, H. Obloh, and K. Bachem, Appl. Phys. Lett. **71**, 1837 (1997).
[8] D. G. Thomas and J. J. Hopfield, Phys. Rev. **128**, 2135 (1962).
[9] P. J. Dean and D. C. Herbert, in: Excitons, Ed. K. Cho, Springer-Verlag, Berlin 1979 (p. 55).
[10] B. Monemar, U. Lindefelt, and W. M. Chen, Physica B **146**, 256 (1987).
[11] B. Monemar, CRC Crit. Rev. Solid State Mater. Sci. **15**, 111 (1988).
[12] W. E. Carlos, J. A. Freitas, Jr., M. Asif Khan, D. T. Olson, and J. N. Kuznia, Phys. Rev. B **48**, 17878 (1993).
[13] K. P. Korona, J. P. Bergman, B. Monemar, J. M. Baranowski, K. Pakula, L. Grzegory, and S. Porowski, Mater. Sci. Forum **258–263**, 1125 (1997).
[14] K. Kornitzer, T. Ebner, K. Thonke, R. Sauer, C. Kirchner, V. Schwegler, M. Kamp, M. Leszczynski, I. Grzegory, and S. Porowski, Phys. Rev. B **60**, 1471 (1999).
[15] M. Leroux, B. Beaumont, N. Grandjean, P. Lorenzini, S. Haffouz, P. Vennegues, J. Massies, and P. Gibart, Mater. Sci. Eng. B **50**, 97 (1997).
[16] M. Leroux, N. Grandjean, B. Beaumont, G. Nataf, F. Semond, J. Massies, and P. Gibart, J. Appl. Phys. **86**, 3721 (1999).
[17] A. K. Viswanath, J. I. Lee, D. Kim, C. R. Lee, and J. Y. Leam, Phys. Rev. B **58**, 16333 (1998).
[18] R. A. Mair, J. Li, S. K. Duan, J. Y. Lin, and H. X. Jiang, Appl. Phys. Lett. **74**, 513 (1999).
[19] U. Kaufmann, M. Kunzer, C. Merz, I. Akasaki, and H. Amano, Mater. Res. Soc. Symp. Proc. **395**, 633 (1996).
[20] R. Stepniewski, A. Wysmolek, M. Potemski, J. Lusakowski, K. Korona, K. Pakula, J. M. Baranowski, G. Martinez, P. Wyder, I. Grzegory, and S. Porowski, phys. stat. sol. (b) **210**, 373 (1998).
[21] M. Kunzer, U. Kaufmann, K. Maier, J. Schneider, N. Herres, I. Akasaki, and H. Amano, Mater. Sci. Forum **143–147**, 87 (1994).
[22] E. R. Glaser, T. A. Kennedy., K. Doverspike, L. B. Rowland, D. K. Gaskill, J. A. Freitas Jr., M. Asif Khan, D. T. Olson, J. N. Kuznia, and D. K. Wickenden, Phys. Rev. B **51**, 13326 (1995).
[23] A. Wysmolek, V. F. Sapega, T. Ruf, M. Cardona, M. Potemski, P. Wyder, R. Stepniewski, K. Pakula, J. M. Baranowski, I. Grzegory, and S. Porowski, in: Proc. Internat. Workshop Nitride Semiconductors, Nagoya, 2000, IPAP Conf. Series **1**, 579 (2000).
[24] R. Stepniewski, M. Potemski, A. Wysmolek, K. Pakula, J. M. Baranowski, J. Lusakowski, I. Grzegory, S. Porowski, G. Martinez, and P. Wyder, Phys. Rev. B **60**, 4438 (1999).

Exciton Diffusion in GaN Epitaxial Layers

Yu. Rakovich [1]) (a, b), J. F. Donegan (a), A. Gladyshchuk (b),
G. Yablonskii (c), B. Schineller (d), and M. Heuken (d)

(a) Physics Department, Trinity College, Dublin 2, Ireland

(b) Physics Department, Brest State Technical University, Moskowskaja Str. 267, 224017 Brest, Belarus

(c) Stepanov Institute of Physics, National Academy of Science of Belarus, F. Skaryna Ave., 220072 Minsk, Belarus

(d) AIXTRON AG, Kackertstr. 15–17, D-52072 Aachen, Germany

(Received July 12, 2001; accepted July 12, 2001)

Subject classification: 71.35.Cc; 78.20.Bh; 78.40.Fy; 78.55.Cr; S7.14

The photoluminescence (PL) and reflection excitonic spectra of GaN epitaxial layers grown on sapphire substrate by MOVPE were measured in the temperature region from 50 to 200 K. A three-layer model of the crystal was used for fitting the reflection spectrum. In this way the dead layer thickness, resonance energies and the broadening parameters of the free excitons were obtained. These parameters were then used for fitting the PL spectra assuming a thermal equilibrium for excitons. A decrease in the diffusion coefficient and increase in the radiative lifetime of excitons was found with increasing temperature.

Introduction For the technological development of GaN based optoelectronic devices a detailed knowledge of GaN optical properties is necessary. Analysis of the line shape of free exciton photoluminescence and reflection spectra (RS) can provide a way for the determination of energy structure and estimation of the basic parameters of radiative recombination processes [1, 2]. Most of the works have reported on identification of the free exciton states in GaN epitaxial layers in reflection, photoreflection and PL measurements [2–5]. However many fundamental parameters of exciton states in GaN are still insufficiently investigated. This includes the temperature dependence of the diffusion coefficient and diffusion length of the free excitons, which are connected with the radiative lifetimes. Picosecond time-resolved PL spectroscopy has been used for investigation of recombination processes in GaN [6] yielding the lifetimes of excitons of the order of several tens of picoseconds. However, the study of exciton diffusion processes in GaN encounters experimental difficulties [7]. It was demonstrated in our previous work [2] that analysis of the exciton PL and reflectance line shape in ZnSe epitaxial layers can provide us with this data. This method was applied in the present work to the investigation of the optical spectra of GaN epitaxial layers.

Growth and Experimental Procedure The 2 μm thick undoped GaN layer was grown at 1180 °C in an AIXTRON multiwafer reactor with H_2 as carrier gas and a V/III ratio of 1000. The layers show n-type conductivity with an electron mobility of 406 cm^2/Vs and the electron concentration was 1×10^{15} cm^{-3}. PL was excited by He–Cd laser radiation with $\lambda_{exc} = 325$ nm. The RS were recorded using a xenon light source.

---

[1]) Corresponding author; Phone: +353-1-608-1987; Fax: +353-1-6711759;
e-mail: Rakovich@lycos.com

Modeling of the Reflection and Luminescence Spectra Evaluation of the excitation levels, used in our experiments, shows that at a non-equilibrium carrier concentration of about 10^{12} cm^{-3} at $I_{exc} = 0.5$ W/cm^2 estimated by the generation rate, a degeneracy does not take place. This allows us to apply a phenomenological description of the excitonic PL line shape, which has been developed by Agranovich and Galanin [8].

The radiation line shape can be described by the following equation:

$$I_{PL}(\omega) = \varrho(\omega)\,(1 - R'(\omega)) \int_0^h n_{ex}(x) \exp(-\alpha(\omega)\,x)\,\mathrm{d}x, \qquad (1)$$

where h is the epilayer thickness, $\varrho(\omega)$ is the probability of generation of a photon of frequency ω per unit time owing to the exciton decay, $R'(\omega)$ is the reflection coefficient for the light incident at the internal surface of the epilayer, $n_{ex}(x)$ is the exciton concentration depending on the lifetime (τ), the diffusion coefficient (D_{ex}) of the excitons and the surface recombination rate (S) [8, 9], $\alpha(\omega)$ is the absorption coefficient.

In accordance with (1) only three exciton parameters $(S, \tau$ and $D_{ex})$ are varied during the fitting of the experimental PL if the complex dielectric function $\tilde{\varepsilon}(\omega)$ near the A and B free-exciton resonance is known [2]. Function $\tilde{\varepsilon}(\omega)$ can be described by the sum of the terms corresponding to contributions of the A and B free-exciton states [4].

To calculate the RS, taking into account the interference in the dead layer, we used the methods described in Ref. [10]. The resonance frequency ω_0, damping parameters $\Gamma_{A,B}$ and the width of the dead layer d were varied during the fitting of the experimental RS.

Results and Discussion The major difficulty from the standpoint of quantitative analysis of the free-exciton line shape of GaN epilayers is the overlapping of PL band A$_{n=1}$ with the I$_2$ line belonging to a neutral donor bound exciton. However, at elevated temperatures thermal quenching of the I$_2$ line takes place. Therefore to avoid this problem, PL spectra measured in the temperature region above 50 K were used in the analysis.

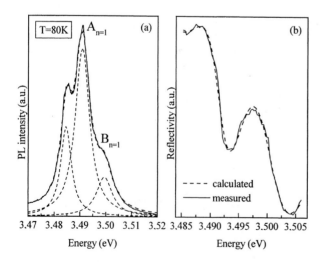

Fig. 1. Experimental (solid curves) and calculated (dashed curves) a) PL and b) reflection spectra of a GaN sample for $T = 80$ K. To calculate the PL spectra parameters $I_{exc} = 1.6 \times 10^{18}$ photons/(cm^2 s), $\alpha_{ex} = 10^5$ cm^{-3} and $S = 100$ cm/s were employed. The Lorentz model was used for modeling the emission bands located on the low energy side from the A exciton peak position

Table 1
Values of diffusion coefficients D_{ex}^A and D_{ex}^B and lifetimes τ_A and τ_B of A and B excitons as estimated from fit to experimental PL spectra

T (K)	D_{ex}^A (cm²/s)	D_{ex}^B (cm²/s)	τ_A (ps)	τ_B (ps)
50	0.38	0.13	26	11
60	0.35	0.12	26	13
70	0.32	0.14	29	13
80	0.30	0.10	36	17
90	0.33	0.10	38	22
100	0.29	0.15	41	16
120	0.27	0.12	36	24
140	0.33	0.13	49	31
160	0.33	0.17	53	39
180	0.26	0.14	53	31
200	0.22	0.10	57	44

The simulations of RS give an agreement with the experiment (Fig. 1b). To calculate $I_{PL}(E)$, the parameters ω_0, Γ and d obtained from results of the fitting of experimental RS were used. The experimental and calculated PL spectra for $T = 80$ K are given in Fig. 1a.

By fitting the experimental PL spectra, the exciton diffusion coefficient D_{ex} and the exciton lifetime τ were estimated (Table 1). The diffusion coefficient of the A exciton decreases from $D_{ex}^A = 0.38$ cm²/s at 50 K to $D_{ex}^A = 0.22$ cm²/s at 200 K. At the same time the diffusion coefficient of the B exciton stays nearly constant up to 200 K. The corresponding exciton lifetime increases slightly towards $T = 200$ K.

The observed temperature dependence of the exciton lifetimes agrees qualitatively with data of time-resolved PL measurements of GaN epilayers [6]. In our sample the dependence of the relation of intensities of free- and bound-exciton PL bands on inverse temperature is characterized by a Boltzmann factor $\exp(-\Delta E/kT)$ with an activation energy $\Delta E = 5.6$ meV (Fig. 2). The latter coincides with the energy distance between lines. This behaviour testifies to a thermal equilibrium in the excitonic system and the observed increase of the lifetime of free excitons can be explained by the thermal release of bound excitons.

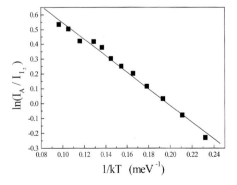

Fig. 2. Dependence of logarithm of the intensity ratio of free-exciton (I_A) and bound-exciton (I_{12}) PL bands on inverse temperature. Squares represent the experimental data. The solid line is a linear fit, with slope $\Delta E = 5.6$ meV

Conclusion We have analyzed the line shape of the free-exciton PL spectra of a GaN epitaxial layer. By fitting the experimental PL spectra, the exciton diffusion coefficient D_{ex} and the lifetime τ were estimated in the temperature region 50–200 K. The lifetimes of the free excitons were found to increase with increasing temperature because of the release of bound excitons. It is supposed that the scattering of free excitons by impurities and defects is responsible for a slight decrease of the diffusion coefficient with increasing temperature.

Acknowledgement This work was carried out within the framework of the project BELARUS-INTAS-98-0995.

References

[1] U. Behn, A. Thamm, O. Brandt, and H.T. Grahn, phys. stat. sol. (a) **180**, 381 (2000).
[2] A.L. Gurskii, Yu.P. Rakovich, E.V. Lutsenko, A.A. Gladyshchuk, G.P. Yablonskii, H. Hamadeh, and M. Heuken, Phys. Rev. B **61**, 10314 (2000).
[3] R. Stepniewski, K.P. Korona, A.V. Wysmolek, J.M. Baranowski, K. Pakula, M. Potemski, G. Martinez, I. Gregory, and S. Porowski, Phys. Rev. B **56**, 15151 (1997).
[4] D.K. Nelson, Yu.V. Melnik, A.V. Selkin, M.A. Yakobson, V.A. Dmitriev, K.J. Irvine, and C.H. Carter Jr., Phys. Solid State **38**, 455 (1996).
[5] S. Chichibu, T. Azuhata, T. Sota, and S.J. Nakamura, J. Appl. Phys. **79**, 2784 (1996).
[6] J.S. Im, A. Moritz, F. Steuber, V. Härle, F. Scholz, and A. Hangleiter, Appl. Phys. Lett. **70**, 631 (1997).
[7] J.Y. Duboz, F. Binet, D. Dolfi, N. Laurent, F. Scholz, J. Off, A. Sohmer, O. Briot, and B. Gil, Mater. Sci. Eng. B **50**, 289 (1997).
[8] V.M. Agranovich and M.D. Galanin, Electronic Excitation Energy Transfer in Condensed Matter, North-Holland Publ. Co., Amsterdam 1982 (pp. 262–273).
[9] V.V. Travnikov and V.V. Krivolapchuk, Sov. Phys. – Solid State **24**, 547 (1982).
[10] F. Evangelisti, A. Frova, and F. Patella, Phys. Rev. B **10**, 4253 (1974).

Excitonic Transitions in Homoepitaxial GaN

G. Martínez-Criado[1]) (a), C. R. Miskys (b), A. Cros (a), A. Cantarero (a), O. Ambacher (b), and M. Stutzmann (b)

(a) Materials Science Institute and Department of Applied Physics, University of Valencia, Dr. Moliner 50, E-46100-Burjasot, Spain

(b) Walter-Schottky-Institut, Technische Universität München, Am Coulombwall, D-85748 Garching, Germany

(Received June 25, 2001; accepted July 4, 2001)

Subject classification: 71.35.Gg; 78.55.Cr; S7.14

The photoluminescence spectrum of a high quality homoepitaxial GaN film has been measured as a function of temperature. As temperature increases the recombination of free excitons dominates the spectra. Their energy shift has successfully fitted in that temperature range by means of the Bose-Einstein expression instead of Varshni's relationship. Values for the parameters of both semi-empirical relations describing the energy shift are reported and compared with the literature.

Introduction Photoluminescence (PL) spectroscopy in GaN results in rich and defined excitonic structures which provide useful physical information like the band gap energy [1]. Accordingly, the study of excitons by this technique occupies a peculiar position in the physics of group III nitrides. However, in contrast to what happens for other III–V compounds such as GaAs, the agreement between some important parameters derived by different optical experiments is still poor [2]. In a significant part, this is due to the fact that factors such as the strain caused by lattice mismatch and differences in the thermal expansion coefficients between film and substrate (thermal strain), crystalline defects, residual impurities, etc., give rise to significant differences from the ideal bulk crystal. However, significant advances have been made in GaN homoepitaxial growth [3, 4]. Here we present high-resolution PL measurements of a GaN film grown homo-epitaxially on a free-standing GaN substrate.

Experimental Details The investigated sample of 2 µm thickness was grown by metal-organic chemical vapor phase deposition (MOCVD) on a free-standing high quality GaN substrate obtained by a laser-induced lift-off process of 300 µm thick films grown by hydride vapor phase epitaxy (HVPE) [3]. For the PL measurements the sample was mounted in a continuous He-flow cryostat and as exciting source the 333.6 nm line of the Ar^+ laser was used. The luminescence light was detected with a 0.8 m DILOR triple spectrometer equipped with a cooled charge-coupled device (CCD) detector. The Raman experiments were carried out in the same system using the 514.5 nm laser line in backscattering geometry.

Results and Discussion First, we have employed Raman spectroscopy in order to investigate the crystalline quality of the GaN layer. Two spectra are shown in Fig. 1; one corresponds to a clean GaN substrate, i.e. with no layer; the other one corresponds to

---

[1]) Corresponding author; Phone: +34 96 398 3606; Fax: +34 96 398 3633; e-mail: gmc@uv.es

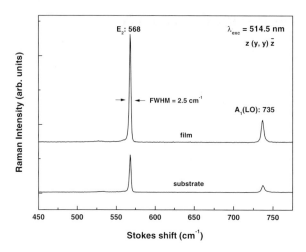

Fig. 1. Raman spectra taken in backscattering configuration. The assignment of the phonon modes is indicated in the graph

the same substrate with a deposited homoepitaxial film. Both measurements were taken in backscattering configuration, with the laser beam along the c-axis of the wurtzite structure (z-direction). The allowed E_2 and $A_1(LO)$ phonon modes at 568 and 735 cm^{-1}, respectively, are in good agreement with previous observations in strain-free GaN samples [5, 6], and exhibit a full width at half maximum (FWHM) of about 2.5 cm^{-1}. Therefore, the obtained linewidth clearly corroborates the high crystal quality of the homoepitaxial growth.

Figure 2 shows the temperature variation of the PL for the homoepitaxial GaN layer. There are three well-defined emission peaks in each spectrum. The higher energy peak is assigned to the free exciton X_A, the strongest and very narrow peak at 3.474 eV is assigned to the donor bound exciton recombination (D^0-X_A), whereas the low energy peak has been identified as an acceptor–bound-exciton transition, (A^0-X_A). The X_A emission is relatively weak at low temperatures, but its relative intensity increases with T, becoming the dominant transition at 55 K. A similar but not so clear behavior is shown for the lowest energy band, the free exciton X_B, which appears with T forming a shoulder at the right side of X_A. The weakening and quenching of the bound exciton recombinations, (D^0-X_A) and (A^0-X_A) is observed due to thermal dissociation of the respective complexes, contributing to free exciton emission. In order to evaluate accurately the energy position of the different recombination processes each spectrum was fitted using a multi-Gaussian procedure. In Fig. 3 the obtained energy shift of X_A with temperature is plotted (solid circles). (D^0-X_A) and (A^0-X_A) behave in a similar way, hence only the X_A data points were fitted with the empirical Varshni expression [7],

$$E(T) = E(0) - \frac{\alpha T^2}{T + \beta}, \tag{1}$$

$E(0)$ being the transition energy at 0 K, α and β are constants. Alternatively, we have used an average Bose-Einstein-statistics expression [8],

$$E(T) = E(0) - \frac{2a_B}{e^{\Theta_B/T} - 1}, \tag{2}$$

where a_B represents of the average exciton phonon coupling constant, and Θ_B is the mean phonon temperature. The effects of the thermal expansion and of the electron-

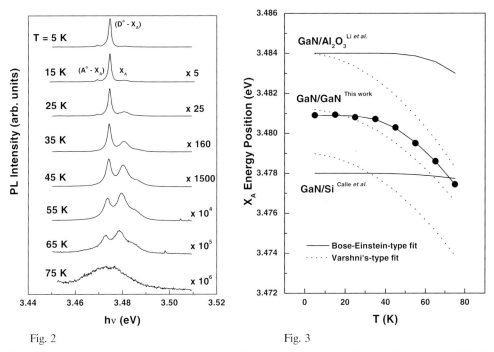

Fig. 2

Fig. 3

Fig. 2. Photoluminescence spectra measured at different temperatures with an exciting power of 70 μW

Fig. 3. Energy position vs. temperature. The dashed and solid lines are the least-squares fits to Eqs. (1) and (2), respectively

phonon interaction are not explicitly considered or separately taken into account in the two relations; therefore, the physical meaning of the parameters appearing in them is not simple. However, it has been demonstrated that the phonon contribution to the energy shift coincides with Eq. (2), and that the Varshni equation (1) can be considered as second-order approximation of Eq. (2), valid for high temperatures ($\Theta_B/T \ll 1$) [9]. In addition, the fitting curves proposed in the literature [10, 11] have been included in Fig. 3. In Table 1 our parameters are compared to those obtained by Li et al. [10] and

Table 1

Parameters used to describe the free exciton energy shift with temperature (Eqs. (1) and (2)), obtained from GaN samples grown on different substrates

sample	$E(0)$ (eV)	α (10^{-4} eV/K)	β (K)	a_B (meV)	Θ_B (K)	T (K)
GaN/GaN[a]	3.4812[d] 3.4809[e]	11.3	1307	17.5	181	5–75
GaN/Al$_2$O$_3$[b]	3.484[d] 3.484[e]	12.8	1190	110	405	15–475
GaN/Si[c]	3.479[d] 3.478[e]	8.70	884	47	271	4–300

[a] This work. [b] Ref. [10]. [c] Ref. [11]. [d] Corresponding value for Varshni's expression. [e] Corresponding value for Bose-Einstein relation.

Calle et al. [11] for GaN layers grown on sapphire and Si, showing the significant shift of $E(0)$ due to different residual strains. A closer insight into our fitting curves shows that in this temperature region there are small but clear differences between the trend of both calculated curves, getting an excellent fit with the Bose-Einstein relationship instead of the Varshni's law. This result is consistent with the fact that the former expression gives essentially the dynamic part of the energy shift which dominates at low temperatures. For higher temperatures the thermal expansion contribution becomes important, which follows the latter empirical relation (Eq. (1)). Evidently, in order to get expressions valid in a wider temperature range we think that a more detailed determination is needed taking into account that strain of thermal origin may modify these dependencies.

Conclusions Photoluminescence and Raman spectroscopies were used to investigate a homoepitaxial GaN film. Raman spectra exhibit a very narrow E_2 phonon peak evidencing the high crystal quality of the homoepitaxial growth. The temperature dependence of excitonic transitions were studied in detail. Using two semi-empirical relations, Varshni's and Bose-Einstein expressions, the thermal behavior of our data was fitted. At low temperatures, where the effect of the thermal dilatation on the energy shift is negligible, the fit of the data with Bose-Einstein relation is in much better agreement. This shows that the values of the parameters with which a given empirical law describes the experimental data depend on the temperature interval in which the data have been fitted and are valid only in that interval.

Acknowledgements G. M.-C. would like to thank the Agencia Española de Cooperación Internacional (AECI) for granting her fellowship, as well as the Generalitat Valenciana for continuous support. In the same way, A.C. acknowledges the financial support under Grant No. GV00-080-15 and MAT2000-0772.

References

[1] Z. X. Liu, S. Pau, K. Syassen, J. Kuhl, W. Kim, H. Morkoç, M. A. Khan, and C. J. Sun, Phys. Rev. B **58**, 6696 (1998).
[2] K. Reimann, M. Steube, D. Fröhlich, and S. J. Clarke, J. Cryst. Growth **189/190**, 652 (1998).
[3] C. R. Miskys, M. K. Kelly, O. Ambacher, G. Martínez-Criado, and M. Stutzmann, Appl. Phys. Lett. **77**, 1858 (2000).
[4] M. Schauler, F. Eberhard, C. Kirchner, V. Schwegler, A. Pelzmann, M. Kamp, K. J. Ebeling, F. Bertram, T. Riemann, J. Christen, P. Prystawko, M. Leszczynski, I. Grzegory, and S. Porowski, Appl. Phys. Lett. **74**, 1123 (1999).
[5] M. S. Liu, L. A. Bursill, S. Prawer, K. W. Nugent, Y. Z. Tong, and G. Y. Zhang, Appl. Phys. Lett. **74**, 3125 (1999).
[6] A. Link, K. Bitzer, W. Limmer, R. Sauer, C. Kirchner, V. Schwegler, M. Kamp, D. G. Ebling, and K. W. Benz, Appl. Phys. Lett. **86**, 6256 (1999).
[7] Y. P. Varshni, Physica **34**, 149 (1967).
[8] L. Viña, S. Logothetidis, and M. Cardona, Phys. Rev. B **35**, 9174 (1984).
[9] A. Manoogian and J. C. Woolley, Can. J. Phys. **62**, 285 (1984).
[10] C. F. Li, Y. S. Huang, L. Malikova, and F. H. Pollak, Phys. Rev. B **55**, 9251 (1997).
[11] F. Calle, F. J. Sánchez, J. M. G. Tijero, M. A. Sánchez-García, E. Calleja, and R. Beresford, Semicond. Sci. Technol. **12**, 1396 (1997).

Donor and Donor Bound Exciton Spectroscopy in Wurtzite GaN Heterostructures

M. Teisseire, G. Neu[1]), and C. Morhain

Centre de Recherche sur l'HétéroEpitaxie et ses Applications, CNRS,
Rue Bernard Grégory, Parc Sophia-Antipolis, F-06560 Valbonne, France

(Received June 30, 2001; accepted July 12, 2001)

Subject classification: 71.35.Gg; 71.55.Eq; 78.55.Cr; S7.14

Neutral donor bound excitons (I_2) and donor related electronic Raman scattering (ERS) in wurtzite GaN epilayers deposited onto Si, 6H-SiC, Al_2O_3 and GaN substrates are studied at low temperature by high-resolution selective photoluminescence spectroscopy (SPL). Due to the presence of a large strain distribution in the heterostructures, the observation of the I_2 excited states under selective laser excitation is possible from tensile to compressive strain along the [0001] direction. It is shown that the I_2 spectra strongly depend on which valence band (A or B) the bound exciton hole belongs to. Resonant ERS and resonantly excited two-electron spectra reveal that two main different residual donor species are present both in MBE and MOCVD epitaxial samples. The donor binding energies are shown to vary noticeably with the biaxial strain field.

Introduction Wurtzite GaN has become a major wideband gap semiconductor mainly because of its importance in the fabrication of various commercial optoelectronic devices [1] and because many other new industrial applications based on GaN heterostructures are reasonably expected in a near future [2]. GaN's breakthrough is mainly due to the significant improvements in heteroepitaxial growth and to the subsequent decrease in density of structural defects, in particular in epitaxially laterally overgrown (ELO) layers [3]. However, low temperature microscopic photoluminescence [4] and cathodoluminescence [5, 6] experiments establish that partial local strain relaxation still occurs in such layers and broadens the luminescence lines [6]. In more usual GaN/Al_2O_3 and GaN/SiC heterostructures, it is demonstrated that the Raman line broadening mechanism of the E_2^{high} phonon is closely related to the presence of inhomogeneous strain, inherent to the use of substrates with large lattice mismatch [7]. Lateral fluctuations of the local emission wavelength where evidenced by cathodoluminescence studies over a micrometer scale. Lateral stress distribution was also supposed to be at the origin of the resonant features observed under selective excitation over the whole bound exciton emission spectral range [8].

In this paper, the neutral donor bound excitons (NDBEs) are studied by high-resolution SPL spectroscopy at low temperature in GaN epilayers deposited onto Si, 6H-SiC, Al_2O_3 and GaN substrates. The NDBE excited state spectra are determined in a large range of strain, from tension (GaN/Si) to compression (GaN/Al_2O_3), allowing to precise the A and B valence band crossing energy. Beside, by using ERS and two-electron (TET) spectroscopies, the main donors are identified. Finally, the strain dependence on the donor binding energy is experimentally determined.

---

[1]) Corresponding author; Phone: +33 49 395 42 15; Fax: +33 49 395 83 61;
e-mail: gneu@crhea.cnrs

Experiment The heterostructures were grown in our laboratory by metal organic vapor phase epitaxy (MOVPE) on 6H-SiC and on sapphire (0001) as well as by molecular beam epitaxy (MBE) on Si(111) substrates. Homoepitaxial layers were grown on high quality GaN(0001) substrates by MBE and MOVPE. Photoluminescence (PL) and SPL measurements were performed at 1.8 K. The excitation light was provided by the second harmonic wave of a tuneable Ti:sapphire ring laser pumped by a cw Ar-laser.

Excited States of the Neutral Donor Bound Exciton The PL spectra of homoepitaxial layers display a narrow NDBE recombination line I_{20}, followed at higher energy by four well defined peaks [9, 10] attributed to the NDBE excited states (I_{2i}, i = a, c, d and e) [9]. Unfortunately, the broadening of the emission lines generally prevents the direct observation of the NDBE excited states on the PL spectra of the heterostructures. However, it was shown that the SPL spectra of GaN grown on sapphire display narrow resonant NDBE emission lines superimposed on I_2 [8]. Stress distributions induce distributions of the bound exciton ground and excited state energies, allowing for the possibility within the excitation laser spot of populating distinct bound exciton states with a given excitation energy [9]. After thermalization in their respective ground states, the excitons recombine, displaying a picture of the neutral donor bound exciton excited state spectrum. The PL spectra of the samples studied in this work and the energy difference between the NDBE ground and excited states measured by SPL spectroscopy of these samples is reported in Figs. 1a and b, respectively. The resonant peaks, labelled a, c, d and e in Fig. 1b, are observed at a nearly constant energy from that of the laser, when the excitation energy is varied over I_2. The linear extrapolation of the energies of the a, c, d and e resonant peaks down to the homoepitaxial layer NDBE excited states (triangles in Fig. 1b), demonstrates that the same electronic states are involved in the SPL spectra of compressively strained layers and in the PL spectra of strain-free layers. Such a conclusion rules out previous interpretations of homoepitaxial PL spectra, that identified some of the recombination peaks between I_2 and X_A as NDBEs related to the B valence band [10, 11] or to excitons bound to donor complexes in different structural environments [11].

Conversely, owing to the tensile strain field generally present in GaN/SiC and GaN/Si heterostructures, the recombination spectra are expected to involve the B valence band related NDBEs, especially near the A–B valence band crossing. It is obvious that, far

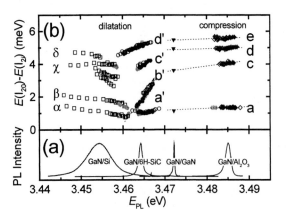

Fig. 1. a) PL spectra of GaN under tensile strain (GaN/Si and GaN/6H-SiC), compressive strain (GaN/sapphire) and unstrained (GaN/GaN). b) Energy difference between the ground and the excited states of the NDBEs versus the emission line energy

from the A–B crossover, the rotational excited state spectra are deeply modified depending on whether the involved hole is in the A (a, c, d and e resonant transitions) or in the B valence band (α, β, χ and δ resonant transitions). This result clearly illustrates that the nature of the hole state is crucial in the kinetics of the NDBE rotational excited states [12]. While the excited states are well defined and roughly at constant energy from I_{20} for the compressively strained GaN/sapphire and the GaN/Si in extension, rapid variations of the energy difference between the ground and the excited states of NDBEs are observed in the intermediate energy range corresponding to the A–B crossing point [13]. Among the a', b', c' and d' resonant transitions observed in this spectral range on SPL spectra recorded for GaN/SiC, the b'-line exhibits the steepest variation and moves linearly with respect to the PL energy. Its variation extrapolated at low energy could induce a crossing with the A related NDBE near 3.459 eV, i.e. very close to the expected crossing between the A and B valence bands [13]. The extrapolation performed on the high energy side shows that the b-transition is likely to be the ground or the first excited state of the NDBEs associated with the B valence band.

Donor Binding Energy in GaN Heterostructures As the recombination process of NDBEs via the two-electron mechanisms was shown to involve NDBEs excited states [9, 14], the two-electron transitions (TET) spectra are compared with resonant ERS in Fig. 2 in order to avoid a misinterpretation of the replicas. The samples used in this study are GaN (ELO) heterostructures grown on sapphire with two different MOVPE kits. The excitation laser was tuned on the B free exciton to record the resonant ERS spectra or tuned on I_2 to detect the 1s–2s donor TET's and the resonant Raman signal involving the low energy E_2 phonon. Only one donor (d_1: probably Si [14]) is detected in the G973 and G730 samples grown in the same kit. As previously mentioned [8], the resonant ERS signal is coincident with only one TET replica. The second one is attributed to the recombination of the NDBE first rotational excited state leaving the neutral donor in the 2s sate. Both resonant ERS and TET spectra show that a second donor (d_2) is present in the sample grown in the second MOVPE kit.

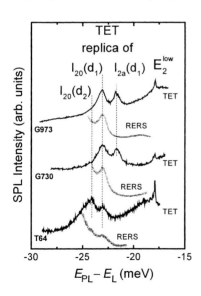

The 1s–2s transition energies of the samples exhibiting an A valence-band related NDBEs are reported in Fig. 3 as a function of the exciton ground state recombination energy. The transition energies are shown to vary with the strain. The 1s–2s donor transition energy is shown to increase from GaN/SiC to GaN/sapphir, evidencing an increase in the d_1 and d_2 donor binding energies in compressively strained heterostructures. On the contrary, the

Fig. 2. 2 K resonant electronic Raman scattering (RERS) and two-electron transition (TET) spectra of ELO GaN heterostructures. The G973 and G730 samples are grown in the same MOVPE kit

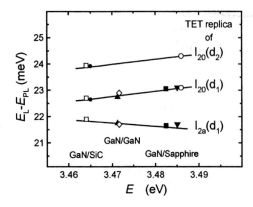

Fig. 3. Two-electron transition energies deduced from SPL measurements at 2 K. The symbols used for data are different for each studied sample

TET transition energy associated with the first NDBE excited state (I_{2a}) decreases as the strain is increased. This effect is related to the strain dependence of the energy difference between I_{20} and I_{2a} (see Fig. 1b). This is a new confirmation that the first NDBE excited state is involved in the low temperature two-electron spectra of GaN.

Conclusion The neutral donor bound exciton excited states of GaN have been studied over a large domain of tensile and compressive strains. It is shown that the I_2 spectra strongly depend on which valence band (A or B) the bound exciton hole belongs to. The observation of A and B valence band related NDBE states allow us to give an accurate evaluation of the A and B crossing energy. Resonant ERS and resonantly excited two-electron spectra reveal that two main different residual donor species are observed both in MBE and MOCVD epitaxial samples. The donor binding energies are shown to vary noticeably with the biaxial strain field.

Acknowledgements The authors are grateful to Dr. N. Grandjean and Dr. F. Semond for the growth of GaN/Si and GaN/GaN samples by MBE and to Dr. B. Beaumont and Dr. H. Lahreche for providing us with the samples grown by MOVPE.

References

[1] S. Nakamura, M. Senoh, S. Nagahama, N. Iwasa, T. Yamada, T. Matsushita, H. Kiyoku, Y. Sugimoto, T. Kozaki, H. Umemeto, M. Sano, and K. Choto, Jpn. J. Appl. Phys. **37**, L309 (1998).
[2] J.-Y. Duboz, phys. stat. sol. (a) **176**, 5 (1999).
[3] P. Venneguès, B. Beaumont, V. Bousquet, M. Vaille, and P. Gibart, J. Appl. Phys. **87**, 4175 (2000).
[4] M. Yoshimoto, J. Saraie, and S. Nakamura, Jpn. J. Appl. Phys. **40**, L386 (2001).
[5] A. Kaschner, A. Hoffmann, C. Thomsen, F. Bertram, T. Riemann, J. Christen, K. Hiramatsu, T. Shibata, and N. Sawaki, Appl. Phys. Lett. **74**, 3320 (1999).
[6] F. Bertram, T. Riemann, J. Christen, A. Kaschner, A. Hoffmann, C. Thomsen, K. Hiramatsu, T. Shibata, and N. Sawaki, Appl. Phys. Lett. **74**, 359 (1999).
[7] M. Giehler, M. Ramsteiner, P. Waltereit, O. Brandt, K. H. Ploog, and H. Obloh, J. Appl. Phys. **89**, 3634 (2001).
[8] G. Neu, M. Teisseire, B. Beaumont, H. Lahreche, and P. Gibart, phys. stat. sol. (b) **176**, 79 (1999).
[9] G. Neu, M. Teisseire, P. Lemasson, H. Lahreche, N. Grandjean, F. Semond, B. Beaumont, I. Grzegory, S. Porovski, and R. Triboulet, Physica B **302/303**, 39 (2001).
[10] K. Kornitzer, M. Grehl, K Thonke, R. Sauer, C. Kirchner, V. Schwegler, M. Kamp, M. Leszczynski, I. Grzegory, and S. Porowski, Physica B **273/274**, 66 (1999).
[11] A. Wismolek, V. F. Sapega, T. Ruf, M. Cardona, M. Potemski, P. Wider, R. Stepniewski, K. Pakula, J. M. Baranowski, I. Grzegory, and S. Porowski, IAP Conf. Series **1**, 579 (2000).
[12] W. Rühle and W. Klingenstein, Phys. Rev. B **18**, 7011 (1978).
[13] B. Gil, F. Hamdani, and H. Morkoç, Phys. Rev. B **54**, 7678 (1996).
[14] G. Neu, M. Teisseire, E. Frayssinet, W. Knap, M. L. Sadowski, A. M. Witowski, K. Pakula, M. Leszczynski, and P. Prystawko, Appl. Phys. Lett. **77**, 1348 (2000).

Playing with Polarity

M. Stutzmann (a), O. Ambacher (a), M. Eickhoff (a), U. Karrer (a),
A. Lima Pimenta (a), R. Neuberger (b), J. Schalwig (b), R. Dimitrov (c),
P. J. Schuck (d), and R. D. Grober (d)

(a) Walter Schottky Institut, Technische Universität München, Garching, Germany

(b) EADS Germany GmbH, München, Germany

(c) Nova Crystals, Inc., San José, CA, USA

(d) Department of Applied Physics, Yale University, New Haven, CT, USA

(Received August 22, 2001; accepted August 31, 2001)

Subject classification: 68.35.Ct; 68.37.Ps; 68.55.Jk; 73.40.Kp; S7.14

We review the influence of GaN crystal polarity on various properties of epitaxial films and electronic devices. GaN films grown on sapphire by MOCVD or HVPE usually exhibit Ga-face polarity. N-face polarity is obtained either on the backside of such layers after removal from the substrate, or by turning the crystal polarity in MBE growth via a thin AlN buffer layer. In addition to rather obvious differences in their structural and morphological features, Ga- and N-face samples differ also in their electronic properties. Thus, different Schottky barrier heights are observed for both polarities, the position and detailed properties of spontaneously formed two-dimensional electron gases vary with polarity, and the adsorption of gases and ions also show an influence of the two different surfaces. A particular interesting possibility is the growth of lateral polarity heterostructures with predetermined macroscopic domains of different polarity separated by inversion domain boundaries. These structures make use of the crystal polarity as a new degree of freedom for the investigation of electronic properties of III-nitrides and for novel devices.

Introduction The fact that the III-nitrides InN, GaN, and AlN are essentially ionic solids with a strong charge transfer between the very electronegative nitrogen atoms and the less electronegative metal atoms (In, Ga, or Al) has important consequences for many physical properties of this material system.

The thermodynamically stable wurtzite structure of III-nitrides has a polar axis parallel to the c-direction of the crystal lattice. Deviations of the real atomic charge distributions from the point charge model of the ideal wurtzite lattice give rise to a macroscopic spontaneous polarization **P** of the III-nitrides, which can reach values up to 0.1 C/m^2. Electrostatically, such macroscopic lattice polarizations are equivalent to two-dimensional fixed lattice charge densities $\sigma = \mathbf{P}$ with values between 10^{13} and 10^{14} e/cm^2 located at the two surfaces of a sample (about 1–10% of a monolayer) [1, 2]. A direct consequence of this large macroscopic polarization and the corresponding fixed surface charges is the appearence of large internal electric fields,

$$\mathbf{E} = \mathbf{P}/\varepsilon_0 \, (\varepsilon - 1), \tag{1}$$

which in the case of AlN with $P = 0.09$ C/cm^2 can reach values of up to 10^7 V/cm, or 1 V/nm. Corresponding values for P in GaN and InN are 0.034 and 0.042 C/cm^2, respectively. In the zincblende structure of III-nitrides, this spontaneous polarization is absent due to the higher lattice symmetry.

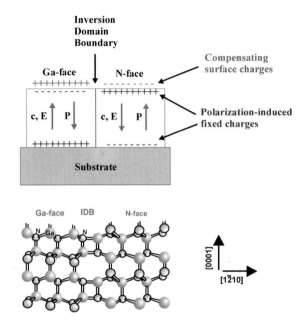

Fig. 1 (colour). Crystal structure and orientation of the c-axis as well as of the macroscopic spontaneous polarization **P** and the corresponding electric field **E** for Ga-face and N-face GaN grown on a heterosubstrate. Also shown are the polarization-induced terminating fixed charges at the substrate interface and the sample surface, the compensating surface charges due to adsorbed ions, and the inversion domain boundary (IDB) separating the two crystal domains laterally. The atomic structure of the IDB shown is based on the results of Ref. [3]

Because of their polar axis, III-nitride crystals grown on a heterosubstrate can have two inequivalent orientations as shown in Fig. 1[1]). In the specific case of GaN, these two orientations are refered to as Ga-face (N-face) polarity, depending on whether the Ga atoms (N atoms) of the Ga–N bilayers forming the crystal are facing towards the sample surface. In Ga-face samples, the crystallographic c-axis and the internal electric field point away from the substrate towards the surface, whereas the polarisation has the opposite direction. The polarisation-induced fixed lattice charges are negative at the surface and positive at the substrate interface. For N-face material, all charges and directions are inverted.

Whereas the presence of the macroscopic polarization (both spontaneous and piezoelectric) in III-nitrides and its effect on many material and device properties by now is well established and has attracted a lot of attention, several important details still remain elusive. As also shown in Fig. 1, positive or negative ions from the environment can be adsorbed at the free surface, thus causing a partial compensation of the polarization charges. Nothing is known in detail about this compensation. The same is true for the substrate interface, where a high density of structural defects and charge transfer to the substrate can easily give rise to significant deviations from the simple situation shown in Fig. 1. Finally, Ga-face and N-face domains can coexist side by side on the same substrate, leading to a further complication of the overall situation. Although the microscopic nature of the inversion domain boundaries (IBD) separating adjacent domains of different polarity has been studied theoretically [3], not much is known about their real structural and electronic properties.

Setting and Analyzing Lattice Polarity In order to "play" with crystal polarity, it is of course important that we are able to, both, influence and analyze the polarity in a reli-

---

[1]) Colour figures are published online (www.physica-status-solidi.com), where indicated.

able way. The most obvious approach is to use the two sides of opposite polarity of a free-standing III-nitride sample e.g. obtained by high pressure–high temperature synthesis [4] or by laser-induced lift-off from a heterosubstrate [5]. A second approach is the use of polar substrates such as ZnO or 6H-SiC, which can be prepared as bulk crystals with defined polarity and thus can set the polarity of heteroepitaxially grown III-nitride layers accordingly. For an authoritative review of this issue see [6] and references therein. However, most of the epitaxial layers today are grown on sapphire (Al_2O_3) by MOCVD or MBE, and we, therefore, briefly review polarity in this heteroepitaxy system.

High quality III-nitride MOCVD-layers on sapphire invariantly turn out to have Ga-face polarity, whereas similar layers grown by MBE exhibit N-face polarity. A simple chemical argument rationalizing this general observation is shown in Fig. 2. Usually, "epiready" sapphire substrates are oxygen-terminated. Under deposition conditions typical for MBE growth (substrate temperatures around 850 °C, low flux of nitrogen radicals towards the substrate), this termination is found to be stable against nitridation for a considerable amount of time (in hours). Therefore, it is reasonable to assume that the first monolayer of the deposited epitaxial film consists of group III metals, as shown in Fig. 2 for Ga (left) and Al (right) as the first metallic monolayer. The difference between these two cases comes from the different chemical strength of the III–N versus the III–O bonds. Since Ga–N bonds are stronger than Ga–O bonds, the first Ga monolayer deposited on the sapphire substrate already belongs to the GaN epilayer, which consequently has N-face polarity (left side of Fig. 2). In contrast, if Al is deposited on sapphire, the first metal monolayer still belongs to the substrate, since Al–O bonds are stronger than Al–N bonds. Instead, the subsequently deposited N-layer is the first monolayer belonging to the growing III-nitride crystal which, therefore, now has the opposite polarity (Al-face or Ga-face). Thus, by depositing a thin AlN buffer layer at the beginning of MBE growth, the polarity can be inverted from N-face to Ga face (right-hand side of Fig. 2). Indeed, this has been observed experimentally [7] and has also been discussed more rigorously by a theoretical study of thin AlN films on sapphire [8].

In the case of MOCVD growth, the N-rich growth conditions together with the higher substrate temperatures and the usually employed nitridation step at growth start always give rise to the formation of a thin AlN layer on the sapphire substrate, thus explaining the Ga-face polarity of MOCVD material on sapphire.

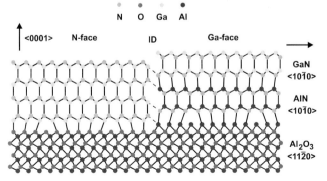

Fig. 2 (colour). Schematic view of the III-nitride/sapphire interface with (right part) or without (left part) an AlN buffer layer. This causes a polarity inversion, as discussed in the text

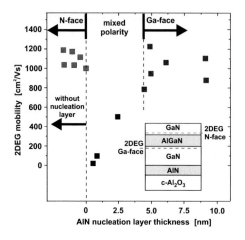

Fig. 3 (colour). Influence of a thin AlN buffer layer on the Hall mobility of two-dimensional electron gases formed at the upper (N-face polarity) or lower (Ga-face polarity) interface of a GaN/AlGaN/GaN heterostructure grown by molecular beam epitaxy on sapphire substrates. In accordance with the schematic model of Fig. 2, the epilayer has N-face polarity without an AlN buffer and is inverted to Ga-face polarity by an AlN buffer with a thickness of 5 nm or more. Thinner AlN buffers lead to mixed polarity epilayers

Figures 3 and 4 show in more detail the pronounced effect of a thin AlN buffer layer on the polarity and the structural properties of MBE-grown GaN/AlGaN/GaN heterostructures. Such heterostructures are a very sensitive probe for polarity because of the spontaneously formed two-dimensional electron gases (2DEGs) at the GaN/AlGaN interface. Determined by the sign of the polarization-induced fixed charges, for N-face heterostructures such a 2DEG is formed at the upper AlGaN/GaN-interface, as shown by the inset in Fig. 3. On the other hand, a 2DEG is formed at the lower interface for a Ga-face layer [2]. The position of the 2DEG can be conveniently probed by $C-V$ profiling. The existence of inversion domains has a strong influence on the macroscopic mobility of the electrons in the respective 2DEGs determined from Hall measurements. As seen from Fig. 3, the room temperature mobility in N-face heterostructures without an AlN buffer layer is about 1000 cm^2/Vs. AlN buffers with a thickness of less than 3 nm are too thin to cause a large scale polarity inversion. Instead,

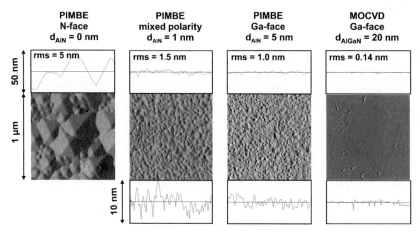

Fig. 4. Atomic force micrographs and surface contours of MBE-grown AlGaN/GaN epilayers on sapphire substrates with different AlN buffer layer thicknesses, d_{AlN}. A Ga-face sample grown by MOCVD on an AlGaN buffer is included for comparison. Rms values of the respective surface roughnesses are also given

the corresponding epilayers exhibit mixed polarity, with a large density of inversion domains limiting the lateral size of coherent 2DEG regions and causing a pronounced drop in Hall mobility. For AlN layers with a thickness exceeding 5 nm, however, the polarity of the entire epilayer has changed to Ga-face, and the mobility reaches again values of 1000 cm^2/Vs.

Despite of the similar electron mobilities for the two polarities in Fig. 3, the structural properties of the respective epilayers are quite different, as depicted by the AFM topographs in Fig. 4. The N-face polarity MBE-grown material has a noticeable surface roughness due to the existence of quite large hexagonal crystallites. These disappear when a thin AlN buffer is introduced, leading to a much smoother surface in the case of mixed polarity and for Ga-face samples. This observation is in agreement with the "standard framework" formulated by Hellman [6]. Note, however, that absolute values of surface roughness also depend sensitively on the growth method employed. Thus, there is little resemblance between the surface morphologies of the two Ga-face layers in Fig. 4 which were grown by plasma-induced MBE and MOCVD, respectively.

For practical applications of GaN/AlGaN heterostructures, e.g. in high electron mobility transistors, it is important to conclude that there is little difference between the room temperature mobilities of 2DEGs for both polarities, despite their significantly different interface roughnesses. However, cases with mixed polarity on a microscopic scale definitely have to be avoided.

Determining the crystal polarity of GaN thin films via the formation of a 2DEG at an epitaxial AlGaN barrier is a reliable, but rather indirect method. On the other hand, the apparently easier polarity analysis using surface morphology is not very reliable and sometimes even misleading. Thus, a case of mixed polarity is difficult to distinguish from a Ga-polar epilayer, as shown in Fig. 4. Moreover, the difference between N and Ga-face morphology is just opposite in bulk GaN samples, where the Ga-face is rough, whereas the N-face is smooth [9]. Direct structural analysis methods using X-ray analysis or electron diffraction have been successfully employed, but require a significant sample preparation effort and/or highly specialized equipment [9–11]. So far, the easiest albeit destructive way of polarity determination for epitaxial III-nitride films is based on the different chemical properties of both kind of surfaces: whereas surfaces with Ga-polarity are resistant to most chemical etchants except at structural defects, N-polar surfaces are easily etched in KOH/H_2O solutions or KOH/NaOH eutectics [9, 12]. These characteristic chemical properties of Ga- versus N-polar III-nitride surfaces are also important for a better understanding of surface adsorption processes or Schottky contact formation, which recently have been found to be quite different for the two crystal polarities [13, 14].

Lateral Polarity Heterostructures The possibility to locally set the lattice polarity by using a thin AlN buffer layer in conjunction with MBE growth on sapphire substrates has opened up an entirely new degree of freedom for the design of novel device structures in the III-nitrides. These lateral polarity heterostructures (LPHs) are interesting, both, as far as basic physics as well as possible applications are concerned, but their structural and electronic properties still remain to be determined in detail.

Figure 5 schematically outlines the basic technological steps involved in the preparation of LPHs [15]. First, a thin AlN buffer layer with a thickness of 5 nm or more is deposited on the sapphire substrate. This nucleation layer is then laterally structured

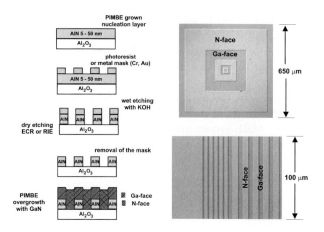

Fig. 5 (colour). Preparation of lateral polarity heterostructures (left part, see text for details), and optical micrographs of such structures, showing adjacent regions of N-face and Ga-face material (right part)

by conventional lithography in conjunction with wet chemical or reactive ion etching. Finally, the substrate is reinserted into the MBE machine and overgrown with GaN and AlGaN epilayers. The optical micrographs on the right-hand side of Fig. 5 clearly can distinguish the resulting LPH structures. This is partly due to the different surface morphologies of different crystal polarities as shown in Fig. 4. In addition, the MBE growth rates for N and Ga-polar surfaces are slightly different, giving rise to further contrast in optical microscopy.

We finally turn to the electronic properties of LPHs, which in principle should be governed by the different arrangement of bulk and surface charges (cf. Fig. 1) and the density of structural defects associated with the inversion domain boundaries. In the ideal case of defect free IDBs [3] and no surface charge compensation, one would expect a lateral band alignment as depicted in Fig. 6. For N-face material, the positive polarization-induced fixed charges give rise to an electron accumulation layer at the free surface, fixing the Fermi level close to the conduction band edge. The inverse situation with E_F close to E_V should occur for Ga-face material. For practical purposes, however, these extreme cases will be modified by adsorbed ions at the sample surface and charged defects at the IDB, both of which will have a strong but yet unknown influence on the details of the complex potential landscape perpendicular or parallel to the LPH growth direction. Anyhow, we expect electronic transport across inversion domain boundaries to be quite complex, and indeed a rectifying behaviour of IDBs in

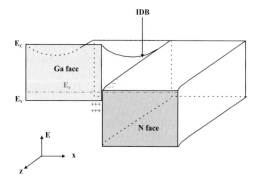

Fig. 6 (colour). Idealized band diagram of the local potential surface in the vicinity of a lateral polarity inversion domain boundary, IDB. Coordinates x and z denote directions perpendicular and parallel to the growth direction, respectively. E_C, E_F, and E_V are the conduction band edge, the Fermi level, and the valence band edge, respectively. The lateral band edge shift from Ga-face to N-face should occur on the scale of several μm, unless the local charge density is increased by a high density of IDB-related structural defects

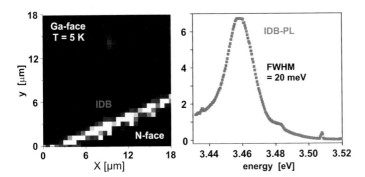

Fig. 7 (colour). Spatially resolved low temperature (5 K) photo-luminescence of an inversion domain boundary between adjacent N and Ga-face regions. Radiative recom-bination occurs predomi-nantly at the IDB (left part), with an energy close to that of the N or Ga-face regions (right part)

lateral polarity heterostructures has been observed recently [16]. A more detailed study is currently under way.

A little more is already known about the optical properties of IDBs in lateral polarity heterostructures, due to a recent investigation of spatially resolved luminescence in such structures [17]. As shown in Fig. 7, the boundary between two adjacent domains with different polarity is a very efficient radiative recombination center, with luminescence intensities exceeding those of the "bulk" Ga-face or N-face regions by more than an order of magnitude. The IDB-related luminescence peaks at 3.46 eV, very close to the peak at 3.49 eV of the Ga-face material. This suggests that, indeed, the IDBs are characterized by a defect free local reconstruction of the inverted wurtzite crystals (e.g. such as shown in Fig. 1) and, moreover, that excitons are attracted towards the IDB because of local strain and/or electric fields. Again, a more detailed understanding has to wait for future studies, but the properties of LPHs seen so far appear quite promising.

Conclusions and Outlook We have reviewed some basic properties of group III-nitrides related to the crystal polarity of the wurtzite structure. Whereas the important consequences of the polarization-induced effects e.g. for the formation of 2DEGs at heterointerfaces or for radiative recombination in quantum wells are now widely recognized, the specific use of lattice polarity as a new degree of freedom is just beginning to be explored. Using MBE deposition in conjunction with a lithographically structured AlN buffer layer, lateral polarity heterostructures (LPHs) with interesting new electronic and optical properties can be prepared on sapphire substrates. In particular, the inversion domain boundaries in such LPHs seem to be free of structural defects, which is a major prerequisite for the use of LPHs in novel devices. As an example for such a device, we show in Fig. 8 a lateral tunnel diode, which uses the different locations of the spontaneously formed 2DEGs at an AlGaN barrier in N-face versus Ga-face samples for the definition of a one-dimensional tunnel junction between two 2DEGs. Also, the strong radiative recombination observed at inversion domain boundaries of LPHs can be of interest for novel light emitting devices. Thus, we believe that it may be worthwhile to have a closer look at the electronic and optical properties of polarity

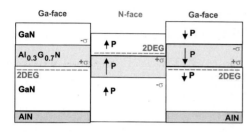
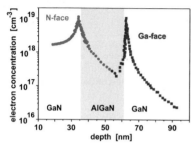

Fig. 8 (colour). Schematic outline of a simple device making use of a lateral polarity heterostructure. 2DEGs are formed at opposite interfaces of a GaN/AlGaN/GaN heterostructure, depending on the lattice polarity, which in turn is determined by a laterally structured AlN buffer layer. The inversion domain boundary thus defines a 1D tunnel junction with a barrier determined by MBE overgrowth only, without the need for postgrowth processing. The lower part of the figure shows experimentally determined electron concentrations of the 2DEGs in the two regions with opposite lattice polarity

heterostructures, and to extend such investigations also beyond the GaN/AlGaN heterosystem. If successful, what started out as "playing with polarity" may eventually develop into serious business.

Acknowledgement Part of this work was supported by the Deutsche Forschungsgemeinschaft (Schwerpunktprogramm "Gruppe III-Nitride").

References

[1] A. ZORODDU, F. BERNARDINI, P. RUGGERONE, and V. FIORENTINI, Phys. Rev. B **64**, 045208 (2001).
[2] O. AMBACHER et al., J. Appl. Phys. **85**, 3222 (1999).
[3] J. E. NORTHRUP, J. NEUGEBAUER, and L. T. ROMANO, Phys. Rev. Lett. **77**, 103 (1996).
[4] H. TEISSEYRE et al., J. Appl. Phys. **76**, 2429 (1994).
[5] M. K. KELLY et al., Jpn. J. Appl. Phys. **38**, L217 (1999).
[6] E. S. HELLMAN, MRS Internet J. Nitride Semicond. Res. **3**, 11 (1998).
[7] R. DIMITROV et al., J. Appl. Phys. **87**, 3375 (2000).
[8] R. DI FELICE and J. E. NORTHRUP, Appl. Phys. Lett. **73**, 936 (1998).
[9] J. L. ROUVIERE et al., Appl. Phys. Lett. **73**, 668 (1998).
[10] F. A. PONCE, C. G. VAN DE WALLE, and J. E. NORTHRUP, Phys. Rev. B **53**, 7473 (1996).
[11] A. KAZIMIROV et al., J. Appl. Phys. **84**, 1703 (1998).
[12] J. L. WEYHER et al., J. Cryst. Growth **182**, 17 (1997).
[13] M. EICKHOFF et al., Proc. 4th Internat. Conf. Nitride Semiconductors, July 16–20, 2001, Denver (USA); phys. stat. sol. (b) **228**, 519 (2001).
[14] U. KARRER, PhD Thesis, Technische Universität München, 2001, to be published.
[15] R. DIMITROV et al., Mater. Res. Soc. Symp. **622**, T4.6.1 (2000).
[16] C. R. MISKYS et al, to be published.
[17] P. J. SCHUCK et al., Appl. Phys. Lett. **79**, 952 (2001).

Investigation of Defects and Polarity in GaN Using Hot Wet Etching, Atomic Force and Transmission Electron Microscopy and Convergent Beam Electron Diffraction

P. Visconti (a, b, c), D. Huang (a), M. A. Reshchikov (a), F. Yun (a),
T. King (a), A. A. Baski (a), R. Cingolani (b), C. W. Litton (d),
J. Jasinski (e), Z. Liliental-Weber (e), and H. Morkoç (a)

(a) Department of Electrical Engineering and Physics Department,
Virginia Commonwealth University, Richmond, VA 23284, USA

(b) NNL – National Nanotechnology Laboratory of INFM,
Unit of Lecce and Department of Innovation Engineering, University of Lecce,
Via per Arnesano, I-73100, Lecce, Italy

(c) Istituto per lo Studio di Nuovi Materiali per l'Elettronica, CNR-IME,
Via per Arnesano, I-73100, Lecce, Italy

(d) Air Force Research Laboratory, Wright Patterson AFB, OH 45433, USA

(e) Lawrence Berkeley National Laboratory, Berkeley, CA 94720, USA

(Received July 2, 2001; accepted July 18, 2001)

Subject classification: 61.72.Ff; 68.37.Lp; 68.37.Ps; 68.55.Ln; S7.14

Availability of reliable and quick methods to investigate defects and polarity in GaN films is of great interest. We have used photo-electrochemical (PEC) and hot wet etching to determine the defect density. We found the density of whiskers formed by the PEC process to be similar to the density of hexagonal pits formed by wet etching and to the dislocation density obtained by transmission electron microscopy (TEM). Hot wet etching was used also to investigate the polarity of MBE-grown GaN films together with convergent beam electron diffraction (CBED) and atomic force microscopy (AFM). We have found that hot H_3PO_4 etches N-polarity GaN films very quickly resulting in the complete removal or a drastic change of surface morphology. On the contrary, the acid attacks only the defect sites in Ga-polar films leaving the defect-free GaN intact and the morphology unchanged. The polarity assignments, confirmed by CBED experiments, were related to the as-grown surface morphology and to the growth conditions.

Introduction Successful fabrication of GaN-based devices depends on the ability to grow epitaxial films on sapphire with low defect density. The poor match in both lattice parameter and thermal expansion coefficient results in a high dislocation density (DD). Wurtzite GaN is a polar material that has two different planes along its c-axis. The (0001) plane is the Ga-terminated face while the ($000\bar{1}$) plane is the N-face [1]. It is known that the surface and bulk properties of GaN layers depend greatly on the polarity [2]. Therefore the availability of reliable and quick methods to determine the defect density and the polarity of GaN films is of great importance.

We have investigated defects in GaN films grown by MBE using PEC method [3] and hot wet etching [4]. We found the whisker density to be similar to the etch pit

---

[1]) Corresponding author; Tel.: +39-0832-320231; Fax: +39-0832-326351;
e-mail: paolo.visconti@unile.it

densities (EPD) for samples etched under precise conditions. TEM studies confirmed the DDs obtained by etching which increased our confidence in the validity of the methods used. Additionally, we have demonstrated that hot H_3PO_4 can be used to determine the polarity of GaN films. Already, NaOH and KOH based aqueous solutions at different temperatures have been demonstrated to attack N-polar surface whereas the morphology of Ga-polar GaN films remains unchanged [5–7]. Similarly, we found that a few seconds of etching in hot H_3PO_4 leads to the complete removal or a drastic change in the morphology of N-polar GaN films. On the contrary, a defect free Ga-polar film is hardly affected by the hot acid. Only hexagonal pits associated with the surface defects are formed after several minutes of etching. The polarity assignments, confirmed by CBED experiments, were related to the as-grown morphology and to the growth conditions. We found that the GaN films grown by MBE on high temperature (HT) AlN ($> 890\,°C$) and GaN (770–900 °C) buffer layers show Ga- and N-polarity, respectively [8].

Experimental Details The GaN samples consisted of unintentionally n-doped GaN layers grown by MBE on c-plane of sapphire using radio-frequency activated N. Some samples utilized GaN buffer layers grown at 800 °C (HT). Others utilized AlN buffer layers grown at 890–930 °C (HT). Following the buffer layers, ≈1 μm thick GaN layers were grown at a temperature between 720 and 850 °C with growth rates in the range of 0.3–1 μm/h under N-limited (Ga-rich) conditions. PEC etching was carried out in a standard electrochemical cell at room temperature using a 0.02M KOH solution and a He–Cd laser. Further details of the PEC experimental set-up can be found elsewhere [9, 10]. The morphology of samples etched by PEC and wet etching was investigated using AFM and scanning electron microscopy (SEM). Additionally, some samples were observed by TEM to estimate the DD. The polarity was determined by hot H_3PO_4 etching, CBED experiments and AFM investigation of their as-grown morphology. Additionally, in-situ RHEED patterns during the MBE growth supported the polarity assignments.

Results and Discussion PEC etching has been demonstrated to be suitable for DD estimation in n-type GaN films. Freestanding whiskers were obtained by selectively etching GaN between dislocation sites under precise etching conditions. With XTEM analysis, the presence of the dislocations in the whiskers was demonstrated and the whisker density was found close to the DD [3]. We used slightly carrier-limited conditions to etch Ga-polar GaN selectively, leaving threading vertical wires on the surface.

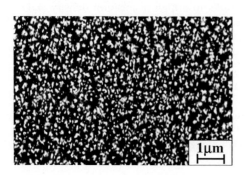

Fig. 1. Plan-view SEM image of the PEC etched sample. The density of the whisker-like features (white dots in the image) is ≈$1–2 \times 10^9$ cm^{-2}

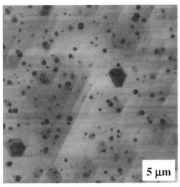 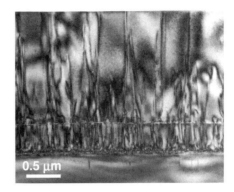

Fig. 2   Fig. 3

Fig. 2. AFM image of the Ga-polar GaN surface morphology after etching in H_3PO_4 for 5 min at 160 °C. The EPD is $\approx 1 \times 10^9$ cm^{-2}. The vertical scale ranges from 0 to 10 nm

Fig. 3. Cross-sectional TEM image of a Ga-polar GaN layer. The total DD, near the top surface, is about 4×10^9 cm^{-2} (edge dislocations about 95% of the total)

The density of features (white dots in the SEM image of Fig. 1) formed by the PEC process is $\approx 2 \times 10^9$ cm^{-2}, the same value obtained from the AFM investigation.

Wet etching is a suitable technique to determine in Ga-polar GaN the density of defects propagating to the surface. We have used hot H_3PO_4 as a defect etchant, which produces hexagonal pits at the defect sites. During etching, a careful balance must be struck to ensure that every defect is delineated, but not over-etched to cause merging which would lead to an underestimation of the defect density. H_3PO_4 for 5 min at 160 °C was used to determine the defect density in Ga-polar GaN films grown by MBE. The AFM image of Fig. 2 shows the etched surface with EPD of $\approx 1 \times 10^9$ cm^{-2}. TEM analyses were carried out in order to determine the effective DD. From XTEM observation (Fig. 3), the threading dislocations were observed starting from the buffer/GaN interface and propagating within the GaN layer. The total DD, near the surface, including screw ($<1 \times 10^7$ cm^{-2}), mixed ($\approx 1 \times 10^8$ cm^{-2}) and edge dislocations (about 95% of the total) is $\approx 4 \times 10^9$ cm^{-2} close to the values obtained by defect revealing etches.

It is known that the electrical and optical properties of GaN layers depend greatly on the polarity. We have investigated the polarity of GaN films grown by MBE on HT ($T > 770$ °C) GaN buffer layers or HT ($T \approx 900$ °C) AlN buffer layers in order to obtain N- and Ga-polar GaN films, respectively. The polarity was determined by etching in hot H_3PO_4 and AFM imaging of as-grown surface morphologies. The polarity assignments were supported by in-situ RHEED patterns and confirmed by CBED experiments. As reported in the literature, the surface of a Ga-polar film is either very flat or shows stepped terraces. The surface of a N-polar film grown by MBE shows tall columns or terraces separated by deep troughs. Similar to NaOH- and KOH-based etchings, we have found that hot H_3PO_4 etches N-polar GaN films very quickly resulting in the complete removal or a drastic change of the morphology as revealed by AFM or optical microscopy. The etching rate is from 0.3 to 0.7 µm/min. On the contrary, a defect free Ga-polar film is hardly affected by the hot acid. As shown in Fig. 2, only hexagonal pits associated with the surface defects are formed after several minutes of etching of Ga-polar films.

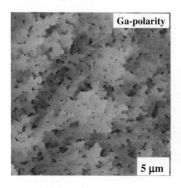
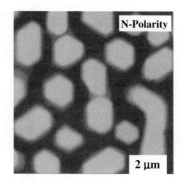

Fig. 4

Fig. 5

Fig. 4. AFM image of Ga-polar GaN sample. The rms roughness is ≈1.3 nm. The vertical scale ranges from 0 to 20 nm

Fig. 5. AFM image of N-polar GaN sample. The rms roughness is ≈20 nm. The vertical scale ranges from 0 to 80 nm

The surface morphology of an as-grown Ga-polar GaN film is shown in Fig. 4. A HT AlN buffer layer leads to Ga-polar film with a smooth, but pitted surface morphology with a rms roughness of ≈1.3 nm. According to the literature, the Ga-polar film shows a 2×2 RHEED pattern upon cool-down at temperatures between 280 and 650 °C. Shown in Fig. 5 is the typical surface morphology of a N-polar film obtained using a HT GaN buffer layer. The rough morphology presents 50–100 nm high non-coalesced columns with a rms roughness as high as 20 nm. The RHEED pattern upon cooling shows only the bulk 1×1 structure. Dipping of the N-polar GaN layer in hot (160 °C) H_3PO_4 for only 20 s results in a drastic change of the morphology. Longer dipping in hot H_3PO_4 for just few minutes instead results in the complete removal of the film, obtaining the clean surface of sapphire substrate by AFM imaging. In the AFM image of Fig. 6, we show the surface morphology of a N-polar film after etching in hot H_3PO_4 for 20 s. Etching of the c-plane with an etch rate of ≈0.5 μm/min produces 300–400 nm high features on the sapphire substrate (rms roughness ≈ 150 nm).

In order to confirm the polarity assignments, we carried out CBED experiments. For a GaN layer grown on HT AlN buffer layer, the CBED study confirmed the Ga-polarity, whereas from XTEM study the density of inversion domains (ID) was estimated $\approx 1 \times 10^7$ cm^{-2}.

For a GaN layer grown on HT GaN buffer layer, the CBED study again confirmed the assigned N-polarity, whereas from TEM study the density of IDs was estimated $\approx 1 \times 10^{11}$ cm^{-2}.

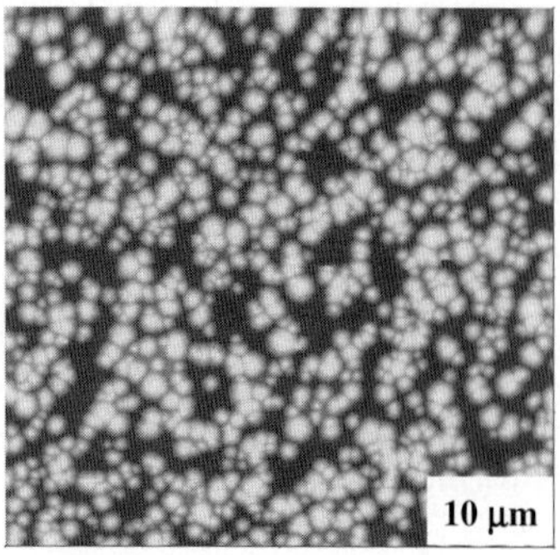

Fig. 6. AFM image of N-polar GaN morphology after etching in H_3PO_4 for 20 s at 160 °C. The rms roughness is ≈150 nm. The vertical scale ranges from 0 to 1200 nm

Conclusion PEC and hot wet etching experiments were carried out to determine the defect density in GaN films. The density of whiskers formed by the PEC process was similar to the density of pits formed by wet etching and close to the DD obtained by TEM. Hot wet etching has been used also to determine the polarity of MBE-grown GaN layers. We found that few seconds of etching in H_3PO_4 produce a drastic change of the morphology whereas few minutes result in the complete removal of N-polar GaN film. On the contrary, the hot acid attacks only the defect sites in Ga-polar films producing pits, but leaving the defect-free GaN intact and the morphology unchanged after several minutes of etching.

Acknowledgements The authors would like to thank Prof. K.J. Wynne and Ms. J. Uilk for the use of large area AFM. The VCU portion was funded by grants from AFOSR (Dr. G.L. Witt), NSF (Drs. L. Hess and G. Pomrenke), and ONR (Drs. C.E. C. Wood and Y.S. Park). LBNL portion of this work was supported in part by Air Force Office of Scientific Research, through the U.S. Department of Energy under Order No.AFOSR-ISSA-00-0011. J.J. and Z.L.W. thank NCEM at Berkeley for the use of TEM facility.

References

[1] H. MORKOÇ, Nitride Semiconductors and Devices, Springer Verlag, Berlin/Heidelberg/New York 1999.
[2] E. S. HELLMAN, MRS Internet J. Nitride Semicond. Res. **3**, 11 (1998).
[3] C. YOUTSEY, L. T. ROMANO, R. J. MOLNAR, and I. ADESIDA, Appl. Phys. Lett. **74**, 3537 (1999).
[4] P. VISCONTI, K. M. JONES, M. A. RESHCHIKOV, R. CINGOLANI, H. MORKOÇ, and R.J. MOLNAR, Appl. Phys. Lett. **77**, 3532 (2000).
[5] M. SEELMANN-EGGEBERT, J. L. WEYHER, H. OBLOH, H. ZIMMERMANN, A. RAR, and S. POROWSKI, Appl. Phys. Lett. **71**, 2635 (1997).
[6] A. R. SMITH, R. M. FEENSTRA, D. W. GREVE, M.-S. SHIN, M. SKOWRONSKI, J. NEUGEBAUER, and J. E. NORTHRUP, Appl. Phys. Lett. **72**, 2114 (1998).
[7] J.L. ROUVIERE, J.L. WEYHER, M. SEELMANN-EGGEBERT, and S. POROWSKI, Appl. Phys. Lett. **73**, 668 (1998).
[8] D. HUANG, P. VISCONTI, K. M. JONES, M. A. RESHCHIKOV, F. YUN, A. A. BASKI, T. KING, and H. MORKOÇ, Appl. Phys. Lett. **78**, 4145 (2001).
[9] P. VISCONTI, M. A. RESHCHIKOV, K. M. JONES, F. YUN, D. F. WANG, R. CINGOLANI, and H. MORKOÇ, Proc. MRS Fall Meeting, Boston, Nov. 2000, Vol. 639, no. G3.14.
[10] P. VISCONTI, M. A. RESHCHIKOV, K. M. JONES, D. F. WANG, R. CINGOLANI, H. MORKOÇ, R. J. MOLNAR, and D. J. SMITH, J. Vac. Sci. Technol. B **19**, 1328 (2001).

Wetting Behaviour of GaN Surfaces with Ga- or N-Face Polarity

M. Eickhoff[1]) (a), R. Neuberger (b), G. Steinhoff (a), O. Ambacher (a), G. Müller (b), and M. Stutzmann (a)

(a) Walter Schottky Institute, Technical University Munich, Am Coulombwall, D-85748 Garching, Germany

(b) EADS-Germany GmbH, D-81663 München, Germany

(Received July 31, 2001; accepted August 2, 2001)

Subject classification: 68.08.Bc; 68.65.Ac; 73.40.Kp; 79.60.Bm; S7.14

The wetting behaviour of GaN surfaces with N-face and Ga-face polarity and the influence of different surface treatments is studied by measuring the wetting angle of highly purified water by microscopic imaging. We found that wet thermal oxidation of the surface leads to a decreased wetting angle indicating an improved wetting behaviour. The presence of Al in AlN or AlGaN leads to a further reduction of the wetting angle, which is attributed to the presence of Al_2O_3 on the surface. In addition the comparison of Ga- and N-face material revealed a lower wetting angle for all N-face samples. XPS analysis showed the enhanced formation of native oxide on the surface with N-face polarity.

Introduction Due to their outstanding material properties, wide bandgap semiconductors like silicon carbide or diamond have been demonstrated to bear a great potential for sensor applications at high temperatures and under harsh environmental conditions, as for example in the case of gas sensitive devices for combustion monitoring or high temperature pressure sensors [1–3].

Less widely recognized up to now are the possible sensor applications of HEMT devices based on heterostructures of pyroelectric GaN/AlGaN layers. In those devices the carrier density of the confined two-dimensional electron gas is highly sensitive towards manipulation of the surface charge and of the mechanical strain in the sample. These effects have been successfully exploited for the detection of ion-flux [5], mechanical strain and gas detection [6, 7] and the detection of polar fluids [8]. In the latter case the current–voltage characteristics of a HEMT device is influenced by electronic interaction of the polar device surface with the molecular dipole moments in the fluid which is covering the non-metallized, non-passivated gate area. Figure 1 shows the variations of the current–voltage characteristics of a GaN-face AlGaN/GaN heterostructure due to exposure of the gate area to different polar fluids [4]. A decrease of the saturation current at constant source–drain voltage compared to the case of exposure to air is observed for all investigated polar fluids. The amplitude of this change is determined by the dipole moment of the fluid, the geometric structure of the molecules and by the wetting behaviour of the fluid on the gate area. The time dependence of the channel current, especially during the transition from a homogeneously covered gate to the formation of droplets or the evaporation of the fluid from the gate area is also strongly influenced by the latter effect.

---

[1]) Corresponding author; Phone: +49 89 289 12889; Fax: +49 89 289 12737; e-mail: eickhoff@wsi.tum.de

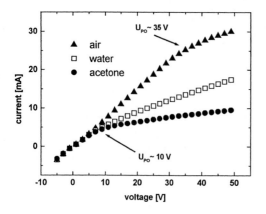

Fig. 1. Current–voltage characteristics of AlGaN/GaN HEMT structure upon exposure of the gate area to different polar fluids [4]

Therefore the analysis of the wetting behaviour of GaN surfaces and its manipulation is inevitable for understanding and optimising the sensor performance. The present work shows a comparative analysis of the wetting behaviour for N-face and Ga-face GaN and the influence of different surface manipulations by measurements of the wetting angle of highly purified water. The observed difference between both types of polarity with respect to the wetting behaviour as well as the influence of wet thermal oxidation, which turns out to significantly reduce the wetting angle, was analysed by means of X-ray photoelectron spectroscopy (XPS).

Wetting Angle Measurements The GaN and AlGaN layers were deposited by plasma induced molecular beam epitaxy on c-plane sapphire substrates as described in detail elsewhere [8]. Whereas deposition on the bare sapphire substrate results in GaN layers with N-face polarity, Ga-face polarity is achieved by initial deposition of an AlN nucleation layer with a thickness of 10 nm. The wetting behaviour was measured by microscopic imaging of the contact angle of a 5 µl drop of highly purified water on the surface. Figure 2 shows the measured contact angle for the investigated GaN, AlN and AlGaN layers after different surface treatments.

Each surface treatment leads to a decreasing wetting angle corresponding to an improvement of the wetting behaviour in comparison to the samples stored in air. As the investigated surface treatments result also in cleaning of the surface (prior to the thermal oxidation step a 30 s dip in 10% hydrofluoric acid (HF) was performed) we attribute that effect to the presence of adsorbates on the surface of the samples "stored in air". This is confirmed by the fact that after successful reduction of the wetting angle it

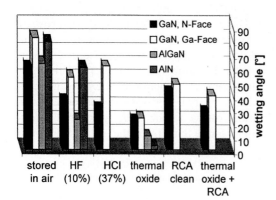

Fig. 2. Wetting angle of highly purified water on GaN and $Al_{0.28}Ga_{0.72}N$ surfaces after different surface treatments (HF, HCl: 60 s; thermal oxidation: 2 h at 800 °C; RCA-clean: 60 min)

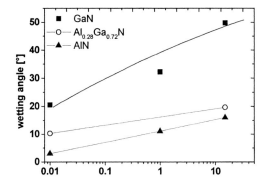

Fig. 3. Increase of the wetting angle with increasing time after wet thermal oxidation for N-face GaN, $Al_{0.28}Ga_{0.72}N$ and AlN

increases again when the sample is stored in air again with no further chemical treatments, which is shown in Fig. 3 for the case after wet thermal oxidation. We also attribute this effect to a coverage of the surface by adsorbed species. Any kind of oxidation of the surface leads to a significant reduction of the wetting angle. We observed this for the case of H_2O_2 treatment as well as thermal oxidation. In case of the latter treatment, which gave rise to the lowest contact angles, it is shown that the incorporation of Al into the GaN layer leads to a further reduction, whereas pure AlN shows even smaller wetting angles. These effects can be attributed to the formation of Al_2O_3 at the surface, which is known to result in a hydrophilic behaviour. Furthermore a systematic difference between GaN of N-face and Ga-face polarity is observed. For all surface treatments the wetting angle on N-face GaN is significantly lower than in the case of Ga-face. As the surface roughness of the investigated samples was similar this difference can only be attributed to inherent material properties. To study this we investigated both types of samples by XPS analysis.

XPS Analysis Taking into account that the formation of an oxidic layer on the surface is beneficial for the wetting behaviour, the smaller wetting angle in the case of N-face GaN could be caused by preferential formation of a native oxide. Therefore the oxygen-related peak structures in the corresponding XPS spectra were analysed and the difference was compared to the influence of wet thermal oxidation. XPS analysis was performed using Mg K_α excitation (1253.6 eV) and a hemispherical electron energy analyser. A comparison of the Ga-face and N-face samples after storage in air revealed a peak shift of ≈0.3 eV to higher binding energies for the most surface sensitive Ga 2p core level in the case of N-face material, whereas the Ga 3d core level peak was broadened and an increase of the high energy part was detected. The oxygen 1s core level also showed a broadening, however due to increasing intensity of the low binding energy part [9]. The most striking difference occurred in analysing the GaLMM Auger levels, as shown in Fig. 4. The sample with N-face polarity shows an additional peak forming a high binding energy shoulder as shown in Fig. 4a. Etching the sample in HF for 30 s leads to the disappearance of this peak (Fig. 4b), giving rise to the conclusion that the presence of this peak is caused by the enhanced formation of native oxide on the N-face surface. As shown in Fig. 4b oxide formation by wet thermal oxidation results in the same spectral characteristics as in the case of the N-face sample before etching but to a much higher extent: The high energy part is strongly enhanced. In addition thermal oxidation significantly reduces the intensity of the low binding energy peak. The Ga 2p core level is shifted by 0.5 eV to higher binding energies. Also, wet thermal oxidation results in an exchange of nitrogen by oxygen on the surface. The

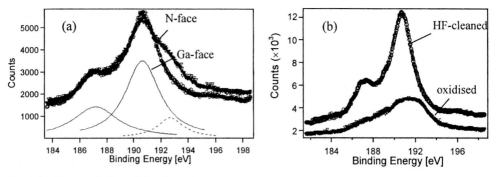

Fig. 4. a) Deconvolution of the GaLMM peak structure showing an additional high binding energy peak for the N-face polarity sample after storage in air. b) GaLMM peak structure for N-face GaN after etching in HF and after wet thermal oxidation

N 1s core level peak nearly disappears after the oxidation process, whereas a drastic increase of the O 1s core level is observed, accompanied by an increased contribution of the low energy part (not shown). This gives evidence for the conclusion that oxygen, which is present on the surface before oxidation due to adsorption of different species exhibits different bonding states, whereas during thermal oxidation preferentially Ga_2O_3 is formed, characterized by an O 1s core level of 532.3 eV [10].

Conclusions Oxidation was shown to improve the wetting behaviour of N-face and Ga-face GaN drastically. The incorporation of Al is also beneficial due to the formation of Al_2O_3 on the surface. A systematically lower wetting angle on N-face GaN surfaces in comparison to Ga-face surfaces is also observed. In addition to that, surfaces with N-face polarity show a higher amount of oxygen on the surface forming a Ga_xO_y-like layer after storage in air, as proved by XPS analysis. This observation is in agreement with theoretical predictions [11]. This effect is shown to a much higher extent if the surface is subject to thermal oxidation, proposing the more oxide-like behaviour of the N-face surface to be beneficial for the wetting behaviour.

References

[1] A. LLOYD SPETZ, A. BARANZAHI, P. TOBIAS, and I. LUNDSTRÖM, phys. stat. sol. (a) **162**, 493 (1997).
[2] J. SHOR, D. GOLDSTEIN, and A. D. KURTZ, IEEE Trans. Electron Devices **40**, 1093 (1993).
[3] M. EICKHOFF, H. MÖLLER, G. KRÖTZ, J. V. BERG, and R. ZIERMANN, Sens. Actuators A **74**, 56 (1999).
[4] R. NEUBERGER, G. MÜLLER, O. AMBACHER, and M. STUTZMANN, phys. stat. sol. (a) **185**, 85 (2001).
[5] R. NEUBERGER, O. AMBACHER, G. MÜLLER, and M. STUTZMANN, phys. stat. sol. (a) **183**, R10 (2001).
[6] M. EICKHOFF, G. KRÖTZ, O. AMBACHER, and M. STUTZMANN, J. Appl. Phys., in print (2001).
[7] J. SCHALWIG, G. MÜLLER, M. EICKHOFF, O. AMBACHER, and M. STUTZMANN, Proc. E-MRS Spring Meeting, Strasbourg, June 2001.
[8] O. AMBACHER, J. Phys. D **31**, 2653 (1998).
[9] S. D. WOLTER, B. P. LUTHER, D. L. WALTEMYER, C. ÖNNEBY, and S. E. MOHNEY, Appl. Phys. Lett. **70**, 2156 (1997).
[10] K. PRABHAKARAN, T. G. ANDERSSON, and K. NOZAWA, Appl. Phys. Lett. **69**, 3212 (1996).
[11] T. K. ZYWIETZ, J. NEUGEBAUER, and M. SCHEFFLER, Appl. Phys. Lett. **74**, 1695 (1999).

Kinetic Process of Polarity Selection in GaN Growth by RF-MBE

K. Xu[1] (a), N. Yano (b), A. W. Jia (a, b), A. Yoshikawa (a, b), and K. Takahashi (c)

(a) Center for Frontier Electronics and Photonics, Chiba University,
1-33 Yayoi-cho, Inage-ku, Chiba 263-8522, Japan

(b) Department of Electronic & Mechanical Engineering, Chiba University, Japan

(c) Department of Media Science, Teikyo University of Science & Technology, Japan

(Received July 3, 2001; accepted August 4, 2001)

Subject classification: 68.35.Fx, 68.49.Sf; 81.15.Hi; S7.14

Extensively nitridated and non-nitridated sapphire substrate, Al insertion layer and AlN intermediate layer were used as the platforms to investigate polarity selection processes of GaN grown by rf-MBE, aimed at giving a comprehensive understanding to the issues of GaN polarity. GaN growth was started on these platforms with different surface stoichiometry by controlling the order of source supplies. The results showed that GaN tended to grow with Ga polarity which was kinetically favorable on thermally cleaned sapphire substrate and Al covered surfaces. N polarity could be reversed to Ga polarity by Al insertion layers or AlN intermediate layer. It is suggested that the polarity conversion of GaN by AlN or Al insertion layers relies on the fact that they provide an Al platform on which the subsequent epilayer prefers to grow with Ga polarity.

Introduction GaN polarity is of major importance in the epilayer growth and the resulting surface morphology and intereface quality, as well as the device performance [1, 2]. Ga polarity was favored both in the sense of the film growth and the resulting electrical and optical properties. However, GaN grown by RF-plasma assisted MBE (rf-MBE) was reported to predominantly exhibit N polarity. High temperature AlN buffer layer and intermediate layer has been used to realize the Ga-polar epilayer in rf-MBE growth [3, 4]. However, the related mechanisms were not understood well so far. There is a lack of general idea how the polarity is determined, how the polarity can be changed, both from N polariy to Ga polarity and from Ga polarity to N polarity. The present work is devoted to give a comprehensive understanding to the polarity selection process in GaN growth by rf-MBE through investigating the polarities of GaN grown on sapphire substrate, Al layers, and AlN intermediate layers.

Experimental GaN was grown by rf-MBE. Active N was supplied from Applied EPI Uni-bulb rf plasma cell. After thermal cleaning at 890 °C for 30 min, sapphire substrate was nitridated for 40 min at the substrate temperature of 200 °C with rf power of 500 W, and N_2 flow rate of 1.2 sccm. Then, a GaN buffer layer was grown at 650 °C with different surface stoichiometry by timing Ga and N supplies. For the GaN buffer layer growth, the N plasma condition was 200 W rf power and 0.4 sccm N_2 flow rate, and Ga deposition flux was 0.5 Å/s. The epilayer growth temperature was 820 °C, with

---

[1]) Corresponding author; Phone: +81 43 290 3988; Fax: +81 43 290 3993;
e-mail: xuke@vbl.chiba-u.ac.jp

the N plasma condition of 300 W rf power and 0.8 sccm N_2 flow rate. The growth rate was about 0.7 µm/h. Three groups of experiments were carried out to understand the GaN polarity selection process:

Group 1, GaN epilayer was grown on sapphire substrate with and without nitridation, and the growth was started with different surface stoichiometry, i.e. Ga rich or N rich by timing the order of source supply.

Group 2, GaN epilayer was grown on the nitridated sapphire substrate, and an Al layer of 1 or 2 ML was introduced during the epilayer growth at 820 °C. Al deposition was 0.5 Å/s.

Group 3, AlN intermediate layer was introduced during GaN epilayer growth. For AlN intermediate layer growth, the Al beam flux was 0.5 Å/s, N plasma condition was the same as that for GaN epilayer growth stated above. But AlN layer was grown by two methods. One is starting the growth by closing the Ga cell shutter and opening Al cell shutter, thus AlN was grown in N rich condition. The other is adjusting the Al/N ratio to above unity by modulating the N shutter operation periodically open and close.

Polarities of GaN films were characterized by using RHEED and Coaxial Impact Collision Ion Scattering Spectroscopy, i.e. CAICISS (Shimadzu, TALIS-9700). The polarity of GaN as well as the Ga/N polar ratio was determined by comparing the experimental spectra and the theoretically simulated ones [5, 6].

Results and Discussions CAICISS spectra of GaN grown on nitridated and non-nitridated sapphire substrate are shown in Figs. 1a and b. By comparing with the simulated results, we knew the film grown on extensively nitridated sapphire substrate was not 100% N polarity, and the estimated Ga polarity was below 5%; while the GaN epilayer grown on non-nitridated sapphire substrate was well-defined Ga polarity. This result showed that, even that the chemical activity of N supplied from rf plasma was very high, Ga polarity was favourable on thermally-cleaned sapphire surface regardless of source supply order. Therefore, we could not simply refer to rf-MBE grown GaN as N polarity.

Thermally-cleaned sapphire substrate was usually terminated by Al in ultra-high vacuum [7]. In the following, we investigated the effect of Al deposition on GaN polarities.

Figure 2a shows the CAICISS spectrum of a GaN film which was grown on nitridated sapphire substrate but with 1 ML Al insertion during the epilayer growth. The signal is very complicated at low angle. This may be caused by the excess Ga coverage on the real surface. According to the peak shape from 60° to 80° in Fig. 2a, the film was mixed-polarity, and the spectrum is close to the simulated one assuming 30% Ga polarity. For the sample grown with 2 ML Al deposition, the measured CAICISS spectrum is similar to Fig. 2a, but the Ga polarity percentage increased to about 70%. Following this, we tried the 2 ML Al deposition twice during the epitaxy. After 20 nm GaN growth on the first 2 ML Al, the second deposition of 2 ML Al was done. The corresponding CAICISS spectrum is shown in Fig. 2b, it is clear that the polarity of the epilayer was completely changed to Ga one. Thus it has been confirmed that 2 ML Al coverage is necessary and also enough to reverse N polarity to Ga polarity. Logically, if 2 ML is not enough, the polarity will not be changed regardless of the deposition times. The yield of around 30% Ga polarity by 1 ML Al deposition is thought due to 2 ML Al islands formation. The yield of about 70% Ga polarity was attributed to the fact

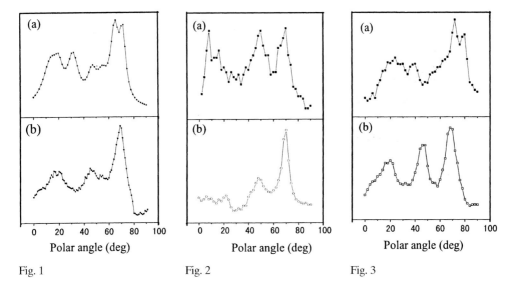

Fig. 1. CAICISS spectra of GaN grown by rf-MBE on a) extensively nitridated and b) non-nitridated sapphire substrate

Fig. 2. CAICISS spectra of GaN, a) 1 ML Al deposition during epitaxy, b) 2 ML Al deposition by twice during epilayer growth

Fig. 3. CAICISS spectra of GaN grown on AlN, a) without N shutter modulation, i.e. under N rich condition, b) with N shutter modulation

that the surface was not fully covered by 2 ML Al uniformly even the mean thickness of the deposited Al layer was 2 ML.

In the third-group experiment, GaN polarity selection on AlN intermediate layer was investigated. N-polarity GaN epilayer was grown at 820 °C with N plasma condition of 300 W rf-power and 0.8 sccm N_2 flow rate, corresponding GaN growth rate was 0.7 µm/h. Under this N plasma condition, Ga flux as high as 2.5 Å/s was needed to realize the pseudo-two-dimensional growth and this was also the upper limit for Ga flux to avoid Ga droplet. The AlN intermediate layer was grown by different methods. As stated above, the Al flux was fixed at 0.5 Å/s for all experiments in this paper. We could switch GaN growth to AlN intermediate layer growth by simply closing the Ga shutter and open Al shutter while keeping the N plasma condition the same. Accordingly, AlN was grown under N rich condition by this method. The CAICISS spectrum of GaN with 20 nm thick AlN intermediate layer grown under N rich condition is shown in Fig. 3a; it is a well-defined N-polarity spectrum. In-situ RHEED monitoring confirmed that 20 nm AlN can completely cover the GaN surface under above growth condition. This result shows that AlN intermediate layer grown under N rich condition will not change the polarity of GaN epilayer.

However, if N shutter was modulated during AlN growth while Al shutter was kept open so that the Al/N ratio was adjusted to above unity, the polarity of the GaN epilayer could be perfectly changed to Ga polar, as shown in Fig. 3b. In this growth method, N shutter operation was modulated as 4 s open and 6 s close. The 6 s closing of N shutter allowed about 1.1 ML Al deposition. We also examined the effect of AlN sur-

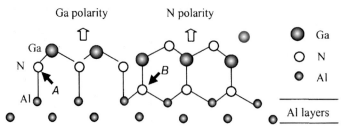

Fig. 4. Polarity selection process of GaN on Al layer

face stoichiometry on GaN polarity. In the case that AlN intermediate layer was grown under nitrogen rich condition, we did excess Al deposition before starting GaN growth on AlN at 820 °C. The Al deposition time was 12 s, corresponding to 2.2 ML coverage. The resulting GaN epilayer was proven to be Ga polarity. The experimental results suggest that excess Al coverage on AlN intermediate layer was crucial to reverse N polarity to Ga polarity by AlN high temperature intermediate layer.

The GaN polarity selection process is schematically shown in Fig. 4. Since the bonding energy between Al and N is higher than that between Ga and N, the polarity of GaN grown on Al covered surface depends on bonding configuration among Al and N when Ga and N species arrive at the surface. If N atom takes position A, Ga polarity will be obtained; if N atoms take position B, N polarity will be obtained. STM characterization and theoretical calculation showed that Ga atoms in GaN surface would form metallic bonds which were only slightly weaker than those formed in bulk Ga [8, 9], coherently, we think Al adlayer on GaN surface may form much stronger metallic bonds than Ga atoms. For N atoms entering position B, it needs the cooperation among three Al atoms, therefore, N polarity is kinetically unfavorable on Al covered surface. However, due to the high ionicity of III-nitrides, the growing epilayer tends to keep its polarity. To reverse Ga polarity to N polarity, much thicker Al layer coverage and stronger N plasma are necessary, which will be reported separately.

Conclusions The polarity selection process of GaN grown on nitridated and non-nitridated sapphire substrate, Al layer, and AlN intermediate layer was investigated. Results showed that polarity of rf-MBE grown GaN depended on nitridation condition of sapphire substrate; GaN tended to grow with Ga polarity which was kinetically favorable on thermally cleaned sapphire substrate and Al covered surfaces. N polarity could be reversed to Ga polarity by Al insertion layers or AlN intermediate layer. It is suggested that the polarity conversion of GaN by AlN or Al relies on the fact that they provide an Al platform on which the subsequent epilayer preferred to grow with Ga polarity.

Acknowledgement This work was partly supported by the "Research for the Future Program", Japan Society for the Promotion of Science (JSPS-RFTF 96R16201).

References

[1] F.A. PONCE, D.P. BOUR, W.T. YONG, M. SAUNDERS, and J.W. STEEDS, Appl. Phys. Lett. **69**, 337 (1996).
[2] R. HELD, G. NOWAK, B.E. ISHUG, S.M. SEUTTER, A. PARKHOMOVSKY, A.M. DABIRAN, P.I. COHEN, I. GRZEGORY, and S. POROWSKI, J. Appl. Phys. **85**, 7697 (1999).

[3] X.Q. Shen, T. Ide, S. H. Cho, M. Shimizu, S. Hara, H. Okumura, S. Sonoda, and S. Shimizu, J. Cryst. Growth **218**, 155 (2000).
[4] A. Kikuchi, T. Yamada, S. Nakamura, K. Kusakabe, D. Sugihara, and K. Kishino, Jpn. J. Appl. Phys. **39**, L330 (2000).
[5] D.H. Lim, K. Xu, Y. Taniyasu, K. Suzuki, S. Arima, B.L. Liu, K. Takahashi, and A. Yoshikawa, IPAP Conf. Ser. **1**, 150 (2000).
[6] K. Xu, N. Yano, A.W. Jia, A. Yoshikawa, and K. Takahashi, 13th Internat. Conf. Crystal Growth, Kyoto (Japan), July 29–August 4, 2001, to be published in J. Cryst. Growth (2002).
[7] D.L. Medlin, K.F. McCarty, R.Q. Hwang, S. E. Guthrie, and M.I. Baskes, Thin Solid Films **299**, 110 (1997).
[8] T. Zywietz, J. Neugebauer, M. Scheffler, J. Northrup, and C.G. Van de Walle, MRS Internet J. Nitride Semicond. Res. **3**, 26 (1998).
[9] A.R. Smith, R.M. Feenstra, D.W. Greve, J. Neugebauer, and J.E. Northrup, Phys. Rev. Lett. **79**, 3934 (1997).

V/III Ratio Dependence of Polarity of GaN Grown on GaAs (111)A-Ga and (111)B-As Surfaces by MOMBE

O. Takahashi (a), T. Nakayama (a), R. Souda (b), and F. Hasegawa (a)

(a) Institute of Applied Physics, University of Tsukuba, Tsukuba, Ibaraki 305-8573, Japan

(b) Advanced Materials Laboratory, National Institute for Materials Science, Tsukuba, Japan

(Received June 19, 2001; accepted August 4, 2001)

Subject classification: 68.55.Jk; 81.15.Hi; S7.14

In order to investigate the influence of the substrate polarity and growth conditions on the polarity, GaN was grown on GaAs (111)A-Ga and B-As surfaces with different V/III ratios by metalorganic molecular beam epitaxy (MOMBE). It was found that for GaN grown on GaAs (111)A-Ga surface polarity was dominant independent of the V/III ratio, but the polarity of GaN grown on GaAs (111)B depends on the V/III ratio; N polarity was dominant for high V/III ratios (N rich) and Ga polarity was dominant for low V/III ratios (Ga rich).

Introduction It is well known that the polarity of hexagonal GaN influences greatly the crystal quality of the grown layer, but it is not understood well how the polarity of the grown layer is controlled. GaN grown by metalorganic vapor phase epitaxy (MOVPE) on a sapphire substrate is known to have usually Ga polarity [1]. Polarity of GaN grown by gas source molecular beam epitaxy (GSMBE) is much more complicated than the MOVPE case and depends on the reactive gas and the buffer layer. GaN grown by radio-frequency plasma-assisted molecular beam epitaxy (rf-MBE) is usually a mixture of Ga and N polarities, and generally the N polarity is dominant [2, 3]. Layers grown with NH_3 as the N source is reported to have Ga polarity [4]. Ga polarity layers can be obtained by AlN buffer layer or AlN/GaN superlattice [5, 6]. The quality of GaN grown by rf-MBE is improved very much by controlling the Ga polarity [5, 6].

When GaN is grown on a substrate having its own polarity, it should depend on the substrate polarity. However, it is not made clear yet how the polarity of the grown layer is governed by the substrate and growth conditions such as the V/III ratio.

The purpose of this work is to clarify how the polarity of GaN is governed by the substrate polarity and by the V/III ratio.

Experimental GaN was grown by a conventional metalorganic molecular beam epitaxy (MOMBE) apparatus, using metal Ga as Ga source. Di-methyl hydrazine (DMHy) was used as nitrogen source [7]. The beam equivalent pressure (BEP) of Ga was changed from 2×10^{-8} to 8×10^{-8} Torr (Ga cell temperature: 866–935 °C) to grow layers with different V/III ratios with constant DMHy flux of 2×10^{-4} Torr. The growth temperature was 700 °C.

(111) GaAs substrates were etched in an etchant of $NH_4OH:H_2O_2:H_2O = 1:1:5$ at 0 °C for 2 min and dipped into HF solution for 15 min. A thermal cleaning was performed at 600 °C for 10 min in vacuum prior to the growth to remove the surface oxide. GaAs (111)A and B substrates were set together in the same Mo holder, and GaN was grown on the both wafers in the same run under the same growth conditions.

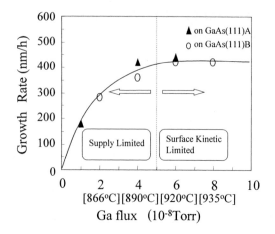

Fig. 1. Ga flux dependence of growth rate for GaN/GaAs(111)A and B

The crystal quality was measured by reflection high-energy electron diffraction (RHEED) and X-ray diffraction (XRD), and the polarity was characterized by coaxial impact collision ion scattering spectroscopy (CAICISS) [8].

Results and Discussion The growth rate of GaN increased from less than 200 nm/h for Ga BEP of 1×10^{-8} Torr (Ga cell temperature: 830 °C) to more than 400 nm/h for Ga BEP of 4×10^{-8} Torr (Ga cell; 890 °C) with increase of apparent BEP, and saturated at above 6×10^{-8} Torr (Ga cell temperature: 920 °C) as shown in Fig. 1. (The BEPs shown here are the values measured in our MOMBE system, and look slightly smaller than those expected from the Ga cell temperatures.) This result indicates that the growth with Ga BEPs at less than about 5×10^{-8} Torr is supply limited and that with Ga BEPs larger than about 5×10^{-8} Torr is surface kinetic limited. Therefore, the former is N rich growth condition and the latter a Ga rich one.

RHEED patterns of the grown layers with different Ga BEPs are shown in Fig. 2. It can be seen that RHEED patterns for the layers grown on GaAs (111)A-Ga surface are always streaky and better than those for the layers grown on GaAs (111)B-As surface. The RHEED pattern of the layer on GaAs (111)B at near stoichiometry (Ga cell temperature: 920 °C) is the worst with spotty ring and the XRD intensity of the grown layer was very low. This is probably due to the fact that the grown layer is a mixture of Ga and N polarities as shown below. The RHEED pattern of the layer grown on GaAs (111)B became better for Ga rich condition as shown in Fig. 2 (right bottom part).

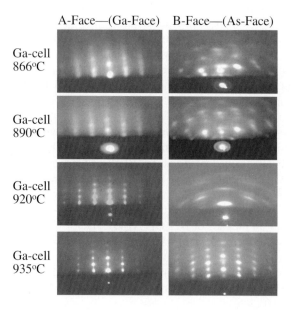

Fig. 2. Ga flux dependence of RHEED pattern for GaN/GaAs(111)A and B

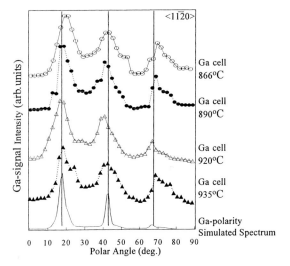

Fig. 3. V/III ratio dependence of CAICISS spectra of GaN/GaAs(111)A. GaN/GaAs(111)A always exhibited Ga-polar spectrum

Figure 3 shows CAICISS spectra of GaN layers grown on GaAs (111)A with different V/III ratios or Ga beam fluxes in comparison with a simulated spectrum of Ga polar surface.

As can be seen, they are all Ga polar independent of the V/III ratio or Ga beam flux. On the other hand, GaN grown on GaAs (111)B in N rich condition was N-polarity dominant as shown by the upper two curves of Fig. 4. They have peaks at 14°, 55°, and 70°, which are features of the N-polar CAICISS spectrum as shown in the simulated one. For nearly stoichiometric V/III ratio, the CAICISS spectrum showed no specific features as shown in the second curve from bottom, indicating that the grown layer is a complete mixture of Ga and N polarities. GaN grown in Ga rich condition (Ga cell temperature: 935 °C) became Ga-polarity dominant because it has peaks at 20°, 42°, and 67°. These results indicate that the polarity always inherits the substrate one for GaAs (111)A but depends on the V/III ratio for GaAs (111)B and inherits the substrate one only under N rich growth condition.

Summary The dependence of GaN polarity on the substrate polarity and growth conditions using GaAs (111) substrates and MOMBE growth has been investigated. It was found that for GaN grown on GaAs (111)A always Ga polarity was dominant independent of the V/III ratio, but the polarity of GaN grown on GaAs (111)B depends on the V/III ratio. The crystal quality was the worst for the layer grown on GaAs(111)B under nearly stoichiometric condition.

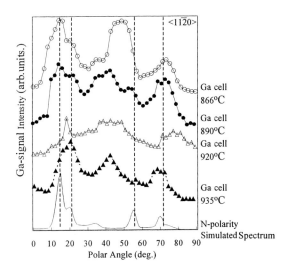

Fig. 4. V/III ratio dependence of CAICISS spectra of GaN/GaAs(111)B. Under N-rich growth condition, the CAICISS spectra showed an N-polarity spectrum. Otherwise, GaN grown under Ga rich condition became Ga-polarity dominant

Acknowledgements The author would like to express their sincere thanks to Prof. M. Sasaki for his kind help on the maintenance of the MOMBE system, and to Dr. T. Suemasu, Mr. M. Namerikawa and Mr. T. Sato for their support to the experiments and for discussions of the experimental results.

References

[1] F. A. PONCE, D. P. BOUR, W. T. YOUNG, M. SAUNDERS, and J. W. STEEDS, Appl. Phys. Lett. **69**, 337 (1996).
[2] E. S. HELLMAN, MRS Internet J. Nitride Semicond. Res. **3**, 1 (1998).
[3] S. SONODA, S. SHIMIZU, Y. SUZUKI, K. BALAKLISHNAN, J. SHIRAKASHI, and H. OKUMURA, Jpn. J. Appl. Phys. **39**, L73 (2000).
[4] S. SONODA, S. SHIMIZU, X. Q. SHEN, S. HARA, and H. OKUMURA, Jpn. J. Appl. Phys. **39**, L202 (2000).
[5] X. Q. SHEN, T. IDE, S. H. CHO, M. SHIMIZU, S. HARA, H. OKUMURA, S. SONODA, and S. SHIMIZU, Jpn. J. Appl. Phys. **39**, L16 (2000).
[6] A. KIKUCHI, T. YAMADA, S. NAKAMURA, K. KUSAKABE, D. SUGIHARA, and K. KISHINO, Jpn. J. Appl. Phys. **39**, L330 (2000).
[7] M. SASAKI, T. NAKAYAMA, N. SHIMOYAMA, T. SUEMASU, and F. HASEGAWA, Jpn. J. Appl. Phys. **39**, 4869 (2000).
[8] S. SHIMIZU, Y. SUZUKI, T. NISHIHARA, S. HAYASHI, and M. SHINOHARA, Jpn. J. Appl. Phys. **37**, L703 (1998).

Electron Backscattered Diffraction Patterns from Cooled Gallium Nitride Thin Films

F. Sweeney[1]) (a), C. Trager-Cowan (a), J. Hastie (a), D.A. Cowan (a),
K.P. O'Donnell (a), D. Zubia (b), S.D. Hersee (b), C.T. Foxon (c),
I. Harrison (c), and S.V. Novikov (c)

(a) Department of Physics and Applied Physics, John Anderson Building, University of Strathclyde, 107 Rottenrow, Glasgow G4 0NG, UK

(b) Centre for High Technology Materials, University of New Mexico, 1313 Goddard, SE Albuquerque, NM, USA

(c) University of Nottingham, Nottingham NG7 2RD, UK

(Received July 10, 2001; accepted August 14, 2001)

Subject classification: 61.14.Lj; 68.55.Jk; S7.14

The acquisition of electron backscattered diffraction (EBSD) (or Kikuchi diffraction) patterns in the scanning electron microscope is proving to be a useful technique with which to probe the structural properties of nitride thin films. In this paper we show that if a sample is cooled the patterns improve dramatically, an increase in intensity of the Kikuchi lines and a decrease in the intensity of the diffuse background is observed. Kikuchi lines from higher order planes become visible and the HOLZ rings become better defined. Such cooled patterns yield more information on the sample, particularly on non-centrosymmetric planes, from which the polarity of the nitride thin film under investigation may be deduced.

Introduction The acquisition of electron backscattered diffraction (EBSD) (or Kikuchi diffraction) patterns in the scanning electron microscope is a very powerful method for the microstructural characterisation of crystalline materials. EBSD is at present predominantly used in metallurgy for the measurement of texture, i.e., the mapping of the orientation of individual grains in polycrystalline samples [1], and to identify different crystalline phases. However, Wilkinson [2] and Troost et al. [3] applied EBSD to the measurement of strain in SiGe epilayers while Baba-Kishi [4] has used EBSD to investigate crystallographic polarity of non-centrosymmetric structures. These results strongly suggest that EBSD may be a useful technique to apply to the characterisation of nitride thin films.

In previous work [5] we have used EBSD to reveal the relative orientation of a nitride thin film with respect to its substrate; determine the tilt of a GaN thin film; detect improved surface quality in As-doped GaN films grown under high As_4 flux; and observe more detail in EBSD patterns from cooled GaN thin films. In this paper we show that EBSD patterns obtained from cooled GaN thin films can provide information on non-centrosymmetric planes, from which the polarity of the nitride thin film under investigation may be deduced.

Electron Backscattered Diffraction Technique In electron backscattered diffraction an electron beam is incident on a sample which is tilted at an angle typically $\geq 70°$. The impinging electrons are scattered elastically through high angles forming a spherical

---
[1]) Corresponding author; Tel.: +44 (0) 141 548 3458, e-mail: acu96168@strath.ac.uk

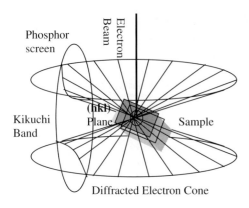

Fig. 1. Formation of Kikuchi Bands

source of diffracted electrons. Electrons that satisfy Bragg's condition for a given plane emanate in diffraction cones from both the upper and lower surfaces of the plane [1]. When these cones intersect the phosphor screen Kikuchi lines are observed (Fig. 1). The Kikuchi lines appear as almost straight lines because the cones are very shallow, as the Bragg angle θ_D is of order $1°$ ($\lambda \approx 0.008$ nm for a 25 keV electron beam). Each Kikuchi band is effectively the trace of the plane from which it is formed, the EBSD pattern is therefore a 2D projection (in fact the gnomic projection) of the crystal structure. Each Kikuchi band subtends an angle of $2\theta_D$ with the sample [3].

Our home built EBSD system (see Fig. 2) is assembled around a Cambridge 600S scanning electron microscope (SEM). Our EBSD system comprises a Cohu CCD low light camera, a P22 phosphor screen and a Dell Personal Computer, equipped with a Data Translation framegrabbing card to record the EBSD patterns. Samples can be mounted on a cold stage incorporating an Oxford instruments closed cycle helium cryorefrigerator allowing samples to be cooled down to 30 K from room temperature.

Results and Discussion The sample under investigation is a 3 μm thick GaN thin film grown by metalorganic chemical vapour deposition (MOCVD) on a sapphire substrate. It was nucleated at 480 °C, using a 30 nm GaN buffer layer, then the main GaN thin film was grown at 1050 °C using trimeythylgallium and ammonia as described previously [6].

The EBSD patterns shown in Fig. 3 were obtained from the GaN thin film at approximately 220 K and 200 K, respectively. As observed in Fig. 3, as the temperature decreases, the pattern quality dramatically increases. In the patterns from cooled samples contrast is greatly enhanced, Kikuchi lines from higher order planes become visible (as indicated by the arrows in Fig. 3b) and HOLZ (Higher order Laue zones) rings (see Fig. 3b) become more clearly defined. This result is expected because as the sample is cooled, the displacement of the atoms from their equilibrium positions is reduced [7]. (Note: the Debye temperature for GaN is 570 K [8].)

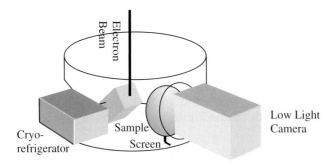

Fig. 2. Schematic of EBSD System

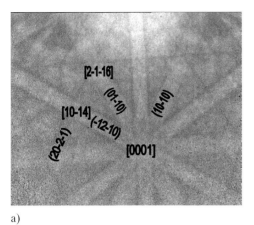 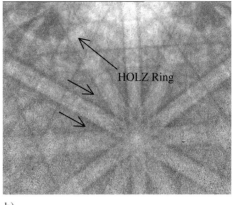

Fig. 3. EBSD Patterns from GaN thin film obtained at a) ≈220 K and b) ≈200 K

On examination of the pairs of Kikuchi lines corresponding to the $(10\bar{1}0)$ and $(\bar{1}2\bar{1}0)$ planes (labelled in Fig. 3a), it is clear that these pairs of lines are of the same intensity (this is clearest in Fig. 3b). This is not unexpected because, as illustrated in Fig. 4 (produced using CaRIne Crystallography), for the $(\bar{1}0\bar{1}0)$ plane (and similarly for the $(\bar{1}2\bar{1}0)$ plane), the N atoms are in-line with the Ga atoms. This means that the intensity of the electrons diffracted from the upper and lower surfaces of these planes are the same, leading to the pairs of Kikuchi lines being of equal intensity. However, on inspection of the pair of Kikuchi lines corresponding to the $(20\bar{2}\bar{1})$ plane, the line closest to the edge of the pattern is observed to be darker than the line closer to the centre of the pattern. We attribute this difference in intensities to the non-centrosymmetry of the atoms in this plane and the violation of Friedel's law due to dynamical scattering between simultaneously excited diffracted beams [4]. The non-centrosymmetry of the $(20\bar{2}\bar{1})$ plane is illustrated in Fig. 4 which shows that for the $(20\bar{2}\bar{1})$ plane the N atoms are offset from the Ga atoms. Modelling of the electron diffraction intensities from the non-centrosymmetric $(20\bar{2}\bar{1})$ plane will allow us to determine which face of the $(20\bar{2}\bar{1})$ plane each Kikuchi line corresponds [9]. Once this is determined, it will be possible to ascertain the polarity of the GaN epilayer under investigation. Such modelling is the subject of future work. Our conclusions will be verified by the use of calibration samples.

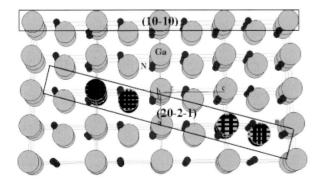

Fig. 4. Schematic diagram of a wurtzite GaN crystal showing that the N atoms are in-line with the Ga atoms for the $(10\bar{1}0)$ plane (see upper rectangle) and the N atoms are offset from the Ga atoms for the $(20\bar{2}\bar{1})$ plane (lower rectangle)

MOCVD grown GaN thin films are generally found to be of Ga polarity [10], while GaN thin films grown by molecular beam epitaxy (MBE) are generally found to have N polarity [10]. Preliminary room temperature measurements on an MBE grown As-doped GaN thin film produced patterns where the ($20\bar{2}1$) Kikuchi lines show the opposite contrast to that observed in Fig. 3, i.e., the line closest to the centre of the pattern is observed to be darker than the line closer to the edge of the pattern. From this observation we may deduce that the relative intensities of the Kikuchi lines for the ($20\bar{2}1$) plane are indeed a marker for the polarity of GaN films. We are presently comparing EBSD patterns from cooled MOCVD and MBE grown GaN thin films.

Summary and Future Work We have illustrated that if a sample is cooled the EBSD patterns improve dramatically. The contrast and detail in the patterns increases and the intensity of the diffuse background decreases. Cooling is therefore useful when high quality EBSD patterns are required. Future work includes the modelling of the electron diffraction intensities from individual planes, along with the study of calibration samples to allow the polarity of nitride thin films to be unambiguously determined. Thus introducing a simple non-destructive technique for polarity measurement on nitride thin films.

Acknowledgements Thanks to David Dingley of TSL and John Graham of Cryophysics for useful discussions on sample cooling. Thanks to David Clark for constructing the EBSD system and thanks to Ged Drinkwater for the electronic modifications of the SEM.

References

[1] D.J. Dingley and V. Randle, J. Mater. Sci. **27**, 4545 (1992).
[2] A.J. Wilkinson, J. Electron Microsc. **49**, 299 (2000).
[3] K.Z. Troost, P. van der Sluis, and D.J. Gravesteijn, Appl. Phys. Lett. **62**, 1110 (1993).
[4] K.Z. Baba-Kishi, J. Appl. Cryst. **24**, 38 (1991).
[5] C. Trager-Cowan, F. Sweeney, J. Hastie, S.K. Manson-Smith, D.A. Cowan, D. Mccoll, A. Mohammed, K.P. O'Donnell, D. Zubia, S.D. Hersee, C.T. Foxon, I. Harrison, and S.V. Novikov, submitted to J. Microsc.
[6] S.D. Hersee, J. Ramer, K. Zheng, C. Kranenberg, K. Malloy, M. Banas, and M. Goorsky, J. Electron. Mater. **24**, 1519 (1995).
[7] M.H. Loretto, Electron Beam Analysis of Materials, Chapman & Hall, London 1994.
[8] J.C. Nipko, C.K. Loong, C.M. Balkas, and R.F. Davis, Appl. Phys. Lett. **73**, 34 (1998).
[9] D.M. Bird and A.G. Wright, Acta Cryst. A **45** 104 (1989).
[10] E.S. Hellman, MRS Internet J. Nitride Semicond. Res. **3**, 11 (1998).

Influence of Polarity on Surface Reaction between GaN{0001} and Hydrogen

M. Mayumi[1]) (a), F. Satoh (a), Y. Kumagai (a), K. Takemoto (b), and A. Koukitu (a)

(a) Department of Applied Chemistry, Faculty of Technology, Tokyo University of Agriculture and Technology, Koganei, Tokyo 184-8588, Japan

(b) Itami Research Laboratories, Sumitomo Electronic Industries Ltd., 1-1-1 Koya-kita, Itami, Hyogo 664-0016, Japan

(Received June 25, 2001; accepted July 11, 2001)

Subject classification: 68.55.Jk; 81.05.Ea; 81.10.Bk; 81.15.Kk; S7.14

The influence of polarity on GaN decomposition has been investigated by an in situ gravimetric monitoring (GM) method using freestanding GaN(0001). The decomposition rate of the GaN was measured as a function of P_{H_2} in the temperature ranging from 800 to 950 °C. In the low-temperature region, the decomposition rate of GaN(0001) is faster than that of GaN(000$\bar{1}$), where the decomposition rates of both surfaces are proportional to $P_{H_2}^{3/2}$. On the other hand, the decomposition rate of GaN(000$\bar{1}$) is faster than that of GaN(0001) in the high-temperature region. In this case, the decomposition rates of both surfaces are proportional to $P_{H_2}^{1/2}$. These results indicate that the rate-limiting reactions of GaN decomposition can be written as follows: N(surface) + 3/2H$_2$(g) → NH$_3$(g) at lower temperatures, and Ga(surface) + 1/2H$_2$(g) → GaH(g) at higher temperatures.

Introduction Because wurtzite GaN is polar, there are two surface structures along the c-axis direction: the (0001) Ga face and the (000$\bar{1}$) N face, where a Ga atom combines with three N atoms in the bulk and a N atom combines with three Ga atoms, respectively. A few groups of scientists reported the characterization of the polarity of GaN grown on sapphire (0001) substrate. Also the determination of the polar direction of GaN growth was reported using several methods such as coaxial impact-collision ion scattering spectroscopy (CAICISS) analysis [1], and convergent beam electron diffraction (CBED) [2]. However, which reaction occurs on each polarity surface, (0001) and (000$\bar{1}$) has been unclear yet.

The GaN films can be grown using metalorganic vapor phase epitaxy (MOVPE) and halogen-transport VPE (HVPE), where the reactions governing the GaN deposition are Ga(g) + NH$_3$(g) → GaN(s) + 3/2H$_2$(g) and GaCl(g) + NH$_3$(g) → GaN(g) + HCl(g) + H$_2$(g), respectively. In general, H$_2$ is used as the carrier gas in these VPE methods. Additionally, H$_2$ is formed as by-product in GaN growth. From the viewpoint of understanding the GaN growth mechanism in VPE, it is very important to investigate the reaction between the GaN surface and hydrogen.

The in situ GM method provides direct information on the growth or etching rate in real time with an accuracy of atomic layer level [3–5]. Recently, we reported the decomposition process of GaN surface grown on sapphire (0001) using an in situ gravimetric monitoring (GM) method [6]. It was found that the decomposition rate was proportional to the 3/2 power of H$_2$ partial pressure (P_{H_2}) in the carrier gas at 850 °C,

---

[1]) Corresponding author; Phone: +81-42-388-7469; Fax: +81-42-386-3002; e-mail: mayumiho@cc.tuat.ac.jp

which indicates that the rate-limiting reaction for the GaN decomposition is GaN(surface) + 3/2H$_2$(g) → Ga(surface) + NH$_3$(g). In this paper, the polarity dependence of GaN decomposition is studied by the in situ GM method using GaN(0001) substrate.

Experimental The process of GaN decomposition on its surface was investigated under atmospheric pressure using an in situ GM system [6]. This system consists of a vertical quartz reactor and a recording microbalance has a sensitivity of 0.004 μg (about 6.5×10^{-6} μm of GaN per 1.0 cm^2 area). We used a freestanding GaN(0001) substrate ($1.0 \times 1.0 \times 0.03$ cm^3) [7]; to measure the dependence of the decomposition rate on surface polarity, one side of the GaN substrate was covered by a protective SiO$_2$ mask. The substrate was suspended from the microbalance with a fused quartz fiber. The GM system provides direct information on the substrate weight change due to the decomposition of GaN under a certain condition. The carrier gases used were H$_2$ and He as an inert gas. NH$_3$ was introduced over the GaN substrate while heating the furnace to prevent GaN decomposition before measurements. The decomposition rate of the GaN under each fixed set of conditions was monitored by switching off the NH$_3$ flow. By monitoring the weight change of the GaN substrate, the decomposition rates of both GaN(0001) and GaN(000$\bar{1}$) faces were measured in the temperature range from 730 to 950 °C with $P_{H_2} = 1$ atm. In addition, the decomposition rate in the carrier gas with various hydrogen partial pressures P_{H_2} was monitored by changing the ratio of H$_2$ to H$_2$ + He.

Results and Discussion Figure 1 shows the decomposition rates of GaN(0001) and GaN(000$\bar{1}$) faces as a function of temperature (730–950 °C) and at a constant H$_2$ carrier gas pressure $P_{H_2} = 1$ atm. The decomposition rates of both GaN(0001) and GaN(000$\bar{1}$) faces increase markedly with increasing temperature regardless of the polarity. This result agrees with that using an epitaxial GaN film grown on sapphire (0001) reported by us previously [3]. In Fig. 1, each surface has two slopes, depending on the

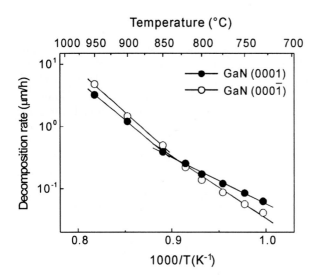

Fig. 1. Decomposition rates of the GaN(0001) face (open circles) and the GaN(000$\bar{1}$) face (filled circles) as a function of temperature at $P_{H_2} = 1$ atm

temperature: in the temperature range from 850 to 950 °C, GaN(0001) has an activation energy of 242 kJ/mol and that for GaN(000$\bar{1}$) is 259 kJ/mol, in the temperature range from 730 to 850 °C, GaN(0001) and GaN(000$\bar{1}$) have the activation energies of 143 and 191 kJ/mol, respectively. These results indicate that the rate-limiting reactions for decomposition on both surfaces change with temperature.

Furthermore, it is seen that the decomposition rate depends on the polarity of the GaN surface. In the temperature range below about 820 °C, the decomposition rate of GaN(0001) is faster than that of GaN(000$\bar{1}$). Conversely, the decomposition rate of GaN(000$\bar{1}$) is faster than that of GaN(0001) at temperatures ranging from about 850 to 950 °C. These results suggest that the difference of the value of decomposition rate on the lattice polarity is due to the different bonding configuration of the GaN surfaces as discussed later.

Next, the reaction between the GaN surface and hydrogen was investigated in detail. The decomposition rates of GaN were measured by varying the P_{H_2} from 0 to 1 atm at temperatures ranging from 800 to 950 °C. Figure 2 shows the P_{H_2} dependence of decomposition rates for GaN(0001) and GaN(000$\bar{1}$) in the temperature range from 800 to 950 °C.

The decomposition rates of both GaN(0001) and GaN(000$\bar{1}$) increased with increasing P_{H_2} at a given temperature. Therefore, it is clear that H_2 plays an important role in the decomposition of GaN. The rate equation for GaN decomposition can be written as $r = k P_{H_2}^n$, where r, k, and n are the decomposition rates, rate constants, and order of reaction, respectively. In Fig. 2, the values of n are also drawn at each temperature. The values of n for GaN(0001) are 1/2 at temperature ranging from 900 to 950 °C, hereafter high-temperature region, and 3/2 at 800 and 850 °C, hereafter low-temperature region, namely, the decomposition rates of GaN(0001) are proportional to $P_{H_2}^{1/2}$ in the high-temperature region, whereas the decomposition rates are proportional to $P_{H_2}^{3/2}$ in the low-temperature region. On the other hand, the decomposition rates of GaN(000$\bar{1}$) are proportional to $P_{H_2}^{1/2}$ at 925 and 950 °C (high-temperature region), whereas the decomposition rates are proportional to $P_{H_2}^{3/2}$ between 800 and

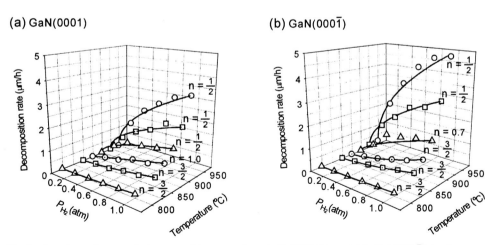

Fig. 2. P_{H_2} dependences of decomposition rate for a) GaN(0001) and b) GaN(000$\bar{1}$) in the temperature range from 800 to 950 °C

875 °C (low-temperature region). From these results, the rate-limiting reactions for decomposition of both GaN(0001) and GaN(000$\bar{1}$), in each temperature region are as follows:

$$\text{Ga(surface)} + \tfrac{1}{2}\text{H}_2(\text{g}) \rightarrow \text{GaH(g) (high-temperature region)} \tag{1}$$

and

$$\text{N(surface)} + \tfrac{3}{2}\text{H}_2(\text{g}) \rightarrow \text{NH}_3(\text{g}) \text{ (low-temperature region)} . \tag{2}$$

Consequently, the rate-limiting reactions for the decomposition of both GaN(0001) and GaN(000$\bar{1}$) likely shift from N(surface) + 3/2H$_2$(g) → NH$_3$(g) to Ga(surface) + 1/2H$_2$(g) → GaH(g) with increasing temperature.

We summarize our findings from Fig. 2. As described above, the decomposition of the GaN is limited by the Ga surface in the high-temperature region. The topmost Ga atoms on the GaN(0001) surface each combine with three N atoms (three back-bonds) in the bulk, whereas Ga atoms on the GaN(000$\bar{1}$) surface each form one back-bond with only one N atom in the bulk. This would explain why the decomposition rate of GaN(000$\bar{1}$) is faster than that of GaN(0001) in the high-temperature region. On the other hand, the limiting surface is the N surface in the low-temperature region. The decomposition rate of GaN(0001) is faster than that of GaN(000$\bar{1}$) because each N atom on the GaN(0001) surface combines with one Ga atom in the bulk (one back-bond), whereas each N atom on GaN(000$\bar{1}$) forms back-bonds to three Ga atoms in the bulk. Furthermore, the rate-limiting reaction for GaN decomposition is shifted from N(surface) + 3/2H$_2$(g) → NH$_3$(g) to Ga(surface) + 1/2H$_2$(g) → GaH(g) at temperatures ranging from 850 to 900°C for GaN(0001) and at that from 875 to 925 °C for GaN(000$\bar{1}$). This difference is also due to the different bonding configuration of the GaN surfaces.

Conclusions The dependence of the GaN decomposition on the lattice polarity has been investigated by an in situ GM method using a freestanding GaN(0001) substrate. At temperatures below about 820 °C, the decomposition rate of GaN(0001) was faster than that of GaN(000$\bar{1}$). Conversely, in the temperature range from about 850 to 950 °C, the decomposition rate of GaN(000$\bar{1}$) was faster than that of GaN(0001). The decomposition rates of both surfaces drastically increased with increasing temperature.

The decomposition rate of GaN as a function of P_{H_2} showed that the decomposition rates for both polarities were proportional to $P_{\text{H}_2}^{1/2}$ and $P_{\text{H}_2}^{3/2}$ in the high- and the low-temperature regions, respectively. The relation between the decomposition rate and P_{H_2} indicates that the rate-limiting reaction for GaN decomposition shifted from N(surface) + 3/2H$_2$(g) → NH$_3$(g) to Ga(surface) + 1/2H$_2$(g) → GaH(g) with increase of temperature. Based on these results, the relations between the lattice polarity and the decomposition of GaN can be explained clearly by considering the rate-limiting reactions and the bonding configurations on GaN surfaces.

Acknowledgements The authors would like to express their sincere thanks to Y. Matsuo for preparation of our GaN samples. This work was partly supported by the Foundation for Promotion of Material Science and Technology of Japan.

References

[1] M. Sumiya, M. Tanaka, K. Ohtsuka, T. Ohnishi, I. Ohkubo, M. Yoshimoto, H. Koinuma, M. Kawasaki, and S. Fuke, Appl. Phys. Lett. **75**, 674 (1999).
[2] J. L. Rouviere, J. L. Weyher, M. Seelmann-Eggebert, and S. Porowski, Appl. Phys. Lett. **73**, 668 (1998).
[3] A. Koukitu, H. Ikeda, H. Yasutake, and H. Seki, Jpn. J. Appl. Phys. **30**, L1847 (1991).
[4] A. Koukitu, Y. Kumagai, T. Taki, and H. Seki, Jpn. J. Appl. Phys. **38**, 4980 (1999).
[5] Y. Kumagai, M. Mayumi, A. Koukitu, and H. Seki, Appl. Surf. Sci. **159**, 427 (2000).
[6] M. Mayumi, F. Satoh, Y. Kumagai, K. Takemoto, and A. Koukitu, Jpn. J. Appl. Phys. **39**, L707 (2000).
[7] K. Motoki, T. Okahisa, N. Matsumoto, M. Matsushima, H. Kimura, H. Kasai, K. Takemoto, K. Uematsu, T. Hirano, M. Nakayama, S. Nakahata, M. Ueno, D. Hara, Y. Kumagai, A. Koukitu, and H. Seki, Jpn. J. Appl. Phys. **40**, L140 (2001).

A Comparative Study of MBE-Grown GaN Films Having Predominantly Ga- or N-Polarity

F. Yun[1]) (a), D. Huang (a), M. A. Reshchikov (a), T. King (a), A. A. Baski (a),
C. W. Litton (b), J. Jasinski (c), Z. Liliental-Weber (c), P. Visconti (a),
and H. Morkoç (a)

(a) Department of Electrical Engineering and Physics Department,
Virginia Commonwealth University, 601 W. Main St., Richmond, VA 23284, USA

(b) Air Force Research Laboratory (AFRL/MLPS), Wright Patterson AFB,
OH 45433, USA

(c) Lawrence Berkeley National Laboratory, Berkeley, CA 94720, USA

(Received June 21, 2001; accepted August 1, 2001)

Subject classification: 61.72.Dd; 61.72.Ff; 68.37.Ps; S7.14

Wurtzitic GaN epilayers having both Ga and N-polarity were grown by reactive molecular beam epitaxy (MBE) using a plasma-activated nitrogen source on c-plane sapphire. The polarities were verified by convergent beam electron diffraction (CBED). High-resolution X-ray diffraction, atomic force microscopy (AFM) imaging and transmission electron microscopy (TEM) were employed to characterize the structural defects present in the films. The different topographic features of Ga and N-polarity samples and their appearance after wet etching were correlated to the measured X-ray rocking curve peak widths for both symmetric [0002] and asymmetric [10$\bar{1}$4] diffraction. For Ga-polarity samples, the [0002] diffraction is narrower than the [10$\bar{1}$4] diffraction, while for N-polarity ones the [0002] peaks are broader than [10$\bar{1}$4]. The half width of [10$\bar{1}$4] peaks for both polarity types were in the range of 5–7 arcmin indicative of, among possibly other defects, a high density of pure edge threading dislocations lying parallel to the c-axis. The 1–2 arcmin [0002] linewidths of Ga-polarity samples suggest a low density of screw dislocations, which corresponds with the TEM observations where the screw dislocation density is less than 10^7 cm^{-2}. In N-polarity samples, however, the [0002] diffraction peak was typically wider than 5 arcmin, suggesting either a higher density of edge dislocations and inversion domains in N-polarity samples, or the columnar structural features in AFM images, where the effective coherence length for X-ray diffraction is drastically reduced.

Introduction The potential application of III-nitride materials for electronic devices has generated extensive research on GaN grown by MBE [1, 2]. GaN is polar along the c-direction, which is the most commonly used orientation for MBE growth. The quality of GaN films and performance of related device structures depend critically on the polarity of the epilayers [3]. Study of the freestanding GaN template grown by hydride vapor phase epitaxy (HVPE) has demonstrated distinct features for Ga-polarity and N-polarity GaN [4]. X-ray rocking curve linewidth of both the symmetric and asymetric reflections are narrower on the Ga-polarity side than on the N-polarity side. The overall defect density on the Ga-polarity side is more than an order of magnitude lower than that of N-polarity side. Moreover, the dependence of X-ray peak width on the dislocation structure has been studied on GaN films grown by metalorganic vapor phase deposition (MOCVD) [5], together with quantitative defect analysis by transmission electron microscopy (TEM) [6, 7].

---

[1]) Corresponding author; Fax: (804)828-4269; e-mail: fyun@mail1.vcu.edu

Recently, we have established the topological growth pattern for Ga-polarity and N-polarity GaN films by MBE through control of buffer layer parameters [8]. This allows us to assess the structural properties specific to GaN polar structure. In this work, we examine both Ga- and N-polarity GaN films using atomic force microscopy (AFM), X-ray rocking curves (ω-scan), and high-resolution TEM.

Experimental The GaN thin films were grown by MBE using a rf-plasma nitrogen source. The substrate used was c-plane sapphire. Both GaN and AlN buffer layers were optimized [9] and used to obtain Ga-polarity and N-polarity films through control of growth temperature and growth rate [8]. In particular, the Ga-polarity sample for TEM analysis was grown with an AlN buffer at 915 °C for 20 min and the GaN epilayer at 740 °C for 2 h. The N-polarity sample for TEM was grown with GaN buffer layer at 500 °C for 1 h, followed by GaN epilayer grown at 725 to 790 °C for 4.5 h.

The polarity of the GaN films was monitored by in-situ RHEED pattern and was established through wet chemical etching experiments, details of which can be found elsewhere [10]. It was also confirmed by convergent beam electron diffraction (CBED) on the two samples for TEM measurements. High-resolution X-ray measurements were performed on a Philips X'Pert MRD system equipped with a four-crystal Ge(220) monochromator. The instrumental broadening was ~10 arcsec. CBED, conventional and high-resolution TEM studies were peroremed using an 002B TOPCON microscope operated at 200 keV. AFM images were taken on Ga-polarity and N-polarity GaN samples before and after wet chemical etching by hot H_3PO_4.

Results and Discussions The surface morphology of Ga- and N-polarity GaN films was imaged by tapping mode AFM and is shown in Fig. 1[2]), both as-grown and after chemical etching by H_3PO_4 at ~160 °C. Distinct features of surface morphology can be observed between Ga- and N-polarity samples. The surface of as-grown Ga-polarity GaN

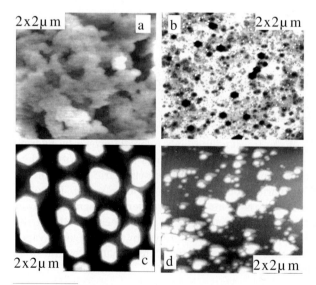

Fig. 1 (colour). AFM images of a) as-grown Ga-polarity GaN, b) Ga-polarity after etching for 5 min, c) as-grown N-polarity GaN, and d) N-polarity after etching for 30 s. Vertical scales for a) and b) are 10 nm, and for c) and d) 70 nm

[2]) Colour figure is published online (www.physica-status-solidi.com).

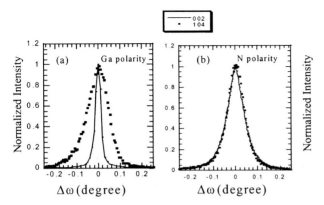

Fig. 2. Symmetric and asymmetric X-ray rocking curves for a) Ga-polarity and b) N-polarity GaN grown by MBE

(Fig. 1a) is smooth, terraced, and well-coalesced, with a root-mean-square (rms) roughness of ~3.5 nm. After exposure to etchants for 5 min, the surface of GaN remains relatively smooth, except for the presence of many hexagonal etch pits (Fig. 1b). The as-grown surface of the N-polarity film (Fig. 1c) shows isolated columns with feature size of ~0.2 µm, and a roughness (rms) of ~50 nm. The N-polarity film is etched quickly by hot H_3PO_4 leaving only sporadic spots (Fig. 1d) of GaN after etching for only 30 s.

Symmetric [0002] and asymmetric [10$\bar{1}$4] ω-scans are plotted in Fig. 2 for the Ga- and N-polarity samples used for TEM measurements, respectively. The Ga-polarity sample has a full-width at half-maximum (FWHM) of 1.98 arcmin for the [0002] peak, and 5.0 arcmin for the [10$\bar{1}$4] peak. The N-polarity sample shows a much larger FWHM of 5.26 arcmin for the [0002] peak, and 5.24 arcmin for the [10$\bar{1}$4] peak. Through accumulative statistics of more than 20 samples, we found that the FWHM figures of [0002] direction were between 1 and 2 arcmin for Ga-polarity GaN, and 5 and 7 arcmin for N-polarity. The FWHMs for [10$\bar{1}$4] direction were 5–7 arcmin for both Ga- and N-polarity films. This suggests that the nature and density of dislocations dominating Ga- and N-polarity films are different.

Three types of dislocations can be identified in GaN by their Burgers vectors. The edge dislocation (with Burgers vectors of ±[100], ±[110], or ±[010]) is parallel to the growth plane (c-plane) and will produce an in-plane strain which can only affect the asymmetric diffraction [10$\bar{1}$4]. The screw dislocation (with Burgers vectors of ±[001], parallel to c-axis) will produce a shear strain that can broaden the symmetric diffraction [0002]. There are also dislocations with mixed Burgers vectors (±[101], ±[011], ±[111]) which can contribute to both the symmetric and asymmetric diffractions. In order to differentiate between different types of dislocations and estimate their densities we applied conventional TEM methods since application of the diffraction contrast under two-beam condition allows for observation of a specific type of dislocations.

Shown in Figs. 3a and b are cross-sectional multi-beam bright field TEM images taken from Ga- and N-polarity GaN, repectively showing all types of dislocations. Also shown are the bright field images with a **g**-vector being parallel (**g** = 0002, Figs. 3c and d) or perpendicular (**g** = 11$\bar{2}$0, Fig. 3e and **g** = 01$\bar{1}$0, Fig. 3f) to the c-axis. For both Ga- and N-polarity films, very high dislocation density was observed (Figs. 3a and b) at the

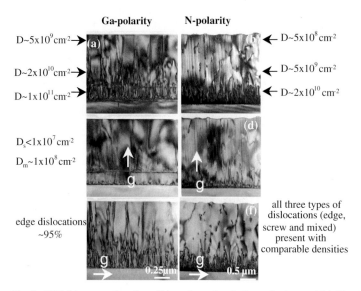

Fig. 3. TEM images showing dislocations in a) Ga-polarity and b) N-polarity GaN films. Multiple-beam dark-field images of Ga and N-polarity films with $\mathbf{g} = 0002$ are shown in c) and d), respectively, and with $\mathbf{g} = 11\bar{2}0$ in e) and $\mathbf{g} = 01\bar{1}0$ in f)

initial stage of growth (buffer layer). Some of these dislocations terminate within the buffer layer, others thread through the epilayers. The density gradually decreases with continuous growth. The total density of dislocations for the Ga-polarity sample (Fig. 3a) decreases from $D = 1 \times 10^{11}$ cm^{-2} at the buffer region to 5×10^9 cm^{-2} near the growth front. The total density for the N-polarity sample (Fig. 3b) decreases from $D = 2 \times 10^{10}$ cm^{-2} at the bottom to 5×10^8 cm^{-2} near the top surface. It should be noted that the total dislocation density of N-polarity GaN is an order of magnitude lower than that of Ga-polarity samples.

With the $\mathbf{g} = 0002$ reflection, only pure screw dislocations and mixed dislocations are visible. With $\mathbf{g} = 11\bar{2}0$, only pure edge dislocations and mixed dislocations can be observed. Therefore, it can be concluded (from Figs. 3, c, d, e and f) that for the Ga-polarity sample, edge disloactions are dominant (~95%), while the density of screw dislocations is very low ($D_s < 10^7$ cm^{-2}) and the density of mixed dislocations ($D_m = 1 \times 10^8$ cm^{-2}) is also much lower than the edge dislocations. The N-polarity sample, on the contrary, shows no dominance by any dislocation type. Instead, it exhibits an equivalent amount of all three types of dislocations. Also pronounced in N-polarity film is the large density ($D_{ID} = 1 \times 10^{11}$ cm^{-2}) of inversion domains (IDs) which occur in the Ga-polarity regions (see Fig. 3d). The density of IDs in the Ga-polarity film, however, is found to be very low (less than 10^7 cm^{-2}).

The features of Ga and N-polarity films in X-ray rocking-curve measurements can be well correlated with dislocation characteristics observed in TEM images. The narrow [0002] FWHM for Ga-polarity GaN is consistent with the very low density of screw and mixed dislocations. For N-polarity GaN, the much wider FWHM of [0002] peak agrees with the existence of a high density of screw dislocations, together with even higher density of IDs which may contribute to the broadening in a way similar to mixed dislocations. The asymmetric [10$\bar{1}$4] rocking curve peaks for both Ga- and N-polarity films have a comparable FWHM, reflecting the TEM observation of comparable edge-dislocation densities in Ga- and N-polarity films.

The symmetric X-ray diffraction peak linewidth can also be influenced by the feature size of grains or columns with respect to Ga- and N-polarity. The well-coalesced surface features of the Ga-polarity sample (in Fig. 1a) indicates less influence of broadening effects due to its large coherent length. For the N-polarity film in Fig. 1b, the columnar structure with lateral feature size of $\sim$0.2 μm will centainly contribute to the broadening of the symmetric [0002] peak due to the size effect predicted by the Scherrer formula [11].

Summary Distinct structural features from X-ray rocking curve measurements were found in Ga-polarity and N-polarity GaN films grown by MBE. Ga-polarity samples show much narrower [0002] linewidth than N-polarity films, while both types of films show comparable linewidths of the asymmetric [10$\bar{1}$4] diffraction. TEM images reveal the different nature and density of dislocations in Ga and N-polarity films, and are in good agreement with X-ray data. The surface morphology measured by AFM also shows different features for Ga and N-polarity films.

Acknowledgements The authors would like to thank Prof. K. J. Wynne and Ms. J. Uilk for the use of a large area AFM. The VCU portion of this work was funded by grants from AFOSR (Dr. G. L. Witt), NSF (Dr. L. Hess and Dr. G. Pomrenke), and ONR (Dr. C. E. C. Wood and Dr. Y. S. Park). The TEM group (J. J. and Z. L.-W.) supported by Air Force Office of Scientific Research, through the U.S. Department of Energy under Order No. AFOSR-ISSA-00-0011 would like to thank W. Swider for her excellent TEM sample preparation and would like to acknowledge the use of the facilities at the National Center for Electron Microscopy at Lawrence Berkeley National Laboratory.

References

[1] H. MORKOÇ, Nitride Semiconductors and Devices, Springer-Verlag, Berlin 1999.
[2] S. J. PEARTON, J. C. ZOLPER, R. J. SHUL, and F. REN, J. Appl. Phys. **86**, 1 (1999).
[3] H. MORKOÇ, R. CINGOLANI, and B. GIL, Solid-State Electron. **43**, 1909 (1999).
[4] F. YUN, M. A. RESHCHIKOV, K. JONES, P. VISCONTI, H. MORKO, S. S. PARK, and K. Y. LEE, Solid-State Electron. **44**, 2225 (2000).
[5] B. HEYING, X. H. WU, S. KELLER, Y. LI, D. KAPOLNEK, B. P. KELLER, S. P. DENBAARS, and J. S. SPECK, Appl. Phys. Lett. **68**, 643 (1996).
[6] D. KAPOLNEK, X. H. WU, B. HEYING, S. KELLER, U. K. MISHRA, S. P. DENBAARS, and J. S. SPECK, Appl. Phys. Lett. **67**, 1541 (1995).
[7] W. QIAN, M. SKOWRONSKI, M.. DEGRAEF, K. DOVERSPIKE, L. B. ROWLAND, and D. K. GASKILL, Appl. Phys. Lett. **66**, 1252 (1995).
[8] D. HUANG, P. VISCONTI, K. M. JONES, M.. A. RESHCHIKOV, F. YUN, A. A. BASKI, T. KING, and H. MORKOÇ, Appl. Phys. Lett. **78**, 4145 (2001).
[9] F. YUN, M. A. RESHCHIKOV, P. VISCONTI, K. M. JONES, D. WANG, M. REDMOND, J. CUI, C. W. LITTON, and H. MORKOÇ, Mater. Res. Soc. Symp. Proc. **639**, G3.17 (2001).
[10] P. VISCONTI, K. M. JONES, M. A. RESHCHIKOV, F. YUN, R. CINGOLANI, H. MORKOÇ, S. S. PARK, and K. Y. LEE, Appl. Phys. Lett. **77**, 3743 (2000).
[11] B. E. WARREN, X-Ray Diffraction, Dover Publ. Co., New York 1990.

Polarity Inversion by Supplying Group-III Source First in MOMBE of GaN/AlN or GaN on GaAs (111)B (As Surface)

F. Hasegawa (a), O. Takahashi (a), T. Nakayama (a), and R. Souda (b)

(a) University of Tsukuba, Institute of Applied Physics, Tsukuba, Ibaraki 305-8573, Japan

(b) Advanced Materials Laboratory, National Institute for Materials Science, Tsukuba, Japan

(Received June 19, 2001; accepted August 4, 2001)

Subject classification: 68.35.Bs; 81.15.Hi; S7.14

In order to investigate the influence of the Ga or Al and N supply sequence on the polarity, GaN/AlN or GaN was grown on a GaAs (111)B (As surface) from Ga/Al source or N source in MOMBE. 300 nm GaN/10 nm AlN/GaAs (111)B grown with N source supply first in N rich condition exhibited N polarity dominance, but it became Ga polarity dominant when about five monolayers Al were deposited prior to N source supply. The same phenomenon was observed when GaN/GaAs (111)B was grown in N rich condition by supplying Ga source first. These results indicate that polarity of GaN is controlled not only by substrate polarity but also by the surface condition of the substrate.

Introduction It is well known that the polarity of hexagonal GaN influences greatly the crystal quality of the grown layer, but it is not understood well how the polarity of the grown layer is controlled. GaN grown by MOVPE on a sapphire substrate is known to have usually Ga polarity [1]. Polarity of GaN grown by GSMBE is much more complicated than the MOVPE case and depends on the reactive gas and the buffer layer. GaN grown by rf plasma-assisted MBE is usually a mixture of Ga and N polarities, and generally the N polarity is dominant [2, 3]. Layers grown with NH_3 as the N source are reported to have Ga polarity [4]. Ga polarity layers can be obtained by an AlN buffer layer or AlN/GaN superlattice [5, 6]. The quality of GaN grown by rf plasma-assisted MBE is improved very much by controlling the polarity to be the Ga one [5, 6].

When GaN is grown on a substrate having its own polarity, it should depend on the substrate polarity. However, it is not made clear yet how the polarity of the grown layer is governed by the substrate, surface and growth conditions. Influence of an AlN intermediate layer is not known well.

The purpose of this work is to clarify how the polarity of GaN is governed by the surface condition, an AlN intermediate layer or the substrate polarity itself, by growing GaN or GaN/AlN on GaAs (111)A and B surfaces.

Experimental GaN and AlN were grown by a conventional MOMBE apparatus, using metal Ga as the Ga source and Di-methyl-aluminum-hydride (DMAH) as the Al source. Di-methyl-hydrazine (DMHy) was used as the nitrogen source for both growths [7]. The beam equivalent pressure (BEP) of Ga was kept at 2×10^{-8} Torr (Ga cell temperature; 866 °C) so that the growth condition was N rich. The growth rate was 200 nm/h compared to the normal growth rate of 400 nm/h in stoichiometry condition, and the grown GaN normally exhibits N polarity dominant. The BEPs of DMAH and

DMHy were fixed at 1×10^{-6} Torr and 2×10^{-4} Torr, respectively. The growth temperature was 700 °C for both AlN and GaN layers.

(111) GaAs substrates were etched in an etchant of $NH_4OH:H_2O_2:H_2O = 1:1:5$ at 0 °C for 2 min and dipped into HF solution for 15 min. A thermal cleaning was performed at 600 °C for 10 min in vacuum prior to the growth to remove the surface oxide. GaAs (111)A and B substrates were set together in the same Mo holder, and GaN/AlN was grown on the both wafers in the same run with the same growth conditions.

Crystal quality was measured by RHEED and XRD, and the polarity was characterized by CAICISS (Coaxial Impact Collision Ion Scattering Spectroscopy) [8].

Results and Discussion GaN grown on GaAs (111)A (Ga surface) always exhibits Ga polarity independently of V/III ratio, but polarity of GaN grown on GaAs (111)B (As surface) depends on the V/III ratio and it becomes N polar for large V/III ratio, i.e., N rich growth condition. By insertion of 20 nm AlN layer, it became Ga polar even on GaAs (111)B with N rich growth condition. However, when the AlN thickness was reduced to 10 nm, GaN grown on it was N polarity dominant for usual growth procedure with DMHy supply first.

In order to see the effect of DMAH pre-flow, DMAH corresponding to 2 ML, 5 ML and 10 ML of Al was supplied on GaAs (111)B before supply of DMHy. Although the RHEED pattern was spotty ring for GaN grown with DMHy first, a spotty streaky RHEED pattern was obtained for GaN grown on AlN with the 5 ML Al pre-flow as shown in Fig. 1. The RHEED pattern became spotty ring again with the 10 ML Al pre-flow.

As to the polarity characterized by CAICISS, N polarity was dominant when DMHy was supplied first as shown by the top curve of Fig. 2. Simulated CAICISS spectra of Ga polar and N polar GaN are also shown in Fig. 2 for reference; the best Ga polarity

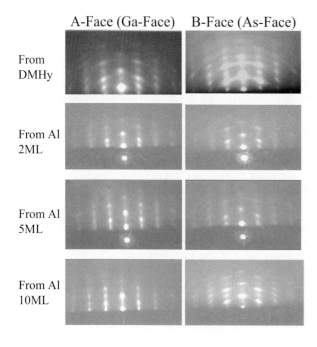

Fig. 1. Dependence of RHEED pattern on supply sequence and amount of Al for GaN/10 nm AlN/GaAs (111)A&B. Left: on A-face, right: on B-face

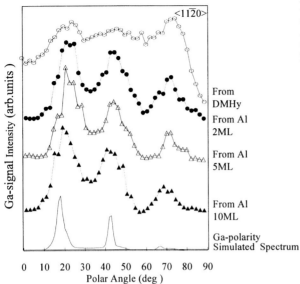

Fig. 2. Dependence of the polarity on supply sequence and amount of Al for GaN/10 nm AlN/GaAs (111)B. Ga polarity became dominant by 2 ML Al pre-flow

spectrum was obtained for 5 ML Al as can be seen in Fig. 2. The CAICISS spectrum deteriorated slightly for 10 ML Al pre-flow, but it was still Ga polarity dominant. GaN grown on GaAs (111)A (Ga surface) showed always Ga polarity independently of the growth condition and of the sequence of the source material supply as shown in Fig. 3.

An almost similar result was obtained for Ga metal pre-deposition as shown in Fig. 4. When GaN was directly grown on GaAs (111)B in a usual way with DMHy first in Ga deficient condition, a CAICISS spectrum corresponding to a layer including a lot of N

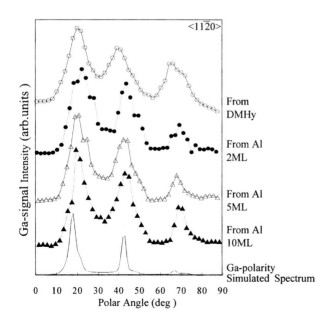

Fig. 3. Dependence of CAICISS spectra on supply sequence and amount of Al for GaN/10 nm AlN/GaAs (111)A. The grown layers have always Ga-polarity independently of the supply sequence on GaAs (111)A surface

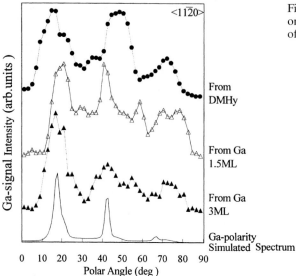

Fig. 4. Dependence of CAICISS spectra on supply sequence and pre-deposition of Ga for GaN/GaAs (111)B

polarity columns was observed as shown in the top curve of Fig. 4. Ga polarity dominant GaN layer was obtained with 1.5 ML Ga pre-deposition. These results indicate that group-III materials supplied first are very effective to obtain Ga polarity dominant GaN layers.

Summary In order to investigate influence of the supply sequence of source materials on the polarity, GaN/AlN or GaN was grown on GaAs (111)B (As surface) from Ga/DMAH or DMHy in MOMBE. It was found that the polarity greatly depended on which source was supplied first; when a group-III source of five monolayers was supplied first, a Ga polarity layer was always obtained independently of the growth condition.

Acknowledgements The authors would like to express their sincere thanks to Prof. M. Sasaki for his kind help on the maintenance of the MOMBE system, and to Dr. T. Suemasu, Messrs. M. Namerikawa and T. Sato for their support to the experiments and for discussions on the experimental results.

References

[1] F. A. PONCE, D. P. BOUR, W. T. YOUNG, M. SAUNDERS, and J. W. STEEDS, Appl. Phys. Lett. **69**, 337 (1996).
[2] E. S. HELLMAN, MRS Internet J. Nitride Semicond. Res. **3**, 1 (1998).
[3] S. SONODA, S. SHIMIZU, Y. SUZUKI, K. BALAKLISHNAN, J. SHIRAKASHI, and H. OKUMURA, Jpn. J. Appl. Phys. **39**, L73 (2000).
[4] S. SONODA, S. SHIMIZU, X. Q. SHEN, S. HARA, and H. OKUMURA, Jpn. J. Appl. Phys. **39**, L202 (2000).
[5] X. Q. SHEN, T. IDE, S. H. CHO, M. SHIMIZU, S. HARA, H. OKUMURA, S. SONODA, and S. SHIMIZU, Jpn. J. Appl. Phys. **39**, L16 (2000).
[6] A. KIKUCHI, T. YAMADA, S. NAKAMURA, K. KUSAKABE, D. SUGIHARA, and K. KISHINO, Jpn. J. Appl. Phys. **39**, L330 (2000).
[7] M. SASAKI, T. NAKAYAMA, N. SHIMOYAMA, T. SUEMASU, and F. HASEGAWA, Jpn. J. Appl. Phys. **39**, 4869 (2000).
[8] S. SHIMIZU, Y. SUZUKI, T. NISHIHARA, S. HAYASHI, and M. SHINOHARA, Jpn. J. Appl. Phys. **37**, L703 (1998).

Charge Screening of Polarization Fields in Nitride Nanostructures

A. Di Carlo and A. Reale

INFM and Dipartimento di Ingegneria Eletteronica, Universita' di Roma "Tor Vergata", Via di Tor VerGata 110, I-00133 Roma, Italy

(Received July 3, 2001; accepted July 18, 2001)

Subject classification: 77.65.–j; 78.47.+p; 78.55.Cr; 78.67.De; S7.14

Spontaneous and piezoelectric polarization play a crucial role in nitride-based nanostructures and devices. The effects of polarization fields can, however, be reduced by free and fix charge screening. Optical transition energy, oscillator strength and recombination times are typical physical parameters that crucially depend on the degree of screening of polarization fields. In the following we will review the topic of polarization field screening and we will show how the screening can influence both static and dynamic properties.

1. Introduction The recent developments in the field of GaN-based blue-UV optoelectronic devices, for a review, see [1] have stimulated several experimental and theoretical studies on GaN/AlGaN MQWs. Most of the attention has been paid to the existence of built-in electric fields in such heterostructures. The theory of Bernardini and Fiorentini [2–4] establishes that the internal electric field is originated by the difference in the spontaneous polarization between the barrier and the well, and by the piezoelectric polarization, due to the pseudomorphic growth of the well or of the barrier, depending on the buffer layer.

Several experiments have indeed shown that the ground level emission of thick AlGaN/GaN QWs falls below the energy gap of the bulk GaN [5–12]. Though this is a confirmation of the presence of an internal electric field, there is still a quantitative disagreement between the experimental ground level energy deduced by the optical experiments and the calculated energy in the presence of the built-in electric fields. To circumvent this problem, the experimental data are usually fitted by using either the value of the electric field as a free parameter [7–12], or in more refined self-consistent models [5, 6, 13] by adjusting the actual carrier density in the well to screen the built-in field. In both cases, either the adjusted electric field (which is different from the real value) or the rather large injected carrier density (somewhat unphysical for a cw optical excitation) indicate that a quantitative understanding of the optical properties of GaN quantum wells is still somehow unresolved.

In this paper, we present a new approach to the problem and we show that accumulation of the charge in both steady state and time dependent conditions screens the built-in field. However, this accumulation is strongly related to both radiative and non-radiative channels. Thus, even in nominally identical systems difference in the non-radiative recombination time can induce different optical properties.

2. Model and Results In order to describe the static and dynamic screening of the polarization fields we need to properly account for the process of generation and recombination of the electron–hole pairs. We will assume a very simple model for the time evolution of the carrier density in the quantum well, that is

$$\frac{dn}{dt} = G - R_{sp} - R_{nr}, \qquad (1)$$

where G is the generation rate, R_{sp} is the spontaneous (radiative) recombination rate and R_{nr} the non-radiative recombination rate.

The spontaneous recombination rate R_{sp}, is given by [14]

$$R_{sp} = \int_{-\infty}^{+\infty} \frac{e^2 \mu E_p \omega n_r}{\hbar^2 c^3 \pi^3 \varepsilon_0 m_0} \sum_{i,f} f_{c,i}(1 - f_{v,f}) I_{i,f}^2 \theta(\omega - \omega_{if}) \, d\omega \qquad (2)$$

where all the symbols have their usual meaning [14], n_r is the refractive index, f_c, f_v are the Fermi distributions, θ is the Hevyside function, $I_{i,f}$ is the wavefunctions overlap and ω_{if} is the energy transition from the initial level i in conduction band to the final level f in valence band. In Eq. (2) both quasi-Fermi levels, overlap matrix elements, and transition energy are all charge density dependent. Overlap and transition energy are related to the density through the coupling between the Poisson and Schrödinger equations. Equation (2) represents the microscopic expression of the bimolecular recombination term which is widely used in rate equation models describing carrier dynamics in QWs.

The non-radiative recombination rate accounts for all the non-radiative recombination channels such as defects assisted and surface recombination [15]; for the sake of simplicity, we will assume a very simple expression for R_{nr}, that is

$$R_{nr} = \frac{n}{\tau_{nr}} \qquad (3)$$

with the time constant τ_{nr} equal for all the wells.

2.1 CW excitation Under steady state conditions, and assuming a constant generation rate, Eq. (1) becomes

$$G = R_{sp} + R_{nr}, \qquad (4)$$

which has to be solved for all samples by accounting for the dependence of R_{nr} and R_{sp} on the carrier density. A double iteration is performed in order to determine the actual carrier density and energy levels. Assumed an initial photogenerated carrier concentration, the energy of quantized levels needed for the evaluation of the spontaneous recombination rate is obtained by self-consistently solving Schrödinger and Poisson equations. Equation (4) is then solved for n, and this value is put again into the Schrödinger-Poisson solver. The two-step procedure is repeated until convergence is achieved. In the calculations the value of the non-radiative carrier lifetime τ_{nr} has been assumed to be of the order of 5 ns [16]. The reference structure we consider in the following is a $Al_{0.15}Ga_{0.85}N/GaN$ MQW with a barrier width of 15 nm and several well widths.

In Fig. 1 we show the calculated fundamental PL transition as a function of the well width for three values of the non-radiative recombination time. The central trace is obtained with $\tau_{nr} = 5$ ns, which is a typical value that can be extracted from time-resolved measurements for AlGaN/GaN systems [16]. In the absence of non-radiative recombinations (i.e. $\tau_{nr} = \infty$), the screening of the polarization field due to charge accumulation (*mesoscopic capacitor effect* [16]) is dominant, especially for wide wells. On

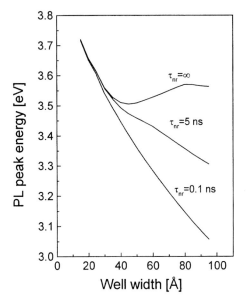

Fig 1. Theoretical values of the emission energy of the samples. The calculations are performed using Al barrier concentration of 15% and considering three different values of the non-radiative lifetime τ_{nr}: 0.1 ns, 5 ns and ∞. The curve with $\tau_{nr} = 0.1$ ns is equivalent to the usual calculation accounting for the built-in field without charge accumulation effect

the contrary, a strong non-radiative rate ($\tau_{nr} = 0.1$ ns) depletes the ground level, preventing charge accumulation and thus leaving the field unscreened. Indeed, the result coincides perfectly with that obtained in the absence of free carriers by using the Bernardini-Fiorentini fields. The flattening of the PL peak energy at increasing well width is therefore due to the competition between radiative and non-radiative recombination processes. For the larger wells, the built-in field is effectively screened due to the mesoscopic capacitor effect, that is to the charge accumulation favored by the long radiative lifetime (which is, in turn, a consequence of the spatial separation of the electron and hole wavefunctions). On the contrary, non-radiative recombinations tend to deplete the ground level energy, thus reducing the effect of screening and red-shifting the PL spectra.

We should point out that, according to the results of Fig. 1, the extraction of a mean electric field from the optical transitions is non-radiative lifetime dependent.

In Fig. 2 we show the mean electric field in the well as a function non-radiative lifetime for three MQWs under the same illumination condition as in Fig. 1. For systems where non-radiative channel is very efficient (low τ_{nr}) the field is unscreeneed and the Bernardini-Fiorentini field values are recovered. On the opposite, in systems where the non-radiative channel is negligible (high τ_{nr}) we observe a reduction in the average field induced by the charge accumulation screening in the well. This field reduction is more evident for wider wells where the radiative channel is less effective.

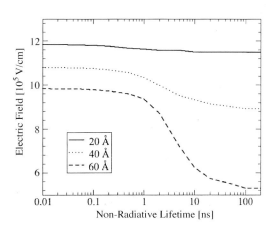

Fig. 2. Mean electric field as a function of non-radiative lifetime for three different MQWs

2.2 Time-resolved photoluminescence We have evaluated the time dependent role of carrier screening on internal polarization fields in time-resolved PL. Equation (1) is solved as following: at each time step t_i we calculate the right-hand side of Eq. (1) by solving self-consistently the Schrödinger and Poisson equation for a charge density $n(t_{i-1})$. This defines the new charge density $n(t_i)$ and the procedure is iterated up to a desired time.

It must be noted that the integration of Eq. (1) allows also for the calculation at any given time of the number and spectral components of photons re-emitted by radiative recombination. This permits the calculation of time-resolved PL spectra.

In Fig. 3 we plot the time-resolved PL (in steps of 200 ps) after Gaussian optical pulse excitation for three reference MQWs. We consider 2 ps pulses having two different powers.

In case of low excitation power, the induced charge screening is low, and in the three samples no significant spectral shift can be observed (lower traces). Moreover, from the first to the last trace the change of intensity is uniform between the three samples.

At high excitation power (upper traces) a significant spectral shift takes place. The shift is more evident for the wider MQW, where the screening of photogenerated carriers induces a more pronounced band flattening. Another important feature of high excitation conditions can be observed for the narrower MQW system, where a much faster decay in PL intensity takes place. The reason of this different behavior is that radiative recombination rate is much more effective than non-radiative channels. We assume 5 ns for the non radiative lifetime.

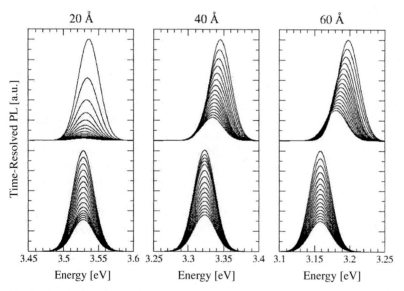

Fig. 3. Time-resolved PL spectra for three different MQWs, for high energy (10 pJ, upper traces) and low energy (10 fJ, lower traces) pulse. The spectra are normalized to the maximum PL intensity. For any set of curves, the upper trace is referring to the first 200 ps; traces are in steps of 200 ps up to 3.2 ns

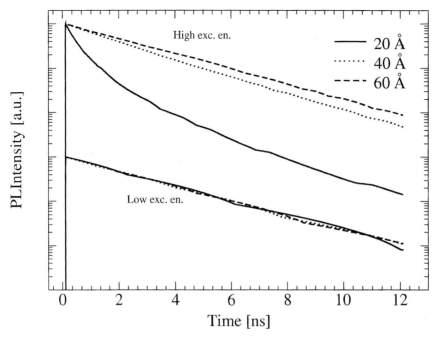

Fig. 4. Normalized PL decay rates for three different MQWs, for two different pulse energies. The upper traces are shifted for sake of clarity

The time dependent decay of PL intensity in the three MQW systems considered above is shown in Fig. 4. For low excitation energies (lower traces), the time decay is governed by the same non-radiative time constant in all the three samples. In high excitation conditions (upper traces), the decay rate is no more uniform for the narrower MQWs. The induced carrier screening strongly increases the oscillator strength and consequently the spontaneous recombination rate (Eq. (3)) is enhanced. Few nanoseconds after the excitation all the samples recover the same decay rate, indicating that screening is no more effective, and the dominant recombination rate is non-radiative.

3. Conclusions We have shown how non-radiative channels can strongly affect the charge screening in nitride based nanostructures. A new model for cw and time dependent photoluminescence has been developed which fully accounts for the self-consistent evaluation of all the recombination channels in the nitride system. Moreover, non-radiative channels can influence the evaluation of the mean polarization field evaluated from optical measurements.

Acknowledgements This work has been partially supported by Italian MURST and EU Project CLERMONT.

References

[1] S. NAKAMURA and G. FASOL, The Blue Laser Diode, Springer-Verlag, Berlin 1997.
[2] F. BERNARDINI, V. FIORENTINI, and D. VANDERBILT, Phys. Rev. B **56**, R10024 (1997).

[3] F. BERNARDINI, V. FIORENTINI, and D. VANDERBILT, Phys. Rev. Lett. **79**, 3958 (1997).
[4] F. BERNARDINI and V. FIORENTINI, Phys. Rev. B **57**, R9472 (1998).
[5] R. CINGOLANI et al., Phys. Rev. B **61**, 2711 (1999).
[6] A. BONFIGLIO et al., J. Appl. Phys. **87**, 2289 (2000).
[7] M. LEROUX et al., Phys. Rev. B **58**, R13371 (1998).
[8] M. LEROUX et al., Phys. Rev. B **60**, 1496 (1999).
[9] T. TAKEUCHI et al., Appl. Phys. Lett. **73**, 1691 (1998).
[10] T. TAKEUCHI et al., Jpn. J. Appl. Phys. **36**, L382 (1997).
[11] S. H. PARK and S. L. CHUANG, Appl. Phys. Lett. **72**, 3103 (1998).
[12] J. S. IM et al., Phys. Rev. B **57**, R9435 (1998).
[13] F. DELLA SALA et al., Appl. Phys. Lett. **74**, 2002 (1999).
A. DI CARLO et al., Appl. Phys. Lett. **76**, 3950 (2000).
[14] K. EBELING, Integrated Opto-Electronics, Springer-Verlag, Berlin 1992.
[15] P. BHATTACHARYA, Semiconductor Optoelectronic Devices, Prentice-Hall Inc., Upper Saddle River (New Jersey) 1994.
[16] A. DI CARLO et al., Phys. Rev. B **63**, 235305 (2001).

Polarization Effects and UV Emission in Highly Excited Quaternary AlInGaN Quantum Wells

E. Kuokstis[1])*) (a), Jianping Zhang (a), J. W. Yang (a), G. Simin (a), M. Asif Khan (a), R. Gaska (b), and M. Shur (b)

(a) Department of EE, University of South Carolina, Columbia, SC 29208, USA

(b) Sensor Electronic Technology, Inc., Latham, NY 12110, USA

(Received June 22, 2001; accepted June 24, 2001)

Subject classification: 73.63.Hs; 77.65.–j; 78.55.Cr; S7.14

We report on UV photoluminescence (PL) dynamics in highly excited quaternary $Al_xIn_yGa_{1-x-y}N$ epilayers and multiple quantum wells (MQWs). At low temperature, in MQWs we have observed new PL band which appeared at low excitation on the long-wave side of the spectrum and we show it to arise from localized carriers (excitons). The strong blue-shift of PL maximum with excitation intensity in MQWs is caused by localized state filling and screening of piezoelectric and spontaneous polarization electric field.

Introduction Recently III-nitride materials are intensely studied for light emission over most of the visible and near UV spectrum. Remarkable progress in the fabrication of ternary InGaN or AlGaN based multiple quantum wells (MQWs), light emitting diodes (LEDs) and laser diodes (LDs) has been realized [1, 2]. Novel quaternary AlInGaN compounds are ideally suited for band engineering in UV region for the development of UV emitters, as well as for solar blind detectors [3]. We previously reported on the advantages of using quaternary AlInGaN materials for fabrication of high quality quantum structures for efficient UV sources for pumping of phosphors for white light emission [4, 5]. However, AlInGaN UV luminescence dynamics has not been analyzed in a wide excitation and temperature range. In this study, we investigate PL of AlInGaN-based MQWs and epilayers and discuss the mechanisms responsible for emission peculiarities.

Experimental The structures were grown using novel pulsed atomic layer epitaxy (PALE) technique. The advantage of this method against conventional metalorganic chemical vapor deposition (MOCVD) arises from growth temperatures being approximately 200–300 °C lower. The use of reduced growth temperatures allowed us to significantly increase the In incorporation in high Al-fraction quaternary AlInGaN layers with transmission cut off in the 250–350 nm range. This improves optical emission, the structural properties of the lattice, reduces the number of band states and forms smoother quantum well interfaces [5]. The structures were grown over sapphire substrates. Prior to the quaternary layers a low temperature (450 °C) 250 Å thick AlN buffer layer was used to initiate the growth on the sapphire substrate, then a 1.5 µm thick intrinsic i-GaN layer was grown at 1000 °C and 76 Torr using conventional low-

---
[1]) Corresponding author; Phone: 803-777-8921; Fax: 803-777-2447; e-mail: koukstis@engr.sc.edu
*) On leave from Department of Semiconductor Physics, Vilnius University, Sauletekio al. 9, 2040 Vilnius, Lithuania.

pressure MOCVD. This was followed by depositing either $Al_xIn_yGa_{1-x-y}N$-based MQW structures or 0.1 μm thick $Al_xIn_yGa_{1-x-y}N$ epilayers with composition corresponding to the well materials. These active layers were grown at 760 °C by PALE technique. Al molar fraction in the quantum wells and barrier layers was close to 20% and 40%, respectively. The indium content in both wells and barriers was 1% and 2%. The MQW structures consisted of four 4 nm thick wells separated by 5 nm thick barriers.

The PL spectra were measured in a wide temperature range (7–300 K) using pulsed excimer laser ($\lambda = 193$ nm, $\tau = 8$ ns) excitation. The laser beam was focused to the surface of the sample to a spot of about 0.1 mm. The laser light maximum power density could reach $\sim$5 MW/cm^2. The samples were mounted onto the cooled finger of a closed-cycle He cryostat. Luminescence was analyzed in backscattering geometry using monochromator SPEX550 with UV-enhanced charge coupled device (CCD) array.

Results and Discussion The PL spectra of $Al_{0.22}In_{0.02}Ga_{0.76}N/Al_{0.38}In_{0.01}Ga_{0.61}N$ MQWs at 10 K under different excitation power densities are shown in Fig. 1. Two clearly resolved PL bands can be seen. The long-wave band dominates below excitation power density of $\sim$3 kW/cm^2, whereas the short-wave one appears under higher excitation and dominates the spectrum until the maximum possible excitation. Investigation of PL in AlInGaN epilayers revealed quite different properties. The PL spectrum consisted of a broad band at low, as well as at high excitation. In well material epilayers under the excitation below $\sim$5 kW/cm^2 its maximum locates at 3.780 eV. The further increase of pumping leads to a slight ($\sim$14 meV) red-shift of emission maximum. It is shown in Fig. 2 by triangles. Also shown in Fig. 2 is the different behaviour of the two bands in MQW emission. As can be seen a total of $\sim$300 meV blue-shift of emission maximum was observed when excitation was increased from 40 W/cm^2 to 2 MW/cm^2.

The PL temperature dependence was similar both in AlInGaN epilayers and MQWs except at lower excitations. In the latter case the long-wave band position in MQWs slightly shifts towards the high energy side when temperature is increased, whereas the emission maximum at high excitation undergoes a long-wave shift. The PL spectrum of AlInGaN epilayer at various temperatures and $\sim$1 MW/cm^2 excitation is demonstrated in Fig. 3.

The temperature change from 7 to 300 K also leads to the decrease of PL intensity about three times in both epilayers and MQWs. Note that the long-wave band was not resolved at room temperature even under the lowest excitation.

At room temperature in AlInGaN well and barrier material the epilayer PL maximum remains unchanged

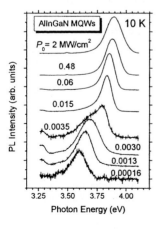

Fig. 1. PL spectra of $Al_{0.22}In_{0.02}Ga_{0.76}N/Al_{0.38}In_{0.01}Ga_{0.61}N$ MQWs at 10 K under different excitation power densities

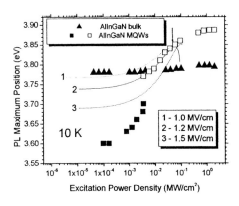

Fig. 2. Position of PL maximum as a function of laser excitation power density in AlInGaN well material epilayer (triangles) and MQWs (squares) at 10 K. Solid and open squares correspond to long-wave and short-wave bands, respectively. Curves 1–3 are theoretical dependences (see text)

up to ~100 kW/cm^2, and shifts towards low energies about 30 meV when excitation is increased to 2 MW/cm^2. For MQWs the PL emission maximum position depends non-monotonously on excitation power. This dependence on excitation power density is shown in Fig. 4 by squares. At first, the maximum undergoes a blue-shift (it shifts about 60 meV when excitation is increased up to 100 kW/cm^2), whereas further increase of excitation leads to a red-shift.

The observed difference in the PL properties of MQWs and epilayers arise from the quantum nature of the structures. First we analyzed the possible built-in electric field.

As the buffer layer is thick GaN epilayer (~1.5 μm) we assume that it is completely relaxed and thin quantum AlInGaN structures grown on the top undergo tensile in-plain strain due to lattice mismatch and remain in a pseudomorphic state. In the calculations we have used the parameters of AlN, GaN and InN from Ref. [6, 7] and linearly extrapolated them according to the fraction of each compound in quaternary alloy. Our calculation has shown that barriers and wells undergo tensions of 0.815% and 0.314%, respectively. It corresponds to piezoelectric charges at interfaces induced by this mismatch, –0.0484 C/m^2 for the well and –0.0134 C/m^2 for the barrier [8]. Meanwhile, polarization charge was calculated as –0.041 and –0.049 C/m^2 for wells and barriers, respectively [7]. The total electric field in the well for an alternating sequence of barriers and wells is found to be 1.2 MV/cm. Note that 50% of the field was due to piezoelectric effect and the rest 50% was caused by spontaneous polarization both having the same direction. Under high excitation the generated nonequilibrium carriers accumulate at the well–barrier interface and produce opposite electric fields and lead to a screening of built-in-field and thus a blue-shift of

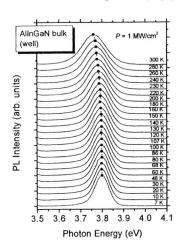

Fig. 3. Normalized PL spectra of Al$_{0.22}$In$_{0.02}$Ga$_{0.76}$N epilayers at different temperatures

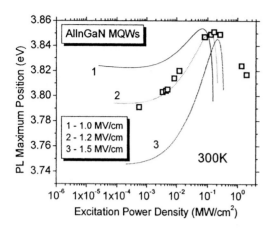

Fig. 4. Experimental data (squares) and theoretical dependences (curves 1–3, see text) of PL maximum position as a function of excitation power density at room temperature

emission maximum. Assuming square recombination as the main radiative channel for band-to-band transitions and using well material parameters from [6], we determined the PL maximum position as a function of excitation power density. The dependences are shown as curves 2 in Figs. 2 and 4 for a field of 1.2 MV/cm. As can be seen, quite satisfactory agreement is achieved with experimental data.

At low temperature and low excitation field screening is negligible and blue-shift, as well as the low energy band have different origin. The most probable mechanism is the localization of carriers (presumably excitons) at imperfections in MQWs, namely, quantum structure interface disorders. In this case, density of state tails occur below the forbidden gap and emission band red-shift with respect to absorption edge is observed. The increase of excitation leads to the filling of tail states and hence blue-shift of emission. The PL band also should vanish with increasing temperature. Detailed theoretical and experimental investigations of PL spectra and kinetics of disordered II–VI solid solutions from localized states at low temperature in Ref. [9] show a similar behavior of emission including strong red-shift of PL band, its dependence on excitation and band shape.

Acknowledgements This work at USC was supported by the Ballistic Missile Defense Organization (BMDO) under Army SMDC contract DASG60-98-1-0004, monitored by Mr. Tarry Bauer, Dr. Brian Strickland and Dr. Kepi Wu. The work at SET, Inc. was supported by the Office of Naval Research under Small Business Technology Transfer Program monitored by Dr. Y.-S. Park.

References

[1] S. J. Pearton, J. C. Zolper, R. J. Shul, and F. Ren, J. Appl. Phys. **86**, 1 (1999).
[2] S. Nakamura and G. Fasol, The Blue Laser Diode, Springer-Verlag, Berlin 1997.
[3] M. E. Aumer, S. F. LeBoeuf, F. G. McIntosh, and S. M. Bedair, Appl. Phys. Lett. **75**, 3315 (1999).
[4] M. A. Khan, J. W. Yang, G. Simin, R. Gaska, M. S. Shur, G. Tamulaitis, A. Zukauskas, D. J. Smith, D. Chandrasekhar, and R. Bicknell-Tassius, Appl. Phys. Lett. **76**, 1161 (2000).
[5] Jianping Zhang, J. W. Yang, G. Simin, M. Shatalov, M. A. Khan, M. S. Shur, and R. Gaska, Appl. Phys. Lett. **77**, 2668 (2000).
[6] D. W. Palmer, www.semiconductors.co.uk, 2001.04.
[7] F. Bernardini, V. Fiorentini, and D. Vanderbilt, Appl. Phys. Lett. **56**, R10024 (1997).
[8] T. Takeuchi, S. Sota, M. Katsuragawa, M. Komori, H. Takeuchi, H. Amano, and I. Akasaki, Jpn. Appl. Phys. **36**, L382 (1997).
[9] A. Klochikin, A. Reznitsky, S. Permogorov, T. Breitkopf, M. Grü, M. Hetterich, C. Klingshirn, V. Lysenko, W. Langbein, and J. M. Hvam, Phys. Rev. B **59**, 12947 (1999).

Photoluminescence Study of Piezoelectric Polarization in Strained $Al_xGa_{1-x}N/GaN$ Single Quantum Wells

V. Kirilyuk[1]) (a), P. R. Hageman (a), P. C. M. Christianen (a), F. D. Tichelaar (b), and P. K. Larsen (a)

(a) Research Institute for Materials, University of Nijmegen, Toernooiveld 1, NL-6525 ED Nijmegen, The Netherlands

(b) National Center for HREM, Laboratory of Materials Science, Delft University of Technology, Rotterdamseweg 137, NL-2628 AL Delft, The Netherlands

(Received June 25, 2001; accepted July 19, 2001)

Subject classification: 77.65.Ly; 78.55.Cr; 78.67.De; S7.14

We report a low temperature photoluminescence study on two identical $Al_{0.13}Ga_{0.87}N/GaN$ single quantum wells (QWs), which are pseudomorphically grown on either a GaN or an AlGaN buffer layer. The red shift of the QW emission due to the quantum confined Stark effect, is found to be strongest in the QW deposited on AlGaN, in contrast to what is to be expected form the estimated built-in electric fields due to spontaneous and piezoelectric polarization fields. Screening of the built-in electric field by a relatively high sheet charge is one of the possible reasons for the observed discrepancy.

Introduction The optical properties of $Al_xGa_{1-x}N/GaN$ quantum wells (QWs) strongly depend on the built-in electric field that originates from both spontaneous polarization and strain-induced piezoelectric fields. In particular, QW transition energies vary in a non-trivial manner with well and barrier width, excitation intensity, doping profile and built-in strain. Theoretically [1, 2] and experimentally [3–8] obtained values of the piezoelectric and spontaneous polarization show considerable discrepancies, which are primarily due to uncertainties in the material constants, the high sensitivity of the QW structures to the growth conditions and the large variety in the layer compositions of the heterostructures used. Most of the reported studies have been performed on multiple $Al_xGa_{1-x}N/GaN$ QWs with a relaxed QW region and strained barriers. In order to determine the relative importance of strain-induced piezoelectric effects, we have investigated two identical single $Al_xGa_{1-x}N/GaN$ QWs deposited by metal-organic chemical vapor deposition (MOCVD) on either a GaN (sample I) or an $Al_xGa_{1-x}N$ (sample II) buffer layer. Assuming that the QW structures are grown coherently on the buffer, either the barrier (I) or the QW (II) is strained. Low temperature photoluminescence (PL) measurements reveal a significantly larger red shift of the ground state energy for the QW deposited on AlGaN as compared to that on GaN. The change in the transition energies is discussed in terms of the differences in piezoelectric and spontaneous polarization fields and screening due to the presence of an excess background charge density.

Experimental Details The two single QW structures, with a (2.8 ± 0.3) nm well width and a (55 ± 1) nm barrier thickness, were grown in a horizontal MOCVD reactor on

---

[1]) Corresponding author; Phone: +31 24 365 30 75; Fax: +31 24 365 26 20; e-mail: vika@sci.kun.nl

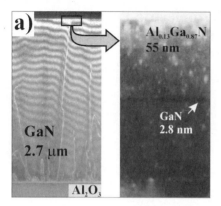 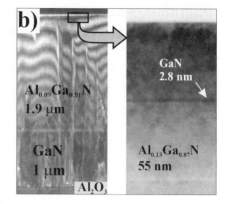

Fig. 1. TEM images of samples a) I and b) II

(0001) sapphire substrates [9]. The QWs were deposited either on a 2.7 μm thick GaN or on a 1.9 μm thick $Al_{0.09}Ga_{0.91}N$ buffer layer. In the latter case, a 1 μm GaN intermediate layer was used between the $Al_{0.09}Ga_{0.91}N$ layer and the sapphire substrate. The growth of the GaN and $Al_{0.09}Ga_{0.91}N$ buffers was performed at 1170 and 1190 °C, respectively, resulting in a growth rate of ∼1.7 μm/h. Ammoniac, trimethylgallium and trimethylaluminium were used as precursors. The $Al_{0.13}Ga_{0.87}N$ cladding layers were grown at conditions identical to those of the GaN deposition, i.e. a growth temperature of 1170 °C, to avoid long growth stops at the QW interfaces. Both the active and cladding layers were grown using a 50% reduced growth rate in order to increase the interface sharpness. Transmission electron microscope (TEM) cross-sectional images were obtained on a CM30T Philips instrument using an accelerating voltage of 300 kV. The samples were mechanically ground, polished to a ∼15 μm thickness and subsequently thinned to electron transparency by a Gatan PIPS 691 ion mill, using Ar at 3.5–4.5 kV. The TEM images (Fig. 1) were used to determine the thickness of the layers and to monitor the continuity of the QWs, the sharpness of the interfaces, and the dislocation densities.

Low temperature photoluminescence (PL) spectra were measured in a cold-finger helium-flow cryostat using the 325 nm line of a 50 mW CW HeCd laser. The PL signal was dispersed by a 0.6 m single grating monochromator and detected by a cooled GaAs photomultiplier, resulting in a spectral resolution of about 0.25 meV.

Built-in Electric Fields Figure 2 shows a schematic representation of the spontaneous and piezoelectric polarizations in the QWs under investigation. Spontaneous polarization depends only on the material composition and has a fixed direction in the crystal. Since the structural parameters of both QWs are the same, the electric fields induced by the spontaneous polarization in the active layer must be identical in samples I and II [1]. A tensile strained barrier with negative piezoelectric polarization has basically the same effect as a compressively strained well with positive piezoelectric polarization. Moreover, because the thickness of both QWs is the same, the difference in the induced electric fields does only depend on the piezoelectric polarization. The piezoelectric constants of AlGaN are larger than those of GaN [2], while the change in the lattice parameters is identical. This implies that the polarization induced electric field

a) Tensile-strained barrier

b) Compressively-strained QW

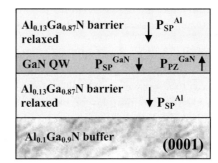

Fig. 2. Spontaneous P_{SP} and piezoelectric P_{PZ} polarizations in pseudomorphic $Al_{0.13}Ga_{0.87}N/GaN$ single QWs grown on a) thick GaN or b) AlGaN buffer. In case a) the barrier is tensile strained and the active layer is relaxed, in b) the barrier is relaxed and the active layer is compressively strained. The spontaneous polarization is identical in both structures

and, therefore, the red-shift of the QW-related PL peak due to the quantum confined Stark effect is expected to be larger in sample I, which is grown on GaN.

Photoluminescence Results Figure 3, curves a and b display PL spectra measured at $T = 4$ K using an excitation power of 30 W/cm^2, on samples I and II, respectively. Pronounced PL peaks are observed from the QW layers (FWHM $\sim$30 meV), the barriers and the buffer layers. From the energy positions of the barrier and buffer related peaks we can accurately determine the Al composition in the different layers and the amount of stress in the thick GaN layer of sample I. The energy position of the free A exciton in the bulk-like GaN buffer shows a slight compressive strain in comparison with homoepitaxial GaN layers [9]. The resulting strain in the active layer of sample I can, however, be neglected because of its much smaller value as compared to that of sample II. The Al compositions in the barriers and the thick AlGaN buffer of sample II are estimated to be $\sim(13 \pm 1)\%$ and $\sim(9 \pm 1)\%$, respectively (Ref. [7]). Secondary Ion Mass Spectroscopy (SIMS) measurements have confirmed this slightly lower Al fraction in the buffer layer, which was caused by the difference in the growth rates. The most striking result in Fig. 3 is the fact that the red shift of the QW emission appears to be largest in sample II, grown on the $Al_{0.09}Ga_{0.91}N$ buffer, in contradiction with the predic-

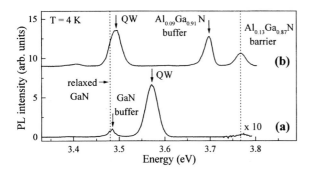

Fig. 3. Photoluminescence spectra ($T = 4$ K) of the $Al_{0.13}Ga_{0.87}N/GaN$ single QWs grown on a (a) thick GaN or (b) AlGaN buffer layer

tions mentioned above. It should be mentioned that the PL peaks of both QWs do not shift with varying the excitation power by two orders of magnitude (not shown), which proves that the experiments correspond to the low power regime, meaning that screening of the internal electric fields by photo-excited carriers is absent.

Discussion Following the reasoning above the electric field induced in the GaN single QW grown on a GaN buffer (sample I), should be slightly larger as compared to that of the identical QW grown on AlGaN (sample II). However, the PL data (Fig. 3) show the largest red-shift (meaning larger built-in electric field) for sample II. From our estimations, the possible experimental errors and uncertainties in the determined parameters of the QWs cannot account for the observed discrepancy. The only reason for the unexpected small red shift of the QW peak in sample I may be the high sheet charge density in this sample. The sheet charge densities found in our HEMTs structures, grown under similar conditions are varying in the range $(3-6) \times 10^{12}$ cm^{-2}, which is high enough to partially screen the polarization-induced electric fields in the QWs. Indeed, the position of the PL peak in QW I is considerably blue-shifted as compared to similar QWs grown by MBE, suggesting a high carrier concentration [5]. The reason why both samples, grown under similar conditions but on different buffer layers, would contain such different background concentrations is not clear at present, and further investigations are necessary.

Conclusions Low temperature experiments of Al$_{0.13}$Ga$_{0.87}$N/GaN single QWs grown on either a GaN or an AlGaN buffer are presented. Both structures show pronounced PL emission of the QW, the peak position of which reflects the presence of polarization-induced electric fields. The PL peak in the QW grown on AlGaN is stronger shifted to lower energy as compared to that of the QW grown on GaN. This effect is attributed to screening of the built-in electric field in the latter sample by the presence of sheet charges at the well/barrier interfaces.

References

[1] V. Fiorentini, F. Bernardini, F. Della Sala, A. Di Carlo, and P. Lugli, Phys. Rev. B **60**, 8849 (1999).
   P. Bigenwald, A. Kavokin, B. Gil, and P. Lefebvre, Phys. Rev. B **61**, 15621 (2000).
[2] F. Bernardini, V. Fiorentini, and D. Vanderbilt, Phys. Rev. B **56**, R10024 (1997).
[3] M. Smith, J. Y. Lin, H. X. Jiang, A. Salvador, A. Botchkarev, W. Kim, and H. Morkoç, Appl. Phys. Lett. **69**, 2453 (1996).
[4] M. Leroux, N. Grandjean, M. Laugt, J. Massies, B. Gil, P. Lefebvre, and P. Bigenwald, Phys. Rev. B **58**, R13371 (1998).
[5] N. Grandjean, J. Massies, and M. Leroux, Appl. Phys. Lett. **74**, 2361 (1999).
[6] R. A. Hogg, C. E. Norman, A. J. Shields, M. Pepper, and N. Iizuka, Appl. Phys. Lett. **76**, 1428 (2000).
[7] G. Steude, B. K. Meyer, A. Goldner, A. Hoffmann, F. Bertram, J. Christen, H. Amano, and I. Akasaki, Appl. Phys. Lett. **74**, 2456 (1999).
[8] R. Cingolani, A. Botchkarev, H. Tang, H. Morkoç, G. Traetta, G. Coli, M. Lomascolo, A. Di Carlo, F. Della Sala, and P. Lugli, Phys. Rev. B **61**, 2711 (2000).
[9] V. Kirilyuk, A. R. A. Zauner, P. C. M. Christianen, J. L. Weyher, P. R. Hageman, and P. K. Larsen, Appl. Phys. Lett. **76**, 2355 (2000).

GW Self-Energy Correction to the Band Mass of Nitride Semiconductors

M. Oshikiri[1]) (a) and F. Aryasetiawan (b)

(a) National Institute for Materials Science, 3-13 Sakura, Tsukuba, Ibaraki 305-0003, Japan

(b) Research Institute for Computational Sciences, 1-1-1 Umezono, Chuo 2, Tsukuba, Ibaraki 305-8568, Japan

(Received June 21, 2001; accepted July 4, 2001)

Subject classification: 71.15.Mb; 71.20.Nr; S7.14

The effective masses of nitride semiconductors obtained by the local density approximation have been corrected by taking into account the dynamical screening effect, which has been calculated by the so-called GW self-energy calculated within the GW approximation. The obtained correction factors would be very useful for device design technology.

Introduction Most ab initio calculations on band structure so far have been based on the density functional theory (DFT) [1, 2] within the local density approximation (LDA) [1, 2]. They are intended to describe the ground state and cannot describe the dynamically excited states in principle. In the conventional DFT-LDA, the important properties of non-locality and energy dependence in the exchange–correlation potential are neglected and the self-interaction problem for the occupied states is present. These defects bring the smaller band gap than experiment in semiconductors and sometimes give wrong effective mass. The mass obtained by the DFT-LDA is not in general the mass of the quasiparticle (QP) formed by the dynamical many-body interaction. One of the first-principles methods, the so-called GW approximation (GWA) [3–5] is known as a relatively simple way to remove these defects in the conventional DFT-LDA since the self-energy (SE) correction is taken into account. It can take into account both the non-locality and dynamic (i.e. energy dependent) features of correlation in the many-body system within random phase approximation (RPA; i.e. screening effect) [6]. The self-interaction problem is also removed and good band gap value is predicted. But electron-phonon coupling is not included. This work is intended to obtain the factors to correct the LDA band mass of the nitride semiconductors, which are in the focus of recent technological interest, by the GWA approach. An early work in this field may be found in Ref. [7].

Theory and Computational Framework The theoretical framework is as follows. The standard linear muffin tin orbital (LMTO) basis [8] within the atomic sphere approximation (ASA) is used to obtain the LDA eigen values and wavefunctions by solving the following equation:

$$(E_j^{LDA} - H_0)\, \varphi_j - V_{XC}\varphi_j = 0, \qquad V_{XC} = V_{XC}^{LDA}. \tag{1}$$

---

[1]) Corresponding author; Phone: +81 298 59 5030; Fax: +81 298 59 5010; e-mail: OSHIKIRI.Mitsutake@nims.go.jp

These LDA eigen values and wavefunctions are employed to make an initial Green function for the GWA scheme. H_0 includes the kinetic energy operator, the potential due to the ions and the Hartree potential of the electrons. V_{XC}^{LDA} is the exchange correlation potential within the LDA. j denotes the band index. The 3d orbitals of Ga and In are treated as valence states explicitly in this study. The polarization function P is computed with the Green function which is constructed from the LDA eigen energies and functions, $P = -iGG$, where $G = G_0(E_j^{LDA}, \varphi_j^{LDA})$. The dynamically screened potential is calculated within the RPA using the dielectric function ε as $\varepsilon = 1 - vP$, $W = \varepsilon^{-1}v$ where v is bare Coulomb potential, $v = 1/(|r'' - r'|)$. The SE is obtained by the convolution of the Green function and the screened potential by

$$\Sigma(k;\omega) = i \int \frac{d\omega'}{2\pi} G(k;\omega+\omega') W(k;\omega') e^{i\delta\omega'}, \qquad (2)$$

where $G(k;\omega+\omega')$ is the full Green function and $\delta = 0^+$. The many-body exchange and relatively long-range correlation (i.e. screening effect) corrections are taken into account by this nonlocal and energy-dependent SE operator. The LDA eigen energy is corrected by the obtained SE, and the QP energy structure is obtained at arbitrary k-point as follows:

$$E_j^{QP} = E_j^{LDA} + \langle j | \Sigma - V_{XC}^{LDA} | j \rangle. \qquad (3)$$

The procedure of the effective mass correction is as follows. The effective mass of the quasiparticle is defined as [9]

$$\frac{dE_j^{QP}}{dE_j^{LDA}} = \frac{m_j^{LDA}}{m_j^{QP}}, \quad \text{where} \quad E_j^{QP} = E_j^{LDA} + \Delta \operatorname{Re} \Sigma(\mathbf{k}, E_j^{QP}). \qquad (4)$$

From Eq. (4), the effective mass at around the Γ-point of the QP in the $\mathbf{k}_\alpha$ direction is obtained as follows:

$$\frac{m_{j,\mathbf{k}_\alpha}^{LDA}}{m_{j,\mathbf{k}_\alpha}^{QP}} = \lim_{|\mathbf{k}_\alpha| \to 0} \frac{1 + \frac{m_j^{LDA}}{\hbar^2 |\mathbf{k}_\alpha|} \frac{\partial \Delta \operatorname{Re} \Sigma(\mathbf{k},\omega)}{\partial \mathbf{k}}\Big|_{\mathbf{k}=\mathbf{k}_\alpha}}{1 - \frac{\partial \Delta \operatorname{Re} \Sigma(\mathbf{k}_\alpha,\omega)}{\partial \omega}\Big|_{\omega=E_j^{QP}}}. \qquad (5)$$

Therefore, the LDA effective mass is corrected by both the energy dependence (the denominator of Eq. (5)) and the k-dependence (the numerator of Eq. (5)) of the SE. We define the part of the energy dependence as Zfac as follows:

$$\text{Zfac} = \left[1 - \frac{\partial \operatorname{Re} \Delta\Sigma(\omega)}{\partial \omega}\Big|_{\omega=E_j^{QP}}\right]^{-1}. \qquad (6)$$

The convergence of the energy dependence of the SE is much better than that of the k-dependence. To obtain sufficiently high accuracy for the k-dependence of the SE, one needs a large computational task at present. Fortunately, the simpler case for AlN shows that the k-dependence accounts for only less than a few percents, which is much smaller than the correction by the energy dependence of the SE. So we could regard the numerator of Eq. (5) to be unity, approximately. For the conventional AlN, GaN, InN, we calculate the Zfac by using the experimental lattice constants for more meaningful comparison with the experimental mass data. The experimental lattice constants

Table 1
Experimental lattice constants (in Å) of AlN, GaN and InN in the wurtzite structure

compound	a	c	u
AlN	3.110	4.980	0.382
GaN	3.190	5.189	0.377
InN	3.5446	5.7034	0.375

are shown in Table 1. On the other hand, the lattice constants of AlGaN$_2$, GaInN$_2$ and AlInN$_2$, have been obtained by minimizing the total energy obtained by DFT-LDA calculation with assumption that those structures have ideal wurtzite structure. The predicted lattice constants are shown in Table 2. The spin–orbit interaction is not included and the spin direction is not distinguished here. In this work, we simply present the k-dependence of the LDA eigen energy $\varepsilon(k)$ as a parabolic feature around the conduction band minimum (cross terms of the k-dependence of the energy are neglected):

$$\varepsilon(k) = \frac{\hbar^2}{2}\left(\frac{k_x^2}{m_x} + \frac{k_y^2}{m_y} + \frac{k_z^2}{m_z}\right). \tag{7}$$

The directions x, y and z correspond to the directions of Γ–M, Γ–K and Γ–A in the first Brillouin zone, respectively. The k-dependence around the valence top is not discussed at present, however, the Zfac of the valence top was calculated.

Results Table 3 indicates the Zfac of nitride compound semiconductors for the conduction-band bottom and the valence-band top. Since the QP mass parameter, m_a^{QP}, is obtained as $m_a^{QP} = m_a^{LDA}/\text{Zfac}$ from Eqs. (5) and (6), the tendency that the corrections

Table 2
Lattice constants (in Å) of AlGaN$_2$, GaInN$_2$, AlInN$_2$ obtained by first-principles method of DFT-LDA under the condition of $c/a = \sqrt{8/3}$, $u = 3/8$

artificial compound	a	c	u
AlGaN$_2$	3.13	5.11	0.375
GaInN$_2$	3.37	5.50	0.375
AlInN$_2$	3.35	5.47	0.375

Table 3
The Zfac of nitride compound semiconductors. C.B. and V.T. mean conduction-band bottom and valence-band top, respectively

nitride compound	Zfac of C.B.	Zfac of V.T.
AlN	0.853	0.827
GaN	0.838	0.805
InN	0.813	0.766
AlGaN$_2$	0.831	0.806
GaInN$_2$	0.821	0.775
AlInN$_2$	0.841	0.800

Table 4

The effective QP (LDA) mass parameters for conduction bands of the nitride compounds. The unit is the electron rest mass m_0

nitride compound	$m_{\Gamma-M}$	$m_{\Gamma-K}$	$m_{\Gamma-A}$
AlN	0.37 (0.32)	0.37 (0.32)	0.38 (0.32)
GaN	0.21 (0.17)	0.21 (0.17)	0.19 (0.16)
AlGaN$_2$	0.27 (0.23)	0.27 (0.23)	0.32 (0.26)
GaInN$_2$	0.064 (0.052)	0.064 (0.052)	0.085 (0.070)
AlInN$_2$	0.16 (0.13)	0.16 (0.13)	0.17 (0.15)

for around the valence top are larger than those for conduction band has been found. The dynamical screening effect of the RPA enhances the mass of the conduction band bottom by about 16–23% and the valence-band top by 20–31%. Table 4 shows the QP mass parameters compared with the LDA mass parameters. The LDA mass parameters are enclosed by parentheses in the table.

For example, in the case of GaN, the experimental effective mass is reported to be $0.22m_0$, which was obtained by cyclotron resonance and therefore can be regarded as a polaron mass. If the polaron effect is subtracted, the bare mass is about $0.20m_0$ [10]. The QP mass of the conduction-band electron in the plane perpendicular to the c-axis, which corresponds to the alignment of the experiment, is regarded to be $0.21m_0$ approximately. They are in good agreement. Similar correspondence was confirmed in the ZnO system, previously [11]. Since it is known that the LDA calculation for InN system does not work very well, we do not describe the k-dependence of the LDA bandstructure of InN system in detail.

Conclusion The correction factors for the LDA mass obtained by the GW self-energy have been calculated. We have found that the dynamical screening effect of the RPA enhances the mass of the conduction-band bottom by about 16–23% and the valence-band top by 20–31%. The experimental conduction-band mass of GaN well corresponds to our calculation. Since we have confirmed a good agreement also for the conduction-band mass of ZnO in our previous work, the correcting method by Zfac seems promising. We have predicted the masses of AlN and of related artificial nitride semiconductors. They would be convenient for device designing since reliable experimental data do not always exist for those materials.

References

[1] P. HOHENBERG and W. KOHN, Phys. Rev. **136**, B864 (1964).
[2] W. KOHN and L. J. SHAM, Phys. Rev. **140**, A1133 (1965).
[3] L. HEDIN and S. LUNDQVIST, Solid State Phys. **23**, 1 (1969).
[4] M. S. HYBERTSEN and S. G. LOUIE, Phys. Rev. Lett. **55**, 1418 (1985).
[5] R. W. GODBY, M. SCHLUETER, and L. J. SHAM, Phys. Rev. B **37**, 10159 (1988).
[6] F. ARYASETIAWAN and O. GUNNARSSON, Rep. Prog. Phys. **61**, 271 (1998).
[7] X. ZHU, M. S. HYBERTSEN, and S. G. LOUIE, Mater. Res. Soc. Symp. Proc. **193**, 113 (1990).
[8] O. K. ANDERSEN, Phys. Rev. B **12**, 3060 (1975).
[9] G. D. MAHAN, Many-Particle Physics, 2nd ed., Plenum Press 1993 (p. 157).
[10] M. DRECHSLER, D. M. HOFMANN, B. K. MEYER, T. DETCHPROHM, H. AMANO, and I. AKASAKI, Jpn. J. Appl. Phys. **34**, L1178 (1995).
[11] M. OSHIKIRI, Y. IMANAKA, T. TAKAMASU, and G. KIDO, Physica B **298**, 472 (2001).

Group-III Nitrides Hot Electron Effects in Moderate Electric Fields

E. A. Barry[1]) (a), K. W. Kim (a), and V. A. Kochelap (b)

(a) Department of Electrical and Computer Engineering, North Carolina State University, Raleigh, North Carolina 27695-8617, USA

(b) Institute of Semiconductor Physics, National Academy of Sciences of Ukraine, Kiev-28, 252650, Ukraine

(Received June 25, 2001; accepted July 9, 2001)

Subject classification: 72.10.Di; 72.20.Ht; S7.14

We studied the distribution function and basic characteristics of hot electrons in InN, GaN and AlN under moderate electric fields, and found that in relatively low fields (of the order of kV/cm) the optical phonon emission dominates the electron kinetics. This strongly inelastic process gives rise to a spindle-shaped distribution function and an extended portion of quasi-saturation of the current–voltage characteristics (the streaming-like regime). We prove that this hot electron regime holds for all three nitrides. We suggest that the effects can be detected by the measurement of the $I-V$ characteristics, or the thermopower of hot electrons in the transverse direction.

Introduction Recent intensive studies of the electron kinetics in group-III nitride materials are mostly focused on two subjects: the problem of the low-field mobility and the problem of the peak (saturation) velocity in extremely high electric fields (hundreds of V/cm, see Refs. [1–5]). Meanwhile, such basic properties of the nitrides such as relatively low effective masses, high optical phonon energies, strong electron–optical phonon interaction and large energy separations of the upper valleys bring about a number of new hot electron effects in moderate electric fields. These effects can be of interest for both the understanding of fundamentals of the electron kinetics in the nitrides and their applications. This paper addresses the hot electron kinetics in group-III nitrides at moderate electric fields.

Model and Results We analyzed the electron kinetics in the nitrides by the Monte-Carlo method. The electron bands for InN, GaN and AlN were considered in the isotropic and parabolic approximation. Scattering by ionized impurities, acoustic phonons, and polar optical phonons were taken into consideration with all material parameters given by Ref. [6]. Corresponding scattering rates as functions of the electron energy are presented in Fig. 1.

As examples of calculations of the electron distribution functions, in Figs. 2a and b we depict the contour plots of these functions for GaN at two values of the electric field. There V_l and V_t are the longitudinal and transversal field projections of the electron velocity, respectively, $V_O \equiv \sqrt{2\hbar\omega/m}$, with m and $\hbar\omega$ being the effective mass and the optical phonon energy, respectively. Using the distribution functions we found the basic characteristics of the electron kinetics for three materials, InN, GaN and AlN. In Figs. 3a and b we depict the average drift velocity, the average kinetic energy asso-

---
[1]) Corresponding author; Phone: 919 515-5402; Fax: 919 515-3027; e-mail: eabarry@eos.ncsu.edu

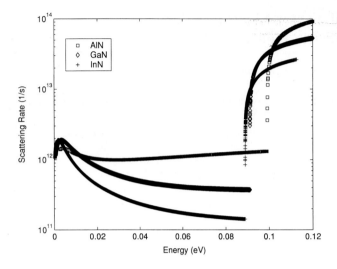

Fig. 1. Energy dependence of the total scattering rate at $T = 77$ K and $N_D = 10^{16}$ cm^{-3}

ciated with the electron motion transverse to the field and the mean-square deviation of the longitudinal component of the electron velocity,

$$v_{dr} = \frac{\langle V_l \rangle}{V_0}, \quad \epsilon_t = \frac{\langle V_t^2 \rangle}{V_0^2}, \quad \epsilon_t = \frac{\langle V_l^2 - V_{dr}^2 \rangle}{V_0^2},$$

respectively. Here $\langle \ldots \rangle$ means averaging over the distribution function, the indexes l and t indicate directions along and transverse to the field, respectively. Results are presented for 77 K.

According to Fig. 1, the rates of optical phonon emission are indeed much higher than other scattering rates. This results in pronounced hot electron effects under electric fields of several kV/cm. At higher fields the optical phonon emission processes begin to dominate the kinetics. A kind of streaming regime is formed: the spreading of the carriers over the "transverse" velocities decreases with increasing field. This manifests itself in decreasing of the transverse average energy as shown in Fig. 2b. The mini-

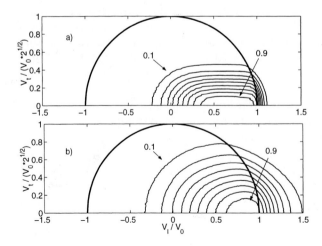

Fig. 2. Velocity space distribution functions at $T = 77$ K, $N_D = 10^{16}$ cm^{-3} and a) $E = 10$ and b) 60 kV/cm. Each contour line represents an increase of 0.1 from 0.1 to 0.9

mum values of ϵ_t are close to the lattice temperature (≈ 77 K for InN). As the drift velocities v_{dr} approach the high value $\approx 1/2$, as seen in Fig. 3, the distribution function takes on a spindle shape as seen in Fig. 2a. Notice that almost all electrons have a positive velocity component V_t and the penetration of the electrons into the region with the energy larger than $\hbar\omega$ is quite small. In this transient field range, the field dependences of ϵ_t, ϵ_l and v_{dr}, are slightly different for the three nitrides. The electric fields corresponding to the lowest ε_t and the maximum ratio ϵ_l/ϵ_t can be interpreted as optimal fields for the streaming regime (most close to the ideal streaming). These fields are found to be 2.5, 10 and 20 kV/cm for InN, GaN and AlN, respectively.

As the electric field is increaced further, the electron kinetics become entirely dominated by optical phonons. Under this regime, in a wide range of the electric field all characteristic parameters, ϵ_t, ϵ_l and v_{dr} depend very slowly on field. Small variations in ϵ_t, ϵ_l mean that the distribution function approximately preserves its shape. The drift velocity demonstrates a linear portion with small slope (almost quasi-saturation). At the end of this 'quasi-satura-

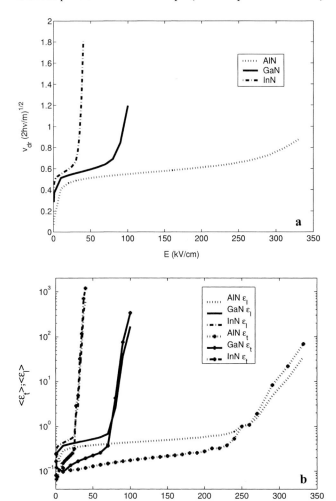

Fig. 3. a) Electric field dependence of the drift velocity. b) Electric field dependence of the longitudinal and transverse components of the average electron kinetic energy $T = 77$ K and $N_D = 10^{16}$ cm^{-3}

tion' portion, both the penetration of the electrons into the active region and subsequent optical phonon emission destroy the spindle shape of the distribution. The new shape of the electron distribution is depicted in in Fig. 2b. Still, the greatest portion of the electrons have a positive longitudinal component of the velocity V_t.

At higher fields the electron distribution spreads quickly over the velocity space, the electron energy sharply increases, as seen from Fig. 3b. The system approaches the regime of the runaway of the electrons [7], where our model fails. For the considered model of the electron spectra the threshold electric fields of the runaway effect in polar materials are: 50, 129 and 415 kV/cm for InN, GaN and AlN, respectively [8].

Conclusion In conclusion, we revisited the problem of hot electrons in polar materials under moderate electric fields. The analysis was applied to the group-III nitrides for which we studied the distribution functions and field dependences of basic hot-electron characteristics. We established that the optical phonon scattering becomes dominant in relatively low electric fields (the kV/cm range). As a result of this strongly inelastic scattering, a spindle-shaped electron distribution occurs. The formation of the streaming-like transport regime is supplemented by a nonmonotoneous dependence of the electron transverse energy ϵ_t. The regime manifests itself in an extended linear portion of the $I-V$ characteristic with very small differential mobility (almost saturation regime). This is shown to be a universal character of this behavior for all three nitrides (it is valid also for their alloys). We suggest that the discussed regimes can be experimentally detected not only through the measurements of the $I-V$-characteristic, but also through the study of transverse hot electron effects. Specifically, the thermopower of hot electrons measured in directions transverse to the current gives directly the transverse energy ε_t (see Ref. [9] and references therein). The thermopower measurements would allow the determination of an optimal streaming regime. A spindle-shaped electron distribution is favorable for the practical application – generation of microwave emission in sub-TGz range [10] (see also recent calculations of high-frequency differential mobility for GaN in Ref. [11]).

Acknowledgements This work was supported by the ONR under MURI-program, Grant No. N00014-98-1-0654 (Project Monitor Dr. J. C. Zolper) and by the ARO. V.A.K. would like to acknowledge the support from ERO of US Army (Contract N68171-01-M-5166).

References

[1] S. J. PEARTON, J. C. ZOLPER, R. J. SHUL, and F. REN, J. Appl. Phys. **86**, 1 (1999).
[2] B. GELMONT, K. KIM, and M. SHUR, J. Appl. Phys. **74**, 1818 (1993).
[3] N. S. MANSOUR, K. W. KIM, and M. A. LITTLEJOHN, J. Appl. Phys. **77**, 2834 (1995).
[4] U. V. BHAPKAR and M. S. SHUR, J. Appl. Phys. **82**, 1649 (1997).
[5] B. E. FOUTZ, L. F. EASTMAN, U. V. BHAPKAR, and M. S. SHUR, Appl. Phys. Lett. **70**, 2849 (1997).
   B. E. FOUTZ, S. K. O'LEARY, M. S. SHUR, and L. F. EASTMAN, J. Appl. Phys. **85**, 7727 (1999).
[6] M. LEVINSHTEIN, S. RUMYANTSEV, and M. SHUR (Eds.), Properties of Advanced Semiconductor Materials: GaN, AlN, InN, BN, SiC, SiGe, John Wiley & Sons, Inc., New York 2001.
[7] E. M. CONWELL, High Field Transport in Semiconductors, Academic Press, New York 1967.
[8] S. M. KOMIRENKO, K. W. KIM, V. A. KOCHELAP, and M. A. STROSCIO, Phys. Rev. B **64**, 113207 (2001).
[9] Y. K. POZHELA, in: Hot-Electron Transport in Semiconductors, Vol. 58, Ed. L. REGGIANI, Springer-Verlag, Berlin/Heidelberg/New York 1985 (p. 113).
[10] A. A. ANDRONOV and V. A. KOZLOV, Zh. Eksp. Teor. Fiz. Pisma **17**, 124 (1973).
[11] E. STARIKOV, P. SHIKHTOROV, V. GRUZINSKAS, L. REGGIANI, L. L. VARANI, J. C. VAISSIERE, and J. H. ZHAO, J. Appl. Phys. **89**, 1161 (2001).

Temperature Dependent Transport Parameters in Short GaN Structures

A. F. M. Anwar[1]) (a), Shangli Wu (a), and R. T. Webster (b)

(a) Electrical and Computer Engineering Department, University of Connecticut, CT 06269-2157, USA

(b) Electromagnetics Technology Division, Air Force Research Laboratory, Hanscom AFB, MA 01730, USA

(Received June 26, 2001; accepted July 25, 2001)

Subject classification: 72.10.Bg; 72.80.Ey; 73.50.Dn; S7.14

We present the mobility and diffusion constant in short GaN structures. Differential mobility decreases with increasing temperature while it increases with decreasing sample size. Theoretically calculated transconductance of an AlGaN/GaN HEMT, using the short channel mobility, is in good agreement with experimental data.

Introduction GaN-based bipolar and field effect transistors are recently being vigorously pursued for applications in high power and high temperature applications. Zolper [1] has reported power density as high as 9.2 W/mm at 8 GHz from an AlGaN/GaMN HEMT fabricated on SiC substrate. GaN based FETs have also been successfully demonstrated by Khan et al. [2] to operate at elevated temperatures. They have reported f_T and f_{max} of 22 GHz and 70 GHz, respectively at 25 °C, which decrease to their corresponding values of 5 GHz and 4 GHz at 300 °C. The use of SiC as a substrate for GaN FETs may extend the operating temperature up to 750 °C, as has been demonstrated by Daumiller et al. [3]. Fabrication and growth techniques are also being perfected to increase the operating frequency as has recently been demonstrated by Wu et al. [4], where f_T and f_{max} of 30 GHz and 100 GHz, respectively, have been obtained using a 0.12 μm × 100 μm $Al_{0.2}Ga_{0.8}N$/GaN FET. The modeling of these devices therefore requires the evaluation of the temperature dependence of the transport parameters namely, mobility and diffusion constant.

The theoretical investigation of the carrier transport in bulk GaN has recently been reported by Gelmont et al. [5]. The temperature dependence of the transport properties, such as the velocity-field characteristic, has also been reported [6]. However, these results are only valid for bulk material and their use becomes questionable in devices where the length of the active region is of the order of 1 μm or less. In devices with very small active regions such as narrow base HBTs or sub-micron gate length FETs, carriers transit the active region before a steady state distribution can be achieved. The transit time of electron is comparable to the mean free flight time between collisions, and the electron transport becomes non-stationary and quasi-ballistic. In this paper the temperature dependence of low field mobility and diffusion constant is reported for short GaN structures.

---

[1]) Corresponding author; Phone: 860-486-3979; Fax: 860-486-2447; e-mail:anwara@engr.uconn.edu

Table 1
Materials Parameters [6]

parameter	symbol (units)	GaN	InN	AlN
dielectric constant	ε_∞	4.2	8.4	4.77
dielectric constant	ε_0	8.1	15.3	8.5
mass density	ϱ (10^3 kg/m)	6.1	6.81	3.23
phonon energy	$\hbar\omega_{LO}$ (meV)	91.2	89.0	99.2
piezoelectric constant	h_{pz} (C/m^2)	0.55	0.3	0.5
acoustic deformation potential	E_1 (eV)	12.5	8.5	9.5
effective mass	m^*	$0.22m_0$	$0.11m_0$	$0.48m_0$
lattice constant	a (Å)	5.125	5.7	4.980
band gap	E_g (eV)	3.38	1.89	5.94
sound velocity	v (10^3 m/s)	5	5	10
nonparabolicity parameter	$a = (1 - m^*/m_0)^2/E_g$ (eV^{-1})	0.187	0.135	0.147
valley separation	ΔE (eV)	1.5	1.1	2.5
intervalley deformation potential	D_n (10^9 eV/cm)	1.0	1.5	0.8

Analysis An ensemble Monte Carlo simulation is carried out by taking into account the scattering due to polar optical phonon, acoustic phonon, equivalent and non-equivalent intervalley scattering, impact ionization, ionized impurity, alloy, interface and self-scattering to investigate transport in short structures [6]. The simulation accounts for the Γ–L–X valleys. The valleys are considered to be spherical and the non-parabolicity in energy is modeled as $\gamma = \hbar^2 k^2/(2m^*) = E(1 + \alpha E)$, where $\alpha = (1 - m^*/m_0)^2/E_g$, with E_g being the band gap. The simulation uses about 20000 electrons where each electron suffers 10000 real scattering events. The position of each electron is tracked along the channel, and the calculation proceeds until the electron leaves the channel region. The material constants used in the simulation are listed in Table 1.

Results and Discussion The velocity-field characteristic for GaN is plotted in Fig. 1 with channel length as a parameter. For reference the bulk velocity-field characteristic is also shown. With decreasing channel length the peak in the velocity-field characteristic moves toward higher electric field, or in other word transport becomes ballistic.

In Fig. 2, the differential mobility ($\mu = v/\varepsilon$) is plotted as function of temperature with sample length as a parameter. The mobility is determined at $E = 20$ kV/cm. Mobility decreases with increasing temperature and decreases with increasing channel length. We found that the temperature dependent low field mobility approaches a constant value for channel lengths greater than 1 μm.

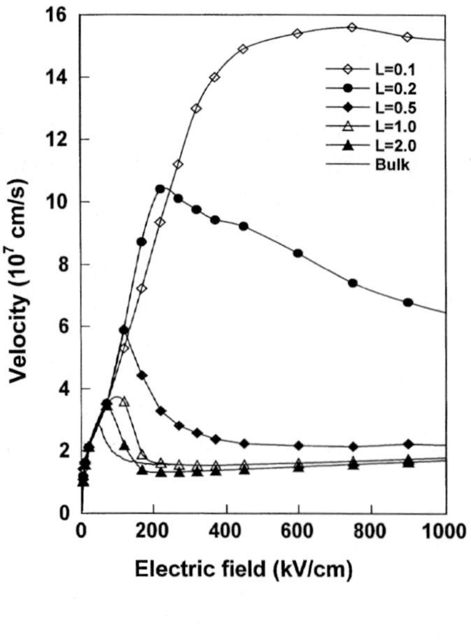

Fig. 1. Drift velocity plotted as a function of the applied electric field with channel length as a parameter (L in μm)

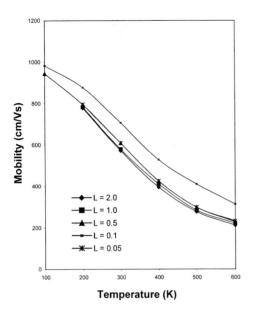

Fig. 2. Differential mobility plotted as a function of temperature for varying sample length (L in μm). The electric field is held constant at 20 kV/cm

The differential mobility is a strong function of the longitudinal electric field as evident from Fig. 3, where μ is plotted as a function of temperature for an applied electric field of 60 kV/cm. This applied field translates roughly into an applied drain bias of 24 V for a GaN/AlGaN HEMT with 4 μm source–drain spacing. The present formulation is validated by comparing theoretically calculated results with experimental transconductance data as shown in Fig. 4. The device under consideration is a 1 μm × 500 μm AlGaN/GaN HEMT reported by Wu et al. [7]. Theoretical results are obtained from the self-consistent solution of the Schrödinger and Poisson equations as a function of temperature [8] to determine 2DEG carrier concentration. For the given device structure, the charge control

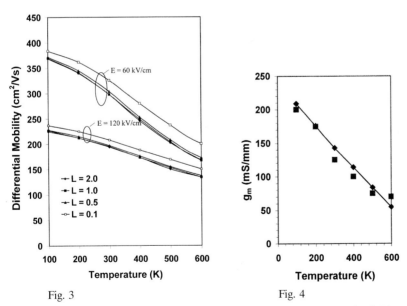

Fig. 3

Fig. 4

Fig. 3. Differential mobility as a function of temperature for electric fields of 60 kV/cm and 120 kV/cm for different sample size (L in μm)

Fig. 4. Comparison of calculated (diamonds) and experimental (squares) transconductance of a 1 μm × 500 μm AlGaN/GaN HEMT [7]

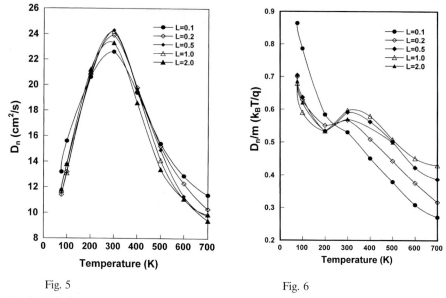

Fig. 5.

Fig. 6.

Fig. 5. Diffusion constant plotted as a function of temperature for varying sample size (L in μm)

Fig. 6. Ratio of diffusion coefficient to mobility normalized to $k_B T/q$ as a function of temperature for different sample size (L in μm)

model is obtained by relating the 2DEG concentration to the applied gate bias. Charge control and temperature dependent mobility are used to compute transconductance.

In Fig. 5, the temperature dependent diffusion coefficient is shown with varying channel length. The diffusion coefficient is computed as one-half of the derivative of the variance of the carrier position

$$D(\varepsilon) = \lim_{t \to \infty} \frac{1}{2} \frac{\delta Z(t) - \delta Z(t_0)}{t - t_0},$$

where $Z(t)$ and $Z(t_0)$ are positions of the electron at time t and t_0, respectively.

The temperature dependence of the ratio of diffusion constant to differential mobility for varying channel length is shown in Fig. 6. For the short channel considered here, Einstein's relationship, $D/\mu = k_B T/q$, no longer applies. The availability of the diffusion constant and mobility presented in this paper enables one to use the drift-diffusion model for transport to analyze devices.

References

[1] J.C. Zolper, in: 1999 IEDM Tech. Dig. (p. 389).
[2] M.A. Khan, M. Shur, J.N. Kuznia, Q. Chen, J. Burm, and W. Schaff, Appl. Phys. Lett. **66**, 1083 (1995).
[3] I. Daumiller et al., in: 1998 Device Research Conf. Dig. (p. 114).
[4] Y.-F. Wu, D. Kapolnek, J. Ibbetson, N.-Q. Zhang, P. Parikh, B. P. Keller, and U. K. Mishra, in: 1999 IEDM Tech. Digest (p. 925).
[5] B. Gelmont, K. Kim, and M. Shur, J. Appl. Phys. **74**, 1818 (1993).
[6] A.F.M. Anwar, Shangli Wu, and R.T. Webster, IEEE Trans. Electron Devices **48**, 567 (2001).
[7] Y.-F. Wu, B.P. Keller, S. Keller, D. Kapolnek, P. Kozodoy, S.P. DenBaars, and U.K. Mishra, Solid-State Electron. **41**, 1569 (1997).
[8] R.T. Webster and A.F.M. Anwar, Mater. Res. Soc. Symp. Proc. **482**, 929 (1998).

Simultaneous Impurity-Band and Interface Conduction in Depth-Profiled n-GaN Epilayers

C. Mavroidis[1]) (a), J. J. Harris[*]) (a), K. Lee (a), I. Harrison (b), B. J. Ansell (b), Z. Bougrioua (c), and I. Moerman (c)

(a) Department of Electronic and Electrical Engineering, University College, London WC1E 7JE, UK

(b) Department of Electronic and Electrical Engineering, Nottingham University, NG7 2RD, UK

(c) INTEC-IMEC, University of Gent, B-9000 Gent, Belgium

(Received July 3, 2001; accepted July 8, 2001)

Subject classification: 71.55.Eq; 73.20.Hb; 73.50.Dn; S7.14

We report on temperature-dependent differential Hall-effect and resistivity measurements, between 10 and 300 K, on two Si-doped GaN epitaxial layers grown by MOCVD on sapphire substrates. These experiments indicate parallel conduction paths in our layers, while depth profiling using plasma-etching shows that two paths are simultaneously present: an impurity band in the Si-doped region and a conducting layer at the GaN/sapphire interface.

Introduction The interpretation of the behaviour of the carrier density and mobility in bulk GaN layers at low temperatures is still unclear: the characteristic fall and then increase of the Hall carrier density, n_H, as the temperature (T) decreases has been proposed to arise from parallel conduction either in a thin degenerate region near the sapphire substrate [1], or alternatively via an impurity band [2, 3]. Recent scanning probe microscopy experiments [4] and depth-profiled temperature-dependent Hall measurements [5] have independently suggested the existence of a compensated donor impurity band near the sapphire-substrate interface. The donor-like impurities were attributed to oxygen atoms diffusing from the Al_2O_3 substrate and the compensating acceptors to Ga vacancies (V_{Ga}) [3, 4, 6]. These results concerned only unintentionally doped samples; in this work, we investigate the transport properties of two silicon-doped layers, by depth-profiling experiments. We see evidence of a conducting region at the interface, but also demonstrate the existence of an impurity band in the Si-doped region which when highly populated, dominates over transport at the interface.

Experimental The GaN layers used in this study were grown on sapphire substrates by low-pressure (100 Torr) MOCVD, as previously described [7]. Following a low temperature GaN nucleation layer, and a ~0.2 μm undoped GaN buffer layer, the samples were doped with SiH_4 at donor concentrations of 2.5×10^{18} and 5×10^{17} cm^{-3} for samples #1 and #2, respectively. Hall effect and resistivity measurements were performed between room temperature and 10 K in a magnetic field of 0.28 T, on square samples with Ti/Al ohmic contacts. The contacts were then masked with photoresist, and the

---

[1]) Corresponding author; Phone: +44 20 767 93 326, Fax: +44 20 738 74 350, e-mail: c.mavroidis@ee.ucl.ac.uk
[*]) Present address: Thermo VG Semicon, East Grinstead, W. Sussex RH19 1TZ, UK.

samples were reactive ion plasma etched using SiCl$_4$, with ~0.25–0.5 μm being removed in each etch. Electrical characterisation between each of the etch steps gave the depth dependence of carrier density and mobility.

Results and Discussion Figures 1a, b and 2a, b plot the Hall sheet carrier density, n_{Hs}, and mobility, μ_H, of samples #1 and #2, respectively, against temperature after each etch step.

These curves are characteristic of material in which conduction occurs both in the conduction band and in a parallel path which is metallic at low temperature. Initially, there is a reduction in sheet density with etching which depends linearly on thickness. This is shown in Fig. 3a (from 1.2 to ~0.3 μm) and the first three etch steps in Fig. 3b (from 2.17 to ~0.3 μm), and corresponds to the uniform Si-doped regions in the bulk of the GaN. If an interface layer were responsible for the parallel conduction, we would expect the plots in Fig. 3 to tend to the (non-zero) interface sheet carrier density as we approach the substrate. Such behaviour is seen in Fig. 3b, where below 0.3 μm the sheet carrier density saturates at a value of ~3 × 10^{13} cm^{-2}. We have seen similar flat characteristics in an undoped GaN layer [5], and thus attribute the flat part of Fig. 3b to the undoped buffer region of sample #2. Unfortunately, we were unable to measure sample #1 below 0.3 μm, but our results put an upper limit on the interface carrier density in this layer of ~1 × 10^{13} cm^{-2}.

The mobilities in Figs. 1b and 2b tend to decrease with decreasing layer thickness, probably as a result of the removal of better quality material in which the conduction band mobility is higher. After the fourth etch, μ_H for sample #2 is almost temperature independent at ~30 cm^2/Vs, which is indicative of degenerate Fermi statistics in the interfacial region [6].

By rearranging the equations for the Hall carrier density and mobility in multilayer systems [1], we can characterise the sample in layers of thickness t, equal to the etching thickness, to obtain the average carrier density, n_r, and mobility, μ_r, of each removed layer:

$$n_r = (n_{sb}\mu_b - n_{sa}\mu_a)^2 / t(n_{sb}\mu_b^2 - n_{sa}\mu_a^2), \tag{1}$$

$$\mu_r = (n_{sb}\mu_b^2 - n_{sa}\mu_a^2)/(n_{sb}\mu_b - n_{sa}\mu_b), \tag{2}$$

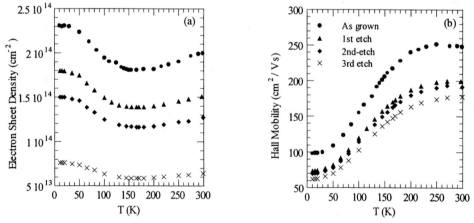

Fig. 1. Temperature dependence of a) sheet carrier density n_{Hs} and b) mobility μ_H after each etch for sample #1. (●) As grown, 1.18 μm thick; (△) 0.94 μm; (♦) 0.82 μm; (×) 0.56 μm

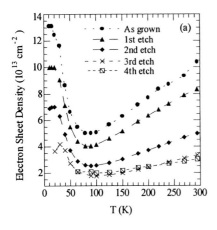 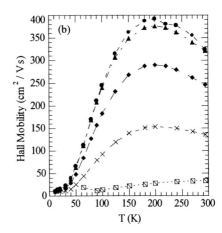

Fig. 2. Temperature dependence of a) sheet carrier density n_{Hs} and b) mobility μ_H after each etch for sample #2. (●) As grown, 2.17 μm thick; (□) 1.74 μm; (♦) 0.81 μm; (×) 0.31 μm; (⊠) 0.08 μm

where n_{sb} and μ_b are the Hall sheet carrier density and mobility of the sample measured before an etch, and n_{sa} and μ_a are the same parameters measured after the etch.

Figure 4a plots the temperature dependence of n_r for each etch step of sample #1. For all three layers, the volume density is $\sim$2.4–2.5 × 10^{18} cm^{-3} at 300 K, falls to a shallow minimum at $\sim$150 K, and approaches $\sim$3 × 10^{18} cm^{-3} at 10 K; this indicates that all three regions are similarly doped, and still contain two conduction paths. Since this analysis has removed the effect of any parallel conduction at the interface, the observed behavior in $n_H(T)$ must originate from a high density ($\geq 3 \times 10^{18}$ cm^{-3}) donor band in the Si-doped region. This data allows accurate impurity-band parameters (activation energy, mobility, compensation, etc.) to be extracted [1, 2, 8] without errors from interface effects, and this work is in progress.

Figure 4b plots these quantities for sample #2; the first three layers correspond to the Si-doped region, with 300 K carrier densities of 4–5 × 10^{17} cm^{-3}, and an impurity band

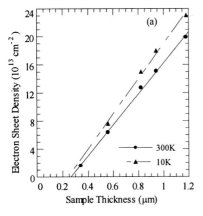 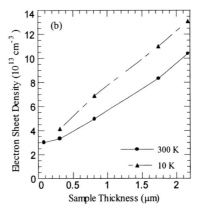

Fig. 3. Sheet carrier density n_{sH} against thickness for a) sample #1 and b) sample #2; lines are a guide to the eye

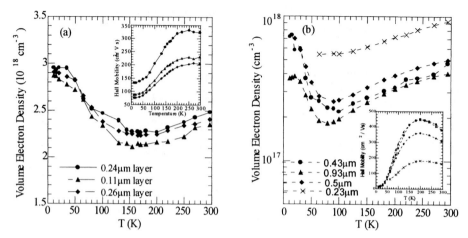

Fig. 4. Temperature dependence of volume carrier density n_{Hr} and mobility μ_{Hr} (inset) of removed layers using the analysis of Eqs. (1) and (2), for a) sample #1 and b) sample #2. Lines are a guide to the eye

with $4–7 \times 10^{17}$ cm^{-3} net donors. The fourth step, with $n_r > 9 \times 10^{17}$ cm^{-3} carriers, is presumably an average over the undoped spacer and part of the highly doped interface region. In both these doped samples, before etch, the sheet density of carriers in the impurity band exceeds that of the interface region, and hence constitutes the dominant parallel conduction process.

The donor concentration, N_D, in sample #1 exceeds Mott's limit [9] of 1×10^{18} cm^{-3} for the transition to a metallic impurity band in GaN; however, in sample #2 the net donor density is below this limit. Nevertheless, impurity band formation is still possible, for two reasons: (a) a compensation ratio of ≥ 0.5 would allow the total donor density to exceed the Mott limit [10] and (b) disorder should bring the limit for banding to lower than 1×10^{18} cm^{-3} [11, 12].

The insets of Figs. 4a, b plot the mobility of the Si-doped regions in the bulk of the samples. The 300 K mobility of sample #2 is almost unchanged over the first ~1 μm of removed material, but then falls rapidly over the remaining 1 μm, as it does in layer #1. This supports our earlier interpretation of Figs. 1b and 2b, and agrees with published work showing that increasing layer thickness improves crystalline quality, and hence mobility [13].

Our depth-profiled Hall data thus show that these Si-doped GaN samples contain two parallel conduction paths: an impurity band in the doped region of the structure, and a highly-doped interfacial layer. The likely simultaneous presence of both these processes must therefore be considered to achieve an accurate analysis of conduction mechanisms in all GaN layers.

Acknowledgement This work is supported in part by EPSRC Grant No GR/M89263.

References

[1] D.C. Look and R.J. Molnar, Appl. Phys. Lett. **70**, 3377 (1997).
[2] R.J. Molnar, T. Lei, and T.D. Moustakas, Appl. Phys. Lett. **62**, 72 (1993).

[3] D.C. Look and J.R. Sizelove, Phys. Rev. Lett. **82**, 1237 (1999).
[4] J.W.P. Hsu, D.V. Lang, S. Richter, R.N. Kleiman, A.M. Sergent, and R.J. Molnar, Appl. Phys. Lett. **77**, 2873 (2000).
[5] C. Mavroidis, J.J. Harris, M.J. Kappers, N. Sharma, C.J. Humphreys, and E. J. Thrush, Appl. Phys. Lett. **79**, 1121 (2001).
[6] D.C. Look, C.E. Stutz, R.J. Molnar, K. Saarinen, and Z. Liliental-Weber, Solid State Commun. **117**, 571 (2001).
[7] W. van der Stricht, I. Moerman, P. Demeester, J.A. Crawley, E.J. Thrush, P.G. Middleton, and K.P. O'Donnell, Proc. MRS Fall Meeting, Boston 1995 (p. 231).
[8] E. Arushanov, Ch. Kloc, and E. Bucher, Phys. Rev. B. **50**, 2653 (1994).
[9] N.F. Mott and W.D. Twose, Adv. Phys. **10**, 107 (1961).
[10] E.M. Conwell, Phys. Rev. **103**, 51 (1956).
[11] W. Baltensperger, Philos. Mag. **44**, 1355 (1953).
[12] P. Aigrain, Physica **20**, 978 (1954).
[13] W. Gotz, L.T. Romano, J. Walker, N.M. Johnson, and R.J. Molnar, Appl. Phys. Lett. **72**, 1214 (1998).

Band Structure Effects on the Transient Electron Velocity Overshoot in GaN

M. Wraback[1]) (a), H. Shen (a), E. Bellotti (b), J.C. Carrano (c), C.J. Collins (d), J.C. Campbell (d), R.D. Dupuis (d), M.J. Schurman (e), and I.T. Ferguson (e)

(a) U.S. Army Research Laboratory, Sensors and Electron Devices Directorate, AMSRL-SE-EM, 2800 Powder Mill Road, Adelphi, MD 20783, USA

(b) Electrical and Computer Engineering Department, Boston University, 8 Saint Mary's Street, Boston, MA 02215-2421, USA

(c) DARPA, 3701 N. Fairfax Dr., Arlington, VA 22203-1714, USA

(d) Microelectronics Research Center, Department of Electrical and Computer Engineering, The University of Texas at Austin, Austin, TX 78712, USA

(e) EMCORE Corporation, Somerset, NJ 08873, USA

(Received June 21, 2001; accepted June 30, 2001)

Subject classification: 78.47.+p; S7.14

Time-resolved electroabsorption measurements on an AlGaN/GaN heterojunction p–i–n photodiode have been used to study the transient electron velocity overshoot for transport in the c-direction in wurtzite GaN. The velocity overshoot increases with electric field up to ~320 kV/cm, at which field a peak velocity of 7.25×10^7 cm/s is attained within the first 200 fs after photoexcitation. However, theoretical Monte Carlo calculations incorporating a GaN full-zone band structure show that the majority of electrons do not attain sufficient energy to effect intervalley transfer until they are subjected to higher fields (> 325 kV/cm). Insight into this behavior can be gleaned from the band nonparabolicity deduced from the constant energy surfaces in the Γ valley, which shows that the effective mass in the c-direction can be viewed as becoming larger at high k values.

The use of high electric fields applied across sub-micron distances for III-nitride semiconductor device applications such as high power, high frequency electronics and ultraviolet avalanche photodiodes suggests that transient electron velocity overshoot and negative differential resistivity effects may play a critical role in their realization. In direct bandgap semiconductors, these phenomena have been primarily associated with scattering from a high mobility central valley to satellite valleys with higher effective masses. Transient electron velocity overshoot has been observed in III–V materials such as GaAs and InP [1], and has been predicted to occur in III-nitride semiconductors as well [2].

In this paper, we report time-resolved electroabsorption measurements of the transient electron velocity overshoot for transport in the c-direction in wurtzite GaN. In these experiments 80 fs pump and probe pulses with photon energy near the bandgap of GaN passed through the $Al_{0.1}Ga_{0.9}N$ p-type window layer of an AlGaN/GaN heterojunction p–i–n diode [3] and excited carriers directly in the i-region of the diode, where the electric field E is approximately uniform. The time-resolved bleaching of the probe transmission associated with the screening of the field by transport of the pump-

---
[1]) Coresponding author; Phone: 301-394-1459; Fax: 301-394-5451; e-mail:mwraback@arl.army.mil

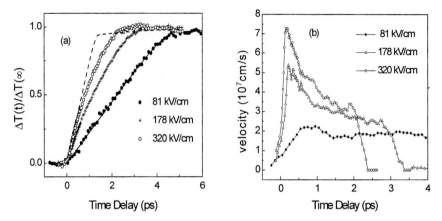

Fig. 1. Experimental data, a) normalized bleaching of the p–i–n diode electroabsorption as a function of time delay for three representative electric fields. The solid and dashed lines are constant electron velocity fits to the data; b) transient electron velocity obtained from the experimental $\Delta T(t)/\Delta T(\infty)$ data

excited electron–hole pairs enabled us to monitor the average position of the electron distribution as a function of time to obtain the transient velocity [4].

Figure 1a shows the normalized photoinduced bleaching $\Delta T(t)/\Delta T(\infty)$ as a function of time delay in this sample at three representative electric fields. The normalization factor $\Delta T(\infty)$ refers to the change in transmission after all the carriers have traversed the sample. All of the curves are characterized by an increase in $\Delta T(t)/\Delta T(\infty)$ that begins at $t = 0$ and eventually reaches a plateau at $\Delta T(t)/\Delta T(\infty) = 1$, at which point the electrons and holes drifting in the electric field cease to move apart. The fact that the sharp rise in transmission normally observed at $t = 0$ in pump–probe experiments is not present in this case indicates that the pump-generated carrier density is small enough to avoid the creation of any change in the probe transmission in the absence of field-induced transport. It has been shown [4] that the leading term in $\Delta T(t)/\Delta T(\infty)$ is $\int v_e \, dt$, so the slope of the normalized bleaching provides a measure of the electron velocity. For an electric field of 81 kV/cm, $\Delta T(t)/\Delta T(\infty)$ exhibits a quasi-linear increase to a plateau at ~5 ps that is well described using a single, constant electron velocity fit to the data (solid line). In contrast to the 81 kV/cm curve, the 178 and 320 kV/cm data exhibit multiple slopes, as evidenced by the fact that the constant electron velocity fit (dashed line) diverges significantly from the experimental results at times greater than 0.25 ps. This fit predicts a transit time (~1.5 ps) far shorter than those experimentally observed, thus indicating that the velocity of the electron distribution becomes slower at longer times.

Such behavior, in which the electron velocity at short times exceeds its steady-state value, is referred to as transient velocity overshoot. Figure 1b shows the velocity obtained by solving for v_e [4] using the experimental $\Delta T(t)/\Delta T(\infty)$ data from Fig. 1a as an input. While the velocity obtained from the 81 kV/cm data exhibits a step-like rise within experimental error to a constant velocity of $\sim 2 \times 10^7$ cm/s, the velocities obtained from the other two data sets are characterized by a peak attained within the first 200 fs that decays on a subpicosecond time scale. The 320 kV/cm data shows the largest peak velocity measured, 7.25×10^7 cm/s. The velocity decays toward a steady-state value with a ~0.67 ps time constant, but does not appear to reach the steady-state regime before the final drop associated with the completion of electron transport across the depletion region. At higher electric fields we are unable to measure higher peak velocities because the initial slopes of the $\Delta T(t)/\Delta T(\infty)$ curves are limited by the 80 fs duration of our pulses. The 178 kV/cm data represents an intermediate case, for which the velocity exhibits a lower peak (~5.5×10^7 cm/s), possesses an initial decay rate that is somewhat slower, and reaches a quasi-equilibrium value of $\sim 3 \times 10^7$ cm/s before transport is completed.

Our new results are in reasonable agreement with an ensemble Monte Carlo (EMC) calculation employing a multi-valley analytic band structure [2], which predicts a transient peak electron velocity of 6.5×10^7 cm/s at ~100 fs after transport is initiated for an electric field of 280 kV/cm. Although the driving mechanism for this calculated transient velocity overshoot is intervalley scattering and related mass changes associated with carrier transfer, it is important to note that transient overshoot effects may also occur in other situations for which the electron momentum relaxation rate is larger than the energy relaxation rate [5, 6]. It has been shown [7] that in direct gap semiconductors with a large energy separation between the Γ valley and the satellite valleys the negative differential resistance observed may be associated with transport in the Γ valley. Monte Carlo calculations indicate that polar optical scattering acting in a nonparabolic Γ valley alone can yield a peak steady-state velocity and a negative differential resistivity region of the steady-state velocity–field characteristic [7]. The separation between the Γ valley and the nearest satellite valley in GaN is ~2 eV. Figure 2 shows the calculated constant energy surface 2 eV above the minimum in the Γ valley of GaN. The shape of this surface suggests that this valley becomes nonparabolic in this energy range, with a larger "local effective mass" in the c-direction. The implications of this result may be further understood through examination of the calculated electron number density $f(E) g(E)$, where $f(E)$ is the distribution function and $g(E)$ is the density of states, as a function of energy E above the Γ valley minimum for two representative electric fields. At a field of 100 kV/cm, the electron density peak occurs at an excess

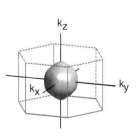

2.0 eV energy surface above the conduction band edge in GaN

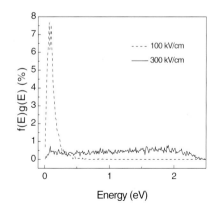

Fig. 2. Calculated isoenergy surface in the Γ valley of the lowest conduction band in GaN at 2 eV above the band minimum (left); Calculated electron number density as a function of energy above the Γ valley minimum for two representative electric fields (right)

energy of less than 200 meV, with negligible population above 0.5 eV. At 300 kV/cm, the low energy peak vanishes and is replaced by a plateau which indicates that the electron density is approximately equally distributed with respect to excess energy up to ~2 eV. These results show that the vast majority of electrons remain in the Γ valley at fields up to 300 kV/cm, due in large part to the greater effective mass that these carriers possess at higher energies in this valley. This larger effective mass and the concomitant shorter momentum relaxation time with increasing field in the high field regime are instrumental to the onset of a negative differential resistivity region of the steady-state velocity–field characteristic primarily associated with transport in the Γ valley of GaN. Moreover, the fact that a large transient velocity overshoot occurs for fields at which the electron density is spread out in energy such that it experiences the nonparabolicity of the Γ valley suggests that this overshoot may be related to both a decrease in the transient velocity at longer times due to the larger effective mass at higher excess energies, and a reduction in momentum relaxation time due to scattering into a greater density of final states associated with the larger effective mass.

In conclusion, time-resolved electroabsorption measurements on an AlGaN/GaN heterojunction p–i–n photodiode have been used to study the transient electron velocity overshoot for transport in the c-direction in wurtzite GaN. The velocity overshoot increases with electric field up to ~320 kV/cm, at which field a peak velocity of 7.25×10^7 cm/s is attained within the first 200 fs after photoexcitation. However, theoretical Monte Carlo calculations incorporating a GaN full-zone band structure show that the majority of electrons do not attain sufficient energy to effect intervalley transfer until they are subjected to higher fields (>325 kV/cm). Insight into this behavior can be gleaned from the band nonparabolicity deduced from the constant energy surfaces in the Γ valley, which shows that the effective mass in the c-direction can be viewed as becoming larger at high k values, thus, leading to a decrease in the transient velocity at longer times and a reduction in momentum relaxation time due to scattering into a greater density of final states associated with the larger effective mass.

References

[1] A. Leitenstorfer, S. Hunsche, J. Shah, M. C. Nuss, and W. H. Knox, Phys. Rev. B **61**, 16642 (2000), and references therein.
[2] B. E. Foutz, S. K. O'Leary, M. S. Shur, and L. F. Eastman, J. Appl. Phys. **85**, 7727 (1999).
[3] C. J. Collins, T. Li, A. L. Beck, R. D. Dupuis, J. C. Campbell, J. C. Carrano, M. J. Schurman, and I. T. Ferguson, Appl. Phys. Lett. **75**, 2138 (1999).
[4] M. Wraback, H. Shen, J. C. Carrano, T. Li, J. C. Campbell, M. J. Schurman, and I. T. Ferguson, Appl. Phys. Lett. **76**, 1155 (2000).
[5] C. G. Rodrigues, V. N. Freire, A. R. Vasconcellos, and R. Luzzi, Appl. Phys. Lett. **76**, 1893 (2000).
[6] S. Tiwari, Compound Semiconductor Device Physics, Academic Press, San Diego 1992.
[7] J. R. Hauser, T. H. Glisson, and M. A. Littlejohn, Solid-State Electron. **22**, 487 (1979).

Photoconductivity in Porous GaN Layers

M. Mynbaeva (a), N. Bazhenov (a), K. Mynbaev[1]) (a), V. Evstropov (a),
S. E. Saddow (b), Y. Koshka (c), and Y. Melnik (d)

(a) Ioffe Physico-Technical Institute, 194021 St. Petersburg, Russia

(b) EE Dept. & Center for Microelectronics Research, USF, Tampa, FL 33543, USA

(c) Mississippi Center for Advanced Semiconductor Prototyping,
Mississippi State University, MS 39762, USA

(d) Technologies and Devices International, Inc., MD 20877, USA

(Submitted July 17, 2001; accepted July 19, 2001)

Subject classification: 73.50.Pz; 73.61.Ey; 78.55.Mb; S7.14

The first observation of photoconductivity (PC) in porous GaN layers fabricated by surface anodization of 0.6 μm thick GaN layers grown by HVPE on 6H-SiC substrates is reported. PC was studied at room temperature and a strong photoresponse was observed in anodized GaN layers, whereas no photoconduction in initial (not anodized) epitaxial layers was detected. Both the build-up and the decay of the PC in anodized GaN demonstrated exponential behavior with two time constants of 1.8 and 100 s, respectively. PC spectra of anodized GaN showed a steady increase of the photoresponse as the photon energy was increased from 2.5 to 3 eV. Room and low temperature (15 K) photoluminescence studies were also performed on initial and anodized structures. EBIC measurements showed formation of a barrier on the GaN/SiC interface after anodization. Photoresponse in anodized GaN layers is ascribed to a decrease in carrier concentration in the anodized semiconductor.

Introduction Recently, there has been much interest in the properties of porous wide band-gap semiconductors, such as GaN and SiC. Porous materials have been already successfully used as buffer layers for epitaxial growth of SiC [1, 2] and GaN [3] layers. The properties of porous GaN have been also reported [4].

At the same time, GaN currently attracts attention as a potential light detector, especially in the ultraviolet part of the spectrum [5, 6]. Porous SiC has been already used for fabricating efficient photodetectors [7, 8], but the photoresponse in porous GaN has not been studied.

In the present work, we report on the first study of photoconductivity (PC) in porous GaN layers. The layers were fabricated by surface anodization of 0.6 μm thick epitaxial GaN layers grown on 6H-SiC substrates. Both anodized and initial (not anodized) GaN layers were also characterized by photoluminescence (PL) measurements at low (15 K) and room temperature and Scanning Electron Microscopy (SEM). The obtained data were used to interpret the origin of the photoresponse observed in the anodized GaN layers.

Experimental Porous GaN layers were prepared by surface anodization of epitaxial n-type GaN layers grown by hydride vapour-phase epitaxy (HVPE) on (0001) oriented

---

[1]) Corresponding author; Phone: +7 812 247 91 82, Fax: +7 812 247 1017,
e-mail: mynbaev@ieee.org

6H-SiC substrates [9]. The concentration $N_D - N_A$ in the substrates was 5×10^{18} cm^{-3}. The thickness of the layers was ≈0.6 μm, and the carrier concentration after growth was ∼5×10^{17} cm^{-3}. The anodization was carried out in aqueous solutions of HF under UV illumination as described elsewhere [10]. As the result of anodization, the whole GaN layer was converted into porous material.

Photoresponse was studied at room temperature with excitation by a broad spectra light source. The anodized GaN/SiC structure exhibited a noticeable photovoltaic (PV) effect with '+' on the GaN metal contact. In order to study the PC, a +3 V bias was applied to this contact (at such 'forward' bias the PV effect could be ignored). PC spectra were recorded using a MDR-23 grating spectrometer with radiation being modulated by a mechanical chopper with frequency variation. The PC signal was detected by a lock-in amplifier. The kinetics of the PC was studied in dc mode. PL measurements were carried out using photoexcitation from a 305 nm wavelength Ar$^+$ laser (9 mW). The PL signal was detected by a photon counter.

SEM studies on the GaN/SiC structure were performed on its cleft edge in Electron Beam Induced Current (EBIC) and Back Scattered Electron (BSE) modes.

Results and Discussion The initial GaN epitaxial layer showed no photoresponse within the limits of the detection capability of our setup, whereas anodized GaN demonstrated a strong photoresponse when illuminated by a broad spectra light source. Figure 1 shows a typical buildup and decay of a PC signal from the anodized GaN sample. Both curves could be easily fitted with an exponential function with two time constants of 1.8 and 100 s, respectively.

A typical PC spectrum of the anodized GaN sample is presented in Fig. 2. It was recorded at a chopper frequency of 8 Hz. Test measurements at higher chopper frequencies showed that the PC signal became much weaker, as could be expected from the kinetics of the PC response in Fig. 1. The spectrum presented in Fig. 2 demonstrates a steady increase of photoresponse as the photon energy was increased from 2.5 to 3 eV. The anodized layer was also sensitive to the UV light from a Hg lamp with photon energy up to ≈4.0 eV, but the high-energy part of the spectrum was not recorded because of experimental setup limitations.

We attempted to find correlation between the PC data and the physical properties of the layers determined by other methods. Of those, PL studies are usually most helpful, as PC and PL phenomena are often related to each other. Figure 3 presents PL spectra of anodized and initial GaN layers at room temperature. In both cases the PL spectrum demonstrates

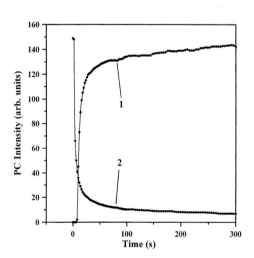

Fig. 1. Typical buildup (1) and decay (2) of PC in anodized GaN layer

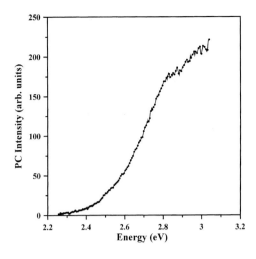

Fig. 2. PC spectrum of anodized GaN/SiC structure at 300 K

a near-band-gap peak and a broad band of yellow luminescence. For anodized GaN layer, the near-band-gap peak experienced an ≈11 meV blue shift, which is not unusual for porous semiconductors [11], and a slight broadening. The peak also became weaker as related to the yellow luminescence band. Low-temperature PL spectra (not shown) for both initial and anodized GaN layers exhibited near-band-gap peaks due to bound excitons and three peaks at lower energies due to donor–acceptor pairs and their phonon replicas. Again, for the anodized layer the near-band-gap peak was weaker as related to the other peaks, slightly blue-shifted and broadened. EBIC measurements indicated the existence of a potential barrier in the anodized GaN/SiC structure, while the BSE signal showed that the barrier was positioned at the GaN/SiC interface. No barrier was detected at this interface in the initial structure.

A study of I–V characteristics showed that the resistivity of the anodized sample was about two orders of magnitude higher than that of the initial GaN layer. Similar effects have been observed in anodized 4H-SiC [12]. We believe this effect can be related to a decrease in the carrier concentration in the anodized semiconductor. However, the origin of this effect is not clear yet and is the subject of continuing study.

On the basis of the data obtained, we can consider a possible origin of the PC and formation of potential barrier after anodization. High carrier concentration in the initial GaN layer makes the photocurrent contribution negligible and no PC signal is detected. After anodization, the carrier concentration in GaN is substantially reduced. Then the photocurrent value becomes noticeable compared to that due to dark current and a PC signal can be detected. The low-energy part of the PC spectrum in Fig. 2 shows that the PC signal emerges when the anodized structure is illuminated with photons with energy

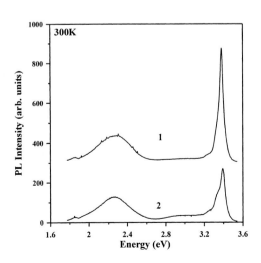

Fig. 3. PL spectra of initial (1) and anodized (2) GaN/SiC structure at 300 K

less than that corresponding to band-to-band transitions in either GaN or SiC. It may be that anodization of GaN layers introduces acceptor levels within the band gap of the material. At 300 K these levels are occupied and therefore seem not to contribute to the room temperature PL response (see Fig. 3). However, they do contribute to the PC response, as their appearance results in some compensation and a decrease in the free carrier concentration.

As to the PV effect, close electron affinities and carrier concentrations in GaN and SiC before anodization result in nearly the same positions of the Fermi level in GaN and SiC as related to the bottom of their conduction bands. Thus, the work functions of both materials are quite similar and no barrier at the interface is formed, whereas the holes experience a 'step' of ≈ 0.5 eV, mostly due to the difference in band gaps. The decrease in electron concentration after anodization results in increasing work function of GaN. A potential barrier appears at the GaN/SiC interface, and it is this barrier that causes the observed PV effect.

Conclusions A strong photoresponse was observed in epitaxial GaN layers grown on 6H-SiC substrates and anodized in aqueous solutions of HF. Both the buildup and the decay of the PC in anodized GaN demonstrated exponential behavior with two time constants of 1.8 and 100 s, respectively. PC spectra of anodized GaN showed a steady increase of the photoresponse as the photon energy was increased from 2.5 to 3 eV. Formation of a potential barrier was also observed at the GaN/SiC interface after anodization. Both the generation of the PC signal and formation of the barrier are attributed to the effect of decreasing carrier concentration in GaN as a result of anodization.

Acknowledgements We would like to thank ONR (Contract Monitor Colin Wood) and TDI for support. Thanks are due to V. Soloviev for performing the SEM studies.

References

[1] M. MYNBAEVA, S. E. SADDOW, G. MELNICHUK, I. NIKITINA, M. SCHEGLOV, A. SITNIKOVA, N. KUZNETSOV, K. MYNBAEV, and V. DMITRIEV, Appl. Phys. Lett. **78**, 117 (2001)
[2] J. SPANIER, G. DUNNE, L. ROWLAND, and I. HERMAN, Appl. Phys. Lett. **76**, 3879 (2000)
[3] M. MYNBAEVA, A. TITKOV, A. KRYZHANOVSKI, I. KOTOUSOVA, A. ZUBRILOV, V. RATNIKOV, V. DAVYDOV, N. KUZNETSOV, K. MYNBAEV, D. TSVETKOV, S. STEPANOV, A. CHERENKOV, and V. DMITRIEV, MRS Internet J. Nitride Semicond. Res. **4**, 14 (1999)
[4] M. MYNBAEVA, A. TITKOV, A KRYGANOVSKII, V. RATNIKOV, K. MYNBAEV, H. HUHTINEN, R. LAIHO, and V. DMITRIEV, Appl. Phys. Lett. **76**, 1113 (2000)
[5] E. MONROY, F. CALLE, J. PAU, E. MUNOZ, F. OMNEZ, B. BEAMONT, and P. GIBART, phys. stat. sol. (a) **185**, 91 (2001)
[6] C. QUI and J. PANKOVE, Appl. Phys. Lett. **70**, 1983 (1997)
[7] S. KIM, J. SPANIER, and I. HERMAN, Jpn. J. Appl. Phys. **39**, 5875 (2000)
[8] K. WU, Y. FANG, W. HSIEH, J. HO, W. LIN, and J. HWANG, Electron. Lett. **34**, 2243 (1998)
[9] Y. MELNIK, I. NIKITINA, A. ZUBRILOV, A. SITNIKOVA, Y. MUSIKHIN, and V. DMITRIEV, Inst. Phys. Conf. Ser. **142**, 863 (1995)
[10] M. MYNBAEVA and D. TSVETKOV, Inst. Phys. Conf. Ser. **155**, 365 (1997)
[11] J. SHOR, L. BEMIS, A. KURTZ, M. MACMILLAN, W. CHOYKE, I. GRIMBERG, and B. WEISS, Inst. Phys. Conf. Ser. **137**, 193 (1993)
[12] S. ZANGOOLE, P. PERRSON, J. HILFIKER, L. HULTMAN, H. ARWINAND, and Q. WAHAB, Silicon Carbide and Related Materials, Mater. Sci. Forum **338–342**, 537 (1999).

High-Field Electron Transport in Nanoscale Group-III Nitride Devices

S. M. Komirenko[1]) (a), K. W. Kim (a), V. A. Kochelap (b), and M. A. Stroscio (c)

(a) Department of Electrical and Computer Engineering, North Carolina State University, Raleigh, NC 27695-7911, USA

(b) Institute of Semiconductor Physics, National Academy of Sciences of Ukraine, Kiev-28, UA-252650, Ukraine

(c) U.S. Army Research Office, Research Triangle Park, NC 27709-2211, USA

(Received June 23, 2001; accepted August 4, 2001)

Subject classification: 72.10.Bg; 73.50.Fq; S7.14

Focusing on the short-size group-III nitride heterostructures, we have developed a model which takes into account main features of transport of electrons injected into a polar semiconductor under high electric fields. The model is based on an exact analytical solution of Boltzmann transport equation. The electron velocity distribution over the device is analyzed at different fields and the basic characteristics of the high-field electron transport are obtained. The critical field for the runaway regime, when electron energies and velocities increase with distance which results in the average velocities higher than the peak velocity in bulk-like samples, is determined. We have found that the runaway electrons are characterized by a distribution function with population inversion. Different nitride-based small-size devices where this effect can have an impact on the device performance are considered.

Introduction Unique fundamental properties of group-III nitride semiconductors make them very attractive for a number of optoelectronic and high-power, high-frequency applications. The steady-state high-field transport in group-III nitrides has been studied in a number of works. In particular, it was found [1] that the onset of the velocity overshoot takes place in the subpicosecond time scale and ten nanometer spatial scale regimes at fields which correspond to the peak velocities. It is expected that the maximum velocities of about $(6-8) \times 10^7$ cm/s can be achieved for InN and GaN. In the present paper, we report the results of detailed investigation of the electron runaway effect in short group-III nitride devices based on exact analytical solution of the Boltzmann transport equation.

The runaway effect arises in crystals with predominant carrier scattering by polar optical phonons [2]. Above a critical field, the momentum and energy gained by the electrons from the field cannot be relaxed to the lattice. The carriers then runaway to higher energies. In a *bulk-like* sample the electron runaway has to be stabilized by a breakdown due to impact ionization, a nonparabolicity or transfer to upper valleys, an additional scattering actual at high energies, etc. In a *short sample* the previously-mentioned mechanisms are not important and the electron transport can occur in the runaway regime. Below we present the results obtained for the latter case.

The Basic Model and Results We consider a short sample with the electron injection from the cathode at $z = 0$. Injection and reflection of the electrons from the anode are

---

[1]) Corresponding author; Phone: 919 515 5402, Fax: 919 515 3027, e-mail: smkomire@eos.ncsu.edu

neglected. Only the electron–optical-phonon interaction is taken into account. It is assumed that $\exp(-\hbar\omega/k_B T) \ll 1$, where ω is the optical phonon frequency, T is the lattice temperature. Under these conditions, we found *exact* solution for the distribution function. Then, the electron characteristics were calculated as function of the coordinate along the sample, z, at different electric fields, E. Our model has the following characteristic parameters: the energy $\hbar\omega$, the velocity $V_0 \equiv \sqrt{2\hbar\omega/m}$, the length $l_0 \equiv \hbar^2 \kappa_\infty \kappa_0 / e^2 m (\kappa_0 - \kappa_\infty)$, and the field $E_0 \equiv \hbar\omega/el_0$. Here, $\kappa_{0,\infty}$ are the low frequency and high frequency permittivities; e, m are the elementary charge and effective mass, respectively. The electron characteristics rescaled to these parameters show universal behavior. In particular, we found that at the dimensionless electric field $\varepsilon = E/E_0 \approx 0.5$ the distribution function changes qualitatively. Instead of the distribution function concentrated at zero kinetic energy, we obtained the distribution for which the number of high-energy electrons progressively increases with the distance from the injection plane. Thus, $\varepsilon \approx 0.5$ can be defined as the *runaway threshold field*. The evolution of the runaway effect can be illustrated by analyzing the average electron velocity as function of $\zeta = z/l_0$ for different electric fields. These results are shown in Fig. 1a in terms of the dimensionless average velocity $v = V_z/V_0$. All dependences $v(\zeta)$ are oscillating due to sequential single-optical-phonon emissions. For the small field $\varepsilon = 0.2$, the velocity–distance dependence has a regular behavior, oscillating around the value 0.6. No runaway is observed. Note, this value is close to the average velocity of the *one-dimensional* electron motion in an electric field under optical phonon emission $v = 0.5$ [3]. This is due to the strong anisotropy of the distribution function even at small fields. At the field $\varepsilon = 0.5$, the magnitude of velocity oscillations is smaller. The tendency of velocity increasing with distance is observed. This result indicates the onset of runaway. At $\varepsilon = 0.6, 0.8$, and 1.2 the increase in v with distance becomes more pronounced. When the field increases well above the threshold, $\varepsilon = 3$, the velocity oscillations are almost suppressed and the velocity magnitudes grow considerably with the distance. Although under the well developed runaway effect the velocity magnitude is below that of the ballistic case, the transport can reasonably be interpreted as *quasiballistic*.

Another remarkable property of the well-developed runaway effect is the extremely nonequilibrium distribution of the electrons over the momentum and energy. In Fig. 1b

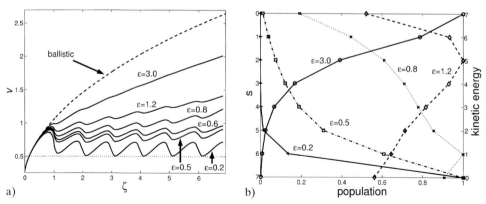

Fig. 1. a) The average velocity–distance dependences for different fields. b) Formation of population inversion

Table 1

The characteristic parameters for group III-nitrides and GaAs

material	l_0 (Å)	E_0 (kV/cm)	V_0 (10^7 cm/s)	$\hbar\omega$ (meV)
InN	90	99	5.3	89
GaN	35	257	4	91
AlN	12	828	2.7	99
GaAs	525	6.9	4.4	36

we show the results of the calculations of populations of the energy stairs s separated by $\hbar\omega$ at different electric fields for $\zeta = 7$. These results demonstrate the formation of a *population inversion* at high fields. At small fields, the carrier distribution is a decreasing function of the kinetic energy. When the field increases up to the threshold value, the distribution becomes wider and all energy stairs become populated. At $\varepsilon = 0.8$, a population inversion between the two lowest energy stairs is formed. Above this value, the population inversion becomes more pronounced as illustrated in Fig. 1b for $\varepsilon = 1.2$ and 3.

Characteristic parameters calculated for the nitrides are collected in Table 1. Then, taking into account all restrictions associated with our model we consider different nanoscale devices where the runaway effects can strongly impact the device performance.

The Runaway Effect in Nanoscale Devices

The runaway-effect diode The simplest vertical device operating in the runaway effect regime is an n^+–i–n^+ heterostructure diode with doped narrow-gap n^+-contact regions and a wide-gap i-base. An example of such a heterostructure is n^+-InGaN–i-GaN–n^+-InGaN. Our estimates show that if the base length is 590 Å and the applied voltage is 0.9 V, the electron velocity averaged over the base can reach 4×10^7 cm/s and the transit time is 0.17 ps (the corresponding cut-off frequency $\nu = 940$ GHz). If the length of the base is decreased by a factor of 2, the average velocity and the transit time are 7.2×10^7 cm/s and 0.06 ps ($\nu = 2.7$ THz). For a device with a 140 Å AlN base, and applied voltage 0.7 V, we estimate the average velocity and the transit time to be 2.5×10^7 cm/s and 0.04 ps ($\nu = 4$ THz), respectively.

The runaway-effect electron spectrometer In thin III–V-compound layers, the ballistic electrons can be observed and studied by using the hot-electron spectroscopy method at low temperatures [4, 5]. The method exploits a three-terminal device where the emitter and the base are separated by a thin barrier and a relatively wider barrier is placed between the base and collector electrodes. The emitter barrier serves for the hot electron injection controlled by the base voltage. The collector barrier is designed to control the energy of electrons traversing the device by the base–collector voltage. Using the principle of the injection of ballistic electrons, a variety of monopolar hot-electron devices (transistors) has been proposed for ultra-fast operation [3].

Similar three-terminal devices can be exploited to observe the runaway effect in the nitrides. For this case, the emitter barrier has to be wide and the base and the collector barrier have to be as thin as possible to avoid perturbation of the electron distribution, and to facilitate control of the emitter–base and collector–base currents. For example, an InGaN–GaN–InGaN–GaN–InGaN heterostructure can be used. In such a heterostructure, three doped narrow-gap n^+-InGaN layers are contacts. An i-GaN emitter barrier has width of about 700 Å. The collector barrier is an i-GaN layer of the width below 100 Å and designed to isolate electrically the base and collector (in general, it can be an InAlGaN narrow layer, etc.). In such a device, the runaway electron transport will be realized in the biased emitter barrier. The collector barrier will control the energy of collected electrons. Using this spectrometer, one can observe the runaway effect by measuring the distribution of collected electrons at different emitter–base biases. When the conditions of the effect are fulfilled, the devices with identical layers, but wider emitter barriers, should demonstrate higher energies of collected electrons at the same emitter-barrier field.

The runaway effect in other heterostructure devices Since group-III nitrides have a potential for high power devices which have to operate under high biases, it is expected that the runaway effect may occur in active regions of these devices. Examples of such devices include induced base transistor [6], the heterostructure bipolar transistor [7], etc.

Consider, for example, the induced base transistor proposed in Ref. [6]. In this device, a heterostructure with AlGaN emitter barrier of the width of about 100 Å operating under electric fields up to 1 MV/cm is assumed. From the considerations given above, it follows that under these conditions, the runaway effect should determine the emitter–base transit time. Then, the collector InAlGaN barrier is designed in such a way that only high energy electrons can be collected. The energy distribution of the electrons traversing the emitter barrier is critically important for the current gain of the device. Thus, the runaway effect will impact the overall device performance.

Conclusions In this paper we report the results of an investigation of the electron runaway under strong electric fields in short group-III nitride heterostructures. We have determined the electron distribution as a function of the electron momenta and the distance from the plane of injection, the critical field for the runaway regime, and the electron velocity distribution over the device as a function of the field.

For InN, GaN and AlN, we have obtained the basic parameters and characteristics of the high-field electron transport. We have estimated the mean path of the electrons between two sequential optical phonon scatterings to be of the order of tens angstroms. Thus, generally, the transport in the nitrides is *always dissipative*. However, for the runaway transport regime we have found that the electrons progressively gain energy and velocity. As the result the average velocity can reach higher values than the peak velocity in bulk-like samples. We have shown that the runaway electrons are characterized by the extreme distribution function with population inversion.

We suggested a three-terminal heterostructure to observe and measure the runaway effect and considered briefly different nitride-based small-size devices where this effect may have an impact on the device performance. Finally, we have established that the runaway regime is the main high-field transport regime in the group-III nitride nanoscale devices.

Acknowledgements This work was supported by the ONR, Grant No. N00014-98-1-0654, ARO, and ERO of US Army (Contract N68171-01-M-5166).

References

[1] B. E. FOUTZ, S. K. O'LEARY, M. S. SHUR, and L. F. EASTMAN, J. Appl. Phys. **85**, 7727 (1999).
[2] E. M. CONWELL, High Field Transport in Semiconductors, Academic Press, New York 1967.
[3] V. V. MITIN, V. A. KOCHELAP, and M. STROSCIO, Quantum Heterostructures for Microelectronics and Optoelectronics, Cambridge University Press, New York 1999.
[4] A. F. J. LEVI, J. R. HAYES, P. M. PLATZMAN, and W. WEIGMAN, Phys. Rev. Lett. **55**, 2072 (1985).
[5] M. HEIBLUM, M. L. NATHAN, D. C. THOMAS, and C. M. KNOEDIER, Phys. Rev. Lett. **55**, 2200 (1985).
[6] M. S. SHUR, A. D. BYKHOVSKI, R. GASKA, M. A. KHAN, and J. W. YANG, Appl. Phys. Lett. **76**, 3298 (2000).
[7] L. S. MCCARTHY, I. P. SMORCHKOVA, H. XING, P. KOZODOY, P. FINI, J. LIMB, D. L. PULFEY, J. S. SPECK, M. W. RODWELL, S. P. DENBAARS, and U. K. MISHRA, IEEE Trans. Electron Devices **48**, 543 (2001).

Electrical Characterization at Cubic AlN/GaN Heterointerface Grown by Radio-Frequency Plasma-Assisted Molecular Beam Epitaxy

T. Kitamura[1])[*]) (a), Y. Ishida (a), X. Q. Shen (a), H. Nakanishi (b), S. F. Chichibu (c), M. Shimizu (a), and H. Okumura (a, b)

(a) National Institute of Advanced Industrial Science and Technology (AIST), Power Electronics Research Center, central 2, 1-1-1, Umezono, Tsukuba, Ibaraki 305-8568, Japan

(b) Department of Electrical Engineering, Science University of Tokyo, 2641, Yamazaki, Noda, Chiba 278-8510, Japan

(c) Institute of Applied Physics, University of Tsukuba, 1-1-1, Tennoudai, Tsukuba, Ibaraki, 305-8577, Japan

(Received June 22, 2001; accepted August 4, 2001)

Subject classification: 73.40.Kp; 81.05.Ea; 81.15.Hi; S7.14

We successfully fabricated high quality cubic AlN/GaN heterostructures on 3C-SiC substrates by radio-frequency plasma assisted molecular beam epitaxy, and characterized their electrical properties for the first time. The Hall mobility value of 1290 cm^2/Vs was obtained at room temperature, and it drastically increased to 7330 cm^2/Vs at 100 K. These values as well as other electrical characterization results suggest the generation of a two-dimensional electron gas.

Introduction III-nitrides and their alloys have been demonstrated to be very important materials for high power and high temperature electronic devices. III-nitrides usually crystallize in hexagonal wurtzite structure as a thermodynamically stable phase. On the other hand, they also crystallize in cubic zincblende structure under certain conditions [1]. Cubic III-nitrides have more prospective advantages due to their higher crystallographic symmetry, which brings about lower carrier scattering, higher doping efficiency, etc. Thus, cubic (c-) GaN and its alloys also have future potential for electronic devices. For electrical properties of hexagonal GaN and its alloys, a number of studies have been reported. However, the study of electrical properties for cubic GaN and its alloys has been limited due to the difficulty of pure cubic GaN growth and the lack of appropriate semi-insulating substrates. For example, in the case of c-GaN epilayers on GaAs substrates, it is difficult to measure only the c-GaN layers precisely due to the formation of unusual conductive layers at the c-GaN/GaAs interface, which may be caused by the thermal decomposition of GaAs. Even for the case of temperature-resistive 3C-SiC substrates, the conductive nature of 3C-SiC prevents precise characterization. Thus, for the electrical characterization, the insulation between epilayers and substrates is important. In recent years, we have obtained high quality c-GaN, c-AlN

---

[1]) Corresponding author; Phone: +81 298 61 3386; Fax: +81 298 61 5434; e-mail: t-kitamura@aist.go.jp
[*]) On leave from: Science University of Tokyo.

epilayers by using 3C-SiC substrates [2–4]. Cubic AlN layers were found to be essential for the electrical insulation from substrates, as well as the fabrication of cubic heterostructures. On the other hand, in the case of hexagonal (h-) GaN, a two-dimensional electron gas (2DEG) is known to exist at h-AlGaN/GaN and h-AlN/GaN interfaces. These 2DEG systems are important ones for the application to GaN-based high-speed electronic devices, and many characterization results for the 2DEG have been reported [5–7]. However, for cubic nitrides, any electrical characterization on AlN/GaN, AlGaN/GaN heterostructures has not been reported.

In this study, we attempted to fabricate c-GaN/AlN heterostructures by using 3C-SiC substrates and investigate their electrical properties.

Experimental Procedure Cubic AlN/GaN heterostructures were grown on chemical vapor deposition (CVD)-grown 3C-SiC(001) substrates by plasma-assisted molecular beam epitaxy (rf-MBE) using metallic Al, Ga as III-group elements and rf N_2 plasma source (SVT model-4.5) as a N source. The flux intensities of Al and Ga were measured by an ion gauge and adjusted through the effusion cell temperatures. Sample surfaces during the growth were monitored by reflection high energy electron diffraction (RHEED) technique in-situ, in order to optimize the effective III/V ratio. For all the samples, a 100 nm thick c-AlN layer was firstly grown at 800 °C following the thermal cleaning. Next, a 800 nm thick c-GaN epilayer was grown at 760 °C. Subsequently, a c-AlN/GaN heterostructure was deposited at the same temperature. The plasma condition was 340 W, 3.0 sccm during the growth of c-GaN and c-AlN. The growth rates were 400 nm/h for the c-GaN layers and 300 nm/h for the c-AlN layers, respectively. Hall mobility and sheet carrier density were measured as a function of the temperature by Van der Pauw method. Ti/Al electrodes were fabricated for ohmic contacts by e-beam evaporation at each corner on the square samples and annealed at 600 °C for 40 s by rapid thermal annealing (RTA).

Results and Discussion First, we describe the growth of c-AlN/GaN heterostructures. Figure 1 shows the RHEED patterns during the growth of c-AlN and c-GaN epilayers. These streak patterns indicate the excellent flatness of the sample surfaces during the growth. In addition, these patterns indicate that the mixing of the hexagonal phase did not occur. The (4×1) pattern for c-GaN and the (2×1) pattern for c-AlN were also clearly observed [8]. These results strongly indicate that the c-GaN and c-AlN layers and their heterointerfaces have good structural quality.

The electrical insulating between the epilayers and substrates was checked for a 100 nm thick c-AlN on 3C-SiC substrate. The c-AlN layer was confirmed to be highly

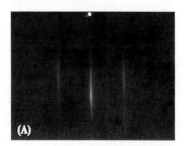 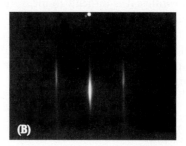

Fig. 1. The RHEED patterns during the c-AlN growth (A) and c-GaN growth (B), respectively

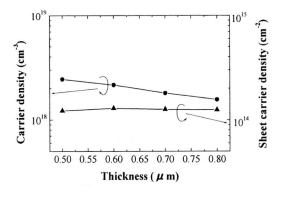

Fig. 2. The c-GaN thickness dependence of apparent carrier concentration (circles) and sheet carrier density (triangles). The Hall mobilities are almost constant for all measured regions

resistive enough to separate epilayers from substrates electrically. Next, the Hall measurement was carried out for a 800 nm thick c-GaN layer on the c-AlN layer aiming at the characterization of c-GaN bulk layers. The Hall mobility at room temperature (RT) showed values as high as around 1100 cm^2/Vs. This value may be too high as a c-GaN bulk mobility. Then, for a 800 nm thick c-GaN on 100 nm thick c-AlN heterostructure, we examined the c-GaN thickness (d) dependence of the carrier concentration (n) and the sheet carrier density (n_s) by chemical etching, in order to reveal whether the dominant current path is the c-GaN/AlN interface or the c-GaN layer. The etching for the c-GaN layer was done with hot phosphoric acid. Figure 2 shows the variation of the sheet carrier density and the carrier concentration; where $n = n_s/d$. The sheet carrier density is almost the same regardless of the c-GaN thickness and the apparent carrier concentration increases with the decrease of the c-GaN thickness. This result indicates that the dominant current path is localized near the c-GaN/AlN interface, which strongly suggests the 2DEG generation at the c-GaN/AlN interface similar to the hexagonal AlN/GaN heterostructures. In order to confirm this assumption, we attempt to fabricate c-AlN/GaN heterostructures on 800 nm thick c-GaN under-layers, expecting the improvement of the c-AlN/GaN interfaces. Figure 3 shows the temperature dependence of Hall mobility and sheet carrier density for the c-AlN/GaN heterostructure. The sheet carrier densities were almost the same in the measured temperature region. Neither the decrease of Hall mobility nor that of sheet carrier density is observed at lower temperature region. These results are typical characteristics of 2DEG. The Hall mobility shows values as high as 1287 cm^2/Vs and 7330 cm^2/Vs at RT and 100 K, respectively. These mobility values are comparable to those for hexagonal 2DEG systems. In addition, the sheet carrier density values reached as high as 10^{14} cm^{-2}, which is much higher than that of hexagonal systems, although there is no piezoelectric field. For the similar structure by AlAs/GaAs, it is well

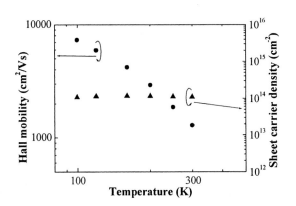

Fig. 3. The Hall mobility (circles) and sheet carrier density (triangles) as a function of temperature for the c-AlN (15 nm)/GaN (800 nm) heterostructure

known that a 2DEG system is generated, however its typical sheet carrier density is in the range of 10^{12} cm^{-2}. Considering the situation of AlAs/GaAs, the present value may be too high. However, carriers from deep traps in AlN layers can move to the GaN channel layer due to the large band offset between AlN and GaN, if the AlN layer contains much of deep traps. For AlAs/GaAs case, the level of equivalent deep traps should be located below the GaAs channel level energetically. Thus, carriers cannot be accumulated in the GaAs channel layer.

Summary We successfully fabricated c-AlN/GaN heterostuctures having excellent quality interfaces on 3C-SiC substrates by rf-MBE. The mobility as high as 1290 cm^2/Vs at RT and the sheet carrier density as high as 10^{14} cm^{-2} were observed at c-AlN/GaN heterointerface. These characteristics strongly suggest the generation of a 2DEG system, however further study on its existence is necessary for confirmation.

References

[1] H. Okumura, S. Misawa, and S. Yoshida, Appl. Phys. Lett. **59**, 1058 (1991).
[2] H. Okumura, K. Ohta, G. Feuillet, K. Balakrishnan, S. Chichibu, H. Hamaguchi, P. Hacke, and S. Yoshida, J. Cryst. Growth **178**, 113 (1997).
[3] H. Okumura, H. Hamaguchi, T. Koizumi, K. Balakrishnan, Y. Ishida, M. Arita, S. Chichibu, H. Nakanishi, T. Nagatomo, and S. Yoshida, J. Cryst. Growth **189/190**, 390 (1998).
[4] T. Koizumi, H. Okumura, K. Balakrishnan, H. Harima, T. Inoue, Y. Ishida, T. Nagatomo, S. Nakashima, and S. Yoshida, J. Cryst. Growth **201/200**, 341 (1999).
[5] I.P. Smorchkova, C.R. Elsass, J.P. Ibbetson, R. Vetury, B. Heying, P. Fini, E. Haus, S.P. DenBaars, J.S. Speck, and U.K. Mishra, J. Appl. Phys. **86**, 4520 (1999).
[6] I.P. Smorchkova, S. Keller, S. Heikman, C.R. Elsass, B. Heying, P. Fini, J.S. Speck, and U.K. Mishra, Appl. Phys. Lett. **77**, 3998 (2000).
[7] R. Gaska, M.S. Shur, A.D. Bykhovski, A.O. Orlov, and G.L. Snider, Appl. Phys. Lett. **74**, 287 (1999).
[8] H. Okumura, T. Koizumi, Y. Ishida, S.H. Cho, X.Q. Shen, and S. Yoshida, Mater. Sci. Forum **338–342**, 1545 (2000).

Transport Properties of Two-Dimensional Electron Gases Induced by Spontaneous and Piezoelectric Polarisation in AlGaN/GaN Heterostructures

A. Link[1]) (a), T. Graf (a), R. Dimitrov (a), O. Ambacher (a), M. Stutzmann (a), Y. Smorchkova (b), U. Mishra (b), and J. Speck (b)

(a) Walter Schottky Institute, Technical University Munich,
Am Coulombwall, D-85748 Garching, Germany

(b) Electrical and Computer Engineering Department and Materials Department,
College of Engineering, University of California, Santa Barbara, California 93106, USA

(Received June 21, 2001; accepted June 30, 2001)

Subject classification: 72.10.–d; 73.40.Kp; 77.65.–j; S7.14

We have performed Shubnikov-de Haas (SdH) and Hall effect measurements to investigate the electronic transport properties of polarisation induced 2DEGs in $Al_xGa_{1-x}N$/GaN heterostructures with alloy compositions between $x = 0.10$ and 0.35 and sheet carrier concentrations of up to $n_s = 1.05 \times 10^{13}$ cm^{-2}. From SdH measurements of 2DEGs with sheet carrier concentrations of 2.1×10^{12} and 4.6×10^{12} cm^{-2}, effective electron masses were determined to be 0.24 and $0.207 m_0$, respectively. In addition, angle resolved SdH measurements were performed to evaluate the effective g-factor from the angle of zero oscillation amplitude. Including the measured electron masses, the g-factors were calculated to be 2.11 and 2.47, respectively. In order to identify the main electron scattering mechanism we determined the ratio between transport- and quantum scattering times as a function of sheet carrier concentration. We observe a significant decrease of the τ_t/τ_q ratio with increasing 2DEG carrier concentration, indicating a transition from a dominant small angle to large angle scattering mechanism when n_s exceeds 7×10^{12} cm^{-2}.

Introduction AlGaN/GaN-based polarization induced high electron mobility transistors (PI-HEMTs) have been a subject of intense recent investigation and have emerged as attractive candidates for high voltage, high power operation at microwave frequencies [1–3]. Detailed studies of these heterostructures have shown that two-dimensional electron gases (2DEGs) forming the device channel are generated by positive polarization induced interface charges [4]. The identification of electronic scattering mechanisms which limit the low temperature mobility of these polarization induced 2DEGs in undoped $Al_xGa_{1-x}N$/GaN heterostructures is of fundamental importance for optimising the performance of these electronic devices. Significant progress has been made in the growth of high mobility two-dimensional electron gases [5, 6], but the mobility limiting scattering mechanisms are still under discussion [7, 8].

In this paper, we present detailed electron transport studies of high mobility 2DEGs with sheet carrier concentrations between 2.1×10^{12} and 1.05×10^{13} cm^{-2} by measuring quantum as well as transport scattering times, effective masses and effective g-factors. From the dependence of the ratio between transport τ_t, and quantum scattering time τ_q, on carrier density, we aim to identify the nature of the dominant electron scattering mechanism.

---

[1]) Corresponding authors; e-mail: link@wsi.tum.de

Experimental The investigated, undoped AlGaN/GaN based heterostructures were grown by plasma induced MBE and MOCVD or by a combination of both methods, discussed in more detail in Refs. [6, 9]. The AlGaN barriers with a thickness of (25 ± 5) nm and alloy compositions between $x = 0.1$ and 0.35 are deposited on insulating GaN buffer layers with a thickness of about 2 µm. Hall effect and SdH measurements were made in van der Pauw geometry, using In ohmic contacts and typical currents of about 10 µA. The samples were mounted in a ^{3}He cryostat with a base temperature of 330 mK and equipped with a revolving sample holder. The angle dependent transport measurements that were used to find the "spin zero" of the SdH oscillations were performed at 450 mK and fields of up to 14 T. All magnetotransport measurements were performed by standard low-frequency lock-in technique.

Results and Discussion The effective electron mass, m^*, is a fundamental parameter in electronic transport, as quantum scattering time, effective g-factor and transport scattering time are mass dependent. Applying the theory developed by Ando et al. [10], the oscillatory part of the longitudinal resistivity, ΔR_{xx}, can be written as

$$\frac{R_{xx}}{R_0} = 4 \frac{X(T)}{\sinh X(T)} \exp\left(-\frac{\pi}{\omega_c \tau_q}\right), \qquad (1)$$

where $X(T) = 2\pi^2 k_B T / \hbar \omega_c$, R_0 is the 2DEG resistance at zero magnetic field B, and $\omega_c = eB/m^*$ is the cyclotron frequency. Thus, the effective mass can be deduced from the decay of the SdH amplitude A with increasing temperature as follows:

$$\ln\left(\frac{A}{T}\right) \approx C - \frac{2\pi^2 k_B m^*}{e\hbar B} T, \qquad (2)$$

where C is a temperature independent constant. In Fig. 1 SdH oscillations of a sample with $n_s = 4.6 \times 10^{12}$ cm^{-2} measured at temperatures between 0.5 and 8.5 K are shown. From fits of the SdH amplitudes measured at a fixed magnetic field but at different temperatures we obtain $m^* = (0.24 \pm 0.02) m_0$ and $(0.207 \pm 0.013) m_0$ for 2DEGs with $n_s = 2.1 \times 10^{12}$ cm^{-2} (sample A) and 4.6×10^{12} cm^{-2} (sample B), respectively. These results are in good agreement with effective masses determined for 2DEGs with similar sheet carrier concentrations confined in AlGaN/GaN heterostructures [11, 12].

Based on these values for the effective mass, the effective g^*-factor was determined by performing magnetotransport measurements in tilted magnetic fields [13]. By rotating the sample inside the magnetic field we can take advantage of the fact, that the landau level splitting is proportional only to the field perpendicular to the 2DEG plane $B_\perp$, while the spin splitting is dependent of the total field B,

$$E = \left(n + \frac{1}{2}\right) \frac{\hbar e}{m^*} B_\perp \pm \frac{1}{2} g^* \mu_B B, \qquad (3)$$

where E is the energy of an electron, μ_B is the Bohr magneton, $B_\perp = B \cos(\Theta)$ and Θ is the angle between B and the surface normal of the sample. There exists an angle Θ_0, where the spin splitting is equal to half of the cyclotron energy. At this angle the SdH amplitude goes to zero and the phase changes by π [13]. In Fig. 2 normalized SdH oscillations are plotted for different tilt angles. With a zero angle of about 59.3° we can calculate an effective g^*-factor of 2.47 ± 0.14 (sample A: $\Theta_0 = 59.5°$ and $g^* = 2.11 \pm 0.23$), using the

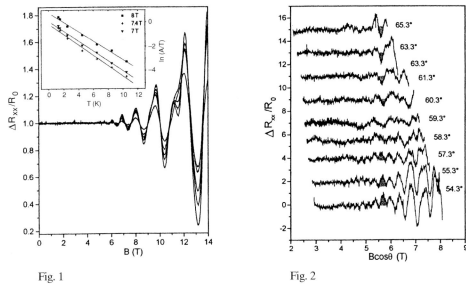

Fig. 1

Fig. 2

Fig. 1. SdH oscillations of R_{xx} measured at $T = 0.5, 1.3, 2.5, 4.5$ and 8.5 K. The inset shows the linear fit to the decay of SdH amplitude versus temperature, from which the effective mass is derived

Fig. 2. Normalized longitudinal resistivity for different tilt angles between sample and magnetic field, showing phase shift and zero amplitude at an angle of about $60°$

expression

$$g^* = \frac{\hbar e}{2\mu_B m^*} \cos \Theta_0 . \quad (4)$$

In addition, we found an exact $\cos \Theta$ dependence of the SdH oscillations at lower tilt angles, which is a proof for the two-dimensional nature of the accumulated electrons at the AlGaN/GaN interface.

Important physical properties which can be deduced from magnetotransport measurements are the electron transport and quantum scattering time. The quantum scattering time, which takes into account all scattering events, can be extracted from the SdH amplitude dependence on the magnetic field. According to Eq. (1), the logarithm of $\Delta R_{xx} \sinh(X(T))/4R_0 X(T)$ is linearly dependent on the inverse magnetic field (Dingle plot, Fig. 3). The slope of the fit to the measured data results in a quantum scattering time of $\tau_q = 0.93 \times 10^{-13}$ s for sample B. By increasing n_s from 2.1×10^{12} to 7.5×10^{12} cm^{-2} we observe a decrease of τ_q from 1.7×10^{-13} s to 0.5×10^{-13} s. A further increase in n_s up to 1.05×10^{13} cm^{-2} results in an increase τ_q to 1.3×10^{-13} s (Fig. 4a). The measured transport scattering time $\tau_t = \mu m^*/e$, is an order of magnitude higher in comparison to the quantum scattering time, e.g. $\tau_t = 4.45 \times 10^{-12}$ s for sample A. The ratio between transport and quantum scattering time is about 25 for 2DEGs with $n_s < 5 \times 10^{12}$ cm^{-2} (Fig. 4b), indicating that the dominant scattering mechanism is related to long range potential scattering, such as alloy scattering or Coulomb scattering (caused by impurities, charged defects or dislocations). It is significant that for higher sheet carrier concentrations the ratio τ_t/τ_q is reduced. This can be explained by the increasing influence of interface roughness scattering. In undoped AlGaN/GaN heterostructures, n_s is enlarged by increasing the Al

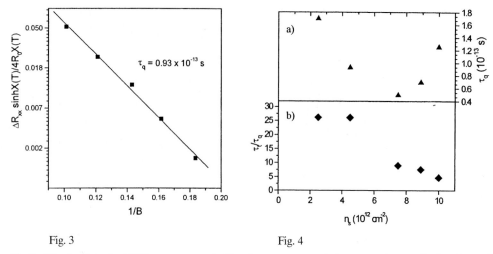

Fig. 3.

Fig. 4.

Fig. 3. Dingle plot of a SDH measurement. The slope corresponds to a quantum scattering time of 0.93×10^{-13} s

Fig. 4. a) Quantum scattering times and b) ratios of transport and quantum scattering times versus 2DEG sheet carrier concentration

concentration of the barrier. This leads to a reduction in the structural quality of the interface as well as of the average distance of the electrons towards the interface. At the same time charged point defects or dislocations are screened more efficiently. In any case it becomes obvious that the dominant scattering mechanism for polarisation induced 2DEGs with low sheet carrier concentrations causes small angle scattering, whereas for $n_s > 7 \times 10^{12}$ cm^{-2} large angle scattering becomes more significant.

In conclusion we have determined important physical properties relevant to understand the electronic transport of polarisation induced 2DEGs in AlGaN/GaN heterostructures. The effective electron mass, effective g-factor, transport and quantum scattering time were obtained by a combination of Hall effect and SdH measurements. The results indicate different dominant scattering mechanisms for 2DEGs with low and high sheet carrier concentrations.

References

[1] L. F. EASTMAN et al., IEEE Trans. Electron Devices **48**, 479 (2001).
[2] Y. F. WU et al., IEEE Electron Device Lett. **18**, 290 (1997).
[3] O. AMBACHER, J. Phys. D **31**, 2653 (1998).
[4] O. AMBACHER et al., J. Appl. Phys. **85**, 3222 (1999).
[5] C. R. ELSASS et al., Appl. Phys. Lett. **74**, 3528 (1999).
[6] I. P. SMORCHKOVA et al., J. Appl. Phys. **86**, 4520 (1999).
[7] E. FRAYSSINET et al., Appl. Phys. Lett. **77**, 2551 (2000).
[8] M. J. MANFRA et al., Appl. Phys. Lett. **77**, 2888 (2000).
[9] R. DIMITROV et al., J. Appl. Phys. **87**, 3375 (2000).
[10] T. ANDO, A. B. FOWLER, and F. STERN, Rev. Mod. Phys. **54**, 437 (1982).
[11] Y. J. WANG et al., J. Appl. Phys. **79**, 8007 (1996).
[12] W. KNAP et al., Appl. Phys. Lett. **70**, 2123 (1997).
[13] W. KNAP et al., Appl. Phys. Lett. **75**, 3156 (1999).

Energy Relaxation by Warm Two-Dimensional Electrons in a GaN/AlGaN Heterostructure

N. M. Stanton[1]) (a), A. J. Kent (a), S. A. Cavill (a), A. V. Akimov (a), K. J. Lee (b), J. J. Harris (b), T. Wang (c), and S. Sakai (c)

(a) School of Physics and Astronomy, University of Nottingham, University Park, Nottingham, NG7 2RD, United Kingdom

(b) Department of Electronic and Electrical Engineering, University College London, London, WC1E 7JE, United Kingdom

(c) Satellite Venture Business Unit, Department of Electrical and Electronic Engineering, University of Tokushima, 2-1 Minami-Josanjima, Tokushima 770-8577, Japan

(Received June 26, 2001; accepted July 30, 2001)

Subject clasification: 63.20.Kr; 73.40.Kp; S7.14

The rate of energy loss per electron, P_e, by a two-dimensional electron gas in an GaN/AlGaN heterostructure has been measured as a function of electron temperature, T_e, in the range 0.4–35 K. A combination of zero and high magnetic field electrical transport measurements were used to determine T_e as a function of the power dissipated in the device. It was found that $P_e \propto T_e^n$, with $n \approx 5$ at the lowest temperatures, $T_e \ll 2$ K, while for higher temperatures, $T_e > 10$ K, $n \to 1$. The experimental results are compared with numerical calculations of the energy relaxation rate. In the range of temperatures studied, emission of piezoelectrically coupled acoustic phonons was found to be the dominant energy relaxation mechanism.

Introduction The process of hot carrier energy relaxation is of fundamental importance to the performance of all semiconductor electronic and optoelectronic devices. Measurement of energy relaxation rates provides information about the electron–phonon interaction which is of use in modelling device behaviour. Energy relaxation by hot two-dimensional (2D) carriers has been extensively studied in GaAs heterojunctions and quantum wells and Si MOS devices [1]. However, to date, there has been much less work with the aim of examining this process in GaN and its alloys.

Recently Lee et al. [2] have reported measurements of the energy relaxation by warm 2D electrons in a GaN/AlGaN heterojunction. The temperature dependence of the amplitude of Shubnikov-de Haas (SdH) magnetoresistance oscillations was used as a thermometer for the electron temperature, T_e. However, these measurements were limited by the technique to the electron temperature range $T_e < 4$ K. Stanton et al. [3] studied the energy relaxation rate in bulk (3D) GaN samples using zero-field transport measurements to determine T_e. In the low mobility samples used, this technique was applicable to a wide range of electron temperature, 1.5 K $< T_e <$ 300 K. However, in higher mobility samples, where phonon scattering makes a significant contribution to the mobility in the temperature range of interest, the technique must be used with caution [4].

The main aim of the work described in this paper was to use the zero-field transport techniques to extend the measurements of Lee et al. to higher electron tempera-

---
[1]) Corresponding author; e-mail: Nicola.Stanton@nottingham.ac.uk

tures (up to $T_e \sim 35$ K) on the same sample. The results of the different techniques are compared in the range of overlap and also compared with the results of numerical calculations, allowing details of the 2D electron–phonon interaction in GaN to be determined.

Experimental Details The undoped GaN/AlGaN heterostructure used in this work (J401) was grown by MOCVD on a sapphire substrate [5]. A 2 μm undoped GaN layer was grown on a 25 nm GaN buffer layer and capped with 120 nm $Al_xGa_{1-x}N$ ($x = 0.18$). A Hall bar geometry was defined using standard photolithographic techniques and the mesa formed by reactive ion etching. Indium contacts were made by thermal evaporation and annealing. The sample was characterized by measuring its longitudinal magnetoresistance, see Fig. 1. At 1.4 K the sheet carrier density was $N_s = 5.8 \times 10^{12}$ cm^{-2} and the mobility $\mu = 10000$ cm^2 V^{-1} s^{-1}.

To experimentally determine the energy relaxation rate requires accurate determination of T_e as a function of the average power dissipated per electron, P_e. Two different methods were used, high field magnetotransport for $T_e < 4$ K and zero field transport for 2 K $< T_e <$ 35 K:

$T_e \sim 0.4$–4 K Low temperatures measurements were made in a helium dilution refrigerator at the High Magnetic Field Laboratory, Grenoble [2]. The amplitude of SdH oscillations was used as a thermometer of the electron temperature. With the sample held at base temperature, $T_l = 50$ mK, the magnetoresistance oscillations were measured for different values of current through the device. Comparison of the amplitude of these oscillations (in the situation where $T_e > T_l$) with those obtained under equilibrium conditions (where $T_e = T_l$) for different values of T_l allowed the electron temperature to be extracted.

$T_e \sim 2$–35 K In this range, the measurements were made in a liquid helium cryostat with a base temperature of $T_l = 1.3$ K. Measurements were made of the device resistance as a function of the power dissipated in the device. To obtain T_e, these measure-

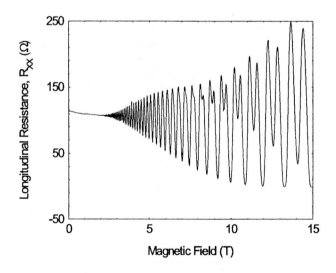

Fig. 1. The device resistance as a function of magnetic field. SdH oscillations are clearly visible from 2 T onwards

ments were compared with a calibration of the device resistance as a function of temperature. The calibration was made at low power so that electron heating was negligible ($T_e \approx T_l$). In this relatively high mobility device, the resistance was dependent on both T_e and T_l at all temperatures in the range of interest. It was therefore necessary to use a novel technique to separate the contributions of impurity and phonon scattering [6].

Results and Discussion Figure 2 shows the energy loss rates per electron as a function of the electron temperature. The two sets of data obtained by the different methods are in excellent agreement in the range of overlap $2\ K < T_e < 4\ K$. It is known that application of a strong magnetic field can have an significant effect on the phonon emission by a 2D electron gas owing to the formation of Landau levels in the electron density of states [7]. However, the agreement between the two sets of data indicates that no such effects are present in the low temperature measurements.

The energy loss rates are calculated using the same approach as for GaAs, see e.g. [1], except that the material parameters including the piezoelectric coupling are different for the case of GaN. The solid and dashed lines in Fig. 2 are given by numerical evaluation of Eq. (1).

$$P_e = \sum_\lambda 2 \int_0^{\frac{\pi}{2}} d\theta \int_0^\infty d\omega \, \frac{m^{*3/2} |Z|^2 \omega^{2+\gamma} C_\gamma^2\left(\frac{\omega}{v_\lambda},\theta\right)}{4\sqrt{2}\,\pi^3 \hbar^2 \rho v_\lambda^{3+\gamma}} \, [n(T_e) - n(T_l)] \int_{E_0}^\infty \frac{(f_{E'} - f_E)\,dE}{\sqrt{(E-E_0)}}. \tag{1}$$

Here ω is the phonon frequency, θ the angle between the direction of the emitted phonon and the normal to the 2D gas, λ represents the phonon polarization (mode), $\gamma = +1$ for deformation potential (DP) coupling and -1 for piezoelectric (PE) coupling, ρ is the crystal density, v_λ is the speed of mode λ, $|Z|^2 = [1 + (\omega a_0/v_\lambda)^2 \cos^2\theta]^{-3}$ is the bound-state form factor where a_0 is the Fang-Howard parameter, $n(T) = [1 - \exp(\hbar\omega/k_B T)]^{-1}$ is the phonon occupation number, $f_E = [1 + \exp((E - E_F)/k_B T_e]^{-1}$ is the electron distribution

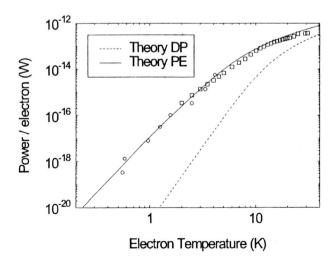

Fig. 2. P_e as a function of T_e. The square symbols are experimental points obtained from the zero-field transport measurements. The circles are experimental points from the SdH measurements. The solid and dashed lines are from numerical evaluation of Eq. (1)

function; $E_0 = (\sin^2\theta/2m^*v_\lambda^2)[\hbar\omega/2 + m^*v_\lambda^2/\sin^2\theta]^2$, $C_\gamma(\omega/v_\lambda, \theta)$ is the (quasi-statically) screened electron–phonon coupling parameter and the other symbols have their usual meanings.

For DP coupling to longitudinal polarized (LA) modes only,

$$C_{+1}^2\left(\frac{\omega}{v_\lambda}, \theta\right) = \Xi_d^2 S_\lambda^2, \tag{2}$$

where Ξ_d is the deformation potential constant and S_λ the screening parameter, given by

$$S_\lambda = \left(\frac{r_s\left(\frac{\omega}{v_\lambda}\right)\sin\theta}{1 + r_s\left(\frac{\omega}{v_\lambda}\right)\sin\theta}\right). \tag{3}$$

Here $r_s = 2\pi\varepsilon\hbar^2/m^*e^2$ is the Thomas-Fermi screening radius and $v_\lambda = v_{LA}$.

For PE coupling, the coupling constant C_{-1} is given by

$$C_{-1}^2 = e^2\left(\frac{e_{15}(q_x^2 + q_y^2)\lambda_z + e_{33}q_z^2\lambda_z + (e_{15} + e_{31})q_z(q_x\lambda_x + q_y\lambda_y)}{\varepsilon_{11}(q_x^2 + q_y^2) + \varepsilon_{33}q_z^2}\right)^2 S_\lambda^2, \tag{4}$$

where e_{15}, e_{31} and e_{33} are the piezoelectric parameters, ε_{11} and ε_{33} the dielectric constants and q_i are the phonon wavevector components.

We used the following values in the calculation: $v_{LA} = 6560$ ms^{-1}, $v_{TA} = 2680$ ms^{-1}, $\rho = 6150$ kg m^{-3}, $\Xi_d = 8.3$ eV, $e_{15} = -0.3$ Cm^{-2}, $e_{31} = -0.49$ Cm^{-2}, $e_{33} = 0.73$ Cm^{-2}, $\varepsilon_{11}/\varepsilon_0 = 10.4$, $\varepsilon_{33}/\varepsilon_0 = 9.5$, and $m^* = 0.22m_e$.

At the lowest temperatures, the Bloch-Grüneisen regime, and for $T_\ell \ll T_e$ it is found that DP coupling gives $P_e \propto T_e^7$ and for PE coupling $P_e \propto T_e^5$. Comparison with the experimental data suggests that PE coupling is dominant at low temperatures. For both mechanisms the temperature dependence approaches linear for $T_e > 10$ K. However, using the above values of the DP and PE constants we find that, throughout the temperature range of interest, emission of DP coupled phonons is much weaker than emission of PE coupled phonons. This result is in stark contrast to similar measurements of the AlGaAs/GaAs system, where deformation potential coupling is dominant above $T_e \sim 3$ K. The difference in behaviour is attributed to the larger piezoelectric coefficients in GaN. These results could be confirmed by heat pulse measurements of the emitted phonons, which also allows angular resolution of the modes.

Conclusions In summary, the energy relaxation rate of warm electrons in an AlGaN/GaN heterostructure has been measured over the temperature range 0.4–35 K. This wide range of temperatures was achieved using a combination of high magnetic field and zero-field transport measurements to determine T_e. The latter employed a novel technique for separating the contributions from phonon and impurity scattering. Excellent agreement between the two independent measurement methods and numerical calculations was found. The results show that the power loss is dominated by the emission of piezoelectrically coupled acoustic phonons in this range of temperature.

References

[1] A. J. Kent, in: Hot Electrons in Semiconductors: Physics and Devices, Ed. N. Balkan, Clarendon Press, Oxford 1998.
[2] K. J. Lee, J. J. Harris, A. J. Kent, T. Wang, S. Sakai, D. K. Maude, and J. C. Portal, Appl. Phys. Lett. **78**, 2893 (2001).
[3] N. M. Stanton, A. J. Kent, A. V. Akimov, P. Hawker, T. S. Cheng, and C. T. Foxon, J. Appl. Phys. **89**, 973 (2001).
[4] P. Hawker, A. J. Kent, O. H. Hughes, and L. J. Challis, Semicond. Sci. Technol. B **7**, 29 (1992).
[5] T. Wang, Y. Ohno, M. Lachab, D. Nakagawa, T. Shirahama, and S. Sakai, Appl. Phys. Lett. **74**, 3531 (1999).
[6] F. F. Ouali, H. R. Francis, and H. C. Rhodes, Physica B **263**, 239 (1999).
[7] G. A. Toombs, F. W. Sheard, D. Neilson, and L. J. Challis, Solid State Commun. **64**, 577 (1987).

2DEG Characteristics of AlN/GaN Heterointerface on Sapphire Substrates Grown by Plasma-Assisted MBE

K. Jeganathan[1]), T. Ide, S. X. Q. Shen, M. Shimizu, and H. Okumura

Power Electronics Research Center, National Institute of Advanced Industrial Science and Technology, Central 2, 1-1-1 Umezono, Tsukuba, Ibaraki 305-8568, Japan

(Received June 22, 2001; accepted July 5, 2001)

Subject classification: 73.40.Kp; 77.65.–j; 81.15.Hi; S7.14

We report on the growth and transport properties of a two-dimensional electron gas (2DEG) confined at the AlN/GaN heterointerface grown by plasma-assisted molecular beam epitaxy on c-plane sapphire substrate nucleated with an AlN buffer. The sheet carrier density of the 2DEG formed at the interface was characterized with respect to the AlN barrier thickness on doped and undoped GaN channels. The strong variation in the sheet carrier density was observed from the two types of grown structures. As the AlN barrier thickness grown on semi-insulating GaN increases from 15 to 25 Å, the carrier sheet density monotonously increased from 0.8×10^{12} to 1.1×10^{13} cm^{-2} due to the existence of spontaneous and piezoelectric polarization. Further, the channel conductivity was increased by inserting a thin n-GaN layer on the SI-GaN template, which favoured to enhance the sheet carrier concentration to 4.3×10^{13} cm^{-2}, a 35 Å thick AlN barrier layer was grown. The surface depletion could be responsible for the lowest sheet carrier density in the case of thin AlN barrier.

Introduction The effects of spontaneous and piezoelectric polarization make AlN/GaN heterostructures a unique and effervescent semiconductor system [1–3]. The high drift carrier velocity, large conduction band discontinuity and strong electrostatic field at the interface of the AlN/GaN heterostructure have led to the study of a variety of new phenomena, including the formation of a high density ($>10^{13}$ cm^{-2}) two-dimensional electron gas (2DEG) at the interface by the influence of macroscopic polarization without any doping [2, 4]. In the recent past, nitride-based heterostructures pace a remarkable interest to the electronic devices such as HFET, MISFET and IGFET [5, 6].

Generally, AlGaN/GaN HEMT structures have been realized with relatively moderate sheet carrier density with barrier Al content of 20–40%. Further, the sheet carrier density can be improved by the increase of Al content in the ternary barrier layer with respect to the strength of polarization [7]. However, the increasing of Al mole fraction in the ternary layer posed several problems, the growth of tensile-strained AlGaN barrier layer is problematic due to the large lattice mismatch, layer critical thickness and alloy scattering disorder in the heterostructure [8–11]. Theoretical and recent experimental studies show that AlN/GaN heterostructures of strained nature results in best sheet carrier density and 2D mobility in which the macroscopic polarization induced electric field is much stronger [2, 4].

Smorchkova et al. [4] have achieved the highest two-dimensional sheet carrier density (3.65×10^{13} cm^{-2}) by growing tensile-strained thin AlN barrier on MOCVD grown GaN template using plasma-assisted molecular beam epitaxy. There are only few reports available so far on the transport properties of AlN/GaN heterostructures. Binari

---

[1]) Corresponding author; Phone: +81-298-61-3374; Fax: +81-298-61-5434;
e-mail: k.jeganathan@aist.go.jp

et al. [12] have reported a sheet carrier density of 4.8×10^{12} cm^{-2} for MOCVD grown AlN/GaN heterostructures and also Alekseev et al. [13] measured a room temperature sheet carrier density of 2×10^{13} cm^{-2} (μ = 320 cm^2/Vs) with a barrier thickness of 11 nm grown by low-pressure MOCVD technique. Some of the recently appeared papers have reported AlN as a suitable optimistic gate material for the fabrication of high power devices [5, 12].

In this paper, we report the growth and transport properties of two-dimensional electron gas formed by an AlN/GaN heterointerface. The samples grown with different barrier thicknesses were subjected to investigations, and interestingly, we found that with the increase of AlN thickness the 2D sheet carrier density increases monotonously. A 35 Å thick AlN layer grown on doped GaN channel layer resulted at the highest value of 2D sheet carrier density ever achieved. Further, it starts to decline due to the barrier relaxation at which the effect of strain-induced piezoelectric polarization vanishes.

Experimental Low temperature nitrided sapphire substrates were used to grow high crystalline quality GaN layers by plasma-assisted MBE with an optimized III/N flux ratio [14]. It has been found that N-stable (N-rich) grown AlN buffer layers show a highly rough surface accumulated with a large number of pits. Nevertheless, an intermediate III/N regime growth maintains 2D streaky RHEED pattern, the surface morphology shows some irregularities due to the large difference in thermal expansion coefficients. The templates described in this context consisted of 2 μm thick high structural quality GaN layers grown on two step AlN buffer process [14]. A 100 nm thick Si-doped (2×10^{17} cm^{-3}) GaN layer was deposited on undoped GaN template as the channel for the heterostructure. The transport measurements were carried out by van der Pauw Hall technique using In as an Ohmic contact to reveal 2DEG density formed at the AlN/GaN interface with respect to different AlN barrier thickness.

Results and Discussion The schematic representation of the heterostructures used in our studies is depicted in Figs. 1a and b.

In both cases, the thickness of the barrier varied from tensile strained to relaxed state, to perform an explicit transport study of the structure. It is well understood that the formation of 2DEG in III–V system (AlGaAs/GaAs), the modulation doping of the barrier close to the heterointerface induces charge carriers in the undoped channel layer and forms a two-dimensional electron gas. In the case of n-type doped barrier, the depleted electrons in the barrier close to the interface accumulate in a triangle-shaped potential box at the bottom of the barrier which is close to the interface at the channel layer. The two-dimensional electron gas is formed at the interface due to charge accumulation in the potential well. This me-

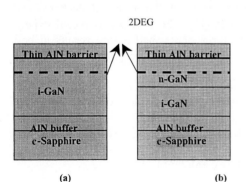

Fig. 1. Schematic diagram of two different AlN/GaN based heterostructures grown by MBE on c-Al$_2$O$_3$ substrates

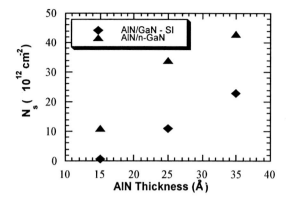

Fig. 2. Two-dimensional electron sheet density as a function of AlN barrier thickness grown on undoped and doped GaN channel by MBE

chanism is responsible for the improvement of sheet carrier density and mobility of the structure in comparison with the bulk mobility. However, in the case of AlN/GaN heterointerface, the formation of 2DEG is possible even at the interfaces of undoped layers with high structural perfection due to the free carrier concentration in the active channel layer and barrier layer, where the activation of two-dimensional charge carriers is stimulated by the macroscopic polarization. Hence, we have taken into account these two structures for the evaluation of transport properties. In order to increase the free charge carrier density in the active GaN channel layer, the channel was slightly doped and a dramatic increment in the two-dimensional sheet carrier density and the channel conductivity was found.

For AlN/GaN heterostructure, the barrier thickness kept far below the critical layer thickness in order to avoid the formation of cracks and surface irregularities to achieve high sheet carrier concentration values. The two-dimensional electron density as a function of the AlN barrier thickness grown on doped and undoped GaN channels is shown in Fig. 2. The sheet carrier density and mobilities were strongly influenced by the AlN barrier thickness. The insertion of a thin doped GaN channel layer reduces the resistivity of the heterostructures as compared with those without doped layers. The accomplishment of tensile-strained thin AlN barrier (15 Å) on a GaN layer causes a low sheet carrier density due to the surface depletion of the accumulated charge carriers at the interface. Further, with the increase of AlN thickness, the sheet carrier density increases monotonously. In our case, for a thickness of 35 Å, the sheet carrier density is 2.3×10^{13} and 4.3×10^{13} cm^{-2} for undoped and doped GaN channel layers, respectively. It is obvious that the barrier should be thick enough for the carrier injection to the interface and carrier confinement. Although these MBE grown structure resulted in higher sheet carrier concentration, the mobility was rather quite low as compared with the barrier and GaN template grown by hybrid growth methods, plasma-assisted MBE and MOCVD [4]. We strongly guess that the collapse of the mobility in the structure, both GaN and barrier grown by MBE, is due to the high interface roughness and increased alloy scattering disorder [15].

Here, we speculate that the reason for this high 2DEG density in the structure is due to the existence of non-zero spontaneous polarization caused by the symmetry of the wurtzite crystal structure and strain induced piezo-electric field by the barrier and well [1–3, 16, 17]. The partial or complete relaxation of AlN barrier may prevent to reach theoretical value because of the annihilation of strain-induced piezoelectric component. The 2DEG sheet carrier density has reached a maximum of 4.3×10^{13} cm^{-2} for the AlN/GaN heterostructure with a barrier thickness of 35 Å, but the measured mobility was quite low 173 cm^2/Vs. The structure fabricated with a thick AlN barrier (240 Å) resulted in a large decrease in the 2DEG density (1.1×10^{12} cm^{-2}) due to the barrier relaxation in which the piezoelectric-induced charge is completely absent and its strength is determined only by the spontaneous polarizations. The 2DEG mobility was quite low with respect to the barrier thickness changes on neither doped nor undoped GaN channel layers. Although in AlN/GaN interfaces the segregation effects are very weak, it should not affect their surface morphology [15], the scattering of free carriers by interface roughness is an important factor that limits 2D electron mobility and lifetimes in quantum structures [3]. To explain the consequences of 2D mobility deterioration for the AlN/GaN structure, further experimental studies on the interface roughness need to be carried out in detail.

Summary In summary, high quality AlN/GaN heterostructures of different AlN barrier thickness have been grown by plasma-assited MBE on doped and undoped GaN channel layers. The sheet carrier density is found to be a function of AlN barrier width. 2DEG sheet densities of 4.3×10^{13} cm^{-2} have been achieved for AlN barrier thickness of 35 Å due to strong built-in electrostatic field originated from the change in polarization field.

References

[1] F. BERNARDINI and V. FIORENTINI, Phys. Rev. B **57**, R9427 (1998).
[2] F. BERNARDINI, V. FIORENTINI, and D. VANDERBILT, Phys. Rev. B **56**, R10024 (1997).
[3] M. B. NARDELLI, K. RAPCEWICZ, and J. BERNHOLC, Phys. Rev. B **55**, R7323 (1997).
[4] I. P. SMORCHKOVA, S. KELLER, S. KEIKMAN, B. HEYING, P. FINI, J. S. SPECK, and U. K. MISHRA, Appl. Phys. Lett. **77**, 3998 (2000).
[5] S. IMANAGA and H. KAWAI, J. Appl. Phys. **82**, 5843 (1997).
[6] H. KAWAI, M. HARA, F. NAKAMURA, T. ASATSUMA, T. KOBAYASHI, and S. IMANAGA, J. Cryst. Growth **189/190**, 738 (1998).
[7] Y. ZHANG, Y. SMORCHKOVA, C. ELSASS, S. KELLER, J. IBBETSON, S. DENBAARS, U. K. MISHRA, and J. SINGH, J. Vac. Sci. Technol. B **18**, 2322 (2000).
[8] J. Z. LI, J. Y. LIN, H. X. JIANG, M. A. KHAN, and Q. CHEN, J. Vac. Sci. Technol. B **15**, 1117 (1997).
[9] H. K. CHO, J. Y. LEE, S. C. CHOI, and G. M. YANG, J. Cryst. Growth **222**, 104 (2001).
[10] C. KIM, I. K. RABINSON, J. MYOUNG, K. H. SHIMAND, and K. KIM, J. Appl. Phys. **85**, 4040 (1999).
[11] O. AMBACHER, J. SMART, J. R. SHEALY, N. G. WEIMANN, and K. CHU, J. Appl. Phys. **85**, 3222 (1999).
[12] S. C. BINARI, K. DOVERSPIKE, G. KELNER, H. B. DIETRICH, and A. E. WICKENDEN, Solid-State Electron. **41**, 177 (1997).
[13] E. ALEKSEEV, A. EISENBACH, and D. PAVLIDIS, Electron. Lett. **35**, 2145 (1999).
[14] K. JEGANATHAN, X. Q. SHEN, T. IDE, M. SHIMIZU, and H. OKUMURA, to be communicated.
[15] P. BOGUSAWSKI, K. RAPCEWICZ, and J. J. BERNHOLC, Phy. Rev. B **61**, 10820 (2000).
[16] M. LEROUX, N. GRANDJEAN, J. MASSIES, B. GIL, P. LEFEBVRE, and P. BIGENWALD, Phys. Rev. B **60**, 1496 (1999).
[17] B. E. FOUTZ, O. AMBACHER, M. J. MURPHY, V. TILAK, and L. F. EASTMAN, phys. stat. sol. (b) **216**, 415 (1999).

Electron Transport in III–V Nitride Two-Dimensional Electron Gases

D. Jena[1]) (a), I. Smorchkova (b), A. C. Gossard (a, c), and U. K. Mishra (a)

(a) Department of Electrical and Computer Engineering, University of California, Santa Barbara, CA 93106, USA

(b) TRW, One Space Park R6/1563B, Redondo Beach, CA 90277, USA

(c) Materials Department, University of California, Santa Barbara, CA 93106, USA

(Received June 26, 2001; accepted July 4, 2001)

Subject classification: 72.20.Dp; 72.80.Ey; S7.14

We present a study of electron scattering processes in AlGaN/GaN two-dimensional electron gases. A theoretical study of the effect of deformation potential scattering from strain fields surrounding dislocations is presented. The most important scattering mechanisms limiting electron transport are identified. We find that for AlGaN/GaN 2DEGs, mobility is limited by alloy scattering at high 2DEG densities. For AlN/GaN 2DEGs, interface roughness scattering limits mobility at high densities; there is a large improvement by the removal of the alloy barrier. At low 2DEG densities, dislocation scattering from charged cores and strain fields are the dominant scattering mechanisms.

Introduction The effect of dislocations on electron transport is generally assumed to occur by coulombic scattering from a charged core. Here, we show that deformation potential scattering from the strain fields surrounding dislocations is as important as the charged core in carrier scattering. To study the effect of strain field scattering on electron transport in AlGaN/GaN 2DEGs, we evaluate the contribution of all scattering mechanisms to electron mobility.

Dislocation Scattering in 2DEGs The effect of dislocation scattering on electron transport in 2DEGs has been a relatively unexplored area. Recently, we derived the scattering rate for a 2DEG electron in the presence of a charged dislocation line [1]. In addition to the possibility of the core being charged, there exist strain fields around dislocations. These strain fields cause a perturbation of the potential for the effective mass electrons, and will cause scattering.

The strain field around an edge dislocation is given by

$$u(r, \theta) = -\frac{b_e}{2\pi} \frac{1 - 2\gamma}{1 - \gamma} \frac{\sin \theta}{r} . \tag{1}$$

The strain field leads to a compression/dilatation of the unit cells around the dislocation. The deformation potential theorem of Bardeen and Shockley tells us that the shift in the conduction band edge due to this is given by

$$\Delta E_C(r, \theta) = \Xi_{ij} \operatorname{Tr}[u_{ij}] = -\frac{b_e \Xi_d^{CB}}{2\pi} \frac{1 - 2\gamma}{1 - \gamma} \frac{\sin \theta}{r} , \tag{2}$$

---

[1]) Corresponding author; Phone: +1 805 893 5404; Fax: +1 805 893 5714; e-mail: djena@engineering.ucsb.edu

where Ξ_{ij} are the components of the conduction band deformation potential tensor in GaN. For the conduction band (CB) perturbation, the CB deformation potential Ξ_d^{CB} at the Γ-point of bandstructure is used. Using this as the scattering potential, we derive the momentum scattering rate for 2DEG electrons in the Born approximation, using Fermi's golden rule. We derive the 2DEG electron mobility limited by strain field scattering of threading edge dislocations to be

$$\mu_{disl}^{strain} = \frac{2eh^3 \pi k_F^2}{N_{disl} m^{*2} b_e^2 \Xi^2} \left(\frac{1-\gamma}{1-2\gamma}\right)^2 \frac{1}{I_{disl}^{strain}(n_s)}. \qquad (3)$$

where $\hbar$ is the reduced Planck constant, $k_F = \sqrt{2\pi n_s}$ is the Fermi wavevector for the 2DEG, N_{disl} is the dislocation density in cm^{-2}, γ is the Poisson ratio for the crystal, m^* is the conduction band effective mass, $\Xi_d^{CB} = -8.0$ eV is the conduction band deformation potential, and b_e is the Burgers vector for the dislocation. $I_{disl}^{strain}(n_s)$ is a dimensionless integral depending only on the 2DEG carrier density[2]).

All Scattering Mechanisms How strong is the effect of deformation potential scattering from the strain fields around dislocations? To evaluate this, we calculate the effect of all scattering processes affecting electron mobility at low temperatures. We consider scattering by background unintentional donors ($N_{back} \approx 10^{17}$ cm^{-3}), scattering by the alloy disorder due to finite penetration of the wavefunction into the alloy AlGaN barrier (this is absent for AlN barriers), interface roughness at the Al(Ga)N/GaN heterojunction ($\Delta = 2.5$ Å, $L = 10$ Å), scattering from remote surface donors ($N_{surf} \approx n_s$), acoustic phonon scattering, and dislocation scattering.

Low temperature mobility reveals important facts about impurities in the heterostructure and gives valuable clues for the design of better devices. We plot the calculated low temperature 'mobility map' against 2DEG sheet density in Fig. 1. The highest reported experimental electron mobilities [2] are also plotted in the same figure for both alloy (AlGaN) and binary (AlN) barriers.

From the plot, it is clear that alloy scattering is severe, and is the main low temperature mobility limiting scattering mechanism. At high carrier densities ($n_s \geq 10^{13}$ cm^{-2}), it starts affecting even room temperature electron transport. The reason is that electron mobility limited by polar optical phonon scattering at room temperature is $\mu_{300K}^{2DEG} \approx 2000$ cm^2/V s, and as is evident from Fig. 1, alloy scattering reaches that value when n_s exceeds 10^{13} cm^2.

Removal of the alloy barrier by growing thin AlN layers causes a marked jump in the mobility for high sheet densities [3]. This is depicted by a jump from the curve of alloy scattering limited mobility to interface roughness limited mobility. In Fig. 1a, the empty circles represent data for AlGaN barriers and the filled circles for AlN barriers. Since this translates to a higher conductivity, introduction of a thin AlN layer at the AlGaN/GaN interface is an attractive technique for high-electron mobility transistors.

As is evident, dislocation scattering dominates electron transport properties at low sheet densities. Figure 1b shows the effect of different dislocation densities on maximum achievable electron mobilities. Note that deformation potential scattering from

---

[2]) $I_{disl}^{strain}(n_s) = \int_0^1 \frac{u^2 \, du}{(u + q_{TF}/2k_F)^2 \sqrt{1-u^2}}$, where q_{TF} is the Thomas-Fermi wavevector for 2DEGs, and $k_F = \sqrt{2\pi n_s}$ is the Fermi wavevector.

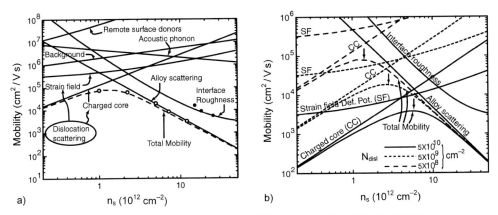

Fig. 1. a) Scattering processes limiting electron mobility in the Al(Ga)N/GaN 2DEG system evaluated for $N_{disl} = 5 \times 10^8$ cm^{-2}; experimental data is also shown (circles). b) Effect of dislocation scattering for three different dislocation densities

the strain fields of dislocations is as important as charged core scattering. Inverted arrows show the peak mobilities achievable for the respective dislocation densities.

In conclusion, we studied the effect of strain fields or dislocations on electron transport properties. We have identified the chief mobility limiting scattering mechanisms for various 2DEG densities. The identification of the chief scattering process in HEMTs has led us to propose a method for an improvement of conductivity.

References

[1] D. JENA, A.C. GOSSARD, and U.K. MISHRA, Appl. Phys. Lett. **76**, 1707 (2000).
[2] I. P. SMORCHKOVA et al., J. Appl. Phys. **86**, 4520 (1999).
[3] I. P. SMORCHKOVA et al., Appl. Phys. Lett. **77**, 3998 (2000).

Investigation for the Formation of Polarization-Induced Two-Dimensional Electron Gas in AlGaN/GaN Heterostructure Field Effect Transistors

H. W. Jang (a), C. M. Jeon (a), K. H. Kim (a), J. K. Kim (a), S.-B. Bae (b), J.-H. Lee (b), J. W. Choi (c), and J.-L. Lee[1] (a)

(a) Department of Materials Science and Engineering, Pohang University of Science and Technology (POSTECH), Pohnag, Kyungbuk 790-784, Korea

(b) Department of Electric and Electronic Engineering, Kyungpook National University, Daegu 702-701, Korea

(c) Center for Advanced Microstructures and Devices, Louisiana State University, Baton Rouge, LA 70806, USA

(Received June 23, 2001; accepted August 27, 2001)

Subject classification: 73.20.At; 85.30.Tv; S7.14

The formation of two-dimensinal electron gas (2DEG) in AlGaN/GaN heterostructure field effect transistors was investigated using synchrotron radiation photoemission spectroscopy and high-resolution X-ray diffraction. The surface band bending, interfacial strain and 2DEG density were evaluated as a function of AlGaN barrier thickness and composition. The 2DEG density increased with the thickness, but the surface Fermi level was independent of the thickness of AlGaN layer. This suggested that the polarization-induced 2DEG originated from the AlGaN barrier, which is unintentionally doped due to the nonstoichiometry of nitrogen deficiency and donor-like impurities.

Introduction AlGaN/GaN heterostructure field effect transistors (HFETs), where the polarization-induced charge at the AlGaN/GaN interface allows very high electron densities ($> 1 \times 10^{13}$ cm^{-2}) with no intentional doping, have recently been attracting much attention because of their promising uses for high power and high temperature microwave applications [1, 2]. Previous works on piezoelectric doping, polarization effects coupled with thermal generation, unintentional impurities in AlGaN, and surface states have explained the high 2DEG density in AlGaN/GaN HFETs [3–6]. However, the formation of the 2DEG is still unclear and under debate, especially for understanding the origin of the 2DEG in AlGaN/GaN HFETs, which is important for the improvement and optimization of device performances.

In the present work, we examined experimentally the surface band bending of AlGaN/GaN HFETs and interfacial strain as a function of the AlGaN barrier thickness and composition using sychrotron radiation photoemission spectroscopy (SRPES) and high-resolution X-ray diffraction (HRXRD). These studies demonstrate the mechanism for the formation of the 2DEG in AlGaN/GaN HFETs, focusing on the origin of the 2DEG.

Experimental Procedure AlGaN/GaN films used in this work were grown on (0001) sapphire substrate using metal organic chemical vapor deposition. An undoped GaN

---

[1]) Corresponding author; Phone: +82 54 279 2152; Fax: +82 54 279 2399; e-mail: jllee@postech.ac.kr

layer with a thickness of 1 µm was grown, followed by the growth of an undoped AlGaN cap layer. Two types of samples were prepared. The first set are samples with the only variation in the AlGaN thickness. The second set of specimen are samples with the only variation in the Al composition of the AlGaN barrier. The samples of one set were grown in sequence and with no time gap, to maintain the growth condition as same as possible. After sample growth, carrier concentration and mobility were measured using Hall measurements.

SRPES and HRXRD analysis were performed in the Pohang Accelator Laboratory. The exact barrier thicknesses and compositions of grown AlGaN layer were determined using HRXRD.

Results and Discussion In strained films, the lattice constants c and a are connected via the Poisson ratio $v(x)$,

$$[c - c_0(x)] + v(x) \frac{c_0(x)}{a_0(x)} [a - a_0(x)] = 0, \tag{1}$$

where $c_0(x)$ and $a_0(x)$ are lattice constants of relaxed AlGaN alloyed with Al content of x. Using the definition $\xi(x) = v(x)[c_0(x)/a_0(x)]$ and assuming a linear dependency of $\xi(x)$ on the composition between GaN and AlN, AlGaN composition x could be obtained by solving a quadratic equation for x [7],

$$(\Delta\xi \Delta a) x^2 + (\Delta\xi \delta a - \Delta c - \xi_{GaN} \Delta a) x - (\delta a \, \xi_{GaN} + \delta c) = 0, \tag{2}$$

where $\Delta a = a_{AlN} - a_{GaN}$, $\delta a = a - a_{GaN}$, $\Delta c = c_{AlN} - c_{GaN}$, $\delta c = c - c_{GaN}$, $\Delta\xi = v_{AlN}[c_{AlN}/a_{AlN}] - v_{GaN}[c_{GaN}/a_{GaN}]$. AlGaN (100–450 Å) barriers of all samples showed fully strained structure with no relaxation ($\delta a = 0$).

Figures 1a and b show the change of 2DEG density with AlGaN barrier thickness and Al content in AlGaN, respectively, determined from HRXRD analysis. The 2DEG density was increased with the $Al_{0.24}Ga_{0.76}N$ thickness, as shown Fig. 1a. It also increased with Al content. This means that the 2DEG density is dependent on both the thickness and Al content in the barrier layer, in good agreement with previous results [6].

Figures 2a and b display the SRPES spectra of Ga 3d for the $Al_{0.24}Ga_{0.76}N$/GaN HFETs with various thicknesses and AlGaN (450 Å)/GaN HFETs with various Al con-

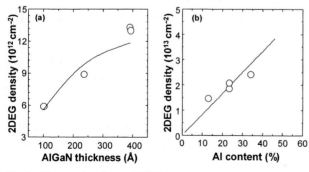

Fig. 1. Measured values of 2DEG density as a function of a) $Al_{0.24}Ga_{0.76}N$ barrier thickness and b) Al content in AlGaN (450 Å)/GaN HFETs

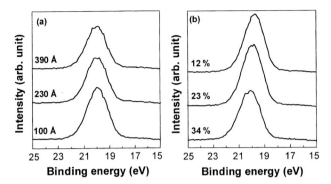

Fig. 2. SRPES spectra of Ga 3d for the a) $Al_{0.24}Ga_{0.76}N$/GaN HFETs with various thicknesses and b) AlGaN (450 Å)/GaN HFETs with various AlGaN compositions

tents, respectively. No peak shift was observed with the change of thickness, as shown in Fig. 2a. When we examined the energy levels of valence band maximum (VBM) from valence band spectra, no change was also found, showing the Fermi level pinned at the surface level ~1.6 eV below the conduction band minimum. This suggests that the surface Fermi level was not changed with the variation of AlGaN barrier thickness. Meanwhile, for the AlGaN/GaN HFETs with various Al contents, the binding energy of Ga 3d peak was shifted toward higher binding energies with the increase of Al content. The peak shift for the Ga–N bond between $Al_{0.12}Ga_{0.88}N$ and $Al_{0.34}Ga_{0.66}N$ was about 0.4 eV and the surface Fermi level shifted by about 0.5 V between them. The binding energy increase is nearly the same as the surface Fermi level one. This means that the surface band bending was not changed, considering the increase in bandgap energy of AlGaN barrier with increase of Al content.

The surface Fermi level pinning observed at the $Al_{0.24}Ga_{0.76}N$/GaN HFETs with various thicknesses can be explained as follows. Figure 3 shows the band diagram of AlGaN/GaN HFETs with different thickness. In an ideal case of undoped AlGaN barrier, the band rises straight toward the surface due to the charge-induced electric field. It is expected that the surface Fermi level of the AlGaN barrier is located closer to the VBM than that of a thinner AlGaN barrier, as shown in Fig. 3a. However, in real case, the AlGaN barrier could not be an undoped layer because of the unintentional doping with donor-like impurities and bulk and/or surface states, which acts as a source for 2DEG at the AlGaN/GaN interface. The substantial band bending occurs in the AlGaN barrier, as shown in Fig. 3b. It is well known that GaN films are slightly nonstoichio-

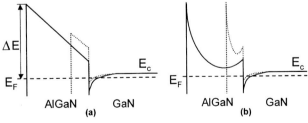

Fig. 3. Schematic band diagram of AlGaN/GaN HFETs with different thickness; for the case of a) an ideal undoped AlGaN barrier and b) a real undoped AlGaN barrier

metric compounds including an abundance of N vacancies and O donor impurities because an undoped GaN film shows n-type conducting property. Therefore, it is suggested that the surface Fermi level of the AlGaN barriers was pinned to the energy level due to the donor-like surface states, which could be the primal source of 2DEG in AlGaN/GaN HFETs even without intentional doping. The higher carrier concentration in the thicker AlGaN barrier, shown in Fig. 1a, originated from a larger amount of the surface states. Based on this postulation, the fact that the doping of AlGaN barrier results in the increase of 2DEG density can be well understood.

Conclusion The mechanism for the formation of a 2DEG in AlGaN/GaN HFETs was investigated using synchrotron radiation photoemission spectroscopy and high-resolution X-ray diffraction analysis as a function of AlGaN barrier thickness and composition. The surface Fermi level of the AlGaN barriers was pinned to the energy level of ~ 1.6 eV below the conduction band minimum due to donor-like surface states, which should be the major source of 2DEG in AlGaN/GaN HFETs with no intentional doping.

Acknowledgement This work was supported by the Korean Institute of Science and Technology Evaluation and Planning (KISTEP) through the NRL project.

References

[1] M. A. KHAN, Q. CHEN, M. S. SHUR, B. T. MCDERMOTT, and J. A. HIGGINS, IEEE Electron Device Lett. **17**, 325 (1996).
[2] Y. F. WU, B. P. KELLER, S. KELLER, D. KAPOLNEK, P. KOZODOY, S. P. DENBAARS, and U. K. MISHRA, Appl. Phys. Lett. **69**, 1438 (1996).
[3] P. M. ASBECK, E. T. YU, S. S. LAU, G. J. SULLIVAN, J. VAN HOVE, and J. REDWING, Electron. Lett. **33**, 1230 (1997).
[4] O. AMBACHER, J. SMART, J. R. SHEALY, N. G. WEIMANN, K. CHU, M. MURPHY, W. J. SCHAFF, and L. F. EASTMAN, J. Appl. Phys. **85**, 3222 (1999).
[5] L. HSU and W. WALUKIEWICZ, Appl. Phys. Lett. **73**, 339 (1998).
[6] J. P. IBBETSON, P.T. FINI, K. D. NESS, S. P. DENBAARS, J. S. SPECK, and U. K. MISHRA, Appl. Phys. Lett. **77**, 250 (2000).
[7] L. GÖRGENS, O. AMBACHER, M. STUTZMANN, C. MISKYS, F. SCHOLZ, and J. OFF, Appl. Phys. Lett. **76**, 577 (2000).

2DEG Mobility in AlGaN–GaN Structures Grown by LP-MOVPE

Z. Bougrioua[1]) (a), J.-L. Farvacque (b), I. Moerman (a), and F. Carosella (b)

(a) INTEC, IMEC-Ghent University, Sint-Pietersnieuwstraat 41, B-9000 Ghent, Belgium

(b) LSPES, Université des Sciences et Technologies de Lille, F-59655 Villeneuve d'Ascq, France

(Received July 9, 2001; accepted August 26, 2001)

Subject classification: 68.55.Ac; 72.10.Di; 73.61.Ey; 81.15.Kk; S7.14

Two-dimensional electron gas (2DEGs) could be tailored through the growth by LP-MOVPE of intentionally undoped AlGaN–GaN heterostructures with 6% < x_{Al} < 36%. The carrier density (n_s) is shown to be controlled by polarisation effects. Large carrier mobilities as high as 1710 cm^2 V^{-1} s^{-1} at $n_s \sim 9 \times 10^{12}$ cm^{-2} can be measured at 300 K. For larger n_s, the mobility drops first smoothly, then sharply when $n_s > 1.4 \times 10^{13}$ cm^{-2}. A two-subband model for 2DEG transport, taking into account phonons and impurity scattering mechanisms is proposed to explain semi-quantitatively the soft decay regime.

Introduction HEMTs based on AlGaN–GaN heterostructures grown on sapphire or on SiC are today better controlled and extra improvement will lead to increase further their reliability and microwave power performance [1]. The behaviour of these transistors is based on the 2DEG located in the quantum well (QW) at the AlGaN–GaN interface. Cut-off frequency and other characteristics can be anticipated through the knowledge of the sheet carrier density and carrier mobility. In this paper, we present the transport properties obtained on different intentionally undoped AlGaN–GaN HEMT-like structures grown by MOVPE, amongst which some were processed and led to excellent transistors [2]. The 2DEG mobility measured at room temperature and 77 K decreases significantly as n_s increases. A modeling of the 2DEG low field transport is proposed to explain one part of the evolution.

MOVPE Growth The AlGaN–GaN structures are grown on c-plane sapphire, in a close-coupled showerhead MOVPE reactor (Thomas Swan) using ammonia, TMGa and TMAl as precursors. The epitaxy is carried out at 100 Torr using a so-called two-step procedure: i.e., a thin nucleation layer (NL) is first deposited at 490–550 °C, then the wafer is brought to a high temperature and maintained for a few tens of seconds, before GaN and AlGaN are finally deposited at 1040–1070 °C. For the realisation of HEMT-like structures the key step corresponds to the growth of a semi-insulating GaN buffer that will allow a good inter-device insulation. The insulating character is realised through the use of a new growth process ("i.3" [3]) based on results described in [4]. The thickness of the highly resistive GaN buffer is between 2.4 and 4.5 μm. The threading dislocation density is reduced through the NL optimisation: it is as low as a few 10^9 cm^{-2}. AlGaN growth is carried out at 50 Torr and aluminum composition (Al%) is

---

[1]) Corresponding author: Phone: 329 2643341; Fax: 329 2643593; e-mail: Zahia.Bougrioua@intec.rug.ac.be

controlled by varying the TMGa flux and keeping TMAl flux constant. For this study: 6% < x_{Al} < 36% and AlGaN layer thickness is chosen between 11 and 54 nm. The morphology of the AlGaN surface is perfectly smooth in any case. TEM analysis on a few structures has shown clear planar and misfit-dislocation-free AlGaN–GaN interfaces.

2DEG Transport Properties Room temperature and 77 K resistivity Hall electrical measurements have been carried out on several AlGaN–GaN heterostructures, using van der Pauw square-shaped specimens. As underlined before, the GaN buffer layers are highly resistive. Provided that there is no parallel conduction in the AlGaN, this means that the 300 K and 77 K transport properties are exactly the 2DEG ones. Figure 1 depicts the evolution of the 300 K sheet carrier density as a function of Al composition for three sets of AlGaN thickness. As the structures are fully undoped, the very dense 2DEG is essentially generated by spontaneous and piezoelectric polarisations [5, 6]. An empirical relationship between n_s and the Al composition is given in Fig. 1. It suits quite well the theoretical predictions proposed in [6]. The evolution of n_s versus AlGaN thickness should saturate after a few 10 nm [7, 8]. However, for thick AlGaN, n_s seems to diverge (example in Fig. 1 for 28%Al).

In Fig. 2, the evolution of the 300 K Hall mobility is plotted as a function of the Hall carrier density measured on several 2DEG specimens corresponding to about some 30 2-inch wafers. For instance, the triangles correspond to structures with an Al content of 28% and where the AlGaN thickness is varied. Three results can be underlined: (i) for low n_s, the 2DEG presents pretty good mobilities, most of them in the range of 1000–1710 cm^2 V^{-1} s^{-1} with a maximum at $n_s = 9 \times 10^{12}$ cm^{-2}, (ii) the mobility starts to decrease slowly for $n_s > 9 \times 10^{12}$ cm^{-2}, (iii) then it drops dramatically for $n_s > 1.4 \times 10^{13}$ cm^{-2}. For a fixed Al composition, the lowest 2DEG mobility corresponds to structures with the thickest AlGaN, for which it was already mentioned that n_s was higher than expected. For these structures it is still unclear why the 2DEG behaviour seems to be lost. Spillover of the carriers into the GaN buffer region close to the interface may be the explanation [9]. The 77 K transport properties were measured for some of the specimens (not shown here). It is observed that: (i) the mobility presents a maximum as high as 6030 cm^{2}V^{-1}s^{-1} at $n_s = 9 \times 10^{12}$ cm^{-2} and (ii) for high carrier density (high Al content and/or thick AlGaN) the mobility quenches as observed for the transport at 300 K.

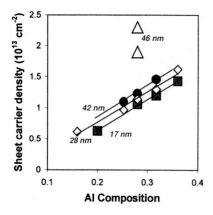

Fig. 1. Sheet carrier density versus Al composition: $n(\text{cm}^{-2}) \sim 4.9 \times 10^{13} x_{Al} + \text{const}$. No 2DEG can be detected for Al% as low as 6%. For too thick AlGaN (here 46 nm), n seems too high

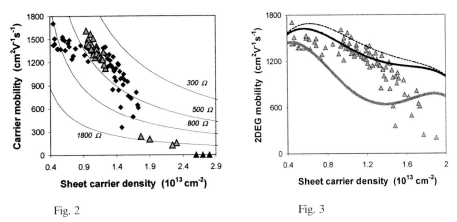

Fig. 2.

Fig. 3.

Fig. 2. 2DEG mobility versus sheet carrier density. The triangles on the x-axis correspond to structures expected to display high n_s but for which sheet resistance was huge. The continuous lines represent sheet resistance contours

Fig. 3. Modeling of the mobility versus 2DEG density at 300 K assuming that one (grey line) or two subbands (dark line) are occupied. Subband separations: 70 meV (dotted line), 80 meV (continuous line)

Interpretation of the Mobility The 2DEG mobility can be limited by several unavoidable solid-state or material related scattering mechanisms. In order to explain one part of our experimental findings, we have developed an original modelling for 2DEG transport based on the dynamical theory [10] and on a specific multi-subband screening tensor [11]. The scattering centers, taken into account for this evaluation, correspond to ionised impurities, acoustic (deformation and piezoelectric potential) phonons and polar optical phonons (POP). We assumed 1.5×10^{18} cm^{-3} ionised impurities in order to fix the low-n_s mobility around a value consistent with the experimental values and in order to symbolise the effect of all the other scattering processes not taken into account here. If the simulation is done with a single occupied subband, the model predicts that the mobility will first increase with n_s (screening of ionised impurity scattering), then decreases severely for carrier densities of the order of or larger than 6×10^{12} cm^{-2} (Fig. 3) because the Fermi level gets larger than the optical phonon energy (91 meV), allowing therefore their emission. The model shows also that as n_s increases further, the screening becomes again strong enough to permit some recovery of the mobility. Owing to the huge density of carriers present in nitride 2DEGs, at least two subbands are occupied in the QW at 300 K. Considering two occupied subband for the transport modelling, POP emission results in a less severe mobility drop. The QW is supposed to be an infinite triangular QW and the separation between the two first subbands is varied from 40 to 150 meV. Orthogonalised wave functions are considered for the calculation using a Fang-Howard parameter which varies with n_s. Two results of the modelling at 300 K for a band separation of 70 and 80 meV are shown in Fig. 3. A smooth drop can be obtained, however, at very high n_s the theoretical decay is not large enough to explain the experimental data. Also, at 77 K the experimental trend could not be fitted (not shown here).

If the low-n_s mobility decay could be due to the POP emission, how to interpret the mobility quench for high n_s? Some authors have already reported lower mobility for higher n_s with limited set of experimental points (for instance in [7, 9, 12]). In [7], the

13 K mobility was seen to drop significantly as a function of the Al composition and barrier thickness. Alloy disorder or interface roughness was proposed to be responsible for the trend. A model for this experimental result was proposed in [8], assuming that only one QW subband was occupied which is a very restrictive assumption. Very recently, a theoretical prediction of the room temperature 2DEG transport was also proposed using the "ladder technique" in [13], but it is also limited to one occupied subband and one scattering center: the polar optical phonon. We are developing further the two-subband model including extra scattering mechanisms not taken into account in this study. As the growth process is the same for all the structures, the interface roughness should be equivalent in all the cases. Nonetheless, as the 2DEG mean localisation is shifted toward the interface when n_s gets higher, the interface roughness scattering would have a greater impact on the mobility [6, 14]. It will be included in the model. The effect of other scattering mechanisms associated to dislocations, alloy disorder or carrier–carrier interaction and the addition of extra-occupied subbands is under progress. On the experimental point of view, further characterisations, as C–V and TEM investigations, are being carried out for a set of samples corresponding to the quenching regime (variable AlGaN thickness). Though the morphology of the AlGaN supply-layers used for the HEMT structure is perfect [15], one must also consider the possibility of a significant change in its microstructure as for instance inhomogeneous distribution of the Al (may result into a heterogeneous distribution of the polarisations and therefore to a non-uniform n_s).

Conclusions Undoped AlGaN–GaN 2DEG-like structures have been grown by LP-MOVPE using a new process for the semi-insulating templates. The sheet carrier density could be tuned from 4×10^{12} to 1.7×10^{13} cm^{-2}. For most of the structures, the mobility was between 1000 and 1710 cm^2 V^{-1} s^{-1}. A transport modelling taking into account the occupancy of two subbands in the QW and considering POP emission could explain the soft mobility decay observed from 9×10^{12} to 1.4×10^{13} cm^{-2}. Other phenomena are responsible for the mobility quench at higher carrier density.

Acknowledgements This work is partially supported by the European Space Agency under contract 14205/00/NL/PA.

References

[1] S. J. PEARTON, F. REN, A. P. ZHANG, and K. P. LEE, Mater. Sci. Eng. R **30**, 129 (2000).
[2] R. A. DAVIES, R. H. WALLIS, Z. BOUGRIOUA, I. MOERMAN, and W. A. PHILLIPS, unpublished.
[3] Z. BOUGRIOUA et al., to be submitted.
[4] Z. BOUGRIOUA et al., J. Cryst. Growth **230**, 373 (2001).
[5] F. BERNARDINI, V. FIORENTINI, and D. VANDERBILT, Phys. Rev. B **56**, R10024 (1997).
[6] O. AMBACHER et al., J. Appl. Phys. **87**, 334 (2000).
[7] I. P. SMORCHKOVA et al., J. Appl. Phys. **86**, 4520 (1999).
[8] L. HSU and W. WALUKIEWICZ, J. Appl. Phys. **89**, 1783 (2001).
[9] R. GASKA et al., Appl. Phys. Lett. **74**, 287 (1999).
[10] J.-L. FARVACQUE, Phys. Rev. B **62**, 2536 (2000).
[11] J.-L. FARVACQUE, F. CAROSELLA, and Z. BOUGRIOUA, submitted to Phys. Rev. B (2001).
[12] C. J. EITING et al., phys. stat. sol. (b) **216**, 193 (1999).
[13] D. R. ANDERSON et al., J. Phys.: Condens. Matter **13**, 5999 (2001).
[14] Y. ZHANG and J. SINGH, J. Appl. Phys. **85**, 587 (1999).
[15] R. CAMPION and I. HARRISON, AFM Analysis, unpublished.

Author Index

ABBAR, B. B457
ABERNATHY, C.R. A239, B337
ADACHI, M. B5, B239
ADAM, D. A325
ADELMANN, C. A575, A673, A711
ADERHOLD, J. A255
AFIFUDDIN, ... A667
AKAHANE, T. A415
AKASAKI, I. A117, A293, A799, A895,
. . . . B157, B461
AKIMOTO, K. A375, B319, B395
AKIMOV, A.V. B107, B607
AKITA, M. A207
ALAM, A. A155, A199, A647
ALDERIGHI, D. A851
ALLUMS, K.K. A239
ALVES, E. B59, B173
ALVES, H.R. A425, A453
AMANO, H. A117, A293, A799, A895,
. . . . B157, B461
AMBACHER, O. B497, B505, B519, B603
AMIMER, K. A515
ANDERSON, T.J. A407, A467
ANDO, H. B21
ANDO, Y. A191
ANDROULIDAKI, M. A515
ANSELL, B.J. A279, B579
ANTOINE-VINCENT, N. A519
ANWAR, A.F.M. B575
AOKI, D. B269, B273
AOKI, T. B475
AOURAG, H. B457
AOYAGI, Y. A83, B195, B461
ARAKAWA, Y. . . A37, A877, B187, B191, B433
ARAKI, T. A677, B13, B17, B395
AREHART, A.R. B309
ARYASETIAWAN, F. B567
AS, D.J. A699, B437
ASAHI, H. A601, A605
ASAMI, K. A601, A605
ASANO, T. A55, A69, A101
ASATSUMA, T. B45

BADER, S. A59, A65, A109, B407
BAE, S.-B. B621
BAHIR, G. A345
BALDANZI, A. B425
BALMER, R. A195
BANG, H. B319
BANNAI, R. A187
BARANOWSKI, J.M. B179
BARDWELL, J.A. . . . A233, A271, A389, A715
BARGHOUT, K. A783
BARJON, J. A673, A695
BARRIÈRE, A.S. A171
BARRY, E.A. B571
BARSKI, A. A867
BASAK, D. A147
BASKI, A.A. A571, B513, B543
BATES, S. A239
BAY, H. A647
BAZHENOV, N. B589
BEAUMONT, B. . A531, A733, A747, A899, B475
BECCARD, R. A751
BECK, A.L. A283
BEDAREV, D.A. A73, A91
BELL, A. B207
BELLET-ALMARIC, E. A695
BELLOTTI, E. B425, B585
BENAMARA, M. B345
BENEYTON, R. A711
BENNDORF, G. B279
BENYOUCEF, M. A747
BEN-YAACOV, I. A775
BER, B.JA. A433, B227
BERGMAN, J.P. B157, B485, B489
BERTRAM, F. B35, B41
BESULKIN, A.I. A91, A885
BHUIYAN, A.G. B27
BIDNYK, S. A857
BIGENWALD, P. B111, B471
BIMBERG, D. A91
BIRBECK, J. A195
BIRKHAM, R. A289
BLÄSING, J. A155, A425, A453
BLISS, D. A477
BÖTTCHER, T. B379, B403
BOETTINGER, W.J. A407
BORISOV, B. A881
BOSZE, E.J. A179
BOUGRIOUA, Z. . A255, A307, A367, B579, B625
BOUHAFS, B. B457
BOUR, D.P. B115
BOUTHILLETTE, L. A477
BOUWHUIS, P. A389
BRAGA, G.C.B. A457
BRANDT, O. B49
BRAULT, J. A575, A673
BRETAGNON, T. B65
BRILLSON, L.J. B441
BROWN, A.S. A491, A561
BROWN, J. B199
BRÜDERL, G. A59, A65, A109
BRUNO, G. A561
BUGAJSKI, M. B111
BUTCHER, K.S.A. A667, B179, B365
BYUN, D. B315

CAI, J. A833
CALLE, F. A307, A367, A899

CAMPBELL, J.C. A283, A333, B585
CAMPION, R.P. . . . A663, B203, B213, B219,
. . . . B223, B227
CANTARERO, A. B497
CAO, H.-J. B263
CAPEZZUTO, P. A561
CARLO, A. DI A251, A851, B553
CARLSTRÖM, C.F. A447
CAROSELLA, F. B625
CARRANO, J.C. B585
CARTWRIGHT, A.N. B115
CAVILL, S.A. B107, B607
CHANG, K.M. A175
CHANG, L.. . . . . A811
CHANG, W.-H. B77
CHAUDHURI, JOY A757
CHAUDHURI, J. A757, A783
CHEN, F. B115
CHEN, F.R. A811
CHEN, J. B411
CHEN, LU A135
CHEN, L.F. B353
CHEN, P.P.-T. A667
CHEN, T.T. A383
CHEN, W.M. B489
CHEN, Y. A653
CHENG, C.C. A175
CHENG, YUNG-CHEN B121, B357
CHEUNG, M.C. B115
CHEUNG, N.W. B91
CHI, JIM Y. A73, B95
CHICHIBU, S.F. A705, B599
CHIERCHIA, R. B403
CHITNIS, A. A147
CHO, A.Y. A825
CHO, HYUNG KOUN B183, B235
CHO, H.K. A163, B165, B231
CHO, S. B315
CHO, YONG-HOON A815
CHO, Y. B91
CHOCHO, K. A1
CHOI, HYUN-CHUL A247
CHOI, H.W. A393, A399
CHOI, J.W. B621
CHOI, K. A881
CHOI, S. A857
CHOI, W.-K. B315
CHOWDHURY, U. . . A283, A289, A301, A333,
. . . . A807
CHOYKE, W.J. A591
CHRISTEN, J. A155, A425, A453, A733,
. . . . A751, B35, B325, B419
CHRISTIANEN, P.C.M. A473, B563
CHU, S.N.G. A825, B337
CHUA, SOO JIN A329, B449
CHUA, S.J. . . . A393, A399, A421, B341
CHUNG, YI-YIN B121

CHUO, C.-C. B77
CHYI, JEN-INN B121, B357
CHYI, J.-I. B77
CIMALLA, V. A567
CINGOLANI, R. B513
CLAUSEN, B. B73
CLAYTON, N.W. A275
COFFIE, R. A355
COLLINS, C.J. A283, A333, B585
COLOCCI, M. A851
CONSTANTINIDIS, G. A259
CONTRERAS, O. A179
CORBEEK, W.H.M. A473
CORDIER, Y. A501
CORNET, A. A515
CORREIA, M.R. B59
CORREIA, R. B173
COWAN, D.A. B533
CROS, A. B497
CUI, Y. A583
CULLIS, A.G. A641, A871

DADGAR, A. A155
DAMILANO, B. . . . A171, A325, B65, B129
DAMILANO, D. A501
DANG, LE SI . . . A575, A673, A695, A711
DARAKCHIEVA, V. A447
DAUDIN, B. A575, A673, A695, A711
DAVIS, C.S. B203, B213, B223, B227
DAVIS, R.F. A729
DAVYDOV, A.. . . . A407, A411, A429, A467
DAVYDOV, V.YU. . . A615, A863, A885, B9
DAWSON, M.D. A743, B91, B169
DAWSON, P. B137, B453
DEATCHER, C.J. A743, B129
DEELMAN, P.W. A317
DELL, J.M. A311
DEMANGEOT, F. A511
DENBAARS, S.P. A213, A297, A355,
. . . . A775, B309, B353
DENYSZYN, J.C. A301, A333
DETCHPROHM, T. A293, A799
DEVATY, R.P. A591
DIAGNE, M. A105, A139
DIESSELBERG, M. B403
DIEZ, A. A155
DIMAKIS, M. A515
DIMITROV, R. B505, B603
DMITRIEV, V. A411, A429, A463, A881
DÖHLER, G.H. A131
DOGHÈCHE, E. A537, B429
DONEGAN, J.F. B493
DOOLITTLE, W.A. A491, A561
DUBOZ, J.Y. A171, A325
DUGGAN, G. B137
DUPUIS, R.D. . . . A283, A289, A301, A333,
. . . . A807, B345, B585

Author Index

EASTMAN, L.F. A203
EDDY, JR., C.R. A289
EDGAR, J.H. A757, A769, A783
EDWARDS, P.R. A743, B91, B169
EGAWA, T. A151
EICKHOFF, M. B505, B519
EINFELDT, S. A729, B379, B437
EITING, C.J. B345
ELLENS, A. A143
ELSASS, C. B199
EL-EMAWY, A.A. B263
EMTSEV, V.V. A863
ENJALBERT, F. A695
EVANS, N.D. A757
EVERITT, H.O. A793, B85
EVSTROPOV, V. B589

FAREED, Q. A95, A147
FARVACQUE, J.-L. B625
FAURIE, J.P. A733
FEDLER, F. A255
FEENSTRA, R.M. A595
FEICK, H. B147
FELTIN, E. A531, A537, A733, A899
FENG, SHIH-WEI B121, B357
FERGUSON, I.T. A289, B585
FERNÁNDEZ, S. A899
FERRO, G. A695
FIGGE, S. B379, B403
FLYTZANIS, N. A259
FOMIN, A.V. A433
FOX, A. A199
FOXON, C.T. . . . A279, A663, A691, B203, B207,
 . . . . . B213, B219, B223, B227, B283, B533
FRANCO, N. B59
FRANCOEUR, S. B287
FRANDON, J. A511, B173
FRASER, J. A715
FREITAS, JR., J.A. A457
FUJIOKA, H. A497, B391
FUJITA, SG. A543, B81, B153
FUJITA, SZ. A543
FUKUI, K. A337, B461
FUNAOKA, C. B125
FUNATO, K. A69
FUNATO, M. A543
FURUKAWA, K. B103

GAINER, G.H. A815, A857
GALLART, M. B65
GARBER, V. A345
GARCIA, R. A179
GARDNER, N.F. A15, B73, B147
GASKA, R. A95, A147, A219, B559
GENG, L. A803, B35
GEORGAKILAS, A. A259, A515, A567
GERMAIN, M. B385

GESSMANN, TH. A359
GHERASOIU, IU. A881
GIBART, P. A531, A537, A733, A747,
 . . . . A899, B429, B475
GIL, B. B65, B111, B471
GILA, B.P. A239
GIRKIN, J.M. A743
GLADYSHCHUK, A. B493
GLEIZE, J. A511, B173, B195
GMACHL, C. A825
GOANO, M. A425
GODFREY, M.J. B137, B453
GODLEWSKI, M. A447, B179, B365
GÖSLING, I. A109
GÖTZ, W.K. A15, B73, B147
GOLDYS, E.M. . . . A447, A667, B179, B365
GONCHARUK, I.N. A863
GOSS, S.H. B441
GOSSARD, A.C. B617
GOTO, S. A55
GOTTSCHALCH, V. B259, B279
GRAF, T. B603
GRAFF, J.W. A359, A889
GRAMLICH, S. A439
GRANDJEAN, N. . . A171, A325, A501, A519,
 . . . . A839, A851, B65, B111, B129
GRAUL, J. A255
GREEN, B.M. A203
GREGOR, R. A425
GROBER, R.D. A729, B505
GRZEGORCZYK, A. A523, A659
GRZEGORY, I. B345
GU, WENHUA. A329
GURSKII, A.I. B361
GURSKII, A.L. A79
GUTOWSKI, J. A65, A109, B379
GWO, S. A383

HABEL, F. A751
HÄRLE, V. A59, A65, A109, B407
HAFFOUZ, S. A523, A659
HAGEMAN, P.R. . . . A473, A523, A659, B563
HAGEMAN, W. A783
HAGIO, Y. A375
HAGIWARA, E. A485
HAHM, SUNG-HO A267, A341
HAIRSTON, A. A289
HAMAMURA, Y. A337
HAN, J. A105, A135, A139
HAN, MYUNG-GEUN B375
HANGLEITER, A. A59, B407
HANSEN, M. A297, B309, B353
HAO, MAO SHENG A329
HAO, M. A421
HAO, M.S. B341
HARADA, Y. B21
HARAFUJI, K. A635

HARDTDEGEN, H. A199, A647
HARIMA, H. B1, B103
HARRIS, J.J. B579, B607
HARRISON, I. . . A275, A279, A663, B203, B207,
 . . . . B213, B219, B223, B227, B533, B579
HASEGAWA, F. A443, B529, B549
HASEGAWA, H. A371
HASHIMOTO, A. . A691, B1, B5, B27, B239, B283
HASHIMOTO, P. A31
HASHIMOTO, T. A587
HASHIZUME, T. A371
HASTIE, J. B533
HAYAMA, N. A191
HE, YIPING A105
HEBARD, A.F. B337
HEIKMAN, S. A355, A775
HEINKE, H. B403
HEPPEL, S. A59
HERBERT, D. A195
HERSEE, S.D. B533
HEUKEN, M. . A79, A155, A199, A425, A453,
 . . . . A647, A751, A845, B361, B385, B493
HEYING, B. B309, B485
HIERRO, A. B309
HIJIKATA, Y. B269, B273
HILL, G. A227
HILSENBECK, J. A263
HINO, T. A55, A69, A101, B45, B395
HIRAI, R. B461
HIRAKO, A. A621
HIRAMATSU, K. A337, A725, A739
HIRANO, A. A293
HIRATA, A. A83
HIRATA, G.A. A179
HIRAYAMA, H. A83, B461
HIRSCH, L. A171
HITZEL, F. B407
HOFFMANN, A. A751
HOFMANN, D.M. A425, A453
HOLTZ, M. A881
HOLTZ, P.O. B467
HOMMEL, D. B379, B403, B437
HONDA, T. A587
HONG, CHANG-HEE B183, B235
HONG, C.H. B315
HONG, C.-H. B231
HONG, M.H. A371
HONG, YOUNG KUE B183, B235
HONG, Y.K. B231
HORIBUCHI, K. A739
HORIO, N. B125
HOSHINO, K. A877, B433
HOUSTON, P.A. A227
HSU, J.W.P. A595
HSU, T.M. B77
HU, X. A219
HUANG, D. A571, B513, B543
HUANG, S. A491
HUGHES, B.T. A195
HUMG, T.-V. A163
HUMPHREYS, J. B165
HURST, P. B137
HUSSAIN, T. A31
HWANG, S.W. A163, A167
HWU, Y. A379

IDE, T. A351, B99, B613
IKEDA, M. A55, A69, A101, B395
IKENAGA, M. A579
IKUTA, K. B239
IM, J.-S. A135
IMADA, Y. A121
IMANISHI, Y. A601, A605
INOUE, K. B81
INUSHIMA, T. B9, B31
IRISAWA, T. A553
ISHIBASHI, H. B55
ISHIDA, K. A687
ISHIDA, M. B103
ISHIDA, S. B187
ISHIDA, Y. A705, B599
ISHII, H. A243
ISHIKAWA, H. A151
ITO, Y. B239
ITOH, Y. A159
IVANOV, S.V. A615, B9, B481
IVANTSOV, V. A411
IWASA, N. A1
IWATA, M. B133
IWAYA, M. A117, A293, B157
IWAYAMA, S. A9
IYECHIKA, Y. A725

JAGGANATHAN, S. A783
JALABERT, D. A695
JANG, H.J. A167
JANG, H.W. B621
JASINSKI, J. A571, B345, B513, B543
JAVORKA, P. A199
JE, J.H. A379
JEFFS, N.J. A275
JEGANATHAN, K. B613
JENA, D. B617
JEON, C.M. B621
JEON, S.-R. A163, A167
JEONG, HWAN-HEE A247
JIA, A.W. B523
JIMBO, T. A151
JMERIK, V.N. A615
JO, M.S. A163
JOHNSON, J.W. A239
JOHNSON, N.M. A23, A131
JOKERST, N.M. A491

Jones, S.K.	A195	Kim, H.S.	B169
Jung, Byung-Kwon	A341	Kim, H.-S.	A743, B91
Jung, Hung Sub	B235	Kim, Jong-Wook	A267
Jung, Young-Chul	A247	Kim, J.	B315
Jursenas, S.	A95	Kim, J.K.	A379, B621
Jyouichi, T.	A121	Kim, K.H.	B621
		Kim, K.S.	A167, A301, A333
Kai, J.J.	A811	Kim, K.W.	B571, B593
Kaisei, K.	A543	Kim, K.Y.	A403
Kalashyan, A.	A857	Kim, K.-S.	A743, B91, B169
Kalliakos, S.	B65	Kim, Min Hong	B183, B235
Kamata, N.	B433	Kim, Sun-Ki	A247
Kaminska, E.	B365	Kim, Y.-W.	B231
Kamiyama, S.	A117, A799	Kimura, R.	A687
Kaneta, A.	B81, B153	King, T.	A571, A591, B513, B543
Kang, J.H.	A527	Kinoshita, A.	A83
Kang, S.	A491	Kipshidze, G.D.	A317, A881
Kang, X.J.	A399	Kirilyuk, V.	A473, A523, A659, B563
Kanie, H.	A481	Kishimoto, K.	A207
Kano, K.	B17	Kishino, K.	A187, A321
Karakostas, Th.	A567	Kishor, G.K.	A415
Karpov, S.Yu.	A611, A763	Kitamura, T.	A705, B599
Karrer, U.	B505	Kitano, T.	B283
Kasahara, K.	A191	Kiyoku, H.	A1
Kaschner, A.	A751	Klausing, H.	A255
Kasic, A.	B437	Klein, A.	A439
Kasu, M.	A779	Klochikhin, A.A.	A863
Kato, M.	A121	Kneissl, M.	A23, A131, B115
Kattner, U.R.	A407	Kobayashi, K.	A69
Katz, O.	A345, A789	Kobayashi, N.	A113, A183, A223, A363, A779
Kaufmann, U.	A143	Kobayashi, T.	B45, B395
Kavokin, A.	B471	Kobusch, M.	A143
Kawakami, Y.	B81, B153	Kochelap, V.A.	B571, B593
Kawanishi, H.	A587, B283	Köhler, K.	A143
Kayambaki, M.	A259	Köhler, U.	A699, B437
Kehagias, Th.	A567	Koide, K.	B5
Keller, S.	A297, A355, A775	Koike, M.	A9
Kent, A.J.	B107, B607	Kojima, A.	A9
Kent, P.R.C.	B253	Kokolakis, G.	A851
Kern, R.S.	A15	Kokorev, M.F.	A885
Khan, M. Asif	A95, A147, A219, B559	Komirenko, S.M.	B593
Kielburg, A.	B325	Komninou, Ph.	A567
Kiesel, P.	A131	Kordos, P.	A199
Kijima, M.	A677	Kosaki, M.	A895
Kijima, S.	A55, A101, A375	Koshka, Y.	B589
Kikawa, J.	A159	Kostopoulos, A.	A259, A567
Kikuchi, A.	A187, A321	Koudymov, A.	A219
Kikuchi, S.	B273	Koukitu, A.	A549, A553, A557, B537
Kim, A.Y.	A15	Kovalenkov, O.	A429
Kim, Chang-Seok	A267	Kovarsky, A.P.	A433, B227
Kim, Chi Sun	B183, B235	Koynov, S.	A845
Kim, C.C.	A379	Kozaki, T.	A1
Kim, C.S.	B165, B231	Kozin, I.E.	A863
Kim, Dong-Joon	A375	Kozłowski, J.	B415
Kim, G.	B315	Kozodoy, P.	A289
Kim, Hyun-Min	B375	Krames, M.R.	A15, A105
Kim, H.	A203	Krishnamoorty, V.	A239

KROST, A. A155, A425, A453, B325
KRTSCHIL, A. B325
KRYLIOUK, O.M. A467
KUBALL, M. A511, A747, A769, B195
KUDO, H. A121, B55
KUDO, Y. B45
KUDRNA, J. A851
KUDRYASHOV, V.E. B141
KUEK, J.J. A311
KÜMMLER, V. A59, A65
KUHN, B. A59, A629, B437
KUJIMA, S. A69
KULIK, A.V. A763
KUMAGAI, Y. . . . A549, A553, A557, B537
KUMAKURA, K. A183, A363
KUNZE, R. A537
KUNZER, M. A143
KUOKSTIS, E. A95, B559
KURAMOTO, M. A47
KURILCIK, G. A95
KURIMOTO, E. B1, B103
KURODA, T. B125
KURYATKOV, V.V. A317, A881
KUWANO, N. A557, A739
KUWATA-GONOKAMI, M. B475
KUZNETSOV, N.I. A433, A463
KUZUHARA, M. A191
KWON, M.K. A527
KWON, Y. A857
KYHM, K. B475
KYUTT, R.N. A863, A885

LADA, M. A641, A871
LAI, C.Y. B77
LAI, S.K. A811
LAM, J.B. A429, A815, A857
LAMARRE, P. A289
LAMBERT, D.J.H. A807, B345
LANG, C.C. A175
LANGER, R. A867
LAPOINTE, J. A271
LARSEN, P.K. . . . A473, A523, A659, B563
LAÜGT, M. A531, A899
LAVRENTIEV, A. A411
LEBEDEV, A.V. B481
LEDENTSOV, N.N. A91
LEE, C.D. A595
LEE, C.-M. B77
LEE, C.-W. B85
LEE, HYUNG JAE B183
LEE, H.J. A403, B231
LEE, JAE-HOON A267, A341
LEE, JAE-SEUNG A267
LEE, JEONG YONG B183, B235
LEE, JUNG-HEE A247, A267, A341
LEE, J.Y. A163, B165
LEE, J.-H. B263, B621

LEE, J.-L. A379, B621
LEE, K. B579
LEE, K.J. B607
LEE, K.P. B337
LEE, K.Y. A457
LEE, MYOUNG-BOK A341
LEE, SEONGHOON B375
LEE, S.H. B371
LEE, S.J. B315
LEE, S.K. A457
LEE, YONG-HYUN A247, A341
LEEM, SHI-JONG B235
LEFEBVRE, P. B65, B111
LEIBIGER, G. B259, B279
LELL, A. A59, A65, A109
ŁEPKOWSKI, S.P. A839
LEROUX, M. A519, A531
LEVETAS, S.A. B137
LEYMARIE, J. A519
LI, G.H. A653
LI, L. A583
LI, Q. A681
LI, T. . A663, A691, B203, B213, B223, B227
LI, YUN-LI A359
LI, Z. B319
LIAO, CHI-CHIH B121, B357
LIAO, Y. B203, B213, B219
LIERDE, P. VAN A457
LILIENTAL-WEBER, Z. . . . A571, B345, B513,
. . . . B543
LIM, K.Y. A527
LIMA PIMENTA, A. B505
LIMPIJUMNONG, S. B303
LIN, YEN-SHENG B121, B357
LINK, A. B603
LISCHKA, K. A699
LITTON, C.W. A571, B513, B543
LIU, B. A757, A769
LIU, CHIH-WEN B121
LIU, C. A743
LIU, K.-Y. B45
LIU, L. A757, A769
LIU, R. B41
LIU, WEI A329
LIU, YING A271, A389
LIU, Y. A233
LOOK, D.C. B293, B441
LORENZ, K. B331
LOSSY, R. A263
LOSURDO, M. A561
LÜBBERS, M. A699
LÜNENBÜRGER, M. A79
LÜTH, H. A199, A647
LUGAUER, H.-J. B407
LUGLI, P. A251, A851
LUNDIN, W.V. . A73, A91, A863, A885, B95
LUO, B. A239

Author Index

LUTSENKO, E.V. A79, B361
LYNAM, P. A871
LYNCH, R.J. A641

MA, KUNG-JENG B121, B357
MA, T.P. A213
MACELWEE, T.W. A233, A271
MACHT, L. A473, A659
MAEDA, N. A223
MAEDA, T. A725, B269
MAHAJAN, S. B161
MAKARONA, E. A105
MAKAROV, YU.N. A611, A763
MAKIMOTO, T. A183, A363
MALPUECH, G. B471
MANSON-SMITH, S.K. B445
MARIETTE, H. . . . A575, A673, A695, A711
MARKO, I.P. A79, B361
MARSHALL, P. A271, A389
MARSO, M. A199
MARTIN, P.S. A15
MARTIN, R.W. . A227, A743, B91, B129, B169
MARTIN, T. A195
MARTÍNEZ-CRIADO, G. B497
MARTINEZ-GUERRERO, E. . . A575, A673, A695, A711
MARUTSUKI, G. B153
MARUYAMA, T. A375
MASCARENHAS, A. B243, B287
MASSIES, J. . . A171, A325, A501, A511, A519, A839, A851, B65, B111, B129
MASTRO, M.A. A467
MASUMOTO, Y. B395
MATSUMOTO, K. A579
MATSUMOTO, S. B269
MATSUMURA, H. A1
MATSUO, Y. A553
MATSUOKA, T. A485
MATSUSHITA, T. A1
MAVROIDIS, C. B579
MAYUMI, M. B537
MCALISTER, S.P. A233
MCCRAY, L. A31
MCKITTRICK, J. A179
MEHANDRU, R. A239
MELNIK, YU. . . . A411, A429, A463, B589
MENONI, C.S. B73
MEYER, B.K. A425, A453, A845
MEYER III, H.M. A757
MEYLER, B. A345, A789
MICHLER, P. A65, A109, B379
MICOVIC, M. A31
MIERRY, P. DE A531, A899
MIKI, H. B391
MIKROULIS, S. A259, A567
MIMKES, J. A699
MINAMI, T. A677

MINSKY, M.S. B133
MIRAGLIA, P.Q. A729
MISHRA, U. B603
MISHRA, U.K. . . A213, A297, A311, A355, A775, B309, B617
MISKYS, C.R. B497
MISTELE, D. A255
MITATE, T. A557
MIYAJIMA, T. . A69, A101, A375, B45, B395
MIYAKE, H. A725, A739
MIYAMOTO, H. A191
MIYAMURA, M. B191
MIYASHITA, N. A23
MIZUTA, M. A47
MIZUTANI, T. A207
MOCHIZUKI, S. A799, A895
MOERMAN, I. . . . A307, A367, B579, B625
MOISA, S. A715
MOLNAR, R.J. A457, B441
MONEMAR, B. . . A447, B157, B467, B481, B485, B489
MONROY, E. A307, A367
MONTEIL, Y. A695
MONTOJO, M.T. A307
MOON, J.S. A31
MOON, YONG-TAE B375
MOORE, W.J. A457
MORAN, B. A213, A297, A355
MOREL, A. B65
MORETTO, P. A171
MORHAIN, C. B501
MORI, H. A691
MORISHIMA, S. B319
MORKOÇ, H. . . . A251, A571, A591, A793, B513, B543
MOTOGAITO, A. A337
MOTOJO, M.T. A367
MOTOKAWA, M. B9
MOURI, H. B103
MOWBRAY, D.J. A871
MÜLLER, G. B519
MUKAI, T. A1, B81, B153
MULA, G. . . . A575, A673, A695, A711
MUÑOZ, E. A307
MURAKAMI, H. A549, A557
MURAKAMI, K. B55
MURAKAMI, Y. B5
MURANO, K. B31
MURATA, S. A579
MYNBAEV, K. B589
MYNBAEVA, M.G. A433, B589

NAGAHAMA, S. A1
NAGAI, S. A9
NAHM, K.S. A527, B371
NAKAGAWA, Y. . A803, A833, B35, B41, B153
NAKAJIMA, H. A69, B395

NAKAMURA, K.	A579	OKI, K.	A557, A739
NAKAMURA, R.	A117	OKUMURA, H.	A351, A705, B99, B599, B613
NAKAMURA, T.	A895	OKUYAMA, M.	B391
NAKANISHI, H.	A705, B599	OMIYA, H.	A803, A833, B35, B41
NAKAYAMA, T.	A191, B529, B549	OMNÈS, F.	A307, A367, B429
NAMERIKAWA, M.	A443	ONABE, J.	A719
NAMKOONG, G.	A561	ONABE, K.	B269, B273
NANISHI, Y.	A375, A677, B13, B17, B395	O'NEILL, J.P.	A871
NAOI, H.	A725	ONO, Y.	B31
NARUKAWA, M.	A725	ONSTINE, A.H.	A239
NASTASE, N.	A647	OOTOMO, S.	A371
NATALI, F.	A501, A519	OR, C.T.	A681
NEBAUER, E.	A439	OSHIKIRI, M.	B567
NEMOTO, T.	A351	OSHIMA, M.	A497, B391
NENER, B.D.	A311	OSIŃSKI, M.	B263
NEU, G.	B501	OVERBERG, M.E.	B337
NEUBERGER, R.	B505, B519	OZDEN, I.	A139
NG, H.M.	A825		
NGUYEN, N.D.	B385	PADUANO, Q.	A821
NIDO, M.	A47	PAKHNIN, D.V.	A885
NIEHUS, M.	A845	PAKULA, K.	B179
NIKISHIN, S.A.	A317, A881	PALACIOS, T.	A367
NIKITINA, I.P.	A411, A433, A463	PARBROOK, P.J.	A227, A641, A871
NIKOLAEV, A.E.	A433	PARISH, G.	A297, A311
NISHIDA, T.	A113	PARK, SEONG-JU.	B375
NISHIKAWA, N.	A151	PASKOV, P.P.	A447, B157, B467, B489
NISHIMOTO, S.	A739	PASKOVA, T.	A447, B157, B467, B481, B489
NISHIO, Y.	A691	PASZKIEWICZ, R.	B415
NITTA, S.	A895, B157	PATANE, A.	B283
NOH, D.Y.	A379	PATEL, D.	B73
NOMURA, M.	B319	PAU, J.L.	A307
NOSEI, D.	B21	PAVLOVSKII, V.N.	A79, B361
NOUET, G.	B411, B429, B457	PAYZANT, E.A.	A757
NOVIKOV, S.V.	B203, B207, B213, B219, B223, B227, B533	PEARTON, S.J.	A239, B337
		PECHNIKOV, A.	A463
NURMIKKO, A.V.	A105, A135, A139	PEIRO, F.	A515
		PELEKANOS, N.T.	A867
OBLOH, H.	A143, A263	PENDLEBURY, S.T.	A871
OCHALSKI, T.J.	B111	PEREIRA, E.	B59, B173
O'DONNELL, K.P.	B59, B129, B445, B533	PEREIRA, S.	B59, B173
ÖZGÜR, Ü.	A793, B85	PERLIN, P.	A839
OFF, J.	A59, B437	PERNOT, C.	A293
OGURA, M.	B269	PETROFF, P.M.	B199
OH, C.S.	A403	PFISTERER, D.	A425, A453
OH, JAE-EUNG	A267	PHILLIPS, M.R.	B179
OHKAWA, K.	A621	PIDDUCK, A.J.	A195
OHNISHI, K.	A601, A605	PIOTROWSKA, A.	B365
OHNO, Y.	A191	PLETSCHEN, W.	A143
OHOYA, S.	B31	PLOOG, K.H.	B49
OHTA, J.	A497	POBLENZ, C.	B199
OHTA, K.	A337	PONCE, F.A.	A179, A803, A833, B35, B41, B207
OHUCHI, Y.	A121, A337		
OKADO, H.	A481	POPHRISTIC, M.	A289
OKAGAWA, H.	A121	POROWSKI, S.	B345
OKAMOTO, K.	B81, B153	POSCHENRIEDER, M.	A155
OKAMOTO, W.	B433	POZINA, G.	B157, B485, B489
OKAMOTO, Y.	A191	PREBLE, E.A.	A729

Author Index

PROTZMANN, H. A79
PULFREY, D.L. A311

RAJASINGAM, S. A769
RAKOVICH, YU. B493
RAMM, M.S. A763
RAMOS, F. A179
RATNIKOV, V.V. A615
RAUHALA, S. A389
RAUSCHENBACH, B. B325
RAYMOND, S. A715
REALE, A. A851, B553
RECHENBERG, I. A439
REED, M.D. A467
REINE, M.B. A289
REN, F. A239
RENNER, F. A131
RENUCCI, M.A. A511, B173
RESHCHIKOV, M.A. . . . A571, B513, B543
REVERCHON, J.-L. A325
RHO, J.I. A527
RICHTER, E. A439
RIEMANN, T. A155, A425, A453, A733,
. . . . A751, B35, B419
RINGEL, S.A. B309
RÖVER, K.S. A255
RÖWE, M. A65, A109
ROLFE, S. A271, A715
ROSKOWSKI, A.M. A729
ROSSNER, W. A143
ROTTER, T. A255
RUSKE, F. B331
RUTERANA, P. . . A367, A379, B411, B429,
. . . . B457
RYAN, J.F. B475
RYS, A. A783

SACCONI, F. A251
SADDOW, S.E. B589
SAGAR, ASHUTOSH A595
SAITO, Y. B13, B17
SAITOH, T. A223
SAKAI, S. B107, B607
SAKHAROV, A.V. . A73, A91, A863, A885, B95
SAKON, T. B9
SALAMANCA-RIBA, L. A595
SALZMAN, J. A345, A789
SAN ANDRÉS, E. A307
SÁNCHEZ, F.J. A307, A367
SÁNCHEZ-OSORIO, J. A367
SANDS, T. B91
SANGUINO, P. A845
SANO, M. A1
SANO, S. A799
SANO, T. A117
SARNEY, W.L. A595
SASAKI, C. B133

SASAKI, F. B99
SASOU, R. B125
SATAKE, A. B395
SATO, H. B125
SATO, K. A587
SATOH, F. B537
SCHAD, S.S. A127
SCHAFF, W.J. A203
SCHALWIG, J. B505
SCHEGLOV, M.P. A863
SCHENK, H.P.D. A537, A899
SCHERER, M. A127
SCHINELLER, B. . . . A79, B361, B385, B493
SCHLOTTER, P. A143
SCHMEITS, M. B385
SCHMIDT, H. A537
SCHMIDT, R. A143, A647
SCHMITT, J. A783
SCHÖN, O. A79, B361
SCHOLZ, F. A59, A629, B437
SCHUBERT, E. A889
SCHUBERT, E.F. A359
SCHUBERT, M. B259, B279, B437
SCHUCK, J. A729
SCHUCK, P.J. B505
SCHURMANN, M.J. B585
SCHWABE, R. B279
SCHWARZ, R. A845
SCHWARZ, U. A729
SCHWEGLER, V. A127
SEGAL, A.S. A763
SEKI, H. A549
SEMCHINOVA, O.K. A255
SEMENDY, F. A807
SEMOND, F. . . A325, A501, A511, A519, A851
SENOH, M. A1
SEO, J.W. A403
SEO, S.W. A491
SEONG, M.J. B243
SEQUEIRA, A.D. B59
SETIAGUNG, C. A719
SEYBOTH, M. A127, A751
SHAPIRO, A. A429, A467
SHAPIRO, N.A. B147
SHARMA, N. B165
SHATALOV, M. A147
SHEALY, J.R. A203
SHEN, H. A807, B585
SHEN, L. A775
SHEN, S.X.Q. B613
SHEN, X.Q. A351, A705, B99, B599
SHI, Y. A757, A769
SHIGEMORI, A. A687
SHIKE, J. A687
SHIMIZU, M. . . A351, A705, B99, B599, B613
SHIMOGAMI, K. A543
SHIN, JIN-HO A267
SHIN, J. B315

Shin, Moo-Whan	A267	Suscavage, M.	A477
Shin, Sang-Hoon	A341	Suski, T.	A839
Shinohara, M.	A587	Suzuki, A.	A351, B13, B17
Shinomiya, G.	B81, B153	Suzuki, Y.	A705
Shiraishi, T.	B31	Swadener, J.G.	A757
Shiraki, Y.	A719	Sweeney, E.	B533
Shishkin, Y.	A591	Sweeney, F.	B59
Shubina, T.V.	A615, B481	Sweeney, P.M.	B115
Shur, M.S.	A95, A147, A219, B559		
Sia, E.K.	A421	Tachibana, A.	A579
Siegrist, T.	A825	Tachibana, K.	B187, B191
Sikora, P.	A715	Tackeuchi, A.	B125
Simin, G.	A95, A147, A219, B559	Tadatomo, K.	A121, A337
Simon, J.	A867	Tagliente, M.A.	B49
Singh, R.	A289	Taguchi, T.	A121, B55, B133
Skolnick, M.S.	A871	Takahashi, H.	A497
Smart, J.A.	A203	Takahashi, K.	A625, A687, B283, B523
Smirnov, A.N.	A863	Takahashi, M.	B103
Smorchkova, I.	B617	Takahashi, O.	A443, B529, B549
Smorchkova, Y.	B603	Takanami, S.	A117
Södervall, U.	A447	Takemoto, K.	B537
Soltani Vala, A.	B453	Takeuchi, T.	A105, A139, B133
Someya, T.	A877, B187, B191, B433	Takeya, M.	A55
Son, S.J.	A163, A167	Taliercio, T.	B65, B111
Song, I.J.	A457	Tampo, H.	A601, A605
Song, J.	A857	Tamulaitis, G.	A95
Song, J.J.	A429, A815	Tan, W.-S.	A227
Song, Y.H.	A167	Tanabe, Y.	B13
Sonoda, Y.	A557	Tanaka, S.	A803, A833, B35, B41, B195
Sood, A.K.	A289	Tanaka, T.	B461
Souda, R.	A443, B529, B549	Taneya, M.	B103
Soukhoveev, V.	A411	Tang, Haipeng	A271
Speck, J.S.	B147, B199, B309, B353, B485, B603	Tang, H.	A233, A389, A715
		Tanikawa, T.	B239
Srinivasan, S.	B35, B41	Tansley, T.L.	A667
Staddon, C.R.	B203, B213, B223, B227	Tapfer, L.	B49
Stanton, N.M.	B107, B607	Tarakji, A.	A219
Steigerwald, D.A.	A15	Tarsa, E.J.	A289
Steinhoff, G.	B519	Taylor, R.A.	B475
Stemmer, J.	A255	Teepe, M.	A23
Steranka, F.M.	A15	Teisseire, M.	B501
Stiles, T.	A729	Teisseyre, H.	A839
Stockman, S.A.	A15	Temkin, H.	A317, A881
Strassburg, M.	A155	Teraguchi, N.	B13, B17
Strauf, S.	B379	Terao, S.	A117
Stroscio, M.A.	B593	Terazima, M.	B81
Stutzmann, M.	B497, B505, B519, B603	Tezen, Y.	A9
Sugimoto, K.	A481	Theodoropoulou, N.	B337
Sugimoto, Y.	A1	Thomsen, C.	A751
Suh, Eun-Kyung	B183	Tichelaar, F.D.	B563
Suh, E.K.	B371	Tilak, V.	A203
Suh, E.-K.	B231	Tisch, U.	A789
Sun, J.	A15, B73	Tkachman, M.G.	A615
Sun, Wenhong	B449	Tłaczala, M.	B415
Sun, W.H.	B341	Tobin, S.P.	A289
Sun, X.L.	A653, B441	Tojyo, T.	A55, A101
Sung, Changmo	A477	Tomé, C.N.	B73

Author Index

TOMIYA, S. A69, A101, B395
TONG, S.Y. A681
TOROPOV, A.A. B481
TOTTEREAU, O. A899
TOYOURA, Y. A321
TRAGER-COWAN, C. . . . A227, A743, B59,
. . . . B445, B533
TREAT, D.W. A23
TRETYAKOV, V.V. A885
TSAGARAKI, K. A515
TSATSULNIKOV, A.F. . . A73, A91, A885, B95
TSENG, K.-U. B77
TSUNEKAWA, T. A121
TSVETKOV, D.V. . . A411, A429, A433, A463,
. . . . A881
TSVETKOVA, K. A411
TU, C.W. B287
TU, RU-CHIN. A73, B95
TUSUBAKI, K. A223

UCHIDA, S. A55, A101
UKAI, T. A117
ULRICH, S.M. B379
UMEMOTO, H. A1
UMENO, M. A151
UREN, M.J. A195
URUGA, T. B45
USIKOV, A.S. . . A73, A91, A863, A885, B95

VAILLE, M. A537, A899
VALCHEVA, E. A447
VASCHENKO, G. B73
VASSON, A. A519
VECKSIN, V.V. B9
VEHSE, M. A109
VEIT, P. A425, A453
VEKSHIN, V.A. A615
VENNÉGUÈS, P. A531, A733, A899
VERDÚ, M. A307, A367
VÉZIAN, S. A501
VIANDEN, R. B331
VINATTIERI, A. A851
VISCONTI, P. A571, B513, B543

WAGNER, J. A143
WAHAB, Q. A447
WAKI, I. B391
WALDRON, E.L. A359, A889
WALKER, L.R. A757
WALLE, C.G. VAN DE B303
WALTEREIT, P. B49, B147
WANG, H.M. A95
WANG, LIANSHAN B449
WANG, L.S. B341
WANG, SHENG QI A477
WANG, S.Q. A821
WANG, T. B107, B607
WANG, WEN A329
WANG, X.W. A213
WANG, Y.T. A653
WANG, Z.G. A653
WATANABE, S. B133
WATANABE, Y. A895
WATSON, I.M. . . A743, B59, B91, B137, B169
WATT, A. A227
WEBB, J.B. A233, A271, A389, A715
WEBB-WOOD, G. A793
WEBER, E.R. B147
WEBSTER, R.T. B575
WEIHNACHT, M. A537
WEIMAR, A. A59, A65
WENZEL, A. B325
WESTMEYER, A.N. B161
WEYBURNE, D. A821
WEYERS, M. A439
WEYHER, J.L. A473, A659
WHITE, M.E. B129
WHITEHOUSE, C.R. A227
WIERER, J.J. A15
WILSON, R.G. B337
WINSER, A.J. B203, B213, B219, B223,
. . . . B227
WITTE, H. B325
WOJCIK, A. B111
WOLTER, M. A199
WONG, K.K. A289
WONG, M.M. . . . A283, A289, A301, A333,
. . . . A807
WONG, W.S. A23, A31
WOOD, D.A. A227, A641, A871
WOOLLAM, J.A. B437
WRABACK, M. A807, B585
WU, J. A719
WU, SHANGLI. B575
WÜRFL, J. A263

XIE, M.H. A681
XIE, Z. A783
XIN, H.P. B287
XU, J.M. A135
XU, K. B523
XU, S.J. A681

YABLONSKII, G.P. A79, B361, B493
YAGI, E. B319
YAGUCHI, H. B269, B273
YAMADA, K. A601, A605, B433
YAMADA, N. B133
YAMADA, Y. B55, B133
YAMAGUCHI, A.A. A47

YAMAGUCHI, S. A895, B461
YAMAGUCHI, T. B13, B17
YAMAMOTO, A. . . . A691, B1, B5, B27, B239,
. . . . B283, B461
YAMAMOTO, S. A543
YAMANE, H. A415
YAMASAKI, S. A9, A383
YANAMOTO, T. A1
YANASHIMA, K. B395
YANG, B. A333
YANG, CHIH-CHUNG B121, B357
YANG, G. A857
YANG, G.M. A163, A167, A403, B165
YANG, HUI A653, A681
YANG, J. A219
YANG, J.W. . . A95, A147, A403, A527, B559
YANG, W. A815
YANO, N. B523
YASAKA, S. B31
YASUDA, T. A383
YASUI, K. A415
YI, M.S. A379
YIN, AIJUN A135
YIN, SUN BIN A73, B95
YODO, T. B21
YONEMARU, M. A321
YOON, C.J. A403
YOSHIDA, S. A159, A243, B269, B273
YOSHIKAWA, A. A625, B523
YOSHITANI, M. A621
YUKAWA, Y. A895
YUN, F. . . A571, A591, A793, B513, B543
YUNOVICH, A.E. B141

ZAMFIRESCU, M. B471
ZANARDI OCAMPO, J.M. B433
ZAVADA, J.M. B337
ZAVARIN, E.E. A91, A863, A885
ZEIMER, U. A439
ZERVOS, M. . . . A259, A515, A567, A867
ZHANG, B.J. A151
ZHANG, JI A329
ZHANG, JIANPING B559
ZHANG, J. A421
ZHANG, J.P. A95
ZHANG, N.-Q. A213
ZHANG, W. A425, A453
ZHANG, XIN HAI A329
ZHANG, X.B. B65
ZHANG, X.H. B341
ZHANG, YONG B243, B287
ZHENG, L.X. A681
ZHENG, R. B55
ZHMAYEV, E. B263
ZHOU, H. A105
ZHU, J.J. A653
ZHU, T.G. A301
ZHUANG, D. A769, A783
ZIELINSKI, M. A447, A473
ZUBIA, D. B263, B533
ZUBIALEVICH, V.Z. A79, B361
ZUBRILOV, A.S. A411, A433
ZUKAUSKAS, A. A95
ZUNGER, A. B253